For Reference

Not to be taken from this room

COLLEGE MISERICORDIA

TREATISE ON ANALYTICAL CHEMISTRY

A comprehensive account in three parts

PART I

THEORY AND PRACTICE

PART II

ANALYTICAL CHEMISTRY
OF THE ELEMENTS

PART III

ANALYTICAL CHEMISTRY IN INDUSTRY

TREATISE ON ANALYTICAL CHEMISTRY

Edited by I. M. KOLTHOFF
School of Chemistry, University of Minnesota

and PHILIP J. ELVING
Department of Chemistry, University of Michigan

with the assistance of ERNEST B. SANDELL
School of Chemistry, University of Minnesota

PART II

ANALYTICAL CHEMISTRY OF THE ELEMENTS

VOLUME 6

Be • Pb • Nb, Ta • Tc Ac, At, Fr, Po, Pa

INTERSCIENCE PUBLISHERS
a division of John Wiley & Sons, New York–London–Sydney

Copyright © 1964 by
John Wiley & Sons, Inc.

All Rights Reserved
Library of Congress Catalog Card Number 59-12439

PRINTED IN U.S.A. BY MACK PRINTING CO., EASTON, PA.

TREATISE ON ANALYTICAL CHEMISTRY

PART II
ANALYTICAL CHEMISTRY OF THE ELEMENTS

SECTION A
Systematic Analytical Chemistry of the Elements

VOLUME 6

BERYLLIUM · LEAD · NIOBIUM AND TANTALUM · TECHNETIUM · ACTINIUM, ASTATINE, FRANCIUM, POLONIUM, AND PROTACTINIUM

AUTHORS OF VOLUME 6

JAMES W. COBBLE
T. W. GILBERT, JR.
SILVE KALLMANN
B. R. F. KJELLGREN

E. STANLEY MELICK
C. W. SCHWENZFEIER, JR.
JACOB SEDLET

Authors of Volume 6

James W. Cobble

Department of Chemistry, Purdue University, Lafayette, Indiana

T. W. Gilbert, Jr.

Department of Chemical and Metallurgical Engineering, University of Cincinnati, Cincinnati, Ohio

Silve Kallmann

Director of Research, Ledoux & Company, Inc., Teaneck, New Jersey

B. R. F. Kjellgren

The Brush Beryllium Company, Cleveland, Ohio

AUTHORS OF VOLUME 6

E. Stanley Melick

The Brush Beryllium Company, Elmore, Ohio

C. W. Schwenzfeier, Jr.

Vice President, The Brush Beryllium Company, Cleveland, Ohio

Jacob Sedlet

Argonne National Laboratory, Argonne, Illinois

PART II. ANALYTICAL CHEMISTRY OF THE ELEMENTS

CONTENTS—VOLUME 6

SECTION A. Systematic Analytical Chemistry of the Elements

Beryllium. By *B. R. F. Kjellgren, C. W. Schwenzfeier, Jr., and E. Stanley Melick* 1

    I. Introduction 3
        A. History 3
        B. Occurrence 4
           1. Minerals and Ores 4
    II. Industrial Processes 5
        A. Extraction as Beryllium Hydroxide 5
        B. Sulfate Extraction 5
        C. The Complex Fluoride Process 7
        D. Production of Beryllium 8
           1. Pyro-Reduction 8
           2. Electrolytic Reduction 9
           3. Vacuum Melting and Powder Metallurgy 9
           4. Miscellaneous 10
           5. Production of Beryllium Copper 10
    III. Toxicology 10
    IV. Industrial Products and Uses 11
        A. Pure Beryllium Metal 11
           1. Nuclear 11
           2. Structural 11
        B. Alloys 12
           1. Copper-Base Alloys 12
           2. Other Alloys 13
    V. Beryllium Compounds 13
        A. Beryllium Sulfate ($BeSO_4 \cdot 4H_2O$) 13
        B. Sodium Beryllium Fluoride (Na_2BeF_4 or $2NaF \cdot BeF_2$) 14
        C. Beryllium Hydroxide ($Be(OH)_2$) 14
        D. Ammonium Beryllium Fluoride (($NH_4)_2BeF_4$ or $2NH_4F \cdot BeF_2$) 15
        E. Beryllium Fluoride (BeF_2) 16
        F. Beryllium Chloride ($BeCl_2$) 16
        G. Beryllium Iodide (BeI_2) 16
        H. Beryllium Oxide (BeO) 16
        I. Beryllium Nitrate ($Be(NO_3)_2 \cdot 3H_2O$) 17
        J. Beryllium Nitride (Be_3N_2) 17
        K. Beryllium Carbide (Be_2C) 18
        L. Beryllium Carbonate—Basic 18
        M. Basic Beryllium Acetate ($Be_4O(C_2H_3O_2)_6$) 18
        N. Beryllium Phosphate ($Be_3(PO_4)_2$) 18
        O. Beryllium Hydride (BeH_2) 19
        P. Intermetallic Compounds—Beryllides 19
    VI. Properties 19
        A. Physical and Mechanical Properties 19

B. Electrochemical Properties	19
C. Chemical Properties	21
D. Optical Properties	23
E. Radiochemical—Nuclear	24
VII. Sampling	25
A. Ore	25
B. Copper-Base Alloys	26
1. Cast Alloys	26
2. Wrought Alloys	26
C. Aluminum-Base Alloys	27
D. Beryllium Metal	27
VIII. Separation and Isolation of Beryllium	28
A. Sample Dissolution	28
1. Ores	28
2. Copper-Base Alloys	29
3. Aluminum-Base Alloys	30
4. Beryllium Metal	30
B. Separation Methods	31
1. Precipitation Methods	32
a. With Ammonia (Initial Separation with Possible Subsequent Purifications)	33
b. With 8-Hydroxyquinoline	33
c. With Cupferron (Ammonia Salt of Nitrosophenylhydroxylamine)	34
d. With n-Benzoylphenylhydroxylamine	34
e. Miscellaneous	34
2. Extraction Methods	35
a. With Acetylacetonate	35
b. With Butyrate	36
c. With Methyl Isobutyl Ketone	36
d. With 8-Hydroxyquinaldine	36
e. With Tri-n-Octylphosphine Oxide	37
3. Other Methods	37
a. By Electrolysis	37
b. By Ion Exchange	37
IX. Detection and Isolation	38
A. Colorimetric Methods	38
1. p-Nitrophenylazoorcinol (Zenia or p-Nitrobenzeneazoorcinol)	38
2. Quinalizarin (1,2,5,8-Tetrahydroxyanthraquinone)	38
3. Quinizarin (1,4-Dihydroxyanthraquinone)	39
4. Morin (A Tetrahydroxy Flavonol)	39
B. Precipitation Methods	39
C. Emission Spectrography	39
D. Radiochemical Methods	39
X. Quantitative Determination	40
A. Gravimetric Methods	40
B. Volumetric Methods	42
C. Photometric Methods	44
1. Spectrophotometric	44
a. Acetylacetone	44
b. Alkanan or Naphthazarin	44
c. Aluminon	44
d. Berillon	45
e. Eriochrome Cyanine R (Also Called Solochrome Cyanine)	45
f. Naphthachrome Azurine 2B (Also Known as Naphthachrome Green)	45
g. p-Nitrophenylazoorcinol (Zenia)	46
h. 2-Phenoxyquinizarin-3,4-Disulfonic Acid	46
i. Quinizarin-2-Sulfonic Acid (1,4-Dihydroxyanthraquinone-2-Sulfonic Acid)	46
j. Sulfosalicylic Acid	47

CONTENTS

k. Others	47
2. Ultraviolet Fluorescence	47
3. Spectrographic	48
D. Radiochemical Methods	50
XI. Recommended Laboratory Procedures	50
A. Metal Assay	50
1. Gravimetric	50
2. Volumetric	51
B. Determination of Oxygen in Beryllium Metal	53
C. Beryllium in Copper Alloys	58
D. Beryllium in Aluminum Alloys	59
E. Beryllium in Ores and Minerals	59
F. Beryllium in Air	60
References	60

Lead. By T. W. Gilbert, Jr.   69

I. Introduction	71
A. Occurrence	72
1. Production	72
2. Industrial Products	72
B. Extraction and Purification	73
C. The Toxicity of Lead and Its Compounds	74
II. Properties of Lead and Its Compounds	77
A. Physical Properties	77
B. Optical and Electrochemical Properties	77
1. Optical Properties	77
2. Electrochemical Properties	78
C. Chemical Properties of Lead	79
1. Properties of Lead Metal	79
2. Properties of Lead Compounds	81
a. Lead(II) Compounds	81
b. Lead(IV) Compounds	90
c. Mixed-Valence Lead Compounds	91
D. Isotopic Distribution of Lead	92
1. Mass Spectrographic Determinations	92
2. Optical Determinations	93
3. Geological Age Determinations	94
III. Separation and Isolation of Lead	96
A. Preliminary Treatment of Inorganic Materials	96
1. The Attack of Lead Ores and Minerals	96
2. Alloys	96
3. The Decomposition of Organic Samples	97
a. Dry Ashing	98
b. Wet Oxidation	98
B. Selective Separation Methods	99
IV. Detection and Identification of Lead	103
A. Spectrographic Detection	103
B. Chemical Methods	103
1. Reactions of Lead(II) Ion	105
2. Microscale Tests for Lead	106
a. The "Triple Nitrite" Crystal Reaction	106
b. The Potassium Iodide Crystal Reaction	106
3. Spot Tests for Lead	107
a. Tests with Benzidine	107
b. Tests with Sodium Rhodizonate	107
c. Test with p-Tetramethyldiaminodiphenylmethane	107
4. Special Tests for Lead	108
a. Detection of Lead in Bismuth Compounds	108

   b. Separation and Detection of Lead in Water.................. 108
   c. Detection of Lead Tetraethyl and Lead Tetraphenyl in Motor
    Fuels... 108
V. Determination.. 108
 A. Precipitation and Gravimetric Determination...................... 108
  1. As Lead Sulfate... 110
  2. As Lead Chromate.. 112
  3. As Lead Molybdate... 113
  4. By Electrodeposition.. 114
   a. Special Electrolytic Methods.................................. 116
  5. With Organic Precipitants... 117
  6. Other Methods.. 118
 B. Titrimetric Methods... 119
  1. Precipitation Titrations.. 119
   a. Titration with Molybdate..................................... 119
   b. Titration with Chromate...................................... 120
   c. Titration with Phosphoric Acid............................... 120
   d. Titration with Ferrocyanide................................... 121
   e. Titration with Alkali Hydroxide.............................. 121
   f. Titration with Fluoride.. 121
  2. Oxidation-Reduction Titrations.................................... 122
   a. Direct Oxidation to Lead(IV)................................. 122
   b. Determination of Lead(IV) in Lead(IV)–Lead(II) Mixtures... 122
   c. Indirect Determination of Lead by Means of Oxidation-Reduction Titrations.. 123
  3. Titrations Based on Complex Formation........................... 125
   a. Direct Titration.. 126
   b. Back-Titration... 126
   c. Replacement Titration... 127
   d. Alkalimetric Titration... 127
   e. Instrumental Methods of End-Point Detection............... 128
 C. Polarographic Methods.. 128
  1. The Dropping Mercury Electrode.................................. 129
  2. The Platinum Electrode.. 130
  3. Methods for Very Low Concentrations of Lead................... 130
  4. Amperometric Titrations... 132
 D. Photometric Methods.. 133
  1. Spectrophotometry and Colorimetry............................... 133
   a. With Dithizone.. 133
   b. With Tetramethyldiaminodiphenylmethane................. 139
   c. With 4-(2-Pyridylazo)Resorcinol (PAR-4).................... 139
   d. With Diethyldithiocarbamate................................. 139
   e. As Lead Chloride.. 140
  2. Flame Photometry.. 140
  3. Atomic Absorption Spectroscopy................................... 141
  4. Spectrographic Methods.. 141
  5. X-Ray Methods... 142
 E. Activation Analysis.. 143
VI. Determination of Lead in Specific Materials.............................. 143
 A. Rocks, Minerals, and Ores.. 143
 B. Alloys... 144
  1. Iron and Steel... 145
  2. Nonferrous Alloys.. 145
 C. Biological Materials... 147
 D. Air.. 147
 E. Foods... 148
 F. Water.. 148
 G. Petroleum Products... 149
 H. Miscellaneous.. 150
VII. Recommended Laboratory Procedures................................... 150
 A. Detection of Lead... 150

1. Tests for Macro Amounts of Lead............................. 150
   a. Precipitation as Lead Chloride............................ 150
   b. Precipitation as Lead Sulfide............................. 151
2. "Triple Nitrite" Crystal Reaction............................ 151
3. Spot Test with Benzidine..................................... 152
B. Determination of Lead... 152
   1. Gravimetric Determination as Lead Sulfate.................. 152
   2. Gravimetric Determination as Lead Chromate by Precipitation from Homogeneous Solution................................. 153
   3. Complexometric Titration with EDTA......................... 154
      a. In Basic Solution with Eriochrome Black T as Indicator.... 154
      b. In Acidic Solution with Xylenol Orange as Indicator....... 155
      c. In Acidic Solution with Copper–PAN as Indicator........... 156
   4. Electrogravimetric Determination as Lead Dioxide............ 157
   5. Photometric Determination with Dithizone................... 159
   6. Polarographic Determination in Copper-Base Alloys........... 161
   7. Amperometric Titration with Potassium Dichromate............ 163
References.. 164

Niobium and Tantalum. By *Silve Kallmann*........................ 177

I. Introduction... 183
   A. Occurrence... 184
      1. In Igneous Rocks... 184
      2. In Niobium and Tantalum Minerals......................... 185
      3. In By-Products... 187
   B. Industrial Uses of Niobium and Tantalum...................... 188
      1. Niobium.. 188
         a. In the Nuclear Industry............................... 188
         b. In Steels... 188
         c. In High-Temperature Alloys............................ 189
         d. Miscellaneous Uses.................................... 189
      2. Tantalum... 190
         a. As Pure Metal... 190
         b. In Miscellaneous Applications......................... 190
   C. Industrial Processes for the Preparation of Niobium and Tantalum Metals and Their Most Important Compounds.................. 191
      1. Extractive Metallurgy.................................... 191
      2. Preparation of the Pure Metals........................... 193
      3. Production of Ferroniobium and Ferrotantalum–Niobium...... 194
      4. Production of Miscellaneous Products..................... 194
   D. Toxicology and Industrial Hygiene............................ 194
      1. Niobium.. 194
      2. Tantalum... 195
II. Properties of Niobium and Tantalum............................. 195
   A. Physical and Mechanical Properties........................... 195
      1. Niobium.. 197
      2. Tantalum... 200
   B. Isotopes of Niobium and Tantalum............................. 200
   C. Electrochemical Properties................................... 201
      1. Tantalum... 201
      2. Niobium.. 201
   D. Optical Properties... 202
      1. Emission Spectra... 202
      2. X-Ray Spectra.. 202
   E. Chemical Properties.. 203
      1. Introduction... 203
      2. Oxides... 204
      3. Earth Acids.. 205

			4. Niobates and Tantalates	206
			5. Peroxy Salts	207
			6. Halogenoniobates and Tantalates	208
			a. Chlorides	208
			b. Bromides and Iodides	210
			c. Fluorides	210
			7. Refractory Binary Compounds	211
			a. Hydrides	211
			b. Nitrides	212
			c. Carbides	213
			d. Borides	213
			e. Silicides	213

III. Sampling of Niobium and Tantalum … 213
  A. Ores and Concentrates … 213
  B. Intermediate Products … 214
  C. Niobium and Tantalum Metals … 215
    1. Powders … 215
    2. Consolidated Metal … 215

IV. Separation and Isolation of Niobium and Tantalum from Other Elements … 215
  A. Decomposition and Other Preliminary Treatment of High-Grade Ores and Concentrates … 216
    1. General Observations … 216
    2. The Pyrosulfate Attack … 217
       a. The Fusion … 217
       b. Treatment of the Pyrosulfate Melt … 218
    3. Attack by Fusion with Potassium or Sodium Hydroxide and Sodium Peroxide … 219
    4. Attack by Borax … 221
    5. Attack by Hydrofluoric Acid … 222
    6. Attack by Chlorination … 223
  B. Attack of Low-Grade Minerals … 224
  C. Decomposition of Intermediate Products … 225
  D. Treatment of Niobium and Tantalum Oxides … 226
    1. Pyrosulfate Attack … 226
    2. Alkaline Attack … 227
    3. Chlorination of Oxides … 228
    4. Attack by Hydrofluoric Acid … 228
  E. Niobium and Tantalum Metals … 228
  F. Alloys … 229
    1. Ferro Alloys … 229
    2. Steels and High-Temperature Alloys … 229
       a. Isolation of Niobium and/or Tantalum Oxide … 229
       b. Subsequent Treatment of Oxides … 231
    3. Titanium-Base Alloys … 232
    4. Uranium- and Plutonium-Base Alloys … 232
    5. Miscellaneous Alloys … 233
  G. Miscellaneous Precipitation Procedures … 234
    1. Hydrolysis … 234
    2. Precipitation with Cupferron … 235
    3. Precipitation with Tannin … 236
    4. Phenylarsonic Acid Procedure … 239
    5. Other Precipitation Procedures … 239
  H. Separations Based on Ion Exchange … 241
    1. General Observations … 241
    2. The Hydrochloric Acid System … 242
    3. The Hydrofluoric Acid Medium … 244
    4. The Mixed Hydrofluoric–Hydrochloric Acid Medium … 245
       a. Introductory Observations … 245
       b. Chemistry of the Exchange Mechanism Involving the HF–HCl Medium … 246
       c. Separation Principles … 250

d. Practical Applications	252
5. The Hydrochloric–Oxalic Acid System	255
I. Separations Based on Cellulose and Paper Chromatography	258
1. Principles Involved	258
2. Preparation of the Sample	259
3. Typical Procedure	259
4. Shortened Procedure	260
5. Paper Chromatographic Methods	261
6. Separation by Electrophoresis	263
J. Separations Based on Solvent Extraction	263
1. General Observations	263
2. The Hydrochloric Acid System	264
3. The Fluoride Medium	265
a. Extractions Involving Hexone	265
b. Extractions Involving Other Solvents	268
4. Other Miscellaneous Extraction Procedures	269
V. Separation of Niobium and Tantalum	269
A. Precipitation Methods	270
1. The Tannin Procedure	270
a. General Considerations	270
b. The Schoeller Procedure	270
c. Modification of Schoeller's Procedure	272
2. The N-Benzoyl-N-Phenylhydroxylamine Procedure	272
3. The Phenylarsonic Acid Procedure	274
4. Separation of Niobium and Tantalum with Selenous Acid	274
5. Other Gravimetric Separation Procedures	274
B. Separation of Niobium and Tantalum by Ion Exchange	276
C. Separation of Niobium and Tantalum by Cellulose Chromatography	276
D. Separation of Niobium and Tantalum by Solvent Extraction	277
VI. Detection and Identification	277
A. Spectroscopic Examination	277
1. Niobium	277
a. Interference with Arc and Spark Lines	278
2. Tantalum	278
a. Interference with Arc and Spark Lines	279
3. Limit of Detection	280
B. Detection by X-Ray Emission	280
C. Detection by Chemical Procedures	280
1. Joint Detection in Minerals	281
a. Tartaric Hydrolysis	281
b. Tannin Procedure	281
2. Detection of Niobium and Tantalum in Minerals by a Paper Chromatographic Method	282
3. Detection of Niobium with Methylthymol Blue	282
4. Other Detection Methods	283
VII. Determination of Niobium and Tantalum	283
A. Gravimetric Methods	284
1. Cupferron Precipitation	284
2. Tannic Acid Precipitation	284
3. Precipitation by Hydrolysis	285
4. Precipitation with Phenylarsonic Acid	285
5. Other Gravimetric Methods	285
B. Photometric Methods	286
1. Thiocyanate Procedure for Niobium	286
a. Extraction Method	286
b. Homogeneous System	291
2. Hydrogen Peroxide Method for Niobium	297
a. Procedure	298
b. Discussion	298
c. Interference by Other Elements	299
d. Application of Procedure	300

CONTENTS

    3. Hydroquinone Method for Niobium........................... 301
        a. Procedure... 301
        b. Discussion.. 302
        c. Interferences... 303
        d. Application of Procedure.............................. 304
    4. 8-Quinolinol Method for Niobium............................ 305
    5. Molybdenum Blue Method for Niobium......................... 306
    6. Pyrogallol Method for Niobium.............................. 307
        a. Procedure... 308
        b. Discussion.. 309
        c. Interferences... 309
    7. Other Photometric Methods for Niobium...................... 309
    8. Photometric Determination of Tantalum by Pyrogallol........ 310
        a. Procedure... 311
        b. Discussion of Various Conditions...................... 311
        c. Interfering Elements.................................. 313
    9. Other Photometric Tantalum Methods......................... 315
C. Titrimetric Determination of Niobium............................ 317
    1. General Observations....................................... 317
    2. Procedure Based on Reduction in Sulfuric Acid Medium....... 317
        a. Preparation of Solution............................... 317
        b. Reduction and Titration............................... 318
    3. Procedure Based on Reduction in Hydrofluoric Acid Medium... 318
D. Polarographic Determination of Niobium.......................... 320
    1. Nitrate System... 320
    2. Sulfuric Acid System....................................... 320
    3. Hydrochloric Acid System................................... 321
    4. Citrate System... 322
    5. EDTA System.. 323
    6. Other Systems.. 324
E. Spectrographic Determination of Niobium and Tantalum............ 324
    1. High-Grade Ores.. 324
        a. No Preliminary Separations............................ 324
        b. After Preliminary Separations......................... 325
    2. Low-Grade Ores and Residues................................ 327
        a. No Preliminary Separations............................ 327
        b. After Preliminary Separations......................... 328
    3. Tantalum in Niobium and Niobium in Tantalum Metals or Oxides 329
    4. Alloys... 329
        a. Semiquantitative Analysis............................. 329
        b. No Preliminary Separations............................ 329
        c. After Preliminary Separations......................... 330
F. Determination of Niobium and Tantalum by X-Ray Fluorescence.... 331
    1. Instrumental Considerations................................ 331
    2. Effect of Other Elements................................... 331
    3. Effect of Physical Characteristics of the Sample........... 332
    4. Elimination of Interferences............................... 332
        a. Use of Correction Factors............................. 333
        b. Preliminary Fusion of Sample.......................... 333
        c. Other Miscellaneous Applications...................... 334
G. Electron Probe Measurements of Niobium Compounds............... 335
H. Radiochemical Methods... 336
    1. Use of Radioactive Tracers................................. 336
    2. Isotope Dilution Procedures................................ 336
    3. Neutron-Activation Analysis of Niobium and Tantalum........ 336
        a. Determination of Niobium.............................. 337
        b. Determination of Tantalum............................. 338
    4. Radiochemical Separations Involving Niobium-95............. 343
        a. Niobium Activity in Aqueous or Organic Solutions...... 343
        b. In Fission Products of Plutonium Fuels................ 344
        c. In Uranyl Nitrate Solution from Irradiated Uranium.... 345

d. In Uranium Target	346
e. After Cupferron Extraction	346
f. After Ion-Exchange Separations	347
5. Other Radiochemical Procedures	347
VIII. Analysis of Niobium and Tantalum Metals, Alloys, and Compounds	349
A. Determination of Interstitial Elements in Niobium and Tantalum	349
1. Oxygen	349
a. By Neutron Activation	349
b. By Halogenation	350
c. By Vacuum Fusion	351
d. By Inert-Gas Fusion	351
e. By Diffusion Extraction	352
f. By Emission Spectrometry	352
g. By Mass Spectrometry	353
h. By Isotopic Dilution	353
2. Hydrogen	353
a. Vacuum-Fusion Methods	353
b. Diffusion Method	354
c. Spectrographic Methods	354
3. Nitrogen	355
a. Vacuum-Fusion Method	355
b. Inert Gas-Fusion Procedure	355
c. Kjeldahl Procedure	356
d. Alkali-Fusion Method	356
4. Carbon	357
B. Determination of Metallic Impurities in Niobium and Tantalum	357
1. Spectrographic Procedures	357
a. General Considerations	357
b. Outline of Several Techniques	358
c. Other Procedures	360
2. X-Ray Fluorescence Procedures	361
3. Radioactivation Analysis for Trace Elements in Niobium	362
a. Nondestructive Technique	363
b. After Radiochemical Separations	365
c. By Differential-Count Method	365
4. Determination of Impurities in Niobium and Tantalum by Chemical Methods	367
a. Determination of Tantalum in Niobium	367
b. Determination of Niobium in Tantalum	368
c. Determination of Iron	368
d. Determination of Tungsten and Molybdenum	369
e. Determination of Silicon and Phosphorus	371
f. Determination of Zirconium	372
g. Determination of Titanium	372
h. Determination of Cobalt	373
i. Determination of Nickel	373
j. Determination of Copper	373
k. Determination of Boron	373
l. Simultaneous Determination of Lead, Copper, Zinc, and Cadmium	374
m. Determination of Tin and Antimony	374
n. Determination of Lead and Cadmium	375
o. Determination of Sodium and Potassium	375
p. Determination of Arsenic	375
IX. Selected Laboratory Procedures	376
A. Determination of Niobium and Tantalum in Ores and Concentrates Involving Ion Exchange	376
1. Apparatus and Reagents	376
2. Procedure	377
a. Decomposition of Sample	377
b. Elution of Impurities	378

xviii   CONTENTS

        c. Elution of Niobium.................................... 379
        d. Elution of Tantalum................................... 379
        e. Determination of Niobium............................. 379
        f. Determination of Tantalum............................ 380
        g. Regeneration of Resin................................ 380
    B. Determination of Niobium and Tantalum in Steels and High-Tempera-
        ture Alloys... 380
    C. Determination of Niobium and Tantalum in Niobium-, Tantalum-,
        Titanium-, Tungsten-, or Molybdenum-Base Alloys............. 381
    D. Spectrographic Determination of Impurities in Tantalum and Niobium   381
      1. Tantalum... 382
        a. Preparation of High-Purity Tantalum Oxide............. 382
        b. Preparation of Synthetic Standards..................... 383
        c. Preparation of Carrier—Internal Standard.............. 383
        d. Preparation of Samples................................ 384
        e. Spectrographic Conditions............................. 386
      2. Niobium.. 387
        a. Preparation of Synthetic Standards..................... 387
        b. Preparation of Carrier and of Samples.................. 389
        c. Spectrographic Conditions............................. 389
References... 389

Technetium.   By *James W. Cobble*............................... 407

   I. Introduction... 408
  II. Properties... 409
    A. Physical Properties...................................... 409
    B. Nuclear Properties...................................... 411
    C. Optical Properties...................................... 412
    D. Chemical Properties..................................... 413
    E. Electrochemical Properties.............................. 417
 III. Separation and Isolation................................... 418
    A. General Procedures..................................... 418
    B. Separation Methods..................................... 418
      1. Precipitation... 418
      2. Electroplating....................................... 419
      3. Solvent-Extraction Behavior.......................... 420
      4. Ion Exchange.. 421
      5. Selective Volatility.................................. 421
 IV. Detection and Identification............................... 421
    A. Radioactivity... 423
    B. Emission Spectroscopy.................................. 423
    C. Spot Tests... 423
  V. Toxicology and Industrial Hygiene.......................... 423
 VI. Determination of Technetium............................... 424
    A. Gravimetric Determinations............................. 424
      1. As Metallic Technetium............................... 424
      2. As Technetium Dioxide............................... 426
      3. As Tetraphenylarsonium Pertechnetate................. 426
      4. As Nitron Pertechnetate.............................. 426
      5. As Technetium Sulfides............................... 427
      6. As Ammonium Pertechnetate......................... 427
    B. Titrimetric Determinations............................. 427
      1. Neutralization Reactions............................. 427
      2. Redox Reactions..................................... 427
    C. Polarographic Methods................................. 428
    D. Spectrophotometric Methods............................ 428
    E. Radiochemical Methods................................. 430
 VII. Recommended Laboratory Procedures....................... 431

A. Gravimetric Methods.. 431
   1. Nitron Pertechnetate ($C_{20}H_{16}N_3 \cdot HTcO_4$)........................ 431
   2. Tetraphenylarsonium Pertechnetate (($C_6H_5)_4AsTcO_4$)........... 431
B. Spectrophotometric Methods... 431
   1. Ascorbic Acid Reduction and Tc(V) Thiocyanate Complex....... 431
   2. Thioglycolic Acid–Tc(VII) Complex............................... 432
References... 432

Actinium, Astatine, Francium, Polonium, and Protactinium. By *Jacob Sedlet*.. **435**

I. Introduction.. 439
  A. Radioactivity of the Elements...................................... 439
  B. Occurrence and Sources... 441
    1. Occurrence in Nature... 441
    2. Sources of the Elements... 445
  C. Industrial and Research Uses...................................... 447
  D. Toxicology and Industrial Hygiene................................ 447
    1. Radioactivity Hazards.. 447
    2. Maximum Permissible Radiation Exposures................. 448
    3. Safe Handling of Radioactive Materials...................... 449
II. Actinium (Atomic Number 89)... 450
  A. Introduction... 450
  B. Properties.. 451
    1. Physical Properties.. 451
      a. Nuclear Properties... 451
      b. Other Physical Properties................................... 451
    2. Electrochemical Properties...................................... 451
    3. Optical Properties.. 452
    4. Chemical Properties.. 453
      a. Tracer Chemistry... 454
  C. Separation and Isolation.. 456
    1. Dissolution of Samples.. 456
      a. Uranium Ores and Ore-Processing Residues............... 456
      b. Biological Samples... 456
      c. Radium, Uranium, and Thorium............................ 456
    2. Separation... 457
      a. Introduction.. 457
      b. Coprecipitation and Precipitation......................... 457
        (1) Fluorides... 457
        (2) Oxalates... 458
        (3) Hydroxides.. 458
        (4) Sulfates... 459
        (5) Precipitation of Other Elements from Actinium...... 459
      c. Solvent Extraction... 459
        (1) 2-Thenoyltrifluoroacetone (TTA)..................... 459
        (2) Organic Phosphates.................................... 461
        (3) Alcohol.. 462
      d. Ion-Exchange Methods....................................... 463
        (1) Cation-Exchange Systems............................ 463
        (2) Anion-Exchange Systems............................. 465
      e. Paper-Chromatographic Separations...................... 465
  D. Detection and Identification...................................... 466
    1. Emission Spectroscopy... 466
    2. Radiochemical Methods.. 466
      a. Growth of Decay Products into Purified Actinium-227... 467
      b. Separation and Counting of Decay Products.............. 468
      c. Gamma-Ray Counting.. 470
  E. Determination.. 472

  1. Introduction.. 472
  2. Determination by Separating Actinium........................ 472
   a. Separations... 472
   b. Counting... 474
  3. Determination by Separating Actinium Decay Products......... 477
   a. Thorium-227... 478
   b. Active Deposit... 479
   c. Francium-223.. 480
  4. Other Methods... 480
 F. Determination in Specific Materials............................... 481
  1. Radioactive Ores and Residues................................ 481
  2. Irradiated Radium... 482
  3. Biological Samples... 482
 G. Recommended Procedures... 483
  1. Extraction with 2-Thenoyltrifluoroacetone..................... 483
  2. Separation and Counting of Francium-223..................... 483
  3. Extraction with Di(2-Ethylhexyl) Phosphoric Acid.............. 485
III. Astatine (Atomic Number 85).. 487
 A. Introduction.. 487
 B. Properties... 487
  1. Physical Properties... 487
  2. Chemical Properties.. 489
   a. Oxidation States... 489
   b. Volatility.. 491
   c. Solvent Extraction....................................... 492
 C. Separation and Isolation.. 492
  1. Separation from Irradiated Targets............................ 492
   a. Distillation... 492
   b. Chemical Separations.................................... 493
  2. Coprecipitation.. 495
  3. Solvent Extraction... 496
  4. Separation from Biological Materials.......................... 497
 D. Detection... 497
 E. Determination... 498
 F. Recommended Procedures... 500
  1. Determination in Solution or in Precipitates................... 500
  2. Determination by Coprecipitation with Silver or Palladium...... 501
  3. Determination in Biological Materials by Coprecipitation with Tellurium... 501
IV. Francium (Atomic Number 87).. 501
 A. Introduction.. 501
 B. Properties... 503
  1. Physical Properties... 503
  2. Chemical Properties.. 503
   a. Coprecipitation.. 504
    (1) Perchlorates....................................... 504
    (2) Picrates... 504
    (3) Tartrates.. 504
    (4) Chloroplatinates................................... 504
    (5) Other Chloroanions................................ 505
    (6) Cesium Cobaltinitrite and Iodate.................... 505
    (7) Cesium Silicotungstate............................. 505
    (8) Heteropolyacids.................................... 505
   b. Volatility.. 506
 C. Separation and Isolation.. 507
  1. Coprecipitation Methods...................................... 507
  2. Chromatographic Methods.................................... 509
  3. Ion-Exchange Resin Separations............................... 510
  4. Solvent Extraction... 511
  5. Other Separations.. 511
 D. Detection... 512

E. Determination.. 513
F. Separation and Determination in Specific Materials.............. 516
   1. Actinium... 516
   2. Irradiated Thorium and Uranium............................. 517
G. Recommended Procedures... 517
   1. Isolation of Francium from Thorium Targets with Silicotungstic Acid.. 517
   2. Separation of Francium-221 with Silicotungstic Acid........ 519
   3. Separation of Francium-223 from Actinium-227 by Precipitation of Other Elements... 520
V. Polonium (Atomic Number 84)..................................... 520
   A. Introduction.. 520
   B. Properties.. 522
      1. Physical Properties..................................... 522
         a. Nuclear Properties................................... 522
         b. Other Physical Properties............................ 524
      2. Electrochemical Properties.............................. 525
      3. Optical Properties...................................... 526
      4. Chemical Properties..................................... 527
         a. Polonium Metal....................................... 527
         b. Properties of the Oxidation States................... 527
         c. Tracer Chemistry..................................... 530
   C. Separation and Isolation.................................... 531
      1. Dissolution of Samples.................................. 531
      2. Separation.. 532
         a. Coprecipitation...................................... 532
            (1) Tellurium.. 532
            (2) Lead Tellurate................................... 533
            (3) Selenium... 533
            (7) Manganese Dioxide................................ 533
            (5) Hydroxides....................................... 533
            (6) Other Carriers................................... 533
         b. Spontaneous Deposition and Electrodeposition......... 534
            (1) Spontaneous Deposition........................... 534
            (2) Electrodeposition................................ 536
         c. Solvent Extraction................................... 536
            (1) Alcohols... 536
            (2) Chelating Agents................................. 536
            (3) Ethers and Ketones............................... 538
            (4) Tributylphosphate................................ 538
            (5) Tri-n-Benzylamine.............................. 539
         d. Ion-Exchange Resins.................................. 539
         e. Volatility Separations............................... 541
         f. Chromatographic Separations.......................... 542
   D. Detection and Identification................................ 543
      1. Coprecipitation... 544
      2. Electrodeposition....................................... 544
      3. Alpha-Particle Counting................................. 544
      4. Gamma- and X-Ray Counting............................... 545
      5. Detection of Milligram Amounts of Polonium.............. 547
   E. Determination... 547
      1. Preparation of Sample for Measurement................... 547
      2. Counting Techniques..................................... 548
      3. Miscellaneous Methods................................... 549
   F. Determination in Specific Materials......................... 549
      1. Biological Materials.................................... 549
      2. Irradiated Bismuth and Other Heavy-Metal Targets........ 550
      3. Ores, Water, and Air.................................... 551
      4. Analysis of Curie Amounts of Polonium Compounds......... 552
   G. Recommended Procedures...................................... 553
      1. Deposition on Silver Foil............................... 553

CONTENTS

2. Determination of Polonium-210 in Uranium Ores	553
3. Determination of Polonium in Irradiated Bismuth by Solvent Extraction	554
4. Separation of Pure Polonium from Radium-DEF Mixtures	554
5. Paper-Chromatographic Separation of Polonium, Bismuth, Tellurium, and Selenium	554
6. Determination of Polonium in Urine and Tissue	555
VI. Protactinium (Atomic Number 91)	555
A. Introduction	555
B. Properties	557
1. Physical Properties	557
a. Nuclear Properties	557
b. Other Physical Properties	557
2. Electrochemical Properties	557
a. Spontaneous Deposition	557
b. Electrodeposition	559
c. Electrode Potential	559
d. Polarographic Behavior	559
3. Optical Properties	559
a. Emission Spectrum	559
b. Absorption Spectrum	560
4. Chemical Properties	561
a. Protactinium(V)	561
b. Protactinium(IV)	563
C. Separation and Isolation	564
1. Dissolution of Samples	564
2. Separation	566
a. Coprecipitation	566
b. Solvent Extraction	571
c. Ion-Exchange Separations	578
(1) Anion-Exchange Resins	578
(2) Cation-Exchange Resins	580
d. Chromatographic Separations	580
D. Detection and Identification	581
E. Determination	585
F. Determination in Specific Materials	587
1. Ores and Ore-Processing Residues	587
2. Irradiated Thorium	591
3. Geological Samples	593
G. Recommended Procedures	594
1. Determination of Protactinium in Uranium Ore Residues by Alpha Pulse-Height Analysis	594
a. Alpha Pulse-Height Analysis Following Chemical Separation	594
b. Alpha Pulse-Height Analysis Without Chemical Separation	595
2. Determination of Protactinium-231 in Ore Residues by Gamma-Ray Spectrometry	596
3. Determination of Protactinium-231 by Differential Gamma-Ray Spectrometry	597
4. Determination of Protactinium-233 in Neutron-Irradiated Thorium	598
5. Preparation of Carrier-Free Protactinium-233	598
References	599
General References	608
Subject Index for Volume 6	**611**

Part II
Section A

BERYLLIUM

By B. R. F. Kjellgren, C. W. Schwenzfeier, Jr., and E. Stanley Melick, *The Brush Beryllium Company, Elmore, Ohio*

Contents

I. Introduction	3
A. History	3
B. Occurrence	4
1. Minerals and Ores	4
II. Industrial Processes	5
A. Extraction as Beryllium Hydroxide	5
B. Sulfate Extraction	5
C. The Complex Fluoride Process	7
D. Production of Beryllium	8
1. Pyro-Reduction	8
2. Electrolytic Reduction	9
3. Vacuum Melting and Powder Metallurgy	9
4. Miscellaneous	10
5. Production of Beryllium Copper	10
III. Toxicology	10
IV. Industrial Products and Uses	11
A. Pure Beryllium Metal	11
1. Nuclear	11
2. Structural	11
B. Alloys	12
1. Copper-Base Alloys	12
2. Other Alloys	13
V. Beryllium Compounds	13
A. Beryllium Sulfate ($BeSO_4 \cdot 4H_2O$)	13
B. Sodium Beryllium Fluoride (Na_2BeF_4 or $2NaF \cdot BeF_2$)	14
C. Beryllium Hydroxide ($Be(OH)_2$)	14
D. Ammonium Beryllium Fluoride (($NH_4)_2BeF_4$ or $2NH_4F \cdot BeF_2$)	15
E. Beryllium Fluoride (BeF_2)	16
F. Beryllium Chloride ($BeCl_2$)	16
G. Beryllium Iodide (BeI_2)	16
H. Beryllium Oxide (BeO)	16
I. Beryllium Nitrate ($Be(NO_3)_2 \cdot 3H_2O$)	17
J. Beryllium Nitride (Be_3N_2)	17
K. Beryllium Carbide (Be_2C)	18
L. Beryllium Carbonate—Basic	18
M. Basic Beryllium Acetate ($Be_4O(C_2H_3O_2)_6$)	18
N. Beryllium Phosphate ($Be_3(PO_4)_2$)	18
O. Beryllium Hydride (BeH_2)	19
P. Intermetallic Compounds—Beryllides	19

Contents (*continued*)

VI. Properties	19
A. Physical and Mechanical Properties	19
B. Electrochemical Properties	19
C. Chemical Properties	21
D. Optical Properties	23
E. Radiochemical–Nuclear properties	24
VII. Sampling	25
A. Ore	25
B. Copper-Base Alloys	26
1. Cast Alloys	26
2. Wrought Alloys	26
C. Aluminum-Base Alloys	27
D. Beryllium Metal	27
VIII. Separation and Isolation of Beryllium	28
A. Sample Dissolution	28
1. Ores	28
2. Copper-Base Alloys	29
3. Aluminum-Base Alloys	30
4. Beryllium Metal	30
B. Separation Methods	31
1. Precipitation Methods	32
a. With Ammonia (Initial Separation with Possible Subsequent Purifications)	33
b. With 8-Hydroxyquinoline	33
c. With Cupferron (Ammonia Salt of Nitrosophenylhydroxylamine)	34
d. With *n*-Benzoylphenylhydroxylamine	34
e. Miscellaneous	34
2. Extraction Methods	35
a. With Acetylacetonate	35
b. With Butyrate	36
c. With Methyl Isobutyl Ketone	36
d. With 8-Hydroxyquinaldine	36
e. With Tri-*n*-Octylphosphine Oxide	37
3. Other Methods	37
a. By Electrolysis	37
b. By Ion Exchange	37
IX. Detection and Isolation	38
A. Colorimetric Methods	38
1. *p*-Nitrophenylazoorcinol (Zenia or *p*-Nitrobenzeneazoorcinol)	38
2. Quinalizarin (1,2,5,8-Tetrahydroxyanthraquinone)	38
3. Quinizarin (1,4-Dihydroxyanthraquinone)	39
4. Morin (A Tetrahydroxy Flavinol)	39
B. Precipitation Methods	39
C. Emission Spectrography	39
D. Radiochemical Methods	39
X. Quantitative Determination	40
A. Gravimetric Methods	40
B. Volumetric Methods	42
C. Photometric Methods	44
1. Spectrophotometric	44
a. Acetylacetone	44

Contents (*continued*)

	b. Alkanan or Naphthazarin	44
	c. Aluminon	44
	d. Berillon	45
	e. Eriochrome Cyanine R (Also Called Solochrome Cyanine)	45
	f. Naphthachrome Azurine 2B (Also Known as Naphthachrome Green)	45
	g. *p*-Nitrophenylazoorcinol (Zenia)	46
	h. 2-Phenoxyquinizarin-3,4-Disulfonic Acid	46
	i. Quinizarin-2-Sulfonic Acid (1,4-Dihydroxyanthraquinone-2-Sulfonic Acid)	46
	j. Sulfosalicylic Acid	47
	k. Others	47
2.	Ultraviolet Fluorescence	47
3.	Spectrographic	48
D.	Radiochemical Methods	50
XI. Recommended Laboratory Procedure		50
A. Metal Assay		50
1.	Gravimetric	50
2.	Volumetric	51
B. Determination of Oxygen in Beryllium Metal		53
C. Beryllium in Copper Alloys		58
D. Beryllium in Aluminum Alloys		59
E. Beryllium in Ores and Minerals		59
F. Beryllium in Air		60
References		60

I. INTRODUCTION

A. HISTORY

Beryllium was discovered in 1797 by Louis Nicholas Vauquelin (109). He noted the formation of a precipitate as the result of boiling a solution of beryl in potassium hydroxide. Thus, beryllium was first isolated as the hydroxide. This difference in chemical behavior between aluminum and beryllium (beryllium precipitates, whereas aluminum is retained in solution as the aluminate) is utilized today in the sulfate processing of beryl as a means of separation of a major portion of the aluminum from the beryllium.

Nearly 30 years elapsed before the element was first separated in metallic form. A practical reduction technique was not developed until nearly 1900, when the Frenchman Lebeau (123) reduced beryllium oxide with carbon in the presence of metallic copper in an electric-arc furnace. The Lebeau reaction is still the basis for commercial production of beryllium–copper master alloy.

The first commercially feasible process for the production of pure beryllium was developed in Germany in the early 1920's. Beryllium oxide was converted to a fluoride and the metallic beryllium was separated by electrodeposition in the form of flake from a high-temperature, mixed, fused sodium and beryllium fluoride bath (87). In the 1930's a method was developed for electrodeposition of beryllium from a sodium chloride–beryllium chloride bath at a temperature of approximately 350°C. The flake beryllium produced is generally contaminated with its own halide, as well as that of the solvent halide, and it is recovered by washing the soluble halides from the flake beryllium. A description of the electrolysis of beryllium chloride has been presented by Windecker (199). Other experimental electrolytic methods for separation of beryllium are described by Magel (130).

A process for the pyro reduction of beryllium fluoride with metallic magnesium as the reducing agent was developed in the 1930's by B. R. F. Kjellgren (107). This is the only process yet discovered that is satisfactory for large-scale production of pure metallic beryllium.

B. OCCURRENCE

1. Minerals and Ores

Beryllium minerals are found quite widely distributed over the surface of the earth, but may constitute only an average of 4 to 6 p.p.m. in exposed rock (74). The most common beryllium mineral, and the only one of commercial importance at the present time, is beryl, beryllium aluminum silicate ($3BeO \cdot Al_2O_3 \cdot 6SiO_2$) (44), of which the aquamarine and the emerald are gem-stone varieties highly valued as early as 1650 B.C. (73). New geological discoveries in 1959 indicate that the beryllium silicates, phenakite (Be_2SiO_4) and bertrandite ($Be_2(BeOH)_2Si_2O_7$), may soon become of industrial significance. A great many other beryllium-bearing minerals are known, but their occurrence is so rare that they are not of commercial importance.

All beryl deposits of commercial importance are found in pegmatite dikes, with the main constituent generally being feldspar. Pegmatite dikes containing beryl in quantities sufficient to justify mining will rarely exceed 0.75% beryllium oxide content. This beryl must be cobbed to produce a mineral with a minimum of 10% beryllium oxide to be of value commercially. The average beryllium oxide content of commercial beryl ores is generally found to be in the range of 10.5 to 12.5% (4).

In recent years, large quantities of beryl have been found in white granites in the Sheeprock Mountain Range in Utah. The mineral also has been found in quartz veins and in wolframite veins in Arizona. These latter two types of occurrences, although not of commercial importance at the present, may well become so as a result of increasing alertness for the detection of beryllium in the course of general mining exploration and analysis. Such alertness has recently resulted in the discovery of substantial quantities of the minerals phenakite and bertrandite in fluorspar deposits in Colorado, Nevada, and Mexico. Because of the very large quantities of beryllium present in such deposits and the relatively high content, in some cases as high as 0.5%, as beryllium oxide, it is expected that this type of beryllium occurrence may well become of major industrial significance in the near future.

Beryllium Resources, Inc., and United Technical Industries are the first to announce (33) processes for recovery of beryllium as mineral and oxide, respectively, from ores in the Topaz and Spors Mountains of Utah.

Leading known world producers of beryl, other than the United States, are Brazil, Argentina, South Africa, India, Southern Rhodesia, Australia, and Mozambique. Madagascar, Portugal, and Morocco produce appreciable amounts. The United States is about fourth in tonnage production.

Summaries listing minerals, their location, exploration, and ore-dressing processes have been presented (44,54,73,74,135).

II. INDUSTRIAL PROCESSES

A. EXTRACTION AS BERYLLIUM HYDROXIDE

Although only one commercially practical process has been developed for production of pure beryllium metal, a number of processes have been developed and used commercially for extraction of beryllium from beryl and for production of beryllium oxide of purity suitable for direct use in ceramics, as input material for beryllium–copper alloy production, or as input material for production of pure beryllium metal (17,45,109).

Only two of these many processes, however, are currently in use on a commercially significant scale: (*1*) extraction of beryllium as sulfate (171) with subsequent precipitation as beryllium hydroxide; and (*2*) extraction as sodium beryllium fluoride (45,100) with subsequent precipitation as beryllium hydroxide.

B. SULFATE EXTRACTION

This process, described in detail by Schwenzfeier (171), involves melting of the beryl, quenching in water, heat treatment of the quenched ore, and ultimate grinding to a −200 mesh size. This finely ground ore is then heated with concentrated sulfuric acid and the sulfated ore is leached with water. Under properly controlled conditions it is possible to recover 90 to 95% of the beryllium content of the ore. The water leach of the sulfated ore becomes a slurry, since approximately 67% of the ore is insoluble silica. The chemical reaction may be represented as follows:

$$3BeO \cdot Al_2O_3 \cdot 6SiO_2 + 6H_2SO_4 \rightarrow 3BeSO_4 + Al_2(SO_4)_3 + 6SiO_2 + 6H_2O \quad (1)$$

Appreciable amounts of iron, chromium, nickel, magnesium, manganese, and lithium sulfates are also present, as well as some free sulfuric acid.

Ammonium hydroxide equivalent to the aluminum content of the filtered solution is continuously added, the solution is cooled, and the ammonium alum is removed by centrifugation.

$$2NH_4OH + H_2SO_4 \rightarrow (NH_4)_2SO_4 + H_2O \quad (2)$$

$$(NH_4)_2SO_4 + Al_2(SO_4)_3 + 24H_2O \rightarrow (NH_4)_2SO_4 \cdot Al_2(SO_4)_3 \cdot 24H_2O \quad (3)$$

After this removal of the major amount of aluminum, the pregnant solution is treated with a suitable chelating agent, such as EDTA, and dilute sodium hydroxide to an excess of approximately $1.5N$. This solution is maintained at a relatively cool temperature to permit the initial gelatinous hydroxides to be redissolved in the excess of sodium hydroxide and to convert the aluminum and beryllium to the corresponding aluminate and beryllate solutions. The sodium beryllate solution is transferred continuously into a boiler in which the beryllium is precipitated as a granular hydroxide, which can be easily removed by continuous centrifugation and water-washed prior to discharge into suitable containers. The amount of expensive chelate is controlled in order to produce hydroxide of suitable quality for ultimate use in (*1*) high-purity beryllium oxide, (*2*) beryllium metal, or (*3*) beryllium copper. The conversion of beryllium sulfate to granular beryllium hydroxide may be illustrated by the following reactions:

$$BeSO_4 + 2NaOH \rightarrow Na_2SO_4 + Be(OH)_2 \text{ (gelatinous)} \quad (4)$$

$$Be(OH)_2 + 2NaOH \xrightarrow{(1.5N\ NaOH)} Na_2BeO_2 + 2H_2O \quad (5)$$

$$Na_2BeO_2 + 2H_2O \xrightarrow{\Delta} 2NaOH + Be(OH)_2 \text{ (crystalline)} \quad (6)$$

Beryllium may be extracted as the sulfate from a beryl ore–lime fusion (45,171). The finely ground fusion is sulfated and water-leached to remove the soluble sulfates from the insoluble calcium sulfates and SiO_2. Impurities are removed by fractional crystallization and the pH-controlled precipitation is followed by precipitation of the beryllium from the sulfate medium with ammonia gas.

C. THE COMPLEX FLUORIDE PROCESS

The original process developed by Copaux (40) consisted of fusing the ore with a fluorinating agent, sodium fluosilicate, at about 750°C.

$$3BeO \cdot Al_2O_3 \cdot 6SiO_2 + 6Na_2SiF_6 \rightarrow 9SiO_2 + 3Na_2BeF_4 + 2Na_3AlF_6 + 3SiF_4 \quad (7)$$

Sodium fluosilicate is an expensive fluorinating agent and the solution containing the water-soluble sodium beryllium fluoride is generally contaminated with very large amounts of silicon formed with excess fluosilicate. Beryllium hydroxide is precipitated from the sodium beryllium fluoride solution with sodium hydroxide, producing a waste product of sodium fluoride. Disposal of this dilute solution of sodium fluoride presented a waste-disposal problem and also a loss of valuable fluoride. Kawecki (100) precipitated sodium fluoride from the solution with ferric sulfate in the form of sodium ferric fluoride, and has used this fluorinating agent as a substitute for approximately 60% of the sodium fluosilicate in the original reaction. A combination of the reactions then actually takes place in this commercial process used to extract beryllium from beryl. The actual reaction is represented by two equations as follows:

$$60\% \quad 2Na_3FeF_6 + 3BeO \cdot Al_2O_3 \cdot 6SiO_2 \rightarrow 3Na_2BeF_4 + Fe_2O_3 + Al_2O_3 + 6SiO_2 \quad (8)$$

$$40\% \quad 3BeO \cdot Al_2O_3 \cdot 6SiO_2 + 2Na_2SiF_6 + Na_2CO_3 \rightarrow 3Na_2BeF_4 + 8SiO_2 + Al_2O_3 + CO_2 \quad (9)$$

The water-soluble sodium beryllium fluoride is then precipitated with sodium hydroxide, according to the following equation:

$$Na_2BeF_4 + 2NaOH \rightarrow Be(OH)_2 + 4NaF \quad (10)$$

In order to recover the sodium fluoride, Kawecki (100) has treated this with ferric sulfate:

$$12NaF + Fe_2(SO_3)_4 \rightarrow 2Na_3FeF_6 + 3Na_2SO_4 \quad (11)$$

Fewer chemical controls are required from the standpoint of product quality when using the fluoride process, as compared with the sulfate procedure, but it is necessary to handle more than ten times the volume of solution to obtain the same amount of beryllium. Beryllium solutions in the sulfate process contain the equivalent of about 35 g./liter of BeO, whereas those in the fluoride process contain only about 2.7 g. of BeO, which is the theoretical saturation point for a sodium beryllium fluoride solution.

Other proposed methods for fluorination are discussed by Darwin and Buddery (45) and Kawecki (100). Other methods of opening ore, which may be suitable in the laboratory but not on a commercial scale, such as fusion with alkali hydroxide, alkali carbonates, calcium and potassium fluoride, or ammonium fluoride, are discussed by Darwin and Buddery (45) and Mellor (135).

D. PRODUCTION OF BERYLLIUM

1. Pyro-Reduction

The commercial production of beryllium today depends upon the use of an anhydrous halide. Based on thermodynamic information, lithium, sodium, potassium, magnesium, and calcium will reduce beryllium halides (102,130). Considering the properties of the halides and the problems involved in their production, the fluoride offers the most promise for commercial application.

Most of the free world supply of beryllium metal has been made by the magnesium reduction of anhydrous beryllium fluoride (171).

Production of anhydrous beryllium fluoride may be accomplished by thermal decomposition of ammonium beryllium fluoride or by direct fluorination of beryllium oxide with ammonium bifluoride or anhydrous hydrogen fluoride. Of the various methods of producing ammonium beryllium fluoride the simplest is that proposed by Kjellgren (110). Beryllium hydroxide or oxide is dissolved in hydrofluoric acid and sufficient ammonia is added to produce the ammonium beryllium fluoride salt ($BeF_2 \cdot 2NH_4F$), which is crystallized from a purified solution. Anhydrous beryllium fluoride, in excess, and magnesium are added to an externally heated graphite crucible. The reacted melt contains water-soluble beryllium fluoride, coarse beryllium "pebble," and magnesium fluoride. On leaching with water, this melt disintegrates into fine, insoluble magnesium fluoride, lump or pebble beryllium of all sizes, and beryllium fluoride solution. This soluble beryllium re-enters the stream and is again crystallized as the high-purity ammonium beryllium fluoride to repeat the process. The complete procedure for production of nuclear purity has been explained in detail by Schwenzfeier (171,172).

Residual and trapped beryllium fluoride, magnesium fluoride, magnesium, and other low-boiling ingredients are removed from the pebbles by vacuum melting in a beryllia crucible (graphite susceptor) and casting in a graphite mold. Barrett et al. (15) suggest a molybdenum sheet susceptor and an alumina-dressed graphite or mild steel mold.

Kura et al. (120) were successful in reducing magnesium content to a trace by melting under a hygroscopic and toxic flux. Vacuum melting is recommended, however. Corzine and Kaufmann (41) report that beryllium melted in a new alumina crucible is contaminated by 1 to 2% aluminum. Molten beryllium also reacts slowly with zirconia.

2. Electrolytic Reduction

The metal also may be produced by electrolysis of beryllium chloride from a fused bath also containing sodium chloride and potassium chloride (199). Since halides of Li, Na, K, Mg, Ca, and Ba are more stable than beryllium halides, they may be used without interference in electrodeposition of beryllium (102). In early electrolytic work the fluorides received most attention (130). The melting point of beryllium is above the boiling point of its halides and electrolysis, therefore, produces flake, which requires purification to remove trapped halides of the bath. This is most simply accomplished by washing followed by vacuum melting. Barrett et al. (15) state that vacuum sublimation at 95°C. will reduce chlorine to a suitable level (200 p.p.m.). Many combinations of $BeCl_2$ and alkali chlorides have been tried (200), including molten eutectic $BeCl_2$–NaCl into a mercury cathode (86,130), with subsequent separation of the mercury from a very fine-particle beryllium—very susceptible to oxidation.

3. Vacuum Melting and Powder Metallurgy

Vacuum melting of the flake removes volatile constituents to a low level. Vacuum-cast beryllium has a large grain size, poor structural and mechanical properties, and cannot be machined to a fine surface with required tolerances. Powder metallurgy processes have been developed to circumvent these problems. Vacuum-cast metal, reduced to a −200 mesh powder of average particle size 30 to 35 μ, can be consolidated to nearly 100% theoretical density when subjected to pressures up to 700 p.s.i. in vacuum (50 to 300 μ absolute pressure) at about 1050°C. Vacuum hot-pressed blocks as heavy as 11,000 pounds have been successfully produced. Nearly all the metal marketed today is produced in this manner. Some direct warm extrusion of powder to small rods and tubes and forging of powder to shapes is also being practiced. Because of the high cost of the metal, partly due to the cost of recycling machine chips from the parts machined from massive beryllium, more and more effort is being directed to powder forging and other methods for producing rough blanks requiring a minimum of machining work. Sheet produced by warm rolling is also available.

Barrett et al. (15) have described a pressureless sintering procedure for beryllium powders. Much of the metal working is done under argon.

4. Miscellaneous

Kellogg (102) has discussed theoretical considerations in electrolytic reductions, thermal decomposition of BeI_2, and reduction of BeO and halides. Attempts to prepare pure beryllium by decomposition of BeI_2 on a hot wire have been unsuccessful because of a reaction of the BeI_2 with the container.

Wood and Brenner (201) have worked in pure argon or helium and successfully deposited thick, coherent beryllium by electrolysis from organic salts. Other attempts at electrolytic deposition from nonaqueous solutions have been reported (26,53,79,101,202).

Because of a considerable interest in the properties of "pure" beryllium, various methods of additional purification are currently under investigation: (1) vacuum distillation and condensation on hot surfaces (175) to reduce Fe, Si, and Al from 0.65, 0.3, and 0.1%, respectively, to less than 0.012% by two distillations; and (2) zone melting.

5. Production of Beryllium Copper

Production of beryllium copper, the major beryllium consumer, depends upon the reduction of beryllium oxide with carbon in an arc furnace in the presence of copper for direct production of the "master" alloy—4% Be. The original Lebeau (123) process is still practiced commercially (108).

III. TOXICOLOGY

It was not until the middle 1940's that a possible toxic effect of beryllium or its compounds on the human body was recognized. The fundamental biochemistry of beryllium toxicology is still not at all understood. The three recognized toxic manifestations of beryllium are:

1. Acute berylliosis, closely resembling chemical pneumonitis, with the symptoms being of relatively short duration and, unless death results, leaving no permanent after effects.

2. Granulomatosis, or chronic berylliosis, a generally disabling disease of long duration.

3. Skin ulcers and dermatitis, localized to areas or points of direct contact with water-soluble beryllium compounds.

Beryllium is being processed and handled in large quantities in a wide variety of forms and chemical compounds over a broad segment of industry with complete safety; however, anyone not thoroughly familiar with beryllium toxicology is strongly urged to make himself thoroughly familiar with the subject before processing any beryllium in the factory or in the laboratory. A description of health hazards, manifestation of troubles,

U. S. Atomic Energy Commission permissible atmospheric contamination, ventilation to avoid exposure, air sampling, and analysis are given by Eisenbud (58). An excellent report by Breslin and Harris (29) of the U. S. AEC describes the Commission's ten-year experience with controls. Many analytical methods and references are given.

IV. INDUSTRIAL PRODUCTS AND USES

A. PURE BERYLLIUM METAL

1. Nuclear

As the pure metal, beryllium is rather unique with an atomic weight of 9.013 and a density of 1.85. It is reluctant to acquire additional neutrons but is willing to part with one of its neutrons. It has an extremely low thermal neutron absorption cross section and is an excellent scatterer and moderator of neutrons. These properties, coupled with its high melting point and resistance to corrosion, make it very interesting in nuclear applications. A materials testing reactor (MTR) and a submarine intermediate reactor (SIR) were two of the first beryllium thermal-energy reactors produced (145). Because of its high melting point and resistance to corrosion it is also an interesting material as cladding agent for uranium fuel elements in reactors (127).

Beryllium is an excellent source of neutrons when bombarded by deuterons and by alpha particles (180).

One of the first uses of beryllium was as an x-ray tube window. Because of its low atomic weight, it permits passage of x-rays 17 times more readily than the same thickness of aluminum (135,188).

2. Structural

Because of its unique combination of physical and mechanical properties (Tables III and IV), beryllium has long been of great interest to structural engineers. Its low density, combined with high strength, even at elevated temperatures, and its very high Young's modulus, 40 to 44 $\times$ 10^6, makes beryllium a very desirable structural material for high-performance aircraft and missiles.

Muvdi (142) has studied five methods of fabricating plate beryllium for structural applications.

Beryllium has a high reflectivity, up to 55% of white light and greater than 55% of ultraviolet. It therefore has possibilities for use as a mirror.

Because of its high heat capacity, good thermal conductivity, and high melting point, it was selected for one of its most glamorous applications—

TABLE I
Typical Alloys of Beryllium

Be, %	Co, %	Ni, %	Ag, %	Uses
Copper-base alloys				
4				Master
1.80–2.05	0.18–0.30			Electrical terminals, springs
1.60–1.80	0.18–0.30			Welder bars
0.40–0.70	2.35–2.70			Good electrical conductors
0.25–0.50	1.40–1.70		0.90–1.00	Bearings
0.25–0.50		1.40–1.60		Welding equipment
2.00–2.25	0.35–0.65			Casting alloys
2.35–2.55		1.00–1.20		Casting alloys
2.60–2.85	0.35–0.65			Casting alloys
Aluminum-base alloys				
5				Master
0.25				Aircraft sheet
Nickel-base alloys				
1.7–1.9				Springs
2.75				Castings

the heat sink for re-entry protection of some of the Mercury space capsules.

Metallic beryllium is used to produce all alloys except the copper-base. Beryllium oxide is the beryllium source for direct production of copper-base "master" alloys (Table I).

B. ALLOYS

1. Copper-Base Alloys

Industrial use of beryllium, particularly in copper-base alloys (16), has become quite extensive, with current U. S. industrial consumption of these alloys approaching 20,000 pounds of beryllium content per month. Industrial applications of these alloys are many and widespread. They find extensive use in springs of all kinds, especially current-carrying springs, as well as in diaphragms of pressure-sensitive instruments. They are especially adaptable to this type of use because of their fatigue resistance. Large quantities are also used in the manufacture of dies for pressing plastics and for deep-drawing metals. The alloy's high electrical, as well as thermal conductivity, coupled with hardness, make it attractive for electrodes and other components for resistance-welding work. It also finds extensive use in the manufacture of sparkless tools for use in mines, refineries, or similar industries (16,50).

2. Other Alloys

Beryllium is used as a minor alloying agent in aluminum-base alloys. As little as 0.005% beryllium in aluminum-base alloys retards oxidation of the melt (188).

The use of as little as 0.0042% of beryllium in magnesium is claimed to make pouring of this material safe at 760°C. (31).

It has found use as an alloying agent in nickel, especially in Germany, to produce a high-strength alloy that is resistant to corrosion and wear. It also finds use in instrument springs, hypodermic needles, and surgical instruments (16).

Beryllium "steels" (16) (0.15 to 1.5% Be), with varying amounts of Ni, Cr, Co, and Mo, are extensively used for springs of many types (some high-temperature), and where good resistance to corrosion and wear is required (valve disks, pump impellers, and gears).

Alloys with many other metals also have been studied and reported (16,49,50,99).

V. BERYLLIUM COMPOUNDS (109,135,149)

The most important compounds, from an industrial application standpoint (more or less in the order of their occurrence in the manufacturing process—see Section I-C), are beryllium sulfate ($BeSO_4 \cdot 4H_2O$), sodium beryllium fluoride (Na_2BeF_4), beryllium hydroxide ($Be(OH)_2$), ammonium beryllium fluoride (($NH_4)_2BeF_4$), beryllium fluoride (BeF_2), and beryllium chloride ($BeCl_2$). Other compounds of particular interest are beryllium oxide (BeO), beryllium nitrate ($Be(NO_3)_2 \cdot 3H_2O$), beryllium nitride (Be_3N_2), beryllium carbide (Be_2C), beryllium iodide (BeI_2), beryllium oxylate ($Be(C_2O_4)$), beryllium acetate ($Be(C_2H_3O_2)_2$), or the basic acetate ($Be_4O(C_2H_3O_2)_6$). Most beryllium compounds are colorless or white, beryllium carbide being a notable exception (see Section V-K).

A. BERYLLIUM SULFATE ($BeSO_4 \cdot 4H_2O$)

The tetragonal, colorless, beryllium sulfate tetrahydrate is the most common form of this compound. One hundred grams of this colorless tetrahydrate crystal can be dissolved in 100 ml. of water at 100°C. About 60% of this may be recrystallized if the solution is cooled to about 20°C. The sulfate extraction of beryllium from ore depended upon the crystallization of the tetrahydrate from the sulfuric acid medium until about 1949 (171). The salt is only slightly soluble in concentrated sulfuric acid ($10M$), and good recovery of beryllium is suggested by crystallization from a medium containing an excess of sulfuric acid. Single crystallization of the tetrahydrate from a sulfuric acid solution of commercial beryllium hydroxide is possible for ultimate production of reactor-purity beryllia (see Section V-H). Hydrates of 1, 2, 4, 6, and 7 molecules of water have been reported (135), although some doubt has been expressed as to the existence of the heptahydrate. Two molecules of water are lost when the tetrahydrate is subjected to temperatures of 100°C. Evidence of hydrate loss has been noted in dry atmospheres at 85°C. and when confined with phosphoric anhydride. At 400°C. the tetrahydrate is converted to the anhydrous form. Anhydrous beryllium sulfate is insoluble in cold water and is converted slowly to the soluble tetrahydrate in boiling water. It is more readily dissolved by boiling in a dilute sulfuric acid medium. Beryllium compounds are generally converted to the sulfate medium for wet chemical analysis. Anhydrous beryllium sulfate can be thermally decomposed to beryllium oxide by heating at temperatures as low as 700°C. (56).

B. SODIUM BERYLLIUM FLUORIDE (Na_2BeF_4 OR $2NaF \cdot BeF_2$)

Two forms of sodium beryllium fluoride are claimed to exist, one involving single molecules of beryllium fluoride and sodium fluoride, and one involving a molecule of beryllium fluoride with two molecules of sodium fluoride. The salt with two molecules of sodium fluoride is generally considered to be the most likely and readily obtainable, in crystalline form, as rhombohedron prisms (135). Although the theoretical solubility of Na_2BeF_4 is only 14.1 g./liter (2.7 g. of BeO/liter), leach solutions of as high as 20 g. of BeO/liter have been obtained (100). This would suggest that much of the beryllium may be dissolved as BeF_2. Sodium beryllium fluoride is most important, however, as the intermediate compound in the extraction of beryllium from beryl by the fluoride process.

C. BERYLLIUM HYDROXIDE ($Be(OH)_2$)

According to Mellor (135), beryllium hydroxide exists in the alpha or granular form and in the beta or gelatinous form. Schwenzfeier (171) describes the alpha form as precipitated from a strongly basic solution of sodium beryllate by boiling, and the beta form as being precipitated by ammonia from an acid medium. Gelatinous beryllium hydroxide precipitated by ammonia from an acid medium is generally quite reactive, and readily dissolves in an acid medium if this dissolution takes place shortly after precipitation. However, upon standing for an extended period of time this hydroxide becomes less reactive with an acid medium. This suggests a transformation to a different, less reactive, form.

Darwin and Buddery (45) describe three forms: (*1*) gelatinous, (*2*) alpha, and (*3*) beta. The gelatinous form will in time convert to the crystalline alpha, which, in turn, over a long period is transformed into the even less reactive beta form. These writers describe the beta form as the crystalline precipitate produced by boiling the beryllate in caustic.

Beryllium hydroxide in the gelatinous form may be precipitated from chloride, nitrate, or sulfate media with hydroxides of ammonia, sodium, or potassium. Although an excess of ammonium hydroxide does not completely redissolve the beryllium hydroxide, an excess of sodium or potassium hydroxide will redissolve the beryllium hydroxide to produce beryllates. The crystalline beryllium hydroxide may be precipitated from a $1.5N$ sodium hydroxide solution by boiling. At higher caustic concentrations the beryllium hydroxide is more soluble. Concentrated sodium and potassium hydroxides will redissolve beryllium hydroxide. According to Duval and Duval (56), beryllium hydroxide loses its combined water completely to become BeO at a temperature of 951°C. Livey (127) states that the hydroxide loses all combined water at 500°C.

D. AMMONIUM BERYLLIUM FLUORIDE ((NH_4)$_2BeF_4$ OR $2NH_4F \cdot BeF_2$)

This salt is the purest form of beryllium (Table II) in the production process for the reduction of beryllium fluoride to produce beryllium metal. It is quite soluble in water, with an equivalent as high as 22 g. of Be/liter.

TABLE II
Typical Analysis of Ammonium Beryllium Fluoride

Element	%
Beryllium	7.45
Aluminum	0.0025
Chromium	0.0006
Iron	0.0007
Manganese	0.0002
Nickel	0.0007
Silver	0.0001
Cobalt	0.0001

This salt may be recrystallized to produce a compound of very high purity. Its principle use is in thermal decomposition to produce beryllium fluoride. This decomposition is carried out above the melting point of beryllium fluoride (171).

E. BERYLLIUM FLUORIDE (BeF_2)

Beryllium fluoride is an amorphous, colorless, glassy compound that is very hygroscopic and tends to produce oxyfluorides when exposed to moisture. It has a melting point of 803°C. and boils at 1159°C. It has a density of about 2 g./cc. Although it is quite soluble in water, with formation of oxyfluorides (135), it dissolves very slowly in the massive form. It is produced by the thermal decomposition of ammonium beryllium fluoride in the absence of air, and its principal use is for the production of beryllium by reduction of the fluoride with magnesium. It may also be used in high-temperature sodium fluoride–beryllium fluoride electrolysis (87). Its properties would indicate possible application as liquid moderator in nuclear reactors.

F. BERYLLIUM CHLORIDE ($BeCl_2$)

Beryllium chloride is generally produced by high-temperature chlorination of a beryllium oxide–carbon mixture. It also can be produced by chlorination of a beryllium sulfate–carbon mixture and, in some cases, it has been produced by high-temperature chlorination of a beryl–carbon mixture. It melts at about 405°C. and boils at 487°C. Since its boiling point is considerably higher than that of the corresponding chlorides of aluminum, iron, and silicon, the principal elements of beryl except for oxygen, it is possible to produce a relatively high-purity beryllium chloride by direct chlorination of beryl. Its principal use is as the beryllium source in various electrolytic separations (103,130,199). It has been suggested as a catalytic agent in oil-cracking processes (109). Aldehydes and ketones readily coordinate to $BeCl_2$, $BeBr_2$, and BeF_2. The chloride forms the tetrahydrate and the stable tetramine ($Be(NH_3)_4Cl_2$) (81).

G. BERYLLIUM IODIDE (BeI_2)

Beryllium iodide can be produced by reaction of the metal in iodine vapors at high temperature, and also by the reaction of hydrogen iodide on beryllium carbide at 700°C. This material decomposes readily in moist air. The crystal melts at 510°C. and boils at 585 to 595°C. (135). The principal interest in this compound is as a possible method of producing pure beryllium by decomposition on a hot metal wire.

H. BERYLLIUM OXIDE (BeO)

Beryllium oxide is a rather unique compound. It can be produced in a purity suitable for nuclear applications by thermal decomposition of beryllium sulfate tetrahydrate. It is of special interest as a reflector and moderator (20,23,32,60,67,94) in high-temperature nuclear applications because of its resistance to corrosion (127), its melting point of 2570°C., and its boiling point of 4300°C. Beryllium oxide can be hot-pressed or cold-pressed and sintered to its near-theoretical density of 3.02 g./cc. The high temperatures needed to accomplish this densification produce a beryllium oxide that is quite inert to attack by all acids except hydrofluoric. In this dense form it is also finding extensive uses in the electronics industry, because of its excellent thermal conductivity and electrical resistance. In its less pure form, as thermally decomposed commercial beryllium hydroxide, it is used as the beryllium source for the production of master beryllium-copper alloy by reduction in an arc furnace with copper and carbon. Beryllium oxide, because of its excellent refractory properties, also finds extensive applications as crucible material for the melting of many metals. It is used to increase the modulus of glass in fibrous form (17). Beryllium oxide is quite abrasive, especially after being subjected to elevated temperatures (197).

I. BERYLLIUM NITRATE (Be(NO$_3$)$_2$·3H$_2$O)

Beryllium nitrate can be crystallized from a nitric acid solution of the hydroxide to produce a crystal, Be(NO$_3$)$_2$·3H$_2$O. In its pure form it is generally colorless or white. However, because of the free nitric acid frequently remaining in the crystalline material, it also may appear yellow. Beryllium nitrate also may be produced by dissolving the hydroxide in nitric acid and concentrating by evaporation to slightly above its boiling point of 142°C. The crystal melts at 60°C. and, on exposure to moist air, quickly absorbs sufficient moisture to dissolve itself. The material is very soluble in both warm and cold water, and is also quite soluble in alcohol. The principal commercial use of beryllium nitrate is in the production of mantles for gas lights—nitrate impregnation is ignited to the oxide.

J. BERYLLIUM NITRIDE (Be$_3$N$_2$)

Beryllium nitride is produced by heating beryllium in a nitrogen atmosphere at elevated temperatures. Reaction starts at approximately 500°C. and is quite rapid at temperatures in excess of 900°C. Nitride of 97%, based on nitrogen content, can be produced by heating beryllium metal powder of −100 mesh size in a pure nitrogen atmosphere. Purified nitrogen, containing up to 5% hydrogen, appears to catalyze the reaction (19). It also can be produced by heating in an ammonia atmosphere. Surface oxides may retard the reaction, but at temperatures of about 900°C. this oxide layer is easily penetrated and reaction takes place quite rapidly. The nitride has a melting point of 2200°C. Beryllium nitride readily dissolves in caustic and acid media to produce ammonia, in the case of caustic dissolution, and ammonium ion in acid dissolution. It must be protected from moist atmosphere to avoid partial conversion to beryllium hydroxide. There is some interest in beryllium nitride as a refractory material because of its high melting point. Its most interesting use, however, is for the production of radioactive carbon-14, which results when it is submitted to reactor-pile irradiation.

K. BERYLLIUM CARBIDE (Be_2C)

Beryllium carbide is one of the few colored beryllium compounds, and this color may vary from yellow through tan to red or black, depending upon the free carbon present and the crystal type formed (49). It shows varying degrees of reactivity with water, depending upon its thermal history. When submitted to high temperatures the carbide is less reactive. It is of interest as a refractory material because of its high melting point, in the neighborhood of 2250 to 2400°C. It may be formed by heating a mixture of fine beryllium and carbon in an inert atmosphere at temperatures above 1000°C.

L. BERYLLIUM CARBONATE—BASIC

No definite composition has been established for this precipitate from boiling ammonium carbonate solution. The precipitates can be produced at high purity and may be decomposed by heat to the oxide. Possible application to gravimetric analysis is proposed (155).

M. BASIC BERYLLIUM ACETATE ($Be_4O(C_2H_3O_2)_6$) (109)

This compound can be crystallized from a hot glacial acetic acid solution of anhydrous beryllium hydroxide. The salt is soluble in chloroform and other organic solvents. It melts at 283°C. and can be sublimed at 330°C. in a closed system without decomposition, thereby accomplishing purification (81). It has been used to prepare high-purity oxide for spectrographic standards.

N. BERYLLIUM PHOSPHATE ($Be_3(PO_4)_2$) (135)

The principal beryllium phosphate of interest to the analytical chemist is the ammonium beryllium phosphate. This salt may be precipitated from a chloride medium with ammonium phosphate and in the presence of chelating agents to produce a relatively high-purity salt (91) of indefinite composition that is converted to beryllium pyrophosphate on thermal decomposition. The pyrophosphate, of a very definite chemical composition, has a high compound-to-beryllium weight ratio and therefore is of interest to the analytical chemist.

O. BERYLLIUM HYDRIDE (BeH_2)

Barbaras et al. (12) have described the preparation of beryllium hydride, a white powder that starts to decompose at 125°C., by the reaction between lithium aluminum hydride and dimethyl beryllium in an ether solution. They were unable to remove all the ether. According to Coates and Glocking (37), the hydride does not decompose below 250°C.

P. INTERMETALLIC COMPOUNDS—BERYLLIDES

Beryllium forms compounds of the MBe_{12} and MBe_{13} types. These compounds are of interest because of their excellent mechanical properties at elevated temperatures. Summaries with many literature citations are given in References 19, 49, and 99.

VI. PROPERTIES

A. PHYSICAL AND MECHANICAL PROPERTIES

It must be remembered that the beryllium industry is still in its infancy. Very extensive and intensive effort is now being expended to produce very high purity metal for further study of its properties. Some of the values reported in Table III may be revised as a result of these studies. Physical and mechanical properties reported to date have not always categorized the specimen examined, i.e., low or high metallic impurity content, beryllium oxide (oxygen) content, grain size (cast or powder metallurgy product)' or the extent of prior mechanical working, to mention a few examples.

The ranges in values given in Table IV reflect a general pattern. The mechanical properties of vacuum hot-pressed material are enhanced by extrusion and rolling. An increase in beryllium oxide content also generally enhances mechanical properties.

B. ELECTROCHEMICAL PROPERTIES

Beryllium can be separated electrolytically from its halides, but considerable difficulty is experienced when this method is used because of the vapor pressure in the molten condition. This problem is generally somewhat alleviated by dissolving the beryllium halide in the corresponding alkali or alkaline earth halide molten bath. The flake beryllium thus produced is generally contaminated with its own halide, as well as with the solvent halide. Purification steps are necessary to obtain pure beryllium. The procedures cannot be used in analytical methods. A description of the

TABLE III
Physical Properties of Beryllium

Property	Value	Reference
Atomic number	4	(111)
Atomic weight	9.013	(111)
Atomic volume, cm.^{3}/mol at 25°C.	4.877	(188)
Atomic diameter, A.	2.221	(125)
Density, 25°C., x-ray (g./cm.^{3})	1.8477	(188)
Density, 1000°C., x-ray (g./cm.^{3})	1.756	(188)
Melting point, °C.	1283	(125)
	1315 ± 20	(188)
Boiling point, °C.	2970	(125)
Latent heat of fusion, cal./g.	250/275	(125)
Thermal neutron cross section, barns/atom	0.009 ± 0.0005	(180)
Thermal conductivity, cal./cm.^{2}/second/°C./cm.^{a}	0.35–0.45	(47,125,188)
Electrical conductivity, % I.A.C.S.	35–42	(170)
Crystal lattice, 20°C.	Hexagonal close-packed	(111)
Lattice constants, A., a	2.285	(125)
c	3.583	(125)
Velocity of sound, m./second	12,600	(170)
Coefficient of thermal expansion, 25–100°C. (in./in./°C.)^a	11.5 × 10^{-6}	(47,69,125,188)
Specific heat, cal./g./°C.^{b}		
100°C.	0.51	(47)
300°C.	0.61	
500°C.	0.66	
700°C.	0.72	
900°C.	0.77	

^{a} Thermal properties vary with the type of product, depending upon the method of densification of the powder or flake to the solid material and cast metal. Chemical composition, primarily oxide content, also may be important.

^{b} These values for specific heat are from the graph on page 170 of Reference 47, and are credited there to D. G. Ginnings, T. B. Douglas, and A. F. Ball, *J. Am. Chem. Soc.*, **73**, 1236 (1951). The beryllium used for these authors' study is alleged to be of higher purity than that of other investigators.

electrolysis of beryllium chloride has been presented by Windeker (199). Other experimental electrolytic methods of separation of beryllium are described by Magel (130).

Polarograms of beryllium chloride and beryllium sulfate solutions show two waves: at −1.4 v. *vs.* S.C.E. due to the discharge of hydrogen and at −1.8 v. *vs.* S.C.E. due to reduction of beryllium ion. The beryllium wave is not well defined and has been reported to coalesce with that of aluminum (115).

TABLE IV
Mechanical Properties of Beryllium (19a,47a,127)

Property	Vacuum hot pressed	Warm extruded annealed	Warm rolled
Ultimate tensile strength, p.s.i.	40,000–50,000	60,000–90,000	80,000–90,000
Yield strength (0.2% offset), p.s.i.	27,000–40,000	45,000–55,000	60,000–70,000
Elongation, % in 1 inch	1–4	2–5	5–10
Modulus of elasticity, p.s.i.	40–44 × 10^6	42–44 × 10^6	44 × 10^6
Compression yield strength, p.s.i.	25,000–35,000	50,000	—
Hardness, Rockwell B	75–90	85–95	96

Kolthoff and Coetzee (114) have reported that well-defined but incompletely reversible waves are possible for beryllium, as well as for magnesium and aluminum, in an acetonitrile solvent. Diffusion currents are proportional to concentrations over a wide range measured at −2.4 v., at a drop time of 2.3 seconds, and at a mercury flow rate of 1.08 mg./second. A drawn-out wave was obtained with beryllium in 0.1N tetraethylammonium perchlorate.

Gyorbiro (77) claims beryllium salts give two polarographic waves in solution of pH 3.5 to 4.4 which are 0.1 to 0.2M in LiCl or $(CH_3)_4N^+I^-$. In the concentration range 0.8 to 6 × 10^{-3}M the second wave is proportional to the Be concentration. It shifts from −1.85 to 1.90 v. with decreasing pH. Data indicate a two-electron reduction of the $(Be(H_2O)_4)^{+2}$ ion, according to the Ilkovic equation.

C. CHEMICAL PROPERTIES

Beryllium is the first member of Group II in the periodic system. At ordinary temperatures, it is quite resistant to oxidation; highly polished pure beryllium will retain its luster for years. At about 700°C. oxidation is noticeable, and at 1000°C. it is quite rapid (111), with the formation of a white scale.

Gulbransen and Andrews (76) have determined that polished beryllium specimens may be heated in a low-pressure pure-oxygen atmosphere to 825°C. before the weight increase exceeds 40 mg./cm.²/hour. In pure nitrogen at 7.6 cm. Hg the protective coating was seriously penetrated only after heating to 875°C. Weight changes were measured with a vacuum microbalance. The data fit a parabolic law:

$$w^2 = Kt + c$$

where w is the weight gain per unit area, t is the time, and K and c are constants.

Finely divided beryllium may oxidize slowly upon exposure to air. It is possible, however, that this apparent increase in oxygen content is due to absorbed moisture on the beryllium which, during the generally accepted analytical procedure for oxygen, is converted to the oxide and thus determined. At temperatures in excess of 900°C. it combines with nitrogen quite readily, forming the nitride (Be_3N_2). Beryllium reacts with carbon to form the carbide (Be_2C) at temperatures about 100°C. in excess of its melting point (1283°C.) (125). Reactions with the above three elements are exothermic.

According to Kroll (119), fluorine reacts with beryllium powder at room temperature, whereas chlorine, bromine, and iodine react at elevated temperatures. The gases hydrogen chloride and hydrogen bromide attack beryllium metal at elevated temperatures to form the volatile halide, but they do not react with contained beryllium oxide. Beryllium reacts vigorously with warm sulfuric and hydrochloric acid and much less vigorously with dilute nitric acid. Concentrated nitric acid has little effect on beryllium. Phosphoric acid attacks beryllium slowly with liberation of hydrogen. The reaction with hydrofluoric acid is extremely vigorous, which also is the case with bifluoride solutions. As a metal ion, it exhibits one state of oxidation (Be(II)).

It reacts vigorously with hot concentrated alkalis, with the evolution of hydrogen. Beryllium is readily precipitated as the hydroxide with alkalis in weak alkaline medium. However, especially with sodium or potassium hydroxide, the precipitated hydroxide will redissolve as the alkali concentration increases, and it is assumed to form the beryllate, $(BeO_2)^{-2}$. Ammonia in large excess tends to convert some beryllium hydroxide to the beryllate, but it does not approach the nearly complete beryllation or dissolution as with sodium or potassium hydroxide. One of the few chemical differences between beryllium and aluminum is the fact that at a certain alkali concentration (1.5N sodium hydroxide medium), beryllium is nearly completely precipitated as the hydroxide by boiling, whereas aluminum is retained as aluminate. This difference is not utilized analytically because the reactions are not quantitative, but it is utilized in the extraction of beryllium from ore by the sulfate process (171).

Kellogg (102) has prepared a table of thermodynamic constants for beryllium compounds including only those for which reliable information is available (the halides, oxide, nitride, sulfide, and silicate).

Because of its small ionic radius the oxide of beryllium is amphoteric. Other elements of Group II, however, increase in basic strength with atomic weight.

Normal salts of beryllium such as the sulfate, nitrate, and chloride will dissolve additional hydroxide or carbonate. The basic compounds are produced by evaporating aqueous solutions of the salts, generally yielding a thick, syruplike liquid. Salts of beryllium are less hydrolyzed than the corresponding salts of aluminum or iron (135).

Latimer (122) gives the reduction potential of beryllium:

$$Be^{+2} + 2e^- \rightarrow Be \qquad -1.70 \text{ v.}$$

Equilibrium constants for various reactions are (122a):

		K
Halide hydrolysis	$2Be^{+2} + H_2O = Be_2O + 2H^+$	4×10^{-7}
	$H_2BeO_2(I) = 2H^+ + BeO_2^{-2}$	2.8×10^{-36}
Solid and acid ions	$H_2BeO_2(II) = 2H^+ + (BeO_2)^{-2}$	1.4×10^{-30}
	$H_2Be_2O_3(S) = 2H^+ + Be_2O_3^{-2}$	7.3×10^{-30}
	$Be_2O(OH)_2 + H_2O = 2Be^{+2} + 4(OH)^-$	1×10^{-49}
Hydroxide–base	$Be_2O(OH)_2 = Be_2O^{+2} + 2(OH)^-$	4×10^{-19}
Hydroxide–metal	$6(OH)^- + 2Be = Be_2O_3^{-2} + 3H_2O + 4e^-$	$E^°_b = 2.28$

Various methods for the determination of beryllium depend upon the formation of soluble ionic states as a complex with the citrate or tartrate, in combination with a beryllate ion, which is in turn converted to a more stable colored lake with an organic dye (Section X-C-1). The fact that beryllium also forms a complex fluoride ion is utilized (Section X-B). An excess of ammonium carbonate added to a beryllium sulfate solution will retain beryllium in solution, which may be precipitated by boiling (Section X-A).

Most inorganic compounds of beryllium are colorless. (For a thorough discussion of beryllium compounds, see Mellor (135).)

Beryllium forms stable compounds with many other metals (intermetallics) (19). A compilation of much of this work, including equilibrium diagrams, has been prepared (49).

D. OPTICAL PROPERTIES

Massive beryllium, with fractured or pickled surface, is steel grey. Reflectivity of white light has been reported to be 50 to 55% (8,170) and of ultraviolet greater than 55% (188). Emissivity of solid and liquid beryllium at a wavelength of 6500 A. is 0.61. At a wavelength of 5500 A. the solid emissivity is 0.61, but the liquid changes to 0.81.

Beryllium windows equal in thickness to aluminum windows permit x-rays to pass 17 times better (188). The substitution of beryllium in x-ray tube windows permits the transmission of a softer ray, and improved sensitivity is realized for those materials of low radiographic density. Although some radiographic testing procedures specify a sensitivity of 2%, it appears possible to achieve a 1% sensitivity by using a beryllium window, especially on thin samples.

The optical spectrum of beryllium presents relatively few sensitive lines suitable for analytical purposes—arc or spark. The two most sensitive arc lines (Table V) are self-reversing and must be used judiciously in analytical work. This suggests possible application at these wavelengths (2348.61 and 3321.34 A.) of absorption spectroscopy for beryllium determination.

TABLE V
Optical Properties

Wavelength, A.	Intensity (30)	
	Arc	Spark
2348.610	2000R	50
2494.730	30	20
2650.781	25	—
3130.416	200	200
3131.072	200	150
3321.013	50	—
3321.086	100	15
3321.343	1000R	30

	Emissivity coefficient, A.
6500: solid	0.61 (188)
liquid	0.61
5500: solid	0.61
liquid	0.81

	Reflectivity, %
White light	50–55 (170)
Ultraviolet	+55 (188)

E. RADIOCHEMICAL—NUCLEAR PROPERTIES

When beryllium, in its normal state, is subjected to irradiation by gamma rays at an energy level in excess of 1.63 M.e.v. (105), neutrons are liberated according to the nuclear reaction expressed in either of two methods:

$$^{9}_{4}\text{Be} + \gamma \rightarrow {^{8}_{4}}\text{Be} + n$$

$$\text{Be}_9\ (\gamma, n)\ \text{Be}_8$$

By counting the neutrons liberated, it is possible to estimate the beryllium content of a sample (Sections IX-D and X-D). This lower isotope itself spontaneously decomposes to produce two α particles.

Pannell (146) studied this application, using conventional electronic equipment and a portable γ source for determination of beryllium in ore. Interference of elements of high absorption cross section were studied and the effect of boron was measured. Commercial equipment for field and laboratory determination of beryllium in ores is available (21,93). The γ source in these units is currently less than 1 Curie. This principle, with higher curie sources, has very interesting possibilities as a means of precise determination of beryllium content, even in pure metal. Its potential accuracy exceeds that of chemical methods. Pannell and Freyberger (147) reported experiments to determine the application to "picking" beryl. Gaudin et al. (65) describe the application to ore concentration.

Beryllium is an excellent source of neutrons when bombarded by α particles, deuterons, or protons (51,70,180). Very complete discussions of the nuclear properties have been presented (51,180) (see Table VI).

TABLE VI
Isotopes of Beryllium

Isotope	Ref.	Half life	Mode of decay	Energy of radiation, M.e.v.	Method of production
^{6}Be	(188)	0.4 sec.	Indefinite	—	—
^{7}Be	(6,10, 113)	53 ± 0.4 days	Electron capture	0.477γ	^{6}Li$(d,n)^7$Be ^{10}B$(p,\alpha)^7$Be
^{8}Be	(47)	$\sim 10^{-16}$ sec.	2α	0.047α	^{9}Be$(\gamma,n)^8$Be ^{9}Be$(n,2n)^8$Be
^{9}Be		Stable		—	Natural
^{10}Be	(188)	2.5 × 10^6 yr.	β	0.577β	^{9}Be$(n,\gamma)^{10}$Be ^{9}Be$(d,p)^{10}$Be
^{11}Be	(198)	13.57 ± 0.15 sec.	β and γ		^{11}B$(n,p)^{11}$Be

VII. SAMPLING

A. ORE

Beryl, the only beryllium-containing ore of commercial value today, is not entirely homogeneous and, therefore, must be sampled in a manner consistent with good principles and practices for such material (14). The ratio of the gross sample collected to the lot size of the ore will be partially dependent upon the particle- (or lump-) size distribution. Since beryl and other minerals containing beryllium are nonmagnetic, it is highly recommended that equipment used to reduce the gross sample to the final sample for analytical consideration be constructed of magnetic materials for the easy removal of possible contaminants. Because of the potential toxic effects of materials containing beryllium (Section III), extreme care must be exercised to protect the persons preparing the sample, but at the same time caution must be exercised to avoid loss of very fine portions of the sample because of excessive drafts. (The more friable material in commercial ore contains little or no beryllium.) Since it is desirable to use a relatively small sample in the final analysis, it is recommended that the final analytical sample of at least 10 g. be reduced to pass through a USBS 200 sieve. This particle size is especially important if preliminary spectrographic inspection of the sample is to be performed as a guide for later chemical procedures.

B. COPPER-BASE ALLOYS

1. Cast Alloys

A sample representative of the melt (pigged) may be obtained by drilling each sample pig in three positions in a diagonal pattern across the top of the pig (14). If the drilling cannot be accomplished entirely through the height of the pig from one side, drilling must be made from opposite sides to obtain a suitable sample in a manner that will avoid contamination. There should be no fewer than three sample pigs per melt.

Samples of a single casting are to be selected in such a manner as to obtain material representative of the casting, but the method must in no way harm the usefulness of the casting. The sampling is a mutual problem and must be agreed upon by the manufacturer and the purchaser. All samples for wet chemical analysis must be fine chips, taken in a manner that will avoid oxidation, and since these alloys are nonmagnetic, the equipment used for sampling should be of a magnetic nature in order to provide for easy removal of possible tramp contamination by the sampling equipment.

Samples for spectrographic use, pin or disk, must be taken in such a manner as to simulate the matrix of the standards used for comparison.

2. Wrought Alloys

Samples must be representative of the material. The procedure must be agreed upon by the manufacturer and the purchaser. The principles of sampling outlined in ASTM Designation E55 (adopted in 1948 and re-approved in 1955) serve as an excellent guide for the selection of samples for all types of copper-base beryllium alloys.

C. ALUMINUM-BASE ALLOYS

If a true, solid solution of beryllium in aluminum exists, it is of a low beryllium content. The "master" alloy commercially available in pig form (5% nominal beryllium content) must be carefully sampled (14), and a relatively large portion of the lot sample must be used for analytical work to minimize the effect of possible segregation. By quenching the sample pig, relatively fine chips may be obtained with a steel drill having a suitable cutting angle. Not less than three holes are to be drilled through the entire height of the sample pig in a diagonal pattern across the face of the pig. There should be a minimum of three sample pigs per melt. Sampling of wrought alloys containing less amounts of beryllium should be according to ASTM Designation E55.

D. BERYLLIUM METAL

Commercially available beryllium metal may be in the form of "pebbles" (magnesium reduction of beryllium fluoride), flake (electrolysis from a fused-salt bath), vacuum-cast, powder, vacuum hot-pressed block, or other massive form (from powder metallurgy). For the pebble or flake, sampling must follow sound procedures for nonhomogeneous particulate matter, since pebbles may contain up to 3% magnesium and/or magnesium fluoride, and flakes may be contaminated with the salts of the fused bath. Particle (pebble) sizes and ratios will determine the manner in which the gross sample is collected. (Consider Section III, this chapter, in collecting and preparing the analytical sample.)

A high-purity beryllium mortar and pestle (power-driven) should be used for crushing the lump (and flake, if desirable) and ultimate grinding of the analytical sample, which should all pass through a USS 100 sieve. This is especially true if spectrographic analysis by direct burn of the metal powder is practiced as a method for impurity determination. This is also advisable for greater uniformity of sample, should small individual samples be used for a single-element determination. (This grinding operation may increase the oxygen content of the powder, compared with original solid metal, by as much as 0.6%, with a corresponding decrease in total beryllium.) Extreme care must be exercised to avoid any contamination of the sample from any source, especially when examining for nuclear application. (Lowe (127a) reports an increase in cobalt content in a silicon carbide sample from 0.02 p.p.m. before grinding to 780 p.p.m. after grinding in a boron nitride mortar and pestle.) Sandell (163,165) reports measurable contamination from grinding in different media: Plattner mortar (163,165) (Fe, Mn, Nb, and Ta); alumina mortar (165) (Ti and Mg); and jar mill with flint balls (165) (Fe, Cu, Zn, Co, and Na). The authors have found measurable uranium contamination from grinding beryllium containing 0.1 p.p.m. of uranium in a beryllium mortar of 35 p.p.m. U.

A representative sample of a commercially available vacuum-melted beryllium casting may be obtained by drilling axially in the geometric center of the casting's bottom (drill of magnetic material for contaminant removal). The drillings are ground to −100-mesh size. Solid metal from the casting, not drillings or powder, must be used for representative oxygen and beryllium determination.

Commercially available powder (−200 mesh) may contain as much as 50% material that will pass a 400-mesh (USBS) sieve. Samples of blended powder are obtained according to ASTM Designation E215. Equal portions of the complete cross section of the powder stream from the blender, near the beginning, middle, and end of the unloading operation, are in turn blended for the lot sample. A great excess of "coarse" or "fine" powder could give lower- or higher-than-true oxygen content, respectively, with converse effect on total beryllium content.

Vacuum hot-pressed block is produced from blended powder, and any portion of metal, after cleaning the surface, will be representative of the block. Experience has demonstrated that the block analysis is identical with the powder used, within analytical error. Results of analysis of the powder blend are generally accepted for the block produced therefrom. Machine chips or turnings generally contain more oxygen and less beryllium than the original block. A portion of the solid metal must be used for reliable beryllium and oxygen determinations when block analysis is required.

VIII. SEPARATION AND ISOLATION OF BERYLLIUM

A. SAMPLE DISSOLUTION

1. Ores

Beryl contains about 66% silica and can be opened by the conventional method for siliceous material. The finely divided ore sample is mixed and covered with sodium carbonate (1:5), fused in a covered platinum crucible to produce a clear melt, and this clear melt is dissolved in sulfuric acid. Care must be exercised in the dehydration of silica in fuming sulfuric acid because of the relatively large amount of silica present. The ore sample also may be fused with sodium fluoride (1:5) to a clear melt and evaporated to sulfuric fumes. Evolution of silicon as the fluoride greatly reduces the danger of loss by bumping associated with the sodium carbonate fusion.

The effect of the sodium carbonate-to-beryl ratio and the heating time were investigated by Huguet and Bamberger (89). Results indicated a 2.2 to 1 ratio at $820 \pm 10°C$. for $2^1/_2$ hours. The melt (0.25 g. of beryl) can be disintegrated in hot distilled water with the help of a flat-tipped stirring rod. The silica is dehydrated with sulfuric acid and separated by filtration.

Adolfo (2) notes that beryl may be accompanied by quartz, feldspar, apatite, topaz, and lithium minerals, and the choice of method may depend upon the presence of alkali metals, iron, manganese, chromium, titanium, calcium, magnesium, and phosphorus. He recommends fusion of 0.25 g. of -200 mesh beryl in 1.5 g. of potassium bifluoride. Complete disintegration is accomplished in less than ten minutes. The melt is digested with 3 ml. of concentrated sulfuric acid to dense SO_3 fumes.

Athavale et al. (7) have tried bicarbonate, carbonate, sodium fluoride, and sodium carbonate–borax flux fusions. Sodium carbonate fusion, followed by water leach to leave a residue of crude beryllia and other impurities, with these impurities corrected by oxine determination, was found to be subject to error from samples containing apatite. The carbonate–borax flux fusion is taken up with cold dilute hydrochloric acid. Silica is retained in solution.

Whereas the use of hydrofluoric and sulfuric acids may be suitable for opening simple silicates with few impurities, it requires more time and may require a final fusion in either a basic or acid flux to effect complete decomposition of the ore. Our recent work with synthetic "low-grade ores" indicates serious loss of beryllium when this method is used for opening the ore. Sandell (161) also reports incomplete recovery of beryllium by hydrofluoric acid treatment of silicates.

2. Copper-Base Alloys

The copper-base alloys containing beryllium are generally completely soluble in a diluted nitric acid (1:4) without heating (1 g. in 50 ml.). A more concentrated nitric, or sulfuric–nitric acid mixture, as described in ASTM Designation E106, may be used for more rapid dissolution of the sample. However, rapid dissolution of a sample, where a temperature of greater than about 60°C. may prevail, converts varying amounts of the silicon to a form that will not react with molybdate, and the silicon then must be determined by the more tedious gravimetric procedure. The nitric–sulfuric acid mixture is recommended for alloys containing more than

0.40% Si. At such concentrations all the silicon is not in a soluble form suitable for colorimetric determination.

A chloride solution can be prepared by the use of hydrochloric acid and hydrogen peroxide (30%) (recommended when using solution technique for spectrographic determination of impurities and the determination of beryllium with aluminon).

3. Aluminum-Base Alloys

Aluminum-base alloys containing beryllium are generally commercially available at a nominal 5% or at less than 1% beryllium content. It is difficult to prepare a representative sample of less than 5 g. Samples of these alloys generally are soluble in either sulfuric, a mixture of hydrochloric and nitric, or a combination of all three acids. The sample is generally covered with about 25 ml. of water/g. of sample and 15 ml. of 1:1 sulfuric acid/g. of sample, added cautiously, since a violent reaction may result. Any insoluble matter present may be handled in a conventional manner to remove silica and to accomplish complete dissolution of the material to be added to the main portion of solution. The aliquot for the determination of beryllium must be from an acid medium.

In case of high silica and for silicon determination in the alloy the dissolution procedure using sodium hydroxide, outlined in ASTM Designation E34, is recommended. The choice of acid or caustic solvent will be dictated by the analytical requirements of the sample.

4. Beryllium Metal

For the determination of total beryllium, a representative sample is dissolved in sulfuric acid (1 g. of sample in 15 ml. of 1:1 sulfuric acid). The acid may be added directly in small increments to the powdered sample in a relatively large, covered beaker, since the reaction is quite violent, followed by a similar cautious addition of 40 ml. of distilled water/g. of metal until the sample is almost completely dissolved. The fine metal has a tendency to creep up the sides of the beaker and requires frequent washing down to insure complete dissolution of the sample. This method of dissolution is quite rapid and very suitable for the determination of beryllium and of most other elements that can be determined from aliquots of such a solution. These other elements would include iron, manganese, nickel, chromium, magnesium, copper, and aluminum.

For a more cautious approach, cover the sample with 20 ml. of warm distilled water/g. and cautiously add 20 ml. of 1:1 sulfuric acid/g. of sample. (The acid must be added to a warm solution or an excess may be added, resulting in gradual accumulation of heat and "boiling over.") In each case, to effect complete dissolution of the sample the few insolubles that are usually present are filtered, washed in the filter paper, and ashed in platinum, and this residue is treated with hydrofluoric and sulfuric acids, followed by fusion in potassium bisulfate. This potassium bisulfate fusion is dissolved in water and is added directly to the main solution. (In very high-purity metal the impurities in the bisulfate may influence the final determination.) The sample is dissolved in a nickel or stainless steel beaker in caustic for colorimetric silicon determination.

If total free metals are sought, the sample may be dissolved in a closed system in concentrated potassium hydroxide solution (157). Correction must be made for methane, which is liberated by the reaction of caustic on beryllium carbide, and for ammonia, which will be liberated by action between the caustic and the beryllium nitride that may be present, as well as the usual corrections for gasometric procedures.

If spectrographic techniques for impurities involving the use of porous cups are to be used, the sample should be dissolved in hydrochloric acid. If the sample is to be converted to the oxide for spectrographic determination of impurity elements, the sample as dissolved in sulfuric acid is evaporated to dryness and is ignited in a platinum crucible to an oxide form. The sulfuric acid conversion is recommended for comparison to beryllium oxide standards to generate the same matrix in both standard and sample. This oxide will probably give slightly higher than true silicon content, and there may be sufficient sodium in the sulfuric acid used to give an erroneously high sodium content. (Contamination by the acid is of significance in very high-purity material.) If the metal is being used for nuclear application, the sample should be dissolved in silica beakers for ultimate conversion to oxide for spectrographic analysis. Silicon should be determined on a separate sample dissolved in caustic in a nickel or stainless steel beaker.

Dissolution in nitric acid with a trace of hydrochloric, using all Teflon apparatus with ultimate evaporation to dryness in the Teflon beaker, followed by physical transfer to platinum and ignition at 600°C., is also recommended (71) for preparation of the oxide form for spectrographic analysis.

B. SEPARATION METHODS

No "specific" reagent for beryllium is known. Separations to be used will depend upon the micro or macro amount of beryllium in the sample, the amount of interfering elements present, and the potential interference with the chosen method of determination. The fluorescent reagent, morin, subject to many element interferences, would be recommended for determinations in silicate rock, biological materials, sediments, water, etc., where complete recovery of the beryllium present may not be an absolute necessity. On the other end of the scale, recovery of the beryllium within 0.1% in the "pure" form (98 + %) is desired, and interfering elements may be determined, corrected for, or separated by various means, and the beryllium as a pure compound recovered and measured. Cheng (34) proposes the terms *selectivity ratio* or *masking ratio* for evaluation and prediction of the equilibrium of the masking and principle reactions of a system. EDTA, for example, permits precipitation of beryllium with ammonia without interference from small amounts of magnesium.

1. Precipitation Methods

Beryllium cannot be quantitatively recovered from solutions containing fluoride ion by precipitation methods. Ignition of samples containing phosphate converts the phosphate to a water-soluble pyrophosphate, which cannot be recovered by precipitation (173). Fluoride ion must therefore be eliminated, and any pyrophosphate converted to phosphate by boiling in a relatively strong acid medium prior to precipitation.

For the fluorimetric or colorimetric determination of small amounts of beryllium, elements forming insoluble precipitates in sodium hydroxide, such as iron, magnesium, manganese, chromium, etc., cause serious interference. For solutions of samples of silicate rock, sediment, water, or biological materials, these elements therefore must be separated from the beryllium. Because of a strong tendency to coprecipitate, beryllate ion cannot be separated from iron, etc., by basic medium. Calcium, barium, and lead usually can be reduced to harmless concentrations by processing through a sulfate medium and filtering. Chromium can be oxidized to chromate with periodate or persulfate and the ammonia group, including beryllium, can be precipitated as the hydroxides in the presence of ammonium chloride to retain magnesium in solution. Thioglycollic acid will complex iron and help keep it in solution.

For samples from silicate rocks (at less than 10 p.p.m. of beryllium), Sandell (164) recommends drying of the ammonia precipitate, physical transfer of the dried cake to a nickel crucible, and a double fusion in sodium hydroxide. Beryllium and aluminum (and titanium in large amounts) will be recovered in the water leach from each fusion. Recovery from the first fusion may be only 60%. (Beryllium oxide subjected to temperatures in excess of 850°C. may be only partially converted to soluble form by this fusion.) Titanium may be separated by a following method. Elements such as cobalt, nickel, and the rare earths are retained in the water insolubles.

a. With Ammonia (Initial Separation with Possible Subsequent Purifications)

Separation from the alkali metals may be accomplished by ammonia precipitation. Ammonia gas precipitation (homogeneous solution) produces an easily filterable precipitate. Aqueous ammonia produces a gelatinous precipitate that may occlude other metal ions, requiring more than a single precipitation to recover only the beryllium. Phosphate ion will precipitate with the hydroxide.

Aluminum will act as a gatherer for beryllium and does not interfere with most color agents used, thereby assisting retention of beryllium in the ammonium hydroxide precipitates. According to Sandell (164), 3 mg. of aluminum will assist in the recovery of all but 0.1 γ of beryllium in 25 ml. by ammonia precipitation (pH 7 to 8).

Beryllium may be separated from a preponderance of copper, nickel and cobalt by an excess of ammonia, forming the soluble, highly colored, ammonia complexes. A double or triple separation may be necessary, when the beryllium-to-metal ratio is less than 1:25 for complete separation and gravimetric determination of beryllium. A single separation is suitable for volumetric beryllium determination, utilizing the formation of the $(BeF_4)^{-2}$ complex. The precipitate is washed with $2N$ ammonium nitrate at pH 8.5.

b. With 8-Hydroxyquinoline

According to Prodinger (154), many metals can be separated from beryllium with 8-hydroxyquinoline at a pH of 5.7 in an ammonium acetate–acetic acid buffered solution (151). These elements, in addition to aluminum, include the members of the ammonium hydroxide group, with the exception of some rare earths and chromium, which may be incompletely precipitated. Manganese and vanadium may be incompletely precipitated. Elements such as Mo, W, U, Cu, Ni, Co, Zn, Cd, Hg, and Bi are completely precipitated. An excess of quinolate, and careful control of temperature, pH, and buffer concentration, are important in the separation. If the ratio of other metals to beryllium is high the precipitate should be redissolved and reprecipitated to permit complete recovery of beryllium. The precipitates are best coagulated at a temperature of about 60°C. At a pH of less than 5.7 incomplete separations are probable. Above this pH beryllium is more likely to be occluded in the precipitate.

Sandell (164) recommends solvent extraction of the precipitates in chloroform instead of filtration as a means of separation from beryllium. There is evidence that some beryllium may be extracted in the chloroform. This possibility should be investigated under proposed operating conditions to determine whether a possible loss will seriously affect the beryllium recovery.

c. With Cupferron (Ammonia Salt of Nitrosophenylhydroxylamine)

Many elements may be precipitated from a mineral acid medium containing beryllium at low pH (164). These elements include Zr, Ti, V(V), Fe(III), W, and Ta, among others. Solubility in chloroform and other organic solvents as a function of acidity and acid medium also has been considered by Sandell (166).

d. With n-Benzoylphenylhydroxylamine

Titanium (2 to 17 mg. of Ti^{+4}) may be precipitated from a $0.1M$ HCl medium with 10 ml. of 4 to 7% wt./vol. solution of n-benzoylphenylhydroxylamine in ethyl alcohol (52). Precipitation of iron and aluminum is complete at pH 4 in a sodium acetate buffer from a nitrate–sulfate medium with 4 to 5% wt./vol. solution in ethyl alcohol. Beryllium starts to precipitate at pH 4.6 and is complete at pH to 6 to 8 as $Be(C_{13}O_{10}O_2N_2)_2$. The precipitate may be dried at 110°C. and weighted as such (2.08% Be). Other interesting applications of this organic reagent are given by Sinha and Schome (177).

e. Miscellaneous

Beryllium may be separated from a preponderance of aluminum (high-aluminum mineral or aluminum-base alloy, with low beryllium) by the method of Gooch and Havens (36). Aluminum is precipitated as the chloride from a hydrochloric acid–ether mixture saturated with HCl gas. The ammonium hydroxide group precipitate is dissolved in hydrochloric acid and is evaporated to a syrup. The HCl-saturated hydrochloric acid–ether mixture is added, is cooled to 0°C., and the insoluble aluminum chloride is separated from the soluble beryllium chloride by filtration.

Sandell (164) reports that Fisher and Wernet were successful in separating 60 γ (± 5) of beryllium from a solution containing 3.0 g. of aluminum and 0.18 g. of TiO_2 by a similar method not including ether but including ammonium chloride. The aqueous solution with ammonium chloride (4 g. in 75 ml.) is saturated with anhydrous HCl gas, precipitating aluminum as $AlCl_3 \cdot 3H_2O$ and titanium as $(NH_4)_2TiCl_6$ at 0°C.

A number of methods have been used to separate phosphate from beryllium. In siliceous beryllium ores, removal of phosphate may be accomplished by precipitation with bismuth from a $0.5M$ nitric acid medium, or as the phosphomolybdate. Bismuth may then be removed with hydrogen sulfide and molybdenum as quinolate. In biological material in which dry ashing or ignition has taken place, possible pyrophosphate must be converted to phosphate ion or beryllium losses may result (185).

Klemperer and Martin (112) separate beryllium from biological materials by a variety of methods. Calcium is removed as sulfate and beryllium is precipitated with ammonia, along with iron phosphate added as collector. Iron and other metals are removed by electrolysis with a mercury cathode. Aluminum is added and beryllium is collected as phosphate and determined fluorimetrically with morin.

Smith et al. (178) isolate beryllium by phosphate precipitations, using versene as chelating agent, for ultimate spectrographic determination (13,35,152).

Hure et al. (91) use a double precipitation of beryllium as phosphate from beryl, using EDTA to chelate interfering elements. Beryllium was weighed as the pyrophosphate after ignition.

Rare earths (124) may be quantitatively separated from beryllium by precipitation as the oxalates, with thorium as carrier. Although the rare earths are not quantitatively removed for estimation, they are removed to harmless concentrations by precipitation as fluoride from an acid fluoride medium (157,191). Hettel and Fessel (82) experienced incomplete recovery of rare earths from zirconium by fluoride precipitation.

2. Extraction Methods

Solvent extraction of quinolates (Section VIII-B-1-b) and cupferrates (Section VIII-B-1-c) from the beryllium solution has been suggested above. The major amount of iron in beryllium steels, for example, may be easily removed by the well-known extraction of the ferric chloride from a $6N$ hydrochloric acid solution with either ethyl or isopropyl ether.

a. With Acetylacetonate

This reagent forms complexes with many metals (167), some of which may be extracted with beryllium. Beryllium complexes with pyrophosphate, fluoride, and silicic acid are more stable than those with acetylacetone (82). Silicic acid may be dehydrated, the fluoride evolved, and the pyrophosphate converted to phosphate in strong acid medium. This extraction is generally used for trace beryllium determination.

Beryllium forms a complex with acetylacetone in aqueous solutions of pH 7 to 8, and may be extracted with chloroform or benzene (80,90). Addition of EDTA prior to the addition of acetylacetone reduces or eliminates interference from other impurities usually present in ores and air samples. Extraction with benzene as solvent at pH 4 to 5 effects separation from alkaline earths and phosphate, but not from iron or aluminum (185). Time of contact between solvent and solution may have to be established according to anions present in the solutions. The beryllium can be extracted from the benzene phase by shaking with hydrochloric acid. Iron and acetylacetone must be removed before beryllium determination with morin is possible. The aqueous extract may be evaporated in strong acid medium to avoid vapor loss of the beryllium–acetylacetonate.

Toribara and Chen (185) studied the recovery of beryllium from biological materials, involving acetylacetone extraction as one step, and established points of loss by using ^{7}Be tracer. A method was developed for satisfactory recovery. Acetylacetone extraction is used as one of the recovery steps for trace amounts of beryllium by a number of authors (4,136,-173,186).

Adam et al. (1) proposed a procedure for acetylacetone extraction of beryllium, but Merrill et al. (136), through the use of ^{7}Be, found many modifications necessary to adapt this procedure to the determination of very low beryllium in sediments and in sea water at 5×10^{-13} g. of Be/ml. (137). Acetylacetone separation is only one of the many separation steps used in the last three references.

b. With Butyrate

Beryllium can be separated from iron(III) and aluminum as the butyrate with chloroform from a solution containing EDTA at a pH of 9.7 (11,183). Sandell (164) suggests further investigation for trace analysis.

c. With Methyl Isobutyl Ketone

Molybdenum can be separated from beryllium by extraction of its thiocyanate complex with methyl isobutyl ketone (84). Copper as its neocuproene complex can be similarly extracted (83).

d. With 8-Hydroxyquinaldine

Beryllium may be separated from copper (in 4% beryllium alloy) by solvent extraction from a chloride medium (106). A chloride solution containing 5 to 30 γ of Be, 3 ml. of $0.1M$ KCN solution, 5 ml. of 10% NH$_4$Cl, and 3 ml. of 1% 8-hydroxyquinaldine is adjusted to pH 8.0 ± 0.2 with ammonia. The beryllium can be quantitatively extracted with 10 ml. of chloroform.

e. With Tri-*n*-Octylphosphine Oxide

Uranium and thorium can be extracted from beryllium nitrate (88) with tri-*n*-octylphosphine oxide. The extraction method has been used to separate uranium and a fluorimetric method is used to detect 0.1 p.p.m. of uranium when a 5-g. sample of beryllium is used. Ross and White (159) have used this reagent to extract thorium from sulfate and phosphate media.

3. Other Methods

a. By Electrolysis

Copper, nickel, and silver, in copper and nickel-base alloys, may be separated from beryllium by conventional electrolysis on a platinum cathode.

Heavy metals such as Fe, Zn, Co, Sn, Ag, Ni, Cu, Co, Mo, Bi, Cd, Tl, Au, etc., can be separated by electrolysis on a mercury cathode (112,129,185). This can best be accomplished in a water-cooled Melaven cell (133). Best results will be from a sulfuric or perchloric medium at a pH of about 1.5, 3 to 5 amp. d.c., and 6 to 15 v. The anode should be platinum in the form of a gauze or spiral in a plane parallel to and near the mercury surface. Agitation of the mercury will promote a cleaner surface for more efficient removal. A high salt concentration will retard deposition.

Beryllium can be separated from relatively large amounts of uranium (158) in a fused fluoride system by removing fluoride as fluoboric acid and by electrolytic deposition of the sexivalent uranium as hydrous oxide on platinum from an ammonium acetate solution at pH 4.0.

b. By Ion Exchange

Florence (64) has reported very efficient separation of such cations as Co, Cu, Cd, Cr, Fe, Mn, Mo, Zn, Zr, and U from beryllium by passing a $9M$ HCl solution through a strong anion-exchange resin in chloride form. Beryllium and aluminum pass through unchanged. Diallyl phosphate resins (104) are reported to absorb Be and U and to permit such metal ions as Ca, Cd, Cu, Al, Hg, Pb, Mn, La, Fe, Zn, and Zr to pass through. Uranium can be eluted, thus separating it from beryllium.

Beryllium can be separated from aluminum and iron by fixing (160) them as oxalates at pH 4.4 to 5 and passing through cationite (SBS). Beryllium is retained on the resin.

A table of equilibrium distribution coefficients, K_d, for 43 cations including Be, in hydrochloric acid, using cation-exchange resin AG 50W-X8. is presented by Strelow (181). Be can be separated (182) from Al(III), Fe(III), Y(III), Ce(III), Sr(II), Ba(II), Ga(III), La(III), and the rare earths, Zr(IV) and Th(IV).

Athavale et al. (7) separated beryllium from silica by passing a carbonate–borax beryl fusion dissolved in cold hydrochloric acid, containing EDTA and peroxide and adjusted to pH 3.5, through a cation exchanger, IR-120, in the sodium form. Beryllium oxide content of the ore, following this separation, was usually 0.6% higher than by any other procedures.

Toribara and Sherman (186) discuss ion exchange in separation of beryllium from biological materials.

Isolation of beryllium from various alloy metals by use of ion-exchange techniques was studied by Kallmann et al. (96).

Vetejska and Mazacek (190) report a method for separation of aluminum on a column while beryllium passes through.

Various methods of isolation in connection with gravimetric and spectrophotometric methods have been discussed by Hunt et al. (90).

IX. DETECTION AND ISOLATION

A. COLORIMETRIC METHODS

1. p-Nitrophenylazoorcinol (Zenia or p-Nitrobenzeneazoorcinol)

One of the easiest methods for qualitative detection (inexpensive, quick, relatively free of interferences) involves a spot test with Zenia (116). The powdered mineral (1 part), suspected of containing beryllium, is fused with sodium hydroxide (3 parts) in a nickel crucible. A water leach of this fusion product is used as the test solution. A synthetic standard made of a known amount of beryl and quartz should be used for a comparative test. A 0.03% weight/volume solution of Zenia is prepared by mixing the dye in 0.1N sodium hydroxide for a minimum of two hours, allowing this mix to stand one day, filtering through an asbestos mat, and storing in a red (low actinic) glass bottle. This solution is stable at room temperature for one month. A drop of potassium cyanide solution (to mask nickel color) is placed on a thick filter paper. Add two drops of the Zenia solution followed by one drop of the leach of the mineral fusion to the center of this brownish-yellow spot. A red to pink color confirms beryllium. As low as 0.2 mg. of beryllium (in 0.04 ml.) can be detected in this manner (61).

2. Quinalizarin (1,2,5,8-Tetrahydroxyanthraquinone) (62,97,164)

This dye has been suggested for a spot-plate test for beryllium. It forms a cornflower blue color, but is subject to many interfering elements including aluminum.

3. Quinizarin (1,4-Dihydroxyanthraquinone) (54)

This reagent is recommended as a fluorescent method of detecting as low as 0.013% beryllium in field mineral exploration. An alcohol solution (20% of saturation) in a water solution of a sodium hydroxide fusion is pink to orange in ultraviolet light.

4. Morin (a Tetrahydroxy Flavinol) (162)

Silicic rock is fused twice with sodium hydroxide and the water leach is combined as filtrate. A yellow-green fluorescence following the morin procedure verifies beryllium.

B. PRECIPITATION METHODS

Simple precipitation methods (hydroxide pptn.) may be used. In the case of a white gelatinous precipitate, in the presence of ammonium chloride, the presence of beryllium may be determined by adding enough sodium hydroxide to just dissolve the precipitate and by boiling the resulting solution (approximately $1.5N$). Beryllium will separate as a precipitate and aluminum will remain in solution. (The alpha beryllium hydroxide crystal, well formed, will be an octahedron.) A white gelatinous precipitate with ammonia, from a solution after oxine separation of aluminum and removal of the oxine with nitric and sulfuric acid, confirms beryllium.

C. EMISSION SPECTROGRAPHY

Where available, an emission spectrograph can be used for quick, convenient, and qualitative detection of beryllium. The powdered ore or mineral sample may be arced directly or mixed with carbon or some type of carrier, such as tin oxide or nickel oxide (126). This method for qualitative detection should be a good means of estimating the amount of beryllium in different ore samples, provided the matrix is relatively uniform. Very little work has been done on determining the matrix effect and its relation to line intensities of beryllium in ore and mineral analysis (Section X-C-3).

D. RADIOCHEMICAL METHODS

The nuclear reaction $^9Be\,(\gamma,n)\,^8Be$, using ^{124}Sb as the gamma-ray source (76% have an energy level above 1.63 M.e.v. (105), the threshold for ^9Be), is utilized for the qualitative and quantitative determinations of beryllium in ore, and has been proposed as a method of "picking" beryl (65,147). A field instrument (called the Berylometer) utilizing this principle, marketed by Isotope Specialties, Burbank, California (93), is available for field exploration. Other field instruments are being developed (21,22). Users of ^{124}Sb must be licensed by the U. S. Atomic Energy Commission. A satisfactory laboratory unit is now available (21).

X. QUANTITATIVE DETERMINATION

The method selected for the quantitative determination of beryllium will depend upon a number of factors: (*1*) the percent of beryllium in the sample; (*2*) possible interfering elements present; (*3*) accuracy required; (*4*) available equipment; (*5*) the number of determinations to be performed; and (*6*) cost of analysis compared with the time required to obtain such analysis, with due consideration to reliability of results.

When accuracy takes precedence over all other considerations (for macro amounts) in the establishment of the beryllium content of a primary standard (as a means of comparison for other methods of analysis) or in the purchase of ore, etc., the longer, more tedious, and perhaps more accurate gravimetric procedure either must or should be used. In standard products, in which the potential interfering element may be known from both a qualitative and quantitative standpoint, a more rapid volumetric, colorimetric, or other method may be chosen. The interfering element(s) may be easily separated or converted to a noninterfering complex ion or, if it interferes quantitatively and its concentration is known, perhaps correction may be applied.

If acceptable spectrographic standards are available, emission spectrographic methods may prove most rapid for the determination of beryllium in low-beryllium-content materials. For very small amounts of beryllium, fluorescent procedures are recommended. Three publications under the title *The Analytical Chemistry of Beryllium*, including methods for the determination of beryllium as well as of impurities in the metal, have appeared in recent years (134,157,191).

An excellent series of papers were presented at a symposium on the analytical chemistry of beryllium sponsored by the United Kingdom Atomic Energy Authority, on June 23, 1960, in England. These papers involved the quantitative determination of beryllium in various grades of ore (90), in atmospheres (90,139,169), and in other forms by colorimetric (80), fluorimetric (139), spectrochemical (169), and radiochemical techniques (22,138), and the identification of impurities (184,193), especially oxygen (39,148), in beryllium.

A. GRAVIMETRIC METHODS

The most widely accepted gravimetric method for the determination of beryllium involves the precipitation of beryllium from a solution, free from other interfering elements (as well as phosphate and fluoride), with ammonium hydroxide at a pH of 8.5 and ultimate ignition of this hydroxide to the oxide. Many references will suggest that this beryllium oxide should be ignited in platinum to constant weight. However, beryllium oxide has a rather high affinity for a small amount of moisture, and Hutchinson and Malm (92) have demonstrated that beryllium oxide will volatilize when heated in the presence of water. Grossweiner and Seifert (75) express the equilibrium constant for the reaction

$$BeO(s) + H_2O(g) \rightleftharpoons Be(OH)_2(g)$$

as

$$\log p_{Be(OH)_2} - \log p_{H_2O} = 1.63 - (9060/T)$$

It has been the authors' experience that, in the gravimetric determination of beryllium, if the precipitate and paper are completely dried, the temperature is raised slowly to 1100°C., and is maintained for a period of not less than two hours at this temperature, the carbon from the paper is completely oxidized and thermal decomposition of beryllium hydroxide to oxide is completed. The ignited sample is cooled in a desiccator over magnesium perchlorate and is covered and weighed quickly to determine the final weight of BeO. Subsequent ignitions "to constant weight" may give low results.

Duval and Duval (56) have presented data to indicate that decomposition of ammonia-precipitated beryllium hydroxide (in the presence of ammonium chloride in a chloride medium) is complete at 951°C.

The determination of beryllium as pyrophosphate has been suggested by a number of authors (4,90,91,129). Huguet and Bamberger (89) precipitate the beryllium from 0.75 g. of beryl, opened by carbonate fusion, as the phosphate in the presence of 5 g. of ammonium chloride and 7.5 ml. of Versene solution (Na salt of EDTA), with 10 ml. of 20% $NH_4H_2PO_4$. The solution pH is raised to 2.5 with concentrated ammonia and to 5.5 by the addition of about 50 ml. of $2M$ ammonium acetate. After boiling the precipitate is separated on medium paper and is washed with pH 5.5 ammonium acetate. The precipitate is dissolved in 30 ml. of hot 1:1 hydrochloric acid and beryllium is reprecipitated as the phosphate, adding 1 ml. of $NH_4H_2PO_4$, 7.5 ml. of Versene, aqueous ammonia, and ammonium acetate.

Possible silica interference is removed by the first phosphate filtration. The phosphate is decomposed at 900°C. to the pyrophosphate. Adolfo (2) also precipitates beryllium as the ammonium phosphate and ignites to pyrophosphate. This procedure generally gives slightly lower result than the conventional carbonate fusion, oxine removal of heavy mealts, and precipitation as hydroxide and ignition to oxide. Duval and Duval (56) recommend the ammonium–beryllium phosphate procedure. They present data indicating complete conversion of the ammonium–beryllium phosphate to beryllium pyrophosphate at 640°C.

This method has the advantage of elimination of the phosphorus separation (in beryl-ore analysis, for example), and the gravimetric factor for beryllium in the ignited pyrophosphate is about one-fourth that of beryllium in the oxide. The precipitate is more granular in nature and is more easily filtered than the gelatinous beryllium hydroxide. The use of ethylenediaminetetraacetic acid (EDTA) as a sequestering agent (153) to prevent the precipitation of potential interfering elements is, according to Hure (91) and others (65), a means of producing a spectrographically pure beryllium pyrophosphate. Equal weights of beryllium in oxide from the above procedure and from the pyrophosphate in this procedure were compared by emission-spectrographic methods, and the oxide precipitate was found to contain a much lower impurity content than the pyrophosphate.

Other gravimetric methods involve the precipitation of beryllium with 8-hydroxyquinaldine (141), as the complex double salt, beryllium–barium–fluoride (55), as the oxine (168), as the basic carbonate (155) and as the hexamine cobaltic complex (90).

B. VOLUMETRIC METHODS

Since beryllium exhibits only one state of oxidation (BeII), any volumetric determination must be by an indirect method. It can be completely precipitated from a sulfate, chloride, or nitrate medium at a pH of 8.5 (Section XI-A-2). The addition of an excess of sodium fluoride (low bulk density, large exposed area per unit mass) liberates hydroxide ions, which may be determined by titration with a standard acid.

$$M(OH)_z + 2ZF^- = (MF_{2z})^{-z} + Z(OH)^-$$

Although the reaction between beryllium hydroxide and sodium fluoride is not completely understood, it is believed to form a stable, complex beryllium fluoride ion $(BeF_4)^{-2}$ (63), and the reaction is encouraged to go to completion by removal of hydroxide ion by neutralization with a standard acid. The reaction does not appear to be stoichiometric. A high-bulk density sodium fluoride is not suitable for use in this reaction. The sodium fluoride that is used must be uniform in its residual acidic (or basic) constituents. The standard acid, which may be prepared at as low as $0.25N$ to as high as $1.0N$, must be checked daily with a standard beryllium solution to determine its beryllium equivalent. In as much as the reaction is not entirely stoichiometric the amount of standard beryllium solution must be chosen to approximate as nearly as possible the amount of beryllium under consideration in the sample. Interfering elements are aluminum, zirconium, uranium, hafnium, thorium, and the rare earths. In pure beryllium and the beryllium–copper alloys the only interfering element that may be present in quantities large enough to warrant separation or correction is aluminum. As a general rule, aluminum is determined in these materials and judicious correction is applied.

This titration can be performed satisfactorily by a person with proper color perception, using a mixed phenolphthalein–thymolphthalein–methyl orange indicator. The end point for visual detection is at a pH of approximately 8 or below. Magnesium (at 0.3% or more of beryllium present) will cause a very slow reaction at the end point, and may interfere unless sufficient time is allowed for equilibrium to be established in initial and final end-point determination. Although the accuracy is not as good in many cases, in the absence of potentiometric or automatic titrimers this operation can be performed satisfactorily with visual end-point detection. Larger samples (higher beryllium content) and stronger sulfuric acid are recommended for visual work.

In the potentiometric titration, a glass electrode suitable for high sodium or alkali content should be used. A completely automatic titrimeter using a $0.25N$ acid will give good accuracy. A simple arrangement, however, utilizing a pH meter, mechanical stirrer, and manually controlled acid buret for addition of titrant will give nearly comparable results. There is little choice between the two procedures from a time standpoint. Although the mixed indicator is not necessary in a potentiometric titration, it does serve as a safety factor in case trouble develops with the potentiometer.

In a series of titrations being compared to a standard, the sodium fluoride source used must not be changed. If the titrating conditions are rigorously maintained, this titrimetric method will result in good accuracy and, when a series of determinations must be performed, a relatively fast method of determining beryllium. Given the sample solution, one technician can make about ten determinations per hour. (Although the first published information describing this method appeared in 1950 (118,131), this procedure has been used for years.)

Singh et al. (176) have developed an acidimetric potentiometric method suitable for the determination of Be, Al, and Th. They use $1M$ KF solution as the fluoride ion source. They obtain an accuracy of $\pm 1.0\%$ for beryllium in the range 1.05×10^{-3} to $4.2 \times 10^{-2}M$, using $0.05084M$ HCl and quinhydrone and calomel electrodes. "Neutral" point for their work is pH 7.1 to 7.2 with a pH meter.

Athavale recommends an indirect volumetric method for beryllium in beryl. Following essentially the phosphate precipitation of Hure (91), the beryllium–ammonium phosphate is dissolved and the phosphate is titrated with standard bismuth perchlorate, using diallyldithiocarbamidohydrazine as an extraction indicator. He believes the gravimetric procedure of Hure (91) is subject to silica interference.

A survey of other volumetric methods has been made (134).

C. PHOTOMETRIC METHODS

1. Spectrophotometric

Methods in general and spectrophotometric methods in particular for the determination of beryllium have been closely scrutinized, and many new methods have been developed in recent years, especially for small amounts of beryllium. Methods for the determination of beryllium by using various dyes have been published recently (179), and surveys with many references are included in other publications (29,80,134,157,191). The methods chosen will depend upon interfering ions present, available equipment, the amount of beryllium in the sample, etc.

a. Acetylacetone (80,90)

In aqueous media at a pH of 7 to 8, in the presence of sufficient EDTA to complex interfering elements, beryllium will form a complex with acetylacetone that may be extracted with chloroform or benzene (Section VIII-B-2-a). In the absence of other impurities that absorb in the ultraviolet region, absorbance may be determined on the chloroform extract at 295 mμ. Three extractions are necessary for complete Be recovery (90). Loss of Be may occur in various steps, as demonstrated by use of a ^{7}Be tracer (136,173,185,186).

b. Alkanan or Naphthazarin (187,189)

A blue color develops quickly in the presence of beryllium in a buffered solution at a pH of about 6.4. The color is stable and pH control is not critical. The beryllium must first be isolated, since aluminum, copper, iron, and fluoride ions interfere. Gum arabic is necessary to prevent flocculation of the beryllium color lake. Problems may be encountered in obtaining or preparing these two compounds in required purity.

c. Aluminon (80,85,117,128)

Beryllium forms a red complex with aluminon (aurintricarboxylic acid). Aluminum can also be determined by the use of aluminon and must be absent or rendered inactive by a chelating or sequestering agent. An intense red complex is formed with a relatively small amount of beryllium (suitable range 0.005 to 0.100 mg. of Be in 100 ml. of solution). Elements such as copper, aluminum, iron, cobalt, and nickel found in commercial copper-base alloys, as well as elements such as zirconium, titanium, and zinc, can be rendered inactive by the use of EDTA. Fluoride and phosphate have a bleaching effect, but magnesium does not interfere. The rate of color development is dependent upon temperature. The pH must be maintained between 6.2 and 6.6. Color also develops continually over an extended period of time. A 20-minute standing time is recommended, with absorbance measured at 530 mμ.

d. Berillon (80,90,98)

Berillon (the tetrasodium salt of 8-hydroxynaphthalene-3,6-disulphonic acid-(1-azo-2')-1',8'-dihydroxynaphthalene-3',6'-disulphonic acid) is colored magenta to violet in solution (pH 12 to 13), whereas beryllium forms a blue complex. Ascorbic acid may be used to reduce iron to the ferrous state in the acid medium, and EDTA is used to complex other possible interfering elements. Absorbance is measured at 630 mμ and calibration curves are determined simultaneously with samples. The calibration curve is smooth over the range 4 to 16 γ of beryllium per 50 ml. of solution, but it shows an appreciable flattening at higher concentrations. The method has been used largely for swab samples and low-grade ore. The order of addition of reagents is very important, and the absorbance must be measured within 30 minutes or excessive color fading may take place.

e. Eriochrome Cyanine R (Also Called Solochrome Cyanine) (80,85)

This dye, which forms a red complex with beryllium, is recommended for use without separation for determination of beryllium in aluminum, steel, copper, titanium, and mixed oxide. The color is developed at a pH of 9.8, has its maximum absorbance at 512 mμ, and Versene and cyanide are used to eliminate interference from other ions. The method claims accuracy at 0.002% Be. Time factors and order of addition of reagents are also important for reliable results.

f. Naphthachrome Azurine 2B (Also Known as Naphthachrome Green) (5)

This method is generally limited to small amounts of beryllium and particularly for the determination of beryllium in biological materials.

The beryllium must be separated by precipitation as the phosphate, with aluminum phosphate as the carrier. The precipitate is then dissolved in trisodium phosphate at a pH between 11.0 and 11.3. Dye concentration is critical, and color develops in about 20 minutes at 30°C. Maximum absorbance is at 650 mμ.

g. p-Nitrophenylazoorcinol (Zenia) (42,191,196)

In a slightly alkaline medium maintained by a borate–citrate–sodium hydroxide buffer, this dye forms a red-brown lake with beryllium. Although this dye is not as sensitive as many others, it is relatively specific for beryllium, being subject to less interference from foreign ions than most of the other dyes. Many possible interfering ions, such as copper, iron, nickel, magnesium, calcium, etc., can be rendered inactive by the buffer (17.8 g. of sodium citrate dihydrate, 8.5 g. of sodium borate decahydrate, and 14.35 g. of sodium hydroxide in 100 ml. of water), or by the use of chelating agents such as EDTA. The pH must be rather rigorously controlled. If alkalinity (pH 5.5), dye strength (75 mg. in 250 ml. of 0.1N sodium hydroxide), and temperature are carefully standardized in the sample and the standards used to prepare the calibration curve, very satisfactory results are obtainable. The dye does decrease in strength with time, and standards must be carried through the determination along with the reagent blank and sample. The sample solution may be as the fluoride, sulfate, or nitrate, and 10 minutes are required for color development. The maximum relative difference in absorption between dye and the colored beryllium complex is at about 515 mμ. A 20-mm. optical path is recommended.

h. 2-Phenoxyquinizarin-3,4-Disulfonic Acid

Owens et al. (144) claim detection of one part Be in 1.25×10^8 parts of solution. The stable violet complex develops in one hour and is measured at 550 mμ. Of 65 elements investigated, Cr(III), Mg(II), Zr(IV), Th(IV), F^-, and PO_4^{-3} interfere seriously. The pH must be controlled to 6.0 ± 0.1.

i. Quinizarin-2-Sulfonic Acid (1,4-Dihydroxyanthraquinone-2-Sulfonic Acid) (43)

This dye is suitable for small quantities of beryllium, to 20 γ in 20 ml., using a Beckman DU spectrophotometer at 575 mμ in a 1-cm. cell. Fluoride, aluminum, magnesium, and phosphate interfere, with copper, nitrate, calcium, and zinc interference being rather serious. Gum arabic is used as a stabilizer. The color is developed in boiling water for 10 minutes; pH is maintained at 6.5. The color is stable for several hours.

j. Sulfosalicylic Acid (132)

This method can be used satisfactorily to determine beryllium (0.002 to 0.25% Be) in aluminum at a pH of 9.2 to 10.8. The aluminum is partially complexed with EDTA, and standards used to prepare the calibration curve must contain the same amount of aluminum as the sample. Iron can be eliminated by extraction with β'-β'-dichlorodiethylether. Copper interferes, but can be eliminated by electrodeposition. Nitrate and fluoride interfere seriously, acetate slightly. The work can be carried out successfully in either sulfuric, perchloric, or hydrochloric acid media. The maximum absorption is at 315 mμ.

k. Others

Chrome Azurol S (174), also known as Solochrom Brilliant Blue B, Polytrop Blue R, and British color index No. 723, is suggested as a color agent for beryllium and fluoride. A lake is produced in a neutral medium with beryllium.

2. Ultraviolet Fluorescence

Because of the tremendous increase in the industrial use of beryllium and its compounds in recent years, and in an effort to better study its toxic effects, more exact methods for the determination of small amounts of beryllium were needed. The fluorimetric method using morin (2',4',-3',5',7-pentahydroxyflavone) has been studied extensively in recent years and is the most promising method for the determination of submicrogram quantities of beryllium in air and in urine, bone, and other biological materials. This material has the highest inherent sensitivity of any known reagent for beryllium (173). Although, according to Toribara and Sherman (186), a spectrographic method may be more sensitive, a fluorimetric method is generally more reproducible and reliable at very low concentrations and equipment for fluorimetric methods of determination is generally available at a much lower cost. A number of photo fluorimeters are available as such on the market, and many colorimetric instruments have fluorimetric attachments. The methods of excitation of fluorescence, fluorescence detectors, and instruments (some modified) for this work differ among various authors.

It is generally agreed that for best sensitivity, commercially available morin must be purified (24,78,140,150). One recent publication (192) gives available commercial suppliers of high-purity morin, and another (173) recommends additional purification of the morin for greatest potential sensitivity. Very complete studies have recently been reported for the dissolution of the sample, isolation of beryllium, and determination of factors affecting the use of morin for the determination of beryllium in air samples (192), in air samples and biological materials (173), in biological materials (112,178,185), and in the determination in general of microgram quantities of beryllium (193,186). Morin concentrations varying from 2×10^{-5} (195) to 2×10^{-3} (121) have been suggested. In general, a higher sensitivity is attained at a lower morin concentration. It has been found that, as the sodium hydroxide content increases above the optimum concentration, there is a quenching effect on the fluorescence.

Another important conditions that must be controlled is the temperature especially at higher beryllium concentration. As morin concentration increases the temperature effect becomes less. A 1°C. change in operating temperature may change the fluorescence by as much as 5 or 6%. The presence of copper may cause air oxidation of morin and low results (173). The use of an alkaline stannite reducing solution is recommended to prevent oxidation by air. Standards and samples must be handled in exactly the same manner to produce similar salt concentration, pH, temperature, and time, all of which are important factors in the determination of beryllium by the use of morin. Strong ultraviolet sources for excitation of fluorescence are to be avoided, as they may also cause rapid fading of the fluorescence. An automobile headlight as a means of fluorescence excitation is recommended (173,186). Fixed diaphragm openings between the ultraviolet light source and the sample are used to diminish light intensity (121).

3. Spectrographic

Relatively speaking, beryllium has few spectral lines from which to choose for analytical purposes. Whereas many elements have a large number of lines of varying intensity in a wide spectral region, the lines from beryllium are very intense or very weak by comparison. The more intense lines do, however, appear in the spectrum in areas relatively free of line interferences from other elements.

Although volumetric or colorimetric methods are generally more suitable and reliable for the determination of beryllium in copper-base alloys, spectrographic methods are frequently used at beryllium content of 2% and under. A common matrix can be established for the standards and samples, but variations in beryllium content of 0.10% at the 2% content level, due to segregation, can cause serious problems in its spectrographic determination. The above statements are generally true for a point-to-plane technique.

Even more serious segregation problems are experienced in aluminum-base master alloys, making spectrographic determination of beryllium on solid samples of this alloy even less reliable.

Spectrographic methods have been utilized in the determination of beryllium in air-dust samples (13,28,35,151,169). In this work, separation of the beryllium is considered necessary in order to establish a uniform matrix and to include an internal standard in the sample. Problems are involved in maintaining a low volume of solution in order to obtain beryllium concentrations that can be read, or in transferring the entire dried sample to an electrode for excitation by arc or spark methods. The spectral lines generally used in the determination of microquantities of beryllium are the 2348.6 and 2650.5 A. lines, when the Al 2367.1 A. line is used as the internal standard (13). The beryllium lines at 3130.4 and 3131.07 A., with the aluminum internal standard line at 3059.9 A., are recommended (178) for the determination of beryllium in urine. Thallium (35), gold, and scandium (169) also have been suggested as internal standards. Baer and Hodge (9) have compared solution techniques in general for spectrochemical determinations.

If spectrographic methods are to be used for air samples or biological materials, the unit should not, of course, also be used for the determination of impurities in pure beryllium, beryllium oxide, or high-beryllium-content alloys. The technique involved would depend upon the need for rapid evaluation of the sample, in which event the solution technique might be advisable; or, if rapid results are not necessary, perhaps a more reliable procedure involving separation of potential interfering ions would be advisable for greatest accuracy (13,35,152). If information is needed immediately to monitor a particular operation in an industrial plant, a continuously operating spectrographic monitor has been suggested (169). Excitation conditions must be established that will not be subject to bias because of the form of beryllium (compound, metal, or refractory oxide) or the particle size. In general the d.-c. arc is recommended, and air samples are automatically selected on filter paper that is mechanically and automatically moved to the arcing position. With proper calibration an almost immediate answer is available for the air sample in this particular area.

Hunt and Martin (90) report that as little as 30 p.p.m. of BeO can be detected in low-grade ores by using a 1:9 ratio of sample to a flux containing sodium borate and graphite, with vanadium as an internal standard. A reasonable facsimile of the sample matrix must be produced,

D. RADIOCHEMICAL METHODS

By counting the photo neutrons resulting from the nuclear reaction ^{9}Be (γ,n) ^{8}Be, a quantitative as well as qualitative (see Section XI-D) estimate of the beryllium content of an ore is possible (66,138,146). This method of determination will provide results in minutes, compared with the hours required for wet chemical methods. Reliable standard samples of a known beryllium-content range must be available for comparison, and high thermal neutron absorbers such as B, Cd, Sm, and Gd (138) may cause a reduction in apparent beryllium content.

Proper handling and shielding of equipment is a must to protect personnel and environs at all times. (The ^{124}Sb must be replaced every six months.) At present, initial equipment cost will be in excess of $2000. Claims of an accuracy of 0.03% or better for beryllium oxide content and of no interference from the presence of other materials have been made (21).

Milner and Edwards (138) have studied the geometry of the necessary equipment, determined potential interfering elements in both solid and liquid state, proposed shielding to eliminate interference, and demonstrated excellent agreement between the photo neutron and acetylacetone method for the determination of beryllium oxide in various grades of ore and ore fractions. Bisby and Hale (22) have described the development of portable (γ, n) electronic equipment for use by field geologists.

Bradshaw et al. (27) have described gamma-ray activation methods for the determination of oxygen, carbon, and the combination of iron, copper, and nickel in beryllium metal. The nitrogen content can be determined by difference when the iron and copper content are known from chemical procedures.

XI. RECOMMENDED LABORATORY PROCEDURE

A. METAL ASSAY

1. Gravimetric

Different grades of beryllium contain varying amounts of oxygen and other impurities, and it is therefore considered advisable, for the purpose of establishing beryllium content in metal for standard use, and for reference purposes, to use the conventional gravimetric procedure by precipitation with carbonate-free ammonium hydroxide and igniting to the oxide. As a general rule vacuum-cast beryllium, or powder produced therefrom, is of sufficient purity that interfering elements may be removed by precipitation as the quinolates at pH of 5.7 from an acetic acid–ammonium acetate buffered solution. The beryllium may then be precipitated from the filtrate as hydroxide after wet oxidation of the excess quinolate with sulfuric and nitric acid. Ammonium hydroxide should be added slowly, with constant agitation, to pH 8.5. The precipitate is washed with 2% ammonium acetate at pH 8.5 and is dried carefully, the filter paper is burned off, and finally it is ignited at 1100°C. for 2 hours.

If the total quinolate weight indicates 0.10% or less of aluminum, after correcting for other heavy metal quinolates such as iron, nickel, etc., it is considered unnecessary for practical purposes to make a correction for equivalent beryllium if the solution is to be used as a standard for volumetric comparison. At a higher per cent of aluminum it is usually advisable first to remove the heavy metals from the beryllium-bearing solution by mercury electrolysis, followed by precipitation of the aluminum as quinolate, so that any equivalent beryllium may be calculated if the solution is to be used for standard volumetric determinations (Section XI-A-2).

Most commercial beryllium available today contains a total of about 98.4% beryllium (minimum), 1% oxygen (equivalent nonvolatile Be (as BeO) as determined by HCl volatilization), 0.12% carbon, 0.15% iron, and approximately 0.3% of other impurities, few of which ever exceed 0.05%.

2. Volumetric

The simplest, most economical, and perhaps most reliable method for determination of beryllium in the pure metal is the volumetric procedure. The amount of beryllium in the sample should be chosen to approximate closely the amount of beryllium in the standard solution, since the reactions do not appear to be stoichiometric.

It is recommended that 0.250 g. of beryllium be used for this work. These solutions should be free from interfering elements or, if interfering elements are present and their amount determined, judicious correction may be applied. As previously stated it is believed that a complex beryllium fluoride ion is formed, with beryllium hydroxide being the basic constituent supplying hydroxide ion. The following three equations are offered as a suggestion or explanation of the reactions that take place. As may readily be seen, acidic or basic constituents present in the slightly soluble sodium fluoride can readily affect the measured hydrogen ion to remove hydroxide ion from the possible equilibrium condition expressed in the second equation

$$Be^{+2} + 2(OH)^{-1} \rightarrow Be(OH)_2$$

$$Be(OH)_2 + 4NaF \rightleftharpoons (BeF_4)^{-2} + 4Na^{+1} + 2(OH)^{-1}$$

$$(OH)^{-1} + H^{+1} \rightarrow HOH$$

The beryllium containing solution should be diluted to approximately 300 ml. and 40% sodium hydroxide solution added dropwise with constant stirring until a precipitate is persistent but not permanent. This solution should then be cooled to approximately room temperature and 40% caustic should again be added to the first permanent precipitate. For visual end-point determination, and as a check on the operation of the automatic titrating equipment, an indicator (12 to 15 drops) made up of a 1% solution of phenolphthalein in ethyl alcohol (two parts), 0.1% solution of thymolphthalein in ethyl alcohol (two parts), and 0.1% solution of methyl orange in distilled water (1 part) is added. Dropwise addition of 40% sodium hydroxide is continued until no change is noted in the intensity of the purple color (pH 10.5 to 11). A mechanical mixer is preferred for the agitation in the addition of the sodium hydroxide, since it is believed that there is a better chance for breakdown of any possible agglomerates of beryllium hydroxide precipitate, which may enclose caustic that later would be titrated as equivalent beryllium. This solution is permitted to stand for five minutes. Waiting or standing time is important and standards for comparison with samples should be permitted to stand exactly the same length of time prior to starting titration to the first end point.

Clean electrodes, mechanical stirrer, etc., are placed in the sample solution, and the standard acid (preferably between 0.55 and 0.60N) is added slowly until pH 8.5. Since the reaction at this pH is rather slow, dropwise addition of the standard acid is continued until the pH is constant for not less than 30 seconds. Twenty g. (± 0.2 g.) of low-bulk density sodium fluoride is added to the sample and standard acid is added at a rate not to exceed 1 ml./second. Evidence of an indicator color change will be noted after the addition of about 90 to 95% of the standard acid. In automatic titrimeters with automatic acid-addition control the pH may fall below 8.5 momentarily as the last few drops of acid are added. If visual end-point detection is practiced, as previously stated, the pH indicator change is at about 8 and, unless the acid addition is carefully controlled manually, the pH may drop below 8 for greater time intervals. For best control of the operation, standard samples should be titrated along with samples in a series of determinations at intervals of not less than 6 minutes. The average beryllium equivalent of the standards should be used to determine beryllium content of samples.

Key points in this determination are the standing time after addition of a sodium hydroxide solution, uniform rate of addition of acid prior to and after addition of sodium fluoride, the use of low-bulk density sodium fluoride, and the use of the same sodium fluoride from the same bottle throughout the series of titrations. A pH of 8.5 must be maintained for a period of not less than 30 seconds after the last addition of standard acid, both on first and second end-point titrations. Very careful work indicates that the values obtained are correct within ±0.25% of the total beryllium content of samples approaching 100% Be.

B. DETERMINATION OF OXYGEN IN BERYLLIUM METAL

A number of methods are now being used for the determination of oxygen in beryllium metal. Vacuum-fusion techniques used on metals such as titanium, zirconium, thorium, and other high-melting metals have been used for beryllium (25,72). It has generally been found that, although as many as 18 samples of zirconium and other nonvolatile metals may be dropped into the platinum bath in graphite, only four to five samples of beryllium may be added prior to changing the platinum in the bath. Because of the higher vapor pressure of beryllium, it appears that it acts as a "gettering" agent after several samples have been fused, and low results are obtained on the samples added to the bath at the end of the series. The addition of tin (148) to the platinum bath is said to reduce the vapor pressure of beryllium and improve reliability of the oxygen values obtained. Another problem in the vacuum-fusion technique is the difference in density between beryllium and platinum. At 1900°C., operating temperature of the bath, beryllium oxide reacts with carbide to produce beryllium and liberates carbon monoxide.

Parker (148) has tried a number of modifications of the standard techniques of vacuum fusion for heavier, high-melting metals and compared these values with those obtained by chemical methods. The chemical methods include HCl and chlorine volatilization and selective solvents, including 10% hydrochloric acid, copper sulfate, and bromine–methanol. He also reports that concentration of the solvent medium must be varied according to physical characteristics of the metal. Satisfactory results can be obtained by the bromine–methanol procedure with flake metal, but high results are likely to be obtained on the powder and cast metal compared with other methods. Hydrogen and nitrogen also can be determined in some of the heavy high-melting metals, but the determination of these two elements in beryllium by the vacuum-fusion technique has been found very unsatisfactory. This procedure does have one advantage in that it is a direct measurement of oxygen in the metal.

It has also been suggested that oxygen content of beryllium metal may be determined by a fusion technique in molten platinum in graphite crucibles, with an inert gas atmosphere instead of vacuum (59). Such a procedure has been successfully used for the determination of oxygen in zirconium and Zircaloy by Elbing and Goward (59), using a 7:1 weight ratio of platinum to sample at 900 to 3700 p.p.m. of oxygen. At 1000 p.p.m. oxygen content the method is accurate to ±6%.

Codell and Norwitz (38) have proposed a method involving high-temperature bromination of a mixture of titanium and carbon, with liberation of the oxygen as CO and oxidation to and determination as CO_2. Although this method may be adaptable to beryllium, it would appear that existing methods are more attractive from the standpoint of practical application.

Various possible applications of activation analysis for the determination of oxygen in beryllium have been proposed. Coleman (39) briefly describes the use of thermal neutrons to produce the unstable nuclide ^{19}O, and suggests the lower limit of detection would be about 0.1% oxygen. He also describes a fast-neutron technique with the use of an accelerator, in which the lower limit may be 0.001% oxygen. Gamma rays and charged particles also are proposed as possible procedures. In the case of charged particles the geometry of the sample used becomes of great importance.

Osmond and Smales (143) have proposed the use of thermal neutron activation of a mixture of beryllium metal powder and lithium fluoride, resulting in the following reactions:

$$^6Li + n' \rightarrow {}^4He + {}^3H$$

$$^{16}O + {}^3H \rightarrow {}^{18}F + n'$$

Since the half life of the ^{18}F generated is 112 minutes, it is possible to chemically separate this fluoride ion by distillation for counting, thereby reducing or eliminating possible interferences. One of the limits of the method is the possible oxygen contamination of lithium fluoride. A blank from lithium fluoride may run as high as 0.1 to 0.2% oxygen. Possible contamination (oxidation) of the beryllium metal powder also may take place unless this powder is generated in an inert atmosphere.

Beard et al. (18) have done extensive work in photon activation of oxygen in beryllium.

Although the above-mentioned techniques are specific for oxygen, other methods, following an indirect approach, are generally used for control and routine analysis. The two most widely used procedures are the hydrogen chloride volatilization and the methanol–bromine selective-solvent technique.

Other selective-solvent methods (148) have been proposed. From the selective solvent using copper sulfate solution containing mercuric chloride, the insoluble beryllium oxide is dissolved in aqueous ammonium bifluoride. Titration procedures are described for the determination of beryllium utilizing the $(BeF_4)^{-2}$ reaction. Selective dissolution by use of 10% HCl in cold solution has also been suggested (148), with the insoluble beryllium oxide eventually determined by a colorimetric method using p-nitrophenylazoorcinol. It is found unsuitable for certain types of beryllium powders unless the powder is first heated in vacuum at 1000°C. for half an hour or more to render the low-temperature oxide less soluble.

The most widely used selective-solvent procedure depends upon dissolution of the sample in a methanol–bromine mixture, as described by Eberle and Lerner (57). (This work was done with vacuum-cast and electrolytic samples—no material from hot-pressed or sintered block.) This method again has the same shortcoming as the 10% HCl selective dissolution, since certain metal powders that are produced in a manner to generate beryllium oxide at a low temperature are likely to show low results compared with HCl evolution or vacuum fusion. The bromine–methanol procedure does not necessarily measure oxygen but rather insoluble beryllium, which is determined as such and then calculated as beryllium oxide or the oxygen equivalent. In vacuum hot-pressed commercial beryllium the bromine–methanol procedure generally gives higher beryllium oxide content than is obtained by other methods. Under carefully controlled conditions of temperature, time of contact of the solvent medium with the metal, and other variables, such as the use of a standard pore-size filter medium for removal of the insoluble oxide, it is possible to duplicate results within rather close tolerances.

Table VII gives dissolution conditions practiced by four laboratories when employing the bromine–methanol method for the determination of insoluble beryllium and calculating it as BeO. Laboratories A, B, and D use the same sample weight, but D uses a larger amount of bromine–methanol. Laboratory A uses a stronger hydrochloric acid–methanol solvent. Laboratory C, however, uses the highest ratio of solvents to sample.

Table VIII gives results of analysis by these four laboratories on eight samples of powder and chip beryllium. Laboratory A has determined the nonvolatile Be from anhydrous HCl evolution and reported this result calculated as BeO. It will be noted that, except for sample 7, the HCl

TABLE VII
Bromine–Methanol Dissolution Conditions

Conditions	Laboratory			
	A	B	C	D
Sample weight, g.	0.50	0.50	0.080	0.50
Methanol, ml.	10	10	0	10
Bromine in methanol (6% vol./vol.), ml.	40	40	25	50
Exposure time, approx. min.	10–15	10–15	8–12	10–15
Hydrochloric acid in methanol				
8% vol./vol., ml.	—	50	25	50
12% vol./vol., ml.	50	—	—	—
Exposure time, min.	30	30	30	30
Insoluble separation-filtration	Fine glass frit	Paper	Paper	Fine glass frit
Determination	Color	Volumetric	Color	Color

result is measurably lower than the bromine–methanol result. Sample 5 is known to contain excessive amounts of Be_3N_2, which react with anhydrous HCl at elevated temperatures of the procedure. Laboratory B determines insoluble Be by the volumetric method.

TABLE VIII
Determination of BeO (%) in Beryllium Metal by Laboratories A, B, C, and D

	By HCl method[a]	By bromine–methanol method			
Sample	A	A	B	C	D
1	0.81	1.33	1.42	1.49	1.45
2	1.72	2.34	—	2.66	2.87
3	1.80	2.14	2.70	2.40	2.55
4	1.74	2.39	2.43	2.59	2.76
5	1.80	10.01	10.10	9.80	10.80
6	2.36	2.77	2.79	2.72	2.51
7	2.69	2.61	2.71	2.74	2.60
8	—	2.50	2.58	2.29	2.41

[a] Anhydrous HCl method.

Examination by x-ray diffraction of bromine–methanol insolubles in commercial metal by laboratory A has detected BeO, Be_2C, Be_3N_2, Si, and free Be. Emission-spectrographic examination of portions of the same insolubles has indicated the presence of Al, Fe, and Mg in the same ratio as in the parent metal and Si at nearly 30 times the concentration in the parent metal. Aluminum may interfere in the volumetric methods used by laboratory *D*. Other laboratories determine the Be by color methods without interference. All laboratories dissolve the residues in sulfuric acid for determination.

The HCl volatilization or evolution technique is actually a method for the determination of nonvolatile beryllium, which is determined as such and calculated as beryllium oxide or the oxygen equivalent. It is apparently not subject to interference by nitride or carbide (191), since at operating temperatures the gas reacts with these compounds but not with beryllium oxide, regardless of manner in which it is formed. Care must be exercised to avoid oxidation of the beryilium metal in the train, thus resulting in values above the true oxygen content. More stringent control of the operation is necessary to avoid possible contamination of the air with beryllium chloride vapors should train blockage occur and vapor evolution escape into the room. Care also must be exercised in the HCl evolution technique to avoid sweeping away the residual BeO, which generally appears as a lacework after volatilization of the solid beryllium or compacted-powder pellet sample. The samples are placed in high-fired beryllium oxide boats, the nonvolatile residue is removed from the boat and dissolved in sulfuric acid, and the beryllium is determined by colorimetric procedure, using *p*-nitrophenylazoorcinol. A standard is included with a series of samples in each furnace load in order to assure proper functioning of the equipment for greater assurance of reliable results (157).

The last-mentioned procedure utilized equipment available in almost any laboratory and requires a 0.5-g. sample (or less) per determination.

TABLE IX
Determination of Breyllium Oxide (%) in Beryllium Metal by Different Methods in Different Laboratories

	Laboratory				
	1	2	3	4	5
Sample	By bromine–methanol method			By HCl method	By vacuum fusion
1	1.37	1.1	1.33	1.45	1.17	
2	1.81	1.6	1.80	1.67	1.58	
3	1.91	1.8	1.83	1.83	1.81	
4	2.37	2.3	2.22	2.34	1.80	
5	1.89	1.7	1.87	2.11	1.71	
6				1.39	1.05	0.97
7				2.22	1.81	1.72
8				3.17	2.60	2.53

This has a considerable advantage over the vacuum-fusion technique in avoiding possible segregation and obtaining a more representative powder sample. The vacuum-fusion technique may use a sample as small as 10 mg. Type of sample, available equipment, etc., will dictate the choice of method used for the determination of the oxygen or beryllium oxide content of beryllium metal.

The results reported in Table IX are intended to demonstrate the unsolved problems concerning the determination of oxygen in beryllium. It is noted, however, that the percentage of BeO by HCl evolution is generally lower than by bromine–methanol. In three cases vacuum fusion compares favorably with the HCl evolution, but the results are lower. Parker (148) verifies this observation. It will also be noted that laboratory *2*, using bromine–methanol methods, is in excellent agreement with laboratory *4*, using HCl evolution, on four of five samples.*

The pelletized powder or solid beryllium (0.5 g.) is placed in a quartz tube with a small outlet in high-fired BeO boats, and the quartz tube, including the HCl generator (hydrochloric and sulfuric acids), is flushed with helium followed by anhydrous HCl. The quartz tube with samples is moved into the hot zone of the furnace at 650°C. for one-half hour. The temperature is raised to 750°C. for an additional half hour and finally to 800°C. for 15 minutes. The nonvolatiles are carefully transferred to a beaker and dissolved in fuming sulfuric acid, and the beryllium is determined colorimetrically (157).

C. BERYLLIUM IN COPPER ALLOYS

A volumetric method is recommended for the determination of beryllium in copper-base alloys at greater than 2% Be. An automatic titrimeter or potentiometric titration procedure is recommended, using a $0.25N$ acid (Section XI-A-2). The beryllium standard used in this procedure should be prepared from beryllium metal of known assay dissolved in nitric acid, and a standard beryllium–copper solution should be prepared that will compare favorably with the type of alloy under consideration in so far as beryllium-to-copper ratio is concerned. There is some evidence of a slight loss in beryllium following the standard separation procedure for copper from beryllium, utilizing the soluble copper–ammonia complex.

The beryllium is separated from the copper in a nitrate medium by addition of ammonium hydroxide, with constant stirring, until no further color change is noted in the intensity of the blue copper–ammonia complex. The insoluble beryllium hydroxide is removed by filtration and is washed

* Gilman and Isserow (68) report that radioactivation gives lower oxygen content than either bromine–methanol or HCl evolution methods.

with ammonium nitrate solution (1% at pH of 7 to 8.5). The major part of the copper may be washed from the precipitate by this operation, as is evidenced by the removal of the blue color. The precipitate is dissolved in warm hydrochloric acid (25%), cooled, and handled as previously described (Section XI-A-2). (Note: beryllium may be separated from nickel-base alloys in this same manner.)

The colorimetric aluminon procedure is recommended for the determination of beryllium in standard copper-base alloys at less than 2% beryllium content. The determinations can be performed according to the method of Luke and Campbell (128), which was slightly modified in the adopted procedure described by ASTM in its Method E 106–54T. Standards and samples must be in the same medium and handled in exactly the same manner, since temperature, time, and order of addition of reagents must be carefully controlled in order to duplicate conditions and results.

D. BERYLLIUM IN ALUMINUM ALLOYS

The colorimetric procedure using p-nitrophenylazoorcinol is recommended for determining beryllium in aluminum-base alloys. Since the beryllium complex does not exactly follow Beer's law, standards utilizing a composition similar to the alloy in question should be prepared and one or two standards carried along with each series of determinations to establish the curve position for reading or determining beryllium content in the samples. These alloys may contain varying amounts of magnesium, and it is recommended that EDTA be used as a complexing agent in standard and sample to avoid interference by magnesium and other possible interfering elements. For very low-beryllium-content aluminum-base alloys, separation by acetylacetone or the method of Gooch and Havens, referenced by Churchill (36), is recommended prior to colorimetric determination of beryllium.

E. BERYLLIUM IN ORES AND MINERALS

For commercial-grade beryl ores the sodium fluoride fusion for decomposition of the ore is recommended, followed by dissolution in sulfuric acid and evaporation to sulfuric fumes. The sample, after fuming, is dissolved in water in the original platinum dish, the solution is transferred to an appropriate volumetric flask, and an aliquot is selected to provide 0.3 to 0.4 mg. of beryllium, removed for colorimetric comparison using p-nitrophenylazoorcinol. Since this dye is nearly specific for beryllium, it is seldom necessary to use any additives or chelating agents to render inactive possible interfering elements.

In lower-beryllium-content minerals the acetylacetone extraction procedure is recommended as the best method of separating beryllium from many potential interfering elements (164). The papers by Hunt and Martin (90) and by Sill and Willis (173) are highly recommended as reference reading prior to a choice of method to be used.

Ahrens (3) recommends the 2348.6 A. line for detection of beryllium in minerals down to 0.0001%, and the 3130.4 and 3131.1 A. lines for greater concentrations.

F. BERYLLIUM IN AIR

Various types of air-sampling equipment depending upon some type of filter medium, are used to remove beryllium from the air steam. In other cases, swab samples from surfaces are used for monitoring beryllium content in air. The methods chosen for determination of beryllium in air or swab samples will depend to a great extent on the potential interfering elements that may be associated with the beryllium, the quantity of beryllium that may be collected in a sample, and the permissible time lapse between collecting the sample and learning the results of the determination. When a relatively large amount of beryllium may be collected in the sample, it may be possible to use p-nitrophenylazoorcinol. This is suggested since it is relatively free of interference from other elements. When high sensitivity is desired or required, the procedure using morin, recommended by Sill and Willis (173) or Walkley (192), is recommended (see also Sections X-C-1, 2, and 3).

REFERENCES

1. Adam, J. A., E. Booth, and J. D. H. Strickland, *Anal. Chim. Acta*, **6**, 462 (1952).
2. Adolfo, A., "Rapid Determination of Beryllium in Beryl," in *Proc. 2nd Intern. Conf. Peaceful Uses At. Energy, Geneva, 1958*, **3**, 598 (No. 1579) (1958).
3. Ahrens, L. H., *Spectrochemical Analysis*, Addison-Wesley, Cambridge, Mass., 1954, p. 169.
4. Airoldi, R., *Ann. Chim. Rome*, **41**, 478 (1951).
5. Aldridge, W. N., and H. F. Lindell, *Analyst*, **73**, 607 (1948).
6. Anderson, W., R. E. Bentley, R. P. Parker, J. O. Crookall, and L. K. Burton, *Nature*, **187**, 550 (1960); through *Nucl. Sci. Abstr.*, **14**, No. 21825 (1960).
7. Athavale, V. T., and 13 other authors, "New and Improved Methods of Analysis of Some Nuclear Raw Materials," in *Proc. 2nd Intern. Conf. Peaceful Uses At. Energy, Geneva, 1958*, **3**, 554 (No. 1671) (1958).
8. Atlee, Z. F., *Mod. Metals*, **1**, 7 (1945).
9. Baer, W. K., and E. S. Hodge, *Appl. Spectroscopy*, **14**, 141 (1960).
10. Balomey, R. A., and L. Wish, *J. Am. Chem. Soc.*, **72**, 4483 (1950).
11. Banerjee, S., A. K. Sundaram, and H. D. Sharma, *Anal. Chim. Acta*, **10**, 256 (1954).

12. Barbaras, G. P., C. Dillard, A. E. Finholt, T. Wartik, K. E. Wilzbach, and H. I. Schlesinger, *J. Am. Chem. Soc.*, **73**, 4385 (1951).
13. Barnes, E. C., W. E. Piros, T. C. Bryson, and G. W. Wiener, *Anal. Chem.*, **21**, 1281 (1949).
14. Barnitt, J. B., "Standard Methods of Sampling," in N. H. Furman, Ed., *Scott's Standard Methods of Analysis*, 5th ed., D. Van Nostrand Co., New York, 1939, Vol. II, pp. 1301–1318.
15. Barrett, T. R., G. C. Ellis, and R. A. Knight, "Pressureless Sintered Beryllium Powder," in *Proc. 2nd Intern. Conf. Peaceful Uses At. Energy, Geneva, 1958*, **5**, 319 (No. 320) (1958).
16. Bass, N. W., "The Role of Beryllium in Industry," in D. W. White, Jr., and J. E. Burke, Eds., *The Metal Beryllium*, The American Society for Metals, Cleveland, Ohio, 1955, Chapt. II-C.
17. Bastian, R. R., and A. C. Ottoson, U. S. Patent 2,978,341 (April 4, 1961) (to Imperial Glass Corporation).
18. Beard, D. B., R. G. Johnson, and W. G. Bradshaw, *Nucleonics*, **17**, 90 (1959).
19. Beaver, W. W., "Cermets and Ceramics," in D. W. White, Jr., and J. E. Burke, Eds., *The Metal Beryllium*, The American Society for Metals, Cleveland, Ohio, 1955, Chapt. XI-A.
19a. *Ibid.*, Chapt. V-B.
20. Benoist, P., "Critical and Subcritical Experiments on U–BeO Lattices," in *Proc. 2nd Intern. Conf. Peaceful Uses At. Energy, Geneva, 1958*, **12**, 585 (No. 1192) (1958).
21. Birminghan, J. M., Boulder Scientific Co., Private communication (1961).
22. Bisby, H., and F. H. Hale, "Alpha and Gamma Irradiation Techniques for beryllium Determination," in Symposium on Analytical Chemistry of Beryllium, United Kingdom Atomic Energy Authority, June 23, 1960.
23. Boettcher, A., "Fuel Elements for Gas-Cooled High-Temperature Reactors," in H. H. Hausner and J. F. Schumar, Eds., *Nuclear Fuel Elements*, Reinhold Publishing Corp., New York, 1959.
24. Bonner, J. F., Jr., *U. S. At. Energy Comm.*, **UR-III** (1950).
25. Booth, E., S. I. Bryant, and A. Parker, *Analyst*, **82**, 50 (1957).
26. Booth, H. S., and G. C. Torrey, *J. Phys. Chem.*, **35**, 3111 (1931).
27. Bradshaw, W., R. Johnson, and D. Beard, "Beryllium Analyzed for Trace Impurities by Gamma-Ray Activation," *U. S. At. Energy Comm.*, **LMSD-288231** (1960).
28. Brash, M. P., *Appl. Spectroscopy*, **14**, 43 (1960); through *Nucl. Sci. Abstr.*, **14**, No. 14753 (1960).
29. Breslin, A. J., and W. B. Harris, "Health Protection in Beryllium Facilities, Summary of Ten Years of Experience," *U. S. At. Energy Comm.*, **HASL-36** (1958).
30. Brode, W. R., *Chemical Spectroscopy*, Wiley, New York, 1956, p. 449.
31. Burns, J. R., *Trans. Am. Soc. Metals*, **40**, 143 (1948).
32. Caillat, R., J. Elston, C. H. Rispal, and M. Roudier, "Corrosion of Beryllium Oxide," in *Proc. 2nd Intern. Conf. Peaceful Uses At. Energy, Geneva, 1958*, **5**, 334 (No. 1147) (1958).
33. *Chem. Week*, **88**, 35 (1961).
34. Cheng, K. L., *Anal. Chem.*, **33**, 783 (1961).
35. Cholak, J., and D. M. Hubbard, *Anal. Chem.*, **20**, 73 (1948).

36. Churchill, H. V., and R. W. Bridges, "Aluminum," in N. H. Furman, Ed., *Scott's Standard Methods of Chemical Analysis*, 5th ed., D. Van Nostrand Co., New York, 1939, Vol. I, p. 12.
37. Coates, G. E., and F. Glocking, *J. Chem. Soc.*, **1954**, 2526.
38. Codell, Maurice, and George Norwitz, *Anal. Chem.*, **27**, 1083 (1955).
39. Coleman, R. F., "The Determination of Oxygen in Beryllium by Activation Analysis," in Symposium on Analytical Chemistry of Beryllium, United Kingdom Atomic Energy Authority, June 23, 1960.
40. Copaux, H., *Compt. Rend.*, **168**, 610 (1919).
41. Corzine, P., and A. A. Kaufmann, "The Melting and Casting of Beryllium," in D. W. White, Jr., and J. E. Burke, Eds., *The Metal Beryllium*, The American Society for Metals, Cleveland, Ohio, 1955, Chap. V-A.
42. Covington, L. C., and M. J. Miles, *Anal. Chem.*, **28**, 1728 (1956).
43. Cucci, M. W., W. F. Neuman, and B. J. Mulryan, *Anal. Chem.*, **21**, 1358 (1949).
44. Darwin, G. C., and J. H. Buddery, *Beryllium*, Academic Press, New York, 1960, Chapt. 1.
45. *Ibid.*, Chapt. 2.
46. *Ibid.*, Chapt. 4.
47. *Ibid.*, Chapt. 6.
47a. *Ibid.*, Chapt. 7.
48. *Ibid.*, Chapt. 8.
49. *Ibid.*, Chapt. 9.
50. *Ibid.*, Chapt. 10.
51. *Ibid.*, Chapt. 11.
52. Das, J., and S. C. Schome, *Anal. Chim. Acta*, **24**, 37 (1961).
53. Dirkse, T. P., and H. T. Briscoe, *Metal Ind.*, **36**, 284 (1938).
54. Dressel, W. M., and R. A. Ritchey, "Field Test for Beryllium" *U. S. Bur. Mines, Inform. Circ. No.* **7946** (1960).
55. Dutta, R. K., and A. K. Sen Gupta, *J. Indian Chem. Soc.*, **33**, 146 (1956).
56. Duval, T., and C. Duval, *Anal. Chim. Acta*, **2**, 53 (1948).
57. Eberle, A. R., and M. W. Lerner, *Metallurgia*, **59**, 49 (1959).
58. Eisenbud, Merle, "Health Hazards from Beryllium," in D. W. White, Jr., and J. E. Burke, Eds., *The Metal Beryllium*, The American Society for Metals, Cleveland, Ohio, 1955, p. 620.
59. Elbing, Phineas, and G. W. Goward, *Anal. Chem.*, **32**, 1610 (1960).
60. Elston, J., and R. Caillat, "Physical and Mechanical Properties of Sintered Beryllia under Irradiation," in *Proc. 2nd Intern. Conf. Peaceful Uses At. Energy, Geneva, 1958*, **5**, 345 (No. 1159) (1958).
61. Feigl, F., *Qualitative Analysis by Spot Test* (Translated by R. E. Oesper), 3rd ed., Elsevier Publishing Co., New York, 1946, p. 149.
62. *Ibid.*, p. 147.
63. Feigl, F., and A. Schaeffer, *Anal. Chem.*, **23**, 351 (1951).
64. Florence, T. M., *Anal. Chim. Acta*, **20**, 472 (1959).
65. Gaudin, A. M., J. Dasher, J. H. Pannell, and W. L. Freyberger, *Min. Eng.*, **187**, 495 (1950).
66. Gaudin, A. M., and J. H. Pannell, *Anal. Chem.*, **23**, 1261 (1951).
67. Gilbreath, J. R., and O. C. Simpson, "The Effect of Reactor Irradiation on the Physical Properties of Beryllium Oxide," in *Proc. 2nd Intern. Conf. Peaceful Uses At. Energy, Geneva, 1958*, **5**, 367 (No 621) (1958).

68. Gilman, A. R., and S. Isserow, *Analysis of Oxygen in Beryllium*, Contract AT-(30-1)-1565, 10 MI 1234 (1960); through *Nucl. Sci. Abstr.*, **14**, No. 13686 (1960).
69. Ginning, D. G., T. B. Douglas, and A. F. Ball, *J. Am. Chem. Soc.*, **73**, 1736 (1951).
70. Glasstone, S., *Sourcebook on Atomic Energy*, D. Van Nostrand Co., Inc., New York, 1950, p. 289.
71. Googin, J. M., "Analytical Methods in the Beryllium Program," *U. S. At. Energy Comm.*, **TID-4500**; *Rept. No.* **Y-1324** (1960).
72. Gregory, J. N., and D. Mapter, *Analyst*, **80**, 230 (1955).
73. Griffith, R. F., "Historical Note on Sources and Uses of Beryllium," in D. W. White, Jr., and J. E. Burke, Eds., *The Metal Beryllium*, The American Society for Metals, Cleveland, Ohio, 1955, Chapt. II-A.
74. Griffitts, W. R., and J. J. Norton, "Occurrence of Beryllium Ores and Their Treatment," in D. W. White, Jr., and J. E. Burke, Eds., *The Metal Beryllium*, The American Society for Metals, Cleveland, Ohio, 1955, Chapt. III.
75. Grossweiner, L. I., and R. L. Seifert, *J. Am. Chem. Soc.*, **74**, 2701 (1952).
76. Gulbransen, E. A., and K. R. Andrews, *Trans. Electrochem. Soc.*, **97**, 383 (1950).
77. Gyorbiro, K., *Acta Chim. Acad. Sci. Hung.*, **22**, 225 (1960) (in German); through *Chem. Abstr.*, **55**, 230i (1961).
78. Haley, T. J., and M. Bassini, *J. Am. Pharm. Assoc.*, **40**, 111 (1951).
79. Harley, F. H., and T. P. Weir, *J. Electrochem. Soc.*, **98**, 703 (1951).
80. Hayes, T. J., "The Absorptiometric Determination of Beryllium," in Symposium on Analytical Chemistry of Beryllium, United Kingdom Atomic Energy Authority, June 23, 1960.
81. Heslop, R. B., and P. L. Robinson, *Inorganic Chemistry*, Elsevier Publishing Co., New York, 1960, p. 250.
82. Hettel, H. J., and V. A. Fassel, *Anal. Chem.*, **27**, 1311 (1955).
83. Hibbits, J. O., W. F. Davis, and M. R. Meinke, *Talanta*, **4**, 101 (1960); in *Nucl. Sci. Abstr.*, **14**, No. 17821 (1960).
84. Hibbits, J. O., W. F. Davis, M. R. Meinke, and S. Kallmann, *Talanta*, **4**, 104 (1960); in *Nucl. Sci. Abstr.*, **14**, No. 17822 (1960).
85. Hill, U. T., *Anal. Chem.*, **30**, 521 (1958).
86. Holden, R. B., M. C. Kells, and C. I. Whitman, "A Continuous Electrolytic Process for the Preparation of Beryllium Metal," in *Proc. 2nd Intern. Conf. Peaceful Uses At. Energy, Geneva, 1958*, **4**, 306 (No. 717) (1958).
87. Hopkins, B. S., and A. W. Meyer, *Trans. Am. Electrochem. Soc.*, **45**, 475 (1929).
88. Horton, C. A., and J. C. White, *Anal. Chem.*, **30**, 1779 (1958).
89. Huguet, J. L., and C. L. Bamberger, "Rapid Determination of Beryllium in Beryl Mineral," in *Proc. 2nd Intern. Conf. Peaceful Uses At. Energy, Geneva, 1958*, **3**, No. 1578, 595 (1958).
90. Hunt, E. C., and J. V. Martin, "Methods of Beryllium Determination Used at the National Chemical Laboratory," in Symposium On Analytical Chemistry of Beryllium, United Kingdom Atomic Energy Authority, June 23, 1960.
91. Hure, J., M. Kremer, and F. LeBerguier, *Anal. Chim. Acta*, **7**, 37 (1952).
92. Hutchinson, C. A., Jr., and J. G. Malm, *U. S. At. Energy Comm.*, **AECD-2345** (1947).
93. Isotope Specialties, Burbank, California (1961).
94. Jain, R. D., and K. R. Srinivasan, "Nuclear Study of a Low-Power Beryllium Oxide Natural Uranium Power Reactor," in *Proc. 2nd Intern. Conf. Peaceful Uses At. Energy, Geneva, 1958*, **12**, 580 (No. 1945) (1958).

95. Kallmann, S., and F. Collier, *Anal. Chem.*, **32**, 1616 (1960).
96. Kallmann, S., R. Liu, and H. Oberthin, "Ion Exchange and Other Chemical Methods for Beryllium-Base Alloys," Wright Air Development Command, *WADC Tech. Rept.* **59-325** (1959).
97. Kanyukova, M. V., *Nauch.-Issledovatel. Inst. i Ministerstva Tsvetnoi Met. S.S.S.R.*, No. 19, 1–13—Indium (1957); through *Chem. Abstr.*, **55**, 229d (1961).
98. Karanovich, G. J., *Zhur. Anal. Chim.*, **11**, 402 (1956); through *Chem. Abstr.*, **51**, 13640i (1957).
99. Kaufmann, A. R., and P. Corzine, "Beryllium-Rich Alloys," in D. W. White, Jr., and J. E. Burke, Eds., *The Metal Beryllium*, The American Society for Metals, Cleveland, Ohio, 1955, Chapt. X-A.
100. Kawicki, H. C., "The Fluoride Extraction of Beryllium from Beryl," in D. W. White, Jr., and J. E. Burke, Eds., *The Metal Beryllium*, The American Society for Metals, Cleveland, Ohio, 1955, Chapt. IV-B.
101. Kawicki, H. C., Thesis, Massachusetts Institute of Technology, 1934.
102. Kellogg, H. H., "Thermodynamic Consideration in the Preparation of Beryllium Metal," in D. W. White, Jr., and J. E. Burke, Eds., *The Metal Beryllium*, The American Society for Metals, Cleveland, Ohio, 1955, Chapt. IV-A.
103. Kells, M. C., R. B. Holden, and C. I. Whitman, *J. Am. Ceram. Soc.*, **79**, 3925 (1957).
104. Kennedy, J., and V. J. Wheeler, *Anal. Chim. Acta*, **20**, 412 (1959).
105. Kern, B. D., D. J. Zaffarano, and A. C. G. Mitchell, *Phys. Rev.*, **73**, 1142 (1948).
106. Kida, K., M. Abe, S. Nishigaki, K. Kobayashi, "Colorimetric Determination of Beryllium in Be-Cu Alloy with 8-Hydroxyquinaldine," *Nippon Gaishi Kaisha Review*, No. 5 (1959).
107. Kjellgren, B. R. F., U. S. Patent 2,831,291 (Aug. 7, 1945) (to the Brush Beryllium Co.).
108. Kjellgren, B. R. F., and C. B. Sawyer, U. S. Patent 2,176,906 (Oct. 24, 1939) (to the Brush Beryllium Co.).
109. Kjellgren, B. R. F., "Beryllium," in R. E. Kirk and D. F. Othmer, Eds., *Encyclopedia of Chemical Technology*, The Interscience Encyclopedia, Inc., New York, 1948, p. 490.
110. Kjellgren, B. R. F., "The Production of Beryllium," *Trans. Electrochem. Soc.*, **93**, 122 (1948).
111. Kjellgren, B. R. F., "Beryllium," in C. A. Hampel, Ed., *Rare Metals Handbook*, Reinhold Publishing Co., New York, 1954, Chapt. 3, p. 42.
112. Klemperer, F. W., and A. P. Martin, *Anal. Chem.*, **22**, 828 (1950).
113. Koch, H. W., "Gamma-Ray Spectroscopy," in J. H. Yoe and H. J. Koch, Jr., Eds., *Trace Analysis*, Wiley, New York, 1957, p. 413.
114. Kolthoff, I. M., and J. F. Coetzee, *J. Am. Chem. Soc.*, **79**, 1852 (1957).
115. Kolthoff, I. M., and J. J. Lingane, *Polarography*, Interscience, NewYork–London, 1952, Vol. II, p. 430.
116. Komarowsky, A. S., and N. S. Poluektoff, *Mikrochem.*, **14**, 315 (1933).
117. Kosel, G. E., and W. F. Neuman, *Anal. Chem.*, **22**, 936 (1950).
118. Kosel, G. E., and W. F. Neuman, "Potentiometric Titrations of Beryllium," *U. S. At. Energy Comm.*, **UR-106** (1950).
119. Kroll, J. W., in H. Uhlig, Ed., *Corrosion Handbook*, Wiley, New York, 1948, Section II, p. 56.

120. Kura, J. C., J. H. Jackson, M. C. Udy, and L. W. Eastwood, *J. Metals*, **1**, No. 10 (1949); *Trans. Am. Inst. Mining, Met. Engrs.*, **185**, 769 (1949).
121. Laitenen, H. A., and P. Kovalo, *Anal. Chem.*, **24**, 1467 (1952).
122. Latimer, W. M., *Oxidation States of the Elements and Their Potentials in Aqueous Solution*, Prentice-Hall, New York, 1938, p. 270.
122a. *Ibid.*, p. 709.
123. Lebeau, P., and H. Moissan, *Compt. Rend.*, **175**, 1172 (1897).
124. Lerner, M. W., and L. J. Pinto, *Anal. Chem.*, **31**, 549 (1959).
125. Lillie, D. W., "The Physical and Mechanical Properties of Beryllium Metal," in D. W. White, Jr., and J. E. Burke, Eds., *The Metal Beryllium*, The American Society for Metals, Cleveland, Ohio, 1955, Chapt. VI-A.
126. Lingard, A. L., *Proc. S. Dakota Acad. Sci.*, **33**, 191 (1954).
127. Livey, D. T., and J. Williams, "Some Aspects of the Fabrication Technology of Beryllium and Beryllia," in *Proc. 2nd Intern. Conf. Peaceful Uses At. Energy, Geneva, 1958*, **5**, No. 311 (No. 319) (1958).
127a. Lowe, L. F., H. D. Thompson, and J. P. Cali, *Anal. Chem.*, **31**, 1951 (1959).
128. Luke, C. L., and M. E. Campbell, *Anal. Chem.*, **24**, 1056 (1952).
129. Lundell, C. E. F., and J. I. Hoffman, *Outlines of Methods of Chemical Analysis*, Wiley, New York, 1938, p. 94.
130. Magel, T. T., "Experimental Production of Beryllium," in D. W. White, Jr., and J. E. Burke, Eds., *The Metal Beryllium*, The American Society for Metals, Cleveland, Ohio, 1955, Chapt. IV-E.
131. McClure, J. H., and C. V. Banks, *U. S. At. Energy Comm.*, **AECU-812** (1950).
132. Meek, H. V., and C. V. Banks, *Anal. Chem.*, **22**, 1512 (1950).
133. Melaven, A. D., *Ind. Eng. Chem., Anal. Ed.*, **2**, 180 (1930).
134. Melick, E. S., "The Analytical Chemistry of Beryllium," in *Analytical Chemistry in Nuclear Reactor Technology*, U. S. At. Energy Comm., **TID-7555** (1958).
135. Mellor, J. W., *A Comprehensive Treatise on Inorganic and Theoretical Chemistry*, Longmans, Green and Co., New York, 1946, Vol. IV, Chapt. XXVIII.
136. Merrill, J. R., M. Honda, J. R. Arnold, *Anal. Chem.*, **32**, 1420 (1960).
137. Merrill, J. R., E. F. X. Lyden, M. Honda, and J. R. Arnold, *Geochim. et Cosmochim. Acta*, **18**, 108 (1960).
138. Milner, G. W. C., and J. W. Edwards, "The Determination of Beryllium by the Photoneutron Method," in Symposium on Analytical Chemistry of Beryllium, United Kingdom Atomic Energy Authority, June 23, 1960.
139. Molineaux, H. F., F. Trowell, G. B. Turnbull, "The Fluorimetric Estimation of Microgram Quantities of Beryllium in Laboratory and Workshop Atmospheres," in Symposium on Analytical Chemistry of Beryllium, United Kingdom Atomic Energy Authority, June 23, 1960.
140. Morris, Q. L., T. B. Gage, and S. H. Wender, *J. Am. Chem. Soc.*, **73**, 3340 (1951).
141. Motojima, K., *Bull. Chem. Soc. Japan*, **29**, 29 (1956).
142. Muvdi, B. B., *Structural Beryllium*, ASTM Preprint, No. 79 (1960).
143. Osmond, R. G., and A. H. Smales, *Anal. Chim. Acta*, **10**, 117 (1954).
144. Owens, I. I., E. Guy, and J. H. Yoe, *Anal. Chem.*, **32**, 1345 (1960).
145. Pahler, Robert E., "The Role of Beryllium in the Atomic Energy Program," in D. W. White, Jr., and J. E. Burke, Eds., *The Metal Beryllium*, The American Society for Metals, Cleveland, Ohio, 1955, Chapt. II-B.
146. Pannell, J. H., *U. S. At. Energy Comm.*, **MITG-227** (1950).

147. Pannell, J. H., and W. L. Freyberger, "Preliminary Experiments on the (γ,n) Reaction for Analyzing and Picking Beryl," *U. S. At. Energy Comm.*, **MITG-214** (1949).
148. Parker, A., "The Determination of Oxygen in Beryllium by Vacuum Fusion and by Chemical Methods," in Symposium on Analytical Chemistry of Beryllium, United Kingdom Atomic Energy Authority, June 23, 1960.
149. Parsons, C. L., *The Chemistry and Literature of Beryllium*, The Chemical Publishing Co., Easton, Pa., 1909.
150. Perkins, A. G., and L. Pate, *J. Chem. Soc.*, **67**, 649 (1895).
151. Peterson, G. E., G. A. Welford, and J. H. Harley, *Anal. Chem.*, **22**, 1197 (1950).
152. Powell, R. J., P. J. Phennah, and J. B. Still, *Analyst*, **85**, 347 (1960).
153. Přibil, R., and J. Kucharsky, *Collection Czech. Chem. Commun.*, **15**, 132 (1950).
154. Prodinger, W., *Organic Reagents Used in Quantitative Inorganic Analysis* (Transl. from 2nd German ed. by S. Holmes), Elsevier Publishing Co., Inc., New York, 1940, pp. 103–112.
155. Read, E. B., private communication (1960).
156. Robert, J., *Ann. Phys.*, **4**, 89 (1959); in *Phys. Abstr.*, **62**, No. 13628 (1959).
157. Rodden, C. J., and F. A. Vinci, "Analytical Chemistry of Beryllium," in D. W. White, Jr., and J. E. Burke, Eds., *The Metal Beryllium*, The American Society for Metals, Cleveland, Ohio, 1955, Chapt. XIII.
158. Rogers, N. E., and D. W. Prather, *Anal. Chem.*, **31**, 1081 (1959).
159. Ross, W. J., and J. C. White, *Anal. Chem.*, **31**, 1847 (1956).
160. Ryabchokov, D. K., and V. E. Bukhtiavov, *Zhur. Anal. Khim.*, **9**, 196 (1954).
161. Sandell, E. B., *Ind. Eng. Chem., Anal. Ed.*, **12**, 674 (1940).
162. Sandell, E. B., *Ind. Eng. Chem., Anal. Ed.*, **12**, 762 (1940).
163. Sandell, E. B., *Anal. Chem.*, **19**, 652 (1947).
164. Sandell, E. B., *Colorimetric Determination of Traces of Metals*, Interscience, New York, 1959, Chapt. IX.
165. *Ibid.*, p. 15.
166. *Ibid.*, p. 61.
167. *Ibid.*, p. 58.
168. Sastri, C. L., G. Sriramulu, and Bh. S. V. Raghava Rao, *J. Sci. Ind. Research*, **14B**, 171 (1955).
169. Sawyer, A. E., "The Spectrochemical Determination of Beryllium in Air, Swabs, Urine, and Biological Tissue," in Symposium on Analytical Chemistry of Beryllium, United Kingdom Atomic Energy Authority, June 23, 1960.
170. Sawyer, C. B., and B. R. F. Kjellgren, *Metals or/and Alloys*, **11**, 163 (1940).
171. Schwenzfeier, C. W., Jr., "The Sulfate Extraction of Beryllium from Beryl," in D. W. White, Jr., and J. E. Burke, Eds., *The Metal Beryllium*, The American Society for Metals, Cleveland, Ohio, 1955, Chapt. IV-C.
172. Schwenzfeier, C. W., Jr., *J. Metals*, **12**, 793 (1960).
173. Sill, C. W., and C. P. Willis, *Anal. Chem.*, **31**, 598 (1959).
174. Silverman, L., and M. W. Shideler. *Anal. Chem.*, **31**, 152 (1959).
175. Sinelnikov, K. D., V. E. Ivanov, V. M. Amonenko, and V. D. Burlakov, "Refining Beryllium and Other Metals by Condensation on Heated Surfaces," in *Proc. 2nd Intern. Conf. Peaceful Uses At. Energy, Geneva, 1958*, **4**, 295 (No. 2051) (1958).
176. Singh, D., B. N. Prasad, and B. B. Prasad, *J. Sci. Res. Banaras Hindu Univ.*, **10**, 67 (1959–60); through *Chem. Abstr.*, **55**, No. 7 (1961).
177. Sinha, S. K., and S. C. Schome, *Anal. Chim. Acta*, **24**, 33 (1961).

178. Smith, R. G., A. J. Boyle, W. G. Frederick, and B. Zak, *Anal. Chem.*, **24**, 406 (1952).
179. Snell, F. C., and C. T. Snell, *Colorimetric Methods of Analysis*, D. Van Nostrand Co., Inc., Princeton, N. J., 1959, Vol. II-A, Chapt. 15.
180. Stehn, J. R., "The Nuclear Properties of Beryllium," in D. W. White, Jr., and J. E. Burke, Eds., *The Metal Beryllium*, The American Society for Metals, Cleveland, Ohio, 1955, Chapt. VI-B.
181. Strelow, F. W. E., *Anal. Chem.*, **32**, 1185 (1960).
182. Strelow, F. W. E., *Anal. Chem.*, **33**, 542 (1961).
183. Sundaram, A. K., and S. Banerjee, *Anal. Chim. Acta*, **8**, 526 (1953).
184. Todd, R., "The Determination of Impurities in Beryllium by Chemical or Neutron-Activation Analysis," in Symposium on Analytical Chemistry of Beryllium, United Kingdom Atomic Energy Authority, June 23, 1960.
185. Toribara, T. Y., and P. S. Chen, Jr., *Anal. Chem.*, **24**, 539 (1952).
186. Toribara, T. Y., and R. E. Sherman, *Anal. Chem.*, **25**, 1594 (1953).
187. Toribara, T. Y., and A. L. Underwood, *Anal. Chem.*, **21**, 1352 (1949).
188. Udy, Murray C., H. L. Shaw, and F. W. Boulger, *Nucleonics*, **11**, 52 (1953).
189. Undewood, A. L. and W. F. Newman, *Anal. Chem.*, **21**, 1348 (1949).
190. Vetejska, K., and J. Mazacek, *Collection Czech. Chem. Commun.*, **25**, 2245 (1960) (in German); through *Chem. Abstr.*, **55**, 231b (1961).
191. Vinci, F. A., *Anal. Chem.*, **25**, 1580 (1953).
192. Walkley, J., *Am. Ind. Hygiene Assoc. J.*, **20**, 241 (1959).
193. Webb, M. S. W., and H. I. Shalgosky, "Physical Methods for the Analysis of Beryllium," in Symposium on Analytical Chemistry of Beryllium, United Kingdom Atomic Energy Authority, June 23, 1960.
194. Webb, S. W., *Atom*, **45**, 20 (1960); through *Nucl. Sci. Abstr.*, **14**, No. 20495 (1960).
195. Welford, G., and J. Harley, *Am. Ind. Hygiene Assoc. Quart.*, **13**, 4 (1952).
196. White, J. C., A. S. Meyer, Jr., and D. L. Manning, *Anal. Chem.*, **28**, 956 (1956).
197. White, J. F., "Refractory Properties of Beryllium Oxide," in D. W. White, Jr., and J. E. Burke, Eds., *The Metal Beryllium*, The American Society for Metals, Cleveland, Ohio, 1955, Chapt. XI-B.
198. Wilkinson, D. H., and D. E. Alburger, *Phys. Rev.*, **113**, 563 (1959).
199. Windecker, C. E., "The Production of Beryllium by the Electrolysis of Beryllium Chloride," in D. W. White, Jr., and J. E. Burke, Eds., *The Metal Beryllium*, The American Society for Metals, Cleveland, Ohio, 1955, Chapt. IV-D.
200. Wong, M. M., R. E. Campbell, and D. H. Baker, Jr., *J. Metals*, **12**, 786 (1960).
201. Wood, G. A., and A. Brenner, *J. Electrochem. Soc.*, **98**, 203 (1951).
202. Wood, G. A., and A. Brenner, *J. Electrochem. Soc.*, **104**, 29 (1957).

Part II
Section A

LEAD

By T. W. Gilbert, Jr., *Department of Chemical and Metallurgical Engineering, University of Cincinnati, Cincinnati, Ohio*

Contents

I. Introduction	71
A. Occurrence	72
1. Production	72
2. Industrial Products	72
B. Extraction and Purification	73
C. The Toxicity of Lead and Its Compounds	74
II. Properties of Lead and Its Compounds	77
A. Physical Properties	77
B. Optical and Electrochemical Properties	77
1. Optical Properties	77
2. Electrochemical Properties	78
C. Chemical Properties of Lead	79
1. Properties of Lead Metal	79
2. Properties of Lead Compounds	81
a. Lead(II) Compounds	81
b. Lead(IV) Compounds	90
c. Mixed-Valence Lead Compounds	91
D. Isotopic Distribution of Lead	92
1. Mass Spectrographic Determinations	92
2. Optical Determinations	93
3. Geological Age Determinations	94
III. Separation and Isolation of Lead	96
A. Preliminary Treatment of Inorganic Materials	96
1. The Attack of Lead Ores and Minerals	96
2. Alloys	96
3. The Decomposition of Organic Samples	97
a. Dry Ashing	98
b. Wet Oxidation	98
B. Selective Separation Methods	99
IV. Detection and Identification of Lead	103
A. Spectrographic Detection	103
B. Chemical Methods	103
1. Reactions of Lead(II) Ion	105
2. Microscale Tests for Lead	106
a. The "Triple Nitrite" Crystal Reaction	106
b. The Potassium Iodide Crystal Reaction	106
3. Spot Tests for Lead	107
a. Test with Benzidine	107

Contents (*continued*)

          b. Test with Sodium Rhodizonate............. 107
          c. Test with *p*-Tetramethyldiaminodiphenyl-
             methane.................................... 107
       4. Special Tests for Lead......................... 108
          a. Detection of Lead in Bismuth Compounds..... 108
          b. Separation and Detection of Lead in Water... 108
          c. Detection of Lead Tetraethyl and Lead
             Tetraphenyl in Motor Fuels................ 108
  V. Determination....................................... 108
    A. Precipitation and Gravimetric Determination....... 108
       1. As Lead Sulfate................................ 110
       2. As Lead Chromate............................... 112
       3. As Lead Molybdate.............................. 113
       4. By Electrodeposition........................... 114
          a. Special Electrolytic Methods................ 116
       5. With Organic Precipitants...................... 117
       6. Other Methods.................................. 118
    B. Titrimetric Methods............................... 119
       1. Precipitation Titrations....................... 119
          a. Titration with Molybdate................... 119
          b. Titration with Chromate.................... 120
          c. Titration with Phosphoric Acid.............. 120
          d. Titration with Ferrocyanide................. 121
          e. Titration with Alkali Hydroxide............. 121
          f. Titration with Fluoride..................... 121
       2. Oxidation-Reduction Titrations................. 122
          a. Direct Oxidation to Lead(IV)................ 122
          b. Determination of Lead(IV) in Lead(IV)–
             Lead(II) Mixtures.......................... 122
          c. Indirect Determination of Lead by Means of
             Oxidation-Reduction Titrations............. 123
       3. Titrations Based on Complex Formation.......... 125
          a. Direct Titration........................... 126
          b. Back-Titration............................. 126
          c. Replacement Titration...................... 127
          d. Alkalimetric Titration..................... 127
          e. Instrumental Methods of End-Point Detection. 128
    C. Polarographic Methods............................. 128
       1. The Dropping Mercury Electrode................. 129
       2. The Platinum Electrode......................... 130
       3. Methods for Very Low Concentrations of Lead... 130
       4. Amperometric Titrations........................ 132
    D. Photometric Methods............................... 133
       1. Spectrophotometry and Colorimetry.............. 133
          a. With Dithizone............................. 133
          b. With Tetramethyldiaminodiphenylmethane..... 139
          c. With 4-(2-Pyridylazo)Resorcinol (PAR-4)..... 139
          d. With Diethyldithiocarbamate................ 139
          e. As Lead Chloride........................... 140
       2. Flame Photometry............................... 140
       3. Atomic Absorption Spectroscopy................. 141
       4. Spectrographic Methods......................... 141
       5. X-Ray Methods.................................. 142
    E. Activation Analysis............................... 143

Contents (*continued*)

   VI. Determination of Lead in Specific Materials 143
      A. Rocks, Minerals, and Ores 143
      B. Alloys 144
         1. Iron and Steel 145
         2. Nonferrous Alloys 145
      C. Biological Materials 147
      D. Air 147
      E. Foods 148
      F. Water 148
      G. Petroleum Products 149
      H. Miscellaneous 150
  VII. Recommended Laboratory Procedures 150
      A. Detection of Lead 150
         1. Tests for Macro Amounts of Lead 150
            a. Precipitation as Lead Chloride 150
            b. Precipitation as Lead Sulfide 151
         2. "Triple Nitrite" Crystal Reaction 151
         3. Spot Test with Benzidine 152
      B. Determination of Lead 152
         1. Gravimetric Determination as Lead Sulfate 152
         2. Gravimetric Determination as Lead Chromate
            by Precipitation from Homogeneous Solution ... 153
         3. Complexometric Titration with EDTA 154
            a. In Basic Solution with Eriochrome Black T
               as Indicator 154
            b. In Acidic Solution with Xylenol Orange
               as Indicator 155
            c. In Acidic Solution with Copper–PAN
               as Indicator 156
         4. Electrogravimetric Determination as
            Lead Dioxide 157
         5. Photometric Determination with Dithizone 159
         6. Polarographic Determination in Copper-Base
            Alloys 161
         7. Amperometric Titration with Potassium
            Dichromate 163
   References 164

I. INTRODUCTION

Inasmuch as lead is easily extracted from its ores, it was one of the few metals known to ancient man. The earliest specimen is a lead figure dating from 3000 B.C., which was found at the site of the city of Abydos, on the Dardenelles (231). In the days of the Pharaohs of Egypt, lead was used for making solder and ornamental objects, and also for glazing pottery. The famed Hanging Gardens of Babylon were floored with lead sheets to retain the moisture demanded by the vegetation. The Babylonians also used lead for calking purposes and for fastening iron bolts in the masonry of bridges, dams, and other stone structures, in much the same way that it is used today.

One of the most important historical uses of lead was for the manufacture of water pipes. The Romans made lead pipe in standard diameters and in regular ten-foot lengths. Many of these pipes, in almost perfect states of preservation, have been recovered from the ruins of Pompeii and Rome.

A. OCCURRENCE

The most important lead ore is the silvery, metallic-appearing mineral, *galena* (PbS). Other minerals of lesser importance are *cerrusite*, $PbCO_3$; *anglesite*, $PbSO_4$; and *pyromorphite*, $Pb_5Cl(PO_4)_3$.

Lead occurs widely in nature in association with other metals, notably silver and zinc. Not so many years ago lead was mainly considered as a by-product of silver mining.

1. Production

The world produces and uses close to two million tons of lead in a single year. Although the United States is the principal lead-producing country, it only mines about 20% of the world consumption, and practically all of this is used within its borders. In addition, an almost equal quantity is recovered from scrap. Imports from Mexico, Canada, Peru, Australia, Europe, and elsewhere combine to bring the United States consumption up to about one-half of the world's total.

Because lead is easily refined and does not usually become seriously contaminated during service, secondary, or lead alloy scrap, is an important factor in the lead market. One of the major sources of secondary lead is from automobile storage batteries. The bulk of this material, which is an alloy containing small percentages of antimony, is resold to the battery manufacturers, and it has been estimated that 80% of lead used for manufacturing batteries reenters the market. Secondary lead containing tin is most often reused for the manufacture of solder, bearing metals, and other tin–lead alloys.

2. Industrial Products

The manufacture of storage batteries consumes by far the largest single amount of lead used in the United States. Other important consumers are the producers of cable covering, paints, and gasoline. Table I shows the consumption of lead by various industries for the year 1950.

TABLE I
Consumption of Lead[a]

Use	Tons
Storage batteries	406,800
Cable covering	136,800
Building (including chemical construction)	61,500
Tetraethyl lead	111,100
Red lead and litharge (other than for storage batteries)	75,900
Solder	88,100
White lead	34,000
Calking	61,100
Bearing metal	33,100
Ammunition	32,800
Type metal	30,900
Foil	3,300
Other uses	136,600
Total	1,212,000

[a] American Bureau of Metal Statistics.

B. EXTRACTION AND PURIFICATION

Crushed lead ores are concentrated primarily by flotation, which also serves to separate zinc sulfide from the galena. The concentrates obtained generally contain 40% or more of lead. The ore is then roasted in air to convert the sulfide to oxide and sulfate and simultaneously to eliminate much of the sulfur as sulfur dioxide. The roasted ore is mixed with limestone and coke and is smelted in a blast furnace to yield crude lead bullion ("hard" lead). Important impurities in the crude lead are copper, antimony, arsenic, bismuth, gold and silver.

The crude lead is refined by first melting it and then allowing the melt to cool below the freezing point of copper. The copper crystallizes and is then removed by skimming. The molten lead is next heated in a reverberatory or "softening" furnace, where it is stirred and subjected to an air blast that oxidizes impurities of antimony and arsenic. The oxides of these elements are then skimmed off. Antimony and arsenic, when alloyed with lead, harden it; hence the term, "softening" furnace.

Most commonly the softened lead is further refined by the Parkes process. This refining technique is based on the fact that silver and gold are much more soluble in molten zinc than in molten lead. Furthermore, molten zinc and lead are only slightly miscible with each other. Accordingly, when zinc is added to the molten lead in the desilverizing kettles, gold and silver are extracted. Being less dense than lead the zinc rises to the surface, solidifies, and can be skimmed off.

This treatment leaves only a little zinc as an impurity, and it can be removed either through oxidation followed by skimming, or by the more recent process of evaporation (285). These techniques make refined lead one of the purest of the commercial metals. The chemical requirements of "corroding lead," the purest commercial grade available in bulk, follow (17):

	Maximum, %
Silver	0.0015
Copper	0.0015
Silver + copper	0.0025
Arsenic + antimony + tin	0.002
Zinc	0.001
Iron	0.002
Bismuth	0.050
	Minimum, %
Lead (by difference)	99.94

Lead may be refined electrolytically to a purity of 99.995%, which is satisfactory for almost all scientific purposes. Further purification in this manner is not practicable, and recourse must be made to a chemical process such as is described by Brauer (50). By such a procedure lead may be prepared that contains as little as 10^{-12} g. of gold and 10^{-10} g. of silver per gram of lead. Ultrapure lead is remarkably soft and far more resistant to atmospheric oxidation than is ordinary lead.

Lead also may be refined to a high degree by making use of the technique of zone refining. The starting material may be ordinary reagent-grade lead (394).

C. THE TOXICITY OF LEAD AND ITS COMPOUNDS

Lead poisoning has long been known and has been exhaustively studied. Although fatal cases of plumbism are now relatively rare, lead and its compounds still constitute one of the most important industrial hazards.

Harmful quantities of lead may, of course, be ingested by swallowing contaminated food and drink, but cases of lead intoxication arising in this manner are now uncommon. Minute quantities of lead are ingested in this way by the average adult, but they are subsequently eliminated by normal body processes before harmful concentrations accumulate. Serious lead intoxication is most frequently encountered in persons who have inhaled the vapors, fumes, or dusts of lead and lead-containing compounds. For example, the processes of sanding and spraying lead paints are extremely hazardous without adequate protection.

In industrial operations it is not practical to attempt to eliminate completely the possibility of exposure. Accordingly, a maximum allowable atmospheric concentration has been specified, which is 0.15 mg. of lead per cubic meter of air. The equilibrium vapor pressure data for pure lead show that this concentration is not reached until the temperature of lead exceeds 500°C., which is well above the melting point (327°C.). One might conclude that there should be little risk in the handling of molten lead; however, the chief hazard is not the lead vapor itself but is the film of lead oxide that is formed on the surface of the molten metal and that can be dispersed mechanically as a dust. Provision of exhaust systems and scrub-

TABLE II
Physical Properties of Lead Metal

Atomic number	82
Atomic weight (1959 International Value)	207.21
Melting point, °C.	327.4258
Boiling point, °C. at 760 mm. Hg	1725
Latent heat of fusion, kcal./g.-atom	1.224
Latent heat of vaporization, kcal./g.-atom	41.9
Specific heat, cal./g.-atom at 25°C.	6.36
Coefficient of linear thermal expansion, per °C. over temperature range of 10 to 100°C.	2.95×10^{-5}
Thermal conductivity at 0°C., cal./(sec.)(cm.^{2})(°C./cm.)	0.083
Electrical resistivity, ohm-cm. at 20°C.	2.065×10^{-5}
Magnetic susceptibility, c.g.s. units at 18 to 330°C.	-0.12×10^{-3}
Density, g./cm.^{3}	
at 20°C.	11.342
at 327.4°C.	11.00 (solid)
at 327.4°C.	10.686 (liquid)
Surface tension, dynes/cm.	444 (327.4°C.)
Ionization potentials for successive electrons, v.	I, 7.42
	II, 15.03
	III, 32.08
	IV, 42.25
	V, 69.7
Thermodynamic properties (291)	
Heat of formation (kcal.) at 25°C.	$\Delta H° = 0.000$ (solid)
	$\Delta H° = 46.34$ (vapor)
Free energy of formation (kcal.) at 25°C.	$\Delta F° = 0.000$ (solid)
	$\Delta F° = 38.47$ (vapor)
Entropy (cal./°C.) at 25°C.	$S° = 15.51$ (solid)
	$S° = 41.890$ (vapor)
Crystal structure	Face-centered cubic
	$a = 4.9489$ A.
Closest approach of atoms	3.499 A.

bers helps prevent air pollution by lead fumes. Nevertheless, workmen in the lead industries should have regular examinations by qualified physicians and industrial hygienists to detect abnormal lead absorption.

TABLE III
Stable and Radioactive Isotopes of Lead (380)

Mass number[a]	Type of decay[b]	Half life[c]	Relative abundance, %
204	Stable	—	1.40
206	Stable	—	25.1
207	Stable	—	21.7
208	Stable	—	52.3
195	E.C.	17 min.	
196	E.C.	37 min.	
197m	E.C. (80%), I.T. (20%)	42 min.	
198	E.C.	2.4 hr.	
199m	I.T.	12.2 min.	
199	E.C., β^+ (weak)	90 min.	
200	E.C.	21.5 hr.	
201m	I.T.	61 sec.	
201	E.C., β^+ (weak)	9.4 hr.	
202m	I.T. (90%), E.C. (10%)	3.62 hr.	
202	E.C.	ca. 3×10^5 yr.	
203m	I.T.	6.1 sec.	
203	E.C.	52.1 hr.	
204m	I.T.	66.9 min.	
205	E.C.	ca. 5×10^7 yr.	
207m	I.T.	0.80 sec.	
209	β^-	3.30 hr.	
210 (RaD)	β^-	19.4 yr.	
211 (AcB)	β^-	36.1 min.	
212 (ThB)	β^-	10.64 hr.	
214 (RaB)	β^-	26.8 min.	

[a] The letter m following a mass number indicates a metastable excited state.
[b] E.C. = orbital electron capture; I.T. = isomeric transition; β^- = negative beta particle emission; β^+ = positive beta particle emission.
[c] The half lives tabulated are believed by the authors of Reference 380 to be the best available. For other measurements and discussion refer to their compilation.

Lead intoxication can be cured and recovery is usually complete, leaving no disability. The most common treatment is the intravenous injection of the sodium calcium salt of ethylenediaminetetraacetic acid; a 10- to 30-fold increase in the urinary excretion of lead results. Within a short period the clinical symptoms of poisoning are relieved. Reference works on industrial toxicology (97) and on the diagnosis and treatment of lead poisoning (13) should be consulted for further information.

II. PROPERTIES OF LEAD AND ITS COMPOUNDS

A. PHYSICAL PROPERTIES

Some of the outstanding physical properties of metallic lead are its high density, softness, malleability, and flexibility; its low melting point, low strength, and low elastic limit. These characteristics, as well as its high lubricity, low electrical conductivity, high coefficient of expansion, and high corrosion resistance, are the basis of its selection in most of its major applications.

The metal alloys well with certain other metals, notably with antimony and tin. The alloys often have physical properties (such as hardness and melting characteristics) that differ considerably from those of the pure metal.

The high density and ready availability of lead makes its use common for radiation shielding purposes. Gamma and x-rays are efficiently attenuated by lead shields. In addition, the thermal neutron cross section of natural lead is low, and that of the nuclide, ^{208}Pb, is very low (0.0006 barns).

Some of the more pertinent physical properties are collected in Tables II and III.

B. OPTICAL AND ELECTROCHEMICAL PROPERTIES

1. Optical Properties

The most sensitive lines in the arc emission spectrum of lead are at 4057.820, 3683.471, 2833.069, 3639.580, and 2614.178 A. These are suitable for the detection and determination of the element (see Sections IV-A and V-D-4).

In the spark spectrum the most sensitive line is at 2203.505 A. (see Section IV-A for a more complete tabulation).

The oxygen–hydrogen and the oxygen–acetylene flames are capable of exciting the emission lines of lead at 405.8, 368.4, and 364.0 mμ. For further details see Section V-D-2.

The K series of the x-ray emission spectrum of lead is of very high energy and is not of any analytical importance. The L series is by far the more useful. The wavelengths of the K and L lines are given in Tables IV and V (236). For analytical methods that utilize x-ray emission and absorption, see Section V-D-5.

TABLE IV
K and L Emission Lines of Lead

Line	Wavelength, A.	Intensity[a]	Line	Wavelength, A.	Intensity[a]
$K\alpha_1$	0.165	100	L_{II}, β_1	0.982	50
$K\alpha_2$	0.170	50	L_{II}, γ_1	0.840	10
$K\beta_1$	0.146	21	L_{II}, η	1.092	1
$K\beta_2$	0.142	3	L_{III}, α_1	1.175	100
L_I, β_3	0.969	6	L_{III}, β_2	0.983	20
L_I, β_4	1.007	4	L_{III}, α_2	1.186	11
L_I, γ_3	0.815	2	$L_{III}, 1$	1.350	3
L_I, γ_2	0.821	1	L_{III}, β_6	1.021	1

[a] The figures given are only approximate relative intensities, which show the same ranking order for the x-ray spectra of all elements. They are given merely to indicate the lines of most analytical importance. The strongest lines of the K and L series are assigned the value 100.

TABLE V
Critical Absorption Edges and Energies

Edge	Wavelength, A.	Energy, kv.
K	0.1408	88.037
L_I	0.7815	15.858
L_{II}	0.8151	15.205
L_{III}	0.9503	13.041

2. Electrochemical Properties

Standard reduction potentials of some half reactions of lead that are of analytical interest are listed in Table VI (230). All potentials are at 25°C. and are referred to the normal hydrogen electrode.

TABLE VI
Standard Reduction Potentials of Lead Half Reactions

Reaction	$E°$, v.
$Pb^{+2} + 2e^- = Pb$	−0.126
$HPbO_2^- + H_2O + 2e^- = Pb + 3OH^-$	−0.54
$PbCl_2(s) + 2e^- = Pb + 2Cl^-$	−0.268
$PbBr_2(s) + 2e^- = Pb + 2Br^-$	−0.280
$PbI_2(s) + 2e^- = Pb + 2I^-$	−0.365
$PbS(s) + 2e^- = Pb + S^{-2}$	−0.98
$PbSO_4(s) + 2e^- = Pb + SO_4^{-2}$	−0.3563
$Pb^{+4} + 2e^- = Pb^{+2}$	ca. 1.7
$PbO_2(s) + 4H^+ + 2e^- = Pb^{+2} + 2H_2O$	1.455
$PbO_2(s) + SO_4^{-2} + 4H^+ + 2e^- = PbSO_4(s) + 2H_2O$	1.685
$PbO_2(s) + H_2O + 2e^- = PbO(red) + 2OH^-$	0.248

The PbSO$_4$–Pb couple is a commonly used reference electrode, the potential of which has been measured as a function of temperature from 0 to 60°C. Potentiometric data of other half reactions of lead that are pertinent to the lead storage battery have also been obtained over this temperature range (154).

For polarographic half-wave potentials, see Section V-C-1.

C. CHEMICAL PROPERTIES OF LEAD

1. Properties of Lead Metal

Metallic lead is bluish white in color and has a bright luster. In contact with air and water vapor, however, the metal is attacked and a thin surface film of oxycarbonate is formed that protects the underlying metal. The formation of this film is easily observed as a dulling of the luster of the freshly cut metal. Although the massive metal is only superficially attacked by this reaction, finely divided lead reacts vigorously and is, in fact, pyrophoric.

Water, in the absence of oxygen and even at elevated temperatures, does not react with lead metal. When oxygen is present, however, attack occurs with the formation of plumbous hydroxide; the suboxide, Pb$_2$O, has been postulated as an intermediate:

$$4Pb + O_2 = 2Pb_2O$$

$$2Pb_2O + O_2 + 4H_2O = 4Pb(OH)_2$$

This attack is, of course, accelerated in the presence of acids, but if the aerated water is made basic, and particularly if it contains carbonate or silicate, the rate becomes exceedingly slow. These considerations are of great importance in the prevention of lead poisoning, which might result from the consumption of water that has been carried in lead pipes.

The solubility of pure metallic lead in water is, on the other hand, quite low. A direct determination has been made by equilibrating lead with air-free water at 24°C. Analysis of the water by the dithizone method showed the solubility to be 311 ± 18γ/liter (1.50 × 10^{-6}M) (302).

In the vapor state at 1870°C. lead has been shown to be monatomic. This is also true of lead dissolved in mercury.

Lead stands just above hydrogen in the electromotive series, its potential relative to the normal hydrogen electrode being + 0.130 v. One would predict from this that lead should, in general, dissolve in dilute acids. However, the discharge of hydrogen on lead has a very high overpotential, and under most experimental conditions it is greater than that of any other metal with the sole exception of mercury. The action of acids on lead is profoundly affected by this fact and also by the formation of insoluble lead compounds that may coat the metal and protect it from further attack. The actions of some of the common acids on lead are tabulated below.

H_2SO_4. Sulfuric acid (unless practically anhydrous) does not dissolve lead since an insoluble coating of lead sulfate is formed. It is for this reason that lead is widely used for the handling and storing of sulfuric acid in industry. It is not attacked by dilute sulfuric acid even when hot, nor by concentrated acid below 200°C.

HCl. Lead is dissolved very slowly by hydrochloric acid. Again, the formation of sparingly soluble lead chloride slows down the rate.

HNO_3. Because of its strong oxidizing power, nitric acid dissolves lead rapidly, resulting in a solution of lead nitrate and evolving oxides of nitrogen.

$HC_2H_3O_2$. In the presence of atmospheric oxygen, acetic acid rapidly attacks the metal. The formation of soluble lead acetate complexes undoubtedly accelerates the attack.

Metallic lead in contact with acidic solutions is a fair reducing agent, and with alkaline solutions it is quite strong (see Section II-B-2). It has been successfully applied to the quantitative reduction of many metallic ions. The technique employed is similar to that used with the familiar Jones reductor of amalgamated zinc. The advantages of the lead reductor are: (1) lead is obtainable in a state of high purity, thereby eliminating the need for blank determinations in many cases; and (2) hydrogen is not evolved, so amalgamation is unnecessary. The primary difficulty associated with the use of the lead reductor appears to be its application to solutions containing sulfuric acid. Under these conditions an adherent film of lead sulfate forms, which decreases the efficiency of reduction. However, it has been shown that the formation of this film can be prevented by maintaining the solutions at least $2.5M$ in hydrochloric acid. Rapid reduction is then achieved even with continuous use (85).

Although the reductive power of the lead reductor is less than that of the Jones reductor, this can be a decided advantage in some determinations. Chromium, niobium, cobalt, and nickel are not reduced and hence do not interfere in the determination of metals such as uranium, iron, tin, etc. One of its most important applications is for the reduction of uranium(VI) to uranium(IV). In contrast to the mixed oxidation states obtained when the Jones reductor is used for this purpose, uranium(IV) is quantitatively produced (367).

The reduction of metallic ions in alkaline solution with lead metal apparently has not been applied to quantitative procedures.

The volatility of lead at high temperatures may be used for its selective distillation from other metallic elements. For example, pure tin–lead alloys have been successfully analyzed by distilling away the lead at 1000°C. under a vacuum of 0.05 mm. Hg. The residue from this treatment is pure tin and may be weighed as such. A precision of less than 0.1% is obtainable. More complicated systems may be handled in a similar manner (318).

2. Properties of Lead Compounds

Lead forms two series of compounds, corresponding to the oxidation states of $+2$ and $+4$. In contradistinction to the other members of Group IV in the Periodic Table, the most common oxidation state of lead in its compounds is $+2$. This behavior has led Sidgwick (365) to describe lead as having an "inert pair" of electrons, which is attributed to a decrease in the ability of lead to form covalent bonds. A theoretical explanation has been advanced that proposes that the energy required for the promotion of an s electron to the p state in order to form covalent bonds of the sp^3 type is unfavorable (94). The compounds of lead(IV) are accordingly generally regarded as being covalent in nature, whereas those of lead(II) are considered to be primarily ionic.

Lead(IV) compounds are very strong oxidizing agents that yield lead(II) on reduction. The simple cation, Pb^{+4}, being very acidic, does not exist as such in solution even when the solution is very strongly acidified. In concentrated hydrochloric acid lead(IV) is believed to exist as complex ions, e.g., $PbCl_6^{-2}$, $PbCl_5^{-}$, etc. As the acidity is reduced, precipitation of $Pb(OH)_4$ ensues, which subsequently dehydrates to the brown PbO_2. The hydroxide is amphoteric and dissolves in dilute sodium hydroxide with the formation of $Pb(OH)_6^{-2}$ or PbO_3^{-2}. Lead(IV) may be produced by oxidation of lead(II) compounds in alkaline solution by strong oxidizing agents, e.g., NaOCl, $KMnO_4$, H_2O_2, etc. Brown PbO_2 is the product formed. The oxidation also may be accomplished electrolytically (see "Electrolytic Determination of Lead," Section V-A-4).

a. Lead(II) Compounds

Lead(II) oxide, PbO, litharge. This is the common oxide of lead formed when lead is heated in the presence of air. It is prepared industrially by blowing air on molten lead, the temperature being kept above the melting point of the oxide (880°C.). Two forms are known: the yellow orthorhombic litharge, the high-temperature form stable above 600°C.; and the red tetragonal form, stable at ordinary temperatures.

The yellow and red forms may both be prepared by dissolving freshly precipitated lead(II) hydroxide in hot sodium hydroxide solution. The yellow form precipitates on cooling $5M$ sodium hydroxide solution; the red form on cooling $10M$ sodium hydroxide.

A third form, black PbO, has been much discussed in the literature. It has been shown, however, to be merely yellow PbO with an altered surface (probably having a thin surface film of elemental lead) (75).

The solubilities of both forms in water and sodium hydroxide solutions have been determined (132). In water at 25°C. they are: yellow PbO, $4.8 \times 10^{-4}M$; red PbO, $2.2 \times 10^{-4}M$.

Lead(II) hydroxide, $Pb(OH)_2$. The oxide, PbO, is known to take up water, but no definite compound of the composition $Pb(OH)_2$ has been isolated (366). Nevertheless the precipitate obtained on the treatment of lead(II) salts with base is usually assigned this composition. On the other hand, Duval, in his thermogravimetric investigations, claims that lead hydroxide precipitated from aqueous solution with ammonia corresponds exactly to the formula $Pb(OH)_2$ in the temperature region of 155 to 410°C. (95). He recommends lead hydroxide as a convenient weighing form for the automatic thermogravimetric determination of lead.

The precipitated hydroxide is amphoteric. Its basic properties are evidenced by the fact that a suspension in water is alkaline to both phenolphthalein and thymophthalein. The hydroxide also has acidic properties since it dissolves readily in dilute solutions of strong bases with the formation of $HPbO_2^-$ or PbO_2^{-2} ions.

The actual formula of the ion present in basic solutions is uncertain. Polarographic studies indicate that the biplumbite ion, $HPbO_2^-$, is the prevalent species in sodium hydroxide solutions of 0.01 to $1.0M$ (239,298). This conclusion also agrees with the interpretation of the solubility measurements of Garrett et al. (132). Since the solubility of both forms of PbO in aqueous sodium hydroxide solutions increases linearly with the concentration of sodium hydroxide from 10^{-3} to ca. $0.2M$, Garrett et al. postulated the reaction to be $PbO + OH^- = HPbO_2^-$, rather than $PbO + 2OH^- = PbO_2^{-2} + H_2O$. On the other hand, the results of recent ultracentrifugation experiments in 0.25 to $0.95M$ sodium hydroxide suggest the ion, PbO_2^{-2}, as the prevalent species (183). This work also showed that, even with lead concentrations of 0.015 to $0.05M$, the lead is present mainly in a mononuclear form, contrary to the hydrolytic behavior of many other metal ions. These results are in essential agreement with recent extensive studies of the hydrolytic behavior of lead by e.m.f. measurements (62,296,297). In these studies the simple plumbite ion was formulated as $Pb(OH)_3^-$; the solutions investigated were less than $0.5M$ in sodium hydroxide, and no evidence of the ion $Pb(OH)_4^{-2}$ was found. However, at moderate and high lead-ion concentrations in perchlorate media, evidence was obtained for the existence of polynuclear species and the following ions were proposed: $Pb_4(OH)_4^{+4}$, $Pb_3(OH)_4^{+2}$, $Pb_6(OH)_8^{+4}$, and $Pb_2(OH)^{+3}$.

Lead(II) halides, PbX_2. The halides of lead(II) are well-known, slightly soluble salts. The solubility increases on going from fluoride to chloride, but then decreases from chloride to bromide to iodide. Values for the solubility products are given in Table VII.

With the exception of the iodide the halides are all colorless crystalline salts. Lead iodide differs from the others in that it is a bright yellow compound crystallizing in characteristic hexagonal plates, which are useful for the microscopic identification of lead. The solubilities of all the halides increase greatly with increasing temperature. Use is made of this fact in the separation and isolation of lead in the usual scheme of qualitative analysis. The solubilities of the chloride, bromide, and iodide in the presence of excess halide ion decrease at first, due to the common-ion effect. However, with further increases in the halide-ion concentration the solubility rapidly increases, due to the formation of complex ions. Because of this, lead cannot be separated completely from aqueous solutions as these salts, and consequently they are little used in gravimetric procedures for lead.

Ahrland claims that the stabilities of the halide complex ions of lead increase in the order, $F^- \ll Cl^- < Br^- < I^-$ (8). No complexes with fluoride are detectable in aqueous solutions (84). The explanation advanced for this sequence is that the complexes formed with the other halides involve partial double-bond formation. These ions, unlike the fluoride ion, have vacant d orbitals that can accommodate extra bonding electrons from the metal (8). Recent values of the thermodynamic formation constants determined by conductance measurements, however, reverse this order for the chloride and bromide complex ions (286).

Double halide salts are also known. One of these, lead chlorofluoride, PbClF, is widely used for the precipitation and determination of fluoride.

Lead(II) nitrate, $Pb(NO_3)_2$. Lead nitrate is usually prepared by the action of nitric acid on lead oxide or lead metal. It is freely soluble in water and is widely used as a starting material for the preparation of other lead compounds. Contrary to the behavior of many nitrate salts, it is not completely dissociated in aqueous solutions. The formation constant at infinite dilution at 25°C. of $PbNO_3^+$ has been reported to be (286):

$$Pb^{+2} + NO_3^- = PbNO_3^+ \qquad K = 15.1$$

The formation constant of this same reaction at 25°C. and at an ionic strength of 2.0 has been determined by several methods and has been found to be approximately 2.6 (159).

Lead nitrate is insoluble in concentrated nitric acid and can be quantitatively precipitated from this medium. A gravimetric method for lead is based on this fact (418).

Lead(II) sulfate, $PbSO_4$. Lead sulfate is a white crystalline solid and is one of the most widely used weighing forms for lead. It is difficultly soluble in water and in acids. It dissolves to a limited extent in concentrated sulfuric acid, probably due to the formation of lead bisulfate or sulfato-complex ions. Organic acid radicals such as acetate and tartrate, which complex lead readily, bring lead sulfate into solution. The solubility of lead sulfate in ammonium acetate provides an important means of separating it from barium sulfate and other insoluble substances.

The decomposition temperature of lead sulfate is 803°C. (299).

Lead(II) acetate, $Pb(C_2H_3O_2)_2$. Lead acetate is a freely soluble, white, crystalline salt which is important in commerce and is known as "sugar of lead." Its solutions are only partially ionized, thus accounting for the strong solvent action of acetates on lead compounds. Three complex acetate ions are known in aqueous solution and have the following formation constants at 25°C. and at an ionic strength of 2.0 (56):

$$Pb^{+2} + OAc^- = PbOAc^+ \qquad K_1 = 145$$
$$Pb^{+2} + 2\,OAc^- = Pb(OAc)_2 \qquad K_2 = 810$$
$$Pb^{+2} + 3\,OAc^- = Pb(OAc)_3^- \qquad K_3 = 2950$$

Lead(II) sulfide, PbS. Lead sulfide is an extremely insoluble compound that occurs in nature as the mineral, galena. It is formed as a black precipitate on treatment of solutions of plumbous salts with hydrogen sulfide or other soluble sulfides. The precipitate is insoluble in acids unless very concentrated (HCl) or oxidizing (HNO_3). It is insoluble in alkaline polysulfide solutions, but is readily oxidized by hydrogen peroxide to the white sulfate, $PbSO_4$.

Lead sulfide is sometimes used as a weighing form for lead. Although it is suitable for this, the drying temperature is critical—the range being only 97 to 107°C. (95).

The solubility in water (approximately 0.8 mg./liter) is higher than might be expected on consideration of the very low solubility product ($K_{sp} = 7 \times 10^{-29}$). This is due mainly to the hydrolysis of the sulfide ion.

The precipitation of lead as the sulfide from dilute acid solutions serves as a means of separation from the many metal ions of groups III, IV, and V of the customary scheme of qualitative analysis.

Lead(II) cyanide, Pb(CN)$_2$. The treatment of solutions of lead salts with soluble cyanides precipitates white, insoluble Pb(CN)$_2$. In the presence of excess cyanide, lead cyanide is moderately soluble, and it has been generally assumed that the lead is present in the form of a cyanide complex ion. Recent polarographic data, however, show no evidence of complexing between lead and cyanide. The lead is present exclusively as the biplumbite ion, HPbO$_2^-$, due to the strongly basic solution that results from the hydrolysis of cyanide (298).

Lead(II) ethylenediaminetetraacetatoplumbate(II). When an excess of lead(II) nitrate is added to an EDTA solution of pH 3.5 to 4.5, white crystals of Pb[PbC$_{10}$H$_{12}$O$_8$N$_2$]·H$_2$O precipitate. One-half of the lead in the compound shows the properties of being ionically bound, whereas the other half appears to be coordinated by the EDTA. At 25°C. and at an ionic strength of 0.2, the measured solubility product is (384):

$$K_{sp} = [\text{Pb}^{+2}][\text{PbC}_{10}\text{H}_{12}\text{O}_8\text{N}_2^{-2}] = 4.34 \times 10^{-6}$$

Basic lead(II) salts. Bivalent lead shows a pronounced tendency to form basic salts, a fact that must be kept constantly in mind in the preparation of lead salts and in the design of analytical procedures. For example, the gradual addition of sodium hydroxide to a lead nitrate solution does not precipitate lead hydroxide, as might be expected, but instead forms a basic lead nitrate. Nitrate has been shown to be present in the precipitate even when it is in contact with a solution of pH 12. Various investigators differ on the exact composition of the solid phases, but the existence of the basic salts cannot be denied. In the case of lead nitrate, Pauley and Testerman (305) state that the following solids are present in the pH regions given:

Pb(NO$_3$)$_2$·Pb(OH)$_2$ pH 5.0–7.0

Pb(NO$_3$)$_2$·5Pb(OH)$_2$ pH 7.0–> 12

For precipitation titrations of lead it has been recommended that the pH of the solution at the end point should be less than 5.5 if the formation of basic salts is to be avoided (164). The results of Pauley and Testerman suggest that perhaps even more strongly acidic solutions should be used.

Many other examples could be cited, since basic salts may be prepared containing practically any anion. A particularly stable basic lead bromide has been used for the alkalimetric determination of lead. A neutral solution of lead salt containing excess bromide is titrated with sodium hydroxide. The reaction, Pb^{+2} + OH$^-$ + Br$^-$ = PbOHBr(s), takes place stoichiometrically, and the end point is detected with phenolphthalein (214).

TABLE VII
Solubility Products of Lead(II) Compounds (45)

Compound	$-\log_{10}$ (solubility product)[a]	Medium	Method of determination	Ref.
$Pb(OH)_2$	14.93	Cor. to $\mu = 0$	E.m.f. (Pb elec.)	137
	19.96 (22°C.)	Extrap. to $\mu = 0$	Quinhydrone elec.	218
	19.49 (40°C.)	Extrap. to $\mu = 0$	Quinhydrone elec.	218
$PbCrO_4$	13.75 (88°C.)	Variable μ	Solubility	31
$Pb_2Fe(CN)_6$	14.46	Variable μ	Solubility	389
$Pb(SCN)_2$	4.70	Variable μ	Solubility	421
$PbCO_3$	13.48 (18°C.)	Variable μ	Solubility	310
	13.0 (18°C.)	Variable μ	Solubility	23
$Pb_3(PO_4)_2$	42.10	Cor. to $\mu = 0$	E.m.f. (Pb elec.)	269
$Pb_3(AsO_4)_2$	35.39 (20°C.)	Variable μ	Solubility	71
PbS	27.47 (°C.?)	Variable μ	Solubility	54
	29.04 (°C.?)	Cor. to $\mu = 0$	Solubility	321
	29.15	Cor. to $\mu = 0$	Calc. from ΔF data	321
	29.37	Cor. to $\mu = 0$	Solubility	190
	28.15		Calc. from ΔF data	230
	27.10	Cor. to $\mu = 0$	Calc. from ΔF data	138
	28.17	Cor. to $\mu = 0$	Calc. from ΔF data	188
PbS_2O_3	6.40	Variable μ	Solubility	421
$PbSO_4$	7.75	Dilute solution	Conductivity	49
	7.80	Cor. to $\mu = 0$	Conductivity	49
	7.78	Cor. to $\mu = 0$	E.m.f. (Pb amal., elec.)	364
	7.79	Cor. to $\mu = 0$	Solubility	390
$PbSeO_3$	11.5 (20°C.)	Variable μ	Solubility (glass elec.)	72
$PbSeO_4$	6.84	Cor. to $\mu = 0$	Solubility	362
PbF_2	7.43 (26.6°C.)	Cor. to $\mu = 0$	Conductivity	205
	7.57	Extrap. to $\mu = 0$	E.m.f. (Pb amal. elec.)	176
$PbCl_2$	4.67	Cor. to $\mu = 0$	E.m.f. (Pb elec.)	126
	4.79	Extrap. to $\mu = 0$	Pb amal. and Ag elec.	295
$PbBr_2$	4.56	Variable μ	Solubility	408
	5.04	Cor. to $\mu = 0$	E.m.f. (Pb elec.)	126
	4.41	Cor. to $\mu = 0$	E.m.f. (Pb amal. elec.)	60
PbI_2	7.8	Dilute solution	Solubility	408
	7.86	Cor. to $\mu = 0$	Solubility	235
	8.01	Cor. to $\mu = 0$	E.m.f. (Pb elec.)	126
	8.15	Extrap. to $\mu = 0$	E.m.f. (Ag elec.)	287
$Pb(IO_3)_2$	12.86 (20°C.)	Dilute solution	Conductance	49
	12.58 (25.8°C.)	Cor. to $\mu = 0$	Conductance	205
	12.59 (r.t.)	Dilute solution	Solubility	134
PbC_2O_4	9.32 (26°C.)	Extrap. to $\mu = 0$	Solubility	212
$Pb(C_9H_6ON)_2$ (oxine)	22.00	Cor. to $\mu = 0$	E.m.f.	288

[a] Temperature of 25°C., except where noted otherwise.

Other lead(II) salts. Other insoluble lead salts sometimes used for the gravimetric determination of lead are: lead iodate, lead molybdate, lead chromate, and lead oxalate.

The following lead salts are quite soluble: lead nitrite, lead chlorate, lead bromate, lead perchlorate, and lead dithionate.

Selected values of the solubility products of lead(II) compounds are listed

TABLE VIII
Instability Constants of Lead(II) Complexes with Inorganic Ligands at 25°C.

Ligand	$-\log_{10}$ (instability constant)^a	Medium	Method	Ref.
OH$^-$	pK_1 7.82, pK_2 3.06, pK_3 3.06	Cor. to $\mu = 0$	Solubility	132
	pK_1 6.9, pK_2 3.9, pK_3 2.5	1M KNO$_3$	Polarography	143
SCN$^-$	pK_1 0.54, pK_2 0.33 $p\beta_3 - 1$, $p\beta_4$ 0.85	$\mu = 2$ (NaClO$_4$)	Polarography	232
NO$_3^-$	pK_1 0.45	$\mu = 2$ (NaClO$_4$)	Polarography	159
	pK_1 0.52	$\mu = 2$ (NaClO$_4$)	E.m.f. (Pb amal.)	159
	pK_1 0.2	$\mu = 2$ (NaClO$_4$)	Spectrophotometry	159
	pK_1 1.18	Extrap. to $\mu = 0$	Conductimetry	286
	pK_1 0.36	$\mu = 2$ (HClO$_4$)	Spectrophotometry	40
	pK_1 1.15	Cor. to $\mu = 0$	Spectrophotometry	24
P$_2$O$_7^{-2}$	pK_1 11.24	0.1M Na$_2$P$_2$O$_7$	Polarography	333
S$_2$O$_3^{-2}$	$p\beta_2$ 5.13, $p\beta_3$ 6.35	μ varied	Solubility	421
F$^-$	$pK_1 < 0.3$	$\mu = 0.5$ (NaClO$_4$)	E.m.f. (redox elec.)	304
	$pK_1 < 0.8$	Cor. to $\mu = 0$	E.m.f. (redox elec.)	304
Cl$^-$	pK_1 1.60, pK_2 0.18, $pK_3 - 0.1$, $pK_4 - 0.3$	Cor. to $\mu = 0$	Ion exchange	293
	pK_1 0.96, pK_2 -0.09, pK_3 0.50	$\mu = 1$ (NaClO$_4$)	Polarography	202
	pK_1 1.59	Cor. to $\mu = 0$	Conductimetry	286
Br$^-$	pK_1 1.77	Extrap. to $\mu = 0$	Spectrophotometry	40
	pK_1 1.47	Cor. to $\mu = 0$	Conductimetry	286
	pK_1 1.11, pK_2 0.32, pK_3 0.75	$\mu = 1$ (NaClO$_4$)	Polarography	203
I$^-$	pK_1 1.26, pK_2 1.54, pK_3 0.62, pK_4 0.50	$\mu = 1$ (NaClO$_4$)	E.m.f. (Pb elec.) and polarography	204
	pK_3 -0.18, pK_4 0.80	Cor. to $\mu = 0$	Solubility	227
	pK_1 1.92	Extrap. to $\mu = 0$	Spectrophotometry	40

^{a} pK values are stepwise constants, and pβ values gross constants; for example, for the Pb(SCN)$_3^-$ complex, $K_3 =$ [Pb(SCN)$_2$][SCN$^-$]/[Pb(SCN)$_3^-$], $\beta_3 =$ [Pb^{+2}][SCN$^-$]^3/[Pb(SCN)$_3^-$], and p$\beta_3 = pK_1 + pK_2 + pK_3$.

TABLE IX

Instability Constants of Lead(II) Complexes with Organic Ligands

Complex-forming agent	Type[a]	$-\log_{10}$ (instability constant)[b]	Temp., °C	Medium	Method	Ref.
Acetic acid	HL	pK_1 2.70, pK_2 1.5	25	μ ca. 0.5	Soly.	178
		pK_1 2.05	25	$1M$ NH_4ClO_4	Soly.	96
		pK_1 2.22, pK_3 2.40, pK_4 2.10	—	—	Pol.	396
Citric acid	H_4L	pK_1 2.52	30	Cor.	E.m.f.	88
		pK^M_{MHL} 6.50	30	Cor.	E.m.f.	196
		$pK^M_{MH_2L}$ 5.72	25	—	E.m.f.	383
Cysteine	H_2L	pK_1 12.20	25	$0.15M$ KNO_3	Glass	234
		pK_1 12.75	25	—	Pol.	234
1:2-Diaminocyclohexane-$NNN'N'$-tetraacetic acid	H_4L	pK_1 19.68	20	$0.1M$ KNO_3	Pol.	360
Ethylenediamine-NN-diacetic acid	H_2L	pK_1 12.22, pK_2 2.90	20	$0.1M$ KNO_2	Glass	356
Ethylenediamine-NN'-diacetic-NN'-dipropionic acid	H_4L	pK_1 13.2	30	$0.1M$ KCl	Glass	66
Ethylenediamine-$NNN'N'$-tetraacetic acid	H_4L	pK_1 18.3	20	$0.1M$ KCl	Glass	359
		pK_1 17.7	—	μ ca. 0.1	Spec.	255
		pK_1 16.8	30	$0.1M$ KNO_3	Spec.	168
		pK_1 18.04, pK^M_{MHL} 10.61	20	$0.1M$ KNO_3	Pol.	360

Ligand	Form	pK values	T	Medium	Method	Ref.
Glutathione	H_3L	pK_1 10.60	25	0.15M KNO_3	Glass	234
Glycine	HL	pK_1 5.47, pK_2 3.39	25	Cor.	Glass	278
		pK_1 5.17	25	Cor.	Sol.	191
Glycylglycine	HL	pK_1 5.53, pK_2 4.45	25	μ ca. 0.01	Glass	252
		pK_1 3.23, pK_2 2.70	25	Cor.	Glass	277
N-2-Hydroxyethylimino-diacetic acid	H_2L	pK_1 9.41, pK_{ML}^H 8.25	20	0.1M KNO_3	Glass	356
8-Hydroxyquinoline	HL	pK_1 9.45, pK_2 4.17	30	0.1M $KClx_3$	Glass	65
		pK_1 9.02	25	Cor.	Glass	288
8-Hydroxyquinoline-5-sulfonic acid	H_2L	pK_1 8.53, pK_2 7.6	25	Cor.	Glass	290
Methyliminodiacetic acid	H_2L	pK_1 8.02, pK_2 4.10	20	0.1M KNO_3	Glass	361
Nitrilotriacetic acid	H_3L	pK_1 11.8	20	0.1M KCl	Glass	358
		pK_1 10.68	—	0.2M KCl	Pol.	219
		pK_1 11.39	20	0.1M KNO_3	Pol.	356
Oxalic acid	H_2L	pK_1 + pK_2 6.54	26	Cor.	Sol.	212
Phenyliminodiacetic acid	H_2L	pK_1 3.49	20	0.1M KNO_3	Glass	356
Thioglycollic acid	H_2L	pK_1 8.5	—	μ = 0.002	Glass	234

[a] L = ligand.
[b] pK values refer to the stepwise addition of ligands. pK_{MHL}^M refers to the equilibrium between Pb^{+2} and the species PbHL; the symbols $pK_{MH_2L}^M$ and pK_{ML}^H have analogous meanings.

in Table VII. Stability constants of complex ions of lead(II) are given in Tables VIII and IX (45).

b. Lead(IV) Compounds

Aside from the organic compounds of lead(IV), there are relatively few simple compounds known.

Of the halides, only PbF_4 and $PbCl_4$ are known. They are of little importance in analytical chemistry. The hydride, PbH_4, is extremely unstable. On the other hand, many organic compounds that may be considered to be derivatives of PbH_4 are known. The most important of these is, of course, lead tetraethyl, $Pb(C_2H_5)_4$, the antiknock compound used in gasoline. For the determination of lead in these organolead compounds, see Section VI-G.

Lead(IV) oxide, PbO_2. Lead dioxide may be prepared by the electrolytic oxidation of solutions of lead salts or by the action of hypochlorite or other strong oxidizing agents in alkaline solution. It is a dense brown powder that is practically insoluble in water. It is amphoteric dissolving in acids as well as bases; its acidic character is more clearly developed, the action of bases giving rise to a whole series of plumbate compounds. Acids will dissolve lead dioxide with the formation of unstable salts of lead(IV). In concentrated hydrochloric acid, complex anions such as $PbCl_6^{-2}$ are formed. In $1M$ hydrochloric acid the lead(IV) exists mainly as the ion, $PbCl_5^-$ (157). All of these solutions are unstable, however, chlorine being slowly released by the reduction of lead(IV) to lead(II).

Lead dioxide is important in analytical chemistry in two respects: (1) the anodic deposition of PbO_2 is widely used for the separation and determination of lead; and (2) it is a powerful oxidant that is useful for the quantitative oxidation of many substances.

Lead dioxide has a tetragonal crystal structure that, until recently, was the only known form. The tetragonal form is produced by the anodic deposition of PbO_2 from a wide variety of solutions of low pH. An orthorhombic form, however, is deposited on electrolysis of plumbite solutions and of neutral nitrate and acetate solutions (121). The solid in the "dry" condition usually contains small amounts of water, which are very difficult to remove. When heated in air, oxygen begins to be lost at 348°C. Further discussion of the drying and weighing of lead dioxide is given in Section V-A-4.

In a pyrophosphate buffer of pH 2 to 6, lead dioxide oxidizes manganese-(II) to manganese(III), cerium(III) to cerium(IV), vanadium(IV) to vanadium(V), and chromium(III) to chromium(VI). The excess lead dioxide may then be easily removed by filtration.

Lead(IV) tetraacetate, $Pb(C_2H_3O_2)_4$. Lead tetraacetate is well known in organic chemistry as an alkylating and oxidizing agent. As an oxidizing agent it is often used for the selective cleavage of *vic*-glycols to the corresponding aldehydes or ketones. As an analytical reagent, however, it has received little attention because of the superior characteristics of metaperiodic acid for this purpose. Nevertheless, it should be very useful for the determination of water-insoluble glycols (124).

The colorless crystalline solid is freely soluble in glacial acetic acid, which is the solvent generally employed for carrying out its reactions. Recent applications include the determination of tartaric acid and tartrates (35), and the potentiometric titration of thallium(I) (36).

c. Mixed-Valence Lead Compounds

There are some lead compounds that may be considered to contain both lead(II) and lead(IV) in various proportions. The most important of these are the oxides intermediate in composition between PbO and PbO_2.

Lead orthoplumbate, Pb_3O_4, "red lead." This well-defined stoichiometric compound is obtained by heating litharge, PbO, in air to about 450°C. If the Pb_3O_4 that is formed is further heated in air, it begins to lose oxygen at about 550°C. and reverts to PbO. For this reason, PbO_2 cannot be prepared directly by calcining litharge. The Pb_3O_4 is a brilliant red dense powder that is commercially important as a corrosion inhibitor in paints.

Treatment of Pb_3O_4 with dilute nitric acid brings two-thirds of the lead into solution and leaves one-third as insoluble lead dioxide, PbO_2. Accordingly, red lead is given the name lead orthoplumbate and the formula $Pb_2(PbO_4)$. It also may be regarded as the result of the union of an essentially basic oxide, PbO, with a predominantly acidic oxide, PbO_2, to form a salt.

Lead sesquioxide, Pb_2O_3. Oxides of this approximate composition have been prepared by the thermal decomposition of lead dioxide and by the reaction of lead(II) salts with solutions of plumbates. Its existence as a definite stoichiometric compound, however, has not been established. The crystal structures and compositions of the intermediate lead oxides are exceedingly complex. For example, a study of the thermal decomposition of lead dioxide by differential thermal analysis revealed at least three transformations between the compositions PbO_2 and Pb_3O_4 (55).

Lead suboxide, Pb_2O. This oxide is reportedly formed by the thermal decomposition of lead oxalate. A careful investigation has shown that it is in reality an intimate mixture of elemental lead and tetragonal PbO (41). Its existence is accordingly extremely doubtful.

D. ISOTOPIC DISTRIBUTION OF LEAD

The isotopes of lead occupy an important place in chemical history. In fact, the first indication of the existence of isotopes was through the discovery of chemically indistinguishable samples of lead that differed in atomic weight. It was found that lead obtained from thorium-rich minerals had an atomic weight greater than that of "ordinary" lead (obtained, for example, from galena), whereas the atomic weight of lead from uranium-rich minerals was lower. These facts, combined with the demonstrated existence of radioactive lead, resulted in the recognition that all atoms of the same element were not necessarily identical. The nature of isotopy, however, was unambiguously demonstrated later through the classic experiments of J. J. Thompson and F. W. Aston with the mass spectrograph.

1. Mass Spectrographic Determinations

By far the most widely used technique of determining isotopic ratios is through the application of the mass spectrograph. The early work of Nier et al. showed the great potentialities of this method, particularly in regard to the determination of the geological ages of minerals (293a, 293b). The samples analyzed by Nier contained macro amounts of lead, which were easily processed, the lead being isolated as lead iodide which then could be volatilized directly into the mass spectrometer. A method more commonly used now for samples containing relatively large amounts of lead is to convert to the very volatile tetramethyl lead compound, $Pb(CH_3)_4$. The lead is then easily introduced as a gas into the ionization chamber of the spectrometer. No isotopic fractionation has been observed with this method, but the volatilization of lead iodide used by Nier has been questioned in this regard (30a). The lead is usually isolated from the sample as lead chloride, which is then reacted with methyl Grignard reagent to yield tetramethyl lead (30a). Yields of about 50% are obtained. An improved synthesis giving higher yields (136a), and in which the chances of contamination of the sample are reduced, has been recently applied to macrolead (greater than 1%) minerals (328a). In this method the isolated lead chloride is reacted with methyl lithium in ether solution in the presence of excess methyl iodide.

The above techniques, which are quite satisfactory for samples containing more than 1% lead, fail if the lead concentration is of the order of only a few p.p.m. Obviously the range of geological materials that could be successfully dated by isotopic distribution measurements would be enormously increased if methods could be developed to handle such samples.

The technique of isotope dilution is the most promising solution to this problem yet devised. As little as 0.1 p.p.m. of lead in granites has been precisely determined in this manner (394c). Briefly, the method works as follows. The sample is divided into two aliquot portions. To one of them, ^{208}Pb tracer is added, and a blank with ^{208}Pb tracer is also prepared. The two sample portions and the blank are then carried through the procedure for the chemical separation of lead in exactly the same way. The lead is first extracted from citrate solution at pH 9.5 with dithizone in chloroform, washed back into the aqueous phase with nitric acid, and then extracted again with dithizone from cyanide solution. The chloroform extract is evaporated to dryness and oxidized with nitric and perchloric acids. Although the lead recovery may only be of the order of 70%, complete recovery is not necessary since comparison of the isotopic distribution in the spiked and unspiked samples allows the recovery to be calculated. The dried sample is treated with a small amount of a borax solution, and it is then evaporated directly on the tungsten filament of the mass spectrometer. The filament is heated to 600°C. and the isotopic composition of lead emitted as Pb$^+$ is determined. Since the ion currents obtained with such small amounts are extremely small, a sensitive electron-multiplier detector must be used. Instrumental improvements and slight modifications of the chemical separation procedure have since been described (70a). It should be emphasized that contamination of the samples with ordinary lead from the reagents, the apparatus and the air is a serious problem. Elaborate precautions must therefore be taken to prevent it (70a, 394c).

2. Optical Determinations

Relative isotopic concentrations also may be determined from optical spectra independently of the mass spectrograph. Under sufficiently high resolution the emission lines of many elements containing a mixture of isotopes break up into a number of components. For the light atoms, the effect is due a modification of the Rydberg constant by the change in nuclear mass. For heavy atoms, this effect is small, but a nuclear volume effect and other factors are sufficient to cause isotope spectral shifts. Intermediate-weight atoms show vanishingly small isotope separations, but isotopic assay may still be carried out by using shifts in molecular spectra (e.g., in the band spectrum of the CuCl molecule). Although these separations for heavy elements are of the same order of magnitude as the so-called "hyperfine" structure, the latter should be distinguished from isotopic shifts since it is due to coupling of the nuclear moment with the total angular momentum of the extranuclear electrons. Thus, since hyperfine splitting is generally limited to nuclei of odd mass number, examination of the Pb 5201 A. line under high resolution shows single lines for ^{204}Pb, ^{206}Pb, and ^{208}Pb, but the ^{207}Pb line is split into several components (45a).

Because of the extremely high resolutions required for this work, the results obtainable are strongly dependent on the quality of the optical equipment available. Through improvements in the manufacture of grating replicas, large gratings exhibiting nearly theoretical resolution are now fairly readily available. To obtain the resolving power necessary, these are usually used in the higher orders of the spectrum. For example, a 30-ft grating spectrograph used in the tenth order was required to resolve the component lines of the Hg 2537 A. line, which were about 0.01 A. apart (394d).

An alternative approach is to use a moderate dispersion spectrograph coupled with a Fabry-Perot interferometer. The spectrograph serves to select the wavelength region of interest, which is then scanned by the interferometer. The intensities of the lines may be obtained by photoelectric recording. The treatment of the data is very similar to conventional spectrographic analytical methods (i.e., intensity ratios are proportional to concentration ratios). Using the Pb 4058 A. line, the relative concentrations of ^{204}Pb, ^{206}Pb, ^{207}Pb, and ^{208}Pb have been successfully determined and compare very favorably with mass spectroscopic values (52a,52b).

3. Geological Age Determinations

The major reason for the interest in accurate measurement of the isotopic concentrations of lead is for the determination of the ages of rocks and minerals. Lord Rutherford many years ago proved that the radioactive elements uranium and thorium decayed ultimately into helium and lead. Boltwood, as early as 1907, recognized that since the rate of decay of uranium could be measured, it should be possible to estimate the ages of uranium-containing minerals (47a): all that needed to be done was to determine the ratio of lead or helium to uranium and to calculate back to the date when the uranium was formed. Accordingly, he published the first list of geological ages that were based on the chemically determined ratios of lead to uranium and correctly showed the order of magnitude of geological time (47a). These first estimates were very crude, but the principle of modern determinations is exactly the same.

The calculated age of a mineral will equal its true age only if the following factors hold (394b): (*1*) the mineral was formed in a time short compared to its age; (*2*) there have been no gains or losses of the daughter or parent by processes other than decay of the parent; (*3*) correction can be made for the amount of daughter present at the time of formation; and (*4*) the decay constant of the parent is accurately known. Considering that these hold for a uranium mineral of a given age, decay products should thus be found in certain specific ratios to uranium itself. Therefore, four different ways of computing the age become immediately evident: from the ratio of ^{238}U to ^{206}Pb (the decay product of ^{238}U); from the ratio of ^{235}U to ^{207}Pb (the decay product of ^{235}U); from the ratio of ^{206}Pb to ^{207}Pb; and from the ratio of helium to uranium. If the sample also contains thorium the ratio of ^{232}Th to ^{208}Pb may also be used. The helium to uranium ratio is seldom applied, however, since helium is known to leak out of minerals, thereby giving lower age estimates. Convenient tables have been prepared for the calculation of geological ages in the region of 1 to 6000 million years from the remaining four ratios above (376a). All of these ratios require correction for the amount of ordinary (nonradiogenic) lead present in the sample at the time it was formed. This correction is made by determining the amount of ^{204}Pb present, since ^{204}Pb is the only isotope of lead known to be completely nonradiogenic. Accordingly, if a reasonable estimate of the isotopic distribution of ordinary lead at the time of formation can be made, the experimentally determined isotopic ratios may be corrected for the presence of nonradiogenic lead.

Complete agreement of the ages determined by the methods described, however, is seldom achieved. Of these, the age determined by the ^{206}Pb/^{207}Pb ratio is generally regarded as the most reliable since it involves only one analysis for the ratio of the two isotopes. In the other methods, of course, the chemical concentrations of lead and uranium, for example, also must be determined. Consequently, three measurements are required. Also, it would be expected that events occurring in the history of the mineral should not appreciably change the ^{206}Pb/^{207}Pb ratio since the chemical properties of the two isotopes are essentially identical. However, recent acid-washing experiments have shown that even this assumption cannot be made categorically (394a). In general, therefore, the agreement of at least two independent age determinations is regarded as necessary for the establishment of the absolute age of a rock. Although discordant ages were originally very disconcerting to the geologist, they actually convey more information than do concordant ages, since they have been affected by the pre- or postmineralization history of the mineral. See References 8a and 394b for more complete discussions of this topic.

III. SEPARATION AND ISOLATION OF LEAD

A. PRELIMINARY TREATMENT OF INORGANIC MATERIALS

The treatment applied to lead-containing samples depends, of course, on the nature of the sample itself and the quantity of lead it contains. The approach chosen is frequently different for the case of determining large amounts of lead as compared to the problem of determining traces.

1. The Attack of Lead Ores and Minerals

Sulfide minerals that contain relatively large amounts of lead are usually decomposed by heating them first with hydrochloric acid. Insoluble sulfides that remain are then oxidized by adding nitric acid. It may be necessary to add additional hydrochloric acid to dissolve any precipitated lead chloride. The lead may then be precipitated as lead sulfate. Since nitric and hydrochloric acids solubilize lead sulfate, they are first removed by adding excess sulfuric acid and then evaporating the solution to fumes. On dilution of the resulting solution with water, the lead sulfate precipitates and may be collected by filtration.

In silicate rocks, lead is rarely found present in macro quantities. However, if such a sample is encountered, the material may be fused with sodium carbonate and the melt leached with water to separate most of the silica from the lead. Some lead is invariably lost in this treatment and a recovery should be made.

Trace amounts of lead in silicate rocks are isolated by dithizone extraction. The sample is treated with perchloric and hydrofluoric acids, and is evaporated to remove silica. After this treatment there may remain undecomposed silicates, quartz, iron ores, and barium sulfate (344). If the latter substance is present, it will contain most of the lead in the sample. The undecomposed material is filtered off and fused with sodium carbonate, and the melt is leached with water and filtered; the washed residue is treated with hydrofluoric and perchloric acids to remove any remaining silica and it is finally dissolved in hydrochloric acid. This procedure thus gives three solutions, which may then be separately extracted with dithizone after proper pH adjustment.

2. Alloys

Lead is present as a major constituent of the following alloys: white metal bearing alloys, solders, type metals, leaded brasses and bronzes, various zinc-base diecasting alloys, and many others. The method of

TABLE X
Some Decomposition Methods for Lead-Containing Alloys (284)

Alloy	Decomposition method
Antimony metal	Treat with HCl + Br_2; evaporate nearly to dryness; add water, tartaric acid, and more HCl to give a clear solution
Copper metal	Boil with 1:1 HNO_3
Pig lead	Heat with concentrated HNO_3
Ingot tin	Treat with HCl + HBr and add Br_2 a little at a time until the sample is in solution. Evaporate to incipient dryness, and make up to volume with 10% HCl plus a few drops of HNO_3. This procedure volatilizes much of the tin
Aluminum metal	Dissolve in HCl and add a little HNO_3 toward the end of the reaction
White metals	Dissolve in HBr + Br_2. Evaporate to dryness and repeat. Cool, add HNO_3, and boil to expel excess Br_2
Zinc-base alloys	Dissolve in HNO_3
Steels	Dissolve in H_2SO_4 (if lead is to be separated as $PbSO_4$); if not, dissolve in HCl

decomposition applied depends, of course, on the composition of the alloy to be analyzed. For this reason, no universally applicable procedure can be cited. Some typical methods of attack are given in Table X (284).

3. The Decomposition of Organic Samples

Many different procedures have been employed for the destruction of organic material prior to the determination of lead in an organic sample. Decision as to the best method is frequently difficult when the individual procedures in the literature are consulted. Fortunately, a comparative study has recently been carried out which has provided much useful information and has confirmed suspicions about some methods of oxidation (140). The recovery of lead was followed by radiochemical means in this investigation. Most of the tests described below were concerned with the recovery of approximately 20 γ of lead from 2 g. of organic material. Cocoa was generally used as the organic material because it contains several types of organic matter, including fat, which is known to be resistant to wet oxidation.

Almost all methods for the destruction of organic matter fall into one of two main classes, i.e., dry ashing and wet oxidation.

a. Dry Ashing

In the dry ashing procedure the oxidation is accomplished by heating the sample to a relatively high temperature (about 400 to 700°C.) in the presence of air. Other substances are sometimes added as "ashing aids." These methods have the advantages of simplicity and freedom from the blank corrections that may be necessary in wet oxidations because of the addition of large amounts of oxidizing acids, which may contain impurities. Also, large samples may be easily handled. On the other hand, serious losses are sometimes encountered through volatilization and through retention on the crucible.

Losses by volatilization were found to be unimportant even on ignition to 650°C. when the lead was present as the nitrate or sulfate (the usual temperature recommended is 550°C.). However, in the presence of either chloride ion or covalently bound chlorine, serious losses of lead were observed.

The retention of trace elements on crucibles has not been fully appreciated. In the case of lead the retention was found to be highest on new silica crucibles and to increase sharply with increasing temperature. Lead oxide is believed to combine directly with silica to form lead silicate. On older crucibles the retention was less, but still serious. The addition of phosphoric acid or sodium chloride as ashing aids make the retention worse, whereas less retention was observed upon the addition of magnesium and aluminum nitrates and boric acid. In general the use of these ashing aids does not appear to be necessary for ignitions at 500 to 550°C. The addition of sulfuric acid slows the oxidation process, but may be beneficial in preventing volatilization losses in the presence of small amounts of chloride. The use of nitric acid results in a clean ash, but caution must be used to prevent deflagration, which results in serious loss. If platinum instead of silica crucibles are used, much lower retention losses are observed. Hence, their use can be recommended.

b. Wet Oxidation

In wet digestion procedures liquid conditions are maintained throughout, the oxidizing agents are chemicals present in solution, and the maximum temperatures achieved are much lower than in dry ashing. These procedures have the advantages of greater speed of oxidation and greater freedom from loss by retention on solid materials in the system.

Of the various acid mixtures studied (i.e., nitric and perchloric; nitric, perchloric, and sulfuric; nitric and sulfuric; and Middleton and Stuckey's (266) method of evaporation to dryness with nitric and sulfuric acids), the mixture of nitric and perchloric acids was found to be the most satisfactory. Mixtures containing sulfuric acid showed low recoveries of lead, presumably due to the deposition of lead sulfate. If samples containing calcium were treated with sulfuric acid, the losses of lead were very high because of the coprecipitation of lead with calcium sulfate.

B. SELECTIVE SEPARATION METHODS

Many procedures are available for commonly encountered and commercially important substances in which the usual interfering elements are properly accounted for. However, in the analysis of unfamiliar and unusual materials the common methods frequently must be modified and a preliminary separation must be made of the lead from those elements that would interfere with its subsequent determination. For this purpose, Table XI lists some typical methods of separation, along with the elements from which the separation of lead is effective.

TABLE XI
Typical Methods for the Separation of Lead

Separation from	Method	Ref.
1. Precipitation methods		
a. Precipitation of lead sulfate		
Numerous elements having soluble sulfates (elements accompanying Pb: Si, W, Nb, Ta, Ba, and, partially, Sr and Ca)	Fume with H_2SO_4; on dilution $PbSO_4$ precipitates (complete precipitation is never achieved)	
b. Precipitation of lead sulfide		
Ca, Sr, Ba, and many other metals	From dilute HCl medium with H_2S (classical scheme of qualitative analysis)	
c. Precipitation of lead molybdate		
Zn, Ni, Co, Mn	From acetate buffer (Cu and Cd are coprecipitated)	
d. Precipitation of lead chromate		
Cu, Cd, Fe(III), Zn, Ag, Ni, Mn, Al, Sr, Ca, Mg	From acetate medium of pH < 4.5 (if precipitated hot in the presence of 1 ml. 60% $HClO_4$/200 ml., separation from Ba is achieved)	53
e. Precipitation of lead periodate		
Al, Cd, Ca, Cu, Mg, Ni, Zn	From $0.025M$ HNO_3 with $NaIO_4$	419
f. Precipitation of lead iodate		
Al, Cd, Mg, Mn, K, Sr, Na, Zn	From hot 3% HNO_3 with HIO_3	135

(continued)

TABLE XI (continued)

Separation from	Method	Ref.
g. Electrodeposition as lead dioxide		
Cu and many other metals	Electrolysis from 15–20% HNO_3 at about 2 v. applied and 1–2 amp. (for details and interferences, see Section V-A-4)	
h. Electrodeposition as lead metal		
Cd, Sn, Ni, Zn, Mn, Al, Fe (Cu and Bi may be removed by preliminary electrolysis)	Controlled potential electrolysis at −0.60 v. vs. S.C.E. from 0.25M tartrate solution of pH 5.	242
2. Precipitation with organic reagents		
a. Thiourea		
Ag, Hg, Cu, Bi, Cd, As, Sb, Sn, Co, Ni, Fe, Mn, Al, Cr, Zn, Ba	With thiourea from 1–2M HNO_3. Quantitative precipitation as $2Pb(NO_3)_2 \cdot 11CS(NH_2)_2$ (not a weighing form). Only known interference is Tl(I)	250
b. Thionalid		
Ag, Cu, As(III), Zn, Ni, Co, Fe(II), Al, Cr, Cd, Ti	With thionalid from tartrate–cyanide–carbonate medium as $Pb(C_{12}H_{10}CNS)_2$	34
c. Mercaptobenzothiazol		
As, Sb, Mo, W, V, and elements which do not precipitate with H_2S in acid solution	With mercaptobenzothiazol as $PbC_7H_4NS_2(OH)$ from ammoniacal solution	375
d. Salicylaldoxime		
Ag, Cd, Zn	With salicylaldoxime from ammoniacal solution as $PbC_7H_5O_2N$	175, 237
3. Volatilization		
a. With ammonium iodide		
	When lead is present as PbO, PbO_2, $PbSO_4$, or PbS, it may be quantitatively volatilized on ignition of the sample with excess ammonium iodide	108
b. By induction heating		
Fe and silicates	The dry sample (rocks or meteorites) is heated to 1400°C. to volatilize trace amounts of lead, which are collected by condensation	254

TABLE XI (*continued*)

Separation from	Method	Ref.
4. Extraction methods		
a. Dithizone		
Cu, Sn, Bi	Aqueous phase: pH 9–10, KCN; organic phase: dithizone in chloroform or benzene	262, 279
Many elements	Aqueous phase: citrate and KCN; organic phase: dithizone in chloroform or carbon tetrachloride	28, 141, 172, 313, 407
b. Diethylammonium diethyldithiocarbamate		
Bi, Tl	Aqueous phase: 1.5M HCl; organic phase: diethylammonium diethyldithiocarbamate in chloroform	262
c. Sodium diethyldithiocarbamate		
	Aqueous phase: Sodium diethyldithiocarbamate and citrate or tartrate; organic phase: toluene–pentanol, chloroform	28, 131, 334
Large amounts of Ca, Mg, and phosphate	Aqueous phase: citrate, NaCN, and sodium diethyldithiocarbamate; organic phase: carbon tetrachloride	46, 245
d. Diethylammonium dithiocarbamate		
Tl and other metals	Aqueous phase: acidic solution; organic phase: diethylammonium dithiocarbamate in chloroform or trichloroethane	273
e. Sodium rhodizonate		
Many elements	Aqueous phase: HCl, NH$_4$F, sodium rhodizonate; organic phase: chloroform, carbon tetrachloride, or chlorobenzene	67
f. Potassium iodide		
Fe, Cu, Zn, Hg, Au, and Pd are removed by prior extraction with thiocyanate	Aqueous phase: 0.5M HCl, KI; organic phase: methyl isopropyl ketone	415
g. Thenoyltrifluoroacetone		
Tl(I), Ac, Ra	Aqueous phase: pH 4; organic phase: 0.25M thenoyltrifluoroacetone in benzene	150

(*continued*)

TABLE XI (*continued*)

Separation from	Method	Ref.
h. Potassium iodide and Tri-n-butylammonium chloride		
Large amounts of Mn(II), Ni, Fe, and the alkaline earths	Aqueous phase: KI and tri-*n*-butylammonium chloride in acidic medium; organic phase: methylene chloride	425
5. Adsorption methods		
a. Coprecipitation		
Numerous metals	Trace amounts of lead are precipitated as PbS, using CuS or HgS as collectors. Mixed sulfides are formed	89, 107
	Trace amounts of lead are collected by precipitation of $SrSO_4$. Mixed crystals are formed	336
	Trace amounts of lead are collected by precipitation of calcium phosphate	105
b. Ion exchange		
Bi, Fe	From $8M$ HCl using anion-exchange resin (Dowex 1). Bi and Fe are retained; Pb passes through	293
Sn(IV)	From 3% malonic acid solution of pH 4.8 using anion-exchange resin (Amberlite IRA-400). Sn(IV) is retained; Pb passes through	90
Ba, Sr, Al	On cation-exchange column (Dowex 50W). Pb eluted with ammonium acetate; others retained	198
Ce(IV), Zr(IV)	From 5% citric acid solution using Dowex 50W. Ce and Zr are eluted with water; Pb with ammonium acetate	198
Bi(III), Fe(III), Th(IV)	From EDTA solution at pH 2.1 using Dowex 50W (Na form). Pb is retained; others pass through	198
Al (feldspar) and many other metals	From $1M$ HCl solution using Dowex 1. Pb is eluted with $8M$ HCl	61
c. Chromatography		
Zn, Cd, Cu, Fe, Co, Bi, Hg(II), Mn, U	On paper using ether–methanol–water–nitric acid eluting solution. Ag and Tl(I) could not be separated from Pb	312

TABLE XI (*continued*)

Separation from	Method	Ref.
Cu, Bi, Cd, Hg, Fe, Al, Mg, Zn, Co, Ni, Sn	On paper using methanol–HCl eluting solution. A semi-quantitative estimation of Pb was made	169
Fe, Mo, Bi, Sb, Cd	On paper using butanol–HCl eluting solution. V, Cu, U, and Ti accompanied Pb, which was determined polarographically	233
Ni	On $CaCO_3$ column traces of Pb and Ni are separated. Pb is retained on the column	82
Hg, Ag, Cu, Bi	On paper impregnated with CdS. Developing solution: NH_4Ac, HAc, isopropanol, and water, 20 γ Pb was detected in the presence of 20 mg. Hg, 3 mg. Cu, 15 mg. Ag, and 0.6 mg. Bi	424

Among the methods listed the precipitation of lead with thiourea is particularly noteworthy for preliminary separation purposes. The only known interference with this procedure is that of thallium(I).

IV. DETECTION AND IDENTIFICATION OF LEAD

A. SPECTROGRAPHIC DETECTION

The basis of qualitative spectrographic analysis is contained in the following statement: "In the qualitative analysis for a certain element one need only look for the persistent lines of that element. If they are absent the assumption may be made that the element is not present in the sample" (52).

The persistent lines of lead and their spectrographic sensitivities are given in Table XII. In carrying out an analysis complete tables of spectra should be consulted for the elements that have lines in proximity to these lines and, consequently, possibly could be mistaken for lead.

B. CHEMICAL METHODS

No truly specific reagent or test for lead exists. In general, therefore, unless the nature of the sample is well known in advance, preliminary separations of lead are made. Some useful separation procedures have been outlined in Section III.

TABLE XII
The Persistent Lines of Lead[a]

Wavelength, A.	State	Sensitivity[b]	Intensity[c] Arc	Intensity[c] Spark	Excitation potential, v.[d]
5608.80	Pb II	V2	—	[40]	16.9
4057.820	Pb I	U1	2000R	300R	4.4
3683.471	Pb I	U2	300	50	4.3
3639.580	Pb I		300	50h	4.4
2833.069	Pb I		500R	80R	4.4
2614.178	Pb I		200r	80	>4.7
2203.505	Pb II	V1	50W	5000R	14.7
2169.994	Pb I		1000R	1000R	5.7
1682.4	Pb II		—	—	—

Key to symbols: R, wide self-reversal; r, narrow self-reversal; h, hazy, diffuse, nebulous; W, very wide or complex; and [ ], discharge-tube intensity.

[a] Compiled from data in References 7 and 256.

[b] For the neutral atom the most sensitive line (*raie ultime*) is indicated by U1 and the other lines by U2, U3, etc., in order of decreasing sensitivity. For the singly ionized atom the corresponding designations are V1, V2, etc.

[c] The values are based on the following convention: the most intense lines (e.g., the Na line at 5889.95 A.) are assigned values of 9000, and the faintest lines are assigned values of 1. For further information see Reference 256.

[d] The excitation potential for an ion includes the ionization potential for the neutral atom, in order to give an approximate idea of the excitation required to produce the line.

In the classical *scheme of qualitative analysis*, lead falls into the silver group (Group I) due to the slight solubility of lead chloride. The precipitation with hydrochloric acid, however, is never complete and provision must always be made in the scheme for the detection and removal of lead in the hydrogen sulfide group (Group II). The difficulty of having the same element appear in two groups is avoided in some schemes by carrying out the precipitation of Group I from hot solution. Under these conditions, lead chloride is very much more soluble and passes into the filtrate to be detected in Group II.

If lead is precipitated together with silver and mercurous mercury from cold solution, it is separated by boiling the precipitate with water. The silver and mercurous chlorides are removed by filtration, and the presence of lead is confirmed by precipitation with K_2CrO_4 (yellow ppt.), KI (yellow ppt.), or H_2SO_4 (white ppt.).

When a solution 0.3M in acid is saturated with hydrogen sulfide, lead sulfide is precipitated along with the sulfides of Hg(II), Bi, Cu, Cd, As, Sb, and Sn, to make up Group II. Lead (along with Hg, Bi, Cu, and Cd) is separated from As, Sb, and Sn by taking advantage of the insolubility of lead sulfide in ammonium sulfide–ammonium hydroxide solutions.* The lead sulfide is then separated (along with the Bi, Cu, and Cd sulfides) from mercury sulfide by warming the mixture with $2M$ nitric acid; the mercury sulfide is insoluble whereas the others dissolve. The separation of the lead from Bi, Cu, and Cd is then generally accomplished by precipitation of the lead as lead sulfate. Confirmation of the presence of lead may be obtained by dissolving the sulfate in ammonium acetate and adding potassium chromate to precipitate yellow lead chromate.

It is to be emphasized that the separations of the usual qualitative scheme are by no means quantitative. The formation of mixed sulfides frequently alters the behavior expected from observation of the properties of the individual sulfides. As a result, for this and other reasons small amounts of some elements cross over into groups where they may not be expected. For qualitative purposes, generally, this is not serious. Further information on such behavior is given in standard works on qualitative analysis.

1. Reactions of Lead(II) Ion

Sodium hydroxide. White precipitate, soluble in excess reagent:

$$Pb^{+2} + 2OH^- = Pb(OH)_2(s)$$

$$Pb(OH)_2(s) + OH^- = HPbO_2^- + H_2O$$

When plumbous salts are treated with sodium hydroxide, basic salts are frequently formed instead of simple hydrated lead oxide. See Section II-C-2-a.

Ammonia. White precipitate, insoluble in excess reagent (lead does not form ammine complex ions):

$$Pb^{+2} + 2NH_3 + 2H_2O = Pb(OH)_2(s) + 2NH_4^+$$

Basic salt formation is also common under these conditions.

Hydrogen sulfide. Black precipitate from acidic, neutral, or basic solution. In the presence of hydrochloric acid the first precipitate obtained may be red in color due to the formation of the double salt, $PbS \cdot PbCl_2$. From $0.3M$ hydrochloric acid the color changes rapidly from

* As, Sb, and Sn form sulfo salts and are separated in the filtrate. If the precipitate is treated with more strongly basic potassium bisulfide–potassium hydroxide solution, mercury accompanies these elements in the filtrate.

red to brown to black, giving the normal sulfide, PbS. From more concentrated HCl the red double salt is more persistent.

Hydrochloric acid. White platelets of sparingly soluble lead(II) chloride from solutions not less than $0.01 M$ in lead. Soluble in hot water, and also in concentrated hydrochloric acid with complex formation. Tends to form supersaturated solutions.

Potassium iodide. Yellow hexagonal platelets of lead(II) iodide. The solubility greatly increases in hot water; the saturated solution is colorless, whereas the precipitate is yellow.

Potassium chromate. Light yellow precipitate of lead(II) chromate; insoluble in ammonia (distinction from silver); soluble in excess sodium hydroxide (distinction from bismuth); insoluble in acetic acid or ammonium acetate.

Sulfuric acid. White precipitate of lead(II) sulfate, insoluble in acidic and neutral solutions. Soluble in sodium hydroxide, hot concentrated sulfuric acid, and in ammonium acetate and sodium thiosulfate (distinction from barium).

Sodium stannite. Black precipitate of metallic lead from hot basic solutions.

2. Microscale Tests for Lead

When carried out under the microscope, the following two tests can be applied to the detection of very small amounts of lead.

a. The "Triple Nitrite" Crystal Reaction

This excellent test is specific for lead and has a limit of identification of only 0.003γ of lead (33) (see Section VII-A-2 for details of the procedure). The formation of very characteristic jet-black crystals of $2KNO_2 \cdot Cu(NO_2)_2 \cdot Pb(NO_2)_2$ serves to establish the presence of lead, and no other element can be substituted for lead without altering the appearance of the crystals. Barium, strontium, and calcium may replace lead to form similarly shaped crystals. The presence of relatively large amounts of cadmium or bismuth hinder the test.

b. The Potassium Iodide Crystal Reaction

Although many ions react with potassium iodide to give precipitates, this reaction is frequently used as a confirmatory test for lead. Briefly, the test is carried out by placing a drop of the slightly acidified test solution on a microscope slide. A small crystal of potassium iodide is then placed at the edge of the drop and is allowed to dissolve and diffuse into the solution. The presence of lead is determined by the crystallization of lead iodide as bright yellow hexagonal plates. When floating at the proper angle, these plates appear greenish or brownish and sometimes even gray. By reflected light they exhibit the brilliant colors of thin films, a very characteristic feature of this compound.

Many other elements (e.g., Ag, Hg, Tl, Cu, Bi, Cd, Sb, and As) form insoluble iodides under these conditions and may obscure the test.

3. Spot Tests for Lead

The following three tests are widely used "spot tests" for the detection of lead. For further examples, see Reference 109.

a. Test with Bendizine

Benzidine reacts with lead dioxide to form a meriquinoid oxidation product commonly known as "benzidine blue." A number of other oxidizing agents are capable of bringing about this reaction. However, under the conditions of the test (see Section VII-A-3) the interference of Ce, Mn, Bi, Co, Ni, and Ag is eliminated by maintaining the solution in an alkaline condition. It is recommended that the oxidation be effected with alkali hypobromite solution since the excess hypobromite is readily destroyed by reaction with ammonia. A slightly increased sensitivity has been reported by using hydrogen peroxide as the oxidant and by substituting 2,7-diaminodiphenylene oxide for the benzidine (87).

b. Test with Sodium Rhodizonate

Sodium rhodizonate reacts with lead ions to produce colored precipitates of basic lead rhodizonates. In neutral solution, violet $Pb(C_6O_6) \cdot Pb(OH)_2 \cdot H_2O$ is formed, and from weakly acid solution the scarlet red $2Pb(C_6O_6) \cdot Pb(OH)_2 \cdot H_2O$ is produced. The test is very sensitive; positive results are obtained on such slightly soluble materials as PbS, $PbSO_4$, and $PbCrO_4$.

c. Test with p-Tetramethyldiaminodiphenylmethane

p-Tetramethyldiaminodiphenylmethane in strongly basic solution in the presence of lead and hydrogen peroxide is oxidized to the corresponding hydrol, which is blue in color. The reaction is sensitive, but other oxidants may also yield the same product.

4. Special Tests for Lead

a. Detection of Lead in Bismuth Compounds (109)

The following procedure is particularly recommended for the examination of pharmaceuticals containing bismuth. The sample is first ashed and fumed with nitric acid. If the residue is treated directly with hypobromite and the benzidine test is applied, then the bismuth is oxidized and interferes strongly. On the other hand, if one boils the residue first with sodium hydroxide the bismuth is converted to the yellow insoluble BiO(OH), which is not oxidized on treatment with hypobromite. The benzidine test may then be applied directly to the supernatant liquid.

b. Separation and Detection of Lead in Water (109)

Traces of lead may be efficiently collected from water by the precipitation of mercuric sulfide. Sodium mercurisulfide solution and ammonium chloride are added to the sample solution and HgS is precipitated. After treatment with hydrogen peroxide, the collected HgS is ignited and the mercury vaporized. The residue of lead sulfate may then be tested with sodium rhodizonate in the usual manner. Lead in extreme dilution may be detected in this way, e.g., in 100 ml. of water at a dilution of $1:10^8$. This test also may be applied to the examination of certain "high-purity" reagents that do not give precipitates with hydrogen sulfide in basic solution.

c. Detection of Lead Tetraethyl and Lead Tetraphenyl in Motor Fuels

The antiknock compounds, lead tetraethyl and lead tetraphenyl, which are commonly added to gasoline, are very volatile and extremely poisonous. The presence of these substances may be detected in the following way. If the gasoline is colored with a dyestuff, it should be decolorized with activated charcoal. A few drops of the sample are then placed on a filter paper and irradiated under an ultraviolet lamp until the gasoline has evaporated. This treatment decomposes the organolead compounds. Lead is then detected by adding a drop of freshly prepared 0.1% solution of dithizone in chlorofrom. A deep red spot of lead dithizonate appears if the sample contained lead; if lead is absent the dithizone remains green.

V. DETERMINATION

A. PRECIPITATION AND GRAVIMETRIC DETERMINATION

No single gravimetric method of the multitude that have been proposed for the determination of lead is clearly the most satisfactory for the majority

TABLE XIII
Thermal Stability of Weighing Forms of Lead (95)

Precipitant	References	Weighing form	Temperature limits, °C.
Hydrogen	377, 397	PbO	>946
Chlorine + sodium hydroxide or electrolysis	161, 331	PbO	>650
Hydrogen peroxide		PbO_2	100–120
Aqueous ammonia		$Pb(OH)_2$	155–410
Electrolysis	78	PbO_x	<340
Hydrochloric acid	77	$PbCl_2$	In the cold
Iodic acid	135	$Pb(IO_3)_2$	<400
Sodium periodate	419	$Pb_3(IO_5)_2$	151–280
Hydrogen sulfide	247	PbS	97.5–107.2
Sulfite, hydrogen sulfite, or pyrosulfite	133, 153, 179	$PbSO_3$	<60 and >900
Sulfuric acid		$PbSO_4$	271–959
Potassium sulfate	388	$PbSO_4 \cdot K_2SO_4$	40–906
Disodium hydrogen phosphate	280, 307	$Pb_2P_2O_7$	>355
Sodium carbonate	180	$PbCO_3$	<142
Ammonium thiocyanate	374	Pb(OH)SCN	In the cold
Molybdophosphoric acid	38	$Pb_{25}Mo_{25}H_{14}P_2O_{112}$	436
Oxalic acid	48	PbC_2O_4	50–300
Sodium phthalate	426	$C_6H_4(CO_2)_2Pb$	288–320
Gallic acid	263	$C_6H_2O_3CO_2Pb$	<152
Sodium salicylate	283		None
Sodium anthranilate	128	$Pb(C_7H_6O_3N)_2$	<198
Dimethylglyoxime	127	$Pb(OH)_2 \cdot PbC_4H_6O_2N_2$	60–88
Salicylaldoxime	175, 237	$PbC_7H_5O_2N$	45–180
Oxine	165	$Pb(C_9H_6ON)_2$	In the cold
5,7-Dibromo-oxine	423	$Pb(C_9H_4ONBr_2)_2$	In the cold
Picrolonic acid	158	$Pb(C_{10}H_7O_5N_4)_2$	58–112
Thionalide	34	$Pb(C_{12}H_{10}NS)_2$	71–134
Mercaptobenzothiazole	222, 375	$Pb(C_7H_4NS_2)_2$	<120
Mercaptobenzimidazole	221, 222	$PbOH \cdot C_7H_5N_2S$	97–172
7-Nitro-5-sulfo-oxine	276	$Pb(C_9H_5O_6N_2S)_2$	<48
Potassium iodide[a]		PbI_2	60–370
Sodium chlorite[b]	229	$Pb(ClO_2)_2$	<77
Sodium chloride and sodium fluoride[c]	6, 376	PbClF	66–538
Disodium hydrogen arsenate[d]	335	$PbHAsO_4$	81–269
		$Pb_2As_2O_7$	320–950
Ammonium molybdate[e]	404	$PbMoO_4$	>505

[a] Not useful because of incomplete precipitation.
[b] Used for the determination of the chlorite ion. Explodes at about 77°C.
[c] Used for the determination of fluoride ion.
[d] Used for the determination of arsenate.
[e] Not recommended by Duval (95) for the determination of lead.

of samples. The thermal stability of a number of the weighing forms available has been studied and compared through the use of the recording thermobalance (95). The results of this valuable study are presented in Table XIII.

1. As Lead Sulfate

The time-honored precipitation of lead as the sulfate has much to recommend it for the determination of lead in a wide variety of samples. It unfortunately suffers from the fact that the solubility of lead sulfate is higher than what is generally regarded desirable for quantitative separation. For example, the solubility has been measured in sulfuric acid solutions and it is found that it decreases at first with increasing concentrations of sulfuric acid. A minimum is reached at 0.3% (w./w.) of sulfuric acid (ca. 0.03M). The solubility then increases and reaches a maximum at about 10% sulfuric acid (ca. 1M), and decreases once again and passes through a second minimum. The solubility at 25°C. in 0.3% sulfuric acid is 4.55 mg. of $PbSO_4$ per liter; in 10% sulfuric acid it is 6.68 mg. per liter; and it increases very rapidly below 0.1% sulfuric acid (86).

The consequence of these facts is that the method is limited to relatively large amounts of lead unless special steps are taken to reduce the solubility. In careful work the unprecipitated lead is frequently recovered by electrolysis of the solution from which the precipitation was made.

In most procedures the sample is brought into solution with nitric and hydrochloric acids. Sulfuric acid is added, and the sample is evaporated until fumes of sulfuric acid are obtained. This evaporation is generally repeated several times in order to ensure that the nitric and hydrochloric acids, which have a strong solvent action on lead sulfate, are completely removed. Since the lead sulfate is soluble in concentrated sulfuric acid, it does not precipitate until the cold solution is diluted with water. Because the solubility is appreciable over the entire range of concentrations of sulfuric acid, to obtain reproducible results the concentration of acid present should be carefully maintained at a known value. If such a precaution is taken, a correction for the solubility loss may be applied, although actual recovery of the lead by electrolysis is to be preferred.

The lead sulfate is separated by filtration through a fine-porosity porcelain crucible, and it is washed with dilute sulfuric acid. Here again the conditions should be standardized with respect to the concentration of acid in the wash liquid and the volume of wash liquid used. If only lead is to be determined, and the lead in the filtrate is not to be recovered, then the washing is best done with sulfuric acid that has been saturated with lead sulfate.

The solubility losses may be reduced to negligible proportions by the addition of alcohol, and this procedure is to be recommended when lead is to be determined in samples not containing elements that form sparingly soluble sulfates. However, if such elements are present (particularly, Bi, Ca, and Ag), serious contamination of the precipitate will result. Also, the presence of alcohol is often objectionable if further determinations are to be carried out on the filtrate.

The process of fuming with sulfuric acid serves to dehydrate silica, and this separates with the lead sulfate on dilution with water. If only small amounts of silica are present, a separation may be made by dissolving the lead sulfate in a hot solution containing 20 g. of ammonium acetate and 2.5 ml. of acetic acid per 100 ml. If large amounts are present the lead cannot be quantitatively removed in this way, and resort must be made to volatilization of the silica with hydrofluoric acid.

The precipitation of lead as the sulfate in the presence of bismuth has often been criticized on the grounds that bismuth is coprecipitated. Etheridge (101) has found that bismuth is heavily coprecipitated when the mixture is fumed strongly (350 to 400°C.), but when the fuming is carried out at a low temperature (250°C.) there is no coprecipitation and good results are obtained. This low-temperature fuming procedure is claimed to be better than the more commonly used preliminary separation of bismuth as the oxychloride or oxybromide, especially when large amounts of bismuth are present. Etheridge also reports that the digestion of lead sulfate that contains bismuth with ammoniacal ammonium acetate completely dissolves the lead sulfate, leaving the bismuth behind as an insoluble residue.

Other elements that precipitate with lead sulfate and silica are tungsten, niobium, tantalum, and barium. Strontium, calcium, antimony, bismuth, silver, and copper may be partially precipitated, depending on the conditions and the amounts present. If tin is present the solution should be filtered shortly after dilution with water to prevent insoluble hydrolysis products of stannic sulfate from contaminating the precipitate. One way to avoid this problem is to add hydrofluoric acid. Lead sulfate formed under these conditions contains only traces of tin (352). Most commonly, however, interference from tin is avoided by a preliminary separation as metastannic acid, which is formed on treatment of the sample with nitric acid. If relatively large amounts of tin are present, some lead may be occluded in the precipitate.

Barium is quantitatively precipitated along with lead on treatment of the sample with sulfuric acid. The dissolution of the lead sulfate in ammonium acetate has been proposed for the separation of these elements after precipitation of the sulfates. It has been found that the barium sulfate tends to prevent the complete solution of the lead, and is itself somewhat soluble in this medium. The procedure, therefore, cannot be recommended, particularly if the Ba:Pb ratio in the sample exceeds 1:10 (251). In such cases, lead should be separated from the barium by a sulfide precipitation.

Potassium, if present in appreciable concentrations, may interfere by formation of the double salt, $PbSO_4 \cdot K_2SO_4$. This salt has been observed to form if solid lead sulfate is contacted with solutions of potassium sulfate of concentration $0.023M$ or greater (320). The crystal habit obtained when lead sulfate is precipitated in the presence of potassium also has been found to be markedly altered (215).

Duval (95) states that lead sulfate is an excellent weighing form for lead. A horizontal was obtained on the thermogravimetric curve from 271 to 959°C. A more recent study by thermogravimetry and differential thermal analysis gives its decomposition temperature as 803°C. (299).

The technique of precipitation from homogeneous solution has been applied successfully to the determination of lead as the sulfate. Lead in copper-base alloys containing Zn, Sn, Fe, Ni, Al, and Mn has been satisfactorily determined (98). The sulfate was generated through the hydrolysis of dimethyl sulfate in 70 to 80% methanol solution. The precipitate obtained was coarsely crystalline and easy to filter and wash. Good recoveries of 1 to 100 mg. of lead were obtained. The concentration of methanol, the amount of dimethyl sulfate, and the digestion time must be adjusted to the quantity of lead and foreign ions present in the sample.

Sulfate generated through the hydrolysis of sulfamic acid in a solution containing ethanol, nitric acid, and hydrochloric acid also results in the quantitative precipitation of lead sulfate (57). These conditions permitted the determination of lead in alloys containing antimony and tin.

2. As Lead Chromate

Lead chromate is more insoluble than lead sulfate and, in this respect, is superior as a precipitation form for lead. The method has received much criticism in the literature, however, and conflicting reports are encountered.

The difficulty appears to be due to the fact that the precipitate obtained does not correspond exactly to the formula, $PbCrO_4$. High results were obtained by Grote when determining lead as the chromate (147). Direct analysis of the precipitates formed showed that they contained an excess of chromium. It was found that an empirical factor for $Pb/PbCrO_4$ of 0.6378, rather than the theoretical 0.6411, gave results in concordance with the lead sulfate procedure.

Other workers, however, have obtained good results using the theoretical factor (149,189). The difference may be due to the use of different precipitation and drying conditions. It is felt that although results of high accuracy and precision have been obtained in some cases (189), the method should not be recommended for work requiring high accuracy due to the uncertainty of the composition of the precipitate.

The fact that lead chromate is insoluble in dilute mineral acid solution, whereas other chromates (particularly barium) are soluble, allows quantitative separations to be made. For example, the precipitation of 0.04 to 0.4 g. of lead by the slow addition of ammonium chromate to a boiling solution containing free nitric acid serves to separate the lead from Cu, Ag, Ni, Ca, Ba, Sr, Mn, Cd, Al, and Fe (189). Thus the ability of this method to determine lead in the presence of barium represents one of its major advantages over the sulfate method. The temperatures recommended for drying the precipitate vary widely; the following have been proposed: 160 (147), 140 (189), 120 (163), and 600°C. (149).

Lead may be determined very nicely as the chromate by precipitation from homogeneous solution (163) (see Section V-B-2). A much larger particle size is obtained with a consequent improvement in washing and drying characteristics. The sample containing 0.1 to 0.2 g. of lead is neutralized, and potassium bromate, chromic nitrate, and an acetate buffer are added. When the solution is heated to 90 to 95°C., the bromate slowly oxidizes Cr(III) to Cr(VI), which then precipitates as lead chromate. After filtration and washing the precipitate is dried at 120°C. and is weighed as $PbCrO_4$. It is noteworthy that the precipitate formed under these conditions was found to have the theoretical composition. Ammonium ions interfere because they are oxidized by bromate, silver interferes because of precipitation with bromide, and anions that complex Cr(III) interfere because they inhibit oxidation to Cr(VI). Al, Fe, Ni, Cu, Zn, Ba, Cd, Bi, and Hg in moderate amounts do not interfere.

3. As Lead Molybdate

Precipitation of lead as the molybdate from dilute nitric acid is also frequently used. The precipitate is less soluble than either the sulfate or chromate, and has a more favorable gravimetric factor. If, however, other ions are to be determined in the filtrate, the addition of molybdate is often objectionable. The method is applicable in the absence of the alkaline earths, which form insoluble molybdates, and in the absence of chromates, arsenates, and phosphates, which form insoluble lead compounds. It serves to separate lead from zinc, nickel, cobalt, and manganese, but copper and cadmium are coprecipitated.

Precipitation is complete from either neutral solution or from solutions that are slightly acidified with acetic acid. The precipitate obtained from neutral solution tends to be colloidal and difficult to filter. For this reason the precipitation of most of the lead is generally carried out in hot, dilute nitric acid (398,413). The acid is then neutralized with ammonia and a slight excess of acetic acid is added. Under these conditions the lead is quantitatively precipitated. The presence of sodium or ammonium chloride in the solution aids in rendering the lead molybdate granular, dense, and filterable.

Ibbotson and Aitchison (171) recommend that the precipitation be carried out in a solution in which the mineral acid has been neutralized with ammonia and then made slightly acidic with acetic acid. Ammonium molybdate is then added to the boiling solution, which contains several grams of ammonium chloride. A large excess of ammonium molybdate should be avoided because contamination of the precipitate with molybdic oxide was found to be roughly proportional to the excess of ammonium molybdate added. This coprecipitation was regarded as being sufficiently important to recommend purification of the lead molybdate to ensure against high results. The precipitate is dissolved in hydrochloric acid, most of the free acid is neutralized with ammonia, and then ammonium acetate is added until methyl orange shows a yellow tint. Under these conditions the lead molybdate is quantitatively reprecipitated. It is filtered off on a Gooch crucible and is ignited at 600°C.

4. By Electrodeposition

When a nitric acid solution of lead is electrolyzed, the lead is deposited on the anode as lead dioxide. The presence of nitric acid is essential because it acts as a cathodic depolarizer (i.e., it is reduced at the cathode more easily than plumbous ion). Unless a high concentration of nitric acid is maintained (about 10 to 20% vol./vol.), lead may be deposited on the cathode. In addition the high concentration also aids against the simultaneous deposition of manganese dioxide on the anode.

Another important consideration is that the presence of substances more easily oxidized at the anode than plumbous ion must be avoided if complete deposition of the lead is to be achieved. Chloride ion is particularly deleterious in this respect, because it is oxidized at a much lower potential than plumbous ion and with very little overvoltage.

$$PbO_2 + 4H^+ + 2e^- = Pb^{+2} + 2H_2O \qquad E° = 1.456 \text{ v.}$$

$$Cl_2 + 2e^- = 2Cl^- \qquad E° = 1.358 \text{ v.}$$

Frequently several drops of concentrated sulfuric acid are added to the electrolyte since its presence seems to aid in the formation of an adherent deposit.

The electrolytic deposition of lead has been the subject of many investigations and much controversy has existed over the specification of optimum conditions. The trouble lies mainly in the fact that when lead dioxide is deposited on the anode it does not correspond to the formula, PbO_2, even when electrolyzed from a solution containing only lead nitrate and nitric acid. Various methods of drying or igniting the deposit to obtain PbO_2 have been used with variable results. One of the most commonly used procedures is to rinse the anode in alcohol and ether and then dry it at 120°C. This always leads to high results if the theoretical factor, $Pb/PbO_2 = 0.8662$, is used. The empirical factor, 0.864, is recommended under such conditions (213).

TABLE XIV
Empirical Factors for the Anodic Deposition of Lead

Quantity of PbO_2 deposited, g.[a]	Electrolysis temperature, ±2.5°C.		
	20°C.	60°C.	95°C.
Less than 0.1	0.8662	0.8662	0.8662
0.1–0.3	0.8644	0.8651	0.8657
0.3–0.5	0.8626	0.8640	0.8650
0.5–0.7	0.8614	0.8629	0.8640
0.7–1.0	0.8595	0.8618	0.8633

[a] Amounts of lead dioxide less than 0.5 g. were dried for 1 hour; greater amounts were dried for 2 hours at 220°C.

L. Hertelendi (160) has made a thorough comparative study of the effects of current density, concentration of nitric acid, temperature, quantity of lead deposited, and drying conditions on the composition of the deposit obtained. The electrolyses were carried out using a stationary inner platinum gauze electrode, and the solutions were stirred throughout the electrolyses at 500 to 550 r.p.m. with a glass stirrer. Initially each solution contained 0.3 g. of copper and had a volume of 150 ml. The walls of the electrolytic cell were washed down twice with water during the electrolysis to make the final volume 190 to 200 ml. Under these conditions it was found that the current density and the concentration of nitric acid (provided that it was greater than 5%) were essentially without effect on the composition of the deposit. The important variables were found to be: (1) the drying conditions, (2) the temperature at which the electrolysis is carried out, and (3) the quantity of lead dioxide deposited.

The procedure recommended was to rinse the anode with alcohol followed by ether, and to dry it at 220°C. Higher results were found at lower drying temperatures. With this drying technique the theoretical factor was found to apply at all conditions where less than 0.1 g. of lead dioxide is deposited. For larger amounts of lead and various solution temperatures, a table of empirical factors was constructed and is reproduced in Table XIV.

It has been proposed that the use of an empirical factor can be avoided by gently igniting the lead dioxide at 600 to 650°C. to PbO and weighing this (261). Other experiments indicate that low results are obtained on ignition at 700°C. (238). These were attributed to the volatilization of PbO at this temperature. The experiments were done using a platinum dish as the anode, thus avoiding the possibility that any solid would flake off the anode during the ignition process.

The anodic deposition of lead gives good results in the presence of all the common metals with the exception of Ag, Bi, and Mn, which form peroxides on the anode; Tl, Sb, Co, and Sn, which contaminate the deposit; and As, Hg, Se, Te, and P, which prevent or hinder complete deposition. The interference of Sn, Sb, and As may be prevented through the addition of fluoride. In this case some fluoride will contaminate the deposit; however, if the deposit is electrolyzed into a fresh 15% nitric acid solution and is redeposited the concentration of fluoride becomes so small that it does not interfere.

The presence of copper in the electrolyte is beneficial because nitrate is smoothly reduced to ammonia at a copper cathode, whereas at a platinum electrode only partial reduction to nitrous acid is achieved. The nitrous acid produced then can be oxidized back to nitrate at the anode and interfere with the oxidation of lead. In most circumstances, therefore, if copper is not present in the sample, it should be added.

a. Special Electrolytic Methods

A micro method has been developed for the determination of 2 to 10 mg. of lead in tin–lead alloys (73). The electrolysis is done in a Clarke-Hermance cell (76), and has been successfully applied to alloys containing 5 to 100% lead. The electrolysis is carried out at an elevated temperature, and urea is added to the electrolyte to remove nitrous acid. With such small amounts of lead the theoretical factor of Pb/PbO_2 was found to be applicable after drying the electrode only 5 minutes at 120°C. In the presence of Sb and large amounts of Sn, a preliminary separation of lead and copper as sulfides must be made.

Lead has been determined electrolytically in iron and steel (142). It is claimed that as little as 0.01% of lead may be satisfactorily estimated. The concentration of iron in the electrolyte must be less than 1 g. in 150 ml. and the solution must be electrolyzed 6 hours. Manganese codeposits with the lead and must be determined separately and a correction applied.

Small amounts of lead (less than 17 mg.) have been successfully separated and determined by internal electrolysis from potassium hydroxide solutions (243). The cell that is used is:

$$C|S_2O_8^{-2}, SO_4^{-2}|PbO_2^{-2}, PbO_2|Pt$$

Advantages of this procedure are that it is rapid and that no electrical instrumentation is required. The PbO_2 is dried at 220°C. and is weighed as such.

Lead has also been separated cathodically and determined as lead metal by internal electrolysis. The technique, however, is not of great importance. For details see Reference 353.

5. With Organic Precipitants

Many organic reagents have been suggested as precipitants for the gravimetric determination of lead. A number of these are listed in Table XIII. The reader should consult the original literature for details of these procedures.

In general, these compounds offer more favorable gravimetric factors and higher insolubility of the precipitate than the methods described above. Because of this, their chief application has been for the determination of lead in amounts smaller than can be conveniently handled by these procedures. None of the reagents listed is particularly selective for lead, and, consequently, many elements may interfere. Often the procedures have been worked out for application to particular analytical problems, with the result that complete studies of interferences have not been made.

Of the precipitants listed the following are noteworthy: salicylaldoxime in strongly ammoniacal solution serves to separate lead from Ag, Cd, and Zn, and Cu may be removed by a preliminary precipitation with salicylaldoxime in weakly acid solution; thionalide in a solution containing sodium tartrate, sodium carbonate, and potassium cyanide separates lead from Ag, Cu, As(III), Zn, Ni, Co, Fe(II), Al, Cr(III), Cd, and Ti; and mercaptobenzothiazole separates lead from elements that do not form sulfides with hydrogen sulfide in acidic solution. These lead salts are all suitable for direct weighing after drying at moderate temperatures (100 to 120°C.).

6. Other Methods

Lead has been determined gravimetrically as lead iodate (135). By slowly adding iodic acid to the hot sample solution, $Pb(IO_3)_2$ is quantitatively precipitated and, after digestion and cooling, may be filtered on a sintered glass crucible, washed, and dried at less than 400°C. (140°C. is recommended). It is weighed as $Pb(IO_3)_2$. It is important that the precipitation be made from hot solution, since cold lead iodate supersaturates badly. The solution should be 3% in nitric acid, and an excess of 0.75 g. of iodic acid should be present for each 100 ml. of solution.

Lead also may be precipitated as triplumbic paraperiodate, $Pb_3H_4(IO_6)_2$, from dilute nitric acid solution (419). The concentration of nitric acid must be less than $0.025M$, and for very small amounts of lead it must be decreased to $0.006M$. Many other elements form insoluble periodates under these conditions, but lead may be satisfactorily separated from moderate amounts of Ni, Cu, Zn, Al, Ca, and Mg. The lead periodate is precipitated from hot solution by the slow addition of sodium periodate. The solution must then be held at 0°C. for 30 minutes to avoid errors due to supersaturation. The precipitate is filtered and washed at this temperature, and is dried at 110°C. and weighed as $Pb_3H_4(IO_6)_2$. Its composition depends on the precipitation and drying conditions, and for this reason a volumetric finish is preferable (see Section V-B-2-c).

Lead is quantitatively precipitated from 84% nitric acid as $Pb(NO_3)_2$, and may be weighed as such (418). The sample is dissolved in nitric acid or in a minimum of water, and the acid concentration is carefully adjusted to 84%. The precipitation is carried out at room temperature and the lead nitrate is filtered and washed with 84% nitric acid and is dried at 135°C. Barium and strontium interfere, and chloride produces low results and must be absent. Interferences have not been fully investigated, but efficient separations from many elements should be possible.

Lead sulfide may be quantitatively precipitated from up to $0.1M$ hydrochloric acid, from up to $0.3M$ nitric acid, or from aqueous ammonia by boiling with a 400% excess of thioacetamide (117). If the precipitate is dried at 110°C. as PbS, slightly high results are found. Duval (95) found that lead sulfide is a suitable weighing form for lead in the narrow temperature range of 97.5 to 107.2°C.

A separation of lead from many elements and its subsequent gravimetric determination is possible by precipitation as lead chloride from a 2% solution of hydrochloric acid in n-butanol (187). The sample is treated with hydrogen peroxide to oxidize Sn(II) to Sn(IV), and then is repeatedly evaporated with hydrochloric acid to convert the sample to the chloride form. In this process As, Sb, and Sn are separated by volatilization. The 2% hydrochloric acid solution in n-butanol is added to the dry residue and, after boiling 5 minutes, is filtered through sintered glass. The collected lead chloride is dried 30 minutes at 105 to 110°C., and 10 minutes at 250°C., and is weighed as $PbCl_2$. Elements which form soluble chlorides in this medium and do not interfere are: Bi, Cu, Zn, Cd, Fe, Al, Cr, Mn, Ca, Mg, Co, and Ni.

B. TITRIMETRIC METHODS

Lead may be determined titrimetrically by means of precipitation and complex-formation reactions. It also may be determined indirectly by precipitation followed by an oxidation-reduction titration. A number of instrumental procedures also are available.

1. Precipitation Titrations

Although precipitation titrations are in general less accurate than gravimetric methods for lead, they are of importance as rapid control methods.

a. Titration with Molybdate

One of the most widely used precipitation titrations is the titration of lead with standard molybdate solution. The classical procedure uses ammonium molybdate as the titrant and tannic acid as external indicator. It is a laborious procedure and has little to recommend it. It has been applied most frequently to the titration of lead that has been separated as the sulfate, which was then subsequently dissolved in sodium acetate–acetic acid solution (246). A better external indicator for this titration is potassium thiocyanate plus stannous chloride. Red $K_3[Mo(SCN)_6]\cdot 4H_2O$ forms at the end point (224).

Improper pH control has been cited as one of the major causes of error in this titration (164). When ammonium molybdate is used as the titrant the following reaction occurs:

$$(NH_4)_6Mo_7O_{24} + 4H_2O + 7Pb(NO_3)_2 = 6NH_4NO_3 + 7\underline{PbMoO_4} + 8HNO_3$$

Since the titrant initially has a pH of 3 to 4 due to hydrolysis and considerable nitric acid is formed by the precipitation reaction, at the equivalence point the solution will have a very low pH. Holness (164) circumvents this difficulty by using magnesium molybdate as the titrant and Solochrome Red B (C. I. No. 216) as an adsorption indicator.

$$MgMoO_4 + Pb(NO_3)_2 = \underline{PbMoO_4} + Mg(NO_2)_2$$

Under these conditions the pH throughout the titration should be less than 5.5 to avoid formation of basic salts, and at the equivalence point must be greater than 2 for proper functioning of the adsorption indicator.

Accordingly, the pH of the sample solution is adjusted between these values and then the hot solution containing 0.1 to 0.9 g. of lead is titrated with standard magnesium molybdate. At the end point the solution changes from colorless to orange red, and the precipitate from bluish pink to white. The color change is reversible and the error is reported not to exceed 0.2%.

The presence of tin causes difficulty by precipitating as stannic oxide and occluding some lead. By boiling the solution with excess molybdate and back-titrating, lead may still be determined but with poorer precision. Antimony interferes seriously.

Lead also may be accurately titrated with molybdate to the clear point (341). When molybdate is carefully added to lead in hot sodium acetate–acetic acid solution it is observed that the colloidal suspension in the supernatant liquid clears suddenly at a point very close to the equivalence point. It is necessary to add the titrant drop by drop near the end point and to allow one or two minutes' settling time after each addition. Although such methods are capable of high precision, they are seldom useful for anything except pure salt solutions.

b. Titration with Chromate

Macro and semimicro amounts of lead have been titrated directly with potassium chromate in dilute nitric acid solution, using 2,6-dichlorophenolindophenol as an adsorption indicator. With macro amounts (*ca.* 400 mg.) the mean error is about 0.5%; with smaller amounts the error increases (123).

c. Titration with Phosphoric Acid

Small amounts of lead (*ca.* 10 mg.) may be directly titrated with phosphoric acid in neutral or weakly basic solution containing about 20% of acetone (102). The indicator used is an aged solution of diphenylcarbazide. Apparently on standing in solution the carbazide is oxidized to diphenylcarbazone, which reacts with lead to give a red color. The titration is carried out using phosphoric acid that has been standardized against a known lead solution. A reagent blank should be used as a comparison solution, and the end point should be approached slowly. Many other metals also form red colors with the indicator and interfere.

The above procedure has been considerably simplified by Stenger (214)

d. Titration with Ferrocyanide

Direct titration of lead is not practicable when diphenylcarbazone is used as an internal indicator, and the titration should be done in the reverse direction. The standard ferrocyanide solution (10 to 15 ml. of $0.05M$ potassium ferrocyanide) is made weakly alkaline with saturated sodium acetate solution, 1 ml. of 0.3% diphenylcarbazone in ethanol is added, and the mixture is titrated with neutral lead nitrate solution until a red color forms, which indicates the presence of excess lead. At the end point the solution is red and the precipitate is colorless (330).

Direct titration with ferrocyanide is possible by making use of the reversible redox indicator, variamine blue. The sample containing 100 to 1000 mg. of lead is buffered at about pH 3.75 with formate buffer solution. One drop of $0.1M$ potassium ferricyanide and some variamine blue are added to the sample. The blue-violet solution is titrated with standard $0.1M$ ferrocyanide solution until the color disappears. An abrupt change in potential occurs after the precipitation of lead ferrocyanide is complete (100).

e. Titration with Alkali Hydroxide

Lead forms a basic bromide of unusual stability which, unlike the normal hydroxide, is neutral to phenolphthalein. An alkalimetric titration is therefore possible, the reaction being:

$$Pb^{+2} + OH^- + Br^- = \underline{PbOHBr}$$

A fivefold excess of bromide is added to the neutral lead solution and the mixture is titrated with sodium hydroxide to the phenolphthalein end point (406).

Adsorption indicators also have been used for detection of the end point during the titration of lead with sodium hydroxide (414). In a similar titration, standard sodium carbonate solution is used as the precipitant (338).

f. Titration with Fluoride

A potentiometric titration based on the precipitation of lead chlorofluoride, PbClF, has been developed (106). A solution of lead of relatively high concentration (0.05 to $0.1M$) containing excess sodium or potassium chloride is titrated directly with standard sodium fluoride. To each 100 ml. of sample solution are added 40 mg. of ferrous chloride and 0.8 mg. of ferric chloride. The ferric ion forms a stable fluoride complex, which causes an abrupt change in potential when the precipitation of lead chlorofluoride is complete. Interferences were not studied, but the precision is excellent.

2. Oxidation-Reduction Titrations

a. Direct Oxidation to Lead(IV)

Although lead exhibits two oxidation states, the direct oxidation of Pb(II) to Pb(IV) by titration with oxidizing agents is practically never used as a method of determining lead. An old method does exist whereby potassium permanganate in basic solution is used for this purpose, but it is of little importance (47). The difficulty with this approach is that it has not been found possible to quantitatively oxidize all of the lead.

An indirect procedure is also available whereby lead in alkaline solution is oxidized to lead dioxide by means of potassium persulfate (228). The lead dioxide is then used to oxidize a known volume of standard arsenite solution, and the excess arsenite is determined by titration with permanganate. The oxidation of lead is not stoichiometric, and nickel must be added as a catalyst. The empirical factor, 104.6, rather than the theoretical 103.6, is used for the reaction weight of lead. Co, Bi, Sn, Sb, Mn, Ce, halides, phosphate, and arsenate interfere; Cu, Fe, and Zn do not.

b. Determination of Lead(IV) in Lead(IV)–Lead(II) Mixtures

Oxidation-reduction titrations are of importance in the assay of many commercial products containing higher oxides of lead. A substantial literature exists on this subject, and the major approaches to the problem will be briefly reviewed here.

Mrgudich and Clark (281) have compared a number of the methods available by analyzing a single sample of C. P. lead dioxide.

Bunsen Method (399)

This method is based on the oxidation of a strong solution of hydrochloric acid by lead dioxide. The chlorine produced is distilled and carried with a stream of carbon dioxide into a solution of potassium iodide, which serves as collector. The liberated iodine is then titrated with thiosulfate. Mrgudich and Clark (281) found results that were low by about 5%, which they attributed to a side reaction between chlorine and water vapor. On the other hand, it has been found that when especially prepared, constant-boiling hydrochloric acid is used good results are obtained with absolute errors not exceeding 0.5% (55). The low results of Mrgudich and Clark are explained as being due to the presence of volatile impurities in the hydrochloric acid. The method tends to be somewhat slow and cumbersome, but is particularly useful if samples of lead oxide are encountered that are difficult to dissolve.

Diehl-Topf Method (395)

In this procedure the sample of lead oxides is added directly to a solution of potassium iodide, which is buffered with acetic acid and sodium acetate. The liberated iodine is then titrated directly with thiosulfate. Blank determinations indicate that no atmospheric oxidation of iodide occurs (281). Difficulties that may be encountered are incomplete solution of the sample and possible side reactions with the iodine produced.

Schaeffer Method (351)

This procedure utilizes a nitric acid solution containing a known amount of hydrogen peroxide to dissolve the sample of lead oxides. The excess peroxide is then back-titrated with permanganate. A similar procedure also has been developed in which the excess peroxide is titrated potentiometrically with ceric sulfate (129). Good results were obtained with samples of red lead, but the results were low with lead dioxide and the dissolution of the sample was slow.

Lux Method (248)

The method as originally proposed by Lux (248) consisted in dissolving the sample in a mixture of nitric and oxalic acids. The oxalic acid not oxidized by the lead dioxide was then back-titrated with permanganate. Mrgudich and Clark (281) showed it to be unreliable due to a side reaction between the oxalic and nitric acids. By substituting perchloric acid for the nitric acid and titrating with cerium(IV), they obtained good results.

Rupp and Siebler Method (339)

To the sample an excess of standard arsenious oxide solution and hydrochloric acid are added. After boiling under reflux because of the volatility of arsenic trichloride, the excess arsenic(III) is titrated with potassium bromate, using methyl orange as indicator. Anderson and Sterns (21) have modified the method by titrating the excess arsenious oxide potentiometrically with permanganate in the presence of a trace of iodide as catalyst. Under their experimental conditions no loss of arsenic trichloride or of chlorine occurred. The procedure is capable of high precision and is convenient for routine determinations.

c. Indirect Determination of Lead by Means of Oxidation-Reduction Titrations

These methods depend on the precipitation of stoichiometric lead compounds with anions that may be conveniently determined by an oxidation-reduction titration.

Titration of Lead Chromate

Lead may quantitatively precipitated as the chromate from acetic acid–sodium acetate solution under the conditions described for its gravimetric determination (Section V-A-2). Two different approaches are then available. A measured amount of standard chromate solution may be used to precipitate the lead, the lead chromate removed by filtration, and the excess chromate in the filtrate determined iodometrically or by titration with ferrous sulfate. Or, the separated and washed lead chromate may be dissolved in hydrochloric acid, treated with potassium iodide, and the liberated iodine titrated with thiosulfate. In general, the latter course is to be preferred (211).

As in the case of the gravimetric determination, any elements that give precipitates with chromate will interfere. Accordingly, the only difference between the two procedures is that the lead chromate is in one case dried and weighed, and in the other is determined volumetrically. Essentially the same considerations apply to each. One milliliter of $0.1N$ thiosulfate is equivalent to 6.907 mg. of lead.

Titration of Lead Iodate

The gravimetric determination of lead as the iodate has already been discussed (Section V-A-6). Gentry and Sherrington (135) also have used a volumetric finish for this determination. The procedure is the same up to the point at which the lead iodate is dried and weighed. Instead, the filtered and washed precipitate is dissolved in hot sodium hydroxide and potassium iodide is added. On acidification with hydrochloric acid, iodine is liberated and is titrated with thiosulfate. One milliliter of $0.1N$ thiosulfate is equivalent to 1.727 mg. of lead.

Titration of Lead Periodate

The gravimetric method of Willard and Thompson (419) for the determination of lead as the periodate has been mentioned previously (Section V-A-6). Because the composition of the precipitate is somewhat dependent on drying conditions the titrimetric finish in this case is to be preferred. The precipitate, which is filtered on paper in a Gooch crucible, is transferred to a glass-stoppered Erlenmeyer flask and an excess of standard sodium arsenite solution is added. After the addition of concentrated hydrochloric acid the excess arsenite is determined by titration with iodate, according to the well-known method of Andrews. Standard iodate is added until the solution assumes a light brown color. A small amount of chloroform is then added and the titration is continued until the violet color in the chloroform is discharged. Vigorous shaking is required between each addition of titrant. One milliliter of $0.1N$ sodium arsenite is equivalent to 5.18 mg. of lead.

3. Titrations Based on Complex Formation

Lead forms few complexes of sufficiently great stability to allow its determination by complexometric titration. The only reagent widely used for this purpose is ethylenediaminetetraacetic acid (EDTA). However, because of the versatility, convenience, and accuracy of the EDTA titrations they have replaced many of the older precipitation titrations (see Table XV).

Much of the effort in developing satisfactory procedures has been devoted to finding indicators that (*1*) have high extinction coefficients; (*2*) show

TABLE XV
Some Visual EDTA Titrations of Lead

Method	Indicator	Solution conditions	Color change	References
Direct titration	Eriochrome Black T C.I. 14645	NH_3 buffer, pH 10; tartrate or triethanol amine added	Blue-violet to blue	116, 119, 316
	Eriochrome Red B C.I. 18760	NH_3 buffer, pH 10; tartrate added	Red to yellow	411
	Copper–PAN	Acetate buffer, pH 4.6	Red to yellow	113
	Xylenol Orange	Acetate buffer, pH 5	Reddish-purple to yellow	216, 217
	Copper–Naphthyl-azoxine	Acetate buffer, pH 5.5–6.5	Yellow to red	125
	4-(2-Pyridylazo)-resorcinol (PAR)	Weakly acidic medium	Red to yellow	412
	Metomega Chrome Cyanine BLL C.I. 17940	Barbiturate buffer, pH 6.8	Red to blue	3
	Omega Chrome Black Blue G C.I. 18160	NH_3 buffer, pH 10; tartrate added	Red to blue	2
	Dithizone	20% pyridine, pH 3.5–6.4	Red to yellow-green	64
Back-titration	Eriochrome Black T Titrant: Mg, Zn.	NH_3 buffer, pH 10	Blue to red	39
	Zincon Titrant: Zn	NH_3 buffer, pH 9–10	Yellow to blue	199

brilliant color changes when converted from the complexed to the uncomplexed form; and (*3*) have high affinities for lead ions but still release the lead when titrated with EDTA. All of these requirements are not easily met, so a number of techniques, some of which are indirect, have been developed for the determination of lead.

a. Direct Titration

Direct titration with a standard solution of EDTA is, of course, the most attractive in principle. The solution is usually buffered at pH 8 to 10 and contains tartrate or other weak complexing agents to prevent precipitation of metal hydroxides. In this pH region, Eriochrome Black T (C.I. 14645) (116,119), Eriochrome Red B (C.I. 18760)(32,411), and Omega Chrome Black Blue G (C.I. 18160) (2), serve as indicators. A sensitive reversible indicator for lead in neutral solution (pH 6.8, barbiturate buffer) is Metomega Chrome Cyanine BLL (C.I. 17940) (3).

The direct titration of lead also may be carried out with advantage in a weakly acidic medium using pyrocatechol violet (381,409), Xylenol Orange (216,217,317), and pyridine(2-azo-4)resorcinol (PAR) (412) as indicators. Perhaps the best of these is Xylenol Orange. The titration is done at pH 5 to 6 using urotropine as buffer. If *o*-phenanthroline is used as a masking agent, it is possible to titrate 47.5 mg. of lead in the presence of 3 to 30 mg. of Cu, Ni, Co, Zn; 5 to 50 mg. of Cd; and 3 to 15 mg. of Mn, with an average error of only ±0.12 mg. of lead (317) (see Section VII-B-3 for details). Cysteine has also been shown to be an effective masking agent for Hg and Cu under similar conditions (37).

Methylene blue and thionine also can be used as indicators for the direct titration of lead (186). This novel method makes use of the fact that these dyes are photoreduced in the presence of free EDTA, but are not affected by the presence of metal–EDTA complexes. Thus by carrying out the titration under brilliant illumination the end point is signaled by the rapid photoreduction of the dye at the point at which excess EDTA is present in the solution.

b. Back-Titration

Excess EDTA may be added to the lead sample and then the excess may be back-titrated with a standard solution of some metal salt. Solutions of magnesium or zinc sulfate generally serve for this purpose. In the case of lead, these procedures have little advantage over the direct titrations described above, so the direct procedures are usually preferred. However, back-titrations frequently are convenient if insoluble lead salts are to be analyzed (e.g., lead sulfate). The salt may be brought into solution by heating with an excess of EDTA and then determining the excess by titration.

c. Replacement Titration

When a metal–EDTA salt that is less stable than the lead–EDTA complex is added to a solution containing less ions, the following reaction occurs:

$$MY^{-2} + Pb^{+2} = PbY^{-2} + M^{+2} \qquad \text{(where } Y = C_{10}H_{12}O_8N_2^{-4})$$

The displaced metal ions may then be titrated with EDTA using the indicator system and reaction conditions for the metal in question. This approach has been used for determining lead by making use of the Mg–EDTA complex (116,119,314) or, preferably, the Zn–EDTA complex (116,118,120). The addition of some zinc or magnesium in the direct titration of lead using Eriochrome Black T as indicator improves the color contrast at the end point, but does not make it sharper (115).

Because of the great stability of the complex between copper and 1-(2-pyridylazo)-2-naphthol (PAN) the following reaction occurs when PAN is is added to a solution containing the Cu–EDTA complex and lead ions:

$$Pb^{+2} + CuY^{-2} + PAN^{-2} = PbY^{-2} + Cu(PAN)$$

This is a partial replacement reaction that makes possible the direct titration of lead in weakly acid solutions (114). A small amount of a solution containing copper and EDTA in exactly equivalent amounts is added to the sample solution. On addition of PAN the bright red Cu–PAN complex is formed. When this solution is titrated with EDTA, the EDTA first complexes with the free lead ions and then decomposes the Cu–PAN complex, thereby causing a color change to yellow.

d. Alkalimetric Titration

If to a neutral or very weakly acidic solution of lead ions, disodium EDTA is added, the reaction,

$$H_2Y^{-2} + Pb^{+2} = PbY^{-2} + 2H^+$$

takes place. The released hydrogen ions are then titrated with sodium hydroxide, and the procedure is in effect a replacement titration. A large excess of disodium EDTA must be avoided, and for this reason it is added in small increments from a second buret until further addition does not cause the pH to drop below 5. The end point is indicated with methyl red–bromcresol green (357).

This procedure is recommended as a rapid preliminary titration so that one may calculate the smallest possible excess of EDTA to add when the excess EDTA is to be determined by back-titration. Otherwise the method is not of great importance. Suspensions of lead sulfate may be titrated in this way, but weak acids and their salts must be absent.

e. Instrumental Methods of End-Point Detection

Although visual titrations are rapid, convenient, and the most suitable for routine determinations, other means have been employed to detect the end point.

Lead has been titrated photometrically with EDTA without the addition of an indicator by employing the absorption of the lead–EDTA complex 240 mμ (329,400,416). An accuracy of about 1% was obtained when 0.2 to 2 mg. of lead was titrated in a 1.5-cm. cell at pH 2. Under these conditions, lead and bismuth may be determined simultaneously with one titration at 240 mμ. Since bismuth reacts preferentially with EDTA and the bismuth–EDTA complex absorbs less light at this wavelength, two end points are observed (416).

Several successful potentiometric EDTA titrations have been carried out. Lead, in addition to other metals, may be determined by adding excess EDTA to the sample solution at pH 8. The excess is then back-titrated potentiometrically with mercuric nitrate using an amalgamated silver indicator electrode (197). A mercury drop may serve as an indicator electrode for a direct titration since it responds to the activity of the free lead ions in the solution when a small amount of mercury–EDTA complex is added and the pH is properly adjusted (326). The titration is carried out potentiometrically with EDTA in an acetate buffer of pH 4.6. An accuracy of 0.4% is obtained with 25 to 250 mg. of lead. Alkaline earths do not interfere, but halides should be absent since they disturb the response of the mercury electrode (327).

Lead has been titrated coulometrically by generating EDTA electrically through the reduction of the mercury–EDTA complex. The end point of the titration is found potentiometrically (325).

C. POLAROGRAPHIC METHODS

Particularly favorable electrochemical characteristics make the polarographic determination of lead very popular. A wide variety of samples are amenable to the polarographic method, and the behavior of lead in a great number of supporting electrolytes has been investigated. Since only a few applications can be discussed here, the reader should consult other monographs and bibliographies (208,272,350).

1. The Dropping Mercury Electrode

The polarographic characteristics of lead in the presence of 0.01% gelatin in various supporting electrolytes are summarized in Table XVI.

TABLE XVI
Polarographic Characteristics of Lead in Various Supporting Electrolytes at 25°C.

Supporting electrolyte	$E_{1/2}$ vs. S.C.E., v.	$i_d/Cm^{2/3}t^{1/6}$
0.1M KCl (or HCl)	−0.396	3.80
1M KCl (or HCl)	−0.435	3.86
1M HNO$_3$	−0.405	3.67
1M NaOH	−0.755	3.39
0.5M tartrate, pH 4.5	−0.48	2.37
0.5M tartrate, pH 9	−0.50	2.30
0.5M tartrate + 0.1M NaOH	−0.75	2.39

Maxima are usually observed during the reduction of lead ions at the dropping mercury electrode. Gelatin is effective in suppressing these, but the concentration added must be maintained constant in quantitative work since the diffusion current has been shown to depend significantly on its concentration (207).

It is seen from the table that the transformation of the simple lead ion in acidic solution to the biplumbite ion, $HPbO_2^-$, in basic solution is accompanied by a large shift in the half-wave potential to more negative values. This property frequently is utilized in practical analyses for the elimination of interferences. Also, the large changes in $i_d/Cm^{2/3}t^{1/6}$, which is proportional to the square root of the diffusion coefficient in going from a chloride to a tartrate medium clearly indicates that complex formation occurs in the tartrate solution. For the most part the reduction of lead ions, either free or complexed, takes place reversibly under the usual polarographic conditions. That is, the rate of electron transfer, $Pb(II) + 2e^- = Pb(Hg)$, is rapid compared to the rate of mass transfer to the electrode.

Lead is effectively determined in either acidic or basic solution. In acidic solution one of the best supporting electrolytes is 0.4M sodium tartrate and 0.1M sodium hydrogen tartrate (to give a pH of 4.5), and a small concentration of gelatin as maximum suppressor (240). The gelatin concentration should not be greater than 0.005% if bismuth is present. In this medium Cu, Bi, Pb, and Cd may be determined simultaneously. The respective polarographic characteristics of these metals are: $E_{1/2}$ = −0.09, −0.23, −0.48, and −0.64 v. vs. S.C.E.; and $i_d/Cm^{2/3}t^{1/6}$ = 2.37, 3.12, 2.37, and 2.34 µamp./mmole/liter at 25°C.

In basic solution, for example, in $1M$ sodium hydroxide, lead as the biplumbite ion, $HPbO_2^-$, is also reversibly reduced. This medium is particularly useful for the determination of lead in the presence of Sn, Sb, and As. Stannate and arsenite ions do not show reduction waves in this medium and the wave of antimonite is 0.5 v. more negative than that of the biplumbite ion. Also, thallium is reduced about 0.3 v. more positive than lead, so that the presence of an approximately equal amount of thallium does not interfere. In acidic solution the waves of these two metals overlap.

A cyanide-supporting electrolyte is useful for the determination of small amounts of lead in the presence of large amounts of cadmium; the reduction wave of lead precedes that of cadmium cyanide complex by about 0.4 v.

The recently reported reduction wave of the plumbate ion in $5M$ sodium hydroxide, where $E_{1/2}[Pb(IV)] = E_{1/2}[Pb(II)] = -0.78$ v. vs. S.C.E., should have some interesting analytical applications (195).

2. The Platinum Electrode

The reduction of lead using a stationary platinum wire electrode has been studied (225), and lead has been determined by making use of the rotated platinum wire electrode (368). But, since reduction of lead on the mercury electrode is far superior with regard to reproducibility, little analytical application of solid electrodes has been made.

One procedure should be mentioned, however, since it is capable of determining lead in the presence of large amounts of thallium (206). Thallium interference is prevented by first oxidizing it to Tl(III) in acidic solution with chlorine. The excess chlorine is then removed with a stream of nitrogen. Then, when the solution is made basic, thallic oxide precipitates and does not interfere. A precision and accuracy of about 2% was observed for $10^{-4}M$ lead in the presence of $10^{-2}M$ thallium.

3. Methods for Very Low Concentrations of Lead

The optimum concentration range for most polarographic determinations is from 10^{-3} to $10^{-4}M$ and may be extended without compensation for the capacity current to concentrations as low as $10^{-5}M$. Although this range applies also to the determination of lead, through the use of specialized techniques and equipment much lower concentrations may be measured. For example, Kelley and Miller (192) have determined lead in solutions as dilute as $2 \times 10^{-7}M$. In this case the diffusion current measured was of the order of only 10^{-3} μamp.

The newer techniques of oscillographic polarography (378), alternating-current polarography (30), polarography with rotating electrodes (386) or with stirred solutions (294), rapid-scanning polarography with a hanging mercury drop (337), and chronopotentiometry (324) have been applied to the determination of lead at concentrations much lower than usually accessible by the conventional methods.

The technique of anodic dissolution polarography has been found to be useful for the determination of very small amounts of lead (253). By electrolyzing solutions containing as little as $10^{-8}M$ lead, using a mercury-plated platinum cathode in a stirred solution, it is possible to deposit all the lead on the electrode. If the potential of the electrode is then made sufficiently positive the dissolution current corresponding to the oxidation of lead may be recorded as a function of time. Integration of the curve gives the number of coulombs equivalent to the amount of lead present. A precision of about 5% is possible. Another method is to estimate the concentration of lead from the peak height of the stripping curve (292). A precision of about 6% was obtained down to concentrations of about $10^{-8}M$. Very careful purification of the background electrolyte was found to be essential.

Very small amounts of impurities in uranium were determined by depositing them in a hanging mercury drop from potassium carbonate solution (193). Anodic dissolution curves were then recorded and lead at a concentration of $8 \times 10^{-9}M$ was determined.

Metallic bismuth has been analyzed polarographically for very small amounts of lead by making use of an unique electrochemical separation (104). Since bismuth is reduced at more positive potentials than lead, it creates considerable difficulty in the determination of lead by conventional d.-c. polarography if present in considerable excess. With square-wave polarography a 10^5 excess of bismuth over lead may be tolerated. For still greater excesses of bismuth the following procedure was developed. The sample of bismuth metal was dissolved in mercury and the amalgam was contacted with $1M$ hydrochloric acid containing a known, small concentration of bismuth. The activities of bismuth in the amalgam and aqueous phases fixed the potential of the system at a value at which the traces of lead were quantitatively oxidized and transferred to the aqueous phase. The aqueous phase was then easily analyzed by square-wave polarography. An accuracy of better than $\pm 2\%$ was found for the determination of 0.5 p.p.m. of lead in bismuth.

Small concentrations of lead have a pronounced effect on the reduction wave of oxygen that may be used for the detection and approximate determination of lead (379). Neutral solutions of KNO_3, K_2SO_4, or $Ca(NO_3)_2$

show a greatly increased oxygen maximum and also a larger limiting current. The lead is believed to catalyze the reduction of oxygen to water instead of the usual reduction product, hydrogen peroxide. By means of this catalytic action, $10^{-7}M$ lead may be detected. Some other metals show similar behavior, but the effect of lead is most pronounced.

4. Amperometric Titrations

Lead may be accurately titrated with dichromate even in very dilute solution by means of the amperometric method using the dropping mercury electrode (210). If the titration is carried out at -1.0 v. vs. S.C.E., both the lead and dichromate yield diffusion currents. A particularly favorable shape for the titration curve is obtained, and the end point may be precisely located (see Section VII-B-7 for details of the procedure).

At the beginning of the titration the diffusion current of the lead ions is observed. As dichromate is added, lead chromate precipitates and the diffusion current decreases. When the solution contains excess dichromate, reduction of Cr(VI) to Cr(III) occurs and the diffusion current increases once again. The curve is rounded in the neighborhood of the equivalence point due to the solubility of lead chromate. The pH should be maintained less than 5.5 since at higher values basic lead chromates tend to precipitate, causing low results. Very high acidities also should be avoided because of increased solubility of the precipitate.

The branches of the titration curve are straight lines if sufficient inert salt is added to eliminate the migration current (e.g., greater than $0.01M$ KNO_3), and if the observed diffusion currents are corrected for dilution by the added titrant. Extrapolation of the straight-line portions of the titration curve then precisely locates the equivalence point.

When titrations are done at -1.0 v. all substances reducible at this potential should be absent from the test solution (e.g., oxygen). The titration, however, may also be effectively done at zero applied potential, at which many substances that are reduced at -1.0 v. do not interfere. The titration curve consists of a straight horizontal line (since lead is not reduced) followed by an abrupt increase in current at the end point. This has the advantage that dissolved oxygen need not be removed and the titration is done quickly and accurately. A satisfactory supporting electrolyte is an acetate buffer of pH 4.

Although the dichromate titration is by far the most widely used amperometric method, other titrants have been investigated. EDTA has been used in acetate medium at pH 4.2 and at an applied potential of -0.60 v. vs. S.C.E. (387). In ammoniacal tartrate solution of pH 8, lead may be titrated with EDTA at -0.72 v. vs. S.C.E. (315). A rotated platinum electrode has been used at zero applied potential *vs.* a mercuric iodide reference cell for the titration of lead in weakly acidic solution, using ammonium oxalate as titrant (401). Copper is added to provide a reducible ion, and an unusual titration curve is observed that shows a very rapid decrease in the diffusion current at the end point.

Quantities of lead of the order of 50 γ were titrated with a precision of 4%, using a mercury pool electrode. Being free of the charging current associated with the dropping mercury electrode, the mercury pool proved to be much superior for small amounts of lead. The titration was done in stirred solution at pH 6 to 9 with EDTA as the titrant (294).

D. PHOTOMETRIC METHODS

1. Spectrophotometry and Colorimetry

a. With Dithizone

Traces of lead are effectively determined colorimetrically by means of the reaction of lead with dithizone. This method dwarfs all others in importance, and will be the only one discussed in detail in this section.

In neutral and in alkaline solutions lead reacts with diphenylthiocarbazone (dithizone), $C_{13}H_{12}N_4S$ (or, H_2Dz), to form a primary dithizonate, $Pb(C_{13}H_{11}N_4S)_2$ (or, $Pb(HDz)_2$), which is soluble in organic solvents with the formation of a carmine red color. In the usual procedure the dithizone is dissolved in carbon tetrachloride or chloroform to give an intensely green solution. This is then shaken with the aqueous phase containing lead; the reaction that occurs may be represented as:

$$Pb^{+2}\ (w) + 2H_2Dz\ (o) = Pb(HDz)_2\ (o) + 2H^+\ (w)$$
$$\text{(green)} \qquad\qquad \text{(red)}$$

where the symbols (w) and (o) represent the aqueous and organic phases, respectively. Thus the lead is extracted with the formation of a red color, which may be used for its spectrophotometric determination.

The choice between carbon tetrachloride and chloroform, which are practically the only solvents used, depends on the results desired. Lead is extracted into carbon tetrachloride at lower pH's than it is with chloroform as solvent. Also, carbon tetrachloride is less soluble in water and has a lower volatility and a greater density than chloroform. On the other hand, the solubility of lead dithizonate in chloroform is much greater, thereby permitting the extraction of greater amounts of lead. In addition, the solubility of dithizone itself is much greater than in carbon tetrachloride. For this reason the aqueous phase must be maintained at a higher pH for the complete removal of dithizone from chloroform.

The fraction of lead extracted in any one equilibration depends on a number of variables: the solvent, the amount and nature of the anions present, the volume ratio of the two phases, the pH, and the concentration of dithizone in the organic phase. Anions influence the extraction equilibrium by forming complexes with lead. For example, the efficiency of extraction decreases in solutions containing equal concentrations of acetate, tartrate, and citrate at the same pH. This is expected since the stabilities of the lead complexes of these anions increase in the same order. This behavior is of importance since tartrate or citrate are commonly added to the aqueous phase to prevent precipitation of metal hydroxides.

The pH interval over which dithizone extracts lead quantitatively is quite wide although, as would be expected, it depends on a number of factors. When 25 ml. of 0.001% (w./vol.) dithizone in carbon tetrachloride is equilibrated with 25 ml. of $0.01M$ sodium perchlorate containing 62 γ of lead, complete extraction is observed between pH 5.5 and 11.5. With an aqueous phase containing the same amount of lead but which is $0.01M$ in sodium citrate and $0.02M$ in potassium cyanide, the pH interval for complete extraction is found to be 7.5 to 11.5. The effect of the formation of the lead citrate complex is clearly evident. When the concentration of dithizone is doubled the pH interval is found to be widened in both directions by about 0.5 pH unit in each case (257).

Because dithizone is a weak acid ($K_1 = ca.\ 2 \times 10^{-5}$, at an ionic strength of 0.1), it is extracted from the organic phase on equilibration with alkaline solutions, the ion HDz^- being formed. It remains quantitatively in the organic phase only at pH's less than 7, and less than 1% remains when the pH is 10.8. Since the lead dithizonate stays in the organic phase up to pH 11.5, lead may be determined photometrically in the organic phase with little interference from free dithizone if the pH is adjusted between 10.8 and 11.5. A pH of 10.8 is recommended as optimum, since a variation of ±0.4 pH units from this value has very little effect on the result (257). On the other hand, above pH 11.5 lead dithizonate decomposes very rapidly with increasing pH, and very large errors may be encountered if a small error is made in setting the pH.

Interferences

In practice, if the sample contains many interfering ions, the lead is often isolated by a preliminary dithizone extraction. The lead dithizonate may then be decomposed by dilute acid and transferred to the aqueous phase, which is buffered at the proper pH for a second dithizone extraction.

Ag^+, Hg^{+2}, Pd^{+2}, Au^{+3}, Cu^{+2}, Zn^{+2}, Cd^{+2}, Co^{+2}, and Ni^{+2} form extractable dithizonates under the pH conditions that are optimum for the extraction of lead. Interference of all of these may be prevented, however, by the addition of cyanide to the aqueous phase. If large amounts of Ag^+, Hg^{+2}, Pd^{+2}, Au^{+3}, and Cu^{+2} are present (up to about 1 mg. of each), it is best to extract these first with dithizone at a pH less than 0.

Bi^{+3}, In^{+3}, Tl^+, and Sn^{+2} interfere and are not masked by the addition of cyanide. Thus, unless special precautions are taken, they are extracted partially or completely in the pH interval used for the extraction of lead. Of these, bismuth is the most commonly encountered and the most serious interference. Most methods developed for its elimination are based on differences between the extractability of the two elements as a function of pH. Many approaches and combinations of procedures are possible, but mainly they are all based on the following two factors.

1. With a chosen solvent, bismuth is extracted by dithizone at lower pH's than is lead. Hence, if the aqueous phase is made sufficiently acidic (generally pH 2.0 to 3.5), bismuth may be selectively extracted, leaving the lead in the aqueous phase to be extracted later from a more basic solution. Alternatively, since these extractions are reversible, lead and bismuth may be extracted together at a high pH and then the organic phase washed with buffer (pH 2.3 to 2.5 with CCl_4 as solvent (174); pH 3.4 with $CHCl_3$ (25)) to strip out the lead. The former procedure suffers from the fact that bismuth reacts rather sluggishly with dithizone at low pH's and long shaking times may be required. However, 25 γ of lead has been successfully separated from 1 mg. of bismuth by this technique (1). With the latter method, a 5- to 10-fold excess of bismuth may be removed.

2. From basic solutions, lead is extracted more efficiently than is bismuth at pH's greater than 10. Accordingly, the solution containing lead and bismuth may be made very alkaline (pH greater than 12), and the lead fractionally extracted. It is more practical, however, to extract them together at pH 7 to 10, and then to wash the organic phase with an alkaline solution (generally KCN, 0.5 to 1% of pH greater than 11) to preferentially decompose the bismuth dithizonate. In this manner lead may be determined in the presence of 2 to 3 times as much bismuth (111).

A combination of the above procedures may be the most effective. It should be noted that the separation of bismuth from lead is generally better in acidic rather than in basic solution since greater differences exist in their extraction behavior. In either case, rigorous control of the pH is mandatory since the separations are never extremely sharp. The accuracy claimed in some reported procedures is undoubtedly due to a more or less fortunate compensation of errors.

Utilization has also been made of the fact that at pH 8 to 9 lead reacts more rapidly with dithizone than does bismuth. If the solution containing lead and bismuth is extracted successively with small portions of dithizone, it is possible to determine approximately the point at which all of the lead has been extracted. This is done by observing the color of the organic phase. When the fractions assume the red-orange color of bismuth dithizonate, the lead is completely extracted and the greatest portion of the bismuth remains in the aqueous phase. The organic extract may then be washed with dilute acid and the lead dithizonate decomposed. In this way lead can be separated from a 500-fold excess of bismuth (112).

With very large amounts of bismuth, volatilization as bismuth tribromide has been used. The sample solution is treated with bromine and hydrobromic acid and is evaporated to dryness. The residue is heated gradually to 300°C. in an electric oven and is maintained at this temperature for 5 min. It may then be dissolved in hot nitric acid (12).

The interference of bismuth may be detected by an alteration of the color of the organic extract of lead or, more sensitively, by shaking the lead dithizonate extract with a solution that is 10% in sulfuric acid and 10% in thiourea. In the presence of bismuth, a yellow color appears (393).

The interference of indium may be eliminated by extraction at pH greater than 10. The *optimum* pH intervals for the extraction of indium are 5.2 to 6.3 (CCl_4) and 8.3 to 9.6 ($CHCl_3$) (260). With carbon tetrachloride as solvent, lead may be extracted at pH greater than 10 in the presence of a 100-fold (and probably even greater) excess of indium. If very large amounts of indium must be separated, extraction of indium(III) chloride or bromide with ether is probably the best procedure.

Thallium(I) interferes seriously since it is extracted from the alkaline cyanide solutions most commonly used for the determination of lead. One method of separating these metals is to extract the lead with dithizone at pH 6.0 to 6.4 at which pH thallium is not extracted appreciably (79). Through a combination of the processes of direct and retrograde extraction, it is possible to separate small amounts of lead from much thallium. The lead is first extracted twice at pH 4; the extracts are combined and then are back-extracted with buffer of pH 5 (258). Another separation that works reasonably well is fractional extraction from cyanide solution. Lead reacts with dithizone more readily than does thallium. The first extracts therefore contain mainly lead, and the later ones mainly thallium. Since the complexes have practically identical colors, they may not be distinguished as was the case with bismuth and lead. Each extract is shaken with 0.5% KCN solution, which decomposes the thallium but not the lead dithizonate. When the extracts are completely decolorized by this treatment, it is known that all of the lead has been extracted (112). Large amounts of thallium also may be separated by a preliminary ether extraction of thallium(III) chloride from 2 to 4M hydrochloric acid.

Although stannous tin can interfere by forming an extractable dithizonate in alkaline cyanide, it is practically never troublesome since the preliminary sample treatment (ashing, dissolution in nitric acid, etc.) converts it to stannic tin, which does not react with dithizone. Such treatments, however, result in the formation of insoluble metastannic acid, which causes the loss of lead by occlusion. Therefore, if tin is present in more than minute amounts it is best removed by volatilization as stannic bromide (349,417).

Ferric iron in the presence of cyanide results in the formation of ferricyanide, $Fe(CN)_6^{-3}$, which interferes by oxidizing the dithizone. By the addition of reducing agents such as NH_2OH, N_2H_5OH, Na_2SO_3, $Na_2S_2O_4$, and others, the ferricyanide is converted to ferrocyanide, which does not interfere. In the absence of cyanide, ferric iron still interferes slightly by its oxidizing action, even in the presence of citrate or tartrate. Copper has a similar but less pronounced action.

When much iron is present, such as in limonite or steel, a preliminary separation is advisable. The iron may be removed by extraction of the sample solution (which is approximately 1.2M in hydrochloric acid) with excess cupferron and chloroform (410). Under these conditions, lead is not precipitated with cupferron and is not extracted. In addition to iron the extraction also removes copper, bismuth, thallium(III), and tin. The excess cupferron in the aqueous layer is removed by extraction with chloroform. The lead may then be separated from Tl(I) by increasing the acidity to 1.5M and extracting the lead with diethylammonium diethyldithiocarbamate in chloroform. After evaporation of the lead-bearing extracts and destruction of organic matter by wet ashing with nitric and perchloric acids, the lead may be determined with dithizone.

In the presence of some metals the extraction of lead may be hindered. In particular, titanium (5 mg. and more) has been found to prevent the complete extraction of lead from ammoniacal citrate medium at pH 7 to 11 (354). A similar effect has been observed in the presence of a relatively high concentration of aluminum. In such cases it is best to isolate the lead by precipitation as the sulfide, adding a little copper to act as a collector if necessary.

Only a few anions are troublesome in the dithizone extraction of lead. Perhaps the most important is sulfide, which is often found in harmful amounts in reagent-grade potassium cyanide. Anions that strongly complex lead may interfere if present in sufficient concentration. Thus, citrate and tartrate may hinder the extraction of lead if present in large concentrations. Phosphate in high concentrations also prevents complete extraction. Its effect is most pronounced at pH 9.5, but its interference is considerably diminished at higher pH values and by the addition of citrate (80). If large amounts of phosphate are present, a preliminary sulfide separation should be carried out. The presence of colloidal silicic acid makes the extraction of lead difficult. Repeated shaking with strong dithizone solution removes the lead, however, unless a large amount of the colloid is present.

Methods of Determination

A number of different experimental procedures are available. The major ones are the monocolor method (112,344), the mixed-color method (25,81,166,372), and the reversion technique (173).

In the monocolor method the lead is extracted with dithizone and then the organic solvent is washed with an ammoniacal cyanide solution to remove the excess free dithizone from the organic phase. The attempt to remove all of the dithizone invariably leads to some decomposition of the lead dithizonate, and for this reason the method is inferior to the mixed-color method.

The lead is extracted with dithizone in the mixed-color method and then is directly determined by the measurement of the light absorbed by the lead dithizonate formed. Since the spectra of lead dithizonate and dithizone overlap, the free dithizone contributes to the total absorption at the wavelength of maximum absorption of lead dithizonate. Its contribution can be made very small, however, by carrying out the extraction from a sufficiently basic solution so that most of the dithizone passes into the aqueous phase. In the recommended procedure (Section VII-B-5) the extraction is carried out at pH 10.8 and very little dithizone remains in the carbon tetrachloride. The procedure thus becomes almost equivalent to the monocolor method.

The disadvantage of the mixed-color method is that for a given concentration of lead and dithizone the concentration of dithizone remaining in the organic phase must be reproducible. This requires that the pH of the sample solutions be carefully regulated. Also, each time a new reagent solution is prepared, a new calibration curve should be constructed.

These difficulties are avoided if the absorbance of the extract is measured with a spectrophotometer at two wavelengths: at 520 mμ ($\lambda_{max.}$ of Pb(HDz)$_2$) and at 620 mμ ($\lambda_{max.}$ of H$_2$Dz), at which the molar absorptivities are 66,500 and 32,000, respectively. In this way, the absorbance of the free dithizone at 520 mμ may be calculated and subtracted from the measured value to give the absorbance due to the lead dithizonate alone. This involves making two measurements and it generally is not necessary if proper technique is used in the usual method.

In the reversion technique the lead is extracted in the usual way, and the absorbance of the resulting solution is measured at about 600 mμ, at which the free dithizone absorbs strongly. The mixture of free dithizone and lead dithizonate is then shaken with aqueous acid (e.g., 0.01M hydrochloric acid), and the absorbance is again measured. The increase due to the decomposition of lead dithizonate into lead ions and free dithizone is proportional to the amount of lead present. In this procedure the concentration of the dithizone solution need not be known or even held constant (173). Two measurements are again required, however, and it has been shown that there is no increase in precision or sensitivity over the usual method (Section VII-B-5) (259).

b. With Tetramethyldiaminodiphenylmethane

Small amounts of lead are isolated by electrodeposition on a platinum anode as lead dioxide. In the presence of relatively large amounts of iron or phosphate poor recoveries are obtained. Chloride must be absent (26). The lead dioxide may then be determined by allowing it to oxidize tetramethyldiaminodiphenylmethane to an intensely blue diphenylmethane dye (282). The anode is simply immersed and rotated in a solution of 15 mg. of the reagent in 30 ml. of glacial acetic acid for 1 minute. The intensity of the blue color which develops is then measured at 600 mμ. The method will detect 0.05 mg. of lead, but seems to be of limited applicability. Manganese dioxide behaves similarly and interferes.

c. With 4-(2-Pyridylazo)Resorcinol (PAR-4)

This is the only reagent thus far reported that forms a water-soluble, intensely colored complex with lead and that is suitable for its spectrophotometric determination. The sample solution is adjusted to pH 10 and the reagent is added. Lead forms a red complex with $\lambda_{max.}$ = 520 mμ; for the reagent, $\lambda_{max.}$ = 410 mμ. Since the method was developed for the determination of lead after chromatographic separation, no systematic study of interferences was made (311).

d. With Diethyldithiocarbamate

When lead is treated with sodium diethyldithiocarbamate in basic cyanide medium, a lead complex is formed that may be extracted into carbon tetrachloride. After separation of the phases the colorless lead complex may be transformed into the intensely colored copper complex by shaking the extract with an aqueous solution of copper sulfate. The absorbance of the resulting brown solution at 435 mμ is proportional to the concentration of lead. The method has been used for the determination of traces of lead in aluminum and copper alloys as well as in iron and steel (392).

e. As Lead Chloride

The chloro complexes of lead absorb light in the ultraviolet region. This absorption has been used for the spectrophotometric determination of lead (265). The molar absorptivity and the absorption maximum vary with the concentration of hydrochloric acid which, consequently, must be maintained constant. In 6M hydrochloric acid the absorption maximum of lead is at 271 mμ and the optimum concentration range for its photometric determination is from 4 to 10 p.p.m. of lead when 1-cm. cells are used.

Bismuth and thallium in 6M hydrochloric acid also absorb in the ultraviolet. Their absorption maxima are sufficiently separated from the absorption maximum of lead so that all three elements may be determined simultaneously in the same solution. Many other metals and anions interfere, however.

2. Flame Photometry

A few emission lines of lead are capable of excitation in the oxygen–hydrogen or oxygen–acetylene flames. The ones most useful for flame photometry are at 405.8, 368.4, and 364.0 mμ. The line at 405.8 mμ is the most sensitive and the one most generally preferred. For samples containing manganese the line at 368.4 mμ is useful since manganese emits at 403.3 mμ and interferes. Potassium also interferes through its emission at 404.4 mμ.

Because of its very low emission sensitivity, lead is only rarely determined in aqueous solutions by this technique. Although emission sensitivities depend on a large variety of experimental factors, the relative sensitivities may be appreciated by considering values obtained under comparable conditions. For example, Dean (91) reports that the emission sensitivities of lead at 405.8, 368.4, and 364.0 mμ are 14, 21, and 41 p.p.m./%T, respectively, using an oxygen–acetylene flame and a Beckman DU Spectrophotometer. For comparison, the emission sensitivity of the calcium line at 422.7 mμ under similar conditions is 0.07 p.p.m./%T. These figures indicate the concentration of metal in the sample solution required to give a response of 1 unit on the per cent transmission scale of the spectrophotometer.

The most useful application of flame photometry has been for the determination of lead as tetraethyl lead in gasoline (136,185,371). The gasoline sample is aspirated directly into an oxygen–hydrogen flame, or it may be suitably diluted first with isooctane. The emission at 405.8 mμ under these conditions is approximately 10 times greater than is observed when aqueous solutions are used.

3. Atomic Absorption Spectroscopy

The technique of atomic absorption spectroscopy has been applied to the determination of lead in motor fuels (332). The procedure is rapid, very sensitive, and requires a minimum of sample preparation. The determination of lead in the range of 0 to 70 p.p.m. is not affected by variations in the sulfur and nitrogen content and the carbon/hydrogen ratio of the fuel. In addition, Sn, Na, Bi, Cu, Zn, Cr, Fe, and Ni do not interfere. Steels and copper-base alloys containing 0.05 to 5% lead also have been analyzed by this technique, with good results (99). In both of these cases the lead line at 2833.1 A. was used for the absorption measurements.

4. Spectrographic Methods

Unfortunately, the most sensitive emission line of lead occurs at 4057 A., where a cyanogen band with a band head at 4216 A. interferes seriously. Consequently the most frequently used analytical line for lead is at 2833 A., even though it does not give maximum sensitivity. The persistent lines of lead are given in Table XII.

The spectrographic method is widely used for the analysis of rocks, minerals, and ores. The following two lines with their corresponding optimum concentration ranges of lead are recommended for these materials: Pb 2833.07, 0.001 to 0.05%; Pb 2663.17, 0.02 to 0.5%. The precision of the method for igneous rocks containing 0.0013 to 0.0055% lead is 10.2% (coefficient of variation of the averages of duplicate determinations) (29).

For cases in which maximum sensitivity is required the line at 4057 A. may be used, providing that the CN emission is eliminated or minimized. Of course, the obvious methods are effective, such as the use of electrodes other than carbon and the shielding of the arc in a nitrogen-free atmosphere. However, the great convenience of graphite electrodes and the nuisance of using nitrogen-free atmospheres make these methods unpopular. With proper technique, it is possible to use the carbon arc in air and to minimize CN emission. Since lead is a volatile element, it will be distilled early in the arcing period. If alkali metals are added to the sample the arc temperature can be maintained sufficiently low so that CN emission is very small. It is evident that the plate exposure must be terminated before the distillation of the alkalies is complete, so that the arc temperature does not reach the point at which CN emission occurs. In this way it is possible to detect and estimate lead concentrations well below 10 p.p.m. (7). In some cases the line of manganese at 4057.95 A. interferes.

Steels containing 0.05 to 0.50% lead have been successfully analyzed by arcing nitric–hydrochloric acid solutions and by measuring the intensity ratio of the Pb 4057.82 and Fe 4017.15 A. lines (122).

Many other examples of the application of the spectrographic method are given in Reference 7.

5. X-Ray Methods

Since the absorption of x-rays increases rapidly with increasing atomic number of the absorber, lead is readily determined in media containing elements of low atomic number. It is not surprising, therefore, that the major application of x-ray methods for the determination of lead has been for the determination of tetraethyl lead in gasolines. Comparative absorptiometry with polychromatic x-rays was applied to this problem as early as 1929 (4). The difference between the absorptions of a reference sample and the sample containing tetraethyl lead was used as a measure of the concentration. The early work was done with ionization detectors, but with the advent of the Geiger tube the technique has enjoyed increased popularity and is presently capable of precision and accuracy comparable to the chemical methods and, in addition, is very much more rapid (382). It must be remembered, however, that x-ray absorption is nonspecific in nature and is subject to serious errors if absorbing species are present that are not compensated for in the reference sample. In this regard, it has been shown that greater sensitivity and less interference from sulfur are encountered if monochromatic x-rays are used for the absorption measurements (167). Further improvements are claimed by measuring x-ray fluorescence intensities (226). Here, the concentration of lead is determined from a calibration curve of fluorescent intensity vs. concentration of tetraethyl lead, and it is claimed that variations in the gasoline base stock and the presence of heavy elements in antiknock additives, such as chlorine and bromine, have a negligible effect on the results.

The application of x-ray techniques to the determination of lead in other types of samples is still pretty much in its infancy, and the reader should consult specialized monographs on the subject for a complete discussion (42,236).

E. ACTIVATION ANALYSIS

Although lead has such a low activation cross section that its determination by activation analysis is hardly worth considering (369), a method has been developed for the determination of traces of lead in meteorites (322,323). Activation is accomplished by the reactions shown in Table XVII.

TABLE XVII
Reactions Used in Determination of Traces of Lead in Meteorites

Reaction	Thermal neutron cross section, barns	Half life of the product nuclide	Approximate sensitivity[a]
$^{208}Pb(n,\gamma)^{209}Pb$	0.00045	3.3 hr.	0.2
$^{204}Pb(n,2n)^{203}Pb$	0.002[b]	52 hr.	0.3

[a] The sensitivity in this case is defined as the number of micrograms of normal lead required to give 10 disintegrations per minute 18 hours after a three-day irradiation in a thermal flux of $5 \times 10^{13} n/cm.^2/second$.

[b] This figure is an effective thermal cross section since the radioactivity is induced by the action of fast neutrons (323). The sample is irradiated inside a special enriched-uranium fuel rod.

Since a 1- to 2-g. sample develops about 1 curie of activity under these conditions, rapid and extensive radiochemical purification is required. Concentrations of lead in the range of 0.05 to 10 p.p.m. were determined. In addition, Ba, Bi, Tl, and U also were determined in the same sample.

VI. DETERMINATION OF LEAD IN SPECIFIC MATERIALS

Many of the methods described elsewhere in this chapter are suitable for the determination of lead in a variety of materials. In most cases, only minor modifications are usually necessary. However, some samples present problems peculiar to them only, and it is intended that this section serve as a guide to the handling of these special materials.

A. ROCKS, MINERALS, AND ORES

Samples containing lead as a major constituent are best analyzed for lead gravimetrically as lead sulfate or lead molybdate. Smaller amounts, of the order of 1 to 2 mg. of lead, are effectively determined electrolytically. In addition the volumetric determination of lead after its separation as the sulfate, followed by the precipitation of lead chromate, is an excellent method for routine analyses of lead ores (162).

Lead ores and concentrates may be assayed by titration with EDTA (201). A single sample containing mainly lead, copper, iron, and zinc is dissolved in concentrated nitric acid in the presence of bromine. After neutralization with ammonia, ammonium bifluoride is added to complex the iron, and the copper is determined iodometrically. Next the iron also may be determined iodometrically by the addition of boric acid to release it from the fluoride complex. Following this the iron is reduced with ascorbic acid and is complexed with cyanide (copper and zinc also are complexed). The lead may then be titrated in the presence of citric acid in an ammonia–ammonium chloride medium with EDTA, using Eriochrome Black T as the indicator (some manganese–EDTA salt is added to sharpen the end point). Zinc is then titrated with EDTA after destruction of the zinc cyanide complex with formalin. Thus all four elements may be determined by titration in the same solution. When lead is present in low concentration in ores or tailings or large amounts of interferences are present, it is recommended that a preliminary separation of copper, lead, and zinc be carried out by extraction of their diethyldithiocarbamates with chloroform.

A somewhat similar method, which is both rapid and precise, for the determination of lead in sulfide ores and concentrates has been reported (139). The lead is first separated electrolytically and then is determined by titration with disodium dihydrogen 1,2-diaminocyclohexane-N,N,N',N'-tetraacetate. The adjustment of the pH is less critical with this reagent, and sharper end points are obtained than when EDTA is used as the titrant.

Rocks, ores, and minerals that contain 0.001 to 0.5% of lead are frequently analyzed spectrographically (7,29).

Trace amounts of lead in silicate rocks are probably most accurately determined colorimetrically by the dithizone method (344). Decomposition of the sample is achieved by evaporation with perchloric acid and hydrofluoric acid to remove most of the silica. The sample is then taken up with hydrochloric acid and filtered, and the residue is fused with sodium carbonate. Lead is then separated from the filtrate and the sodium carbonate leach by extraction with dithizone. It may then be returned to the aqueous phase and determined in the usual way (see Section V-D-1-a).

B. ALLOYS

Since lead is a constituent of a number of commercially important alloys, standard methods for its determination have been developed in many cases. Full descriptions of these procedures are given in the readily available series of volumes published by the American Society of Testing and Materials and hence will not be repeated here (14). The following discussion will therefore be limited to very brief descriptions and to references to some of the newer methods.

1. Iron and Steel

Because lead improves the machinability of steel, it is becoming a frequently encountered alloying element. Accordingly, a standard method has recently been developed for steels containing 0.05 to 0.40% of lead (44). The lead is separated from the bulk of the iron by precipitation as lead sulfide from dilute hydrochloric acid. After collection by filtration the lead sulfide is dissolved in nitric acid, tartrate is added, and the solution is made ammoniacal. Ammonium molybdate is then added, and the precipitated lead molybdate is filtered off, ignited, and weighed as $PbMoO_4$. The procedure is reproducible to ±0.01% Pb on a 10-g. sample, and its applicable to steels containing Cu, Cr, Mo, Ni, not more than 0.25% Sn, and not more than 2% W.

A rapid electrolytic method for lead and copper has been proposed for steels containing similar amounts of lead (405). A preliminary sulfide separation also is used. Another method in which the lead is deposited directly without a preliminary separation is also available (142). In general, however, the electrolytic method has not been widely applied to iron and steel analyses.

A spectrophotometric method in which lead diethyldithiocarbamate is extracted has been applied to iron and steel containing 0.01 to 2% of lead (392). Steels containing lead in the range of 0.05 to 0.5% also have been analyzed by measuring the absorbance due to precipitated colloidal lead sulfide (58). The dithizone method has been applied to steels containing much nickel and chromium and 0.001 to 0.03% lead (51). A sulfide separation is carried out to avoid large amounts of ferric iron, which oxidizes dithizone in basic solution. Sandell gives a slightly modified procedure (347). A rapid dithizone method not requiring a preliminary separation has also been developed (300).

Steels containing 0.1 to 0.4% lead also have been analyzed polarographically (151). Extremely small concentrations of lead of the order of 0.0001% have been determined polarographically in cast iron. Preliminary separations of iron by extraction as ferric chloride and of lead as the diethyldithiocarbamate are carried out (334).

Because of the heterogeneity of leaded steels containing 0.05 to 0.5% lead, solution techniques are recommended for obtaining representative samples for spectrographic analysis (122,303).

2. Nonferrous Alloys

White metal bearing alloys containing 0.1 to 95% lead are analyzed for lead by the gravimetric lead sulfate method. Preliminary separation of interfering elements often is required (15). Lead in fusible alloys containing bismuth also is determined gravimetrically as lead sulfate (101). The determination of lead in white metals, fusible alloys, and tin-free copper-base alloys is conveniently done using a microgravimetric procedure. The lead is separated as the lead nitrate–thiourea complex, which is then converted to lead chromate and weighed as such (267). A microvolumetric procedure is available for tin-base and lead-base alloys. The lead is separated as the sulfate, which is dissolved in ammonium acetate; it is then precipitated with thionalide. The precipitate is filtered, washed, and redissolved, and the liberated thionalide is titrated iodometrically (74). The titration of lead with EDTA (see Section V-B-3) has been applied to the analysis of bronze and brass (200), lead drosses and lead-base alloys (309), and white metals (301).

The electrolytic method is widely used for the determination of lead in nonferrous alloys. The A.S.T.M. has published methods for electrogravimetric determination in aluminum and aluminum alloys, brass, bronze, copper-nickel-zinc alloys, zinc die-casting alloys, and zinc spelter (14). Microelectrolytic procedures are available for the determination of lead in lead-base and tin-base alloys containing antimony and copper (73) and in bronzes (268). Copper-base alloys containing up to 250 γ of lead also may be analyzed by a microelectrolytic procedure. A sample of only 5 mg. is taken (342). Controlled-potential electrolysis has been used for the determination of lead as well as copper, tin, and antimony in brasses and bronzes (9,10) and in other alloys (11). It has also been used for the separation of traces of lead (0.0006 to 0.05%) in zinc-base alloys. The lead is subsequently determined polarographically (391).

The following references describe the polarographic determination of lead in aluminum and aluminum alloys: (83,177,209,363). The square-wave polarograph has also been used for traces of lead in aluminum (110).

Lead in beryllium metal has been determined in the range of 150 to 800 p.p.m., using 2-g. samples (27). Traces of lead in metallic bismuth have been determined with the square-wave polarograph (104).

Polarographic procedures for lead in copper-base alloys have been frequently reported (241,270,271,328). Less than 0.02% lead in metallic copper has been successfully determined by polarographic means. The lead is separated from a 30-g. sample by electrolysis (103).

Magnesium alloys have been analyzed polarographically (148) as well as tin-base alloys containing 0.4 to 20% lead. In the latter determination a $1M$ sodium hydroxide supporting electrolyte is used, since Sn(IV) and Sb(V) are not reduced in this medium (22).

The polarographic determination of lead in zinc base alloys has been reported (156). Lead in high-purity zinc containing 0.001 to 0.0016% lead has been determined by making use of the high sensitivity of the rotated dropping mercury electrode (385).

Dithizone-extraction procedures have been given for the determination of lead in copper-base alloys, zinc-base alloys, refined tin, and in tin-base white metals (274), and in metallic copper, nickel, and cobalt (422). Copper and aluminum alloys containing 0.001 to 2% lead have been analyzed by a procedure utilizing the extraction of lead diethyldithiocarbamate (392).

Lead is routinely determined spectrographically in aluminum using the high-voltage spark. The optimum concentration range is 0.03 to 0.7% lead (16). Concentrations of lead of 0.01 to 1% have been determined as part of the spectrographic characterization of aluminum alloys. The d.-c. arc is used, and it is recommended that the sample be converted to aluminum sulfate (170).

C. BIOLOGICAL MATERIALS

Because of the toxicity of lead, an extremely large literature exists on its determination in biological materials, and this has been collated and critically reviewed in a monograph by the American Public Health Association (12). Of the chemical methods the extraction of lead with dithizone is preferred because of its great sensitivity and selectivity. The spectrographic method is also widely used and is particularly valuable when the need exists for the simultaneous determination of a variety of trace metals. Both the d.-c. arc (69,70) and the a.-c. arc (308) have been used. The polarographic method is very convenient, especially when only occasional samples are to be analyzed (68).

For the routine determination of lead in urine the dithizone method is probably the most satisfactory (249). In this regard, attention should be brought to a method that eliminates the necessity of time-consuming precipitation and ashing steps (264). The lead is extracted as the iodide from the acidified urine sample with methyl isopropyl ketone. It may then be removed from the organic phase with sodium hydroxide and determined by the dithizone method. The reversion dithizone technique (see Section V. D-1-a) has been applied to a variety of biological materials (urine, blood, bone, etc.) (173).

D. AIR

The methods used for sampling air depend on the type of lead exposure under investigation. Tetraethyl lead is best collected by passing the sample through a scrubber containing a solution of iodine (373). The lead may then be determined with dithizone.

Lead dusts or paint sprays may be collected by filtration or by the use of standard impingers containing water or nitric acid (12). In sampling lead fumes (particles of lead oxide of diameter of less than 1 μ), electrostatic precipitators are best for determining the total lead content of the air (12). However, they offer no discrimination of particle sizes, and it is well recognized that the smaller particles of lead fumes are relatively the most toxic. For this reason a rapid field method has been developed, using filtration collection, which does not suffer from the disadvantage that high readings are obtained when occasional large and relatively nontoxic particles are collected (92). Total air-borne lead may subsequently be determined on the same sample (93).

E. FOODS

The general principles of the analysis of foodstuffs are outlined in the *Official Methods of Analysis* of the Association of Official Agricultural Chemists. The techniques employed for the destruction of organic matter are described, and a preliminary separation of the lead as the sulfide or with dithizone is usually made. Depending on the amount of lead present, the analysis is completed electrolytically or colorimetrically with dithizone.

A rapid and unusual method for the destruction of organic material, involving dry ashing followed by treatment with nitric acid vapor, has been described (130). In this connection, it would seem that the recently described wet oxidation with perchloric and periodic acids might prove useful in the analysis of foods. Difficultly oxidized materials were completely destroyed at temperatures less than 200°C. (370). It should be noted, however, that complete destruction is not always necessary, for example, in the analysis of sugar and sugar products (144) and maple syrup (306) by the dithizone method.

Ion exchange has been used for the concentration of traces of lead in foods. The final determination is made with dithizone (182). The following are references to other applications of the dithizone method to food products: cocoa and syrup (181), tea, coffee, sugar, and butter (130), wheat flour (155), and some 22 different foods (244).

The lead content of canned meat and fish has been determined polarographically (184), and wheat has been analyzed spectrographically (194).

F. WATER

Because large quantities of metals that react with dithizone are not usually present in natural waters, a simple procedure is generally applicable for the determination of lead (348). A dithizone procedure for trade effluents has been described (1). A polarographic method has been applied to natural waters, although the lead was concentrated by a preliminary dithizone extraction (223). Lead has been recovered from natural waters and concentrated 1000-fold by the use of a chromatographic column containing cellulose acetate, which served as a support for a solution of dithizone in carbon tetrachloride. The final determination of lead was made polarographically (63). Lead and many other elements have been determined in water by spark-emission spectroscopy on solutions. A preliminary 10-fold concentration of the sample permitted the determination of as little as 0.5 p.p.m. of lead (420).

G. PETROLEUM PRODUCTS

The standard method for lead in lubricating oils involves its precipitation and gravimetric determination as lead sulfate (19). On the other hand, an electrolytic procedure is used for greases (20).

The standard method for tetraethyl lead (TEL) in gasoline is carried out by first decomposing the TEL by refluxing the sample with concentrated hydrochloric acid, evaporating the mixture to dryness, and oxidizing the remaining organic material with nitric acid. The lead may then be determined gravimetrically as the chromate (18). Although this procedure is precise and accurate, it is time-consuming and requires a large sample. For this reason, many more rapid and more sensitive grocedures have been developed. Traces of TEL and inorganic lead in gasoline in the concentration range of 0.01 to 20 p.p.m. have been determined spectrophotometrically with dithizone (145). Larger amounts, which are normally encountered in leaded gasolines, have been determined by titration with EDTA (275,340) and polarographically (152).

All of the above procedures require the chemical decomposition of TEL into inorganic lead, a process that is frequently lengthy. However, this becomes unnecessary when the lead is determined by the absorption of x-rays by the gasoline sample (59,167). Unfortunately, x-ray absorption is dependent on the composition of the base stock. This is especially true with regard to its sulfur content. Generally the sulfur content and other characteristics of the sample are known and corrections may be applied. However, this problem is practically eliminated by the use of x-ray fluorescence (43).

The flame-photometric determination of lead in gasoline offers much lower instrumental cost than the x-ray methods, as well as high speed. The decomposition of TEL is accomplished by combustion in the flame, and thus sample preparation is minimal. The effects of variations in the base stock may be eliminated simply by making standard additions of TEL (371). Atomic absorption spectroscopy has been applied recently to the determination of lead in motor fuels and appears to be an extremely promising technique (332) (see Section V-D-3).

H. MISCELLANEOUS

Glass. The sample of powdered glass is generally treated with a mixture of hydrofluoric and perchloric acids to volatilize the silica. The residue may be analyzed for lead by a variety of methods, depending on the amount of lead present. Classically, glasses containing of the order of 30% of lead oxide are analyzed by separating the lead as the sulfide and determining it as the sulfate. The procedure is slow, however, and a more rapid gravimetric method has been developed in which the lead is precipitated as the iodate (135). It was later demonstrated that small amounts of iron interfere with this procedure (319). Polarographic methods have been used (402,403), as well as titrations with EDTA (343).

Rubber and polymers. After ashing the rubber sample, the residue is dissolved in hydrochloric acid of exactly known concentration. The absorption of ultraviolet light due to lead chloride is then used as a measure of the concentration of lead (220). A similar procedure has been used for the determination of lead in polyvinyl chloride compositions (146).

Paints. Paints containing 0.04 to 50% lead have been analyzed polarographically (5).

VII. RECOMMENDED LABORATORY PROCEDURES

A. DETECTION OF LEAD

The detection and identification of lead have been discussed in a general manner in Section IV. The following few tests are widely used; however, for unusual applications standard works on qualitative analysis should be consulted.

1. Tests for Macro Amounts of Lead

a. Precipitation as Lead Chloride

If 2 to 20 mg. of lead per ml. is present, the following simple test is specific for lead.

Reagents

1. $6M$ hydrochloric acid.
2. Chlorine water.

Procedure

Make the test solution approximately $2M$ in hydrochloric acid. The absence of a white precipitate at this point indicates the absence of large amounts of lead. Add chlorine water to dissolve Hg_2Cl_2 and $TlCl$. Heat the solution to boiling and separate any precipitate by centrifugation (the precipitate should be mainly AgCl). Pour off the supernatant liquid and allow it to cool. The separation of white crystals of $PbCl_2$ on cooling indicates the presence of lead.

b. Precipitation as Lead Sulfide

This test is applicable to solutions containing greater than 0.2 mg. of lead per ml. Bismuth(III) interferes, as does thallium(I). The latter, however, may be oxidized to thallium(III) before the addition of sodium hydroxide.

Reagents

1. Saturated sodium sulfide in water.
2. $4M$ sodium hydroxide.
3. 20% solution of potassium cyanide.

Procedure

To 1 drop of the test solution add 4 drops of $4M$ sodium hydroxide. Boil the solution and separate any precipitate by centrifugation. The supernatant liquid should be clear; if it is not, repeat the boiling and centrifuging. Add 1 drop of the supernatant liquid to 1 drop of potassium cyanide solution, and then add 1 drop of sodium sulfide solution. A black precipitate of lead sulfide shows the presence of lead.

2. "Triple Nitrite" Crystal Reaction

See Section IV-B-2-a for a general discussion of this test.

Reagents

1. Buffer solution. Dilute 450 g. of sodium acetate and 100 ml. of glacial acetic acid to 1 liter.
2. $5.8M$ potassium nitrite solution.
3. Reagent solution. Mix equal volumes of buffer solution and potassium nitrite solution. Store away from strong sources of heat and light and use within 2 days.
4. Copper solution (10 mg. Cu per ml.).

Procedure

Place by means of a platinum loop several small drops of the solution to be tested near the end of a microscope slide. Bring these drops just to dryness by gently warming the slide over a small Bunsen flame. With a capillary pipet, add to the residue of each drop of a volume of copper solution approximately equal to the original volume of each drop. Evaporate the copper solution in the same manner. With a capillary pipet, add a drop of reagent solution small enough so that it does not completely cover the residue, and inspect the result immediately under the microscope, using strong transmitted illumination and a magnification of 70 to 100 ×. In the presence of lead, crystals of the triple nitrite, $2KNO_2 \cdot Cu(NO_2)_2 \cdot Pb(NO_2)_2$, are rapidly formed except when only very small amounts are present. The salt has a very characteristic appearance; cubes and rectangular prisms are formed that are jet-black in color unless they are very thin, in which case they may be brown or orange colored. In reflected light they are brownish in appearance.

3. Spot Test with Benzidine

See Section IV-B-3-a for a discussion of this test.

Reagents

1. $3M$ sodium hydroxide.
2. Saturated bromine water.
3. Solution of benzidine in acetic acid.
4. Aqueous ammonia (1:1).

Procedure

Place a drop of the test solution on filter paper and treat it successively with a drop of sodium hydroxide and a drop of bromine water. After the bromine water has soaked into the paper, add 2 drops of ammonia, allow it to spread, and then remove the excess ammonia by waving the paper over a small flame. Add a drop of the benzidine to the spot; the presence of lead is indicated by the formation of a deep blue fleck. Very small amounts of lead may give a light blue color, which fades on standing.

The limit of identification is 1 γ of lead, and thallium offers the only interference.

B. DETERMINATION OF LEAD

1. Gravimetric Determination as Lead Sulfate

The discussion in Section V-A-1 should be consulted to determine whether the following procedure is applicable to the sample in question. Preliminary separations should be made so that the sample is free from interfering substances. This procedure may be used following the separation of lead by precipitation as the sulfide from $0.3M$ hydrochloric acid. The sulfide is collected and dissolved in nitric acid.

Procedure

To the sample solution containing up to 0.5 g. of lead, add 5 ml. of concentrated sulfuric acid for every 0.1 g. of lead present. Evaporate the solution to fumes of sulfuric acid, cool, wash down the walls of the beaker with water and repeat the evaporation until dense white fumes are again obtained. Cool, and add 100 ml. of water for each 0.1 g. of lead present (Note 1). Mix the solution well and let it stand for an hour or more. Filter off the lead sulfate on a weighed, fine-porosity, porcelain filtering crucible, and wash it with dilute sulfuric acid (1/99) (Note 2). Ignite the crucible to constant weight at 500 to 600°C. in an electric furnace, and weigh as $PbSO_4$ ($Pb/PbSO_4 = 0.68324$).

Note 1. From the data of D. N. Craig and G. W. Vinal (86) on the equilibrium solubility of lead sulfate in solutions of sulfuric acid, this procedure should result in the loss of approximately 0.45 mg. of lead for each 0.1 g. of lead in the sample. Neglecting additional losses on washing the precipitate, this corresponds to a negative error of 0.45%. Obviously, in precise analytical work the lead in the filtrate should be recovered or corrected for. The above figure of 0.45% is given only to indicate an order of magnitude; correction factors should be obtained by carrying known amounts of lead through the procedure. The lead in the filtrate may be conveniently determined electrolytically (see Section V-A-4) after the addition of nitric acid.

It should be noted that miscible organic solvents (e.g., ethanol, isopropanol, etc.) are often added in many procedures to reduce the solubility of lead sulfate to the point where the loss of lead is negligible. This technique must be applied with great caution, however, since the solubilities of many other substances are also lowered and interferences become legion. Accordingly, such procedures are not of general applicability, and for this reason are not included in this section.

2. In the washing step losses may be made insignificantly small by using sulfuric acid that has been previously saturated with lead sulfate.

2. Gravimetric Determination as Lead Chromate by Precipitation from Homogeneous Solution (163)

The following method is recommended because of its extreme simplicity and great convenience. Refer to Section V-A-2 for a discussion of interferences. The method has been found by the author to give accurate results in the hands of inexperienced analysts. Large, well-formed crystals that require very little washing are produced.

Reagents

1. 0.10M chromium nitrate, $Cr(NO_3)_3 \cdot 9H_2O$ (40 g./liter).
2. 0.12M potassium bromate, $KBrO_3$ (20 g./liter).
3. Acetate buffer solution: 6M in acetic acid, 0.6M in sodium acetate.

Procedure

The sample should contain 0.1 to 0.2 g. of lead. After dissolution, neutralize the solution with sodium hydroxide, using universal indicator paper. Adjust the volume of the solution to about 30 to 50 ml., and add 10 ml. of chromic nitrate solution and 10 ml. of potassium bromate solution. Heat to incipient boiling for about 30 minutes, but do not boil. When the supernatant liquid is clear and yellow (Note 1), add 10 ml. of buffer and heat 5 more minutes. Cool the mixture and filter off the lead chromate on a sintered glass or porous porcelain filter crucible. Wash the precipitate with 2 or 3 small portions of 0.1% nitric acid and dry at 120°C. Weigh as $PbCrO_4$ ($Pb/PbCrO_4 = 0.64108$).

Note 1. The color of the solution serves as a guide to the extent of oxidation of the chromic ion. If this is incomplete, low results will be obtained.

3. Complexometric Titration with EDTA

a. IN BASIC SOLUTION WITH ERIOCHROME BLACK T AS INDICATOR (115,355)

Reagents

Disodium salt of ethylenediaminetetraacetic acid (EDTA). Prepare a 0.01M solution in water. The solution is standardized by titration of a nitrate solution prepared by dissolving reagent lead in nitric acid. Lead nitrate, $Pb(NO_3)_2$, also may be used as a primary standard. For work not requiring the highest accuracy the solution may be prepared by direct weighing of primary standard grade $Na_2H_2C_{10}H_{12}O_8N_2 \cdot 2H_2O$ (mol. wt. = 372.25; dried at 80°C.), or $Na_2H_2C_{10}H_{12}O_8N_2$ (mol. wt. = 336.22; dried at 120 to 140°C.). Store the solution in a polyethylene bottle.

Eriochrome Black T indicator. Dissolve 0.2 g. of the dye in 15 ml. of triethanolamine and 5 ml. of absolute ethanol. The solution is stable at least a month.

Buffer, pH 10. Add 57 ml. of concentrated (15M) ammonia to 6.8 g. of ammonium chloride and dilute to 100 ml. To be certain of freedom from the alkaline earths, prepare the ammonia solution from cylinder gas.

Potassium sodium tartrate, $KNaC_4H_4O_6 \cdot 4H_2O$. Prepare an approximately 1M solution in water.

Potassium cyanide, KCN. To be used as required to suppress interference of other metals. If its addition is necessary, it is well to test its purity as follows: pH 10 buffer and Eriochrome Black T indicator are added to the cyanide solution. A clear blue color should result.

Procedure

The solution should not contain more than about 30 mg. of lead per 100 ml. Five milliliters of tartrate solution is added, and the resulting solution is approximately neutralized with sodium hydroxide. If required, potassium cyanide or other masking agents should be added at this point. Next, 2 ml. of pH 10 buffer (Note 1) and the indicator are added, and the solution is titrated with EDTA. Before the end point the solution is violet in color, whereas at the end point it is pure blue. No indicator correction is necessary. One milliliter of $0.01M$ EDTA is equivalent to 2.0721 mg. of lead.

Note 1. If, on addition of the buffer solution, a turbidity develops, insufficient tartrate has been added. Too large an excess of tartrate should be avoided, however, since the color intensity of the lead–Erio T complex diminishes with large excesses.

Interferences

The element that interferes most seriously with this procedure is iron, which blocks the indicator at this pH. No masking reagent is known to prevent this interference, and a preliminary separation is necessary (e.g., precipitation of the lead as $PbSO_4$). The addition of cyanide successfully masks Co, Ni, Cu, Zn, Cd, Hg, and Pt, which would otherwise interfere. The alkaline earths, the rare earths, Mn, and In also interfere. Bi, Sb, and Al do not interfere, but they may cause a turbidity in the solution if insufficient tartrate is added.

b. IN ACIDIC SOLUTION WITH XYLENOL ORANGE AS INDICATOR (216)

Reagents

Disodium salt of ethylenediaminetetraacetic acid. Prepare a 0.01 or $0.05M$ solution in water, as described in Section VII-B-3-a.

Xylenol Orange indicator solution. Dissolve 0.1 g. of Xylenol Orange in 100 ml. of dilute alcohol. The solution is stable.

Buffer, pH 5. Prepared as in Section VII-B-3-c.

o-Phenanthroline. Prepare a $0.15M$ solution by dissolving 2.7 g. of the free base in sufficient nitric acid such that on dilution to 100 ml. the pH is between 5 and 6.

Urotropine (hexamethylenetetramine). Solid reagent.

Procedure

Neutralize any large excess of acid with ammonia, and add buffer solution (pH 5). Add a few drops of Xylenol Orange indicator solution and titrate with $0.01M$ EDTA from a red-violet to lemon-yellow color.

Interferences

Iron(III) interferes by the irreversible formation of a red-purple coloration. Aluminum reacts very slowly with Xylenol Orange and interferes in high concentrations. Bi, Zr, Th, and Sc interfere. Co, Ni, Cu, Sn(II), and rare earths also probably interfere, although their effects have not been specifically studied. Bismuth may be determined in the same sample by the above procedure by carrying out the titration at pH 1 to 2. Lead is not titrated at this pH.

Alternate Procedure (317)

The following procedure is useful for the determination of lead in the presence of Cu, Co, Ni, Cd, and Zn.

The sample is treated with a moderate excess of $0.15M$ o-phenanthroline and several drops of Xylenol Orange indicator solution. The acidity of the solution is reduced by the addition of solid urotropine. Small portions are added until the red-violet complex of lead and Xylenol Orange appears. The titration is then carried out with $0.05M$ EDTA to a lemon-yellow color (or, if copper is present, the color at the end point is green) (Note 1).

Note 1. After the end point is reached the solution should be checked for the complete masking of Cu, Co, Ni, Cd, and Zn, as follows: back-titrate the solution with $0.05M$ lead nitrate (about 0.05 ml. should be required). On further addition of o-phenanthroline the red-violet color should not disappear. If this happens the titration with lead nitrate should be continued until no further change occurs on the further addition of o-phenanthroline.

c. IN ACIDIC SOLUTION WITH COPPER-PAN AS INDICATOR (113,114)

Reagents

Disodium salt of ethylenediaminetetraacetic acid. Prepare a $0.01M$ solution in water, as described in Section VII-B-3-a.

Copper–EDTA solution. A $0.01M$ solution of copper sulfate is prepared. Two 20-ml. portions are titrated with $0.01M$ EDTA solution at pH 10 (ammonia buffer), using murexide as indicator. Another 20-ml. portion of the copper solution is mixed with exactly the average amount of EDTA solution used in the two titrations. The solution is stored in a dropping bottle.

1-(2-Pyridylazo)-2-naphthol indicator (PAN). Dissolve 50 mg. of the indicator in 100 ml. of ethanol. The solution is stable.

Buffer, pH 5. Dissolve 27.3 g. of $NaC_2H_3O_2 \cdot 3\ H_2O$ in water containing 60 ml. of $1M$ hydrochloric acid and dilute to 1 liter.

Procedure

The sample solution containing 20 to 60 mg. of lead is neutralized if necessary. Five milliliters of pH 5 buffer, 5 drops of copper–EDTA solution, and 3 to 5 drops of PAN indicator solution are added. The mixture is diluted to about 50 ml. with water, and 50 ml. of ethanol is added (Note 1). The solution is then titrated with $0.01M$ EDTA until the color changes from red to clear yellow.

Note 1. If ethanol is not added to the solution the reaction between the copper–PAN complex and EDTA takes place very slowly at room temperature. This is believed to be due to a colloidal form of the copper–PAN complex that is sparingly soluble in water In the presence of ethanol the reaction rate is high and a sharp end point is observed. A good end point also is obtained without the addition of ethanol if the titration is done in boiling solution.

Interferences

The major advantage of this procedure over the titration in basic solution using Eriochrome Black T is that the alkaline earths do not interfere since they are not complexed by EDTA in acidic medium. Zn, Cd, In, and Ga interfere and can, in fact, be determined by titration under the same conditions as above. Hg, Ni, Co, and Bi also interfere. Small amounts of Mn do not.

4. Electrogravimetric Determination as Lead Dioxide (160)

Refer to Section V-A-4 for a general discussion of the electrolytic method and its applications.

In the following procedure, for optimum results the sample should contain not more than 0.087 g. of lead (to give a deposit of 0.1 g. of PbO_2). Under these circumstances the theoretical factor, $Pb/PbO_2 = 0.8662$, may be used at all temperatures of electrolysis (20 to 95°C.).

Apparatus

Although many commercial electroanalyzers are available that differ widely in elaborateness, essentially only very simple equipment is required for this determination. A 6-v. storage battery may be used as the source of direct current. It should be connected to the cell through an ammeter and a rheostat so that the current may be regulated. The electrolysis may be conveniently carried out in 250-ml. tall-form beakers that are covered with split watch glasses. For macro amounts of lead, stationary cylindrical platinum gauze electrodes of the following approximate dimensions are suggested: cathode, 2 inches in diameter and $2^1/_4$ inches in height; anode, 1 inch in diameter and $2^1/_4$ inches in height. The electrodes are arranged concentrically in the electrolysis cell with a blade or propeller-type glass stirrer along the central axis. Since more dense and adherent deposits are obtained when the electrolysis is carried out at elevated temperatures, the cell should rest on an electrical hot plate.

Procedure

To the sample solution add 22.5 ml. of concentrated nitric acid to give a solution 15% vol./vol. of nitric acid on dilution to 150 ml. (Note 1). If the sample does not already contain much copper, add 0.3 g. of copper in the form of a solution of copper nitrate, and dilute the sample to 150 ml. with water. Connect the cleaned and weighed anode and the cathode to the electrolysis apparatus. Bring the beaker containing the sample solution beneath the electrodes and adjust its height so that at least two-thirds of the electrodes are immersed. Turn on the hot plate and the stirrer and bring the solution to the desired temperature (95°C. is recommended). Switch on the current and adjust the rheostat to give a flow of about 2 amp. in the circuit. Toward the end of the deposition wash down the watch glass and the walls of the beaker with about 20 to 25 ml. of water and continue the electrolysis. If after about 15 minutes no fresh deposit is obtained on the newly submerged portions of the anode, the electrolysis may be terminated. If more PbO_2 is deposited the electrolysis should be continued, and then the washing procedure repeated until no further deposition is obtained. Without switching off the current, lower the beaker and rinse off the electrodes with water from a wash bottle. Remove the anode, dip it in alcohol and then in ether, and dry it in an electric oven at 220°C. for 1 hour (Note 2). Cool and weigh the electrode and calculate the amount of lead deposited. Clean the electrode by dipping it in dilute nitric acid containing some hydrogen peroxide.

Note 1. A nitric acid concentration of 15% (vol./vol.) is suitable for the deposition of up to 0.3 g. of PbO_2; if greater amounts are to be deposited the concentration should be increased to 25%.

2. These drying conditions permit the use of the theoretical factor for up to 0.1 g. of deposited PbO_2. If greater amounts are deposited the empirical factors of Hertelendi (see Section V-A-4) should be used. Also, if greater than 0.3 g. of PbO_2 is deposited the drying time should be extended to 2 hours. When very large amounts of lead dioxide

are deposited (0.7 g. and over) the deposit tends to flake off the electrode during drying. However, if the electrode is handled carefully and is placed upright in a small beaker for drying and weighing, this difficulty is eliminated.

5. Photometric Determination with Dithizone (346)

Because of the great sensitivity of the dithizone reaction with lead, the cleanliness of the glassware used and the purity of the reagents are of paramount importance. With careful technique to avoid contamination the following procedure yields excellent results for the determination of traces of lead. Interfering elements and some of the many modifications of the procedure are discussed in Section V-D-1-a.

Reagents

Water (lead-free). Distilled from Pyrex or purified by ion exchange.

Carbon tetrachloride. Redistilled and free of oxidizing substances. Add a few drops of liquid bromine to 1 liter of reagent-grade carbon tetrachloride and allow to stand a few days. Reflux with 10% sodium hydroxide for 4 hours, wash free of alkali, and shake with 10% hydroxylamine hydrochloride for 5 minutes. Wash with water, separate the carbon tetrachloride, dry it with Drierite, and distill.

Dithizone stock solution, 0.01% w./vol., 50 mg. of pure dithizone in 500 ml. of redistilled carbon tetrachloride. Store the solution in the dark, preferably in a refrigerator. Some commercial dithizone preparations may be used directly without purification. The purity of this solution should be checked by shaking a small amount with dilute (1:100) ammonia. If the aqueous phase is colorless or only slightly yellow, the solution may be used. If purification is indicated, follow the procedure given by Sandell (345).

Dithizone solution, 0.001% w./vol. Prepare shortly before use by diluting 0.01% dithizone solution 10-fold with redistilled carbon tetrachloride.

Ammonium hydroxide. Pass ammonia from a cylinder into cold redistilled water until saturated. Metal-free ammonium hydroxide also may be conveniently prepared by isothermal distillation. A polyethylene dish containing redistilled water is placed over concentrated ammonium hydroxide in a desiccator, and the system is allowed to come to equilibrium.

Nitric acid, 1:100. Dilute colorless concentrated acid with 100 times its volume of water. Sometimes high blank determinations may be traced to the presence of lead in reagent-grade nitric acid. In this case, the acid should be distilled in Pyrex apparatus.

Ammonium citrate solutions, 50 g. in 100 ml. of water. Purify by making ammoniacal (pH 9) and shaking with portions of dithizone in carbon tetrachloride until all the lead is extracted. Then remove excess dithizone by shaking with chloroform.

Potassium cyanide solution, 10 g. in 100 ml. of water. Check for the presence of lead by diluting 1 ml. with 3 ml. of water and shaking with 1 ml. of 0.001% dithizone. The carbon tetrachloride should show no pink. If necessary, remove lead by repeated extraction with dithizone. The potassium cyanide should also be free from sulfide.

Hydroxylamine hydrochloride solution, 20 g. in 100 ml. of water. If necessary, the solution may be purified by adjusting to pH 9 to 10 with ammonia and removing lead by extraction with dithizone.

Ammonia–cyanide–sulfite solution. Add 30 ml. of 10% potassium cyanide solution and 1.5 g. of sodium sulfite to the equivalent of 725 ml. of $8M$ metal-free ammonia and dilute to 1 liter.

Standard lead solution, 0.00100% in 1:100 nitric acid. Prepare fresh before use by diluting a 0.100% solution (0.160 g. of dried lead nitrate in 100 ml. of 1:100 nitric acid) with 1:100 nitric acid.

Procedure

Isolation of lead. The sample solution should contain 1 to 20 γ of lead in about 10 to 25 ml. Add 10 ml. of ammonium citrate solution, a few drops of thymol blue indicator, and, if more than small amounts of copper and iron are present, 1 ml. of hydroxylamine hydrochloride solution. After making the solution basic with ammonia, add 5 ml. of potassium cyanide solution and adjust the pH with ammonia to 9 to 9.5 (thymol blue assumes a green-blue to blue color). Extract the lead in a separatory funnel with 25 ml. of 0.01% dithizone in carbon tetrachloride (Notes 1 and 2). Drain the carbon tetrachloride layer into another separatory funnel, and repeat the extraction with a second 25-ml. portion of dithizone. No detectable red color should be present in the extract.

Combine the carbon tetrachloride extracts and wash with 10 ml. of water containing 1 drop of 1:1 ammonia. Draw off the carbon tetrachloride into another separatory funnel and wash the aqueous phase with a milliliter or two of carbon tetrachloride. Add the latter to the main carbon tetrachloride extract.

To remove the lead, shake the carbon tetrachloride extract with 5.0 ml. of 1:100 nitric acid for 30 seconds. Separate the phases and extract the carbon tetrachloride with a second 5.0 ml. of nitric acid. Combine the acid extracts and rinse out the second portion with a milliliter or two of water (Notes 3 and 4). Remove residual droplets of dithizone by washing with a small volume of carbon tetrachloride and drawing it off. Swirl the solution to collect remaining droplets of carbon tetrachloride and draw them off, leaving the stopcock filled with organic solvent. Floating droplets should be allowed to evaporate from the unstoppered funnel. With a roll of filter paper, dry out the funnel stem.

Determination of lead. Add 10.0 ml. of ammonia–cyanide–sulfite solution and 10.0 ml. of 0.001% dithizone solution to the combined acid extracts. Shake well for 30 seconds and allow the phases to separate. If the organic phase is free from aqueous droplets, it may be drained directly into a 1-cm. (or 2-cm.) cell (Note 5). The first few drops should be discarded since they will be diluted with carbon tetrachloride from the stopcock bore. Measure the absorbance at 520 mμ against carbon tetrachloride as soon as possible. Avoid exposure to strong light.

A standard curve should be constructed by taking 0, 1, 2, 5, 10, 15, and 20 γ of lead through the procedure for the determination of lead. Each portion of the standard lead solution should be diluted to 10.0 ml. with 1:100 nitric acid.

Note 1. The purity of the dithizone is not critical at this point, so if specially purified dithizone is used for the final determination, 5 ml. of a 0.05% impure dithizone solution may be used instead.

2. If the amount of lead present is expected to be very small the dithizone may be added 5 ml. at a time to give an approximate idea of the amount present. A smaller volume of dithizone can then be used for the final determination to give a higher, and, therefore, more precisely measured absorbance.

3. If greater than 10 γ of lead is present the combined acid extracts may be diluted to 25 ml. in a volumetric flask, and a 10-ml. aliquot taken.

4. If bismuth is not known to be absent, take a portion of the acid extract, adjust it to pH 2.0 with ammonia, and shake it vigorously for 2 minutes with a small volume of 0.001% dithizone solution in carbon tetrachloride. If the latter does not change in color, bismuth is not present in significant amounts. If an orange or mixed color is obtained, bismuth must be removed by adjusting the whole solution to pH 2.0 and successively extracting it with 2 to 3 ml. portions of 0.005% dithizone until the last portion remains green after shaking 3 minutes. After adding the ammonia–cyanide–sulfite solution, the resulting solution must be adjusted to pH 10.8. For methods for avoiding the interference of thallium, see Section V-D-1-a.

5. Small droplets of the aqueous phase that refuse to separate may be removed by inserting a small plug of glass wool in the steam of the separatory funnel and filtering the carbon tetrachloride through it. Discard the first few drops, and beware of contamination from the glass wool.

6. Polarographic Determination in Copper-Base Alloys (241)

Lead may be easily determined by the polarographic method in a wide variety of supporting electrolytes. For example, 0.1M potassium chloride, 1M hydrochloric acid, 1M sodium hydroxide, and acid tartrate media can all be recommended. The choice should be made on the basis of the composition of the sample to be analyzed (see Section V-C-1 for discussion of a variety of approaches). In the following procedure for the determination of lead in copper-base alloys, the copper is first separated along with antimony and bismuth by controlled-potential electrolysis.

Apparatus

The instrumentation for polarography and controlled-potential electrolysis is discussed elsewhere in this Treatise. In this procedure no unusual equipment is required. It is convenient, but not necessary, to have a polarograph of the recording type.

Reagents

1. Hydrochloric acid, concentrated ($12M$), reagent grade.
2. Nitric acid, concentrated ($16M$), reagent grade.
3. Sodium hydroxide, solution ($5M$), reagent grade.
4. Hydrazine dihydrochloride, solid, lead-free.
5. Gelatin solution, 0.1% solution in water.
6. Nitrogen. A cylinder of nitrogen with a low oxygen content.

Procedure

Separation of copper. Dissolve a 0.5- to 1-g. sample in a warm mixture of 4 ml. of water, 6 ml. of concentrated hydrochloric acid, and 1 ml. of concentrated nitric acid. Dilute the solution to about 50 ml. and boil it briefly to remove oxides of nitrogen and chlorine. After allowing it to cool, transfer the solution to a 250-ml. volumetric flask, add 2.5 g. of hydrazine dihydrochloride, and then dilute it to volume.

Transfer the contents of the flask as completely as possible to a dry 300-ml. electrolysis cell containing large platinum gauze electrodes of the type recommended in Section VII-B-4. Commence the electrolysis and maintain the potential of the cathode, as measured against an external saturated calomel electrode, constant at -0.35 v. This may be done manually, but it is preferable and much more convenient to use a potentiostat. When the electrolysis current decreases to a small constant value (about 0.02 amp.) the deposition of the copper is complete. Take a 25-ml. aliquot of the electrolyte for the determination of lead (Notes 1 and 2).

Determination of lead. Add 12 ml. of $5M$ sodium hydroxide and 2.5 ml. of 0.1% gelatin solution to the 25-ml. aliquot in a 50-ml. volumetric flask, and dilute the solution to volume. Transfer a portion of this solution to a polarographic cell in water bath at $25.0 \pm 0.2°C$. and bubble it with nitrogen. Record the polarogram from -0.5 to -1.0 v. *vs.* S.C.E. and measure the diffusion current of lead. Measure the m value of the capillary, and determine its drop time, t, at the potential at which the diffusion current, i_d, was measured. Compute the concentration of lead, C (millimoles/liter), from the relation

$$C = i_d/3.40\ m^{2/3}t^{1/6}$$

Alternatively, a calibration curve may be established by taking known amounts of lead through the procedure.

Note 1. For the determination of tin, a second 25-ml. aliquot of this solution is also taken. The paper by Lingane (241) gives the complete procedure for the determination of Pb, Sn, Ni, and Zn in copper-base alloys.

2. The aliquot will be about $0.4M$ with respect to hydrochloric acid.

7. Amperometric Titration with Potassium Dichromate (210)

Apparatus

A manual polarograph with a dropping mercury electrode. The polarographic cell should have a volume of about 100 ml. and be stoppered with a rubber stopper that has been bored to admit the following:

1. A dropping mercury electrode (D.M.E.).
2. A salt bridge to an external saturated calomel electrode (S.C.E.).
3. A 10-ml. buret.
4. Two nitrogen inlet tubes connected by a three-way stopcock. One tube extends to the bottom of the cell to stir and deaerate the solution; the other does not dip into the solution, but allows a nitrogen atmosphere to be maintained over it.

Reagents

1. Potassium dichromate, $0.05M$. Prepare by accurately weighing dried primary standard potassium dichromate and dissolving in water.
2. Supporting electrolyte. Dilute a solution containing 11.0 g. of potassium nitrate, 10.0 ml. of glacial acetic acid, and 5.0 g. of sodium acetate, to 500 ml.
3. Methyl red, 1% solution.
4. Nitrogen. A cylinder of nitrogen with a low oxygen content.

Procedure

An aliquot of the neutral or slightly acidic sample solution containing 50 to 150 mg. of lead (Note 1) is pipetted into the polarographic cell, 25 ml. of the supporting electrolyte solution and 4 drops of methyl red are added, and the solution is diluted to 50 ml. Before placing the D.M.E. in the cell, raise the levelling bulb to start the flow of mercury. Bubble nitrogen briskly through the solution for 5 to 10 minutes to remove dissolved oxygen. Set the applied potential to -1.0 v. (the negative lead from the polarograph goes to the D.M.E.; the positive to the S.C.E.) (Note 2). Stop the bubbling by turning the stopcock to redirect the nitrogen flow over the surface of the solution. When the solution has come to rest, record the average current indicated by the polarograph galvanometer. Add some of the potassium dichromate solution from the buret (Note 3), bubble the solution for a minute or two to mix it and remove oxygen introduced by the titrant, and record the new diffusion current. Continue in this manner until four or more points are obtained on each side of the equivalence point (Note 4). Multiply each measured current by the factor, $(V + v)/V$, where V = initial volume and v = volume of titrant added. Plot a graph of the corrected diffusion currents against the ml. of titrant added. Draw one straight line through the points before, and another through the points after, the equivalence point. The intersection of these lines gives the end point of the titration. Calculate the amount of lead present from the titration reaction:

$$2Pb^{+2} + Cr_2O_7^{-2} + H_2O = 2PbCrO_4(s) + 2H^+$$

Note 1. Smaller amounts of lead may be determined by reducing the concentration of the standard solution of potassium dichromate. For example, 10 mg. of lead in 50 ml. of $0.01M$ potassium nitrate may be titrated with $0.005M$ potassium dichromate with an accuracy of $\pm 0.3\%$ (210).

2. A convenient alternative procedure is to carry out the titration at 0.0 v. applied potential (short circuit the S.C.E. to the D.M.E. through a low-resistance microammeter or a shunted galvanometer). Lead ions are not reduced, and a different shape of titration curve is obtained (see Section V-C-4). Also, under these conditions, dissolved oxygen need not be removed.

3. A buret with a capillary tip that may be kept immersed in the solution being titrated is most convenient. With this arrangement, no concern need be given to drops hanging from the buret tip.

4. Points close to the equivalence point are not important, and no special effort need be made to obtain them.

REFERENCES

1. A.B.C.M.-S.A.C., Committee on Methods for the Analysis of Trade Effluents, *Analyst*, **81**, 607 (1956).
2. Abd El Raheem, A. A., and F. A. Osman, *Z. Anal. Chem.*, **169**, 328 (1959).
3. Abd El Raheem, A. A., and M. M. Dokhama, *Anal. Chim. Acta*, **20**, 133 (1959).
4. Aborn, R. H., and R. H. Brown, *Ind. Eng. Chem., Anal. Ed.*, **1**, 26 (1929).
5. Abraham, B. M., and R. S. Huffman, *Ind. Eng. Chem., Anal. Ed.*, **12**, 656 (1940).
6. Adolf, A., *J. Am. Chem. Soc.*, **37**, 2509 (1915).
7. Ahrens, L. H., and S. R. Taylor, *Spectrochemical Analysis*, 2nd ed., Addison-Wesley, Reading, Mass., 1961, p. 242.

8. Ahrland, S., *Acta Chem. Scand.*, **10**, 723 (1956).
8a. Aldrich, L. T., and G. W. Wetherill, *Ann. Rev. Nucl. Sci.*, **8**, 257 (1958).
9. Alfonsi, B., *Anal. Chim. Acta*, **19**, 276 (1958).
10. Alfonsi, B., *Anal. Chim. Acta*, **19**, 389 (1958).
11. Alfonsi, B., *Anal. Chim. Acta*, **19**, 569 (1958).
12. American Public Health Association, Inc., *Methods for Determining Lead in Air and Biological Materials*, New York, 1955.
13. American Public Health Association, Inc., *Occupational Lead Exposure and Lead Poisoning*, New York, 1943.
14. American Society for Testing and Materials, *Methods of Chemical Analysis of Metals*, Philadelphia, 1956, p. 383.
15. *Ibid.*, p. 483.
16. *Ibid.*, p. 536.
17. American Society for Testing and Materials, *A.S.T.M. Standards*, Philadelphia, 1955, Part 2, p. 72.
18. American Society for Testing and Materials, *A.S.T.M. Standards*, Philadelphia, 1958, Part 7, p. 267.
19. *Ibid.*, p. 322.
20. *Ibid.*, p. 669.
21. Anderson, J. S., and M. Sterns, *J. Inorg. Nucl. Chem.*, **11**, 2.72 (1959).
22. Ariel, M., and P. Enoch, *Anal. Chim. Acta*, **18**, 339 (1958)
23. Auerbach, F., and H. Pick, *Arb. Kaiser Gesundh.*, **45**, 113 (1913).
24. Bale, W. D., E. W. Davies, and C. B. Monk, *Trans. Faraday Soc.*, **52**, 816 (1956).
25. Bambach, K., and R. E. Burkey, *Ind. Eng. Chem., Anal. Ed.*, **14**, 904 (1942).
26. Bambach, K., and J. Cholak, *Ind. Eng. Chem., Anal. Ed.*, **13**, 504 (1941).
27. Bane, R. W., *Anal. Chem.*, **27**, 1022 (1955).
28. Baskova, Z. A., *Zh. Analit. Khim.*, **14**, 75 (1959).
29. Bastron, H., P. R. Barnett, and K. J. Murata, *U. S. Geol. Surv. Bull.*, **1084-G** (1960).
30. Barker, G. C., and I. L. Jenkins, *Analyst*, **77**, 685 (1952).
30a. Bate, G. L., D. S. Miller, and J. L. Kulp, *Anal. Chem.*, **29**, 84 (1957).
31. Beck, K., *Z. Elektrochem.*, **17**, 843 (1911).
32. Belcher, R., R. A. Close, and T. S. West, *Chemist-Analyst*, **46**, 86 (1957).
33. Benedetti-Pichler, A. A., *Microtechnique of Inorganic Analysis*, Wiley, New York, 1942, pp. 61–63.
34. Berg, R., and E. S. Fahrenkamp, *Z. Anal. Chem.*, **112**, 161 (1938).
35. Berka, A., *Anal. Chim. Acta*, **24**, 171 (1961).
36. Berka, A., J. Dolezal, I. Nemec, and J. Zyka, *Anal. Chim. Acta*, **25**, 533 (1961).
37. Berndt, W., and J. Sara, *Talanta*, **8**, 653 (1961).
38. Beuf, H., *Bull. Soc. Chim. France*, **3**, 852 (1890).
39. Biedermann, W., and G. Schwarzenbach, *Chimia (Aarau)*, **2**, 1 (1948).
40. Biggs, A. I., H. N. Parton, and R. A. Robinson, *J. Am. Chem. Soc.*, **77**, 5844 (1955).
41. Bircumshaw, L. L., and I. Harris, *J. Chem. Soc.*, **1939**, 1637.
42. Birks, L. S., *X-Ray Spectrochemical Analysis*, Interscience, New York, 1959.
43. Birks, L. S., E. J. Brooks, H. Friedman, and R. M. Roe, *Anal. Chem.*, **22**, 1258 (1950).
44. B.I.S.R.A. Methods of Analysis Committee, *J. Iron Steel Inst. (London)*, **193**, 350 (1959).

45. Bjerrum, J., G. Schwarzenbach and L. G. Sillen, *Stability Constants, Part II; Inorganic Ligands*, The Chemical Society, London, 1958.
45a. Blaise, J., *Ann. Phys.*, **3**, 1019 (1958).
46. Bode, H., *Z. Anal. Chem.*, **144**, 165 (1955).
47. Bollenbach, H., *Z. Anal. Chem.*, **46**, 582 (1907).
47a. Boltwood, B. B., *Am. J. Sci.*, **23**, 77 (1907).
48. Bottger, W., *Pharm. Post.*, **40**, 679 (1907).
49. Bottger, W., *Z. Phys. Chem.*, **46**, 521 (1903).
50. Brauer, G., *Handbuch der Praparativen Anorganischen Chemie*, Ferdinand Enke Verlag, Stuttgart, 1954, p. 564–565.
51. Bricker, L. G., and K. L. Proctor, *Ind. Eng. Chem., Anal. Ed.*, **17**, 511 (1945).
52. Brode, W. R., *Chemical Spectroscopy*, 2nd ed., Wiley, New York, 1943, p. 85.
52a. Brody, J. K., and F. S. Tomkins, *Proc. 2nd Intern. Conf. Peaceful Uses At. Energy, Geneva, 1958*, **28**, 639 (1958).
52b. Brody, J. K., F. S. Tomkins, and M. Fred, *Spectrochim. Acta*, **8**, 329 (1957).
53. Brown, D. J., J. A. Moss, and J. B. Williams, *Ind. Eng. Chem., Anal. Ed.*, **3**, 134 (1931).
54. Bruner, L., and J. Zawadzki, *Z. Anorg. Chem.*, **65**, 136 (1909).
55. Butler, G., and J. L. Copp, *J. Chem. Soc.*, **1956**, 725.
56. Burns, E. A., and D. N. Hume, *J. Am. Chem. Soc.*, **78**, 3958 (1956).
57. Burriel-Marti, F., and M. J. Garate, *Rec. Trav. Chim.*, **79**, 495 (1960).
58. Bush, G. H., *Analyst*, **79**, 697 (1954).
59. Calingaert, G., F. W. Lamb, H. L. Miller, and G. E. Noakes, *Anal. Chem.*, **22**, 1238 (1950).
60. Cann, J. Y., and R. A. Sumner, *J. Phys. Chem.*, **36**, 2615 (1932).
61. Cantanzaro, E. J., and P. W. Gast, *Geochim. Cosmochim. Acta*, **19**, 113 (1960).
62. Carrell, B., and A. Olin, *Acta Chem. Scand.*, **14**, 1999 (1960).
63. Carritt, D. E., *Anal. Chem.*, **25**, 1927 (1953).
64. Celse Costa, A., *Chemist-Analyst*, **47**, 39 (1958).
65. Chaberek, S., Jr., R. C. Courtney, and A. E. Martell, *J. Am. Chem. Soc.*, **74**, 5057 (1952).
66. Chaberek, S., Jr., and A. E. Martell, *J. Am. Chem. Soc.*, **74**, 6228 (1952).
67. Ch'en, N. K., C. H. Chu, and H. H. Ch'in, *Sheng Li Hsueh Pao*, **1958**, 153.
68. Cholak, J., and K. Bambach, *Ind. Eng. Chem., Anal. Ed.*, **13**, 583 (1941).
69. Cholak, J., and R. V. Story, *Ind. Eng. Chem., Anal. Ed.*, **10**, 619 (1938).
70. Cholak, J., and R. V. Story, *J. Opt. Soc. Am.*, **31**, 730 (1941).
70a. Chow, T. J., and C. R. McKinney, *Anal. Chem.*, **30**, 1499 (1958).
71. Chukhlantsev, V. G., *Zh. Analit. Khim.*, **11**, 529 (1956).
72. Chukhlantsev, V. G., and G. P. Thomashevskii, *Zh. Analit. Khim.*, **12**, 296 (1957).
73. Cimerman, C., and M. Ariel, *Anal. Chim. Acta*, **15**, 207 (1956).
74. Cimerman, C., and M. Ariel, *Anal. Chim. Acta*, **12**, 13 (1955).
75. Clark, G. L., and R. Rowan, *J. Am. Chem. Soc.*, **63**, 1305 (1941).
76. Clarke, B. L., and H. W. Hermance, *J. Am. Chem. Soc.*, **54**, 877 (1932).
77. Classen, A., *Ausgewahlte Methoden der analytischen Chemie.*, F. Vieweg und Sohn, Braunschweig, 1901, Vol. I, p. 18.
78. Classen, A., *Ber.*, **27**, 163 (1894).
79. Clifford, P. A., *J. Assoc. Offic. Agr. Chemists*, **26**, 26 (1943).
80. Clifford, P. A., *J. Assoc. Offic. Agr. Chemists*, **20**, 192 (1937).
81. Clifford, P. A., and H. J. Wichmann, *J. Assoc. Offic. Agr. Chemists*, **19**, 130 (1936).

82. Cluett, M. L., and J. H. Yoe, *Anal. Chem.*, **29**, 1265 (1957).
83. Coates, A. C., and R. Smart, *J. Soc. Chem. Ind.*, **60**, 249 (1941).
84. Connick, R. E., and A. D. Paul, *J. Am. Chem. Soc.*, **80**, 2069 (1958).
85. Cooke, W. D., F. Hazel, and W. M. McNabb, *Anal. Chem.*, **22**, 654 (1950).
86. Craig, D. N., and G. W. Vinal, *J. Res. Natl. Bur. Std.*, **22**, 55 (1939).
87. Cullinane, N. M., and S. J. Chard, *Analyst*, **73**, 95 (1948).
88. Das, N. K., S. Aditya, and B. Prasad, *J. Indian Chem. Soc.*, **29**, 169 (1952).
89. Dawson, E. C., and A. Rees, *Analyst*, **71**, 417 (1946).
90. Dawson, J., and R. J. Magee, *Mikrochim. Acta*, **1958**, 330.
91. Dean, J. A., *Flame Photometry*, McGraw-Hill, New York, 1960, p. 239.
92. Dixon, B. E., and P. Metson, *Analyst*, **84**, 46 (1959).
93. Dixon, B. E., and P. Metson, *Analyst*, **85**, 122 (1960).
94. Drago, R. S., *J. Phys. Chem.*, **62**, 353 (1958).
95. Duval, C., *Inorganic Thermogravimetric Analysis*, Elsevier, Amsterdam, 1953.
96. Edmonds, S. M., and N. Birnbaum, *J. Am. Chem. Soc.*, **62**, 2367 (1940).
97. Elkins, H. B., *The Chemistry of Industrial Toxicology*, 2nd ed., Wiley, New York, 1959.
98. Elving, P. J., and W. C. Zook, *Anal. Chem.*, **25**, 502 (1953).
99. Elwell, W. T., and J. A. F. Gidley, *Anal. Chim. Acta*, **24**, 71 (1961).
100. Erdey, L., and L. Polos, *Z. Anal. Chem.*, **153**, 411 (1956); cf. L. Erdey, *Chemist-Analyst*, **48**, 106 (1959).
101. Etheridge, A. T., *Analyst*, **75**, 279 (1950).
102. Evans, B. S., *Analyst*, **64**, 2 (1939).
103. Eve, A. J., and E. T. Verdier, *Anal. Chem.*, **28**, 537 (1956).
104. Faircloth, R. F., *Talanta*, **2**, 135 (1959).
105. Fairhall, L. T., and R. G. Keenan, *J. Am. Chem. Soc.*, **63**, 3076 (1941).
106. Farkas, L., and N. Uri, *Anal. Chem.*, **20**, 236 (1948).
107. Feigl, F., and W. Braile, *Analyst*, **69**, 147 (1944).
108. Feigl, F., *Chemistry of Specific, Selective, and Sensitive Reactions*, Academic Press, New York, 1949, p. 649.
109. Feigl, F., *Spot Tests in Inorganic Analysis*, Elsevier, Amsterdam, 1956.
110. Ferrett, D. J., and G. C. W. Milner, *Analyst*, **81**, 193 (1956).
111. Fischer, H., and G. Leopoldi, *Angew. Chem.*, **47**, 90 (1934).
112. Fischer, H., and G. Leopoldi, *Z. Anal. Chem.*, **119**, 161 (1940).
113. Flaschka, H., and H. Abdine, *Chemist-Analyst*, **45**, 58 (1956).
114. Flaschka, H., *EDTA Titrations*, Pergamon Press, New York, 1959, p. 90.
115. *Ibid.*, p. 76.
116. Flaschka, H., and F. Huditz, *Z. Anal. Chem.*, **137**, 172 (1952).
117. Flaschka, H., and H. Jakobljevich, *Anal. Chem. Acta*, **4**, 606 (1950).
118. Flaschka, H., *Mikrochemie ver. Mikrochim. Acta*, **39**, 38 (1952).
119. Flaschka, H., *Mikrochemie ver. Mikrochim. Acta*, **39**, 315 (1952).
120. Flaschka, H., *Mikrochemie ver. Mikrochim. Acta*, **40**, 42 (1952).
121. Fleischmann, M., and M. Liler, *Trans. Faraday Soc.*, **54**, 1370 (1958).
122. Flickinger, L. C., E. W. Polley and F. A. Galletta, *Anal. Chem.*, **29**, 1778 (1957).
123. Fricke, R., and R. Sammet, *Z. Anal. Chem.*, **126**, 13 (1943).
124. Fritz, J. S., and G. S. Hammond, *Quantitative Organic Analysis*, Wiley, New York, 1957, p. 90.
125. Fritz, J. S., W. J. Lane, and A. S. Bystroff, *Anal. Chem.*, **29**, 821 (1957).
126. Fromherz, H., *Z. Phys. Chem.*, **153A**, 376 (1931).

127. Funihashi, H., and M. Ishibashi, *J. Chem. Soc. Japan*, **59**, 503 (1938); *Chem. Abstr.*, **32**, 5722 (1938).
128. Funk, H., and F. Romer, *Z. Anal. Chem.*, **101**, 85 (1935).
129. Furman, N. H., and J. H. Wallace, *J. Am. Chem. Soc.*, **51**, 1449 (1929).
130. Gage, J. C., *Analyst*, **80**, 789 (1955).
131. Gage, J. C., *Analyst*, **82**, 453 (1957).
132. Garrett, A. B., S. Vallenga, and C. M. Fontana, *J. Am. Chem. Soc.*, **61**, 367 (1939).
133. Gasper y Arnal, T., and J. M. Poggis Mesorana, *Anales Soc. Espan. Fis. Quim.*, **43**, 439 (1947).
134. Geilmann, W., and R. Holtje, *Z. Anorg. Chem.*, **152**, 59 (1926).
135. Gentry, C. H. R., and L. G. Sherrington, *Analyst*, **71**, 31 (1946).
136. Gilbert, P. T., *Am. Soc. Testing Mater., Spec. Tech. Publ.*, **116**, 77 (1951).
136a. Gilman, H., and R. G. Jones, *J. Am. Chem. Soc.*, **72**, 1760 (1950).
137. Glasstone, S., *J. Chem. Soc.*, **119(a)**, 1914 (1921).
138. Goates, J. R., M. B. Gordon, and N. D. Faux, *J. Am. Chem. Soc.*, **74**, 835 (1952).
139. Goetz, C. A., and F. J. Debbrecht, *Anal. Chem.*, **27**, 1972 (1955).
140. Gorsuch, T. T., *Analyst*, **84**, 135 (1959).
141. Goto, H., and K. Hirokawa, *Sci. Rept. Res. Inst., Tohoku Univ., Ser. A*, **10**, 10 (1958).
142. Goto, H., and Y. Kakita, *Sci. Rept. Res. Inst., Tohoku Univ., Ser. A.*, **9**, 131 (1957).
143. Goward, G. W., Thesis, Princeton Univ., 1954; Univ. Microfilms **9414**.
144. Gray, T. D., *Ind. Eng. Chem., Anal. Ed.*, **14**, 110 (1942).
145. Griffing, M. E., A. Rozek, L. J. Snyder, and S. R. Henderson, *Anal. Chem.*, **29**, 191 (1957).
146. Grossman, S., and J. Haslam, *J. Appl. Chem.*, **7**, 639 (1957).
147. Grote, F., *Z. Anal. Chem.*, **122**, 395 (1941).
148. Gull, H. C., *J. Soc. Chem. Ind.*, **56**, 177 (1937).
149. Guzelj, L., *Z. Anal. Chem.*, **104**, 107 (1936).
150. Hagemann, F., *J. Am. Chem. Soc.*, **72**, 768 (1950).
151. Haim, G., and W. C. E. Barnes, *Ind. Eng. Chem., Anal. Ed.*, **14**, 867 (1942).
152. Hansen, K. A., T. D. Parks, and L. Lykken, *Anal. Chem.*, **22**, 1232 (1950).
153. Hanus, J., and V. Hovorka, *Chem. Listy*, **31**, 489 (1937).
154. Harned, H. S., and W. J. Hamer, *J. Am. Chem. Soc.*, **57**, 33 (1935).
155. Hart, H. V., *Analyst*, **76**, 692 (1951).
156. Hawkins, R. C., and H. G. Thode, *Ind. Eng. Chem., Anal. Ed.*, **16**, 71 (1944).
157. Heal, H. G., and J. May, *J. Am. Chem. Soc.*, **80**, 2374 (1958).
158. Hecht, F., and J. Donau, *Anorganische Mikrogewichtsanalyse*, J. Springer, Vienna, 1940, p. 151.
159. Hershenson, H. M., M. E. Smith, and D. N. Hume, *J. Am. Chem. Soc.*, **75**, 507 (1953).
160. Hertelendi, L., *Z. Anal. Chem.*, **122**, 30 (1941).
161. Hertelendi, L., and J. Jovanovich, *Z. Anal. Chem.*, **128**, 151 (1948).
162. Hillebrand, W. F., G. E. F. Lundell, H. A. Bright, and J. I. Hoffman, *Applied Inorganic Analysis*, Wiley, New York, 1953, Chapt. 9.
163. Hoffman, W. A., and W. W. Brandt, *Anal. Chem.*, **28**, 1487 (1956).
164. Holness, H., *Analyst*, **69**, 145 (1944).
165. Hovorka, V., *Chem. Listy*, **31**, 273 (1937); *Chem. Abstr.*, **31**, 6997 (1937).
166. Hubbard, D. M., *Ind. Eng. Chem., Anal. Ed.*, **9**, 493 (1937).
167. Hughes, H., and F. P. Hochgesang, *Anal. Chem.*, **22**, 1248 (1950).

168. Hughes, V. L., and A. E. Martell, *J. Phys. Chem.*, **57**, 694 (1953).
169. Hunt, E. C., A. A. North, and R. A. Wells, *Analyst*, **80**, 172 (1955).
170. Hyman, H. M., and S. Weisberger, *Appl. Spectry.*, **9**, 98 (1955).
171. Ibbotson, F., and L. Aitchison, *The Analysis of Non-Ferrous Alloys*, Longmans, Green, New York, 1915, pp. 54–55.
172. U. K. At. Energy Authority, Ind. Group, Rept. **IGO-AM/W-169** (1958).
173. Irving, H., and E. J. Butler, *Analyst*, **78**, 571 (1953).
174. Irving, H., and E. J. Butler, *Analyst*, **79**, 143 (1954).
175. Ishibashi, M., and H. Kishi, *Bull. Chem. Soc. Japan*, **10**, 362 (1935).
176. Ivett, R. W., and T. DeVries, *J. Am. Chem. Soc.*, **63**, 2821 (1941).
177. Jablonski, F., and H. Moritz, *Aluminium*, **26**, 245 (1944).
178. Jacques, A., *Trans. Faraday Soc.*, **5**, 225 (1910).
179. Jamieson, G. S., *Am. J. Sci.*, **40**, 157 (1915).
180. Jilek, A., and J. Kota, *Collection Czech. Chem. Commun.*, **5**, 396 (1933).
181. Johnson, E. I., and R. D. A. Polhill, *Analyst*, **80**, 364 (1955).
182. Johnson, E. I., and R. D. A. Polhill, *Analyst*, **82**, 238 (1957).
183. Johnson, J. S., and K. A. Kraus, *J. Am. Chem. Soc.*, **81**, 1569 (1959).
184. Jones, F. R., and D. M. Brasher, *Analyst*, **72**, 423 (1947).
185. Jordan, J. H., *Petrol. Refiner*, **32**, 139 (1953).
186. Joussot-Dubien, J., and G. Oster, *Bull. Soc. Chim. France*, **1960**, 343.
187. Kallmann, S., *Anal. Chem.*, **23**, 1291 (1951).
188. Kapustinskii, A. F., *Dokl. Akad. Nauk S.S.S.R.*, **23**, 144 (1940).
189. Karaoglanov, Z., and M. Michov, *Z. Anal. Chem.*, **103**, 113 (1935).
190. Kato, H., *J. Chem. Soc. Japan*, **58**, 972 (1937).
191. Keefer, R. M., and H. G. Reiber, *J. Am. Chem. Soc.*, **63**, 689 (1941).
192. Kelley, M. T., and H. H. Miller, *Anal. Chem.*, **24**, 1895 (1952).
193. Kemula, W., E. Rakowska, and Z. Kublik, *J. Electroanal. Chem.*, **1**, 205 (1960).
194. Kent, N. L., *J. Soc. Chem. Ind.*, **61**, 183 (1942).
195. Kern, D. M., *Collection Czech. Chem. Commun.*, **25**, 3159 (1960).
196. Kety, S. S., *J. Biol. Chem.*, **142**, 181 (1942).
197. Khalifa, H., *Z. Anal. Chem.*, **159**, 410 (1958).
198. Khopkar, S. M., and A. K. De, *Talanta*, **7**, 7 (1960).
199. Kinnunen, J., and B. Merikanto, *Chemist-Analyst*, **44**, 50 (1955).
200. Kinnunen, J., and B. Merikanto, *Chemist-Analyst*, **44**, 75 (1955).
201. Kinnunen, J., and B. Wennerstrand, *Eng. Mining J.*, **156**, No. 4, 94 (1955).
202. Kivalo, P., *Suomen Kemistilehti*, **28B**, 155 (1955).
203. Kivalo, P., *Suomen Kemistilehti*, **29B**, 8 (1956).
204. Kivalo, P., and A. Ekman, *Suomen Kemistilehti*, **29B**, 139 (1956).
205. Kohlrausch, F., *Z. Phys. Chem.*, **64**, 129 (1908).
206. Kolthoff, I. M., J. Jordan, and A. Heyndrickx, *Anal. Chem.*, **25**, 884 (1953).
207. Kolthoff, I. M., and J. J. Lingane, *Chem. Rev.*, **24**, 1 (1939).
208. Kolthoff, I. M., and J. J. Lingane, *Polarography*, 2nd ed., Interscience, New York, 1952.
209. Kolthoff, I. M., and G. Matsuyama, *Ind. Eng. Chem., Anal. Ed.*, **17**, 615 (1945).
210. Kolthoff, I. M., and Y. D. Pan, *J. Am. Chem. Soc.*, **61**, 3402 (1939).
211. Kolthoff, I. M., *Pharm. Weekblad*, **57**, 934 (1920).
212. Kolthoff, I. M., R. W. Perlich, and D. Weible, *J. Phys. Chem.*, **46**, 561 (1942).
213. Kolthoff, I. M., and E. B. Sandell, *Textbook of Quantitative Inorganic Analysis*, 3rd ed., Macmillan, New York, 1952, pp. 669–770.

214. Kolthoff, I. M., and V. A. Stenger, *Volumetric Analysis*, 2nd ed., Interscience, New York, 1947, Volume II, p. 317; *ibid.*, p. 181.
215. Kolthoff, I. M., and B. Van't Riet, *J. Phys. Chem.*, **63,** 817 (1959).
216. Korbl, J., and R. Přibil, *Chemist-Analyst*, **45,** 102 (1956); **46,** 28 (1957).
217. Korbl, J., R. Přibil, and A. Emr, *Collection Czech. Chem. Commun.*, **22,** 961 (1957).
218. Korenman, T. M., F. S. Frum, and V. G. Chebakova, *Zh. Obshch. Khim.*, **22,** 1731 (1952).
219. Koryta, J., and I. Kossler, *Collection Czech. Chem. Commun.*, **15,** 241 (1950).
220. Kress, K. E., *Anal. Chem.*, **29,** 803 (1957).
221. Kuras, M., *Collection Czech. Chem. Commun.*, **11,** 313 (1939).
222. Kuras, M., *Collection Czech. Chem. Commun.*, **11,** 367 (1939).
223. Kuroda, K., *Bull. Chem. Soc. Japan*, **15,** 153 (1940).
224. Kutzelnigg, A., *Z. Anal. Chem.*, **129,** 382 (1949).
225. Laitinen, H. A., and I. M. Kolthoff, *J. Phys. Chem.*, **45,** 1061 (1945).
226. Lamb, F. W., L. M. Niebylski, and E. W. Kiefer, *Anal. Chem.*, **27,** 129 (1955).
227. Lanford, O. E., and S. J. Kiehl, *J. Am. Chem. Soc.*, **63,** 667 (1941).
228. Lang, R., and J. Zwerina, *Z. Anal. Chem.*, **93,** 248 (1933).
229. Laseque, G., *Bull. Soc. Chim.*, **11,** 884 (1912).
230. Latimer, W. M., *Oxidation Potentials*, 2nd ed., Prentice-Hall, New York, 1952.
231. *Lead in Modern Industry*, Lead Industries Association, New York, 1952, p. 1.
232. Leonard, G. W., M. E. Smith, and D. N. Hume, *J. Phys. Chem.*, **60,** 1493 (1956).
233. Lewis, J. A., and J. M. Griffiths, *Analyst*, **76,** 388 (1951).
234. Li, N. C., and R. A. Manning, *J. Am. Chem. Soc.*, **77,** 5225 (1955).
235. Lichty, D. M., *J. Am. Chem. Soc.*, **25,** 469 (1903).
236. Liebhafsky, H. A., H. G. Pfeiffer, E. H. Winslow, and P. D. Zemany, *X-Ray Absorption and Emission in Analytical Chemistry*, Wiley, New York, 1960.
237. Ligett, W. B., and L. P. Biefeld, *Ind. Eng. Chem., Anal. Ed.*, **13,** 813 (1941).
238. Lindsey, A. J., *Analyst*, **60,** 598 (1935).
239. Lingane, J. J., *Chem. Rev.*, **29,** 1 (1941).
240. Lingane, J. J., *Ind. Eng. Chem., Anal. Ed.*, **16,** 147 (1944).
241. Lingane, J. J., *Ind. Eng. Chem., Anal. Ed.*, **18,** 42 (1946).
242. Lingane, J. J., and S. L. Jones, *Anal. Chem.*, **23,** 1798 (1951).
243. Lipcinsky, A., and M. Krstewa, *Z. Anal. Chem.*, **164,** 246 (1958).
244. Lockwood, H. C., *Analyst*, **79,** 143 (1954).
245. Lockwood, H. H., *Anal. Chim. Acta*, **10,** 97 (1954).
246. Low, A. H., *Technical Methods of Ore Analysis*, 9th ed., Wiley, New York, 1922, p. 131.
247. Lowe, J., *J. Prakt. Chem.*, **77,** 73 (1859).
248. Lux, F., *Z. Anal. Chem.*, **19,** 153 (1880).
249. Machata, G., and H. Neuninger, *Wien. Med. Wochschr.*, **110,** 39 (1960).
250. Mahr, C., and H. Ohle, *Z. Anorg. Allgem. Chem.*, **234,** 224 (1937).
251. Majdel, J., *Z. Anal. Chem.*, **83,** 36 (1931).
252. Maley, L. E., and D. P. Mellor, *Australian J. Sci. Res.*, **A2,** 579 (1949).
253. Marple, T. L., and L. B. Rogers, *Anal. Chim. Acta*, **11,** 574 (1954).
254. Marshall, R. R., and D. C. Hess, *Anal. Chem.*, **32,** 960 (1960).
255. Martell, A. E., and R. C. Plumb, *J. Phys. Chem.*, **56,** 993 (1953).
256. Massachusetts Institute of Technology, *Wavelength Tables*, Wiley, New York, 1939.

257. Mathre, O. B., Ph.D. Thesis, University of Minnesota, 1956; see also E. B. Sandell, *Colorimetric Determination of Traces of Metals*, Interscience, New York London, 1959, pp. 565–567.
258. *Ibid.;* see also, Sandell, Reference 345, p. 559.
259. *Ibid.;* see also, Sandell, Reference 345, p. 568–569.
260. May, I., and J. I. Hoffman, *J. Wash. Acad. Sci.*, **38**, 329 (1948).
261. May, W. C., *Am. J. Sci.*, **6**, 255 (1873).
262. Maynes, A. D., and W. A. E. McBryde, *Anal. Chem.*, **29**, 1259 (1957).
263. Mayr, C., *Monatsh.*, **77**, 65 (1947).
264. McCord, W. M., and J. W. Zemp, *Anal. Chem.*, **27**, 1171 (1955).
265. Merritt, C., Jr., H. M. Hershenson, and L. B. Rogers, *Anal. Chem.*, **25**, 572 (1953).
266. Middleton, B., and R. E. Stuckey, *Analyst*, **79**, 138 (1954).
267. Miller, C. C., and L. R. Currie, *Analyst*, **75**, 467 (1950).
268. Miller, C. C., and L. R. Currie, *Analyst*, **75**, 471 (1950).
269. Millet, H., and M. Jowett, *J. Am. Chem. Soc.*, **51**, 997 (1929).
270. Milner, G. C. W., *Metallurgia*, **36**, 287 (1947).
271. Milner, G. C. W., *Analyst*, **73**, 472 (1948).
272. Milner, G. C. W., *The Principles and Applications of Polarography*, Longmans, Green, London, 1957.
273. Milner, G. C. W., J. W. Edwards, and A. Paddon, *U. K. At. Energy Authority Rept.*, **AERE-G/R-2612** (1958).
274. Milner, G. C. W., and J. Townend, *Anal. Chim. Acta*, **5**, 584 (1951).
275. Milner, O. I., and G. F. Shipman, *Anal. Chem.*, **26**, 1222 (1954).
276. Molland, J., *Tidssk. Kjemi Bergvesen*, **19**, 119 (1939); *Chem. Abstr.*, **34**, 1932 (1940).
277. Monk, C. B., *Trans. Faraday Soc.*, **47**, 297 (1951).
278. Monk, C. B., *Trans. Faraday Soc.*, **47**, 1233 (1951).
279. Moore, V. J., *Analyst*, **81**, 553 (1955).
280. Moser, L., and W. Reif, *Mikrochemie, Emich Festschrift*, E. Haim & Co., Vienna, 1930, p. 215.
281. Mrgudich, J. N., and G. L. Clark, *Ind. Eng. Chem., Anal. Ed.*, **9**, 256 (1937).
282. Muller, H., *Z. Anal. Chem.*, **113**, 161 (1938).
283. Murgulescu, I. G., and F. Dobrescu, *Z. Anal. Chem.*, **128**, 203 (1948).
284. Naish, W. A., J. E. Clennell, and V. S. Kingswood, *Select Methods of Metallurgical Analysis*, 2nd ed., Chapman and Hall, London, 1953.
285. Najarian, H. K., U. S. Pat. 2,766,114 (March 13, 1952); to St. Josephs Lead Co.
286. Nancollas, G. H., *J. Chem. Soc.*, **1955**, 1458.
287. Nasanen, R., *Suomen Kemistilehti*, **18B**, 45 (1945).
288. Nasanen, R., *Suomen Kemistilehti*, **26B**, 2, 11 (1953).
290. Nasanen, R., and E. Uisitalo, *Acta Chem. Scand.*, **8**, 112 (1954).
291. *Natl. Bur. Std., Circ.* **500**, F. D. Rossini, D. D. Wagman, W. H. Evans, S. Levine, and I. Jaffé, Eds., "Selected Values of Chemical Thermodynamic Properties," 1952.
292. Neeb, R., *Z. Anal. Chem.*, **170**, 321 (1959).
293. Nelson, F., and K. A. Kraus, *J. Am. Chem. Soc.*, **76**, 5916 (1954).
293a. Nier, A. O., *Phys. Rev.*, **55**, 153 (1939).
293b. Nier, A. O., R. W. Thompson, and B. F. Murphey, *Phys. Rev.*, **160**, 112 (1941).
294. Nikelly, J. G., and W. D. Cooke, *Anal. Chem.*, **28**, 243 (1956).
295. Noble, M. V., and A. B. Garret, *J. Am. Chem. Soc.*, **66**, 231 (1944).

296. Olin, A., *Acta Chem. Scand.*, **14**, 126 (1960).
297. Olin, A., *Acta Chem. Scand.*, **14**, 814 (1960).
298. Olver, J. W., and D. N. Hume, *Anal. Chim. Acta*, **20**, 559 (1959).
299. Ostroff, A. G., and G. Sanderson, *J. Inorg. Nucl. Chem.*, **9**, 45 (1959).
300. Ota, K., and S. Mori, *J. Japan Inst. Metals (Sendai)*, **22**, 290 (1958).
301. Ottendorfer, L. J., *Chemist-Analyst*, **47**, 96 (1958).
302. Pariaud, J. C., and P. Archinard, *Bull. Soc. Chim.*, **19**, 454 (1952).
303. Paterson, J. E., *Anal. Chem.*, **29**, 526 (1957).
304. Paul, A. D., Thesis, University of California, Berkeley, 1955; *U. S. At. Energy Comm.*, **URCL 2926** (1955).
305. Pauley, J. D., and M. J. Testerman, *J. Am. Chem. Soc.*, **76**, 4220 (1954).
306. Perlman, J. L., *Ind. Eng. Chem., Anal. Ed.*, **10**, 134 (1938).
307. Petrashen, V. I., *Chem. Abstr.*, **35**, 1346 (1941).
308. Pfeilstricker, K., *Mikrochim. Acta*, **1–3**, 319 (1956).
309. Pinkston, J. L., and C. T. Kenner, *Anal. Chem.*, **27**, 446 (1955).
310. Pleissner, M., *Arb. Kaiser. Gesundh.*, **26**, 384 (1907).
311. Pollard, F. H., P. Hanson, and W. J. Geary, *Anal. Chim. Acta*, **20**, 26 (1959).
312. Pollard, F. H., J. F. W. McOmie, and G. Nickless, *J. Chromatog.*, **2**, 284 (1959).
313. Powell, R. A., and C. A. Kinser, *Anal. Chem.*, **30**, 1139 (1958).
314. Přibil, R., *Collection Czech. Chem. Commun.*, **18**, 783 (1953).
315. Přibil, R., and B. Matyska, *Collection Czech. Chem. Commun.*, **16**, 139 (1951).
316. Přibil, R., and Z. Roubal, *Collection Czech. Chem. Commun.*, **19**, 1162 (1954).
317. Přibil, R., and F. Bydra, *Collection Czech. Chem. Commun.*, **24**, 3103 (1959).
318. Price, J., *J. Soc. Chem. Ind.*, **64**, 283 (1945).
319. Proffitt, P. M. C., and R. C. Chirnside, *Analyst*, **72**, 205 (1947).
320. Randell, M., and D. L. Shaw, *J. Am. Chem. Soc.*, **57**, 427 (1935).
321. Ravitz, S. F., *J. Phys. Chem.*, **40**, 61 (1936).
322. Reed, G. W., K. Kigoshi, and A. Turkevich, *Proc. 2nd Intern. Conf. Peaceful Uses At. Energy, Geneva, 1958*, **28**, 486 (1958–1959).
323. Reed, G. W., K. Kigoshi, and A. Turkevich, *Geochim. Cosmochim. Acta*, **20**, 122 (1960).
324. Reilley, C. N., G. W. Everett, and R. Johns, *Anal. Chem.*, **27**, 483 (1955).
325. Reilley, C. N., and W. W. Porterfield, *Anal. Chem.*, **28**, 443 (1956).
326. Reilley, C. N., and R. W. Schmid, *Anal. Chem.*, **30**, 947 (1958).
327. Reilley, C. N., R. W. Schmid, and D. W. Lamson, *Anal. Chem.*, **30**, 953 (1958).
328. Reynolds, C. A., and L. B. Rogers, *Anal. Chem.*, **21**, 176, 758 (1949).
328a. Richards, J. R., *Mikrochim. Acta*, **1962**, 620.
329. Ringbom, A., *Svensk Kem. Tidskr.*, **66**, 159 (1954).
330. Ripan, R., *Z. Anal. Chem.*, **123**, 244 (1942).
331. Rivot, Beudant, and Daguin, *Compt. Rend.*, **37**, 126 (1853).
332. Robinson, J. W., *Anal. Chim. Acta*, **24**, 451 (1961).
333. Rogers, L. B., and C. A. Reynolds, *J. Am. Chem. Soc.*, **71**, 2081 (1949).
334. Rooney, R. C., *Analyst*, **83**, 83 (1958).
335. Rose, H., *Poggendorff's Ann.*, **76**, 543 (1849).
336. Rosenquist, I. T., *Am. J. Sci.*, **240**, 356 (1942).
337. Ross, J. W., R. D. DeMars, and I. Shain, *Anal. Chem.*, **28**, 1768 (1956).
338. Roy, S. N., *J. Indian Chem. Soc.*, **13**, 40 (1936).
339. Rupp, E., and G. Siebler, *Chem. Ztg.*, **48**, 241 (1924).
340. Russ, J. J., and W. Reeder, *Anal. Chem.*, **29**, 1331 (1957).

341. Sacher, J. F., *Kolloid Z.*, **19**, 276 (1926).
342. Saint, H. C. J., *Analyst*, **83**, 88 (1958).
343. Sales, R., *J. Soc. Glass Technol.*, **43**, 37T (1959).
344. Sandell, E. B., *Ind. Eng. Chem., Anal. Ed.*, **9**, 464 (1937).
345. Sandell, E. B., *Colorimetric Determination of Traces of Metals*, 3rd ed., Interscience, New York, 1959, p. 170.
346. *Ibid.*, p. 569.
347. *Ibid.*, p. 575.
348. *Ibid.*, p. 579.
349. *Ibid.*, p. 583.
350. Sargent, E. H., and Co., *Bibliography of Polarographic Literature*, Sargent, Chicago, 1956.
351. Schaeffer, J. A., *Ind. Eng. Chem.*, **8**, 237 (1916).
352. Scherrer, J. A., *J. Res. Nat.. Bur. Std.*, **21**, 95 (1938).
353. Schleicher, A., *Electroanalitische Schnellmethoden*, F. Enke, Stuttgart, 1947, pp. 89–95.
354. Schultz, J., and M. A. Goleberg, *Ind. Eng. Chem., Anal. Ed.*, **15**, 155 (1943).
355. Schwarzenbach, G., *Complexometric Titrations*, Interscience, New York, 1957, p. 92.
356. Schwarzenbach, G., G. Anderegg, W. Schneider, and H. Senn, *Helv. Chim. Acta*, **38**, 1147 (1955).
357. Schwarzenbach, G., and W. Biedermann, *Helv. Chim. Acta*, **31**, 459 (1948).
358. Schwarzenbach, G., and E. Freitag, *Helv. Chim. Acta*, **34**, 1492 (1951).
359. Schwarzenbach, G., and E. Freitag, *Helv. Chim. Acta*, **34**, 1503 (1951).
360. Schwarzenbach, G., R. Gut, and G. Anderegg, *Helv. Chim. Acta*, **37**, 937 (1954).
361. Schwarzenbach, G., E. Kamplitsch, and R. Steiner, *Helv. Chim. Acta*, **28**, 1133 (1945).
362. Selivanova, N. M., and R. Ya. Boguslavskii, *Zh. Fiz. Khim.*, **29**, 128 (1955).
363. Semerano, G., and V. Capitano, *Mikrochemie ver. Mikrochim. Acta*, **30**, 71 (1942).
364. Shrawder, J., Jr., and I. A. Cowperthwaite, *J. Am. Chem. Soc.*, **56**, 2340 (1934).
365. Sidgwick, N. V., *Ann. Rept.*, **20**, 120 (1933).
366. Sidgwick, N. V., *The Chemical Elements and Their Compounds*, Oxford Univ. Press, 1950, Vol. I, p. 624.
367. Sill, C. W., and H. E. Peterson, *Anal. Chem.*, **24**, 1175 (1952).
368. Skobets, E. M., and S. A. Kacherova, *Zavodsk. Lab.*, **13**, 133 (1947).
369. Smales, A. A., in *Trace Analysis*, J. H. Yoe and H. J. Koch, Eds., Wiley, New York, 1957, p. 527.
370. Smith, G. F., and H. Diehl, *Talanta*, **4**, 185 (1960).
371. Smith, G. W., and A. K. Palmby, *Anal. Chem.*, **31**, 1798 (1959).
372. Snyder, L. J., *Ind. Eng. Chem., Anal. Ed.*, **19**, 684 (1947).
373. Snyder, L. J., W. R. Barnes, and J. V. Tokas, *Anal. Chem.*, **20**, 772 (1948).
374. Spacu, G., and J. Dick, *Z. Anal. Chem.*, **72**, 289 (1927); *Bull. Soc. Stiint. Cluj.*, **4**, 75 (1928).
375. Spacu, G., and M. Kuras, *Z. Anal. Chem.*, **104**, 88 (1936).
376. Starck, G., *Z. Anorg. Chem.*, **70**, 173 (1911).
376a. Stieff, L. R., T. W. Stern, S. Oshiro, and F. E. Senftle, *U. S. Geol. Surv. Profess. Paper*, **334-A** (1959).
377. Stolba, F., *J. Prakt. Chem.*, **101**, 150 (1867).
378. Streuli, C. A., and W. D. Cooke, *Anal. Chem.*, **25**, 1691 (1953).

379. Strnad, F., *Collection Czech. Chem. Commun.*, **11**, 391 (1939).
380. Strominger, D., J. M. Hollander, and G. T. Seaborg, *Rev. Mod. Phys.*, **30**, 585 (1958).
381. Suk, V., and M. Malat, *Chemist-Analyst*, **45**, 30 (1956).
382. Sullivan, M. V., and H. Friedman, *Ind. Eng. Chem., Anal. Ed.*, **18**, 304 (1946).
383. Suzuki, S., *J. Chem. Soc. Japan*, **73**, 92 (1952).
384. Tanaka, N., M. Kamada, and G. Sato, *Bull. Chem. Soc. Japan*, **34**, 541 (1961).
385. Tanaka, N., and T. Koizumi, *Bull. Chem. Soc. Japan*, **30**, 303 (1957).
386. Tanaka, N., T. Koizumi, T. Murayama, M. Kodama, and Y. Sakuma, *Anal. Chim. Acta*, **18**, 97 (1958).
387. Tanaka, N., M. Kodama, M. Sasaki, and M. Sugino, *Japan Analyst*, **6**, 86 (1957).
388. Tananaeff, I. V., and I. B. Mizetzkaja, *Zavodsk. Lab.*, **12**, 529 (1946).
389. Tananaev, I. V., M. A. Glushkova and G. B. Seifer, *Zh. Neorgan. Khim.*, **1**, 66 (1956).
390. Tananaev, I. V., and I. B. Mizetskaya, *Zh. Analit. Khim.*, **1**, 6 (1946).
391. Taylor, J. K., and S. W. Smith, *J. Res. Natl. Bur. Std.*, **56**, 301 (1956).
392. Tertoolen, J. F. W., D. A. Detmar, and C. Buijze, *Z. Anal. Chem.*, **167**, 401 (1959).
393. Thompsett, S. L., *Analyst*, **81**, 330 (1956).
394. Tiller, W. A., and J. W. Rutter, *Can. J. Phys.*, **34**, 96 (1949).
394a. Tilton, G. R., *Trans. Am. Geophys. Union*, **37**, 224 (1956).
394b. Tilton, G. R., and G. L. Davis, in *Researches in Geochemistry*, P. H. Abelson, Ed., Wiley, New York, 1959, p. 190.
394c. Tilton, G. R., C. Patterson, H. Brown, M. Inghram, R. Hayden, D. Hess, and E. Larsen, Jr., *Bull. Geol. Soc. Am.*, **66**, 1131 (1955).
394d. Tomkins, F. S., and M. Fred, *Spectrochim. Acta*, **6**, 139 (1954).
395. Topf, G., *Z. Anal. Chem.*, **26**, 296 (1887).
396. Toropova, F. V., *Zh. Analit. Khim.*, **4**, 337 (1949).
397. Torossian, G., *Ind. Eng. Chem.*, **8**, 331 (1916).
398. Treadwell, F. P., and W. T. Hall, *Analytical Chemistry*, 9th ed., Wiley, New York, 1942, Volume II, p. 60.
399. Treadwell, F. P., and W. T. Hall, *Analytical Chemistry*, 7th ed., Wiley, New York, 1930, Volume II, p. 563.
400. Underwood, A. L., *J. Chem. Ed.*, **31**, 394 (1954).
401. Usatenko, Yu. I., and M. A. Vitkina, *Zavodsk. Lab.*, **23**, 427 (1957).
402. Vandenbosch, V., *Anal. Chim. Acta*, **2**, 566 (1948).
403. Vandenbosch, V., and R. Breckpot, *Chim. Ind. (Paris)*, **63**, 67 (1950).
404. Van Dyke-Cruser, F., and E. H. Miller, *J. Am. Chem. Soc.*, **26**, 676 (1904).
405. Verhun, J., *Chim. Anal.*, **41**, 332 (1959).
406. Viebock, F., and C. Brecher, *Arch. Pharm.*, **270**, 109 (1932).
407. Volkova, A. I., and N. N. Zakharova, *Ukr. Khim. Zh.*, **23**, 530 (1957).
408. Von Ende, C. L., *Z. Anorg. Chem.*, **26**, 129 (1901).
409. Vrestal, J., and J. Havir, *Collection Czech. Chem. Commun.*, **22**, 316 (1957).
410. Warr, J. J., and F. Cuttitta, *U. S. Geol. Surv. Profess. Papers*, **400-B**, 218, 483 (1960).
411. Wehber, P., *Z. Anal. Chem.*, **153**, 253 (1956).
412. Wehber, P., *Z. Anal. Chem.*, **158**, 10 (1957).
413. Weiser, H. B., *J. Phys. Chem.*, **20**, 659 (1916).
414. Wellings, A. W., *Analyst*, **58**, 332 (1933).
415. West, P. W., and J. K. Carlton, *Anal. Chim. Acta*, **6**, 406 (1952).

416. Whilhite, R. N., and A. L. Underwood, *Anal. Chem.*, **27,** 1334 (1955).
417. Wichman, H. J., and P. A. Clifford, *J. Assoc. Offic. Agr. Chemists*, **18,** 315 (1935).
418. Willard, H. H., and E. W. Goodspeed, *Ind. Eng. Chem., Anal. Ed.*, **8,** 414 (1936).
419. Willard, H. H., and J. J. Thompson, *Ind. Eng. Chem., Anal. Ed.*, **6,** 425 (1934).
420. Wilska, S., *Acta Chem. Scand.*, **5,** 1368 (1951).
421. Yatsimirskii, K. B., *Zh. Fiz. Khim.*, **25,** 475 (1951).
422. Young, R. S., and A. Leibowitz, *Analyst*, **71,** 477 (1946).
423. Zan'ko, A. M., and A. Y. Bursuk, *J. Appl. Chem. U. S. S. R.*, **9,** 2297 (1936).
424. Ziegler, M., *Z. Anal. Chem.*, **174,** 323 (1960).
425. Ziegler, M., *Z. Anal. Chem.*, **180,** 351 (1961).
426. Zombory, L., *Magy. Kem. Folyoirat*, **44,** 160 (1938); *Chem. Abstr.*, **33,** 4157 (1939).

Part II
Section A

NIOBIUM AND TANTALUM

By Silve Kallmann, *Ledoux & Company, Inc., Teaneck, New Jersey*

Contents

I. Introduction	183
A. Occurrence	184
1. In Igneous Rocks	184
2. In Niobium and Tantalum Minerals	185
3. In By-Products	187
B. Industrial Uses of Niobium and Tantalum	188
1. Niobium	188
a. In the Nuclear Industry	188
b. In Steels	188
c. In High-Temperature Alloys	189
d. Miscellaneous Uses	189
2. Tantalum	190
a. As Pure Metal	190
b. In Miscellaneous Applications	190
C. Industrial Processes for the Preparation of Niobium and Tantalum Metals and Their Most Important Compounds	191
1. Extractive Metallurgy	191
2. Preparation of the Pure Metals	193
3. Production of Ferroniobium and Ferrotantalum-Niobium	194
4. Production of Miscellaneous Products	194
D. Toxicology and Industrial Hygiene	194
1. Niobium	194
2. Tantalum	195
II. Properties of Niobium and Tantalum	195
A. Physical and Mechanical Properties	195
1. Niobium	197
2. Tantalum	200
B. Isotopes of Niobium and Tantalum	200
C. Electrochemical Properties	201
1. Tantalum	201
2. Niobium	201
D. Optical Properties	202
1. Emission Spectra	202
2. X-Ray Spectra	202
E. Chemical Properties	203
1. Introduction	203
2. Oxides	204

Contents (*continued*)

		3. Earth Acids	205
		4. Niobates and Tantalates	206
		5. Peroxy Salts	207
		6. Halogenoniobates and Tantalates	208
		a. Chlorides	208
		b. Bromides and Iodides	210
		c. Fluorides	210
		7. Refractory Binary Compounds	211
		a. Hydrides	211
		b. Nitrides	212
		c. Carbides	213
		d. Borides	213
		e. Silicides	213
III.	Sampling of Niobium and Tantalum		213
	A. Ores and Concentrates		213
	B. Intermediate Products		214
	C. Niobium and Tantalum Metals		215
		1. Powders	215
		2. Consolidated Metal	215
IV.	Separation and Isolation of Niobium and Tantalum from Other Elements		215
	A. Decomposition and Other Preliminary Treatment of High-Grade Ores and Concentrates		216
		1. General Observations	216
		2. The Pyrosulfate Attack	217
		a. The Fusion	217
		b. Treatment of the Pyrosulfate Melt	218
		3. Attack by Fusion with Potassium or Sodium Hydroxide and Sodium Peroxide	219
		4. Attack by Borax	221
		5. Attack by Hydrofluoric Acid	222
		6. Attack by Chlorination	223
	B. Attack of Low-Grade Minerals		224
	C. Decomposition of Intermediate Products		225
	D. Treatment of Niobium and Tantalum Oxides		226
		1. Pyrosulfate Attack	226
		2. Alkaline Attack	227
		3. Chlorination of Oxides	228
		4. Attack by Hydrofluoric Acid	228
	E. Niobium and Tantalum Metals		228
	F. Alloys		229
		1. Ferro Alloys	229
		2. Steels and High-Temperature Alloys	229
		a. Isolation of Niobium and/or Tantalum Oxide	229
		b. Subsequent Treatment of Oxides	231
		3. Titanium-Base Alloys	232
		4. Uranium- and Plutonium-Base Alloys	232
		5. Miscellaneous Alloys	233
	G. Miscellaneous Precipitation Procedures		234
		1. Hydrolysis	234
		2. Precipitation with Cupferron	235
		3. Precipitation with Tannin	236
		4. Phenylarsonic Acid Procedure	239

Contents (*continued*)

```
       5. Other Precipitation Procedures..................  239
    H. Separations Based on Ion Exchange................  241
       1. General Observations............................  241
       2. The Hydrochloric Acid System...................  242
       3. The Hydrofluoric Acid Medium...................  244
       4. The Mixed Hydrofluoric–Hydrochloric Acid Me-
            dium........................................  245
          a. Introductory Observations...................  245
          b. Chemistry of the Exchange Mechanism Involv-
               ing the HF–HCl Medium...................  246
          c. Separation Principles........................  250
          d. Practical Applications.......................  252
       5. The Hydrochloric–Oxalic Acid System...........  255
    I. Separations Based on Cellulose and Paper Chromatog-
         raphy..........................................  258
       1. Principles Involved............................  258
       2. Preparation of the Sample.....................  259
       3. Typical Procedure.............................  259
       4. Shortened Procedure..........................  260
       5. Paper Chromatographic Methods...............  261
       6. Separation by Electrophoresis..................  263
    J. Separations Based on Solvent Extraction...........  263
       1. General Observations..........................  263
       2. The Hydrochloric Acid System.................  264
       3. The Fluoride Medium..........................  265
          a. Extractions Involving Hexone...............  265
          b. Extractions Involving Other Solvents........  268
       4. Other Miscellaneous Extraction Procedures.....  269
 V. Separation of Niobium and Tantalum.................  269
    A. Precipitation Methods..............................  270
       1. The Tannin Procedure..........................  270
          a. General Considerations......................  270
          b. The Schoeller Procedure....................  270
          c. Modification of Schoeller's Procedure........  272
       2. The N-Benzoyl-N-Phenylhydroxylamine Procedure  272
       3. The Phenylarsonic Acid Procedure..............  274
       4. Separation of Niobium and Tantalum with Selenous
            Acid........................................  274
       5. Other Gravimetric Separation Procedures.......  274
    B. Separation of Niobium and Tantalum by Ion Exchange  276
    C. Separation of Niobium and Tantalum by Cellulose
         Chromatography................................  276
    D. Separation of Niobium and Tantalum by Solvent Ex-
         traction......................................  277
VI. Detection and Identification..........................  277
    A. Spectroscopic Examination........................  277
       1. Niobium.....................................  277
          a. Interference with Arc and Spark Lines.......  278
       2. Tantalum....................................  278
          a. Interference with Arc and Spark Lines.......  279
       3. Limit of Detection............................  280
    B. Detection by X-Ray Emission.....................  280
    C. Detection by Chemical Procedures.................  280
```

Contents (*continued*)

　　　　　1. Joint Detection in Minerals.................... 281
　　　　　　　a. Tartaric Hydrolysis........................ 281
　　　　　　　b. Tannin Procedure.......................... 281
　　　　　2. Detection of Niobium and Tantalum in Minerals
　　　　　　　by a Paper Chromatographic Method.......... 282
　　　　　3. Detection of Niobium with Methylthymol Blue... 282
　　　　　4. Other Detection Methods...................... 283
　　VII. Determination of Niobium and Tantalum.............. 283
　　　　A. Gravimetric Methods.............................. 284
　　　　　1. Cupferron Precipitation....................... 284
　　　　　2. Tannic Acid Precipitation..................... 284
　　　　　3. Precipitation by Hydrolysis................... 285
　　　　　4. Precipitation with Phenylarsonic Acid.......... 285
　　　　　5. Other Gravimetric Methods.................... 285
　　　　B. Photometric Methods............................. 286
　　　　　1. Thiocyanate Procedure for Niobium............. 286
　　　　　　　a. Extraction Method......................... 286
　　　　　　　b. Homogeneous System....................... 291
　　　　　2. Hydrogen Peroxide Method for Niobium......... 297
　　　　　　　a. Procedure................................ 298
　　　　　　　b. Discussion............................... 298
　　　　　　　c. Interference by Other Elements............. 299
　　　　　　　d. Application of Procedure................... 300
　　　　　3. Hydroquinone Method for Niobium.............. 301
　　　　　　　a. Procedure................................ 301
　　　　　　　b. Discussion............................... 302
　　　　　　　c. Interferences............................. 303
　　　　　　　d. Application of Procedure................... 304
　　　　　4. 8-Quinolinol Method for Niobium............... 305
　　　　　5. Molybdenum Blue Method for Niobium.......... 306
　　　　　6. Pyrogallol Method for Niobium................. 307
　　　　　　　a. Procedure................................ 308
　　　　　　　b. Discussion............................... 309
　　　　　　　c. Interferences............................. 309
　　　　　7. Other Photometric Methods for Niobium......... 309
　　　　　8. Photometric Determination of Tantalum by Pyrogallol.. 310
　　　　　　　a. Procedure................................ 311
　　　　　　　b. Discussion of Various Conditions............. 311
　　　　　　　c. Interfering Elements....................... 313
　　　　　9. Other Photometric Tantalum Methods........... 315
　　　　C. Titrimetric Determination of Niobium.............. 317
　　　　　1. General Observations......................... 317
　　　　　2. Procedure Based on Reduction in Sulfuric Acid
　　　　　　　Medium..................................... 317
　　　　　　　a. Preparation of Solution..................... 317
　　　　　　　b. Reduction and Titration.................... 318
　　　　　3. Procedure Based on Reduction in Hydrofluoric Acid
　　　　　　　Medium..................................... 318
　　　　D. Polarographic Determination of Niobium............ 320
　　　　　1. Nitrate System 320
　　　　　2. Sulfuric Acid System.......................... 320
　　　　　3. Hydrochloric Acid System..................... 321

Contents (*continued*)

4. Citrate System	322
5. EDTA System	323
6. Other Systems	324
E. Spectrographic Determination of Niobium and Tantalum	324
1. High-Grade Ores	324
a. No Preliminary Separations	324
b. After Preliminary Separations	325
2. Low-Grade Ores and Residues	327
a. No Preliminary Separations	327
b. After Preliminary Separations	328
3. Tantalum in Niobium and Niobium in Tantalum Metals or Oxides	329
4. Alloys	329
a. Semiquantitative Analysis	329
b. No Preliminary Separations	329
c. After Preliminary Separations	330
F. Determination of Niobium and Tantalum by X-Ray Fluorescence	331
1. Instrumental Considerations	331
2. Effect of Other Elements	331
3. Effect of Physical Characteristics of the Sample	332
4. Elimination of Interferences	332
a. Use of Correction Factors	333
b. Preliminary Fusion of Sample	333
c. Other Miscellaneous Applications	334
G. Electron Probe Measurements of Niobium Compounds	335
H. Radiochemical Methods	336
1. Use of Radioactive Tracers	336
2. Isotope Dilution Procedures	336
3. Neutron-Activation Analysis of Niobium and Tantalum	336
a. Determination of Niobium	337
b. Determination of Tantalum	338
4. Radiochemical Separations Involving Niobium-95	343
a. Niobium Activity in Aqueous or Organic Solutions	343
b. In Fission Products of Plutonium Fuels	344
c. In Uranyl Nitrate Solution from Irradiated Uranium	345
d. In Uranium Target	346
e. After Cupferron Extraction	346
f. After Ion-Exchange Separations	347
5. Other Radiochemical Procedures	347
VIII. Analysis of Niobium and Tantalum Metals, Alloys, and Compounds	349
A. Determination of Interstitial Elements in Niobium and Tantalum	349
1. Oxygen	349
a. By Neutron Activation	349
b. By Halogenation	350
c. By Vacuum Fusion	351
d. By Inert-Gas Fusion	351

Contents (*continued*)

- e. By Diffusion Extraction ... 352
- f. By Emission Spectrometry ... 352
- g. By Mass Spectrometry ... 353
- h. By Isotopic Dilution ... 353
- 2. Hydrogen ... 353
  - a. Vacuum-Fusion Methods ... 353
  - b. Diffusion Method ... 354
  - c. Spectrographic Methods ... 354
- 3. Nitrogen ... 355
  - a. Vacuum-Fusion Method ... 355
  - b. Inert Gas-Fusion Procedure ... 355
  - c. Kjeldahl Procedure ... 356
  - d. Alkali-Fusion Method ... 356
- 4. Carbon ... 357
- B. Determination of Metallic Impurities in Niobium and Tantalum ... 357
  - 1. Spectrographic Procedures ... 357
    - a. General Considerations ... 357
    - b. Outline of Several Techniques ... 358
    - c. Other Procedures ... 360
  - 2. X-Ray Fluorescence Procedures ... 361
  - 3. Radioactivation Analysis for Trace Elements in Niobium ... 362
    - a. Nondestructive Technique ... 363
    - b. After Radiochemical Separations ... 365
    - c. By Differential-Count Method ... 365
  - 4. Determination of Impurities in Niobium and Tantalum by Chemical Methods ... 367
    - a. Determination of Tantalum in Niobium ... 367
    - b. Determination of Niobium in Tantalum ... 368
    - c. Determination of Iron ... 368
    - d. Determination of Tungsten and Molybdenum ... 369
    - e. Determination of Silicon and Phosphorus ... 371
    - f. Determination of Zirconium ... 372
    - g. Determination of Titanium ... 372
    - h. Determination of Cobalt ... 373
    - i. Determination of Nickel ... 373
    - j. Determination of Copper ... 373
    - k. Determination of Boron ... 373
    - l. Simultaneous Determination of Lead, Copper, Zinc, and Cadmium ... 374
    - m. Determination of Tin and Antimony ... 374
    - n. Determination of Lead and Cadmium ... 375
    - o. Determination of Sodium and Potassium ... 375
    - p. Determination of Arsenic ... 375
- IX. Selected Laboratory Procedures ... 376
  - A. Determination of Niobium and Tantalum in Ores and Concentrates Involving Ion Exchange ... 376
    - 1. Apparatus and Reagents ... 376
    - 2. Procedure ... 377
      - a. Decomposition of Sample ... 377
      - b. Elution of Impurities ... 378
      - c. Elution of Niobium ... 379

Contents (*continued*)

d. Elution of Tantalum	379
e. Determination of Niobium	379
f. Determination of Tantalum	380
g. Regeneration of Resin	380
B. Determination of Niobium and Tantalum in Steels and High-Temperature Alloys	380
C. Determination of Niobium and Tantalum in Niobium-, Tantalum-, Titanium-, Tungsten-, or Molybdenum-Base Alloys	381
D. Spectrographic Determination of Impurities in Tantalum and Niobium	381
1. Tantalum	382
a. Preparation of High-Purity Tantalum Oxide	382
b. Preparation of Synthetic Standards	383
c. Preparation of Carrier—Internal Standard	383
d. Preparation of Samples	384
e. Spectrographic Conditions	386
2. Niobium	387
a. Preparation of Synthetic Standards	387
b. Preparation of Carrier and of Samples	389
c. Spectrographic Conditions	389
References	389

I. INTRODUCTION

As was the case with other pairs of chemically similar elements (Se–Te, Zr–Hf, Zn–Cd), the early history of tantalum and niobium is really that of the niobium–tantalum earth acid group. Since niobium-bearing minerals without tantalum are virtually unknown, the first investigators were actually dealing with mixtures of these two chemically similar elements (295). Titanium and other elements present in the samples added to the confusion to the extent that, before the final identification of tantalum and niobium was achieved, seven names had been ascribed to new elements that were thought to be present in one mineral.

Thus the history of tantalum and niobium dates back to 1801, when Hatchett, a British chemist, after examining a heavy black mineral received a number of years previously from Connecticut, described a new acidic oxide of an element for which he proposed the name *columbium*, in honor of the country from which the ore came (240).

A year later, Ekeberg in Sweden announced the discovery of an element in a mineral he called tantalite, whose oxide also showed acidic properties (149). In reference to the fable of Tantalus, he named the element *tantalum*. In 1844, Rose analyzed certain columbites and reported the presence of two distinct acids, one Ekeberg's tantalum, the other a new element, *niobium*, named after Niobe, the daughter of Tantalus. He later also "discovered" pelopium (from Pelops, brother of Niobe) (42). Another investigator (246) added ilminium (from the Ilmen Mountains) and neptunium (from Neptune). Another "discoverer" completed the list with dianium (from Diane) (552). Marignac (358) finally established the identity of Hatchett's columbium with Rose's niobium.

The touch of romanticism added by Rose to the lighter of the two elements (niobium) has been recognized abroad more than in the United States, where the name columbium has remained popular. However, now that the element has been given two distinguished names, leading to considerable confusion and pages of discussions in scientific literature, the question being continually raised is, What shall we call it? In 1951 the American Chemical Society agreed at the International IUPAC Conference to call it niobium. Many metallurgists, engineers, businessmen, physicists, and probably a few chemists in the United States were reluctant to change from the more familiar "columbium" (213).

In his suggestions for preparing this chapter, Dr. Elving referred to "niobium, née columbium." At the time that this author made up his mind to switch from "columbium" to "niobium" the American Engineering Societies voted to retain the name "columbium." However, to avoid further confusion it was decided, for better or worse, to stick with niobium.

A. OCCURRENCE

1. In Igneous Rocks

The average content of niobium and tantalum in various groups of igneous rocks is presented in Table I, which is based on values given by Rankama (448). A more recent investigation reports the tantalum content of granite at between 1.8×10^{-4} to $5 \times 10^{-4}\%$ and the niobium content at $2.8 \times 10^{-3}\%$ (464).

The data in Table I confirm the fact, well known to mineralogists, that niobium and tantalum are concentrated in late crystallates during the

TABLE I
Average Nb and Ta Content and Nb:Ta Ratio in Igneous Rocks

Rock	Nb, g./ton	Ta, g./ton	Nb:Ta
Monomineralic rocks	0.3	0.7	0.4
Ultrabasic rock	16	1.0	16.0
Diorites	3.6	0.7	5.1
Granites	20	4.2	4.8
Syenites	30	2.0	15.0
Nepheline syenites	310	0.8	387.5
Basic alkali rock	10	1.2	8.3

magmatic differentiation. This concentration is very pronounced in granites, and tantalum shows a maximum in these rocks. Niobium, although concentrated in granite, differs clearly from tantalum, and its highest contents are reached in syenites and nepheline syenites (448).

Niobium is approximately as abundant as nickel and one-third as abundant as copper. The relative abundance of niobium and tantalum in relation to other reactive and refractory metals is given in Table II, using zirconium as a basis. Later estimates, which were made after the new discoveries of pyrochlore, place the niobium:tantalum ratio at about 20:1.

TABLE II
Relative Abundance of Niobium, Tantalum, and Some Reactive and Refractory Metals, According to Several Investigators (508)[a]

Element	Information source			
	Mason	Urey	Goldschmidt	Material Advisory Board
Zr	100	100	100	100
Nb	11	0.5	5	30
Ta	1	0.2	0.3	3
W	31	9	10	—
Mo	3	4	7	3

[a] C. T. Sims, in granting permission to reproduce this table from his published work, voiced his objection to calling element 41 by its foreign "pseudonym."

2. In Niobium and Tantalum Minerals

The presently most important commercial niobium and tantalum minerals (see Table III) are the orthorhombic columbite and tantalite, which are respectively a niobate and tantalate of iron and manganese, $(Fe,Mn)(Nb,Ta)_2O_6$, with complete gradation in the components. Tantalite and columbite are found in granite pegmatites and in residual or alluvial deposits derived from such rock. They are frequently associated with albite, microline, beryl, lepidolite, muscovite, tourmaline, spodumene, lithiophilite, triphylite, amblygonite, samarskite, apatite, and microlite, and very frequently with cassiterite and wolframite. Marked variation in the ratio of Nb to Ta and Fe to Mn is found in material from a single locality and even in a single specimen or crystal (125). A very desirable columbite with a 10:1 or higher Nb_2O_5 to Ta_2O_5 ratio is mined in Nigeria, whereas Brazil and Australia produce tantalite with a 10:1 or higher ratio of Ta_2O_5 to Nb_2O_5. High-grade tantalite usually occurs in the form of stout, more

TABLE III
Niobium- and Tantalum-Bearing Minerals

Name	Composition	% Nb_2O_5	% Ta_2O_5	Specific gravity	Crystal structure	Hardness, Mohs	Color
Columbite (theory)	$FeNb_2O_6$	78.72	—	5.2	Orthorhombic, dipyramidal	6	Brown-black
Tantalite (theory)	$FeTa_2O_6$	—	86.01	7.95	Ortho., dipyr.	6.5	Brown-black
Columbite	$(Fe,Mn)(Nb,Ta)_2O_6$	26–78	1–48	5.1–6.8	Ortho., dipyr.	6	Brown-black
Tantalite	$(Fe,Mn)(Ta,Nb)_2O_6$	2–40	42–84	6–7.8	Ortho., dipyr.	6–6.5	Brown-black
Manganocolumbite	$(Mn,Fe)(Nb,Ta)_2O_6$	30–75	5–40	5.2–6.4	Ortho., dipyr.	6	Brown-black
Manganotantalite	$(Mn,Fe)(Ta,Nb)_2O_6$	4–35	35–82	6–7.8	Ortho., dipyr.	6–6.5	Brown-black
Betafite	$(U,Ca)(Nb,Ta,Ti)_3O_9 \cdot nH_2O$	9–45	0–29	3.7–5	Octahedral	4–5.5	Brown-black
Stibiocolumbite	$SbNbO_4$	47.69	—	5.68	Orthorhombic	5.5	Light to dark brown
Stibiotantalite	$SbTaO_4$	—	60.24	7.53	Pyramidal		
Bismutotantalite	$Bi(TaNb)O_4$	0–14	31–49	8.26	Orthorhombic	5	Black
Simpsonite	$Al_2Ta_2O_8$	—	81.25	5.99	Hexagonal		Colorless
Pyrochlore	$NaCaNb_2O_6F$	73.05	—	4.45	Octahedral	5–5.5	Brown-black
Microlite	$(Na,Ca)_2Ta_2O_6(O,OH,F)$	—	82.14	6.33	Octahedral	5–5.5	Pale yellow
Fergusonite	$(Y,Er,Ce,Fe)(Nb,Ta,Ti)O_4$	2–54	4–55	5.4–7.0	Tetragonal	5.5–6.5	Light brown to dark brown
Yttrotantalite	$(Fe,Y,U,Ca,Mn,Ce,Th)(Nb,Ta,Zr,Sm)O_4$	1–20	37–56	5.7	Prismatic	5–5.5	Black-brown
Samarskite	$(Y,Er,Ce,U,Ca,Fe,Pb,Th)(Nb,Ta,Ti,Sm)_2O_6$	27–47	2–27	5.6–5.8	Prismatic	5–6	Velvet black
Eschynite	$(Ce,Ca,Fe,Th)(Ti,Nb)_2O_6$	23–37	0–7	4.9–5.1	Ortho., dipyr.	5.6	Brown-black
Euxenite	$(Y,Ca,Ce,U,Th)(Nb,Ta,Ti)_2O_6$	15–41	1–6	4.7–5	Ortho., dipyr.	5.5–6.5	Black, vitreous
Polycrase	$(Y,Ca,Ce,U,Th)(Ti,Nb,Ta)_2O_6$	4–20	0–14	4.7–5.9	Ortho., dipyr.	5.5–6.5	Black, vitreous
Tapiolite	Fe, Ta_2O_6	0–7	75–86	7.9–8.1	Tetragonal	6–6.5	Black

nearly equidimensional crystals than those of columbite. It has a compact texture with conchoidal fracture and high luster, is purple-black in color, and has a higher specific gravity than columbite. Crystals of columbite generally have a thinner, bladed, or platy form, and often occur in radiated groups. The mineral has a steel-grey color, lower luster, granular texture, and uneven fracture (277).

Since the reserves of columbite are uncomfortably small and insufficient to support large-scale production of the metal, if this should become desirable, much attention is being paid to pyrochlore, which is widely disseminated throughout the world. This mineral is a complex columbite of cerium, calcium, sodium with some titanium, thorium, and fluorine; its approximate formula is $A_2B_2O_6(O,OH,F)$, with A = Na, Ca, K, Mg, Fe^{+2}, Mn^{+2}, Sb^{+3}, Ce, La, Dy, Er, Y, Th, Zr, or U, and B = Nb, Ti, or Fe^{+3}, in the approximate order of importance. When tantalum fully replaces the niobium, the mineral is microlite. Pyrochlore usually occurs in pegmatite derived from alkalic rock, typically for instance in metamorphosed limestone in the Fen region in Norway, Kaiserstuhl in Germany, and other large deposits in Canada and Brazil. The Brazilian deposits of niobium in the form of pyrochlore are the world's greatest known reserves (508) and are reported to have an unusually high (3.5 to 4%) niobium content (109). The composition of the Canadian Oka pyrochlore mineral has been described as $(Na,Fe,Ca)_{0.8}Ca(NbTi)_2O_6(F,OH)_{1.2}$ (271a). The Brazilian pyrochlore is of an unusual composition in as much as it does not contain any calcium, sodium, and fluorine, but approximately 15% of barium oxide and 2% of thorium oxide (282a). Extraction of niobium from pyrochlore deposits still presents metallurgical difficulties owing to the extreme fineness of the grains and a specific gravity not very different from those of some of the associated minerals. (A close relationship between pyrochlore and betafite has recently been established by Hogorth, who considers betafite a hydrous uranian pyrochlore (253).)

During the 1950's more than 99% of the domestic niobium–tantalum production came from the dredging operations at Bear Valley, Idaho, where euxenite associated with columbite and monazite is found in a placer deposit (386) that contains from 23 to 29% Nb_2O_5, 1.5% Ta_2O_5, 22 to 26% TiO_2, 24 to 28% $(Y,Er)_2O_3$, 8 to 10.5% U_3O_8, and 2.5 to 5% ThO_2.

3. In By-Products

Large quantities of tantalum- and niobium-bearing concentrates have been produced as a by-product of tin mining operations in the Belgian

Congo, Nigeria, and Malaya. Slags from tin mining operations often contain from 4 to 12% tantalum and/or niobium oxide.

B. INDUSTRIAL USES OF NIOBIUM AND TANTALUM
(42,102,277,577)

1. Niobium

a. In the Nuclear Industry

With a few exceptions, large-scale uses of niobium as a pure metal have not yet been developed. By far the biggest user of niobium (taking 70% or more of the total in 1960) was Pratt & Whitney Company, for its CANEL (Connecticut Aircraft Nuclear Engine Laboratory) project, the indirect-cycle nuclear aircraft engine program recently terminated by the U. S. Atomic Energy Commission. The AEC, however, at present is still keenly interested in pure niobium and its alloys. In thermal reactors, because of its strength at high temperatures and moderately low nuclear cross section, niobium is being investigated as a constituent of some fuel alloys. For example, the fuel plates in the Experimental Boiling Water Reactor (EBWR) are an alloy consisting of 1.5% Nb, 5% Zr, and the balance of U. It is reported that the addition of 10% of niobium to uranium stabilizes the gamma, body-centered-cubic structure of uranium at room temperature, leading to better corrosion resistance and improved metallurgical properties (159). A group at Battelle Memorial Institute has studied a niobium alloy containing 20% uranium as a fuel. The alloy exhibits excellent strength at temperatures of approximately 1600°F., whereas currently used fuel elements fail at 1200°F. (109). Niobium also has excellent resistance to attack by coolants such as water, liquid metals, and molten salts, which are considered for reactors, solar heating, and thermoelectric devices. The AEC's principal interest in niobium, however, remains as a fuel element cladding and possibly as a fuel alloy in the cores of fast reactors, since with fast neutrons the absorption cross section of metals decreases such that the slight disadvantage of niobium with respect to the usual reactor materials is minimized. Niobium has been chosen by the United Kingdom Energy Authority as a cladding material for uranium in the NaK-cooled Donnray fast breeder reactor (507).

b. In Steels

Niobium in the form of ferroniobium (50 to 65% Nb) or ferroniobium–tantalum (40% Nb, 20% Ta) is added to austenitic chromium nickel (18–8) type stainless steels as a stabilizer to prevent carbide precipitation in the

800 to 1600°F. temperature range. This accounts for about 70% of the total niobium ore consumption in the United States. Niobium inhibits intergranular corrosion and improves strength at high temperature. In order to prevent precipitation of carbide, an addition of niobium at least ten times the amount of carbon is made in Type 347 and 348 stainless steels.

In 4 to 6% chrome steels, the presence of niobium reduces air hardening, gives high impact strength, reduces creep, shortens the annealing time, and improves oxidation resistance (224). Niobium increases considerably the rate of growth of the nitride layer when steel is surface nitrided in ammonia, thus minimizing oxidation at higher temperature (134).

c. In High-Temperature Alloys

Approximately 30% of the total niobium production is consumed in the manufacture of high-temperature alloys. In these "superalloys," niobium and/or tantalum are among the most effective carbide formers by increasing high-temperature strength and enhancing resistance to creep and deformation. There are some 30 or more superalloys in existence, containing niobium in amounts ranging from 0.44 to 4%. Examples include the well-known S-816 cobalt-base alloy, the S-588, S-590, and gamma-Cb iron-base alloys, all containing 4% Nb, the N-155 iron-base alloy containing 1%, and Inconel X containing about 1.00% of niobium. The principal application at present for the super alloys is in components for aircraft jet engines and turbine wheels. The advent of guided missiles and rockets probably will lead to an ever-increasing niobium consumption.

d. Miscellaneous Uses

Niobium-bearing electrodes are used for welding niobium-stabilized stainless steel. Niobium in the form of carbide is being investigated as an addition to titanium carbide for possible application in gas turbines and blades for jet engines. Tungsten alloys containing 2.5 to 10% Nb oxidize more slowly than unalloyed tungsten at 2000°F. (498). Other minor uses of niobium are as construction material and as a "gas-getter" in electronic tubes. The superconductivity of niobium also affords some intriguing possibilities. A niobium–tin alloy has been described that can be used to make permanent magnets due to its ability to carry 10 times as much current as previously used material (456). The niobium–tin alloy (Nb_3Sn) possesses the highest known transition temperature (18°K.) of any known superconductor (267). Empirical rules suggest materials even better than Nb_3Sn, such as Nb_3In, Nb_3Zr, or Nb_3Sb, but it has not yet been possible to produce these compounds (523).

2. Tantalum

a. As Pure Metal

The chemical inertness of tantalum is the most attractive feature of the metal. Because it is not attacked by most acids, equipment can be made of thin metal—0.013 to 0.020 inch thick. Tantalum is particularly useful for handling chlorine and hydrochloric acid and is therefore widely used for heat exchangers, bayonet heaters, coils, condensers, and coolers. Tantalum heaters are used in sulfuric acid concentrators and recovery plants, as well as in metal pickling tanks and chromium-plating baths. Equipment needed for the manufacture of pharmaceutical products of high purity is frequently made of tantalum metal. Tantalum crucibles have found application in the preparation of rare earth metals at the Ames, Iowa, Laboratory of the AEC, and, more recently, in the preparation of scandium (108).

Surgeons have made use of tantalum metal for human implants because it is immune to corrosion by body acids and is nonirritating to living tissues. Injured or severed nerves are sutured with fine tantalum wire, tantalum plates serve to repair skull injuries, and tantalum wire and plates are used in the repair of broken bones and in plastic surgery. Because of its high melting point, ease of fabrication, remarkable strength at high temperatures, and its low vapor pressure and gettering properties, tantalum is used to a great extent in electronic tubes.

Because tantalum forms an anodic oxide film of unusual stability, this property has been exploited in electrolytic rectifiers which operate railway signals and highway-crossing protective devices. The second application of the anodic film is in the construction of capacitors (528). Foil-type capacitors are composed of two 0.005-inch tantalum ribbons separated by a porous ribbon of insulating material saturated with glycol-base electrolyte. These ribbons are rolled into a cylinder and sealed in a metal container. In the sintered porous capacitor the anode is made by pressing tantalum powder around a tantalum wire core and sintering this compact at a temperature that will weld the particles of powder together, but still allows the anode to remain porous. The advantages of tantalum capacitors are greater life, better idle life, wider temperature range, and higher capacitance-to-size ratio.

b. In Miscellaneous Applications

Ferrotantalum is not as efficient as ferrocolumbium in inhibiting intergranular deterioration in austentic stainless steel, since tantalum has twice the atomic weight of niobium and, hence, more of the ferrotantalum is required to achieve the same effect. Tantalum compounds are used as catalysts, especially in the synthesis of butadiene from ethyl alcohol, and tantalum oxide is the principal component of optical glass of a high refractive index. Tantalum carbide, when added to tungsten carbide, produces a good wear-resistant and shock-resistant tool. A tantalum–tungsten alloy containing 7.5% of tungsten maintains its elasticity when heated to a high temperature, and is used extensively as supporting wire springs in electronic tubes. It is also being investigated for exhaust nozzles and engine parts for solid propellant rockets (107). Other tantalum–tungsten alloys have also been extensively investigated (73).

C. INDUSTRIAL PROCESSES FOR THE PREPARATION OF NIOBIUM AND TANTALUM METALS AND THEIR MOST IMPORTANT COMPOUNDS (5)

1. Extractive Metallurgy

Since many niobium and tantalum products submitted for analysis are the result of industrial processes, a short outline of the metallurgical background appears appropriate.

The extractive processes for niobium and tantalum usually involve two basic steps: (*1*) the separation of both niobium and tantalum from other constituents of the concentrate, and (*2*) the separation of niobium and tantalum from each other. One method commonly used for the attack of niobium and tantalum ores is based on fusion with sodium hydroxide. The pulverized ore is treated with molten caustic at red heat in an iron pot. The resulting product is leached with water, is decanted, and is filtered for the removal of free alkali, silica, and other soluble salts, such as sodium stannate and sodium tungstate. The insoluble sodium tantalate and niobate are digested with hydrochloric acid, which leaves a slurry of the white acids of tantalum and niobium, contaminated with small amounts of silicon, iron, titanium, tin, zirconium, and manganese.

A second method used for the attack of ores and concentrates is the direct treatment with 70% hydrofluoric acid. This procedure is looked upon more favorably today than in the past, because the niobium and tantalum complex fluoride solutions can then be subjected to the solvent-extraction processes that are used by a large number of companies. However, the fluoride solution also can be treated with a potassium salt in order to separate the earth acids as the potassium double fluorides by fractional crystallization. The separation procedure based on fractional crystallization, known as the Marignac process, has been in use since 1866 (358). It is based on the difference in solubility of the double potassium salts, the potassium fluotantalate, K_2TaF_7, being soluble in 150 parts of cold water whereas potassium niobium oxyfluoride is soluble in 12 parts. In a typical procedure, a stoichiometric amount of potassium hydroxide is added to the hydrofluoric acid solution of crude tantalic and niobic acids. On partial evaporation and subsequent cooling, needlelike crystals of K_2TaF_7 separate out. The mother liquid from which K_2TaF_7 has been separated contains niobium and other ore residuals. It is transferred to a sodium hydroxide solution to form the insoluble sodium niobate. After conversion to niobic acid with hydrochloric acid, potassium hydroxide is added to the slurry to form potassium niobate. This salt is purified by repeated fractional crystallizations. Niobium is finally recovered as niobium oxide by treating potassium niobate with hydrochloric acid.

Many organic solvents are capable of preferentially extracting tantalum from a hydrofluoric acid solution containing high concentrations of niobium and tantalum. Usually ketones are most effective for obtaining the separation, but many alcohols, amines, and aldehydes also are useful. Solvent-extraction separations are used today by the majority of niobium and tantalum producers. The feed solutions are usually dilute hydrofluoric acid, the extractants are aliphatic or aromatic ketones. A method developed at the U. S. Bureau of Mines (249,566) preferentially extracts tantalum by methyl isobutyl ketone from a hydrofluoric acid solution. Subsequent addition of hydrochloric acid to the aqueous raffinate, and recontacting with fresh ketone, selectively extracts niobium from the remaining impurities. Both batch and continuous counter-current procedures are in use. The advantages of a hydrofluoric acid–sulfuric acid–methyl isobutyl ketone system have recently been described (532).

Halogenation is also used as a means of treating niobium- and tantalum-bearing substances. The feed material usually consists of the mixed oxides of niobium and tantalum, but ores and low-grade slags also are used. Titanium is volatilized as the tetrachloride, whereas niobium is recovered as niobium pentachloride or niobium oxychloride, depending upon the conditions of the reaction (363,503). A selective chlorination procedure of interest treats the oxides at moderate temperatures with ammonia, causing niobium oxide to be nitrided. The mixture is then treated with chlorine at a slightly lower temperature and niobium nitride is preferentially volatilized as niobium chloride (314,512).

Anion-exchange resins can be used to separate niobium and tantalum as mixed chloro–fluoro–anion complexes from a hydrochloric acid–hydrofluoric acid medium. The excellent separability of these elements results from the fact that ions of different charges have widely differing distribution coefficients under specified conditions and that the complex constants relating the various species differ greatly.

2. Preparation of the Pure Metals

Reduction of the hydrated oxides or fluorides obtained by one of the above processes is usually achieved by: *(1)* the carbide process, *(2)* reduction with a metal such as sodium, magnesium, calcium, or aluminum, or *(3)* electrolysis. In the carbide process (44), which is preferred for the preparation of niobium metal, a mixture of carbon and metal oxide is first heated in a high-frequency furnace in an inert atmosphere to form the metal carbide. Stoichiometric amounts of metal carbide and oxide are then pelletized and are placed in a vacuum induction furnace. The following reaction begins at once at 1600°C.: $Nb_2O_5 + 5NbC = 7Nb + 5CO$, the end product at about 1800°C. being metal in the form of a porous ductile mass.

In the sodium reduction of potassium tantalum fluoride (Siemens & Halske process) (56), a sealed iron vessel containing a mixture of the proper amounts of the tantalum salt and granular sodium is heated at about 700°C. to start the exothermic reaction. After cooling, the container is opened and the reaction product is crushed and leached with water. The resulting fine-grained powder is then leached with acid to remove adhering impurities and residual salt.

In the United States, electrolysis of the fused double fluorides has been used to produce a coarse, relatively pure tantalum powder (143,187). The cell is a cast iron pot with a stout graphite rod as anode. Potassium tantalum fluoride, which serves as electrolyte, is heated in the cell to about 900°C. As the tantalum powder accumulates, frequent additions of tantalum oxide or other oxides are made to the bath to depolarize the anode surface. At the completion of the electrolysis the pot is cooled and the cake of powder metal left in the pot is crushed, air-blown, tabled, and acid-washed to produce a pure powder with a purity in excess of 99.8%. The electrolysis of K_2NbF_7–NaCl mixtures has recently also been investigated (298).

The niobium and tantalum powder produced by one or the other process is purified and densified to give a metal that is ductile and has a very low porosity. This is accomplished by heating the metal *in vacuo* at a temperature that is usually a few hundred degrees below the melting point of the metal. Both induction and resistance heating are practicable for sintering the niobium and tantalum. Arc-melting methods find their best use for procedures in which a relatively pure starting material is available; these methods have the advantage of an enormous output potential. A relatively new method of purification, known as electron-beam melting, produces the metal in vacuum of 10^{-7} atm. by bombardment with electrons from a heated tungsten filament. Impurities are boiled from the melt if the vapor pressure of contaminating compounds exceeds a pressure of about 10^{-5} atm. (131). The furnace purifies the metal so highly that a 3-inch bar can be rolled into foil 0.005 inch thick without annealing (510).

In a typical procedure in the production of ductile tantalum metal, the first step consists of compacting the pure powder into bars in a hydraulic press. The bar is held between watercooled terminals and is heated for several hours by its own resistance in an electric circuit. After being worked under a heavy hammer to close the pores, the bar is again heated in vacuum to remove stresses and complete the sintering. By this process, residual impurities such as oxygen, nitrogen, and hydrogen are removed. Probably the most critical impurity eliminated is oxygen, which causes hardness and brittleness that prevent the successful working of the metal. The resulting ductile bar is then forged and rolled into sheet or is drawn into wire at room temperature. The metal must be worked cold to avoid embrittlement by gases.

3. Production of Ferroniobium and Ferrotantalum–Niobium

These alloys, which are important for the manufacture of stainless steel types 347 and 348, are produced in electric furnaces by aluminum or silicon reduction of columbite ores. They also are prepared by the aluminothermic process.

4. Production of Miscellaneous Products

The carbides are prepared by the direct carburization of tantalum powder with the appropriate amount of lampblack, or by the reaction of the oxides and lampblack at about 1900°C. in an inert atmosphere. Niobium–chromium (60% Nb) and niobium–nickel (60% Nb), far less expensive than niobium metal and frequently used as a source of niobium in high-temperature alloys, are prepared by the aluminothermic process (451).

D. TOXICOLOGY AND INDUSTRIAL HYGIENE

1. Niobium

Niobium is immediately adjacent to vanadium in Group V; because of the toxicity of vanadium it could be assumed that niobium might also have toxic properties (465). However, no experimental data are available that would indicate this to be true and no cases of industrial poisonings have been reported (529).

2. Tantalum

Some industrial skin injuries from tantalum have been reported (465). They must be viewed, however, with suspicion because of the widespread use of tantalum wire and tantalum plate in surgery.

Because of the insolubility of tantalum and niobium oxide in most reagents, these substances are so inert that they scarcely could have toxic properties (80).

The toxicology of niobium and tantalum salts, such as fluorides, chlorides, oxalates, etc., should be considered from the standpoint of the anions involved. The widespread use of ketones in liquid–liquid extraction procedures necessitates proper precautions against harm from the solvent.

II. PROPERTIES OF NIOBIUM AND TANTALUM

A. PHYSICAL AND MECHANICAL PROPERTIES

Some physical and mechanical properties of pure niobium and tantalum are summarized in Table IV.

Niobium and tantalum crystallize with the body-centered cubic structure with $a = 3.299$ and 3.298 A., respectively. They thus resemble the alkali metals in crystal structure but have considerably smaller atomic radii (449). The atomic radii of Nb and Ta are identical. The ionic radii

TABLE IV
Physical Properties

	Niobium	Tantalum	Ref.
Atomic number	41	73	
Atomic weight	92.91	180.95	
Melting point	$2468 \pm 10°C.$	$2996 \pm 50°C.$	(491)
			(577)
Boiling point	$5127°C.$	$ca.\ 5300°C.$	(126)
			(577)
Density (g./cm.^{3})	8.57	16.60	(577)
Vapor pressure (mm. Hg)	1×10^{-5} (2194°C.)	9.52×10^{-11} (1737°C.)	(98)
	1×10^{-4} (2355°C.)	4.05×10^{-4} (2727°C.)	(98)
Rate of evaporation	1.16×10^{-7} (2194°C.)	1.55×10^{-7} (2407°C.)	(98)
(g./cm.^{2}/sec.)	1.08×10^{-6} (2355°C.)	1×10^{-3} (2820°C.)	(98)
Specific heat (cal./g.°C. at 100°C.)	0.06510	0.03364	(271)

(*continued*)

TABLE IV (continued)

	Niobium	Tantalum	Ref.
Heat capacity (cpcal./mol. at 100°C.)	6.087	6.101	(271)
Thermal conductivity (cal./cm.²/cm./°C./sec. at 20°C.)	0.125	0.130	(543) (577)
Lattice type	Body-centered cubic	Body-centered cubic	(577)
Lattice constant (A.)	3.294	3.296	(577)
Heat of sublimation (kcal./g. atom)	170.9	185.5	(577)
Heat of combustion (cal./g.)	2379	1380	(577)
Coefficient of linear expansion per °C.	7.1×10^{-6}	6.5×10^{-6}	(577)
Atomic radius (A.)	1.43	1.43	(449)
Radius of 5-valent ion (A.)	0.69	0.68	(449)
Atomic volume (cm.³/g. atom)	10.83	10.89	(577)
Electrical resistivity (μohm-cm. at 20°C.)	15.22	12.4	(543) (577)
Electrical resistivity (temperature coefficient per °C.)	0.00395	0.00382	(577)
Electrical conductivity (international annealed copper standard at 18°C.)	13.3%	13.0%	(126) (164)
Positive ion emission (e.v.)	5.52	10.0	(126) (577)
Ionization potential (v.)	6.77	7.3 ± 0.3	(577)
Electron work function (e.v.)	4.01	4.10	(577)
Electrochemical equivalent (mg./coulomb) with valence 5	0.1962	0.3749	(126) (164)
Tensile strength (p.s.i.)			
Sheet annealed (0.05% O_2)	50.000	50.000	(164)
Sheet worked	100.000	110.000	(577)
Hardness, Rockwell			
Sheet annealed, E scale	—	60	(164)
Sheet worked, E scale	—	95	(164)
Neutron cross section (barns-abs.)	1.1	21.3	(129) (260)
Heat of formation of pentoxide (kcal./g. equiv. of metal)	46.3	49.3	(449)

also differ little from one another. This is reflected in the close resemblance of their chemical properties.

The effect of impurities in niobium and tantalum is reflected in changes in mechanical properties, such as hardness, yield strength and its temperature dependence, ductility, and brittle-to-ductile fracture transition temperature (301). The physical and mechanical properties of pure niobium and tantalum are particularly affected by the gas content of the metals. Thus it was shown (192) that the hardness of tantalum is increased from 40 to 400 (Vickers) when the oxygen content increases from trace amounts to 0.2%. Similarly, the Vickers hardness of arc-melted niobium, which is 141, is reduced to 55 after zone refining (153). As would be expected, the tensile properties of niobium metal are also extremely susceptible to gas content (29), and the $\sim$22 tons/square inch in ultimate tensile strength of annealed production material (O_2 = 0.015 to 0.03%, N_2 = 0.01%, H_2 = 0.001%) can be reduced to $\sim$9 tons/square inch by sufficient lowering of the gas content. Raising the oxygen content to 0.1%, with the nitrogen and hydrogen contents unchanged, increases the ultimate tensile strength to $\sim$30 tons/square inch (569). Additions of optimum amounts of titanium, molybdenum, vanadium, chromium, and zirconium reduce the oxidation rate of niobium and presumably of tantalum at 1800°F. by a factor of 10, and more at lower temperatures (114,291,509).

Hydrogen dissolves readily in niobium and tantalum at low temperatures and renders the metals extremely brittle. This property is, in fact, utilized for the deliberate embrittlement of scrap metal for recirculation purposes. Vacuum treatment at relatively low temperatures ensures the removal of the gas to a very low level (569).

1. Niobium (224)

The modulus of elasticity at 40°C. is 19,000,000 p.s.i. The Brinnell hardness of annealed pure niobium is 75 and its scleroscope hardness is 10. These values allow impurity limits of 0.20% tantalum, 0.002% titanium, 0.03% silicon, 0.01% iron, and 0.02% carbon. The yield strength of niobium is sometimes listed as being 29,000 p.s.i. at 20°C. The electrical conductivity of niobium is $1/3$ that of copper. Transition of niobium to superconductivity occurs in the range 8.74 to 8.18°K. The ability of niobium to spot weld itself or to other metals, when soft-annealed, is quite outstanding. This is accomplished by the electric spot, butt, and roller processes. The metal also can be resistance-welded or arc-welded under a protective atmosphere (163).

TABLE V
Isotopes of Niobium and Tantalum[a]

Isotope	Natural abundance	Half life	Type of decay	Absorption cross section, σ_a	Cross section detn. by activation, σ_{act}	Activation product	Average scattering cross section, $\bar\sigma_s$
^{89m}Nb		1.0 hr.	β^+				
^{89}Nb		1.9 hr.	β^+				
^{90m}Nb		24 sec.	IT				
^{90}Nb		15.0 hr.					
^{91m}Nb		64 days	IT				
^{91}Nb		Long	EC				
^{92m_2}Nb		5.9×10^{-6} sec.	IT				
^{92m_1}Nb		13 hr.	EC				
^{92}Nb		10.1 days	EC				
^{92}Nb		21.6 hr.	β^-				
^{93m}Nb		4.12 yr.	IT				
^{93}Nb	100%			1.1 ± 0.1	1.0 ± 0.5	^{94m}Nb	5 ± 1
^{94m}Nb		6.6 min.	IT (99%) $\beta^-(\sim0.1\%)$				
^{94}Nb		2×10^4 yr.	β^-		15 ± 4	^{95}Nb	
^{95m}Nb		90 hr.	IT				
^{95}Nb		35 days	β^-				
^{96}Nb		23.35 hr.	β^-				
^{97m}Nb		60 sec.	IT				
^{97}Nb		72.1 min.	β^-				
^{98}Nb		30 min.	β^-				

NIOBIUM AND TANTALUM

^{99}Nb	2.5 min.	β^-			
^{176}Ta	8.0 hr.	EC			
^{177}Ta	53 hr.	EC			
^{178}Ta	2.1 hr.	EC(~97%)			
		β^-(~3%)			
^{178}Ta	9.35 min.	EC(~94%)			
^{179}Ta	600 days	EC			
^{180m}Ta	8.15 hr.	{EC(~79%)			
		β^-(~21%)			
		(No β^+)			
^{180}Ta	>10^{12} yr.				
^{181m_2}Ta	2.2 × 10^{-5} sec.	IT			
^{181m_1}Ta	1.2 × 10^{-8} sec.	IT			
^{181}Ta	100%		21.3 ± 1.0	30 ± 10 mb.	^{182}Ta
				19 ± 7 mb.	^{182}Ta 5 ± 1
^{182m}Ta	0.33 sec.	IT			
^{182}Ta	16.5 min.	IT			
^{182}Ta	115 days	β^-			
^{183}Ta	5.2 days	$^{\beta-}$			
^{184}Ta	9.3 hr.	β^-			
^{185}Ta	48 min.				
^{186}Ta	10.5 min.	β^-			

^{a} Key to symbols: the superscript m indicates metastable excited states. When two isomeric states are known, these are denoted by subscripts m$_1$ and m$_2$; β^- = negative beta-particle emission; β^+ = positive beta-particle emission; EC = orbital electron capture; IT = isomeric transition (transition from upper to lower isomeric state of the same nucleus); No β^+ indicates that investigations have failed to find this particular mode of decay. Cross sections are indicated in barns (10^{-24} cm.^{2}) unless they are marked mb.

2. Tantalum (164)

The strength of unannealed tantalum is comparable to cold-rolled steel, and annealed tantalum may be compared to annealed steel. The Young's modulus is valued at 27,000,000 p.s.i., varying, however, with the thickness of the sheet or diameter of the wire. In ultimate strength, tantalum fine wire and thin sheet are similar to steel. As far as ductility is concerned, tantalum is comparable to mild steel or nickel in its ability to be drawn or otherwise formed into complicated shapes.

With regard to workability, tantalum becomes workhardened at a much slower rate than most metals. Tantalum can be machined, but precautions must be taken against its tendency to gall and tear. Tantalum welds to itself and some other metals by resistance-welding techniques. The expansion coefficient of tantalum is linear and is approximately one-half that of steel, four-tenths that of copper, and three-tenths that of silver. The thermal conductivity of tantalum is approximately the same as that of steel. The characteristic color of tantalum is bluish gray. Roll-polished sheet is much whiter and has a luster similar to platinum. The resistivity of tantalum is seven times that of copper.

Tantalum generates against most other metals an e.m.f. whose value per degree is less than that for platinum against platinum–iridium.

With regard to thermionic emission, the work function of tantalum (4.1 e.v.) is lower than that of tungsten but higher than that of niobium (4.01 e.v.) (379). At approximately 2100°K. the intensity of emission from tantalum is equivalent to that from tungsten at 2500°K.

B. ISOTOPES OF NIOBIUM AND TANTALUM

Information on the isotopes of niobium and tantalum is presented in Table V (379,520). Niobium-95, with a half life of 35 days, is the result of

TABLE VI
Radioactive Fission Products of Niobium (53)

Fission yield from U-235, %	6.3	—
Half life, days	35	—
Radioactivity	β	γ
Decay energy, M.e.v.	0.16	0.75
Disintegration, %	100	100
Curies/gram of fission products (230-day irradiation, 104-day cooling)		
Curies/g.	111	111
Total β	111	—
Total γ	111	—

the decay of zirconium-95, formed by the fission of uranium-235. Its principal characteristics are presented in Table VI. Niobium-94 ($T_{1/2} = 6.6$ minutes) and tantalum-182 ($T_{1/2} = 115$ days), the result of neutron activation of Nb-93 and Ta-181 (n,γ) reactions, are of considerable analytical importance for the determination of the two elements by activation analysis (see Section VII-H-4).

C. ELECTROCHEMICAL PROPERTIES

1. Tantalum

Tantalum occupies a position in the electromotive series close to platinum and gold (164). It assumes negative polarity with all metals above platinum and gold in the series. The value of e.m.f. developed by such couples in electrolyte is above the hydrogen overvoltage on water, resulting in deposit of hydrogen on the tantalum. The electrode potential of tantalum has been calculated from the heat of formation of Ta_2O_5, and a Ta/Ta^{+5} value of $E° = -1.12$ v. has been reported (379). Latimer (329) reports that $Ta_2O_5 + 10H^+ + 10e^- = 2Ta + 5H_2O$, with $E° = 0.81$.

2. Niobium

The electrode potential relationships between the various valence states of niobium in acid solution are reported as follows (329):

$$Nb_2O_5 + 10H^+ + 10e^- = 2Nb + 5H_2O \qquad E° = -0.65$$

$$NbO(SO_4)_2^- + 2H^+ + 2e^- = 2SO_4^{-2} + H_2O + Nb^{+3} \qquad E° = ca. -0.1$$

$$NbO(SO_4)_2^- + 2H^+ + 5e^- = 2SO_4^{-2} + Nb + H_2O \qquad E° = ca. -0.63$$

$$Nb^{+3} + 3e^- = Nb \qquad E_0 = ca. -1.1$$

Summary

Nb_2O_5 $ca. -0.1$ $\qquad\qquad\qquad$ Nb^{+3} $\qquad\qquad\qquad$ $ca. -1.1$ Nb

$$-0.65$$

Niobium is fundamentally an active metal, as is evidenced by its rapid rate of solution in aqueous media such as hydrofluoric acid and strongly alkaline solutions, in which a protective oxide coating is not formed. Its apparent action as a noble metal is due to the formation of an inert, very thin oxide coating in most acid solutions. Attempts to electrodeposit pure niobium from aqueous media have not been successful.

The preparation of niobium metal by means of fused-salt techniques is complicated by the fact that the stable lower-valence compounds are soluble in the electrolyte (126).

D. OPTICAL PROPERTIES

1. Emission Spectra

Some of the most sensitive lines of niobium and tantalum useful for emission spectrographic purposes are presented in Table VII (235,511) (see also Section VI-A).

TABLE VII
Emission Spectra

A.	Intensity		A.	Intensity	
	Arc	Spark		Arc	Spark
Niobium			Tantalum		
4137.095	100	60	3406.664	70	18
4123.810	200	125	3318.840	125	35
4100.923	300	200	3311.162	300	70
4079.729	500	200	2963.322	300	100
4058.938	1000	400	2714.674	200	8
3580.273	100	300	2653.274	200	15
3225.479	150	800	2647.472	200	10
3130.786	100	100			
3094.183	100	1000			

2. X-Ray Spectra

Because of its position in the periodic system, the K lines of niobium are very sensitive and are well-suited for the detection and determination of even small amounts of niobium.

On the other hand, the K lines of tantalum cannot be excited by commer-

TABLE VIII
X-Ray Spectra (444)

Niobium, λ		Tantalum, λ	
$K\alpha_1$	0.747	$L\alpha_1$	1.522
$K\beta_1$	0.665	$L\beta_1$	1.327
$L\alpha_1$	5.726	$L\beta_2$	1.285
$L\beta_1$	5.492	$L\gamma_1$	1.38

cially available, 50-kv. x-ray sources, but several of tantalum's L lines are of analytical importance. However, the intensity of the most sensitive tantalum radiation ($L\alpha_1$ $\lambda = 1.522$ A.) is only about 30% that of the most sensitive niobium radiation ($K\alpha_1$ $\lambda = 0.747$ A.) (see Table VIII).

E. CHEMICAL PROPERTIES

1. Introduction

The electronic configurations of niobium and tantalum are presented in Table IX.

TABLE IX
Electronic Configurations of Niobium and Tantalum

At. No.	K	L		M			N				O			P		
							Subshell									
	1s	2s	2p	3s	3p	3d	4s	4p	4d	4f	5s	5p	5d	6s	6p	6d
Niobium (41)	2	2	6	2	6	10	2	6	4		1					
Tantalum (73)	2	2	6	2	6	10	2	6	10	14	2	6	3	2		

Niobium and tantalum, together with vanadium and protactinium, form the fifth subgroup of the periodic system and accordingly are normally pentapositive. Although niobium and tantalum have decidedly characteristic metallic properties, their normal oxides, the pentoxides, are acidic and are frequently referred to as "earth acids."

Like the elements of the fifth main group, those of the fifth subgroup can form compounds from lower valence states, as well as from the +5 state. However, whereas the tendency to assume a lower positive charge than 5 units increases with increasing atomic weight in the fifth main group, it decreases in that direction in the fifth subgroup (449). Thus, in a sulfuric acid solution zinc reduces vanadium to the bivalent state and niobium only to the trivalent state; tantalum is not reduced at all. The reducibility of the niobium is of analytical importance and has been made the basis of volumetric (see Section VII-C) and polarographic (see Section VII-D) procedures.

Unlike the elements of the fifth main group, niobium and tantalum (also vanadium) do not function as electronegative constituents of saltlike compounds (449). This can be explained by their position in the periodic system, since they are not closely followed (as are N, P, As, Sb, and Bi) by inert gases and thus are unable to form, with the usual electron donors, compounds of electron configurations of the special stability of the inert gases. Their position in the periodic system also explains the fact that niobium and tantalum, unlike members of the fifth main group, do not form gaseous hydrides. They form, however, alloy-like hydrides, as will be described below.

Because of the lanthanide contraction, the atomic radii of niobium and tantalum are identical within the error of measurement. Since the ionic radii also differ little, the compounds of niobium and tantalum are very similar in behavior. Thus, KMO_3, M_2O_5, and MCl_4 (where M = Nb or Ta) form mixed crystals with almost identical properties (467). Although the behavior of most niobium and tantalum compounds is very similar and has for years frustrated the efforts of chemists who attempted their separation, the similarity is not as great as that between the compounds of zirconium and hafnium. This has been explained by the fact that the niobium ion polarizes more strongly than does the tantalum ion and generally behaves as if it had an appreciably smaller radius than the tantalum ion (449).

2. Oxides

Analogously to the other subgroups of the periodic system, particularly Groups IV to VII, the stability of the oxides in the V group increases with increasing atomic weight. Consequently, the formation of niobium(V) compounds proceeds less energetically than that of the corresponding tantalum compounds, as is demonstrated by the heat of formation of the pentoxides: $V_2O_5 = 37.3$; $Nb_2O_5 = 46.3$; and $Ta_2O_5 = 49.3$ kcal./g. equivalent of metal. In addition, the relative stability of intermediate oxidation states decreases from V to Nb to Ta, as is shown in the following list of stable oxides (467):

V	VO	V_2O_3	VO_2	V_6O_{12}	V_2O_5
Nb	NbO	—	NbO_2	—	Nb_2O_5
Ta	—	—	—	—	Ta_2O_5

The above compilation indicates that niobium(V) compounds can be reduced more easily than the corresponding tantalum(V) compounds. However, it should be mentioned that intermediate oxidation states of tantalum that are unstable in pure form can be stabilized, if isomorphically incorporated into the corresponding niobium compounds.

The niobium–oxygen system recently has been described in detail (153). Niobium is resistant to oxygen until the temperature approaches 200°C. Strips of sheet or powder heated in air become faintly yellow at 180°C., distinctly yellow at 230°C., and bright blue at 300°C. At 390°C. the color is nearly black and the white oxide begins to appear (72). The sequence of phases is $Nb-NbO-NbO_2-Nb_2O_5$ (291). Niobium oxides dissolve into the metal when heated in vacuum to temperatures above red heat and below 1200°C. At higher temperatures the oxides are evaporated.

Oxygen is dissolved interstitially in the body-centered cubic lattice of tantalum (296). The solubility increases from 1.5 to 6.4% between 700 and 1650°C. (558).

The oxidation rate of tantalum in air at 300 to 350°C. is similar to that described for niobium. The intermediate formation of Ta_4O, Ta_2O, TaO, and finally Ta_2O_5 has been conjectured by some investigators (492), but has not been established with certainty (72). At 400 to 800°C. the rate of oxidation is initially parabolic, then becomes linear (221). At 900 to 1400°C. the reaction varies between linear and parabolic (4). The reaction of tantalum with pure oxygen follows a linear rate, which varies with the square root of pressure (428). As in the case of niobium, oxygen in tantalum metal may be eliminated by heating in a high vacuum to above 2300°C. when the oxygen is distilled out in the form of a lower oxide of tantalum.

Nb_2O_5 and Ta_2O_5 also can be prepared by dehydration of "niobic" and "tantalic" acids and by ignition of the respective sulfides, nitrides, carbides, cupferrates, and tannates in air. Nb_2O_5 and Ta_2O_5 are insoluble in water and in acids other than hydrofluoric acid. They can be brought into solution by fusion with alkali bisulfates, carbonates, or hydroxides, an indication that they form amphoteric oxides. Their acidic character, however, greatly predominates. Unlike Nb_2O_5, Ta_2O_5 is not reduced to the dioxide when heated in hydrogen, even when Ta_2O_5 is mechanically mixed with Nb_2O_5. If the two oxides are present, however, as is frequently the case, as mixed crystals, $(Nb,Ta)_2O_5$, reduction with hydrogen leads to a niobium-rich MO_2 phase and a tantalum rich M_2O_5 phase (471). The exact composition of the two phases depends on the temperature, the H_2O/H_2 ratio, and the Nb_2O_5/Ta_2O_5 ratio of the initial crystals. If the Nb_2O_5 content is high enough, the whole of the Ta_2O_5 present in the mixed crystals can be reduced to homogeneous NbO_2-TaO_2 mixed crystals.

A study of lower oxides of niobium has been made by sintering NbO_2 and Nb in various ratios. X-ray analysis indicates the identity of NbO. The existence of Nb_2O_3 as a compound has been questioned (126).

3. Earth Acids

When alkali niobate or tantalate solutions are treated with sulfuric acid, niobium or tantalum pentoxide hydrate is formed as a white gelatinous precipitate of variable water content. The hydrates commonly referred to as "niobic acid" and "tantalic acid" also can be prepared by leaching potassium pyrosulfate fusions containing niobium or tantalum in water, by rendering fluoride solutions of the two elements ammoniacal, and by hydrolyzing halides of niobium and tantalum by boiling in water. The composition of the earth acids, conjectured by some investigators to be $H_7Nb_5O_{16}$ and $H_7Ta_5O_{16}$, is still uncertain. The water in these gels is firmly retained up to 500°C. with the tantalic acid giving water off somewhat more readily than the niobic acid (449). When the heating is carried out not too slowly, a sudden incandescence occurs due to the sudden crystallization of the oxides previously present in a quasiamorphous state.

4. Niobates and Tantalates

When niobic acid is dissolved in concentrated hydrochloric acid, species such as $Nb(OH)_2Cl_4^-$, $Nb(OH)_2Cl_3$, and $NbOHCl^{+3}$ result, depending on the hydrogen ion and chloride ion concentrations (324). These complexes are of great importance to the analytical chemist, since they afford separations of niobium and tantalum by ion exchange, solvent extraction, and cellulose chromatography (see Sections IV-H to IV-J).

Complex niobates and tantalates are formed by replacing the oxygen atoms with radicals such as oxalate, tartrate, and citrate, and radicals of other oxyacids, as well as radicals of titanic, stannic, arsenic, chromic, and tungstic acids. (The oxalatoniobate, for instance, has the formula $3M_2O Nb_2O_5 \cdot 6C_2O_3$, and has been described as oxotrioxalatoniobate (449); the notation M represents a monovalent element.) Not only can these radicals be incorporated into the niobate and tantalate molecule, but the latter also can enter into the former.

Organic chelate complexes of niobium and tantalum with tannic, oxalic, tartaric, salicylic, and citric acids, and with pyrogallol and other hydroxy acids, are extremely useful to the analytical chemist.

The formation of complex niobates and tantalates containing titanium, tungsten, tin, and zirconium leads to a loss of chemical individuality not only of niobium and tantalum, but also of other associated elements. Thus, the individual constituents cannot behave quantitatively toward reagents as they would in the pure state, because their molecules are too firmly interlocked with those of other constituents of different chemical reactivity (484). When, for instance, a mixed earth–acid precipitate contains more tantalum, it reacts largely like tantalic acid. If the reverse is the case the mixture behaves like niobic acid. Titanic acid masks most of the normal reactions of niobium to a lesser degree than those of tantalum. On the other hand, niobium and tantalum interfere with some of the reactions of tungstic and stannic oxides, and zirconia modifies the normal behavior of titania and subsequently modifies the effect of titania on niobic and tantalic acid.

When niobium or tantalum pentoxide is fused with sodium carbonate, it replaces carbon dioxide, due to its slightly acidic character, to form the orthoniobate or orthotantalate, $Na_3(Nb,Ta)O_4$. The reaction of sodium carbonate with Ta_2O_5 is much more sluggish than that with Nb_2O_5. Both oxides can be fused more readily with sodium hydroxide. When the melt containing ortho sodium salts is leached in water, a white residue remains, consisting of sodium metaniobate or metatantalate, $Na(Nb,Ta)O_3$. The sodium salts also can be obtained by saturating a solution of the potassium salts with sodium chloride. It is thought that the hydrated crystalline tantalate is a salt of pentatantalic acid, $H_7(Ta_5O_{16})$, or is a hexatantalate of the formula $Na_8Ta_6O_{19} \cdot nH_2O$ (449). Part of the water actually may be replaced by NaOH. In order to avoid confusion in the nomenclature, the above compounds are frequently designated as 7:5 or 4:3 salts. In the case of niobium, only a 7:6 salt, $Na_{14}Nb_{12}O_{37} \cdot 31H_2O$, is known besides the metaniobate. The insolubility of the sodium salts, described above, in weak alkaline solutions containing a large excess of the sodium ion represents a valuable analytical tool, since the corresponding tungsten, tin, and molybdenum salts are soluble.

Both niobium and tantalum form with potassium the following salts that, unlike the corresponding sodium salts, are all soluble in water: potassium hexaniobate or tantalate, $K_8(Nb,Ta)_6O_{19} \cdot 16H_2O$, and the potassium metaniobate or tantalate, $K(Nb,Ta)O_3$. The solubility of the potassium salts of niobium and tantalum in slightly alkaline solution has been used in the separation of niobium and tantalum from zirconium (488) (see Section IV-G-5). Rubidium and cesium hexaniobate and tantalate are also known. They carry 14 molecules of water of crystallization (42). A lithium niobate, $LiNbO_3$, also has been mentioned.

5. Peroxy Salts

Both niobium and tantalum form peroxy salts of the formula $K_3(Nb,Ta)O_8$ or $K_3[\overset{O_2}{\underset{O_2}{}}(Nb,Ta)\overset{O_2}{\underset{O_2}{}}]$ when their 4:3 salts (hexaniobate or tantalate) are added to a solution of potassium hydroxide and hydrogen peroxide. They are derived from the orthoniobate or tantalate by exchanging the four oxygen atoms with four peroxy radicals. Upon the addition of dilute sulfuric acid to the concentrated solution of the peroxy salts, free peroxy acids, $H[\overset{O}{\underset{O}{}}(Nb,Ta)O_2]$, are obtained, of which the niobium salt is remarkably stable (371) (see Section V-A-5). The tantalum peroxy complex readily decomposes, however, into tantalic acid (467). Since the absorption maximum of peroxyniobic acid differs considerably from that of peroxytantalic acid, analytical chemists have determined both elements photometrically without the need of a chemical separation (423) (see Sections VII-B-2 and VII-B-9).

6. Halogenoniobates and Tantalates

The halogenoniobates and tantalates also belong to the class of acidoniobates and tantalates and are derived by exchange of oxygen atoms with halogen atoms. The halides are nonpolar compounds and, as such, are characterized by comparatively low melting and boiling points (see Table X).

TABLE X
Melting and Boiling Points of Some Halogenoniobates and Tantalates
(Data based on observations of various investigators)

	M.P., °C.	B.P., °C.
$NbCl_5$	209–212	243–254
$NbCl_4$	Disproportionates	
$NbCl_3$	400, sublimes	
$NbOCl_3$		332, sublimes
$NbBr_5$	227–267	272–361
$NbBr_3$	400, sublimes	
NbI_5	327, decomposes	347
NbF_5	76	225–233
$TaCl_5$	217	239
$TaCl_4$	300, disproportionates	
$TaBr_5$	280	349
$TaBr_4$	300, disproportionates	
TaI_5	496	543
TaF_5	97	229
K_2TaF_7	720	High temp. dec.

a. Chlorides

Yellow needles of niobium pentachloride, $NbCl_5$, and colorless crystals of tantalum pentachloride, $TaCl_5$, are formed when niobium and tantalum are heated in pure dry chlorine above 200°C. The pentachlorides also can be prepared by heating an intimate mixture of the pentoxides and carbon in a stream of chlorine. Chlorination of the pentoxides with carbon tetrachloride ($Nb_2O_5 + 5CCl_4 \rightarrow 2NbCl_5 + 5COCl_2$) and with octochloropropane ($Nb_2O_5 + 5C_3Cl_8 \rightarrow 2NbCl_5 + 5C_2Cl_4 + 5COCl_2$) also have been described. If the chlorination is carried out under atmospheric conditions it may result in the formation of niobium or tantalum oxytrichloride. The niobium oxytrichloride forms double salts of the types $M(NbOCl_4)$ and

$M_2(NbOCl_5)$ (M = monovalent elements) (449). Since the melting points of niobium pentachloride (243°C.) and tantalum pentachloride (239°C.) differ considerably from those of stannic chloride (113°C.) and titanium tetrachloride (136°C.), several analytical chemists have demonstrated the feasibility of using chlorination-distillation procedures as an analytical tool (35,474) (see Section IV-A-6).

Niobium pentachloride can be reduced with metallic niobium or tantalum to niobium tetrachloride (475).

$$4NbCl_5 + Nb = 5NbCl_4$$

$$5NbCl_5 + Ta = 5NbCl_4 + TaCl_5$$

When hydrogen is used as a reducing agent instead of the metals, part of the niobium pentachloride is reduced all the way to the greenish-black trichloride. The trichloride can be reduced to the dichloride by further reaction with niobium metal (472). Tantalum pentachloride cannot be reduced by the conditions described above to the tetrachloride, but can be reduced to a trichloride (457). Niobium tetrachloride mentioned above tends to disproportionate into the tri- and pentachloride: $2NbCl_{4\,solid} = NbCl_{3\,solid} + NbCl_{5\,gas} - 28.3$ kcal. The four phases $NbCl_{solid}$, $NbCl_{4\,solid}$, $NbCl_{5\,liquid}$, and $NbCl_{5\,gas}$ can coexist in equilibrium at 420°C. under a pressure of 22 atm. (473). The niobium trichloride $NbCl_{2.67-3.1}$ is insoluble in water, in the normal organic solvents, and in nonoxidizing acids and alkali solutions (472).

Tantalum trichloride reacts in the molten state (300°C.) with an excess of tantalum pentachloride, forming tantalum tetrachloride: $TaCl_3 + TaCl_5 = 2TaCl_4$. In the absence of tantalum pentachloride (the latter can be distilled off between 350 and 400°C.), the trichloride again splits off tantalum pentachloride, $3TaCl_3 = 2TaCl_2 + TaCl_5$ (449). The tetra-, tri-, and dichlorides of tantalum are all green. With water they react as follows:

$$TaCl_2 + H_2O = Ta^{+3} + 2Cl^- + OH^- + 1/2\, H_2$$

$$2TaCl_4 + 5H_2O = TaCl_3 + Ta(OH)_5 + 5HCl$$

Tantalum trichloride dissolves in water without change. Upon the addition of OH^- a hydroxide is precipitated that amphoterically dissolves in excess of acids and alkali hydroxides. With hot water it undergoes oxidation as follows:

$$Ta(OH)_3 + 2H_2O = Ta(OH)_5 + H_2$$

It should be remembered that in aqueous solution reduced states of tantalum cannot be produced, whereas niobium can be rather easily reduced from the pentavalent to the trivalent state.

b. Bromides and Iodides

Niobium and tatalum pentabromides, MBr_5, and oxytribromide, MOB_3, resemble the corresponding chlorides and are similarly prepared by reaction of bromine with the metals. The tribromides are formed when the pentabromides are reduced with hydrogen (42,528). Iodine does not attack niobium at moderate temperatures, but niobium wire at a temperature of 1300 to 1500°C. reacts with iodine vapor in an evacuated vessel. The resulting NbI_5 is not stable and decomposes, yielding a lower iodide (224). The iodide also can be prepared by reaction of the bromide with anhydrous hydrogen iodide (577). Tantalum pentaiodide, TaI_5, forms when tantalum metal is heated to a bright red heat in iodine. It is more stable than NbI_5. It hydrolyzes when exposed to air, but it does not decompose below its boiling point when protected. It loses iodine when heated somewhat above 1000°C. (8).

c. Fluorides

Both niobium and tantalum react with fluorine at ordinary temperatures, leading to the formation of the pentafluorides. The pentafluorides also are formed by passing anhydrous hydrogen fluoride over the chlorides at room temperature. Niobium pentafluoride is very hygroscopic, hisses when thrown into water, and is hydrolyzed to form hydrofluoric acid and oxyfluoride, $NbOF_3$, or oxyfluoniobic acid (pentafluoniobic acid), $NbF_5 + H_2O = NbOF_3 + 2HF$; $NbF_5 + H_2O = H_2NbOF_5$. This acid is also formed when niobium or "niobic acid" is dissolved in aqueous hydrofluoric acid. If a high concentration of hydrofluoric acid is present in the solution, heptafluoniobic acid, H_2NbF_7, will be present (42). Tantalum pentafluoride is soluble in water without hydrolysis. Niobium trifluoride has been prepared by the action of a hydrogen–hydrogen fluoride gas mixture on $NbH_{0.7}$ (324).

Both niobium and tantalum have a tendency to form double fluorides. If a solution of niobic acid in hydrofluoric acid is treated with metallic fluorides, either fluoroniobates or oxofluoroniobates are formed, depending on the amount and concentration of the hydrofluoric acid (449). The simplest examples of the fluoroniobates have composition such as NbF_5MF or $M(NbF_6)$ and $NbF5 \cdot 2MF$ or $M_2(NbF_7)$ (M is a monovalent element). In the presence of a large excess of potassium fluoride the compound K_3NbOF_6

forms. Some fluoroniobates have more complicated compositions. Among oxofluoroniobates the following compounds are known: $NbOF_3 \cdot 2MF$ and $NbOF_3 \cdot 3MF$, as well as more complex compositions such as $5MF \cdot 3NbOF_3 \cdot H_2O$ and $4MF \cdot 3NbOF_3 \cdot 2H_2O$ (42). If the alkali metal oxofluoroniobate is recrystallized from solutions containing hydrogen peroxide, a peroxy fluoroniobate such as $M_2NbOF_5 \cdot H_2O$ is formed.

Most of the fluorotantalate complexes are of the type $2MF \cdot TaF_5$. In addition, compounds of the type $MF \cdot TaF_5$ and $3MF \cdot TaF_5$ also exist. The potassium salt of the first-mentioned type, $2KF \cdot TaF_5$, structurally a heptafluorotantalate, $K_2(TaF_7)$, that is only sparingly soluble in HF, forms the basis of the classic tantalum–niobium separation developed by Marignac (359). The niobium salt is considerably more soluble and, only after evaporation, the salt $K_2NbOF_5 \cdot H_2O$ crystallizes out. A basic oxyfluoride complex, $K_4Ta_4O_5F_{14}$, formed by evaporation of the double fluoride with water, also has been mentioned (467). However, the tendency to form oxofluoro compounds is much smaller with tantalum than with niobium. This fact is of considerable importance to the analytical chemist and has been made the basis of the separation of niobium and tantalum by ion exchange (see Section IV-H), by paper and cellulose chromatography (see Section IV-I), and by liquid-liquid extraction (see Section IV-J) procedures.

7. Refractory Binary Compounds

Niobium and tantalum form high-melting refractory compounds with light atoms of small atomic dimensions, such as hydrogen, carbon, boron, silicon, nitrogen, and sulfur. The physical properties of these compounds are typical of those formed by members of Group IVa, Va, and VIa. The type of chemical bonding that exists, i.e., whether it is ionic in nature or nonionic, has not been entirely resolved. The high hardness and brittleness have been attributed to the interference of dissolved light atoms with the plastic deformation of the metal lattice (126).

a. Hydrides

Niobium and tantalum absorb gaseous hydrogen over a wide range of temperature, losing their ductility and becoming extremely brittle due to the formation of niobium and tantalum hydride. When the two metals are used in a corrosive environment, hydrogen may be generated on the metal surface by a corrosion process, by cathodic charging, or by galvanic coupling. The exact mechanism by which atomic hydrogen enters and embrittles niobium and tantalum is not well understood. It is thought that hydrogen dissolves in the metals by dissociating into protons until the solubility is exceeded and hydrides form (61). Thus, at ordinary temperature 1 g. of niobium can absorb 157 ml. and 1 g. of tantalum absorbs 55.6 ml. of hydrogen. At 500°C. the corresponding figures are 47.4 and 14.8 ml., and at 1000°C. they are 2.8 and 1.4 ml. (264,449). When exposed to steam (250 to 300°C.), niobium powder decomposes into oxide and hydride according to one of the following mechanisms (397): $16Nb + 5H_2O = Nb_2O_5 + 14NbH_{0.7}$; $2Nb + 5H_2O = Nb_2O_5 + 5H_2$; or $14Nb + 5H_2 = 14NbH_{0.7}$. On the other hand, tantalum is not attacked by steam up to at least 165 pounds/m.^{2}, provided the pH of the condensate does not exceed 8 (43). It should be noted that a vacuum treatment at relatively low temperatures insures the removal of the hydrogen from hydrides to a very low level. Use has been made of this property not only for the determination of hydrogen in niobium and tantalum by vacuum extraction (see Section VIII-A-2), but also for purposes of refining niobium and tantalum scrap, which can be intentionally hydrided, crushed, and vacuum-treated.

b. Nitrides

The tantalum–nitrogen system has recently been thoroughly investigated (40,193). There also are reports on the niobium–nitrogen system (222). Niobium and tantalum can take up only minimal, if any, amounts of nitrogen into their crystal lattices. They form nitrides, however, which have a stability intermediate between the stable nitrides of Group IV and the unstable nitrides of Group VI. A significant reaction of the metals with nitrogen begins below 700°C., is very rapid at 1000°C., remains constant between 1000 and 1500°C., and decreases at higher temperatures. All the nitrogen absorbed is liberated between 1900 and 2000°C. in a vacuum (257). The nitrides also are formed by the reaction between niobium and tantalum oxide or halide and nitrogen in the presence of a reducing agent such as hydrogen. Ammonia gas is sometimes preferred to nitrogen as a reactant, since the dissociation of ammonia releases nitrogen in the active, monatomic form (65). The following nitrides have been characterized: NbN in three modifications (one cubic, two hexagonal) Nb_4N_3, Nb_2N, TaN, and Ta_2N. The existence of Nb_3N_5 and Ta_3N_5 also has been postulated but not yet proven. The nitrides are essentially insoluble in dilute and concentrated HCl, H_2SO_4, and HNO_3. They decompose in fused caustic or boiling KOH with the evolution of ammonia, although Nb_2N is reported to dissolve with the evolution of nitrogen (493). An interesting and peculiar property of NbN is that it has the highest known transition temperature into a superconducting state, 15.2°K., of any known material (126).

c. Carbides

The carbides of niobium and tantalum have been prepared by a number of reactions, as for instance by heating the pentoxides with hydrogen at 1000°C., adding the necessary amounts of carbon for reduction, and further heating in a hydrogen atmosphere (186). The important carbides are NbC and TaC; Nb_4C_3, Nb_2C, and Ta_2C also have been reported. An important characteristic of NbC and TaC is their ability to form solid solutions with each other and with other refractory metal carbides, e.g., with TiC, ZrC, UC, and VC (410), and with certain nitrides (146). It has been stated that the maximum percentage of carbon that can exist in tantalum carbide corresponds to the composition of approximately $TaC_{0.96}$ (454). It was postulated that the carbon atoms donate electrons to the d shell of the tantalum metal atoms and that the optimum number of bonding electrons in relation to the coordination number of the metal atoms is reached in tantalum carbide before stoichiometry is reached.

d. Borides

Niobium borides (Nb_2B, NbB_2, NbB, Nb_3B_4) and tantalum borides (TaB_2, Ta_2B, TaB, Ta_3B_4) are resistant against reduction by other metals, particularly against an attack by the alkalies and alkaline earths and low-melting metals such as Bi, Sn, Zn, and Pb (75).

e. Silicides

Three silicides of niobium have been identified: Nb_4Si, Nb_4Si_3, and $NbSi_2$ (126). The latter exhibits elevated-temperature oxidation resistance.

III. SAMPLING OF NIOBIUM AND TANTALUM

The sampling of niobium- and tantalum-bearing material involves problems common to substances of similar physical characteristics.

A. ORES AND CONCENTRATES

In as much as most of the ores and concentrates processed at the present time are of the tantalite–columbite type, it should be recalled that the origin of this material usually is granite pegmatite with the possibility of considerable variations in the niobium:tantalum ratio in deposits from neighboring locations. For instance, imports of Nigerian columbite into the United States show variations in the $Nb_2O_5:Ta_2O_5$ ratio between 8:1 and 14:1, whereas the FeO:MnO ratio remains more or less constant.

Shipments of the ore or concentrate usually are small, rarely exceeding 10 tons. In order to obtain a representative sample it is advisable that the material be crushed to about 10 mesh before it is sampled.

In contrast to columbite–tantalite, the composition of pyrochlore ore in a particular locality remains fairly constant. Norwegian pyrochlore from the Fen region, for instance, occurring in a metamorphosed limestone matrix and upgraded by a combination of physical and chemical processes to a concentrate containing about 50% Nb_2O_5, has for many years shown a Ta_2O_5 content of 0.6% ± 0.2%, whereas pyrochlore from another locality consistently contains 0.25% ± 0.1% of Ta_2O_5.

Niobium and tantalum concentrates are shipped into the United States in drums or burlap bags and are sampled as follows (551). Samples are drawn from each container by means of a compartmental trier. This tool is inserted in an inverted position completely through the contents of the containers while they are lying on their sides. The trier is then rotated to an upright position and withdrawn. The contents, representing a cross section of the material, is emptied onto a steel plate or similar clean surface. The material thus drawn is mixed by shoveling and the sample is reduced to a convenient size (10 to 20 pounds) by repeated coning and quartering.

The sample is pulverized so that the entire portion will pass through a 10-mesh screen. The 10-mesh material is then mixed and quartered down and further pulverized to pass a 20-mesh screen. The 20-mesh material is again mixed and riffled to 300 g., which is further pulverized so that the analytical sample will pass through an 80-mesh screen.

B. INTERMEDIATE PRODUCTS

Depending on the industrial processes involved, a great multitude of intermediate products may be encountered, and a detailed description of sampling procedures is beyond the scope of this chapter.

Stable products such as slags, oxides, or potassium fluorotantalite can be sampled simply by reducing representative portions to an analytical sample of suitable size by the usual methods such as coning and quartering, rotatory proportioning, or other appropriate splitting procedures. On the other hand, the sampling of unstable products such as the halides is more difficult. These products may be unhomogeneous and may change their composition if exposed to air. Other products, such as those resulting from liquid-liquid extraction, may be volatile. In the former case, the sampling can be carried out in an inert atmosphere in a "dry-box"; in the latter case, it is usually preferred to precipitate the metallic constituents with ammonia in a sample of adequate size.

C. NIOBIUM AND TANTALUM METALS

1. Powders

It is advisable to store powdered samples, particularly those high in hydrogen, under a protective atmosphere of argon (538). The chemical analysis is usually carried out on riffled samples of the unconsolidated metal from a representative portion of the sample (20). For sampling material of coarser particle size than No. 8 ASTM Sieve Designation, it is suggested to press a green compact from a riffled portion of the sample. The compact is drilled or milled to obtain a suitable quantity of chips for analysis.

2. Consolidated Metal

It is difficult to obtain a representative sample for the determination of the gases since chips obtained by milling must be avoided. A piece of metal can be cut off with a diamond wheel that is cooled with deionized water. Small pieces are subsequently cut off with a hacksaw for the determination of oxygen, hydrogen, and nitrogen. For the determination of metallic impurities the solid metal is drilled under distilled water. The drillings are dried. For the determination of iron they are first rinsed with hydrochloric acid (538). As was pointed out before, chips obtained by drilling or milling in air must be avoided in order to prevent a pickup of oxygen and nitrogen. (It was observed, for instance, that a sample of solid niobium metal indicating a nitrogen content of 104 to 106 p.p.m. showed a nitrogen content of 130 and 134 p.p.m. when chips or turnings were used (367).)

IV. SEPARATION AND ISOLATION OF NIOBIUM AND TANTALUM FROM OTHER ELEMENTS

It should be emphasized that, in analytical separations involving niobium and tantalum, the presence of large amounts of other metals may cause serious difficulties because of their marked influence on the otherwise normal chemical reactions of the two metals (156). These interferences, sometimes explained by "loss of individuality" (see Section II-E-1), are typified by the reactions when a bisulfate melt of a mixture of titanium, niobium, and tantalum is extracted with water or dilute sulfuric acid. The presence of titanium causes some niobium and tantalum to pass into solution, depending on the amount of titanium present; on the other hand, an excess of niobium or particularly tantalum causes some of the titanium to be precipitated. The "loss of individuality" referred to above also may occur in an alkaline medium. When, for instance, alkali carbonate or hydroxide melts containing niobium or tantalum and titanium are leached in hot water and filtered, titanium unexpectedly is partly found in the filtrate.

Members of the hydrogen sulfide group, including copper, tin, mercury, molybdenum, and platinum, can be precipitated, and thus separated from niobium and tantalum as sulfides from acid solution containing tartaric acid. Elements, such as manganese, nickel, and iron, also can be separated from niobium and tantalum by precipitation with hydrogen sulfide from an ammoniacal tartrate solution. (The separation from iron, however, is best carried out by passing hydrogen sulfide into a tartaric acid solution until iron is reduced. Ammonium chloride and an excess of ammonium hydroxide are then added and hydrogen sulfide is again passed through the solution, after which it is allowed to stand for at least half an hour until the ferrous sulfide settles.)

The separation of niobium and tantalum from other elements by specific reagents or reactions is described in more detail in Sections IV-A to IV-J.

A. DECOMPOSITION AND OTHER PRELIMINARY TREATMENT OF HIGH-GRADE ORES AND CONCENTRATES

1. General Observations

In view of the complexity in chemical composition of niobium- and tantalum-bearing minerals (see Table III), it is apparent that no one single approach can be effective for the decomposition of the many existing types of ores and concentrates. This complexity is partly due to the fact that in most minerals niobium and tantalum replace each other freely, as is exemplified by the columbite–tantalite and pyrochlore–microlite systems. In addition, iron and manganese substitute for each other in columbite and tantalite, and calcium, the usual basic constituent of pyrochlore and microlite, can be replaced not only with barium and strontium but also by sodium, various rare earths, notably cerium, and by thorium, iron(II), and manganese(II). The acidic portion of the mineral, Nb_2O_5 and Ta_2O_5, can be partly substituted by titanium and iron(III), and oxygen partly by fluorine and the hydroxyl ion. Among other acidic constituents, silicon is often present and zirconium is common, the latter apparently functioning sometimes as acid and sometimes as base because of its amphoteric character (250). The complexity in composition is even more pronounced in a series of minerals exemplified by samarskite, euxenite, fergusonite, and yttrotantalite, which, in addition to the elements already mentioned, contain significant amounts of titanium and/or uranium with no clear distinction in composition between related or associated minerals listed in books on mineralogy.

To complicate matters even more, the most desired mineral, columbite–tantalite, always is accompanied by more or less of cassiterite, frequently by titanium-bearing minerals such as rutile, ilmenite, or brooksite, and also by zircon, but less frequently by wolframite.

From the foregoing presentation or lamentation concerning the complexity of niobium- and tantalum-bearing minerals, it is apparent that, depending on the chemical composition of the minerals, varying approaches must be used for varying situations. In the case of unknown niobium- and tantalum-bearing minerals, therefore, it is of decisive importance to carry out preliminary qualitative tests to obtain as complete a picture as possible of the composition of the material. This qualitative examination is best carried out spectrographically or by means of x-ray fluorescence, as described in Sections VI-A and VI-B.

2. The Pyrosulfate Attack

a. THE FUSION

All niobium and tantalum minerals, if finely powdered, are attacked more or less readily by a fusion with alkali pyrosulfates. The choice of the flux depends on the subsequent treatment of the melt. Potassium pyrosulfate is handled more conveniently than the sodium salt because it can be obtained as "fused potassium bisulfate" with a low water content. The sodium salt, on the other hand, usually consisting of $NaHSO_4$, because of its higher water content, should be preheated to convert it to the pyrosulfate. Potassium pyrosulfate is very effective for columbite–tantalite type minerals. On the other hand, the sodium salt is preferred for minerals that contain rare earth elements and zirconium, because some of the resulting products are more soluble (250).

If the subsequent treatment consists of an anion-exchange procedure involving the fluoride medium, potassium salts naturally must be avoided because of the insolubility of potassium fluotantalate. In choosing a pyrosulfate attack for the determination of niobium–tantalum, it must also be remembered that certain associated minerals, notably zircon, cassiterite, and quartz, are scarcely attacked by the fusion, thus raising the question whether the insoluble residue may contain earth acids. However, an examination of hundreds of residues in Ledoux & Company's laboratory rarely revealed the presence of any niobium and tantalum, and only in cases in which certain details of the procedure were not strictly followed or when the sample consisted of simpsonite. It also must be remembered that a pyrosulfate attack of pyrochlore or microlite leads to the formation of calcium sulfate, which is only moderately soluble in water.

To carry out a pyrosulfate fusion, 0.5 to 1 g. of the finely ground sample is heated in a covered quartz crucible with 10 to 15 g. of pyrosulfate (sodium bisulfate should be preheated in a separate crucible until foaming ceases, then poured into a quartz crucible) for half an hour just above the melting point of the pyrosulfate. The temperature is then gradually raised and is maintained for half an hour at a dull red heat.

b. Treatment of the Pyrosulfate Melt

Probably the best-known continuation (250) consists in leaching the cooled pyrosulfate melt in 10 to 20% (wt./vol.) tartaric acid. If, after filtration, the remaining residue is large or is suspected to contain earth acids the fusion should be repeated. After the addition of a small amount of sulfuric acid, hydrogen sulfide is passed into the solution for the precipitation of any tin and other members of the hydrogen sulfide group introduced by the pyrosulfate attack. Following filtration, the solution is rendered ammoniacal and is again treated with hydrogen sulfide. The precipitate contains all the iron but only part of the manganese, and can be used for the determination of the former by conventional methods. After the ammonium sulfide filtrate is acidified with hydrochloric acid, the earth acids can be precipitated with cupferron. It should be noted that the cupferron precipitate also contains all of the titanium but only a portion of the zirconium (zircon is only partly attacked by the pyrosulfate fusion). If the sample contains tungsten, only part of it will be precipitated with cupferron in the above medium containing tartrate salts. Vanadium, which rarely is found in niobium–tantalum ores, and gallium, which occurs only in trace amounts in any type of ore (usually sulfide), are the only two other elements that have not been removed by the above treatment and that would be precipitated by cupferron.

Alternate treatments: "tartaric hydrolysis" (484) (precipitation of the bulk of the earth acids after the removal of the H_2S group, by the addition of hydrochloric acid to the boiling solution), has no advantage over precipitation with cupferron. This procedure still leaves a small portion of the earth acids in solution (the actual amount depends to a large degree on the amount of titanium in the sample) and requires some cumbersome steps for its recovery.

In the presence of rare earths, the pyrosulfate melt can be dissolved in oxalic acid, resulting in an insoluble rare earth fraction and a solution containing the earth acids, titanium, iron, manganese, etc. A subsequent cupferron precipitation in the filtrate produces insoluble niobium, tantalum, iron, and titanium cupferrates, which can be ignited, fused in pyrosulfates, and leached in tartaric acid for the treatment outlined before. If preferred, the oxalic acid solution containing the earth acids, after the addition of sulfuric acid, is evaporated to destroy the oxalic acid and is then diluted with a solution of tartaric acid.

Another procedure (250), well suited for minerals containing rare earths, consists in dissolving the pyrosulfate melt in water, adding a slight excess of ammonium hydroxide, filtering, washing the precipitate thoroughly, and dissolving it in hydrofluoric acid. This treatment yields an insoluble rare earth fraction and a solution containing, among others, the niobium and tantalum. After evaporating the bulk of the hydrofluoric acid and complexing the rest with boric acid, the solution can be diluted with tartaric acid and treated as described above, or the hydrofluoric acid solution can be transferred to a column containing cellulose for the separation involving ethyl methyl ketone (374) (see Section IV-I).

For the separation of niobium and tantalum from each other and other constituents of the sample by ion-exchange procedures (see Section IV-H), the sodium pyrosulfate melt (not potassium) can be dissolved in appropriate mixtures of hydrofluoric and hydrochloric acids and the solution directly transferred to a column containing strongly basic anion-exchange resin (282). Depending on the acid concentrations, iron, titanium, tungsten, phosphorus, and the sodium salts can be removed by simple elution while niobium and tantalum are retained by the resin. They can subsequently be separated by the use of different elutriants (see Section IV-H).

For a direct determination of niobium in high-grade ore, the pyrosulfate melt can be leached in $18N$ sulfuric acid and hydrogen peroxide. An aliquot is evaporated to fumes of sulfuric acid and the perniobic acid color is produced in a concentrated sulfuric acid–phosphoric acid medium by the addition of hydrogen peroxide (62,430). If the pyrosulfate melt is leached in dilute tartaric acid, niobium also can be determined rapidly by the thiocyanate procedure, using either the extraction method (266,557) or the homogeneous system (361,362) (see Section VII-B-1).

For the direct determination of tantalum, the pyrosulfate melt can be dissolved in dilute tartaric acid (62,362) or a solution of ammonium oxalate (140). In an aliquot the tantalum color is developed in an ammonium oxalate medium with pyrogallol and is corrected for titanium (62,362) (see Section VII-B-8).

3. Attack by Fusion with Potassium or Sodium Hydroxide and Sodium Peroxide

The pyrosulfate methods described above, although simple, have the disadvantage that samples containing cassiterite, zircon, and certain silicates are only partly, if at all, attacked. For the determination of niobium and tantalum this is not a serious disadvantage since virtually all minerals containing the two elements quantitatively yield their earth acid content. However, if the determination of the insoluble elements is part of the analytical scheme, alternate modes of attack may be more desirable.

Some analysts, therefore, prefer an alkaline fusion of the sample, because it assures a complete attack of all components of the sample.

The choice of alkaline flux depends largely on the subsequent treatment of the melt. Filtration of an aqueous sodium hydroxide or sodium peroxide solution leaves most of the earth acids in the precipitate, whereas a potassium hydroxide fusion yields most of the niobium and tantalum in the aqueous solution. Sodium peroxide is the preferred flux if the sample contains considerable tin. If the alkaline melt is leached in water and acidified with sulfuric acid, the earth acids can be hydrolyzed by boiling after the addition of sulfurous acid. The resulting precipitate not only contains the niobium and tantalum (partly as phosphates $Nb_2O_5 \cdot P_2O_5$ and $2Nb_2O_5 \cdot P_2O_5$ if the sample contains phosphorus (200)), but also more or less of the titanium, tin, tungsten, and silica. Zirconium will remain as phosphate, provided the sample contains sufficient phosphorus.

One typical recent procedure (199) involves the following steps.

1. Fusion with sodium peroxide.
2. Solution of the cold melt in dilute hydrochloric acid and hydrogen peroxide.
3. Boiling to expel H_2O_2.
4. Adjustment of pH to 4, addition of SO_2, and hydrolysis of niobium, tantalum, and titanium and part of the tin, tungsten, and phosphorus by boiling.
5. Ignition of filtered precipitate and fusion in potassium pyrosulfate.
6. Solution of the melt in $1N$ hydrochloric acid and hydrogen peroxide and hydrolysis of the bulk of the earth acids by boiling.
7. Filtration and repetition of steps 5 and 6.
8. Precipitation of tin sulfide in the combined filtrates from step 7.
9. Filtration of SnS.
10. Solution of the SnS in ammonium sulfide, leaving a residue containing a small fraction of earth acids.
11. Expulsion of H_2S and precipitation of the iron, titanium, and earth acid recovery in filtrate from step 9 with ammonia, followed by filtration.
12. Ignition of combined precipitates from steps 10 and 11, fusion in $KHSO_4$, fusion of melt in $0.05N$ hydrochloric acid and hydrogen peroxide, and hydrolysis of earth acids by boiling, followed by filtration.

13. Ignition of combined precipitates from steps *7* and *12* and weighing as $Nb_2O_5 + Ta_2O_5$.

14. Correction for coprecipitated tungsten and phosphorus by standard procedures.

It can be noted that in the above procedure the addition of hydrogen peroxide allows the separation of niobium and tantalum from titanium by a type of precipitation from homogeneous solution (124).

The above procedure has been greatly improved (158,200) by leaching the sodium peroxide melt in half-saturated aqueous sodium chloride. After centrifuging the alkaline solution, the clear supernatant liquid contains all the phosphorus, tungsten, tin, and silicon of the sample (also aluminum from the alumina crucible used for the sodium peroxide fusion). The sodium salts or hydroxides of niobium, tantalum, titanium, zirconium, iron, and manganese are quantitatively found in the residue. It must be pointed out that the complete insolubility of the earth acids and titanium is only assured in a very strong alkaline medium ($\sim 2.5M$), which is half saturated with sodium chloride. Dilution of the alkaline solution to permit filtration of the solution instead of centrifugation was shown to lead to losses of niobium, tantalum, and titanium.

4. Attack by Borax

It was recently shown (274) that minerals of the earth acid and rare earth groups, such as columbite–tantalite, samarskite, euxenite, pyrochlore, and microlite are all readily decomposed by a fusion with borax in a platinum crucible. (Bismuto–tantalite, although completely attacked, should not be fused with borax in platinum as damage to the crucible will result.) By dissolving the borax melt in dilute hydrofluoric acid, the rare earths and thorium are separated as insoluble fluorides, accompanied by the major part of the calcium in such minerals as microlite and pyrochlore. Spectrographic analysis has shown the absence of boron, titanium, niobium, and tantalum in the fluoride precipitates from samples of minerals that are known to contain calcium, thorium, and rare earth elements. In this respect the method compares favorably with the direct HF attack described below.

The filtrate from the fluoride precipitate contains all the titanium, iron, niobium, and tantalum, together with the boron that has not yet volatilized. Evaporation with sulfuric acid serves not only to remove all the fluorine, but also all the boron as the volatile boron trifluoride.

Columbite ore also can be decomposed with a fusion mixture consisting of 2 parts of sodium carbonate and 1 part of sodium tetraborate (452)

(1 part of sample to 20 parts of fusion mixture). The fusion is carried out in a platinum crucible. The melt is dissolved in dilute hydrochloric acid and the solution is evaporated to fumes with perchloric acid. After dilution with water, the niobium, tantalum, etc., are precipitated with ammonia. The precipitate is filtered off and decomposed by heating with nitric and sulfuric acids. Tin is expelled with hydrobromic acid. Subsequently, the solution is diluted with water and the earth acids, zirconium, and tungsten, plus most of the titanium, are precipitated by a treatment with ammonium thiocyanate and p-hydroxy phenylarsonic acid and finally are ignited to the oxides.

Fusion of earth acid minerals with borax is also of value for x-ray fluorescence purposes, since it allows the preparation of synthetic standards and also the introduction of internal standards. Lithium metaborate ($LiBO_2$) has been recommended as a flux in an x-ray procedure that incorporates MoO_3 as an internal standard (256)(see Section VII-F-4-b).

5. Attack by Hydrofluoric Acid

Attack of minerals containing earth acids with hydrofluoric acid has not found widespread acceptance. It has been noted that certain rare earth-bearing columbates and tantalates dissolve with great ease in concentrated hydrofluoric acid, whereas others, such as columbite and tantalite, do not (250). Tests carried out in the laboratory of Ledoux & Company, however, conclusively showed that complete decomposition of all niobium- and tantalum-bearing minerals (with the exception of simpsonite) can be obtained if certain essential details are observed. When, for instance, a powdered sample of columbite or tantalite is heated in a platinum dish with 20 ml. of hydrofluoric acid, 20 or more treatments, with intermittent evaporation on a steam bath, are required before complete decomposition of the sample is achieved. When, on the other hand, a finely powdered sample (-200 mesh) is heated with 25 ml. of hydrofluoric acid in a 200-ml. polyethylene beaker and the escape of the acid is prevented by covering the beaker with a thin sheet of polyethylene wrap held tightly against the beaker by a rubber band, complete decomposition of the sample can be obtained in about 2 hours. If part of the hydrofluoric acid is replaced by hydrochloric acid, decomposition is even more rapid, probably due to the greater solubility of the earth acids in a mixture of the two acids.

The solubility of earth acid minerals in hydrofluoric acid or a mixture of hydrofluoric and hydrochloric acids is of great practical importance if ion-exchange or paper-chromatographic methods are used in subsequent separations. If, for instance, it is intended to use a 5:4:11 mixture of HCl, HF, and H_2O for the separation of niobium and tantalum from iron, manganese, titanium, zirconium, tungsten, phosphorus, etc., the sample can be decomposed in a polyethylene beaker (covered with polyethylene wrap as described above) with 25 ml. of HCl and 20 ml. of HF. When decomposition is complete, 55 ml. of water is added and the solution is ready to be passed through an anion-exchange column, as described in Section IV-H-4 (282).

In using the hydrofluoric acid attack, it should be noted that the residue as a rule contains all of the tin (from cassiterite) and all of the zirconium (from zircon). If no hydrochloric acid is used, the total rare earth group, also thorium, is found in form of fluorides in the insoluble residue, as is part of the calcium. On the other hand, spectrographic examinations in the author's laboratory clearly have shown that the residues from an HF or HF–HCl attack are virtually free from niobium, tantalum, titanium, and other elements forming soluble fluorides.

6. Attack by Chlorination

Chlorination of earth acid minerals, combined with subsequent distillation of the various anhydrous chlorides, is more attractive for applications in extractive metallurgy (see Section I-C-1) than for analytical purposes. In most analytical procedures a preliminary separation from iron is required (251), although it has been reported that the addition of sodium chloride promotes the formation of stable nonvolatile ferric chloride (503). The main value of chlorination consists in affording a separation of niobium and tantalum ($NbCl_5$, b.p. = 241°C.; Ta_2O_5, b.p. = 242°C.), from titanium ($TiCl_4$, b.p. = 136°C.) and tin ($SnCl_4$, b.p. = 113°C.) (36,468, 469,546); its main disadvantage consists in not providing a separation of niobium from tantalum.

Recent work with this technique involves either carbon tetrachloride or carbon and chlorine as the chlorinating agents. Chlorination of the impure earth oxides obtained, for instance, by hydrolysis of a pyrosulfate melt, generally is carried out in the 200 to 350°C. temperature range, using sealed and evacuated bomb tubes as reaction vessels. The chlorination of niobium pentoxide proceeds according to one or both of the following reactions:

$$Nb_2O_5 + 5CCl_4 = 2NbCl_5 + 5COCl_2$$

$$Nb_2O_5 + 5COCl_2 = 2NbCl_5 + 5CO_2$$

No niobium oxychloride is formed as long as an excess of the chlorinating agent is present. Pure tantalum oxide requires a much higher chlorinating temperature (320°C.) than niobium oxide (225°C.). In the presence of niobium, however, tantalum can be chlorinated at lower temperatures, provided the earth acids are precipitated together. It has been noted that this catalytic effect of the niobium is even more pronounced in chlorinating titanium and tin oxides. After the chlorination is completed, tin and titanium are separated from niobium and tantalum (after removing the excess CCl_4 and $COCl_2$ by evacuation at room temperature) by raising the temperature to 110°C., applying mild suction, and distilling off the $TiCl_4$ and $SnCl_4$. Subsequently, the niobium and tantalum are distilled off by raising the temperature to 200°C. and lowering the pressure to about 1 mm. Hg.

Chlorination at atmospheric pressure is even more difficult. If carbon tetrachloride is passed over hot niobium pentoxide with a carrier gas, a large portion of the niobium may be lost due to the formation of the volatile oxychloride (417). This can be avoided if octachloropropane is employed as the chlorinating agent (36,251). To obtain quantitative chlorination, however, it is necessary that the oxides be used in a more or less amorphous form. This can only be achieved by low-temperature drying of the precipitated earth acids. It was also noted that among the various oxides studied, the chlorination rate varies in the order niobium pentoxide > tantalum pentoxide > titanium dioxide > zirconium dioxide, i.e., increasing as their acidity is increased. When elements of Group III are present, their chlorination will either fail completely or proceed only to a slight extent. Iron, in particular, should be absent, because it causes a catalytic decomposition of the reagent.

If the sample is heated around 300°C. in a porcelain boat placed in a combustion tube, in a stream of sulfur dichloride and chlorine, the distillate contains all of the niobium, tantalum, tungsten, tin, molybdenum, antimony, arsenic, and part or all of the iron, leaving behind such elements as the alkalies, calcium, silica, and alumina (250).

B. ATTACK OF LOW-GRADE MINERALS

The decomposition of low-grade minerals depends on the composition of the sample. Usually a preliminary treatment is required to remove the matrix elements. Calcite and phosphorus can be removed by a treatment with dilute acetic acid (160). Dilute hydrochloric may be used to dissolve iron or phosphate minerals (361,374). Dilute nitric acid is preferred to decompose calcium minerals and thus concentrate the niobium and tantalum in low-grade ore containing pyrochlore (319). Since niobium and tantalum are frequently disseminated throughout a silica matrix, many analysts prefer to treat such samples first with hydrofluoric and sulfuric acids (39a,315,383,401,405,437,464,525,584). The subsequent treatment consists either in evaporating the solution to dryness, followed by a potassium bisulfate fusion of the residue (315,383,584), or in precipitating niobium and tantalum in the acid medium with tannin (39a,401,405,437,464, 584), pyrogallol (525), or phenylarsonic acid (446). For the determination of niobium in the parts-per-million range in rocks, a different approach is based on the fusion of the sample in a gold crucible with sodium hydroxide (218). Iron and magnesium help to collect the niobium in the subsequent filtration of the aqueous solution of the melt. An ammonium hydroxide separation removes elements such as tungsten, molybdenum, vanadium, and rhenium, which would interfere in the final thiocyanate photometric procedure (see Section VII-B-1-a).

C. DECOMPOSITION OF INTERMEDIATE PRODUCTS

Halogenated substances should first be hydrolyzed to avoid losses of one or the other constituent of the sample due to volatilization. They can also be fused with alkaline fluxes. A fusion with a bisulfate salt invariably will lead to losses of niobium and tantalum and possibly of other elements.

Niobium- and tantalum-bearing slags resulting from the extractive metallurgy of tin metal are high in silica, calcium, and iron. They are best fused in sodium peroxide. The melt is leached in water and acidified with hydrochloric acid, and niobium and tantalum are precipitated with ammonia. The precipitate can be dissolved in a 5:4:11 mixture of HCl, HF, and H_2O and passed through an anion-exchange column. Alternatively, the precipitate after ignition is fused in sodium bisulfate and the melt is dissolved in HF or an appropriate HF–HCl mixture for a separation of niobium and tantalum by either paper chromatography (see Section IV-I) or ion exchange (see Section IV-H). If only a determination of the combined oxides of niobium and tantalum is required, the bisulfate fusion can be dissolved in tartaric acid and the solution treated as described for minerals.

Niobium- and tantalum-bearing slags also can be decomposed by heating with HF and H_2SO_4. After evaporation to fumes, the salts are dissolved in dilute hydrochloric acid and niobium and tantalum, etc., are precipitated with cupferron or tannin.

Possibly the best mode of attack was recently described (282). It consists of fusing the sample in sodium peroxide, leaching the melt in water acidifying the solution with $18M$ sulfuric acid, and adding sufficient hydrochloric and hydrofluoric acids as to provide a 25% HCl (vol./vol.) and 20% HF (vol./vol.) concentration. This solution is now ready to be transferred to an anion-exchange column for a chromatographic separation of various components of the sample (see Section IX-A-2 for details).

D. TREATMENT OF NIOBIUM AND TANTALUM OXIDES

Samples of niobium and tantalum oxide are analyzed for major components as well as for various impurities. The "combined oxides" of the two elements, more or less contaminated with other elements, also are frequently obtained as the result of preliminary chemical separations and serve as a pivotal point for subsequent steps. Thus, the decomposition and preliminary treatment of ores and concentrates described in Section IV-A often aims to obtain first the 'combined oxides," contaminated sometimes with titanium, zirconium, tungsten, and/or phosphorus (190,199,200,250). Similarly, after the decomposition of steels and alloys by various acid treatments, the combined oxides are obtained as the result of hydrolysis or after precipitation with tannin, cupferron, pyrogallol, or other precipitants.

Niobium and tantalum oxide can be decomposed a number of ways. The selection of a suitable flux or acid depends largely on the chemical properties and concentration of the elements to be determined and also on the choice of a chemical or instrumental method.

1. Pyrosulfate Attack

Probably the most frequently used mode of attack consists of a pyrosulfate fusion in a quartz or Vycor crucible. (Platinum crucibles or dishes are less desirable since platinum catalytically accelerates the decomposition of pyrosulfate and thus reduces the effectiveness of the flux.) Niobium pentoxide dissolves rapidly in fused potassium or sodium bisulfate, whereas the decomposition of tantalum pentoxide is less readily achieved. In view of the tendency of the combined oxides to form mixed crystals, $(Nb,Ta)_2O_5$ (when niobium and tantalum have previously been simultaneously precipitated with cupferron, tannin, ammonia, etc. (see Section IV-G)), it is not surprising that an excess of niobium oxide improves the fusibility of the mixture whereas an excess of tantalum oxide retards the decomposition.

Since the pyrosulfate attack of niobium and tantalum oxides represents only the first stepping stone for a great number of separations, no detailed descriptions of all the various procedures can be given here. A few typical examples should suffice.

In the classic procedure by Schoeller the bisulfate melt is dissolved in ammonium oxalate and tantalum is precipitated at a pH of about 4 as a tannin complex (482). Niobium is only precipitated at a higher pH (570). Since titanium in more than trace amounts interferes, it can be complexed as sodium salicylate while the niobium and tantalum are precipitated by the addition of calcium chloride to the oxalate solution. Small amounts of earth acids can be separated from large amounts of titanium by dissolving the pyrosulfate melt in 5% sulfuric acid containing 1% tannin (487). If the melt is dissolved in dilute hydrochloric and tartaric acids, selenous acid can be used as a precipitant for the niobium and tantalum while titanium stays in solution (16). The separation also can be achieved by complexing the titanium in an oxalic acid medium with EDTA and precipitating the earth acids with ammonia (33). For the separation of niobium from tantalum and titanium, the bisulfate melt can be dissolved in ammonium tartrate and the niobium precipitated at pH 6 with 8-hydroxyquinolate (52). If a chromatographic separation of niobium and tantalum from each other or from titanium, zirconium, tungsten, or molybdenum is desired, the bisulfate melt can be dissolved in various combinations of hydrofluoric and/or hydrochloric acids and passed through a column containing strongly basic anion-exchange resin (226) or activated cellulose (91,573). If the bisulfate melt is dissolved in ammonium oxalate containing a little sulfuric acid, another chromatographic separation is possible using activated alumina as the adsorption medium (537). The bisulfate melt has been a favorable starting point for the spectrophotometric determination of both niobium and tantalum. Thus, the melt can be dissolved in ammonium oxalate and the tantalum determined as the pyrogallol (142,265) or catechol (424) complex. For niobium, a pyrogallol finish is also possible in an alkaline medium (574). Other examples involving a bisulfate fusion of the combined oxides are: the determination of niobium by the thiocyanate method (384), the determination of niobium and/or tantalum by the hydrogen peroxide (106,423) or hydroquinone color systems (265), and the determination of niobium by reduction in a Jones Reductor and titration with potassium permanganate (418).

2. Alkaline Attack

When niobium and tantalum oxides are fused in a platinum crucible with potassium carbonate and the melt is leached in water, soluble potassium hexaniobate or tantalate, $K_8(Nb,Ta)_6O_{19} \cdot 16H_2O$, and potassium metaniobate or tantalate, $K(Nb,Ta)O_3$, are formed. Thus, a separation of niobium and tantalum from zirconium, which forms an insoluble carbonate, is possible (488). However, titanium which in the presence of earth acids frequently modifies its normal behavior, dissolves in part and is found in both filtrate and residue (250). It should be noted that the solution of the potassium niobate in water is stable, but the aqueous tantalate solution clouds up on being boiled or on prolonged standing. The addition of a little potassium hydroxide, however, renders the potassium tantalate solution more stable. A sodium carbonate fusion can be employed to separate the earth acids from tungsten. When the melt is dissolved in a solution of a high sodium chloride concentration, practically all of the niobium and tantalum remain insoluble, whereas the tungsten is quantitatively found in the filtrate (484). The alkaline solution also can be used for a determination of phosphorus or boron (335).

3. Chlorination of Oxides

Chlorination of the oxides, usually with carbon tetrachloride in a closed tube, is only of limited use for analytical purposes. Examples are the separation of niobium from tantalum by the action of aluminum chloride (or iodide) on their oxides (104), the separation of tungsten from the earth acids by extracting the soluble WCl_6 with carbon tetrachloride from the mixture of chlorides (476), and the fractional distillation of titanium, tin, niobium, and tantalum chlorides (470).

4. Attack by Hydrofluoric Acid

As would be expected, the pentoxides of niobium and tantalum are appreciably soluble in hydrofluoric acid. Nb_2O_5 dissolves rapidly; however, the reaction of Ta_2O_5 with HF is sluggish. As was pointed out before, if precipitated and ignited together, the oxides form mixed crystals that behave more or less like the preponderant oxide. Thus, all of a "chemical" mixture containing 70% Nb_2O_5 and 30% Ta_2O_5 dissolves easily in HF, whereas all of a combination of 70% Ta_2O_5 and 30% Nb_2O_5 is attacked only slowly. The reaction can be speeded up, however, by dissolving the oxides in a polyethylene beaker, as described in Section IV-A-5. The hydrofluoric acid medium containing niobium and tantalum plus associated impurities is a favorite starting point for a number of separations involving ion exchange (97,226,307,333), cellulose chromatography (91,374), and liquid-liquid extractions (34,381). These techniques are fully discussed in Sections VI-H to VI-J.

E. NIOBIUM AND TANTALUM METALS

The metals can be ignited in a platinum or quartz crucible to oxides. The oxides can then be treated by one or the other techniques outlined above or can be analyzed spectrographically (see Section VIII-B-1).

If an acid attack is preferred, hydrofluoric acid is the usual choice. Although niobium and tantalum metals are both strongly attacked by hydrofluoric acid, it is necessary to add some oxidizing agent (usually nitric acid) to complete solution of the sample. Since the reaction is vigorous, in order to avoid losses of the sample, decomposition of the sample should be carried out in a covered platinum or polyethylene vessel. If required, the nitric acid can be removed by repeated evaporations with hydrofluoric acid. The HF–HNO$_3$ attack of niobium and tantalum metal is feasible for the determination of one or both of the elements by anion exchange (see Section IV-H-4) or for the determination of the metallic impurities, such as iron (22), bismuth, lead, and cadmium (402), tantalum (404,535), or zirconium (184). If a nitrogen determination is required, nitric acid naturally cannot be used. Hydrogen peroxide in conjunction with hydrofluoric acid has been used for this purpose (24).

Another acid attack that has recently been introduced is based on heating the sample with sulfuric acid containing ammonium bisulfate (23). This method has the attractive feature that the sample can be decomposed in a Pyrex or Vycor beaker.

For the determination of nitrogen in niobium or tantalum metal, the sample can be fused in a nickel boat in a closed system. The ammonia formed during the fusion is absorbed in water or boric acid and is determined by the usual volumetric or photometric techniques (234).

F. ALLOYS

1. Ferro Alloys

Ferroniobium and ferroniobium–tantalum can be decomposed with hydrofluoric and nitric acids (489), by a fusion with bisulfate (198,270) or sodium peroxide (571), or merely by heating with a 5:4:11 mixture of hydrochloric acid, hydrofluoric acid, and water, followed by careful oxidation with a few drops of nitric acid. This last-mentioned acid treatment is particularly attractive, since it provides in a very short time a solution that can be directly transferred to an anion-exchange column for a chromatographic separation of all components (21) (see Section IV-H-4).

If a hydrofluoric–nitric acid attack is used, the nitric acid can be removed by repeated evaporations with the intermittent addition of hydrofluoric acid, followed by solution of the salts in a suitable mixture of HCl and/or HF for ion-exchange or paper-chromatographic separations. Iron also can be removed in an ammoniacal tartrate solution with hydrogen sulfide; the earth acids subsequently are recovered with cupferron (198). If a sodium peroxide or potassium bisulfate fusion is used, the melts can be treated as is described in Section IV-A-3 (83).

2. Steels and High-Temperature Alloys

a. Isolation of Niobium and/or Tantalum Oxide

Niobium and, to a lesser degree, tantalum, form an integral part of the composition of Type 347 and 348 stainless steel. Since the function of the niobium and/or tantalum is to form a solid solution of NbC or TaC (see Section I-B-1), samples usually contain about ten times more niobium or twenty times more tantalum than carbon. A great number of other types of steels, which also may contain tungsten, molybdenum, and titanium, are analyzed for niobium and tantalum and other constituents of the sample.

In most chemical methods for determining niobium and tantalum in steel (17,19,265,286,364,384) the sample is dissolved in a mixture of hydrochloric and nitric acids. After the addition of perchloric acid, the solution is evaporated to white fumes. The salts are dissolved in water and hydrochloric acid. Freshly prepared SO_2 water is added and the solution is boiled to cause quantitative precipitation of the niobium and tantalum, accompanied by all of the tungsten (if cinchonine is used) and part of the molybdenum and titanium. After digestion, the precipitate is filtered off and washed with dilute hydrochloric acid.

In another technique the use of perchloric acid is avoided and complete recovery of niobium and tantalum is achieved by dissolving the sample in a mixture of hydrochloric and nitric acids and evaporating several times to dryness with intermittent additions of hydrochloric acid (39,212,396). The residue is usually "baked" before dissolving the soluble salts in dilute hydrochloric acid and filtering off the earth acids. The fact that both procedures are capable of providing a quantitative recovery of niobium, tantalum, and also tungsten, was recently demonstrated with the aid of radioactive tracers (17,71,95) and by other recovery studies (120). Other acid combinations and techniques that have been used are the following.

1. Dilute sulfuric acid; niobium and tantalum are only partially dissolved (38).

2. Nitric, sulfuric, and hydrofluoric acids; the HF is added to prevent hydrolysis of the earth acids during the subsequent phosphomolybdenum blue photometric niobium procedure (395,440).

3. Hydrochloric acid only, for trace amounts of niobium and tantalum, also titanium and zirconium, in conjunction with cupferron; for a preliminary selective separation from large amounts of iron and also manganese, chromium, nickel, and cobalt (227).

4. Hydrochloric and nitric acids (438,489) or perchloric acid (384) in conjunction with tannin.

5. Nitric, hydrochloric, and hydrofluoric acids; leaving, after expulsion of the nitric acid by repeated evaporations with intermittent additions of hydrofluoric acid, a solution that is suited for ion-exchange purposes (280).

6. N hydrochloric acid, *N* potassium chloride, and 1% sulfuric acid; in a procedure in which the sample is dissolved anodically leaving a residue of niobium and/or tantalum carbide (441). An electrolyte consisting of a mixture of hydrochloric acid, citric acid, and potassium iodide also has been used (425).

b. Subsequent Treatment of Oxides

The oxides obtained as a result of hydrolysis are further purified either by reprecipitation after solution in hydrofluoric acid and evaporation with sulfuric acid (71,122,265) or, in order to remove tungsten (39,78,212), by precipitation of niobium and tantalum as insoluble magnesium salts (see Section IV-G-5). They can also be treated by one or the other of the procedures described in Section IV-D. Since the earth-acid content of steels and high-temperature alloys rarely exceeds a few per cent, photometric or instrumental methods usually are preferable over gravimetric or volumetric procedures. A few examples are given here:

For the determination of the niobium, the oxide is fused in potassium carbonate. An aliquot of the solution of the melt is acidified in the presence of potassium citrate and is examined polarographically (396) (see Section VII-D-4).

The oxides are fused in sodium bisulfate and the melt is leached in a mixture of dilute hydrochloric and hydrofluoric acids for a separation of niobium and tantalum from admixed impurities and from each other by anion exchange (227) (see Section IV-H-4).

The bisulfate melt of the combined oxides is leached in a solution of tartaric acid. Niobium is determined in an aliquot photometrically by the thiocyanate method (see Section VII-B-1), and tantalum in another aliquot by the pyrogallol procedure (see Section VII-B-8) (364).

The bisulfate melt is leached in an ammonium oxalate solution. The niobium is determined photometrically by the hydroquinone method (see Section VII-B-3), and the tantalum in another aliquot by the pyrogallol procedure (19,265) (see Section VII-B-8).

The bisulfate melt is dissolved in a solution of citric acid. In an aliquot the niobium is determined photometrically as the oxinate (286) (see Section VII-B-4).

Tantalum and niobium are both determined photometrically with pyrogallol (147) (see Sections VII-B-6 and 8) in the ammonium oxalate solution of the bisulfate melt at different pH.

For the removal of molybdenum, the bisulfate melt of the oxides is dissolved in tartaric acid. Molybdenum is precipitated as the sulfide. After filtration, tartaric acid is destroyed with nitric acid. In an aliquot of the sulfuric acid solution niobium is determined colorimetrically as the peroxy complex (see Section VII-B-2) and in another aliquot tungsten is determined as the thiocyanate (195).

If a gravimetric finish is preferred, the impure oxides containing no tungsten are fused in bisulfate, the melt is leached in a hot solution of ammonium oxalate, and the niobium, tantalum, and titanium are precipitated with tannin in a feebly acid solution (489) (see Section IV-G-3). If tungsten is present, the potassium carbonate–magnesium sulfate separation (see Section IV-G-5) is carried out. The insoluble magnesium salts are dissolved in dilute hydrochloric acid and the earth acids plus titanium are precipitated by boiling with SO_2 water (39).

3. Titanium-Base Alloys

Titanium-base alloys can be dissolved in nitric and hydrofluoric acids (411,412), hydrochloric acid with the addition of hydrofluoric acid (228), ammonium fluoride (141), or $18M$ sulfuric acid (407). Following oxidation with nitric acid, niobium, and tantalum can be separated from the titanium matrix by anion exchange (see Section VII-H-4) (228) or by precipitation with tannin after complexing the hydrofluoric acid with boric acid (141, 411,412). The final determination of the niobium and tantalum is either gravimetrically (228) or photometrically (141,411,412).

4. Uranium- and Plutonium-Base Alloys

Uranium-base alloys usually are dissolved in nitric acid plus hydrofluoric acid. The solution is then evaporated to fumes with sulfuric acid (47,88, 217,380,381,385,394). A mixture of hydrochloric acid and hydrogen peroxide also has been used (309), but it is less effective than HF–HNO_3 acid combinations. Niobium can be determined in the sulfuric acid medium photometrically by measuring the absorbance of a solution of the complex formed with H_2O_2 at 360 mμ and cancelling out the interference of uranium by using a portion of the sample solution as a reference (47,88). Alternatively, the niobium either can be hydrolyzed in an acid SO_2 medium and converted to Nb_2O_5 by ignition at 900°C. (48,217) or it can be precipitated in an acidic medium with cupferron, extracted with chloroform, and, after evaporation of the organic layer, ignited to the oxide (380). Niobium also can be determined photometrically as the quinolate in a portion of the sample (394).

In an extraction procedure, niobium and tantalum are extracted with ethyl methyl ketone from a medium containing 20% sulfuric acid (vol./vol.), 10% hydrofluoric acid (vol./vol.), and 4% ammonium fluoride (w./vol.); subsequently they are determined as pentoxides after precipitation as tannates (385). Because of the low solubility of isobutyl methyl ketone (hexone) in aqueous solutions, this extractant is preferred in another procedure, which features a $10M$ hydrofluoric acid, $6M$ sulfuric acid, and $2.2M$ ammonium fluoride medium and hydrogen peroxide as a "backwash" for niobium and tantalum (381). The same solvent (hexone) and a $6.3M$ H_2SO_4–$1.6M$ hydrofluoric acid medium is used to determine niobium (560) and tantalum (559) in uranium- or plutonium-base alloys. In these procedures, hydroquinone is used for the photometric determination of both the niobium and tantalum (see Sections VII-B-3 and 9).

5. Miscellaneous Alloys

Carbides, unless cemented with nickel or cobalt, can be ignited to oxides. Cemented carbides are best attacked with nitric and hydrofluoric acids (327) or by prolonged heating with sulfuric acid containing ammonium bisulfate (504). The subsequent treatment of the sample follows closely that of impure earth-acid oxides. In one procedure, the hydrous oxides, obtained by evaporating the sulfuric acid solution, are treated three times with ammonia in the presence of EDTA. The precipitate is dissolved in dilute hydrofluoric acid and the cupferrates of the three metals are precipitated at pH 3.5 in the presence of boric acid. After ignition, weighing, and conversion of oxides into sulfates, the titanium is titrated with DCYTA in the presence of H_2O_2. The niobium is determined photometrically as the thiocyanate and the tantalum is calculated by difference (327).

Niobium–vanadium alloys are soluble in nitric and hydrofluoric acids. After evaporating the solution of the sample to fumes with sulfuric acid, the niobium can be hydrolyzed in the usual way by boiling in the presence of SO_2 water (32). To separate tantalum from vanadium the nitric–hydrofluoric acid solution can be passed through an anion-exchange column (175).

A niobium–palladium alloy has been decomposed by fusion with potassium bisulfate. The separation of the two elements is achieved in an oxalic–sulfuric acid medium by precipitating the palladium with potassium iodide and extracting the palladous iodide with isobutyl methyl ketone. Niobium remains in the aqueous layer and is recovered with cupferron (145).

Binary alloys of niobium or tantalum with zirconium can be decomposed with nitric and hydrofluoric acids. In this medium, most of the niobium and tantalum is separated from the zirconium by extraction of the former with isobutyl methyl ketone from a solution that has been equilibrated with $10M$ HF and $6M$ H_2SO_4. Residual niobium or tantalum is separated from the zirconium by precipitating the latter with ammonia in the presence of hydrogen peroxide (382). The separation of niobium and/or tantalum from zirconium also can be achieved through anion exchange (184).

Tungsten–niobium and tungsten–tantalum alloys, as well as binary alloys with molybdenum, are decomposed with nitric and hydrofluoric acids. The once-difficult separation of niobium, tantalum, molybdenum, and tungsten now can easily be achieved by anion exchange in an HF–HCl medium (226) (see Section IV-H-4). Niobium and tantalum can be determined in beryllium metal and oxide by precipitation with cupferron and subsequent separation by anion exchange in an HF–HCl medium (247).

Tungsten concentrate containing small amounts of niobium and tantalum is decomposed by a sodium peroxide fusion. When the melt is leached in water half saturated with sodium chloride, virtually all the niobium and tantalum remain insoluble and can be filtered off; the tungsten is quantitatively found in the filtrate (279).

G. MISCELLANEOUS PRECIPITATION PROCEDURES

1. Hydrolysis

Niobium and tantalum, in contrast to most elements with which they may be associated, distinguish themselves by not forming any simple cations. As was shown in Section II-E-4, the two elements therefore can exist in aqueous solutions only in the form of complex ions. Since they form only very weak acids, they hydrolyze readily in the absence of complexing agents. As a matter of fact, even the stability of the niobium and tantalum complexes formed with organic agents is not great and possible hydrolysis is of great concern to the analytical chemist unless deliberately sought. On the other hand, the formation of acid-insoluble niobic and tantalic acids through the medium of hydrolysis represents a valuable tool for the separation from a great number of other elements. If a hydrolysis step is incorporated into the analytical scheme, the following points are worth remembering.

Hydrolysis will take place when alkaline solutions of niobium and tantalum are acidified and subsequently boiled, or when metal samples are decomposed with mineral acids other than hydrofluoric acid. Precipitation, however, may not be complete.

Since niobic and tantalic acids are soluble in concentrated sulfuric and hydrochloric acids, the use of strong acid concentrations should be avoided.

If the solution contains hydrofluoric acid the latter must be expelled or complexed.

If, in the course of the analysis, niobic and tantalic acids are complexed with hydrogen peroxide, hydrolysis will take place only when the latter is destroyed, preferably by boiling. If a bisulfate melt is dissolved in hydrogen peroxide and nitric acid, more than 99.8% of the earth acids can be recovered by thermal decomposition, whereas the coprecipitation of the titanium is negligible. Precipitation of niobium, tantalum, and titanium also is possible by thermal decomposition of their peroxides from a slightly alkaline medium (pH 7 to 8) with low tungsten coprecipitation. By alternative precipitation from homogeneous solution in a nitric acid and an alkaline medium, it thus becomes possible to determine the tungsten, niobium and tantalum, and titanium contents of a sample after a titanium and tungsten determination in the respective filtrates (124).

Sulfur dioxide water has been used for more than 50 years to hasten the hydrolysis of niobium and tantalum (563). Its purpose is not well defined and difficult to substantiate. Its function may be based on raising the pH of the solution without simultaneously causing, as would ammonia, the precipitation of various R_2O_3 elements. Thus, in the analysis of steel (see Section IV-F-2), SO_2 water is frequently employed, presumably also to reduce Cr(VI) to Cr(III) and Fe(III) to Fe(II), thus lessening the contamination of the earth-acid precipitate. Although generally used, there are only limited data indicating that boiling with SO_2 water actually increases the recovery of niobium and tantalum in steel, since the major portion of the insoluble earth acids, if not all, may actually be the result of direct oxidation of the metals or carbides.

Tartaric hydrolysis, proposed by Schoeller (485) and based on the precipitation of the earth acids with hydrochloric acid from a tartaric acid solution of a bisulfate melt, does not provide a quantitative yield of niobium and tantalum (about 0.003 to 0.01 g. escaping precipitation). The recovery of the unprecipitated fraction is involved and can be achieved only after the cumbersome destruction of tartaric acid with nitric and sulfuric acids. Oxalic acid interferes strongly and sulfuric acid somewhat less with tartaric hydrolysis. Whereas titanium interference is not great and only positive, zirconium apparently forms a stable complex with niobium and tantalum and inhibits their precipitation (484). When tungsten is present, it will be quantitatively collected by the pentoxides (417).

The once-popular pyrosulfate hydrolysis (prolonged boiling of the solution of a bisulfate melt) is useless. It was formerly assumed that the water extracted all the titanium and left the niobium and tantalum unaffected. It has been shown, however, that not only do tantalum and particularly niobium pass into solution according to the relative amounts of titanium present, but also that tantalum, particularly when in excess, prevents some of the titanium from dissolving (250).

A neat separation of niobium and tantalum, however, can be achieved by prior reduction of the titanium and niobium with amalgam and subsequent boiling to cause the precipitation of the earth acids (420). The coprecipitation of titanium also can be obviated by repeated treatments with H_2O_2 and SO_2 water, as described in Section IV-A-3.

In all hydrolytic procedures, most if not all of the phosphorus and tungsten accompany the earth acids and must subsequently be separated or corrected for.

2. Precipitation with Cupferron

Although cupferron is not a specific reagent for niobium and tantalum, its selectivity makes it an outstanding tool in earth-acid analysis (432). Whereas cupferron has been popular in the United States for many years, a survey of the European, and particularly the English, literature indicates a preference for tannin. (This is not surprising, since Schoeller's monumental contribution to the analytical chemistry of niobium, tantalum, and a host of other elements unfailingly involved tannin as a reagent.)

One of the outstanding features of cupferron consists in its ability to precipitate niobium and tantalum over a wide pH range, preferably, however, from a 1 to $2M$ hydrochloric or sulfuric acid solution. The following elements also are quantitatively precipitated from such a medium: Fe(III), V, Zr, Ti, Sn, Ga, and U(IV). Elements that are partly precipitated with cupferron in a strong acid medium are Pd, Ce, Th, W, and Mo. Although the above list of interfering elements appears large, most of them can be eliminated without difficulties. For instance, if a 5:4:11 mixture of hydrochloric acid, hydrofluoric acid, and water, containing all the metals that are precipitated with cupferron, is transferred to a column containing strongly basic anion-exchange resin, cerium and thorium remain as insoluble fluorides that can be filtered off while all other elements, excepting niobium and tantalum, pass unadsorbed through the column (see Sections IV-H-4 and IX-A-2 for details).

Tartaric and oxalic acids do not interfere with the precipitation of niobium and tantalum by cupferron. Tests carried out in the author's laboratory indicate, however, that complete precipitation cannot be obtained from a solution containing citric acid or citrates. Tungsten is partly precipitated by cupferron from a mineral acid solution containing tartaric acid. In the absence of free mineral acid (when, for instance, a bisulfate fusion is dissolved in 25% (w./vol.) of tartaric acid), precipitation of niobium and tantalum, as well as of titanium and zirconium, is complete; tungsten is not precipitated (279).

Ammonium oxalate and oxalic acid also have been investigated as masking agents for tungsten (437). Since cupferron preferentially precipitates niobium, tantalum, titanium, and zirconium, small amounts of these elements can be separated from large amounts of iron(II) by adding only a limited amount of cupferron (38,227). Smaller amounts of iron can be removed as sulfide from an ammoniacal tartrate solution, prior to the precipitation of niobium and tantalum with cupferron. Tin, palladium, or molybdenum can be removed as sulfides from an acid medium containing tartaric acid. Uranium in the course of the analysis usually is present in the hexavalent state, which is not precipitated with cupferron. As a matter of fact, cupferron is sometimes employed to achieve the separation of niobium and tantalum from uranium (310,380).

3. Precipitation with Tannin

As "tincture of galls," tannin or tannic acid has been used for more than a century as a qualitative reagent, as well as for many years as an outside indicator for the volumetric ammonium molybdate lead method. Schoeller and his associates introduced tannin in 1925 as a valuable tool in earth-acid analysis (442). Its function is said to be based on the fact that its aqueous solution represents a colloidal suspension of negatively charged particles that precipitate positively charged metallic hydroxide particles with the formation of colored absorption complexes (489).

It was shown that the tannin complexes of niobium and tantalum, as well as those of tin, germanium, and tungsten, are insoluble, whereas the titanium complex is soluble in dilute sulfuric acid. Other elements form insoluble tannin complexes in weakly acid oxalate solutions. Since the oxalatotantalate complex is more readily decomposed than the oxalatoniobate complex, the tantalic acid liberated by hydrolysis can be precipitated by the addition of tannin, with which it forms a pale yellow adsorption complex. Niobic acid, which is only precipitated at higher pH values, forms a bright red tannin adsorption compound. (For a discussion of the method in regard to the separation of niobium and tantalum, see Section V-A-1.)

Tannin represented for about 25 years an ubiquitous reagent for earth-acid analysis. However, it has been reduced to a lesser role by more recent developments. Although Schoeller and his associates painstakingly investigated the various media suitable for the separation of numerous metals, an exact definition of pH and other conditions was sometimes not provided. For example, Schoeller's instructions for the precipitation of niobium from oxalate solutions mention the cautious addition of ammonium hydroxide to the boiling sample solution (150 ml.) to the appearance of a faint cloudiness, which is immediately removed with a minimum of dilute hydrochloric acid. This solution, which should be slightly acid to litmus, is now treated with 5 drops of strong hydrochloric acid, after which the solution is treated with an equal volume of saturated ammonium chloride solution. The extent to which the precipitation of niobium–tantalum and titanium with tannic acid from various complex solutions is pH-dependent was recently described (381). The optimum pH values for complete precipitation of 20-mg. portions of niobium, tantalum, and titanium from 200-ml. solutions containing 5 g. of the complexing agent, ammonium chloride, acetate buffer of the desired pH, and 1 g. of tannic acid, are given below:

Solution	Element	Optimum acidity
Ammonium acetate and HCl	Nb	0.1 to 2.5N
	Ta	0.1 to 4N
	Ti	0.1N
Ammonium oxalate	Nb	pH = 4.5
	Ta	pH = 4
	Ti	pH = 4
Ammonium tartrate	Nb	pH > 3
	Ta	pH > 2.5
	Ti	pH > 4
Ammonium citrate	Nb	pH > 5
	Ta	pH > 3.5
	Ti	pH > 3.8
Sulfosalicylic acid	Nb	pH > 0.5
	Ta	pH > 0.5
	Ti	pH > 3
Malic acid	Nb	pH > 4
	Ta	2N to pH 7
	Ti	pH > 3
Mandelic acid	Nb	pH > 2
	Ta	2.5N to pH 7
	Ti	pH > 3.5

The optimum pH for the quantitative precipitation of niobium, tantalum, and titanium from an oxalate medium has been independently verified with strikingly close agreement with the data presented above (58). It should be pointed out that the above study was limited to 20 mg. each of the three elements separately and that, at higher acidities in the case of niobium and tantalum, the precipitate was obtained often by a combination of hydrolysis and precipitation with tannic acid. Of particular interest is the fact that in a sulfosalicylic acid solution at pH 0.5 precipitation of niobium and tantalum is complete while only traces of titanium are precipitated. No definite conclusions, however, can be drawn from the above study concerning the separation of niobium and tantalum from larger quantities of titanium.

An earlier separation procedure, based on treating the potassium bisulfate melt of the mixed oxides with a 2% solution of tannin in 4% (vol./vol.) of sulfuric acid (481), or based on precipitation from a slightly hydrochloric acid solution (141,411,412), appears less attractive. Recovery studies carried out in the author's laboratory and elsewhere (94,211) indicate incomplete recovery of tantalum and, particularly, of niobium.

To prevent the interference of large amounts of titanium, it has been suggested to complex the latter with ascorbic acid prior to the addition of the tannin solution (437). Tests carried out with the aid of Nb-95 proved that precipitation of at least the niobium is virtually complete even when the Nb to Ti ratio is 1:44 (96). Prior reduction of titanium to Ti(III) with metallic cadmium also has been suggested to eliminate the interference of large amounts of titanium (544).

Interesting recent studies involve the use of EDTA in conjunction with tannin (127,464,517). It was shown that niobium, tantalum, and titanium are quantitatively precipitated and separated from most of the associated elements by tannic acid from oxalate solutions at a pH of about 4.5 in the presence of EDTA, and, further, that the following amounts of elements (as oxides) do not interfere with the precipitation of 44 mg. of mixed oxides of niobium, tantalum, and titanium: Ca, 20 mg.; Mg, 23 mg.; Fe(III), 50 mg.; Fe(II), 53 mg.; Mn, 48 mg.; rare earths, 26 mg.; Th, 20 mg.; U, 32 mg.; Zr, 12 mg.; Al, 20 mg.; Bi, 52 mg.; Pb, 10 mg.; and W, 4 mg. Tin, antimony, and larger amounts of tungsten interfere and must be removed prior to the precipitation with tannin, or they must subsequently be corrected for. The above procedure is attractive for the precipitation of the mixed oxides in minerals high in iron (ilmenite) or rare earths (samarskite).

Among other modifications of the tannin procedure it is noteworthy that gelatin has been found valuable for the collection of niobium and presumably tantalum (401), and that complete precipitation of the tantalum can be achieved at a pH as low as 1.6 if alkaloids such as cinchonine, strychnine, and brucine are used in conjunction with tannin (242).

4. Phenylarsonic Acid Procedure

Niobium and tantalum, together with titanium, zirconium, and tin, can be quantitatively precipitated with 5 to 10 g. of phenylarsonic acid from a boiling solution about 2 to $3N$ in hydrochloric or sulfuric acid and containing 5 g. of tartaric acid (11,12,446). This method appears attractive for the collection of traces of the above elements in rocks, after preliminary treatment with hydrofluoric and sulfuric acids, solution of the salts in tartaric acid, and fusion of any residue with sodium carbonate and/or potassium bisulfate (see Section IV-B).

Phenylarsonic acid precipitates tantalum completely from an acidity of approximately 10% sulfuric acid up to a pH of 5.8, when buffered with ammonium oxalate. Niobium in an oxalate medium precipitates only when the pH is higher than 4.8 (346). When the ratio of Ta_2O_5 to Nb_2O_5 is higher than 1:2, a double precipitation of tantalum is necessary. With the addition of EDTA and at a pH of 3, tantalum can be separated from most ions except those of titanium, zirconium, lead, barium, and strontium.

Phenylarsonic acid also has been used for the separation of niobium from tantalum by saturating a column of activated carbon with the reagent and passing a mixture of the oxalate complexes through the column in the presence of hydrochloric acid ($0.65N$). Tantalum is completely retained on the column and niobium passes into the eluate. Tantalum is subsequently eluted with 7% (w./vol.) oxalic acid (6). The separation of niobium and tantalum with phenylarsonic acid based on a prior peroxidation of the niobium (189) appears to be incomplete. In addition to phenylarsonic acid, *p*-hydroxy phenylarsonic acid also has been successfully used (452). The precipitation is carried out from a boiling 10% (vol./vol.) sulfuric acid solution after expulsion of tin as the bromide. To prevent the coprecipitation of any iron, the latter is complexed with ammonium thiocyanate and the filtered precipitate contains all of the niobium, tantalum, zirconium, tungsten, and titanium.

5. Other Precipitation Procedures

It has been reported that niobium, tantalum, and titanium can be separated from tungsten, iron, cobalt, and other elements by precipitation with ammonia in the presence of glycerol and EDTA (sodium salt) (328). This collection of the three elements is useful for the analysis of cemented carbides (327). EDTA is also employed in another procedure that features the separation of niobium and tantalum from titanium by precipitation of the earth acids with ammonia in an oxalate medium (33,517). Although the precipitation of the niobium and tantalum is complete, a reprecipitation is required to remove the titanium. However, even after reprecipitation, a small amount of titanium (0.5 to 1.5 mg.) accompanies the earth acids. If the amount of titania originally present in any mixture is twice that of the earth acids, over 75% of the titania is coprecipitated with the earth acids.

Both niobium and titanium (but not tantalum), when peroxidized, form EDTA complexes, thus preventing a direct complexometric titration of either element. However, only the existence of a complex between diaminocyclohexane-tetraacetic acid and the peroxytitanyl ion has been reported (326). This allows the volumetric determination of titanium in the presence of niobium and/or tantalum by adding an excess of DCYTA and back-titrating with copper solution to a PAN end point.

The separation of niobium and tantalum from tungsten by chemical methods is not easy. The only simple procedure known to this author that works irrespective of the amount of niobium, tantalum, and tungsten present is based on anion exchange (see Section IV-H-4). Also of value is the precipitation of small amounts of earth acids in a tartaric acid medium (no mineral acids!) with cupferron (see Section IV-G-2). Somewhat more cumbersome, but still frequently used, is a procedure more than 50 years old (51) that involves a fusion of the mixed oxides with potassium carbonate and treatment of the aqueous extract of the melt with a slightly ammoniacal solution of a magnesium salt. The earth acids, as well as titania and zirconia, are precipitated while alkali tungstate remains in solution (95,212, 443). The earth-acid fraction is subsequently treated with dilute hydrochloric acid. Niobium, tantalum, titanium, and zirconium are precipitated with cupferron or (minus the zirconium) in feebly acid solution with tannin while the tungsten is recovered in the alkaline carbonate solution by standard procedures. Niobium and tantalum also have been separated from tungsten by fusing the mixed oxides with potassium carbonate and treating the aqueous extract of the melt with an excess of sodium chloride, whereby all but a few milligrams of the earth acids are precipitated as the 7:6 sodium niobate and/or 4:3 sodium tantalate (see Section II-E-4), while tungsten stays in solution. The soluble earth-acid fraction is recovered by hydrolysis (486).

Pyrogallol, which is a very valuable reagent for the photometric determination of both niobium and tantalum (see Section VII-B-6 and VII-B-8), also has been used as a precipitating agent for the two elements. It has found application mainly in the preliminary concentration of niobium and tantalum in rocks following an attack with hydrofluoric and sulfuric acids (12,349,464,525). It should be noted that titanium and zirconium are not precipitated, and also that silica gel has been used as a collecting agent.

Niobium and tantalum, as well as tungsten, can be separated from zirconium (but not titanium) by a fusion of the mixed oxides with potassium carbonate and filtration of the aqueous solution of the melt (488).

The once-popular separation of niobium and tantalum from titanium by Schoeller's salicylate procedure (487) is scarcely used any more. It is based on the precipitation of the earth acids by eliminating the complexing oxalate ion while the titanium remains complexed as salicylate. Recovery of niobium and tantalum is seldom complete and is not too well reproducible. A much better separation of niobium and tantalum from titanium can be achieved by means of selenous acid (16). The separation is carried out in a $3M$ hydrochloric acid–1% (w./vol.) tartaric acid medium after fusion of the oxides with potassium bisulfate. The precipitate obtained after boiling the solution with an excess of selenous acid is filtered off, washed with dilute hydrochloric acid, and redissolved in 50 ml. of concentrated hydrochloric acid prior to reprecipitation. If the solution of the sample also contains oxalic acid, the method is suitable for the semimicro determination of tantalum in the presence of both niobium and titanium (219). Other reagents include hexamethylenetetramine (545), used to precipitate both niobium and tantalum; morin and quercetin (539), which precipitate both metals from oxalo–sulfuric acid solution; salicylhydroxamic acid (347); cinnamohydroxamic acid (351); N-cinnamoyl-N-phenylhydroxylamine (348); quinaldohydroxamic acid (351d); 7,8-dihydroxy-4-methylcoumarin (424a); N-benzoyl N-phenylhydroxylamine (351a); benzene–arsonic acid (289a); and acridine in the presence of SCN^- (556a).

H. SEPARATIONS BASED ON ION EXCHANGE

1. General Observations

Even a superficial examination of the various chemical separation procedures presented in the preceding pages underlines the many difficulties that have beset the analyst in the past in his endeavor to separate niobium and tantalum from a host of elements with which they may be associated, not to mention at all his valiant but frequently vain attempts to separate niobium and tantalum from each other.

The fact that niobium and tantalum can exist in solutions only as one or the other complex salt is exploited in these procedures merely to the extent that homogeneous solutions are prepared prior to the addition of a suitable precipitant. However, even in the best known of these procedures, the one involving tannin (Sections IV-G-3 and V-A-1), the optimum pH for the precipitation of the various elements is too close to achieve cleancut separations.

The newer techniques based on anion exchange, cellulose chromatography, and liquid-liquid extraction take full advantage of the fact that niobium, tantalum, and associated elements can exist in solution as characteristic complex species with distinct distribution coefficients and stabilities that are independent of each other.

This author advances with considerable enthusiasm the opinion that the introduction of ion-exchange methods represents the most significant factor contributing to recent rapid developments and simplifications of earth-acid analysis. Although the amount of literature covering ion-exchange procedures for the separation of niobium and tantalum from each other and from associated elements is still comparatively small, this author is aware of more than 50 installations in this country that either apply ion-exchange procedures to earth-acid analysis routinely or are engaged in some form of research to improve present ion-exchange procedures. The minimum equipment required is quite simple: a plastic tube (to accommodate various acid combinations, including hydrofluoric acid) containing a strongly basic anion-exchange resin of proper cross-linkage and mesh size. For a more elaborate setup see Fig. 18 in Section IX-A-1.

Various complexing agents of niobium, tantalum, and associated elements have been investigated for their suitability in ion-exchange separations. Most work involves hydrochloric acid, hydrofluoric acid, hydrochloric–hydrofluoric acid mixtures, and oxalic acid. Undoubtedly, other media will also be found suitable.

2. The Hydrochloric Acid System

In absence of water, niobium(V) and tantalum(V) form chloro complexes of the same type. In the presence of water, however, even in concentrated hydrochloric acid, niobium and tantalum hydrolyze, but in different ways.

It has been suggested (467) that niobium exists in hydrochloric acid in the form of one or the other of the following ions: $NbOCl_4^-$, $NbOCl_5^{-2}$, $Nb(OH)_2Cl_4^-$, or $Nb(OH)_2Cl_5^{-2}$. The complex $Nb(OH)_2Cl_4^-$ has been independently established and its anionic nature proven (284). Other species such as $NbOHCl_3^+$ and $Nb(OH)_2Cl_3$ also may exist, depending on the hydrogen ion and chloride ion concentration. Undoubtedly, when equilibrium is reached, several of the above species can exist at the same

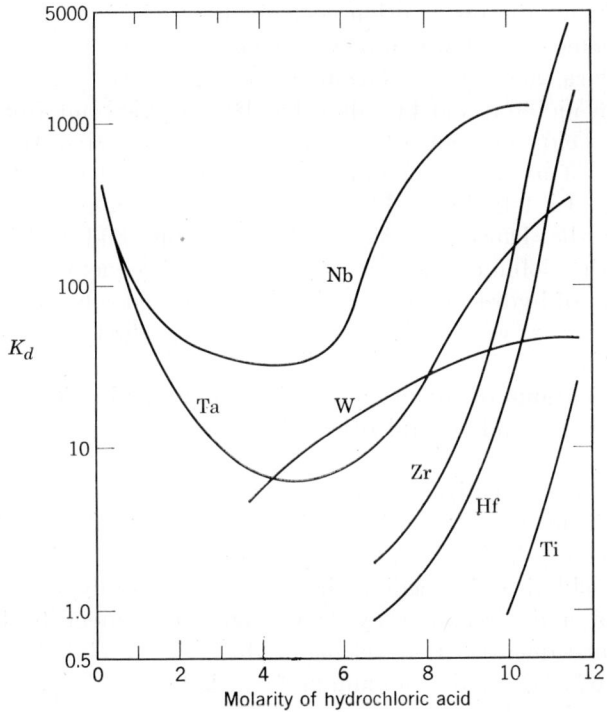

Fig. 1. Dependence of the distribution coefficients at 25°C. on the hydrochloric acid concentration in HCl medium (259).

time. Thus, three separate peaks in the elution curve of niobium were observed when a 10-molar hydrochloric acid solution was passed onto Dowex-2 resin and eluted with $7M$ hydrochloric acid (258).

Less is known about tantalum complexes in strong hydrochloric acid, but the hydrolysis of $TaCl_5$ apparently proceeds further than that of $NbCl_5$. For instance, freshly prepared niobic acid, $Nb_2O_5 \cdot nH_2O$, is soluble in concentrated hydrochloric acid. However, only traces of $Ta_2O_5 \cdot nH_2O$ can be dissolved. Although anionic compounds analogous to those of niobium are formed when $TaCl_5$ is dissolved in 10 to $12M$ hydrochloric acid, they hydrolyze at lower acid concentrations.

When applied to anion exchange, it has been observed that niobium and tantalum adsorb strongly at high concentrations of hydrochloric acid, less so in more dilute solutions, but more at still lower HCl concentrations. Presumably, they are adsorbed as chloro complexes at high-M HCl, and at low acidities are retained by the resin as hydrolyzed species (308).

Figure 1 shows the distribution coefficients of niobium, tantalum, and a few other elements with which they may be associated, as a dependence of the hydrochloric acid concentration. The separation of zirconium and hafnium from niobium and tantalum has been carried out from a strong hydrochloric acid medium (259) employing radioactive tracers. The concentration of niobium and tantalum were small (of the order of 7 to 10^{-6} molar). In 9M hydrochloric acid, niobium and tantalum were adsorbed by the Dowex-2 resin while zirconium and hafnium eluted readily. Both niobium and tantalum were subsequently eluted with dilute solution of hydrochloric acid. In stronger acid solutions (6 to 9M), tantalum, however, was washed from the column first, followed by the niobium.

Other applications of anion exchange in a hydrochloric acid medium involving niobium and several other elements also have been described (90).

Although the mechanism of the retention of niobium in a weakly hydrochloric acid medium by cation-exchange resins is not clear, in view of the comparatively high adsorption by anion-exchange resin (see Fig. 1), a 0.1N hydrochloric acid medium has been successfully used to adsorb both niobium and zirconium by Dowex-50W or Amberlite IR-1 (311, 462,541). Subsequently, both elements have been eluted with dilute oxalic acid (462,541), or the niobium preferentially with dilute hydrochloric acid containing hydrogen peroxide.

It should be emphasized that all ion-exchange separations in a purely hydrochloric acid medium—whether based on an anion- or cation-exchange mechanism—are limited to traces of niobium and/or tantalum because of the ubiquitous dangers of hydrolysis and the inherent difficulties of preparing and maintaining ionic species.

3. The Hydrofluoric Acid Medium

The hydrofluoric acid medium has only recently drawn some attention (165). It was shown that, among others, niobium, tantalum, titanium, zirconium, tungsten, and molybdenum are adsorbed strongly (distribution coefficients >100) from a 1M hydrofluoric acid solution, whereas such elements as iron, manganese, cobalt, nickel, chromium, and copper are adsorbed only slightly or not at all. Anion-exchange separations involving dilute hydrofluoric acid solutions therefore represent an excellent approach for the removal of large quantities of the matrix elements found in steels and high-temperature alloys and for the simultaneous collection of niobium, tantalum, and some other elements that are adsorbed from a weakly hydrofluoric acid medium. A hydrofluoric acid solution of the elements

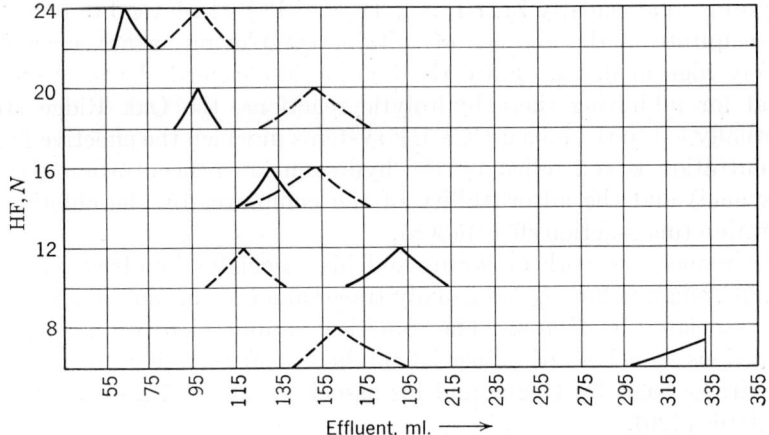

Fig. 2. Elution curves showing separation of niobium and tantalum in HF medium (165). Key: ----, Nb; ——, Ta. Column contains about 1g. of resin.

can be obtained by a solution of the metal in nitric and/or hydrochloric acid and evaporation to dryness with the intermittent addition of hydrofluoric acid (567). After the removal of iron, manganese, cobalt, nickel, chromium, and/or copper, the niobium, tantalum, titanium, zirconium, tungsten, and molybdenum can be separated from each other by a sequence of chromatographic steps involving mixed hydrochloric–hydrofluoric acid solutions, as described in Section IV-H-4. Using a 0.1 to $6M$ HF medium, both zirconium-95 and niobium-95 have been separated from other fission products (Ru, Cs, and Ce) by adsorption on a column containing Dowex-1 (311, 333).

Although the separation of niobium and tantalum is more advantageously carried out from an ammonium chloride–hydrofluoric acid medium (see Section IV-H-4) it should be noted that the separation of the two elements can be achieved from a purely hydrofluoric acid solution (165) (see Fig. 2).

4. The Mixed Hydrofluoric–Hydrochloric Acid Medium

a. Introductory Observations

The outstanding contributions of ion-exchange methods to the analytical chemistry of niobium and tantalum can be largely attributed to the introduction of the mixed hydrochloric–hydrofluoric acid medium. Pioneering research in this area was to a large extent carried out by members of the Oak Ridge National Laboratory, who initiated a comprehensive investigation of separation possibilities of those elements of the 4th and

5th groups (particularly Zr, Hf, Nb, Ta, and Pa) which tend to polymerize or precipitate in the absence of suitable complexing agents, even at extremely high acidities. Since the fluoride ion seemed the most effective ligand for inhibiting these hydrolytic reactions, the Oak Ridge studies eventually led to the mixed HCl–HF systems in which the effective fluoride concentration is controlled by the hydrogen ion concentration (HF is a weak acid) and the adsorbability of the complexes by the chloride concentration (mass-action effect) (308).

The pioneering work of Kraus and his associates involved only small columns, small volumes, and usually tracer amounts of niobium, tantalum, and associated elements. The practical application of ion-exchange separations to a host of industrial products had to wait until it was put on a firmer basis by Hague and his associates at the National Bureau of Standards (226).

b. Chemistry of the Exchange Mechanism Involving the HF–HCl Medium

In a strong hydrochloric acid medium containing no hydrofluoric acid, tantalum passes through the column more rapidly than niobium (see Section IV-H-2). Upon the addition of even small amounts of hydrofluoric acid fundamental changes take place that involve both the niobium and the tantalum (305,306,461).

Niobium. The niobium complex retains oxygen ($NbOCl_4^-$; $NbOCl_5^{-2}$; $NbOCl_3F_2^{-2}$; $NbOF_5^{-2}$) unless the HF concentration becomes high when part or all of the oxygen in the $NbOX_5^{-2}$ complex can be replaced by the fluoride ion.

Tantalum. Upon the addition of hydrofluoric acid, the tantalum changes from a "basic" complex containing more or less O_2 or OH^- to chloride fluoride complexes (such as $TaCl_5F^-$; $TaCl_5F_2^{-2}$; TaF_7^{-2} (467); and others (305)), which are more strongly adsorbed by the anion-exchange resin than is the niobium compound. Since the acidity of niobium and tantalum compounds decreases in the order $H_2MX_7 \ldots H_2MOX_5 \ldots M(OH)_5$, the selective adsorption by the resin can probably be explained as a function of changes in the acidity of the various species. For the elution of the tantalum, elutriants usually are used that contain less acid but more fluoride (for instance, $4M$ NH_4Cl–$1M$ HF), the latter replacing the complex tantalum ion. A better elutriant for tantalum consists of a 3 to $4M$ NH_4Cl–$1M$ NH_4F solution (307). This particular elutriant has subsequently been used by a number of investigators. The rapid elution of the tantalum by the NH_4Cl–NH_4F solution can be explained

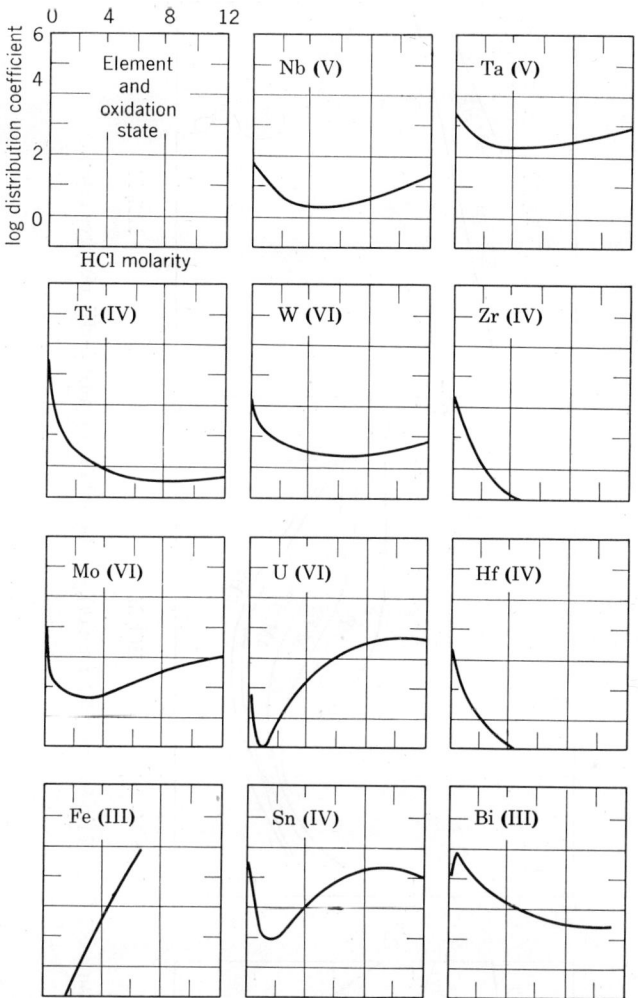

Fig. 3. Adsorption of some elements from HCl–HF solutions (308). Note: 0.5M HF is used with Nb, Ta, Zr, and Hf; 1M HF is used with Ti, W, Mo, U, Fe, Sn, and Bi.

by an increase in the negative charge through the addition of further fluoride ions (307).

It is fortunate to the analytical chemist that all elements (with the exception of bismuth and antimony) with which niobium and tantalum may be associated show dissimilar adsorption characteristics, since their chlorofluoro complexes consist either of different species or exhibit different

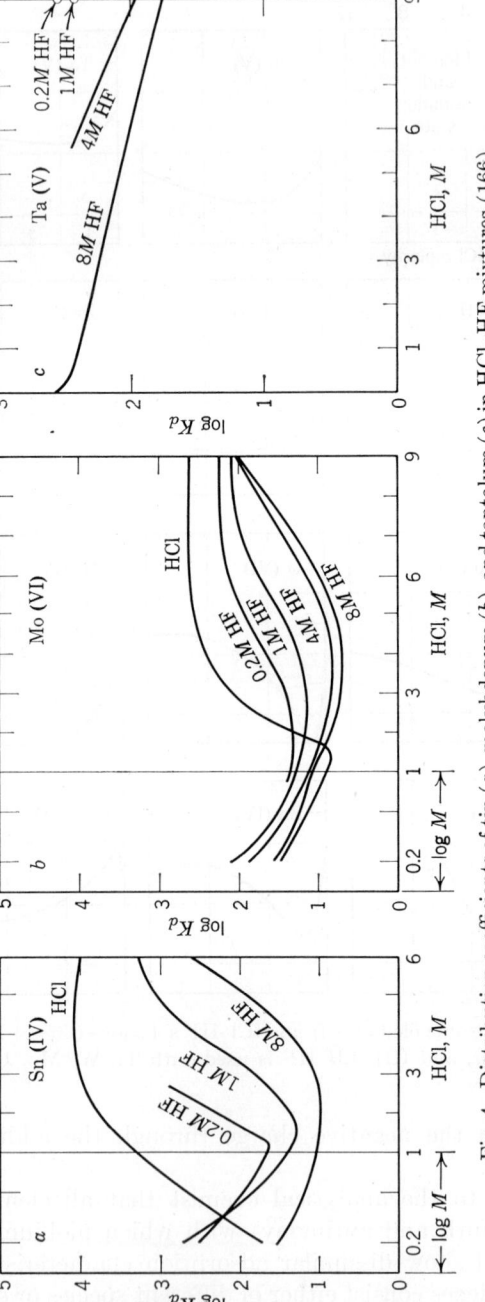

Fig. 4. Distribution coefficients of tin (*a*), molybdenum (*b*), and tantalum (*c*) in HCl–HF mixtures (166).

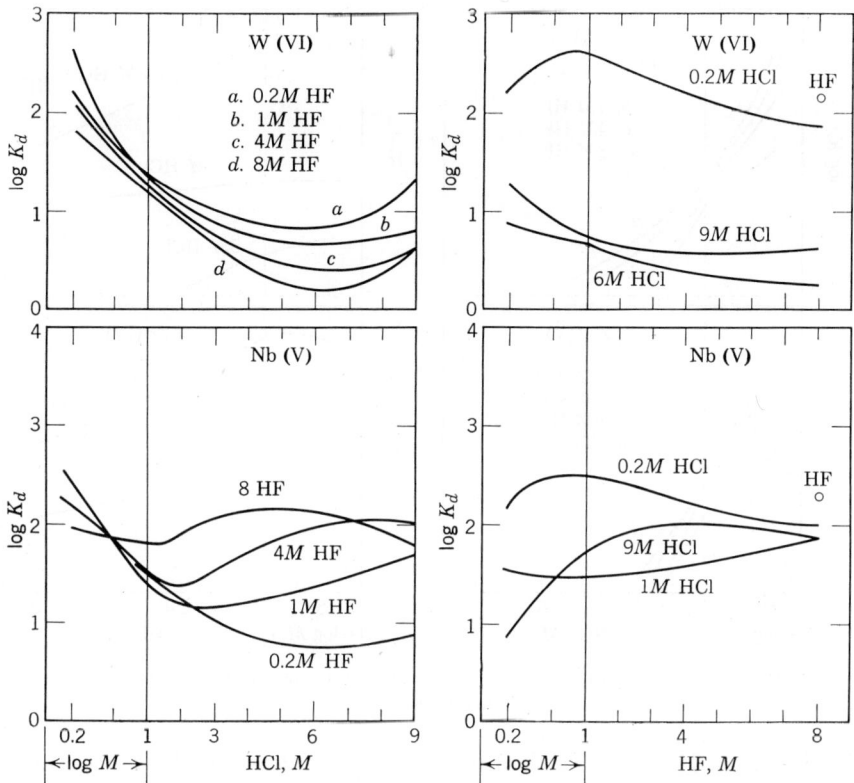

Fig. 5. Distribution coefficients of niobium and tungsten in various HCl–HF mixtures (166).

degrees of stability in various HCl–HF solutions. As a matter of fact, of those elements that frequently accompany niobium and tantalum only the adsorption characteristics of titanium and zirconium are somewhat related; the distribution coefficient of tungsten resembles those of titanium and zirconium.

The adsorption of a number of elements by Dowex-1 resin from HCl–HF solutions recently has been thoroughly studied (166,308) and some of the adsorption curves are here reproduced (see Figs. 3 to 6). Additional adsorption studies involving the HCl–HF system, niobium and/or tantalum, and various fission products (U, Np, Pu, Zr, and Mo) and constituents of complex alloys (Al, Ti, Zr, V, Cr, Mo, W, Mn, Fe, Ni, and Cu) have recently been published (242a,572). An inspection of Fig. 5 would tend to confirm what was stated before, namely that the niobium complexes

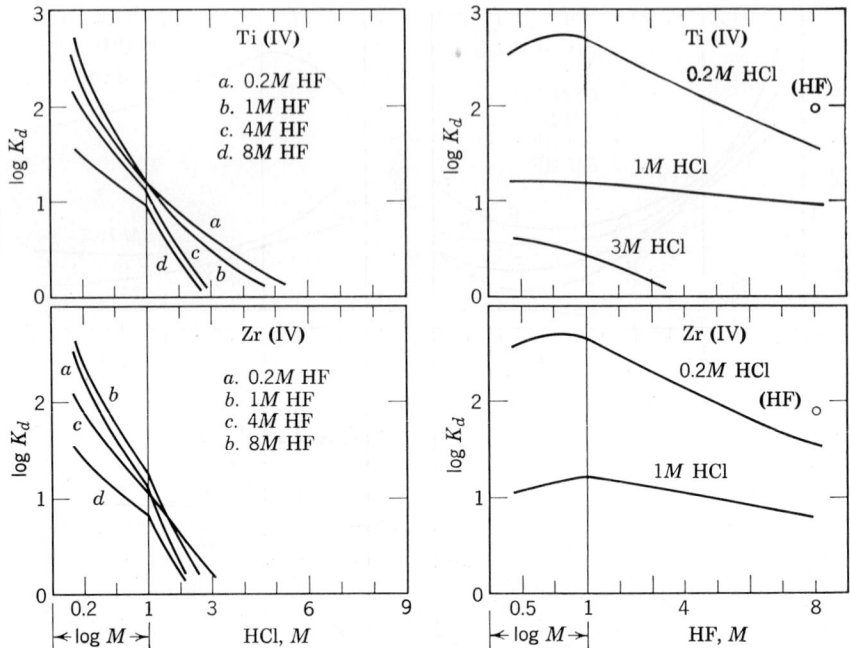

Fig. 6. Distribution coefficients of titanium and zirconium in various HCl–HF mixtures (166).

formed in an HCl–HF solution retain oxygen unless the HF concentration becomes high. It can be seen that in HCl from 3 to 6M the effect of increasing HF leads to an increased adsorption of the niobium, which is the opposite of what has been noted for all other elements. The increased adsorption is explained by the formation of oxygen-free complexes analogous to those of tantalum.

c. Separation Principles

It is desirable to operate under conditions at which the separation factor S, defined in terms of the ratio of distribution coefficients of two ions to be separated, is as large as possible. To achieve separations within a reasonable amount of time, a further condition should be satisfied: the element to be eluted should have distribution coefficients in the neighborhood of unity whereas those of the retained metals are as large as possible. For the separation of several metals, one should first select conditions under which the distribution coefficients of all but one element are high (log K_d = < 1), remove this element, and then progressively

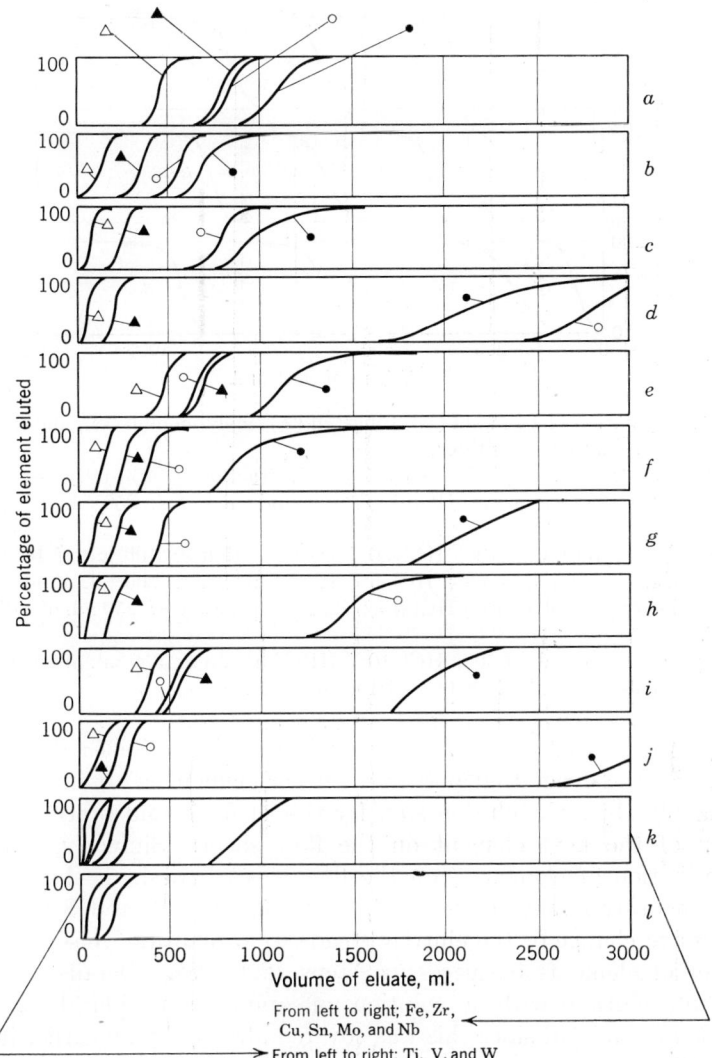

Fig. 7. Elution behavior of titanium (△), tungsten (▲), molybdenum (○), and niobium (●) (226,228). Key to eluate compositions: a, 5% HF, 10% HCl; b, 5% HF, 25% HCl; c, 5% HF, 40% HCl; d, 5% HF, 60% HCl; e, 10% HF, 10% HCl; f, 10% HF, 25% HCl; g, 10% HF, 40% HCl; h, 10% HF, 60% HCl; i, 20% HF, 10% HCl; j, 20% HF, 25% HCl; k, 20% HF, 12.5% HCl, and 7.2% NH$_4$Cl; and l, 20% HF, 12.5% HCl, and 7.2% NH$_4$Cl.

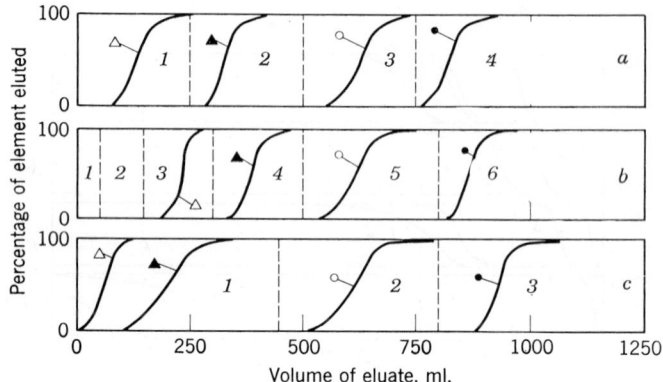

Fig. 8. Separation of titanium (△), tungsten (▲), molybdenum (○), and niobium (●) (225). Key to eluate compositions:

 a. Fraction 1; 250 ml. of 25% HCl–5% HF; fraction 2; 250 ml. of 40% HCl–5% HF; fraction 3: 250 ml. of 40% HCl–10% HF; and fraction 4: 250 ml. of 14% NH₄Cl–4% HF.

 b. Fraction 1: 50 ml. of 10% HCl–20% HF; fraction 2: 100 ml. of 10% HCl–5% HF; fraction 3: 150 ml. of 25% HCl–5% HF; fraction 4: 200 ml. of 40% HCl–5% HF; fraction 5: 300 ml. of 40% HCl–10% HF; and fraction 6: 250 ml. of 14% NH₄Cl–4% HF.

 c. Fraction 1: 450 ml. of 50% HCl–10% HF; fraction 2: 350 ml. of 25% HCl–20% HF; and fraction 3: 350 ml. of 14% NH₄Cl–4% HF.

change the medium so that one metal at a time reaches (log K_d < 1) (308). In the event that one or the other element is absent in a sample, one merely skips the elution step for this element and proceeds with the elution of the next element on the flow sheet. Similarly, if only the determination of element *4* on the flow sheet is desired, elutriant *3* will simultaneously remove elements *1*, *2*, and *3*. Figs. 7 and 8 show how a progressive change of the elutriant leads to a quantitative separation of a number of elements in successive steps (226,228). Details of the final tantalum elution with a solution of ammonium chloride–ammonium fluoride are omitted since this was covered already by an earlier investigation (307).

d. Practical Applications

Based on the principles outlined above, a number of methods have been suggested in the literature. More than 100 mg. each of niobium and tantalum were separated from each other by passing a $3M$ HCl–$0.1M$ HF solution through a column (12 cm. × 1.3 cm.²) containing 120 to 200 mesh De-Acidite FF resin. The eluate contained the niobium and the adsorbed tantalum was subsequently eluted with $4M$ NH_4Cl–$1M$ NH_4F (55,97). Kraus and Moore showed that the separation of niobium and tantalum also can be achieved by means of Dowex-1 (304), however less effectively, in a $9M$ HCl–$0.05M$ HF solution employing a 12.5-cm. column. Protactinium, if present, elutes ahead of the niobium (307). The separation of zirconium from niobium is excellent under a number of conditions (305).

For the determination of small amounts of niobium, tantalum, titanium, and zirconium in steel the following procedure has been suggested (227).

After the solution of the steel (5 to 25 g.) in $4N$ hydrochloric acid, a cupferron separation is carried out to remove the bulk of the iron(II). The ignited precipitate (containing all of the Nb, Ta, Ti, and Zr, some coprecipitated Fe and Cr, and possibly also W, V, Mo, and Sn) is dissolved by a technique that involves treatment with $6N$ hydrochloric and fusion of the insoluble residue with sodium bisulfate. A solution containing (per 100-ml.) 50 ml. of hydrochloric acid and 10 ml. of hydrofluoric acid, and holding the elements mentioned above, is transferred to a polystyrene column (12 × 1 inch) containing a resin bed of 6 to 7 inches of 200 to 400-mesh Dowex-1, 8 to 10% crosslinkage. The first eluate (350 ml. of 50:10 elutriant) contains the titanium and zirconium along with any vanadium, tungsten, or chromium. A second eluate (350 ml. of a mixture containing per 100 ml., 7 g. of NH_4Cl, 12.5 ml. of HCl, and 20 ml. of HF) yields the molybdenum, together with the iron and tin. 300 ml. of a third eluate (14 g. of NH_4Cl and 4 ml. of HF per 100 ml.) contains the niobium, and 300 ml. of a fourth eluate (No. 3 elutriant neutralized with ammonia to pH 6) contains the tantalum.

In a scheme that avoids a preliminary collection of niobium, tantalum, etc., by hydrolysis or cupferron precipitation, a 2.5% hydrofluoric acid solution of steel or a high-temperature alloy (obtained by solution of the sample in an appropriate mixture of HF, HCl, and HNO_3, evaporation to dryness, and solution in 2.5% HF), is passed through Dowex-1 resin. The eluate contains elements such as iron, cobalt, nickel, chromium, manganese, aluminum, and copper. The resin retains elements such as titanium, zirconium, tungsten, molybdenum, niobium, and tantalum. The titanium is eluted with 250 ml. of $8M$ HCl, the tungsten with 300 ml. of a solution that is 10% HF and 60% HCl. The molybdenum is eluted with 300 ml. of a solution that is 20% HF and 25% HCl. The niobium is eluted with 300 ml. of a solution that is 14% NH_4Cl and 4% HF, and the tantalum with 300 ml. of a solution that is 14% NH_4Cl and 4% NH_4F (567).

In the hands of this author and members of his staff certain inherent weaknesses became apparent in the above procedure. The solubility of chromium fluoride (CrF_3) is small in a 2.5% HF medium, thus limiting the amount of high-temperature alloy that can be used for the analysis. The use of $8M$ hydrochloric acid elutriant for the titanium may cause significant losses of other adsorbed elements since the column was previously in the fluoride form. The above procedure has been improved by using $1M$ HF–$0.25M$ HCl solution as the initial medium for introducing the sample onto the column (280). This medium is as suitable for the retention of niobium, tantalum, tungsten, titanium, zirconium, and molybdenum as the 2.5% HF solution previously suggested. It has, in addition, the advantage of accommodating, because of greater solubility, much larger amounts of chromium (up to 0.75 g.) and offers also an excellent means of removing iron, cobalt, nickel, copper, and manganese. As stated above, because of the danger of disturbing the normal elution characteristics of tungsten, molybdenum, niobium, and tantalum, the $8M$ hydrochloric acid elutriant recommended for titanium should be avoided. Instead, titanium and tungsten, together with any zirconium and vanadium present, should be eluted with the 10% HF–60% HF solution.

If the separation of tungsten from titanium and/or zirconium is desired, the 10% HF–60% HCl solution is evaporated to dryness, the three elements are adsorbed on a separate column in a 4% HF medium, and free HF is removed with alcohol, whereupon titanium and tungsten can be eluted without any loss of tungsten with $8M$ hydrochloric acid. Tungsten subsequently can be eluted by the 5:4:11 mixture of HCl–HF and H_2O (280).

Niobium and tantalum have been determined in titanium-base alloy by adsorbing the niobium and tantalum on Dowex-1, using a 7% NH_4Cl (w./vol.)–12.5% HCl (vol./vol.)–20% HF(vol./vol.) medium, which leads to a quantitative removal of the titanium (also vanadium, iron, tin, etc.). Niobium is subsequently eluted with a solution 14% in NH_4Cl (w./vol.) and 4% in HF (vol./vol.); the tantalum is eluted with the same mixture after neutralization to a pH of 5 to 6 (228).

Tantalum is determined in niobium alloys by adsorption on a small column with the 14:4 acid mixture mentioned above and subsequent elution with the neutral 14:4 solution (26). For the determination of zirconium in commercial niobium, both niobium and tantalum are eliminated by adsorption in a 25% HCl–20% HF (vol./vol.) medium on Dowex-1 (184).

The ion-exchange technique is particularly valuable for the determination of one or the other or all constituents of niobium and/or tantalum concentrates (282). After decomposition of the sample with a mixture of 25 ml. of HCl and 20 ml. of HF and dilution to 100 ml. with water, or by fusion with sodium bisulfate and subsequent solution of the melt in the 25:20 HCl–HF mixture (see Section IV-A-5), Dowex-1 resin will retain the niobium and tantalum, whereas titanium, soluble zirconium, iron, manganese, tungsten, molybdenum, and part of the tin are found in the eluate. If in the 25:20 mixture part of the HCl is replaced by NH_4Cl (7% NH_4Cl–12.5% HCl–20% HF) all the tin accompanies the base metals. After removal of the elements listed above the niobium is eluted with the acidic 14:4 mixture mentioned above and, subsequently, the tantalum with the neutral 14:4 elutriant. (For more details see Section IX-A-2.)

If it is desired to retain only tantalum on the column, a 10% nitric acid–10% hydrofluoric acid medium allows the simultaneous removal of niobium with the base metals (281).

5. The Hydrochloric–Oxalic Acid System

A systematic compilation of the various complexes of niobium and tantalum in an oxalate medium is not yet available. It is known, however, that niobium must be in the form of a complex anion that cannot be adsorbed by cation-exchange resin (185). It was found that it forms with oxalic acid complexes of the type $NbO(C_2O_4)_3^{-3}$ (see Section II-E-4). The tantalum complexes contain more oxalic acid and possibly are not oxygenated. If both oxalic acid and hydrochloric acid are present, the complexes may contain both acid radicals.

As to its value as a medium for anion-exchange separations, the HCl–oxalic acid system has been less investigated than the HCl–HF medium, but it offers some interesting possibilities. It was noted fairly early that both niobium and zirconium can be adsorbed on Dowex-1 anion-exchange resin from oxalic acid solution (0.1 to $0.4M$) and preferentially can be eluted (zirconium being removed first) with $1M$ HCl and $0.01M$ oxalic acid (553). Subsequent work with this system was carried out in Belgium (205, 206, 245, 513) and involved mainly the separation of niobium and tantalum in solutions of the same acids (see Fig. 9). Although the separation of niobium and tantalum in a hydrochloric acid–oxalic acid medium is somewhat dependent on the niobium–tantalum ratio, these studies indicated that useful analytical separations for zirconium, titanium, niobium, tantalum, tungsten, and molybdenum possibly could be made through the introduction of additional complexing agents such as peroxide

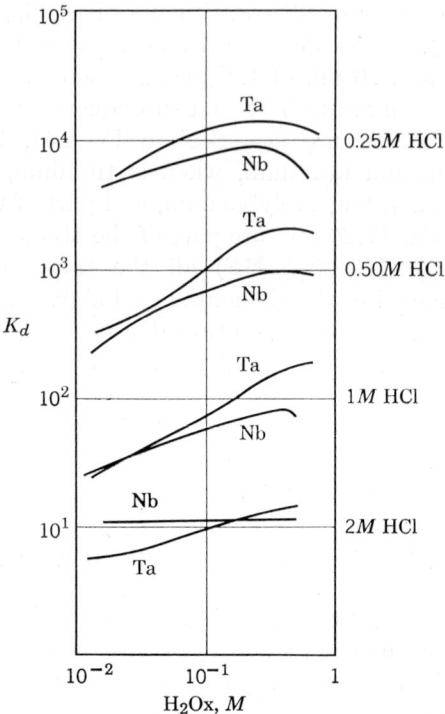

Fig. 9. Distribution coefficients of niobium and tantalum in various hydrochloric–oxalic acid mixtures (513).

and citrate. A method recently published (46) illustrates this point (see Fig. 10). A bisulfate–0.5M oxalate solution of the six elements is introduced onto Dowex-1 resin. Titanium and zirconium are eluted with 200 ml. of a solution that is 1.5M in hydrochloric acid, 0.5M in oxalic acid, and 0.007M in hydrogen peroxide. The passage of an additional 550 ml. of the same elutriant yields the niobium. Tantalum is eluted with 300 ml. of a solution 3M in hydrochloric acid and 0.5M in oxalic acid. The tungsten is subsequently obtained by passing through the column 600 ml. of a solution 4M in hydrochloric acid and 0.1M in citric acid. The molybdenum is finally eluted with 200 ml. of a solution 1.9M in ammonium chloride and 0.44M in ammonium citrate.

The oxalate medium and Dowex-1 resin have been employed also for the separation of mixed fission products (aged for 3 years) in a 0.5% ammonium chloride solution. When such a solution is passed through a column that is in the oxalate form, ^{137}Cs and ^{90}Sr pass through unadsorbed. Rare earth elements (^{90}Y, ^{144}Ce, and ^{147}Pm) as well as ^{95}Zr, ^{95}Nb, and ^{106}Ru are eluted in the following order from the resin: $0.2N$, $0.5N$, $1N$ hydrochloric acid, and $2N$ nitric acid (575).

The separation of niobium and tantalum also has been achieved in a chromatographic procedure involving activated alumina as the adsorption medium (537). An ammonium oxalate solution of the sample (obtained

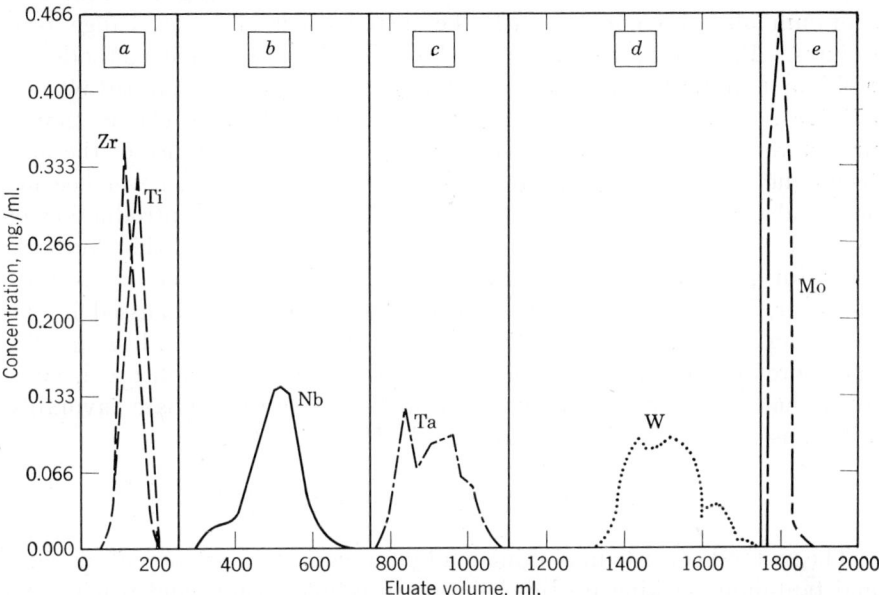

Fig. 10. Anion-exchange separation of 20 mg. each of Ti, Zr, Nb, Ta, W, and Mo using oxalic acid (46). Titanium and zirconium are eluted with 200 ml. of a solution (a) that is $1.5M$ in hydrochloric acid, $0.5M$ in oxalic acid, and $0.007M$ in hydrogen peroxide; the passage of an additional 550 ml. of the same elutriant (solution b) yields the niobium. The tantalum is eluted with 300 ml. of a solution (c) $3M$ in hydrochloric acid and $0.5M$ in oxalic acid, and the tungsten is eluted with the passage of 600 ml. of a solution (d) $4M$ in hydrochloric acid and $0.1M$ in citric acid. Finally the molybdenum is eluted with 200 ml. of a solution (e) $1.9M$ in ammonium chloride and $0.44M$ in ammonium citrate.

by fusion with potassium bisulfate and digestion of the melt with ammonium oxalate solution containing a little sulfuric acid) is transferred onto the column containing the activated alumina, which has been conditioned with ammonium oxalate. The niobium is adsorbed completely in the upper part of the column while the tantalum is only partly adsorbed. The tantalum that is retained is eluted with a solution of ammonium oxalate of pH 7. The niobium can be eluted with a solution of ammonium oxalate containing sulfuric acid.

I. SEPARATIONS BASED ON CELLULOSE AND PAPER CHROMATOGRAPHY

The application of chromatographic separations based on cellulose or paper as adsorbant to metallurgical products containing niobium and tantalum almost coincides with the introduction of ion-exchange procedures. Pioneering research on cellulose chromatographic methods for niobium and tantalum was carried out chiefly in England (91,92,374, 375,573), where up to that time Schoeller and Powell's classic tannin procedure had been reigning supreme. In Europe acceptance of the new technique was general and rapid. The methods, however, never became as popular in the United States, mainly because of the introduction of ion-exchange procedures. In the hands of analysts of this author's laboratory, cellulose chromatographic methods gave satisfactory results. However, in comparison with ion-exchange procedures, discussed in the preceding section, they proved to be somewhat less versatile, less specific, and more cumbersome in manipulation. It should be emphasized, however, that results reported by European laboratories compare favorably with those obtained by ion exchange in the United States.

1. Principles Involved

The method consists in obtaining a hydrofluoric acid solution of niobium and tantalum, soaking up this solution in cellulose pulp, and placing the product so formed at the top of a column of cellulose and extracting it with ethyl methyl ketone containing water or aqueous hydrofluoric acid.

It was found that the use of ethyl methyl ketone containing a low concentration of hydrofluoric acid favors the extraction of tantalum, whereas a high acidity is necessary for the extraction of niobium. The presence of ammonium fluoride in the sample "wad" was shown to be beneficial in assisting the retention of certain other ions present as impurities. The elution of the tantalum is based on the same principles that govern the extraction of tantalum with ketones (see Section IV-J-3). TaF_7^{-2} is extracted by the ethyl methyl ketone while $NbOF_5^{-2}$ is adsorbed by the hydrophile cellulose (467). An increase in the concentration of hydrofluoric acid and water in the solvent favors the extraction of niobium, titanium, tin, zirconium, and small amounts of metal ions, which form soluble fluorides. On the other hand, a low concentration (e.g., 1%) of hydrofluoric acid in the solvent and a very low concentration of water provide unfavorable conditions for extraction of niobium and other metal ions. (For instance, commonly occurring metal ions such as Th, Ce, La, Fe, Al, Mn, Cu, Ni, Ti, Sn, Zr, W, Pb, Ca, and Mg are not extracted under such conditions.) The use of such a solvent serves therefore for the removal of water after extraction of tantalum. A subsequent increase in the H^+ and F^- concentration leads to the formation of NbF_7^{-2} ions, which are extracted by the solvent. Provided, therefore, that water is absent or present in only a very low concentration, as when the intermediate solvent (1% of hydrofluoric acid in ethyl methyl ketone) has previously been used, metal ions other than niobium are not extracted (except tungsten and phosphorus, which are partly extracted).

2. Preparation of the Sample

High-grade concentrates such as pyrochlore can be directly decomposed by heating with hydrofluoric acid and a few drops of nitric acid. After evaporation to dryness, the fluorides are dissolved as described above. Analysis of samples of ferro–tantalum–niobium and columbite by the direct procedure described above sometimes leads to a small retention of tantalum (1 to 2%) of the original concentration, probably due to the presence of certain common ions, e.g., iron. This difficulty can be overcome by making a preliminary chemical separation by "acid hydrolysis" (see Section IV-G-1). In this way, the earth acids can be precipitated, as well as strongly adsorbable ions such as titanium, tin, zirconium, and tungsten, together with minor quantities of common metal ions.

Samples that contain large amounts of silica are treated with hydrofluoric acid in order to remove the silica at the outset of the analysis, followed by evaporation to dryness with the addition of sulfuric acid. If the matrix is iron, as is the case in certain ores or steels, the sample can be decomposed with hydrochloric acid. Such a treatment is particularly useful if the sample contains phosphates. In any case, the earth acids, prior to the chromatographic separations discussed above, are collected by one or the other "acid hydrolysis" technique described in Section IV-G-1. They are then dissolved in hydrofluoric acid and eventually are transferred to the column containing the cellulose.

3. Typical Procedure

The sequence of steps in a typical procedure is as follows (568).

1. A column (65 × 2 cm., tapered to 0.5 ml. at the bottom) is prepared by introducing activated cellulose in the form of a slurry in ethyl methyl ketone. The settled cellulose bed should be about 30 cm. After several washings with ethyl methyl ketone, the solvent is allowed to fall to about 2.5 cm. above the level of the cellulose.

2. The dry fluorides containing the sample constituents are dissolved in 8 ml. of warm $6M$ hydrofluoric acid with the addition of 1 g. of ammonium fluoride.

3. The solution is taken up on 6 g. of cellulose powder and the mixture is introduced onto the top of the cellulose in the column.

4. The sample "wad" and the main body of the cellulose, after the introduction of some Solvent A (ethyl methyl ketone saturated with water), are slightly agitated in such a way as to form a continuous cellulose bed.

5. Sufficient Solvent A is fed into the column until 250 ml. of eluate has been collected. This solution is held for the determination of the tantalum.

6. Solvent B (1% concentrated HF in ethyl methyl ketone) is now introduced into the column and the upper portion of the cellulose is agitated lightly. A total of 400 ml. of Solvent B is passed through the column in two fractions, each of 200 ml. The second 200-ml. portion is held, since it may contain small amounts of niobium.

7. Sufficient Solvent C (12.5 ml. of concentrated HF for each 87.5 ml. of ethyl methyl ketone) is now passed through the column until 500 ml. has been collected. This fraction contains the bulk of the niobium.

8. The tantalum- (step 5) and niobium- (steps 6 and 7) fractions, after the addition of a few drops of sulfuric acid, are evaporated to dryness and ignited to the respective oxides.

9. A portion of the Nb_2O_5 is fused in sodium carbonate and is analyzed for tungsten by the thiocyanate method described by Schoeller (484).

10. If the sample is known or suspected to contain phosphorus, most of it will accompany the niobium (the tantalum fraction is free from phosphorus). Up to 30 mg. of the P_2O_5 can be tolerated, however, if the niobium, after evaporation of the ketone, is precipitated with ammonia.

4 Shortened Procedure

A shortened chromatographic procedure based on the simultaneous extraction of the mixed oxides of niobium and tantalum has been proposed for the analysis of minerals and ores (374). The sample is fused with potassium bisulfate and leached with dilute sulfuric acid, and the hydroxides are precipitated with dilute ammonia and ignited. The oxides are digested with nitric and hydrofluoric acids and, after removal of nitric acid by repeated evaporations with hydrofluoric acid, the dried residue is dissolved in 6 ml. of 25% (vol./vol.) hydrofluoric acid with the addition of 1 g. of ammonium fluoride. Certain minerals can be directly dissolved in hydrofluoric acid, omitting the bisulfate fusion and the subsequent ammonia precipitation (69) (see Section IV-A-5). The sample, after soaking up the solution in 5 g. of cellulose, is transferred to a column 12 × 0.75 cm. I.D., tapered to $1/4$ inch at the bottom and holding 3 inches of cellulose powder, and is washed with ethyl methyl ketone containing 15% (vol./vol.) of hydrofluoric acid. The niobium plus tantalum are eluted with 400 ml. of the same solvent. The solvent is evaporated off, the solution is fumed with sulfuric acid, and the earth acids are precipitated with ammonia. After ignition to the oxides, niobium and tantalum may be determined spectrophotometrically by pyrogallol procedures (262) (see Section VII-B-8), or may be separated by further chromatography along the lines described before. The above procedure may be attractive to those chemists who are merely called upon to determine the combined earth oxides or who prefer to determine niobium or tantalum spectrophotometrically, by x-ray fluorescence, or by emission spectroscopy. A similar chromatographic scheme has been suggested for the separation of niobium and tantalum from titanium (66).

5. Paper Chromatographic Methods

The chromatographic separation of niobium and tantalum from other matrix elements described above, involving the fluoride medium and ethyl methyl ketone as solvent, has been extended to the determination of niobium in low-grade samples by upward diffusion on a paper strip (263). Essentially the method consists of spotting 0.01 ml. of an $8M$ fluoride solution of the sample onto a strip of filter paper and, after drying, chromatographing with a solvent containing 4 ml. of concentrated hydrofluoric acid and 8 ml. of water diluted to 100 ml. with ethyl methyl ketone. After evaporation of the solvent, the strip is made alkaline by exposure to ammonia vapor, and is sprayed with an aqueous solution of tannic acid. The niobium is located as an orange-yellow band near the solvent front. The quantity of niobium present is determined by comparison of the intensity and area of the bands produced on standard strips prepared with known amounts of niobium. Solution of the sample is achieved either by a direct attack with hydrofluoric acid, or by fusion with potassium bisulfate, followed by solution of the melt in dilute hydrofluoric acid. The chromatographic separation is carried out on ten samples simultaneously. Under the recommended conditions few metals move from the original sample spot and of those only tantalum moves as far up the strip as niobium. It gives, however, only a slight yellow color with tannic acid and normally does not significantly interfere with the niobium determination.

The interference of molybdenum is obviated by prior reduction with iron wire.

The method discussed above has been modified to be applicable to the determination of niobium and tantalum in geochemical prospecting (261). Samples are either decomposed with hydrofluoric acid (if organic matter is present they are first ignited) or by a fusion with potassium bisulfate and subsequent treatment with dilute hydrofluoric acid. For the determination of the niobium, the solvent consists of ethyl methyl ketone containing 15% (vol./vol.) of concentrated hydrofluoric acid; the niobium, as in the preceding method, is made visible with tannic acid. For the determination of the tantalum it is necessary to choose a solvent that leaves the niobium behind. A solvent mixture of ethyl methyl ketone containing 2% of hydrofluoric acid and 8% of water was found suitable for this purpose. After chromatography, the tantalum is detected by spraying with a reagent consisting of quinalizarin in a mixture of pyridine and acetone. As little as 0.1 γ of niobium and tantalum can be detected by this method with an accuracy of $\pm 50\%$ or better.

The hydrofluoric acid medium also has been used in conjunction with isobutyl methyl ketone as a solvent (494). This is also a two-phase procedure in which niobium moves with the acid front and tantalum with the ketone front. After loading the hydrofluoric acid solution of the sample (0.1 ml.) onto the paper (Whatman #1) and developing by descending chromatography until the acid front reaches the 10-cm. and the ketone front the 15-cm. marks/(3 hours), the paper is dried in air, sprayed with dilute ammonia (5N), and then with a dilute solution of 8-hydroxyquinoline (5%) in a methanol–chloroform–water mixture (12:12:1). Following drying at 120°C., washing with water to remove the excess 8-hydroxyquinoline, and redrying, the niobium is shown as a brilliant yellow band in ultraviolet light (R_f 0.60), whereas tantalum can be detected by naked eye (R_f 0.90). The lower limit of detection for both metals is 20 γ. For the photometric determination of both elements, the bands can be cut from the paper, the niobium or tantalum dissolved in hot dilute hydrochloric acid, and the quinolates extracted (niobium at pH 8.5, tantalum at pH 10) with chloroform. Subsequent measurement of the niobium quinolate is at 385 mμ; the absorbance of tantalum is measured at 390 mμ (see also Section VII-B-4).

Some attention also has been paid to the oxalate medium. For the paper-chromatographic separation of niobium and tantalum, either acetone–2N HCl–10% oxalic acid (8:1:1) or butanone–8N HCl–10% oxalic acid (8:1:1) have been used as mobile phases (63). The separation of tracer amounts of niobium-95 and tantalum-182 oxalates also has been studied, using ethyl methyl ketone–10N HCl (75:25) as solvent, and R_f values of 0.11 for tantalum and 0.78 for niobium have been reported (85). Since these procedures do not involve hydrofluoric acid, it is interesting to note that the niobium chloride moves more rapidly with the ketone while the stronger-hydrolyzed tantalum chloride moves with the slower acid front. This is in line with the principles discussed in the ion-exchange section, namely that in strong hydrochloric acid solutions the tantalum chloride is stronger hydrolyzed than the niobium chloride. It can be assumed that, in a strong hydrochloric acid solution, the niobium and tantalum are present as chlorides, even in the presence of oxalic acid (467).

6. Separation by Electrophoresis

Niobium and tantalum also have been separated as oxalate complexes by paper electrophoresis for 4 to 8 hours in a citric acid–potassium citrate buffer (pH = 3.42, ionic strength = 0.4) at a current intensity of 4 mamp. (220 v.) (84). A satisfactory separation of sodium niobate and sodium tantalate also can be obtained by paper electrophoresis in a sodium hydroxide–borax buffer (63).

J. SEPARATIONS BASED ON SOLVENT EXTRACTION

1. General Observations

Separations of niobium, tantalum, and associated elements based on solvent-extraction procedures are related in principle to the ion-exchange methods previously discussed. Thus, in a strongly hydrochloric acid solution, and in the absence of hydrofluoric acid, the hydrolysis of tantalum chloride proceeds further than that of niobium chloride, leading to a stronger adsorption of the latter by the anion-exchange resin (see Section IV-H-2). Similarly, when a strong hydrochloric acid solution of niobium and tantalum is treated with suitable extractants, the niobium accumulates in the organic phase while the tantalum remains to a large extent in the aqueous phase (151,249,334). In a dilute hydrochloric acid solution, however, the niobium can be transferred from the organic into the aqueous phase (151,248). In such dilute hydrochloric acid solutions the niobium apparently exists in form of an oxychloride, $HNbOCl_4$ or $H_2(NbOCl_5)$, comparable to the tantalum compound of a strong hydrochloric acid solution.

In a medium containing hydrofluoric acid the niobium is present as an oxyhalogen compound, such as $NbOF_5^{-2}$ and $NbOCl_3F_2^{-2}$, whereas the tantalum exists in solution primarily as TaF_7^{-2}, thus causing a preferential adsorption of the latter by anion-exchange resin or a preferential extraction by ketones (249,515,566). With an increase in the H^+ and F^- concentration the extractability of the niobium becomes more pronounced (467) due to the formation of extractable NbF_7^{-2} ions, whereas many other metallic ions remain unextracted in the aqueous phase (515) (see Section IV-H-4).

Although solvent-extraction techniques were originally introduced as analytical tools, their range of application was soon extended to the production of high-purity niobium and tantalum. In the metallurgical field, therefore, this new technique [see Miller for a detailed discussion of industrial processes (379)] completely replaced, in the span of a few years, the Marignac process, which previously represented the only practical industrial process for the production of tantalum and a rather unsatisfactory process for the production of niobium.

In the analytical field, solvent-extraction procedures are only infrequently used for the separation of niobium and tantalum since partition between their complexes can be more readily achieved by anion-exchange resins (see Section IV-H) or with organic solvents and water by cellulose adsorbents (see Section IV-I). Solvent extraction procedures are, however, extensively used for the separation of niobium and/or tantalum from a host of other elements.

2. The Hydrochloric Acid System

Because it is difficult to prepare hydrochloric acid solutions of niobium and tantalum without causing hydrolysis, the chloride system is suited for solvent extraction only, if the concentration of earth acids is low (as is usually the case when radiochemical separations are carried out), or when the chlorides are obtained as a result of a prior chlorination procedure. Thus, niobium has been separated from tantalum by extraction from strong hydrochloric acid with a solution of methyldioctylamine in xylene. It was finally stripped from the organic phase with nitric, sulfuric, or dilute hydrochloric acids (334). A solution of tribenzylamine in chloroform or methylene chloride has also been suggested to effect the separation of niobium and tantalum from an $11M$ hydrochloric acid medium. Niobium and tantalum sulfate complexes (the exact configuration is in doubt) are also extractable to a degree by the chloroform–tribenzylamine solvent. For example, with a ratio of 15 parts of organic phase—i.e., of methylene chloride–8% tribenzylamine—to 1 part aqueous phase by volume, a quantitative separation of niobium from tantalum apparently can be made in a $4.5M$ sulfuric acid solution (151). Trace amounts of niobium are effectively separated from tantalum by extraction of the former from

6.5M HCl with hexone (86). The extraction behavior of the chloride complexes of protactinium, niobium, and tantalum differ sufficiently under suitable pH conditions to allow an effective separation using diisopropyl carbinol as extractant (103). The addition of a small amount of HF greatly increases the extraction of niobium and tantalum but lowers that of the protactinium. Diisopropyl ketone also has been used to separate tracer amounts of niobium and tantalum in hydrochloric acid solutions. It was shown that niobium is readily extracted from 10M hydrochloric acid solution; the niobium passes back into the aqueous phase when the ketone is equilibrated with 6M hydrochloric acid (248).

3. The Fluoride Medium

The fluoride medium is not particularly suited for the separation of niobium from tantalum. It is, however, extensively used for the separation of one or both metals from a host of other elements.

a. Extractions Involving Hexone

One favorite extractant is methyl isobutyl ketone (hexone). A very extensive study of the hexone–HF–HCl system by the U. S. Bureau of Mines, particularly for industrial applications (249), covers varying concentrations of HF (1.1 to 6.6N) and HCl (0 to 5.72N). Equal volumes of hexone and an aqueous solution of 1.1N HF and 2.86N HCl give optimum conditions for the separation of niobium and tantalum. 99.2% pure tantalum is recovered in the organic phase and 98.4% pure niobium remains in the aqueous phase. Fig. 11 indicates that the tantalum is contaminated by a small amount of niobium and that the unextracted niobium is accompanied by a larger amount ($\sim$2.5%) of tantalum. Although the separation of niobium and tantalum has been perfected for the production of the two elements by the introduction of multiple-batch and continuous-countercurrent procedures, it has found only limited application in the analytical field. In one modification, the separation was applied to the determination of tantalum in niobium metal (535). In this procedure, tantalum is extracted from 2.2N HF and 1N HCl solution of the sample with hexone, followed by its precipitation therein with aqueous NH_3. The precipitated tantalum is purified by evaporating the ammoniacal solution to dryness, adjusting the acidity to 0.9N HF and 2.9N HCl, and carrying out a second extraction and precipitation. The ammoniacal layer is evaporated to dryness, the residue is fused in potassium bisulfate, and the tantalum is determined by the pyrogallol procedure.

266   A. SYSTEMATIC ANALYTICAL CHEMISTRY OF ELEMENTS

For the separation of niobium and/or tantalum from a great number of associated elements it was shown that both elements are quantitatively extracted from 50 ml. of an aqueous solution $10M$ in HF, $6M$ in H_2SO_4, and

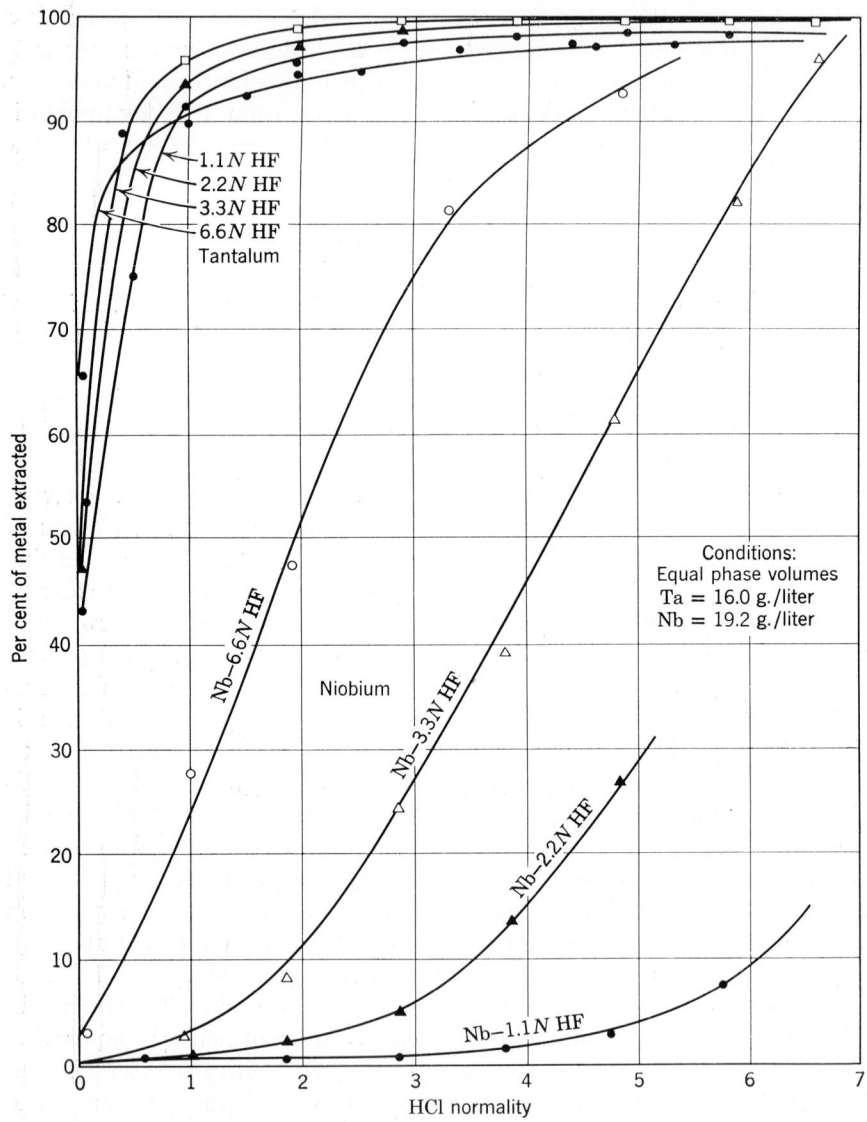

Fig. 11. Extractability of tantalum and niobium from HF–HCl into methyl isobutyl ketone (249).

TABLE XI
Extraction Coefficients for Substances of Importance in Alloy Analysis (381)
(Medium: 10M HF, 6M H$_2$SO$_4$, 2.2M NH$_4$F. Extractant: Hexone)

Element	Extraction coefficient	Concentration used, mg. per 50 ml.
Niobium	2.5×10	60–70
Tantalum	2.15×10^2	80–95
Titanium	1.1×10^{-2}	100
Uranium	7×10^{-3}	100
Molybdenum	1×10^{-1}	100
Zirconium	8×10^{-3}	75
Iron	7×10^{-3}	100
Tungsten	3.5×10^{-1}	100
Aluminum	1.5×10^{-3}	100
Manganese	3.5×10^{-3}	100
Tin	7×10^{-3}	100
Gallium	2×10^{-3}	100

2.2M in NH$_4$F (41,381). Under these conditions the hexone solvent is fairly specific for the earth acid elements (see Table XI).

Starting with 100 mg. of tantalum and 50 ml. of solvent, less than 0.5 mg. of tantalum is left after one, and less than 0.002 mg. after two extractions. The behavior of niobium is only slightly inferior, since starting with 100 mg. of this element, 4 mg. is left after the first extraction, 0.18 mg. after the second extraction, and 0.007 mg. after three extractions. The niobium or tantalum is back-extracted from the combined solvent layers into an aqueous hydrogen peroxide solution and, after destruction of the hydrogen peroxide and residual hydrofluoric acid, finally is precipitated with tannin. Since uranium is not extracted by hexone, this method is well suited for the determination of niobium and/or tantalum in uranium alloys (381).

The separation of niobium or tantalum from uranium (and/or plutonium or "fissium alloy" elements) involving hexone also can be carried out under somewhat different conditions, using for the extraction of niobium a 6.3M sulfuric–1.6M hydrofluoric acid medium (560,561) and for the extraction of tantalum a 6M sulfuric–0.4M hydrofluoric acid medium (559). No interference with the colorimetric hydroquinone determination of niobium and tantalum was caused by the presence of milligram amounts of elements such as Pu, Co, Cr, Tl, V, K, Ce, Nd, Ru, Bi, Ni, Cd, Sn, Pb, Hg, Pt, Ir, Au, and Na. The hexone–HF–H$_2$SO$_4$ system also has been investigated by others (34,382,531), and complete extraction of niobium and tantalum from aqueous solutions 5.6N in HF and 9N in H$_2$SO$_4$

has been reported (531). A $9N$ HF–$12N$ H_2SO_4–hexone system has been suggested for the separation of zirconium from all but traces of niobium or tantalum in binary alloys (382) and for the determination of niobium in intermediate products of titanium manufacture (459a). A $7.0N$ HF–$5.0N$ HNO_3–hexone system produced high-purity niobium and tantalum in the presence of such elements as silicon, iron, aluminum, titanium, magnesium, calcium, lead, and zirconium (172).

b. Extractions Involving Other Solvents

In an HF–H_2SO_4 system (385), methyl ethyl ketone was shown to be less effective than hexone since small amounts of niobium and tantalum escaped extraction. In addition, comparatively large amounts of tungsten, zirconium, chromium, iron, and manganese accompanied the niobium and tantalum.

Cyclohexanone and a $0.4N$ HF–$4N$ H_2SO_4 medium have been suggested for the separation of tantalum from titanium (111) or niobium (112). The fraction containing the solvent is washed with water to strip coextracted niobium and the tantalum is finally re-extracted with a solution containing ammonium oxalate and boric acid. Diisopropyl ketone and HF–H_2SO_4 or HF–HCl media have been investigated for the separation of niobium and tantalum in radiochemical mixtures (515). The reported separation factors (ratios of extraction coefficients) indicate that this system is not suited for separating macro amounts of the earth acids.

Diisobutylcarbinol and a $6N$ HF–$12N$ H_2SO_4 medium have been successfully used to separate niobium from protactinium (388), and the same solvent and a $6N$ HCl–1 to $3N$ HF medium were used for the separation of niobium and tantalum from both zirconium and protactinium (478). Cyclohexane in an H_2SO_4–HF system has been recommended for the simultaneous extraction of niobium and tantalum (215) and for the separation of tantalum from zirconium (152).

Another solvent—tributyl phosphate—has been given considerable attention, primarily, however, for the determination of niobium and/or tantalum in fission products or in mixtures containing only minor amounts of the two elements. By using radioactive isotopes ^{95}Nb and ^{182}Ta, it was found that 90% of the tantalum is extracted from M hydrofluoric acid with tributyl phosphate while niobium remains almost completely in the aqueous phase (458). It was shown that the presence of hydrochloric acid increases the extraction of niobium and that the tributyl phosphate can be diluted with 20% (vol./vol.) of hydrogenated kerosene without significantly affecting the extraction of the tantalum, but facilitating the separation of the layers. The systems TBP–HF–HNO$_3$ and TBP–HF–H$_2$SO$_4$ also have been studied, and valuable information has been derived for the separation and purification of niobium and tantalum in their ores (179). The TBP–HF–H$_2$SO$_4$ system has found some application in the analytical field, as for the isolation of tantalum after neutron activation of rocks (390) or niobium in fission products (369,391). An extraction procedure based on the use of mesityl oxide as solvent in an HCl–HF system has been described (303). Advantages over other systems are claimed due to the low concentration of HF that is required. A HNO$_3$–HF–acetone–isobutyl alcohol system recently appeared in the Russian literature (404) which describes the determination of as little as 0.002% of tantalum in niobium or zirconium metal using only a 1-g. sample.

4. Other Miscellaneous Extraction Procedures

Di-n-butyl ester of phosphoric acid (DBPA), either alone or in a mixture with the monoester (as in commercial "butyl phosphoric acid"), has been investigated as an extractant for the separation of niobium and zirconium (466). A 0.06M solution of DBPA in di-n-butyl ether extracts more than 95% of the zirconium and less than 5% of the niobium. Addition of 3% H$_2$O$_2$ reduces the niobium extraction to less than 1% without affecting the zirconium.

Niobium forms with several organic reagents chelate complexes that are soluble in organic solvents and that may be utilized for analytical purposes, particularly in photometric methods. Oxine (13,15,286), 5,7-dichloro-8-quinolinol (87), and cupferron (380) have been mentioned. These reactions are discussed in more detail in other sections of this chapter.

V. SEPARATION OF NIOBIUM AND TANTALUM

It was already pointed out in Section II-E-1 that, because of the lanthanide contraction, the atomic radii of niobium and tantalum are nearly identical. Since the ionic radii also differ little, the compounds of niobium and tantalum are sufficiently similar to render the separation of the two elements by precipitation methods very difficult. However, advantage may be taken of the fact that niobium and tantalum form a number of chloro, fluoro, and/or oxalato complexes with varying degrees of stabilities, acidities, and tendencies to hydrolyze. Thus, the newer techniques based on anion exchange, cellulose chromatography, and, to a lesser degree, solvent extraction allow clearcut separations of the two elements in a reasonable length of time. These techniques already have been described in some detail in Section IV-H to IV-J, and the discussion in this section is therefore limited to a summary of those features that affect the separation of niobium and tantalum. Since the separation of the two elements is more difficult and less certain by precipitation methods, only the salient points of those procedures are cited here that have not been discussed in previous sections.

A. PRECIPITATION METHODS

1. The Tannin Procedure

a. General Considerations

The basic features of this procedure were already discussed in Section IV-G-3. The separation of niobium and tantalum is best carried out in an oxalic acid solution. Since the order in which elements are precipitated by tannin with increasing pH is Ta > Ti > Nb > V > Fe > Zr > Hf > Th > U > Al, and since neighboring elements are hard to separate (299), it is clear that titanium, if present in more than small amounts, should be removed before the separations of niobium from tantalum is attempted. This can be done by the procedures described in Section IV-G-1, which employ an SO_2 hydrolysis step (199,200). The oxalate salicylate method (487) is more cumbersome. When the tantalum:titanium ratio is 50:1 or larger, titanium does not interfere with the separation of niobium and tantalum. Most of it will be found in the tantalum fraction and can be determined photometrically While niobium is quantitatively precipitated by tannin in an ammonium oxalate medium at pH 4.5, and tantalum at pH 4 (381), it has been shown that an exact pH control is insufficient to prevent the coprecipitation of some niobium with the tantalum (570). If the pH is lowered to 3.2 no niobium is precipitated. The precipitation of the tantalum is, however, only 90% complete (417). Therefore, it is customary to carry out fractional precipitations.

The best known of these procedures is due to Schoeller (442). Since the method involves many steps and has been reproduced elsewhere in detail (299,489), only a short outline is presented here.

b. The Schoeller Procedure

1. The combined oxides of niobium, tantalum, titanium (if Ta:Ti < 50:1, titanium is removed first), and possibly zirconium are fused in potassium bisulfate and the melt is leached in a hot saturated solution of ammonium oxalate. To the boiling solution (300 ml.) 2% tannin solution is added (20 ml.). Either a yellow or orange-to-red coloration will be produced. **Case A** (yellow) is discussed in steps *2* to *7*. **Case B** (orange-red) is described in steps *2a* and *2b*.

Case A: *2.* If the tantalum content is high, a yellow precipitate results. More tannin (20 to 30 ml.) is added in small increments as long as a transient orange tint disappears upon agitation. Complete flocculation is achieved by the addition of a saturated NH_4Cl solution and gentle boiling. The precipitate is filtered, washed, and held (precipitate *a*).

3. The filtrate is evaporated to its original volume and, after the addition of more tannin solution (5 to 10 ml.), is neutralized with N NH_4OH as long as a transient orange tint disappears. The resulting precipitate, if yellow (*b*), is combined with precipitate *a* (step *2*).

4. If the precipitate in step *3* is orange, more ammonia is added until the precipitate appears orange-red. It is then (precipitate *c*) filtered and ignited.

5. If the precipitate in step *3* is yellow, an additional fractionating step is carried out until finally an orange-red precipitate is obtained. This precipitate (*d*) is filtered and ignited.

6. The filtrate from the final orange-red precipitate is further neutralized until a small permanent red precipitate is obtained, which indicates that all of the tantalum has been precipitated (precipitate *e*).

7. The above steps provide three ignited precipitates: *a* and *b* = yellow; *c* and *d* = orange; and *e* = red. They are treated as described in steps *8* and *9*.

Case B: *2a.* The solution is orange-red. It is neutralized, as is described in step *5*. The orange-red precipitate (*c*) is ignited.

2b. The filtrate from step *2a* is further neutralized, resulting in a precipitate (*e*) that is red.

2c. Further treatment of precipitates *c* and *e* is described in steps *8* and *9*.

8. The intermediate fractions *c* = orange and *e* = red are fused in bisulfate, leached in ammonium oxalate solution, and again submitted to fractionation as described in steps *2* to *6*, only on a reduced scale, resulting finally in a yellow precipitate (*f*) and a pale orange precipitate (*g*), which should contain not more than 1 mg. of niobium.

9. Precipitates (*a* and *b*), (*f*), and (*g*) are combined, ignited, and leached with dilute hydrochloric acid for the removal of soluble salts. The oxide is ignited, weighed, fused in potassium bisulfate, and leached in an ammonium oxalate–dilute sulfuric acid mixture. Any precipitate is filtered off (SiO_2) and ignited. In the silica filtrate, titania is determined photometrically. The weights of SiO_2 and TiO_2 are subtracted from the weight of the impure Ta_2O_5.

10. Niobium can be calculated by difference or it can be precipitated in the combined filtrates from the tantalum determination with tannin and ammonia. In the latter case, the weight of Nb_2O_5 must be corrected for soluble salts, silica, titania, and, possibly, for zirconium.

c. Modification of Schoeller's Procedure

As the above outline shows, Schoeller's tannin procedure involves a great number of steps. However, for more than a quarter of a century it represented the only reliable procedure, and the skill of many an analytical chemist was measured by his ability to carry out satisfactory "tannin separations." A great number of modifications of the original procedure have been suggested, and have been discussed elsewhere (299). The best known of these, due to Cunningham (122), has been popular in many United States Laboratories and is based on the intentional coprecipitation of a small amount (0.02 to 0.5 g.) of niobium by precipitating the tantalum at a pH of about 4.3. After weighing, the ignited oxide is fused in potassium bisulfate and the melt is leached in 20% (vol./vol.) of sulfuric acid containing succinic acid and a little hydrogen peroxide. Titanium is determined colorimetrically and the solution is subsequently passed through a Jones Reductor for a volumetric determination of the niobium (see Section VII-C-1).

A similar modification was in use in this author's laboratory for many years (335) prior to the introduction of ion-exchange procedures. In this modification, the bulk of the tantalum and residual titanium are precipitated with tannin free from niobium at pH 3.5. After filtration, more tannin solution is added and the hot solution is further neutralized with ammonia (omitting exact pH measurements) until about 0.02 to 0.05 g. of niobium has been coprecipitated, as recognized by the characteristic color that niobium imparts to the tannin precipitate. The experienced operator has little difficulty in controlling the addition of ammonia. The combined precipitates are treated with hot 10% hydrochloric acid and the precipitation of tantalum plus titanium plus coprecipitated niobium is completed with cupferron. The ignited and weighed precipitate is fused in potassium bisulfate and the analysis is finally concluded with a volumetric (122) or photometric (423) determination of the coprecipitated niobium.

2. The N-Benzoyl-N-Phenylhydroxylamine Procedure

N-Benzoyl-N-phenylhydroxylamine (tantalon), which has been used since 1919 in other analytical procedures (45), was introduced in 1957 by two teams of investigators as a selective reagent that allows not only the separation of niobium and tantalum from each other, but also from a number of other elements (350,393).

Experiments carried out in a $0.15M$ H_2SO_4–$0.1M$ HF medium (393) indicate that this system requires close attention as to acidity, time of standing (2.5 hours), and laboratory ware (polyethylene or platinum required). The maximum amount of tantalum that can be handled is 50 mg. Since approximately 10% of the niobium is coprecipitated under the above conditions, a second and even third precipitation may be required to reduce the niobium contamination to a tolerable level. Titanium and zirconium interfere to a lesser degree and are coprecipitated only to the extent of 3 to 5%. Tungsten and molybdenum must be absent.

It was reported (344,345,350) that, in a tartaric acid medium, tantalum is precipitated by N-benzoyl-N-phenylhydroxylamine at pH values below 1.5. Niobium is precipitated in a pH range of 3.5 to 6.5 while tantalum remains in solution. After the filtration of the niobium compound its composition is:

$$\left(\begin{array}{c} C_6H_5-C=O \\ | \\ C_6H_5-N-O \end{array} \right)_3 NbO \quad (345)$$

containing 17.83% Nb_2O_5. Tantalum can be precipitated by lowering the pH. It is claimed that a single precipitation separates mixtures of niobium and tantalum in ratios of 1:16 and 100:1 (344). At a ratio of 1:100, an effective separation can be achieved by double precipitation. Addition of EDTA inhibits the interference from a wide range of ions, including those of the rare earth elements, but titanium, zirconium, vanadium, and molybdenum interfere (345). It has been observed that the precipitation of the niobium is somewhat dependent on the concentration of the tartaric acid (398). Owing to its stability to heat and its insolubility in hot water, the niobium complex, after being filtered, washed, and dried, can be directly weighed. The tantalum precipitate is of indefinite composition and must be converted to the oxide by ignition (345). It was shown that, at pH 4 to 6, the complex of niobium with the reagent is separated by a single extraction with $CHCl_3$ (98 to 100%), whereas the tantalum complex is not extracted at all (13b).

In a method leading to a quantitative separation of niobium, tantalum, and titanium from each other and from associated elements (323), all three metals are first precipitated with N-benzoyl-N-phenylhydroxylamine at pH 1 in the presence of EDTA and tartaric acid, and are recovered as oxides. The latter are redissolved by fusion with $K_2S_2O_7$ or by heating with HF, and the niobium and tantalum are precipitated by the reagent in the presence of EDTA and H_2O_2. After a reprecipitation by the same process, niobium and tantalum are separated by a modification of the above-described procedure (344) in which niobium is precipitated at pH 4.5 to 5.0. Titanium can be recovered in the filtrate, after destruction of the organic content, with the same reagent in a 5% H_2SO_4 medium.

In another application of the reagent, again involving niobium, tantalum, titanium, and zirconium (349) the two earth acid elements are precipitated with a large excess of pyrogallol. After filtration of the pyrogallates, ignition of the precipitates, and reprecipitation, the mixed oxides of niobium and tantalum are separated by the N-benzoyl-N-phenylhydroxylamine, as discussed above (344). Titanium and zirconium are separated from each other, after destruction of the organic matter, by precipitation of the zirconium with salicylhydroxamine acid in the presence of H_2O_2.

3. The Phenylarsonic Acid Procedure

Phenylarsonic acid (C_6H_5-AsO_3H_2), precipitates tantalum completely from an approximately 10% sulfuric acid solution up to a pH of 5.8, when buffered with ammonium oxalate. Niobium is precipitated in an oxalate medium only at higher pH values (346). The method is discussed in Section IV-G-4.

4. Separation of Niobium and Tantalum With Selenous Acid

Both niobium and tantalum can be precipitated with selenous acid from a $3M$ hydrochloric acid–1% tartaric acid medium (see Section G-5), and thus can be separated from titanium (16). In a tartaric acid (2.5% w./vol)–oxalic acid 6% (w./vol.)–4% sulfuric acid (vol./vol.) medium, in a total volume of 125 ml., the separation of tantalum from both niobium and titanium (219) can be achieved. This method has been applied to the semimicro determination of tantalum in ores.

5. Other Gravimetric Separation Procedures

A great number of additional gravimetric procedures have been suggested. Their usefulness, however, has been reduced to a great extent by more recent analytical techniques, notably by ion exchange. Since these procedures are therefore of lesser interest to the practicing analytical chemist, they are discussed here only briefly.

Niobium can be precipita ed with 8-hydroxyquinoline at pH 6 in a 2% ammonium tartrate medium while tantalum (as well as titanium) stays in solution (52). Even after reprecipitat on the final Nb_2O_5 is only 97.3%

pure. Possibly a better separation can be obtained by extracting the niobium 8-hydroxyquinolate from solutions containing ammonium citrate or tartrate at a pH above 4.5 with various solvents, such as chloroform, dichlorethane, ethyl acetate, cyclohexanone, and isoamyl alcohol. With a ratio of Nb:Ta of 1:1000, isoamyl alcohol extracts 89.1% of the niobium and no tantalum (13). For a photometric determination of the extracted 8-hydroxyquinolate, see Section VII-B-4.

It has been observed recently (352,353) that at a pH between 4.5 and 5.5, and when the ratio of niobium to tantalum is less than 1:2, niobium is completely precipitated with cupferron in a tartaric acid medium while tantalum remains in solution. A more satisfactory separation at a ratio of 30:1 or 1:30 is obtained when tin is used as coprecipitating agent. It is stated that under these conditions niobium can be separated with EDTA (disodium salt) from almost all ions except those of uranium, beryllium, titanium, and phosphate. However, iron and other trivalent elements, when present in 100-fold excess with respect to the niobium, require a double precipitation.

It was already mentioned in Section II-E-4 that niobium exists in acid solutions containing hydrogen peroxide as peroxyniobic acid, $HNbO_4$. The corresponding tantalum peroxy complex is unstable and decomposes readily on boiling. This difference in the stability of the peroxy compounds has been made the basis of a chemical separation (269,270), the reliability of which was confirmed recently (467). In this procedure, the mixed oxides are fused in potassium hydroxide. The melt is leached in water and hydrogen peroxide. After acidifying the solution with hydrochloric acid, it is heated to near 95°C. and is kept at this temperature for about an hour. The precipitated tantalic acid usually includes a small amount of niobic acid and requires reprecipitation.

The greater solubility of potassium oxyfluoniobate (1 part of K_2NbOF_5 is soluble in 12 parts of water), compared to that of potassium fluotantalate (1 part of K_2TaF_7 is soluble in 150 parts of water), is the basis of the once-popular "Marignac procedure" (357). It involves solution of the oxides (preferably free from titanium and zirconium) in hydrofluoric acid, addition of potassium fluoride to the boiling solution, and evaporation to a small volume to cause the crystallization of the K_2TaF_7. After filtration, further evaporations are required that yield additional quantities of the tantalum. The precipitate is fumed with sulfuric acid and the tantalum is recovered with ammonia. The niobium is precipitated with ammonia in the filtrate of the K_2TaF_7, after evaporation with sulfuric acid. Usable results can be obtained if the analysis is carried out with an initial sample weight of 15 to 30 g. (241). A number of modifications of the original procedure have appeared in the literature (83,113,373). However, even the best results are based on compensating errors.

Tantalum also has been separated from niobium by treating the ignited oxides several times with a hot mixture of equal parts of selenium oxychloride and sulfuric acid (376), causing the formation of insoluble tantalum and soluble niobium salts. Although the authors state that the maximum divergence on prepared mixtures of Nb_2O_5 and Ta_2O_5 is 3% of the total, other investigators were unable to obtain satisfactory results (496).

An indirect procedure that avoids an actual separation is based on the chlorination of the pentoxides with carbon tetrachloride in a sealed tube, sublimation and weighing of the chlorides, conversion of the chlorides to oxides, and weighing (474). Because of the large differences in the atomic weights of niobium and tantalum, the method is capable of yielding reasonably good results (error $\pm 0.5\%$).

Other precipitating agents for tantalum that provide more or less quantitative separations from niobium are morin in an oxalic–sulfuric acid medium (312), sodium hypophosphite in an ammonium oxalate medium (10,516), sodium diethyl and sodium tetramethylenedithiocarbonate (354), dihydroquercetin (105), benzohydroxamic acid, and phenylacetohydroxamic acid (351c).

B. SEPARATION OF NIOBIUM AND TANTALUM BY ION EXCHANGE

The separation of niobium from tantalum by ion exchange in a mixed chloride–fluoride medium is straightforward and valid for any ratio of the two metals (see Section IV-H-4). Its introduction into the analytical repertoire represents one of the major advances in earth-acid analysis. Both elements can be adsorbed in a 25% HCl (vol./vol.)–20% HF (vol./vol.) medium by a strongly basic anion-exchange resin, such as Dowex-1, and thus can be separated from virtually all elements with which they may be associated. Subsequently, the niobium is eluted with a solution of 4% HF (vol./vol.)–14% NH_4Cl (w./vol.). After the removal of the niobium, the tantalum is eluted with a 4% NH_4F (w./vol.)–14% NH_4Cl (w./vol.) solution neutralized to a pH of 5.5.

C. SEPARATION OF NIOBIUM AND TANTALUM BY CELLULOSE CHROMATOGRAPHY

The separation of niobium and tantalum by cellulose chromatography has become quite popular in Europe, although less in the United States. In this procedure, niobium and tantalum fluorides are dissolved in dilute hydrofluoric acid, with the addition of ammonium fluoride, which assists in the retention of certain other elements. The solution is soaked up in cellulose pulp, which is placed on top of a cellulose bed. Tantalum is extracted with ethyl methyl ketone saturated with water. To prevent the extraction of other elements with the niobium, water is removed by passing a solvent, consisting of 1% concentrated hydrofluoric acid in ethyl methyl ketone, through the column. Subsequently, the niobium is extracted with a solution containing 12.5 ml. of concentrated HF for each 87.5 ml. of ethyl methyl ketone. For details of the procedure see Section IV-I-3.

D. SEPARATION OF NIOBIUM AND TANTALUM BY SOLVENT EXTRACTION

Methods for the separation of niobium and tantalum by solvent-extraction procedures are somewhat less useful than those based on ion exchange and cellulose chromatography. The hydrochloric acid medium (because of the danger of hydrolysis) is suitable only when dealing with small concentrations of niobium and tantalum, as in radiochemical mixtures. The fluoride or mixed fluoride–chloride medium permits the separation of larger quantities of niobium and tantalum, but it is only capable of producing 99.2% pure tantalum and 98.4% pure niobium. For a more detailed discussion of solvent-extraction procedures see Section IV-J.

VI. DETECTION AND IDENTIFICATION

Since chemical detection and identification tests for niobium and tantalum generally require prior removal of all other metallic elements, spectroscopic procedures are much to be preferred, because they can be applied without the necessity of any chemical separations. X-ray emission tests have the additional advantage that they are nondestructive. They require, however, a somewhat higher concentration of niobium or tantalum.

A. SPECTROSCOPIC EXAMINATION

The intensities of the niobium and tantalum lines described below depend largely on the type of excitation. The extent of interference by other elements depends on the dispersion of the spectrograph.

1. Niobium

Niobium possesses an adequate number of sensitive lines, allowing positive identification of the element by various approaches involving arc, interrupted arc, and spark techniques. In the arc spectrum, the most sensitive line is at $\lambda = 4059.0$ A., followed by $\lambda = 4079.7$ A. The lines $\lambda = 3094.2$, 3580.3, and 4101.0 A. are somewhat weaker. The spark spectrum contains, in the order of decreasing intensities, the lines $\lambda = 3094.2$, 3225.5, 3130.8, and 3163.4 A. (197).

a. Interference with Arc and Spark Lines (273)

$\lambda = 4101.0$ A.: Indium $\lambda = 4101.8$ A. is a strong interfering arc line, as are the arc lines $\lambda = 4102.2$ and 4099.8 A. (vanadium) and $\lambda = 4102.7$ A. (tungsten).

$\lambda = 4079.7$ A.: Strong interference is caused by the arc lines $\lambda = 4079.4$ and 4079.3 A. (manganese); $\lambda = 4081.5$ A. (molybdenum); $\lambda = 4080.6$ A. (ruthenium); $\lambda = 4082.4$ A. (scandium; and $\lambda = 4078.5$ A. (titanium); the spark line 4077.7 A. (strontium) also interferes strongly.

$\lambda = 4059.0$ A.: Strongly interfering arc lines are $\lambda = 4057.8$ A. (lead) and $\lambda = 4058.9$ A. (manganese).

$\lambda = 3580.3$ A.: Particularly interfering are the arc lines $\lambda = 3578.7$ A. (chromium), $\lambda = 3581.2$ A. (iron), $\lambda = 3581.9$ A. (molybdenum), and $\lambda = 3583.1$ A. (rhodium), and the scandium spark lines $\lambda = 3581.0$ and 3576.4 A.

$\lambda = 3163.4$ A.: Strongly interfering spark lines are $\lambda = 3162.6$, 3161.8, and 3161.2 A. (titanium).

$\lambda = 3130.8$ A.: The following arc lines interfere particularly: $\lambda = 3133.3$ A. (iridium), $\lambda = 3132.6$ A. (molybdenum), $\lambda = 3131.8$ A. (mercury), and $\lambda = 3134.1$ A. (nickel). Strong interference also may be caused by the following spark lines: $\lambda = 3130.4$ A. (beryllium), $\lambda = 3132.1$ A. (chromium), $\lambda = 3130.3$ A. (vanadium).

$\lambda = 3094.2$ A.: Strongly interfering lines are: $\lambda = 3092.7$ A. (aluminum arc) and $\lambda = 3093.1$ A. (vanadium spark). Somewhat less interfering are the arc lines $\lambda = 3091.6$ A. (iron) and $\lambda = 3096.9$ A. (magnesium).

2. Tantalum

Although for identical modes of excitation the tantalum spectrum is weaker than that of the niobium, a number of arc and spark lines allow positive identification with adequate sensitivities. However, there are obvious discrepancies in the literature concerning the most sensitive tantalum lines. Some sources (79,235,511) name $\lambda = 3406.7$, 3318.8, and

3311.2 A. as the most sensitive arc lines, whereas other authorities (197, 273) place the most sensitive lines in the lower ultraviolet region at λ = 2714.7, 2653.3, and 2647.5 A. (arc lines) and λ = 2685.1, 2635.6, and 2675.9 A. (spark lines). Scribner, of the National Bureau of Standards, basing his work on a uniform energy scale, considers the lines recorded in Table XII as the most sensitive (370,495).

TABLE XII
Strongest Lines of Niobium and Tantalum (370,495)

	Wavelength	Spectrum	Intensity in arc
Niobium	4058.94	I	1700
	4079.73	I	1200
	4100.92	I	700
	3580.27	I	600
	4123.81	I	550
	4152.58	I	460
Tantalum	2653.27	I	300
	2714.67	I	300
	2647.47	I	280
	3012.54	II	240
	2656.61	I	220
	2850.98	I,II	220

a. INTERFERENCE WITH ARC AND SPARK LINES (273)

λ = 2714.7 A.: Strongly interfering lines are λ = 2714.4 A. (iron), λ = 2713.1 A. (platinum), λ = 2717.3 A. (molybdenum), λ = 2714.6 A. (osmium), λ = 2715.3 A. (rhodium), λ = 2712.4 A. (ruthenium), and λ = 2715.7 and 2711.7 A. (vanadium spark lines).

λ = 2685.1 A.: Strong interference is caused by: λ = 2687.1 A. (chromium), λ = 2684.8 A. (iron), λ = 2688.0, 2684.1, and 2683.2 A. (molybdenum), and λ = 2688.0, 2683.1, and 2682.9 A. (vanadium).

λ = 2675.9 A.: λ = 2676.0 A. is a strongly interfering gold arc line. Additional interfering lines are: λ = 2677.2 A. (chromium), λ = 2677.1 A. (platinum), λ = 2678.7 A. (ruthenium), and λ = 2678.6 and λ = 2677.8 A. (vanadium).

λ = 2653.3 A.: The germanium arc line λ = 2651.2 A. interferes strongly, as do the following lines: λ = 2652.5 A. (aluminum), λ = 2650.8 and 2650.5 A. (beryllium), λ = 2653.6 A. (chromium), λ = 2653.3 A. (molybdenum), and λ = 2650.9 A. (platinum).

$\lambda = 2647.5$ A.: Strong interference is caused by $\lambda = 2650.8$ and 2650.5 A. (beryllium), $\lambda = 2646.5$ A. (molybdenum), $\lambda = 2646.9$ A. (platinum), and $\lambda = 2646.6$ A. (titanium).

$\lambda = 2635.6$ A.: Strongly interfering lines are $\lambda = 2635.8$ A. (iron), $\lambda = 2634.3$ A. (iridium), $\lambda = 2636.7$ A. (molybdenum), and $\lambda = 2637.1$ A. (osmium).

3. Limit of Detection

The intensity of the niobium radiation exceeds that of the tantalum by a factor of about 10. Thus it is possible to detect about 0.05% of niobium and 0.5% of tantalum in steel by a spark technique, with an increase in the limit of detection if an arc technique is employed (480).

Niobium can be detected in minerals following a sodium carbonate fusion. The limit of detection is 1:20,000 (273). The sensitivity of the method can be increased by a prior removal of the silica with HF–H_2SO_4 and collection of the niobium with phenylarsonic acid (445). Another concentration procedure (cupferron extraction) allows the detection of 3 parts of niobium and 40 parts of tantalum in 10^6 parts of sample (uranium ore), employing the "copper spark" technique (455).

B. DETECTION BY X-RAY EMISSION

X-ray emission represents an excellent tool for the detection of small amounts of niobium and tantalum. As in the case of spectroscopy, the intensity of the niobium radiation is considerably greater than that of the tantalum. Tests carried out in Ledoux & Company's laboratory indicate that it is possible to detect, in a matrix of light elements such as silicate rock, 0.01% of niobium and about 0.05% of tantalum (279). The most sensitive niobium line, $K\alpha_1$ ($\lambda = 0.747$ A.), may be interfered with by uranium, $L\beta_4$ ($\lambda = 0.748$ A.), yttrium, $K\beta_1$ ($\lambda = 0.740$ A.), thorium, $L\beta_3$ ($\lambda = 0.755$ A.), and uranium, $L\beta_2$ ($\lambda = 0.755$ A.).

The most sensitive tantalum line, $L\alpha_1$ ($\lambda = 1.522$ A.), may be affected by the presence of large quantities of indium ($K\alpha_1$ (III), $\lambda = 1.536$ A.), barium ($K\alpha_1$ (IV), $\lambda = 1.540$ A.), copper ($K\alpha_1$, $\lambda = 1.542$ A.), uranium ($L\beta_2$ (II), $\lambda = 1.510$ A.), and niobium ($K\beta_1$ (II), $\lambda = 1.500$ A.).

C. DETECTION BY CHEMICAL PROCEDURES

In previous sections the close relationship of the chemical properties of niobium and tantalum compounds was repeatedly emphasized. Since both elements do not form simple cations and usually exist in solution in the form of complexes, only a few specific reactions are available that identify one or both elements. If sufficient niobium and tantalum is present, hydrolysis of the earth acids represents therefore the surest method of identification. Reduction of pentavalent to trivalent niobium and formation of the colored tannin adsorption complexes (after removal of titanium and a few other colored species) represent the only means of differentiating between niobium and tantalum. In the possible presence of interfering elements it is a good practice to proceed first along the line of the quantitative separation procedures outlined in previous section, although this may require almost as much effort as does a quantitative analysis.

1. Joint Detection in Minerals (484)

a. Tartaric Hydrolysis

The finely powdered mineral is fused with bisulfate and the melt is dissolved in hot tartaric acid. The filtered solution is boiled with an excess of hydrochloric acid. Alternatively the bisulfate melt can be dissolved in dilute sulfuric acid containing a little hydrogen peroxide. After filtration, the solution is boiled with an excess of SO_2 water (450). In either case, in the presence of niobium and tantalum a white flocculent precipitate is formed. The sensitivity of the test is approximately 0.03 mg. of $(Nb,Ta)_2O_5$ per ml. and is not affected by five times that amount of titania. Tungsten also is precipitated under the above conditions, but the precipitated tungstic acid is yellow. In the presence of tungsten, the precipitate is filtered off, the residue is fused in potassium carbonate, and the tungsten is removed by the magnesia procedure (see Section IV-G-5).

b. Tannin Procedure (484)

Two to three mg. of the mineral is fused in about 100 mg. of potassium bisulfate and the melt is leached in 1 to 2 ml. of hot saturated ammonium oxalate solution. To the boiling solution 0.1 g. of ammonium chloride and 0.02 g. of tannic acid are added. With minerals high in tantalum, a yellow precipitate is obtained; those containing more niobium give an orange-red precipitate. If the sample contains appreciable titania (hydrogen peroxide test on a separate portion of the sample), it should be removed by the pyrosulfate–tannin method (see Section IV-G-3). According to Schoeller (483), 0.3 mg. of Ta_2O_5 will produce immediately a "heavy" yellow precipitate, whereas 0.1 mg. of Nb_2O_5 can be recognized by a "heavy" red precipitate.

2. Detection of Niobium and Tantalum in Minerals by a Paper Chromatographic Method (261)

Although this procedure is mainly intended for the semiquantitative determination of niobium and tantalum in connection with geochemical prospecting, it is equally suited for the detection of the two elements. One gram of pulverized sample is decomposed by heating with hydrofluoric acid. After evaporation to dryness, the salts are dissolved in 2 ml. of $6N$ hydrofluoric acid.

For the detection of niobium, 0.05 ml. of the solution is transferred to one end of a paper strip on a sheet of Whatman No. 1 filter paper. After allowing the sheet to dry for 1 hour at room temperature, it is rolled and clipped to form a cylinder and is placed into a beaker containing 20 ml. of solvent (85 ml. of ethyl methyl ketone plus 15 ml. of $24N$ hydrofluoric acid). The solvent is allowed to diffuse upward from the sample spots almost to the top of the strips. After removal of the sheet and evaporation of the solvent, the strips are treated for 3 minutes with ammonia vapor and then are sprayed with an aqueous solution of tannic acid. Niobium is detected in a narrow orange-yellow band in the solvent front. As little as 0.1γ of Nb_2O_5 is visible. Larger amounts give more intensely colored bands.

For the detection of tantalum, 0.05 ml. of the sample solution is transferred to the paper strip. After drying at room temperature for 1 hour ($1^1/_2$ hours if niobium is present), the strip is treated in a beaker with 20 ml. of solvent (90 ml. of ethyl methyl ketone, 2 ml. of $24\ N$ hydrofluoric acid, and 8 ml. of water), using the identical technique described for niobium. After removal of the sheet from the solvent beaker, it is sprayed immediately with a solution of quinalizarin (0.05 g. of the salt dissolved in 100 ml. of water containing 10 ml. of pyridine); it is then exposed for 9 minutes to ammonia vapor and then to acetic acid vapor. Tantalum is detected as a narrow mauve-pink band in the solvent front. The limit of detection is 0.1γ of Ta_2O_5. If present, niobium is detected as a similarly colored band between the original spot and the tantalum.

3. Detection of Niobium with Methylthymol Blue

It was found that, at pH 5, peroxoniobate,

$$\left[\begin{array}{c} O \\ | \\ | \\ O \end{array} Nb\!-\!O\!-\!OH \right]^{-2}$$

which is stable in the presence of diaminocyclohexanetetraacetate (DC-YTA) (326), gives an intense blue color with methylthymol blue. Many metal ions that form colored compounds with methylthymol blue (300) can be complexed with DCYTA. It is therefore possible to secure selectivity in the reaction of Nb–H_2O_2 with methylthymol blue via the masking action of DCYTA (325). The rate of the color development is slow; a blue color is attained at room temperature within 30 to 40 minutes, but in a boiling solution the reaction is complete after 1 minute. The detection limit of the test conducted in a test tube is 5 γ and the limit of dilution is 1:800,000. The selectivity of the test is especially favorable. None of the elements that commonly accompany niobium interfere, namely Ta, Ti, Mo, W, Th, and Zr. The following cations show no influence on the test when present in a 100-fold excess: K^+, Na^+, NH_4^+, Ca^{+2}, Sr^{+2}, Ba^{+2}, Mg^{+2}, Mn^{+2}, Al^{+3}, Zn^{+2}, As^{+3}, Sb^{+3}, Hg^{+2}, Bi^{+3}, Cd^{+2}, Pb^{+2}, Ag^+, Ta^{+5}, Zr^{+4}, Hf^{+4}, Th^{+4}, W^{+6}, and Ce^{+4}. Under the test conditions, Mo^{+6}, Ti^{+4}, Fe^{+3}, and UO_2^{+2} cause a yellow color. Cu^{+2}, Ni^{+2}, and Co^{+2} form colored DCYTA chelates; therefore their concentration must not exceed 100-fold that of niobium. Cr and Be interfere.

4. Other Detection Methods

The red color produced by thiocyanates represents a sensitive test for niobium (limit of detection 1γ) (139); however, Ti, Mo, V, and W interfere. In acid solutions, niobium can be reduced by zinc or tin from a colorless pentavalent to a colored (blue or brown) trivalent salt. Other elements (Ti, Mo, W, Cr, and U) are also reduced and produce colored solutions. Reduction with zinc also can be used to differentiate between potassium fluotantalate, K_2TaF_7 (rhombic needles), and potassium oxyfluoride, $K_2NbOF_5 \cdot H_2O$ (thin monoclinic plates). Upon the addition of zinc and hydrochloric acid, as little as 0.005% of niobium will produce a brown-violet coloration. The reduction of niobium(V) salts can also be achieved by fusing the mixed pentoxides (freed from Mo, W, and V) with zinc chloride. As the result of a redox reaction ($Nb_2O_5 + 2\ HCl \rightarrow 2\ NbO_2 + H_2O + Cl_2$) and a catalytic reaction with the molten zinc chloride ($2\ Nb_2O_5 \rightarrow 4\ NbO_2 + O_2$), blue-black niobium dioxide is formed. The test will detect 10γ of niobium in the presence of 100 mg. of tantalum (174).

VII. DETERMINATION OF NIOBIUM AND TANTALUM

After carrying out all necessary separations, niobium and tantalum can be determined gravimetrically, photometrically, volumetrically, or by one or the other instrumental technique.

A. GRAVIMETRIC METHODS

With only a few exceptions, the final weighing form is Nb_2O_5 or Ta_2O_5. The earth acids are first separated from other elements by one or a combination of procedures described in Section IV and then are collected by a suitable precipitating agent.

1. Cupferron Precipitation

Probably the most widely used precipitant is cupferron, which collects niobium and tantalum over a wide pH range, and even in the presence of tartaric or oxalic acids. To improve the filterability of the precipitate, precipitation is best carried out, after the addition of some macerated paper, from a 1 to 1.5M hydrochloric or sulfuric acid medium, cooled to below 10°C. If the solution contains less than 2 mg. of niobium and tantalum, the addition of 10 mg. of zirconium, in the form of zirconium chloride or sulfate, will ensure a quantitative collection of the earth acids. Filtration is best carried out by gentle suction. To facilitate the ignition of the precipitate, a large excess of reagent must be avoided (60 ml. of a 6% aqueous solution is sufficient to precipitate 0.5 g. of niobium or tantalum oxide). The ignition must be very carefully done in the early stages in order to avoid mechanical losses. Final heating should be carried out in a muffle furnace at about 1100°C. Discussions and details of cupferron procedures appear in standard text books (250,299).

2. Tannic Acid Precipitation

Quantitative precipitation of niobium and tantalum can be obtained under various conditions (381). With ammonium oxalate as a buffer, the pH must be raised to 4.5. In hydrochloric acid media, precipitation is complete in 0.1 to 2.5N media. In ammonium tartrate solution, the pH must be raised to 3 to achieve a quantitative precipitation of both niobium and tantalum. The bulky nature of the niobium and tantalum tannate precipitates limit the amount of niobium or tantalum that can be conveniently handled to 100 to 150 mg. Precipitation is carried out by adding 1 g. of tannic acid to the solution after adjustment of the pH, heating to boiling, and then digesting the solution for about 1 hour. The precipitate is filtered, washed with a solution containing ammonium chloride and tannic acid, and ignited in a tared platinum crucible. Final weighing form: Nb_2O_5 or Ta_2O_5. (See Sections IV-G-3 and V-A-1 for a discussion of separation procedures involving tannic acid.)

3. Precipitation By Hydrolysis

This technique is less desirable, since it is usually required to carry out a preliminary fusion with potassium hydroxide or carbonate to obtain a clear solution prior to precipitation. Thus alkali salts are introduced which, to some extent, may be occluded by the precipitate. Precipitation is complete when a potassium hydroxide or carbonate solution (or the insoluble sodium salts) is acidified with a slight excess of hydrochloric or nitric acids and then is boiled. The precipitates become more coarsely flocculent and more readily filterable if the solution is made just ammoniacal and is boiled before filtering (489). (Boiling in the presence of SO_2 water also leads to a quantitative precipitation of niobium and tantalum.) The earth acid precipitate is mixed with filter pulp, collected, washed with a hot 2% ammonium nitrate solution, and ignited in a tared platinum crucible.

4. Precipitation With Phenylarsonic Acid

After the removal of titanium, zirconium, and tin, niobium and tantalum are quantitatively precipitated with phenylarsonic acid (446) or p-hydroxy phenylarsonic acid (452) from a boiling solution about 2 to $3N$ in hydrochloric or sulfuric acid containing tartaric acid. The precipitate, after standing for 48 hours, is filtered, washed with $1N$ hydrochloric acid and then with water, and is ignited to the pentoxides in a tared platinum crucible (see Section V-A-3). (For the determination of small amounts of tantalum, the turbidity developed on adding phenylarsonic acid solution to a solution of the sample in 2 to $3N$ sulfuric acid can be stabilized by the addition of glycerol and the absorbance measured at 350 mμ. An equivalent atomic concentration of niobium can be tolerated if hydrogen peroxide is present (459).)

5. Other Gravimetric Methods

Tantalum can be precipitated in a tartaric acid medium with N-benzoyl-N-phenyl-hydroxylamine at pH values below 1.5. Niobium is precipitated in a pH range of 3.5 to 6.5 (350, 351b) (see Section V-A-2).

Niobium can be precipitated in an ammonium tartrate or an oxalic acid–ammonium acetate medium at pH 6 with 8-hydroxyquinoline. The final weighing form is either $2Nb_2O_5 \cdot 11C_9H_7ON \cdot 8H_2O$, after drying at 105°C. (546), or preferably the pentoxide, after ignition under a protective cover of oxalic acid (52,356). Recent studies would indicate that actually two moles of 8-hydroxyquinoline are bound to one mole of niobium pentoxide, whereas the water content varies from experiment to experiment (138). However, a crystalline precipitate of definite composition can be obtained by homogeneous precipitation (heating a hot M oxalic acid solution containing, in 100 ml., 10 ml. of concentrated sulfuric acid with 5 times the theoretical amount of urea to a pH of 7 to 8). After filtration on a sintered glass crucible and drying at 155°C., the composition of the compound is $NbO(C_9H_6ON)_3$, containing 17.16% of niobium (302,522).

3,3',4',5,7-Pentahydroxyflavone has recently been suggested as precipitating agent for both niobium and tantalum in the absence of titanium. The final weighing form is the pentoxide (105).

B. PHOTOMETRIC METHODS

Much progress has been made during the last 10 years in the development of methods and equipment for the photometric determination of both niobium and tantalum. These methods have the advantage that they are rapid, require only a minimum amount of preliminary separations from associated elements, and that they are capable of differentiating between the two earth acids.

1. Thiocyanate Procedure for Niobium

Niobium salts produce with thiocyanates a yellow color. Not much is known about the composition of the niobium thiocyanate complex. It was once assumed that the product consists of a complex of trivalent niobium (460) produced by the reducing action of the stannous chloride, but it was shown that the yellow color is developed in the absence of any reducing agent (14,331). The main function of the stannous chloride is the elimination of the interference of ferric iron. Since, however, both trivalent and quinquevalent niobium react with thiocyanate (196), it is probable that in acid media two complexes exist (121).

The niobium thiocyanate reaction was first introduced in 1896 as a qualitative test (427). Much later, it was developed into a quantitative colorimetric procedure (14), which now probably is the most widely used method for the determination of small amounts of niobium. Two procedures are used at present. In one the color is developed in a homogeneous water–acetone medium (183), and in the other the yellow niobium thiocyanate complex is extracted into organic solvents, such as ether (14,331) or ethyl acetate (218).

a. Extraction Method

Procedure

The following basic procedure has been recommended. The oxide sample is fused with 50 times its weight of potassium bisulfate and the cold melt is dissolved in $1M$ tartaric acid. A 1-ml. aliquot of this solution is transferred to a separatory funnel and the following reagents are added: 3 ml. of 15% (w./vol.) stannous chloride in $4M$ hydrochloric acid; 5 ml. of $9M$ hydrochloric acid, which is also $1M$ in tartaric acid; and 5 ml. of 20% (w./vol.) of potassium thiocyanate. Within 5 minutes, a 7-ml. portion of diethyl ether is added and the yellow color is extracted. The aqueous layer is drawn into a second separatory funnel and is re-extracted with an additional 7-ml. portion of the ether. With amounts of niobium greater than 25 γ the addition of more thiocyanate reagent and a third extraction are desirable. The combined ether extracts are transferred to a 25-ml. glass-stoppered volumetric flask and are diluted nearly to the mark. The solution is kept for 1 hour, preferably at or below 25°C., and is diluted to volume. A portion of this solution is used to measure the absorbance at 385 mμ.

Discussion

Pentavalent niobium exists as both an unreactive and reactive ion, the latter being probably NbO^{+3} and $Nb(OH)_x^{5-x}$. The unreactive form, possibly consisting of colloidal niobium, can be readily converted to the reactive form by treating the sample solution with concentrated hydrofluoric acid and fuming with sulfuric acid (121). Potassium bisulfate melts of niobium that are dissolved in tartaric acid contain a large part of the niobium in the unreactive form, which is converted to the reactive form by the hydrochloric acid.

One of the drawbacks of the photometric procedure described above consists in the interfering effect of tantalum at ratios of tantalum to niobium greater than 10:1. The two main sources of error have been traced back to an incomplete extraction of the niobium from the aqueous phase and to a loss of niobium due to hydrolysis of the tantalum present. It has been shown that the addition of tartaric acid to the reagents and a change in the order of addition largely eliminates the erratic interference of tantalum (89). In the modified procedure, 1 to 2 ml. of the sample solution containing 1 to 50 γ of niobium is transferred into a separatory funnel, followed immediately by 5 ml. of 20% (w./vol.) potassium thiocyanate and 2 ml. of 15% (w./vol.) stannous chloride containing $1M$ tartaric acid. The funnel is shaken thoroughly after the addition of each reagent. After 5 minutes' standing, 5 ml. of ether is added. The mixture is shaken and allowed to stand for 5 minutes, and the lower aqueous layer is transferred into a second separatory funnel. After the addition of 1 ml. of $9M$ hydrochloric acid and 0.7 ml. of 50% potassium thiocyanate, a second extraction is carried out with 5 ml. of ether. The combined ether extracts are diluted to 25 ml. with ether and the absorbance is measured after 30 minutes' standing at 385 mμ.

When determining small amounts of niobium as found in tantalum metal or oxide, the use of large amounts of tartaric acid is avoided by dissolving the sample in hydrofluoric acid, evaporating to dryness in the presence of potassium bisulfate and oxalic acid, and complexing residual fluorides with boric acid, prior to applying the thiocyanate ether extraction procedure (238).

Effect of Other Cations

The effect of various cations on the niobium thiocyanate–ether extraction procedures is presented in Tables XIII and XIV. It should be further noted that Fe, U, and Ti do not interfere in 100-fold excess. Re,

TABLE XIII

Effect of Various Ions on the Determination of 14.3 γ of Niobium by the Thiocyanate–Ether Extraction Method (331)

Ion	Concentration of ions investigated[a]			Type of interference
	0.125M	0.063M	0.013M	
Cr^{+3}	0	0	0	
Ni^{+2}	+	0	0	Inhibits
Co^{+2}	0	0	0	
Mg^{+2}	0	0	0	
Al^{+3}	0	0	0	
NH_4^+	0	0	0	
Fe^{+3}	+	+	+	Color increased
UO_2^{+2}	+	+	+	Yellow color
Cu^{+2}	+	+	+	Ppt. of CuSCN in ether
W(VI)	+	+	+	Green ppt. and soln. in ether
Pt(VI)	Ppt.	Ppt.	Ppt.	Ppt. of Pt and pink soln.
Hg^{+2}	Ppt.	Ppt.	Ppt.	Ppt. clogs ether
Ti(IV)	+	+	+	Yellow color
$C_2O_4^{-2}$	+	+	+	Strongly inhibits
NO_3^-	+	0	0	Inhibits
F^-	+	+	+	Strongly inhibits
SO_4^{-2}	+	0	0	Inhibits
PO_4^{-3}		+	0	Inhibits
BO_3^{-3}		0	0	
VO_3^-	+	+	+	Red color
MoO_4^{-2}	+	+	+	Red color

[a] Key: + represents interference clearly greater than the normal reproducibility; 0 represents no detectable interference.

TABLE XIV
Detailed Study of Interference of Various Cations and Anions (218)

1. *Do not interfere in the amounts shown* (these amounts are the maximums tested for each oxide)

CaO, MgO, Fe_2O_3, Al_2O_3	50 mg.
Y_2O_3, Ce_2O_3	25 mg.
CoO, Cr_2O_3, ZnO, NiO, Sc_2O_3, Sb_2O_3, PbO, Bi_2O_3, ThO_2	10 mg.

2. *Do not interfere in moderate amounts*

TiO_2: 4 mg., no interference; 10 mg., gave 1.3 and 10.3 γ of Nb_2O_5 when 1 and 10 γ of Nb_2O_5 were taken

ZrO_2: 2.5 mg., no interference; 10 mg., no interference at the 1-γ Nb_2O_5 level; 3 mg., gave 9.8 γ instead of 10 γ of Nb_2O_5

Ta_2O_5: 2.0 mg., no interference if the operations prior to and including the extraction are made within 20 minutes; 3 mg., no interference at the 1-γ level of Nb_2O_5, but gave 9.6 γ of Nb_2O_5 at the 10-γ level

$K_2S_2O_7$: Sulfate diminishes the intensity of the niobium thiocyanate color. Just under 2% error is caused in the range of 1 to 10 γ of Nb_2O_5 when 0.3 g. of potassium pyrosulfate is present in the sample and the calibration curve is made without pyrosulfate. An error of 7% results with 0.6 g. of potassium pyrosulfate. This interference is eliminated by preparing calibration curves from solutions containing the same amounts of added pyrosulfate

3. *Interfere at all levels*

Element	γ equivalent to 1 γ Nb_2O_5
Re(VII)	4.6
W(VI)	9.5
Pt(IV)	11
V(V)	28
Mo(VI)	60
Cu(II)	880
Au(III)	1100
U(VI)	11,700

V, Mo, W, and Pt interfere in 10-fold excess. The interference of vanadium can be prevented by omitting stannous chloride from the aqueous solution and instead shaking the ether extract with stannous chloride to remove ferric thiocyanate (266,557). The modified procedure permits 20 γ of niobium to be determined in the presence of 500 γ of vanadium and 100 γ of Fe, Ti, and U. It has been reported that, in the presence of uranium, low niobium values are obtained if a bisulfate decomposition of the sample is used, whereas a hydrofluoric acid–phosphoric acid attack gives substantially quantitative results (171). A recently published procedure successfully eliminates the interference of Re, W, Pt, V, Mo, Cu, Au, and U in the determination of microgram amounts of niobium in granites and other rocks by a succession of simple separation steps (218) (see Section IV-B).

Effect of Reagents

It was shown (331) that the final hydrochloric acid molarity should be about 4. A higher concentration of hydrochloric acid is undesirable as it promotes incomplete extraction of the niobium thiocyanate complex (89). A too-high acidity causes also polymerization of the thiocyanic acid. The polymer absorbs in the same wavelength region as the niobium complex. Since thiocyanic acid is largely removed by the ether, it is important to add more potassium cyanide before the second extraction (89). Ether itself exerts a bleaching effect on the extracted niobium complex unless it contains sufficient thiocyanide. If it becomes necessary to dilute the extracted solution to a volume larger than 25 ml., ether previously equilibrated with a potassium thiocyanate–hydrochloric acid mixture can be used profitably. On the other hand, although a high thiocyanate concentration in the aqueous phase favors the formation of the complex and its extraction into ether, the thiocyanate concentration must not be raised too high or the blank correction becomes too large. The effect of small variations in the stannous chloride concentration is minor, but it is appreciable over the range of 5 to 25%, and the composition of the reagent should not vary more than 1% absolute of stannous chloride. A number of organic liquids other than diethyl ether have been investigated, such as diisopropylether, β,β-dichlordiethyl ether (331), ethyl and isoamyl acetate, methyl ketone, cyclohexanone and cyclohexanol (14), and butyl alcohol (547). It has been found that the optical stability of the ether extract can be improved by the addition of an equal volume of acetone, which inhibits the polymerization of thiocyanic acid to form yellow products (557). The ether–acetone solution maintains an essentially constant absorbance for at least 20 hours.

Among the more recent studies, the extraction of the thiocyanate complex was investigated by means of the radioactive isotope ^{95}Nb (547). The following optimum conditions were derived from these tests: KCNS, 0.08 to 0.10 g. per ml.; HCl, 0.10 to 0.20 g. of HCl per ml.; SnCl$_2$, 0.01 to 0.05 g. per ml.; extracting agent, butyl alcohol; amount extracted, $\sim$89%. The optimum conditions proposed by another investigator for the extraction are as follows (218): total volume, 25 ml., containing 2 ml. of 25%

(w./vol.) tartaric acid, 0.4 ml. of sulfuric acid, a total of 7 ml. of hydrochloric acid, 0.5 ml. of stannous chloride 40% (w./vol.) in $5M$ hydrochloric acid), and 5 ml. of ammonium thiocyanate 25% (w./vol.); extractant, ethyl acetate, 20 ml. The per cent of niobium extracted is temperature-dependent, at least when ethyl acetate is used as extractant. The decrease in the amount extracted is 1% per degree.

b. Homogeneous System

The pale yellow color that niobium salts form in a $4M$ hydrochloric acid medium upon the addition of stannous chloride and potassium thiocyanate solutions is bleached completely upon dilution with water. Although the thiocyanate complex can be extracted with ether, with the resulting increase in color intensity, the same effect can be achieved by the use of aqueous systems containing various organic solvents miscible with water, such as acetone.

Procedure

The following basic procedure has been recommended (183,383). Ten ml. of concentrated hydrochloric acid, 1 ml. of $2M$ stannous chloride (in $4M$ hydrochloric acid), 5 ml. of water, and 10 ml. of acetone are pipetted into a 50-ml. volumetric flask, mixed, and cooled in a water bath at 20°C. for 15 minutes. 10 ml. of $3M$ potassium thiocyanate and suitable aliquots of the niobium solution (oxide fused with 2.5 g. of potassium bisulfate, dissolved in 200 ml. of $1.2M$ tartaric acid and diluted to 500 ml.) are added, the solution is diluted almost to the mark, and is returned to the water bath for 5 minutes. The flask is removed from the bath, allowed to stand at room temperature, and is adjusted to the mark, and the absorbance is measured. A calibration curve is prepared with known amounts of niobium under the same conditions described above.

Discussion

The color intensity is a function of the acidity, thiocyanate concentration, foreign salt concentration, manner of mixing, and time of standing, and it appears that these factors must be more closely controlled than in the extraction method (461).

Effect of Reagents

The reagents causing an increase in the color intensity are, in the order of decreasing effectiveness, acetone, thiocyanate, hydrochloric acid, and stannous chloride. Tartaric acid and sulfate suppress the color, but this effect is lessened by the action of the acetone in preventing their ionization (361). Although it may be desirable to obtain the maximum intensity of the niobium thiocyanate complex, the amount of reagents required to do so increases the blank and the tendency toward salt formation. Therefore, it is frequently necessary to seek a compromise between opposing effects. (It is convenient but not essential that the order of adding the reagents is stannous chloride in hydrochloric acid and then thiocyanate in water–acetone solution (361,364).)

Acetone. The above procedure (183,383) calls for the use of a 20% (vol./vol.) acetone solution. If the acetone concentration is decreased to 10%, the absorbance of the niobium thiocyanate complex is reduced approximately 50% (364). If the acetone concentration is increased to 40%, the absorbance of the niobium thiocyanate complex is increased about 2.5 times, but the absorbance of the interfering titanium thiocyanate is increased (depending on the wavelength used) 6 to 50 times (361). An even greater intensity of the niobium thiocyanate color can be obtained with a 60% (vol./vol.) acetone concentration (121). This may, however, lead to the precipitation of salts.

Thiocyanates. Both ammonium and potassium thiocyanate can be used. Although an increase in the thiocyanate concentration leads to a corresponding increase in the color intensity, precipitation of various salts will occur when the concentration of the potassium thiocyanate exceeds 15 ml. of $3M$ solution in a volume of 50 ml. (183). In addition, the absorptivity of any titanium present increases much more rapidly with increasing thiocyanate concentration than that of the niobium. It is, therefore, desirable that the thiocyanate concentration be kept below $0.3M$ when the sample contains considerable titanium (399).

Hydrochloric acid. The basic procedure described above (183,383) provides a $2.4M$ hydrochloric acid medium. If the acidity exceeds $3.6M$, salts, presumably potassium chloride, may precipitate. Moreover, a high acidity causes increasing blanks due to polymerization of thiocyanic acid. (However, changing the hydrochloric acid concentration in the final solutions from 3.3 to $5.3M$ causes the absorbance of the niobium–thiocyanate first to increase by about 25% and then to decrease rapidly (364). The increase may be due to the tendency of hydronium ions to decrease the activity of water, thus favoring the formation of the thiocyanate complex. The decrease in absorbance at acidities above $4.8M$ may be due to the formation of hydroxychloro complexes.) While it was once assumed that an increase in the chloride concentration also would cause an increase in the color intensity of the niobium thiocyanate complex

(89), it was later shown that this is not the case. It can be concluded now that the effect of large concentrations of hydrogen ions (or other cations, such as Li, Na, K, and Mg) consists in their reaction with tartrate, thereby freeing the niobium for extraction as the thiocyanate complex (121,183).

Other acids and salts. Sulfates and tartaric acid not only suppress the color slightly but, when present in increased concentrations, cause a slow crystallization of acid tartrates (361). Their quantities, therefore, should be kept to a minimum consistent with an adequate amount of potassium bisulfate for fusion and a tartrate concentration that is sufficient to prevent hydrolysis, particularly if a sample is high in tantalum (hydrolysis of tantalum leads to losses of niobium by adsorption on the basic salt).

Although sulfuric acid exhibits a marked bleaching effect in low acetone media (the bivalent sulfate ion concentration becomes significant and complexes the niobium effectively), good results can be obtained in a 60% (vol./vol.) acetone–$1M$ sulfuric acid medium (121). Because of the greater solubility of sodium salts, sodium bisulfate appears to be superior to potassium bisulfate for fusing the oxide.

Oxalates. In a 60% acetone medium, oxalic acid produces no detectable error at a molar ratio (oxalic acid/Nb) of 100 (121), but in a 20% acetone medium the quenching effect of oxalic acid or oxalate is such that the smallest quantity bleaches completely the niobium thiocyanate complex (362). Ammonium oxalate has even been used to differentiate between the thiocyanate complexes of niobium and the unaffected tungsten (364).

Stannous chloride. Since the intense absorbance peak at 383 mμ is assigned to a niobium(V) thiocyanate complex (121), the stannous chloride serves primarily to reduce traces of iron, thereby preventing the interference due to the ferric thiocyanate complex (183). Although variations in the stannous chloride concentration have little effect on the intensity of the niobium color, they may seriously affect the degree of interference of other colored thiocyanates. Thus it has been observed (299) that, in a method employing 1 g. of stannous chloride (361), the interference of tungsten and vanadium is greater by several magnitudes than in methods that use only half a gram of stannous chloride (183,383).

The interdependence of the various reagents used in the development of the niobium thiocyanate complex were recently re-examined, and the following conclusions were reached (38): the stability of the complex decreases with an increase in the thiocyanate and acid concentration; the absorbance, however, increases (presumably partly due to polymerization of the thiocyanate). With a high concentration of acetone the absorbance of the complex becomes less dependent on the thiocyanate concentration. If the color is developed at optimum acidity ($2M$ HCl), the stability of the color is good, and an increase in the thiocyanate concentration from 0.3 to $6M$, as well as variations in the concentration of the potassium bisulfate (up to $0.18M$) and tartaric acid (up to $0.20M$), have little effect on the absorbance of the complex. The following optimum conditions are suggested: 50 ml. volume, containing 34 ml. of HCl–acetone (147 ml. of HCl + 58.8 ml. of H_2O, diluted to 500 ml. with acetone); 5 ml. of niobium solution (up to 10 γ in $0.05M$ tartaric acid; 0.5 ml. of $SnCl_2$ solution ($2M$ in HCl), and 9 ml. of NH_4CNS solution (6 M). After diluting with water at 23°C. the absorbance is measured at 385 mμ after 5 to 10 minutes. It also should be noted that the absorbance of niobium thiocyanate solutions decreases markedly with increasing temperature in the region of 20 to 35°C. At 25°C. the temperature coefficient is -1.5% per degree rise (364). Hence, all solutions should be brought to constant temperature before absorbance measurements are made.

Interference By Other Elements

The degree of interference of various elements depends to a large extent on the concentration of the reagents used. As was mentioned already, the amount of stannous chloride used significantly affects the percentage error, as is illustrated in Table XV.

The final hydrochloric acid concentration also affects the percentage error due to interfering elements. For instance, with a final hydrochloric acid concentration of $2.4M$, only about 10% of the tungsten(V)–thiocyanate color develops in 30 minutes, whereas with a $4.8M$ hydrochloric acid concentration over 90% of the maximum absorbance is obtained in 15 minutes (364). The acetone concentration significantly affects the degree of interference by titanium, since, for instance, a 40% acetone solution gives an absorbance of titanium 50 times that produced by a 20% solution (405 mμ) (361). The absorbance of interfering elements naturally depends also to a large degree on whether the niobium thiocyanate complex is measured at the wavelength of maximum absorbance (385 mμ) or at some other wavelength (see Fig. 12). For instance, the molar absorbance index of niobium decreases from 11,800 to 10,200 with a wavelength change from 385 to 400 mμ; the same change causes the tungsten index to rise from 10,100 to 13,200 (364).

Elimination of Interference By Other Elements

Methods of separation were discussed in detail in Sections IV and V. As far as individual interfering elements are concerned, the following techniques have been suggested.

TABLE XV

Percentage Error in the Determination of 125 γ. of Niobium Pentoxide Caused by Interfering Elements When Using Different Concentrations of Stannous Chloride (38,183,361)

Interfering element	Error (%) when the ratio of interfering elements to Nb_2O_5 is							
	10:1 in $2M$ $SnCl_2$			1:1 in $2M$ $SnCl_2$			100:1 in $2M$ $SnCl_2$
	0.5 ml.	1 ml.	2 ml.	0.5 ml.	1 ml.	2 ml.	2 ml.
Ta_2O_5	+3.3	+11.2	−15	−1.0	+3.2	Nil	−25
TiO_2	—	+5.6	+1	+120	Nil	Nil	+108
ZrO_2	—	Nil	−1.3	—	—	−0.3	−2.3
MoO_3	+70	+24.8	+8	+3.7	+3.2	+4	+105
WO_3	+14	+20.8	+190	+5.0	+1.6	+38	—
Cr_2O_3	+<1.0	+2.4	−1	Nil	—	−5	+36
U_3O_8	+12	+11.2	−3	−1.2	<+1.0	−15	+42
Co_3O_4	—	<+1.0	Nil	—	—	−2	+7
Fe_2O_3	+2.5	Nil	+0.4	Nil	—	Nil	+22
ThO_2	—	<+1.0	−5	—	—	+6	−13
V_2O_5	—	+13.6	—	30	+4.0	+47.5	—
CuO	—	—	Precipitation	—	—	Precipitation	Precipitation
NiO	—	—	+22	—	—	+9	+174
PtO_2	—	—	+61	—	—	+8	Precipitation
F	—	—	−15	—	−4.0	−4	—
P_2O_5	—	—	Nil	—	—	Nil	−2

Tungsten. The tungsten content of the sample solution can be determined by the dithiol method, referring the resulting reading to a calibration graph prepared by applying the niobium thiocyanate procedure to varying amounts of tungsten (383). Or, alternatively, the tungsten content of the sample solution is determined, after boiling the solution with stannous chloride, by the thiocyanate method. By taking readings at 436 mμ it is possible to obtain the absorbance due to tungsten alone, as niobium does not interfere at this wavelength. A correction then can be made from a calibration graph (361). Probably the best approach is based on a preliminary reduction of tungsten(VI) to tungsten(V) by stannous chloride and measuring the total absorbance of the niobium(V) and tungsten(V) thiocyanates at 400 mμ. Oxalate is then added to bleach the niobium color, and the absorbance due to tungsten alone is measured. The difference in the absorbances gives the niobium concentration (364).

Titanium. The titanium content of the sample solution can be determined by the hydrogen peroxide method, referring the resulting reading to

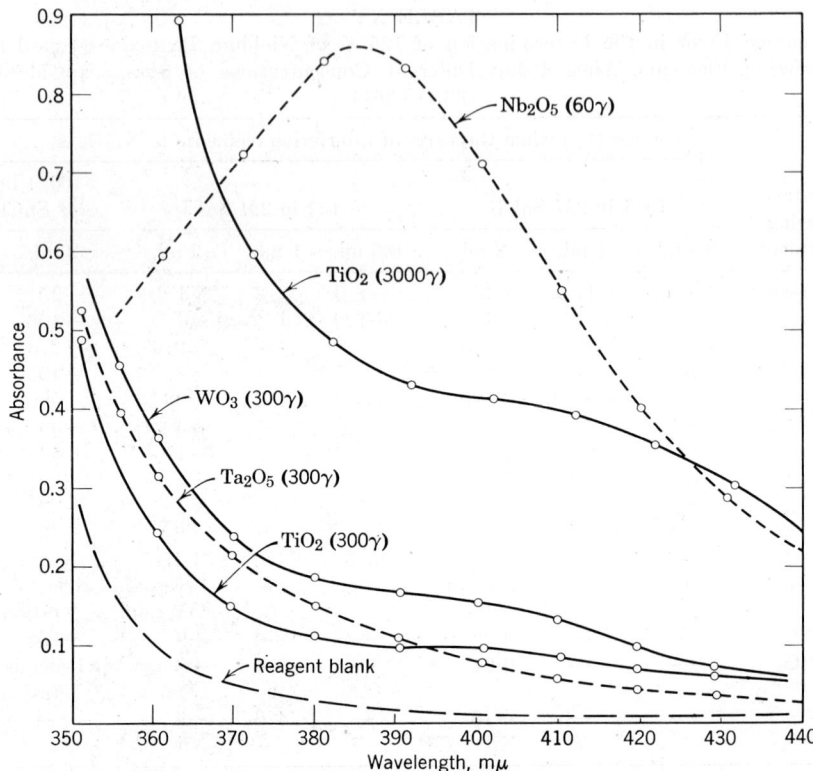

Fig. 12. Variation of absorbance with wavelength for different elements in the thiocyanate procedure (383).

a calibration graph prepared by applying the niobium thiocyanate procedure to varying amounts of titanium (361,383). Although titanium thiocyanate has two absorption peaks (320 mμ and 410 mμ), with low concentration of thiocyanate reagent (< 0.3M) only a single absorption peak (320 mμ) is produced. This condition enables determination of niobium thiocyanate by a ratio procedure, which is based on the fact that at 400 mμ the absorbance is mainly due to niobium whereas at 360 mμ it is mostly due to titanium (399).

Molybdenum. Since its color is about 4% as intense as that of an equal amount of niobium (400 mμ), excessive amounts should be removed by an H_2S separation in an acid tartrate medium (361), or by precipitating the niobium with ammonia (399).

Vanadium. Since its color is about 15% as intense as that of an equal amount of niobium, excessive amounts should be removed by precipitating the niobium either by an hydrolysis procedure or by the carbonate–magnesia procedure described in Section IV-G-5. Uranium depresses the niobium thiocyanate to an extent that equal amounts of niobium and uranium give results about 5% low (399). A cupferron precipitation of the niobium serves to remove excessive amounts of uranium.

Tantalum. When the Ta:Nb ratio is greater than 10:1, as for instance in tantalum metal or oxide, it is advisable to separate the two elements first by ion exchange (55).

2. Hydrogen Peroxide Method for Niobium

Upon the addition of hydrogen peroxide to a strong sulfuric acid solution of niobium(V), a yellow color is produced that is due to a peroxyniobate

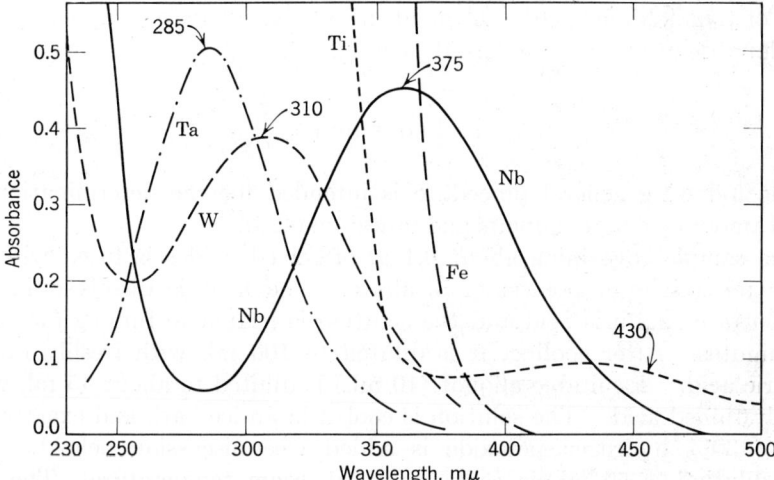

Fig. 13. Absorbance curves of Nb, Ta, Ti, W, and Fe in H_2O_2–H_2SO_4 solutions (477).

complex (371). It has been stated that in a 96% sulfuric acid medium the formation of this peroxyniobate involves niobium and hydrogen peroxide in a 1:1 ratio and that the excess hydrogen peroxide forms mainly H_2SO_5 (477). The following formula has been assigned to the complex (500):

$$HO_4S-Nb\underset{O}{\overset{O}{\diagdown\!\!\diagup}}O-Nb\underset{O}{\overset{O-O}{\diagdown\!\!\diagup}}SO_4H$$

In more recent work however, it has been suggested that in the range of 60 to 100% sulfuric acid two peroxy complexes of niobium (H_2O_2:Nb =

2:1 and 3:2) exist in equilibrium, which is believed (3) to be represented essentially by:

$$\begin{array}{c} O \\ \diagdown \\ \diagup \\ O \end{array}(NbS)\text{—}O\text{—}O\text{—}(NbS)\begin{array}{c} O \\ \diagup \\ \diagdown \\ O \end{array} + H_2O_2 \rightleftharpoons 2 \left[\begin{array}{c} O \\ \diagdown \\ \diagup \\ O \end{array}(NbS)\text{—}O\text{—}O\text{—}H\right]$$

S refers to all coordinated groups except peroxide. The equilibrium of the peroxyniobate complexes is acid-dependent due to an additional equilibrium between hydrogen peroxide and hydroperoxonium ion, $H_3O_2^+$. The decrease in concentration of H_2O_2 in strongly acid solutions favors the formation of the 3:2 niobium complex. Tantalum also forms a peroxy complex, which, however, does not absorb appreciably at the absorption peak of the niobium. Other elements also form colored peroxy complexes or colored solutions in sulfuric acid (see Fig. 13).

a. Procedure

The following general procedure is intended for the determination of small amounts of both niobium and tantalum (423).

The sample containing about 0.1 g. of mixed oxides is fused with 20 times its weight of potassium bisulfate. The melt is dissolved in hot concentrated sulfuric acid and the solution is heated to fuming for 10 to 15 minutes. After cooling, it is diluted to 100 ml. with freshly fumed sulfuric acid. A suitable aliquot (10 ml.) is diluted to about 45 ml. with freshly fumed acid. The solution is cooled in an ice bath and exactly 0.5 ml. of 30% hydrogen peroxide is added (see discussion below). The final dilution to 50 ml. is carried out at room temperature. The absorbances at 285 mμ and 365 mμ are measured against a reference solution of sulfuric acid prepared in the same manner. After determining the absorbance coefficients of the two elements at the two wavelengths under the working conditions, the concentration of niobium and tantalum can be calculated.

b. Discussion

Acid concentration. The absorbances and absorption peaks of the niobium and tantalum complexes vary markedly with the sulfuric acid concentration, as is indicated by the following data (423,461):

	Conc. H_2SO_4, ml. in 100 ml.						
	100	98	95	87.3	70	30	10
Absorption peak, mμ, Nb	365	362	360	355	345	320	—
Ta	287	287	283	281	—	—	255
Relative absorbance, Nb	1.00	1.00	0.98	0.96	0.89	0.12	—
Ta	1.00	0.97	0.93	0.95	—	—	0.99

Other sources report an increase in the absorbance from 0.473 to 0.506 when the sulfuric acid concentration of a niobium solution (7.51 mg. of Nb_2O_5/100 ml.) is raised from 88.7 to 97.8% (477). If the niobium complex is formed in 50 ml. of solution containing 50% of sulfuric acid and 10% of phosphoric acid, the absorption peak lies near 340 mμ and the absorbance is greater than in a medium containing only 50% sulfuric acid (530). The absorbance of the niobium peroxy complex is, however, smaller in sulfuric acid–phosphoric acid mixtures than in concentrated sulfuric acid. In mixtures of 3 parts of sulfuric acid and 2 parts of phosphoric acid, this decrease in absorbance amounts to about 30%; at the same time, the absorbance of the interfering titanium peroxy complex becomes negligible (533).

Hydrogen peroxide concentration. The absorbance of the niobium complex increases with increasing amounts of hydrogen peroxide, but remains virtually constant after the addition of 0.1 ml. of 30% H_2O_2. For 15.2 mg. of niobium in 100 ml. of concentrated sulfuric acid, the addition of 0.4 ml. of H_2O_2 has been suggested (477). Under such conditions, the color of any titanium present is only partly developed.

Beer's law. In a sulfuric acid medium, absorbances are proportional to the element concentration. In a mixed sulfuric acid (50%)–phosphoric acid (10%) system, Beer's law holds up at 342 mμ with niobium concentration between 0 and 125 p.p.m. (530).

c. Interference by Other Elements

If the complex is formed in a concentrated sulfuric acid solution and the absorbance is measured at 365 mμ, the concentration of the following elements in 100 ml. is equivalent to 0.03 mg. of Nb_2O_5 (477): 1.4 mg. of Ta_2O_5, 0.38 mg. of WO_3, 0.32 mg. of TiO_2, or 0.06 mg. of Fe_2O_3. If the complex is measured at 436 mμ (292), 0.442 mg. of titanium is equivalent to 1 mg. of Nb_2O_5. (The amount of titanium in the sample can be determined by developing only the titanium color in a 10% sulfuric acid medium.) Other elements forming peroxy compounds absorbing in the region of the niobium and tantalum are molybdenum and rhenium (461). Vanadium(V) is reduced to vanadium(IV) under these conditions. Ferric iron interferes particularly with the determination of tantalum, by absorbing about 20 times as strongly as peroxytantalic acid at 285 mμ. If only small amounts of iron (or other absorbing substances not reacting with hydrogen peroxide) are present, their effect can be compensated by obtaining the absorbance of an aliquot not treated with peroxide (461).

If a sulfuric acid (50%)–phosphoric acid (10%) medium is used to develop the peroxy niobic acid, the interference of various ions recorded in Table XVI should be taken into account (530).

TABLE XVI

Effect of Diverse Interfering Ions in Sulfuric Acid (50%)–Phosphoric Acid (10%) Medium (530)

Ion	Amount of ion added, p.p.m.[a]	Error, (%) in Nb determination	Permissible amount, p.p.m.
Fe(III)	100	+61	0
VO_2^{-2}	111	+4	25
MoO_4^{-2}	13	+1.3	15
Ti(IV)	10	+4.6	5
NO_3^-	500	+23	100
F^-	24	−2.1	10
SCN^-	1000	+21.4	100
Cl^-	500	+11.1	100
Br^-	1000	+12.6	100
BrO_3^-	1000	+9.3	100
NO_2^-	500	−58.6	0
GeO_3^{-2}	50	+6.0	15
ClO_3^-	100	+100	0
Cr(III)	500	+6.2	200
$Cr_2O_7^{-2}$	575	−12.3	100
UO_2^{+2}	1000	+5.9	500
WO_4^{-2}	140	+13.4	10
Be(II)	1000	+23.8	50

[a] 78-p.p.m. Nb used.

d. Application of Procedure

The method has been applied to the determination of niobium in impure oxides, resulting from the decomposition of steels and ferroniobium by one or the other technique discussed in Section IV-F (195,292). The absorbance is measured at 420 to 436 mμ, where the sensitivity for niobium is rather poor and interference of titanium significant.

The sulfuric acid–phosphoric acid medium appears superior for the determination of niobium in minerals, ores, and concentrates (106,430, 431), since the interference of the titanium is largely repressed by the phosphoric acid. Low-grade samples (Nb > 1%) are decomposed by a sodium peroxide–sodium carbonate fusion and the niobium is collected by a tannin–cinchonine precipitation (431). High-grade samples (50 mg.) are fused in bisulfate (430). The melt can be directly dissolved in $9M$ sulfuric acid. After evaporation to fumes, the clear solution is added to 20 ml. of phosphoric acid (88% w./w.) and the mixture is diluted to 100 ml. with concentrated sulfuric acid. To an aliquot, 0.1 ml. of H_2O_2 (30%) is added and the absorbance is measured at 365 mμ against an unperoxidized aliquot. The procedure is capable of providing results accurate to $\pm 1\%$ relative. The interference from titanium is negligible. In the extreme case of a sample containing 40% Fe_2O_3 and 40% TiO_2, the interference is equivalent to approximately 0.2% of Nb_2O_5. In the above procedure, 90 parts of Ta_2O_5 or 17.5 parts of WO_3 are equivalent to 1 part of Nb_2O_5. Molybdenum also interferes and should be removed. Vanadium, chromium, cobalt, and uranium do not interfere.

The peroxide method also has been successfully applied to the determination of niobium in niobium–uranium alloys (47,88). The concentrated sulfuric acid medium is used and the absorbance is measured at 360 mμ. Interference by uranium is cancelled by using the solution of the sample as a reference or by subtracting the observed uranium absorbance.

The peroxide procedure has even been used for the total assay of high-purity niobium metal (37,297a). After solution of the sample in hydrofluoric and nitric acids and evaporation to fumes with sulfuric acid, the niobium is determined by differential spectrophotometric measurements at 352 mμ in a sulfuric acid (60%)–phosphoric acid (40%) medium.

3. Hydroquinone Method for Niobium

Hydroquinone was introduced into the analytical repertoire in 1946 as a reagent for the determination of niobium, titanium, and tungsten in steel (276). It was assumed that tantalum does not give a color (265,461), but it was recently shown that tantalum actually forms a colored hydroquinone complex with an absorption peak around 355 mμ (559). At this wavelength, however, 90% of the total absorbance (Ta = 0.2 mg. in 10 ml. volume) is due to the reagent. Fig. 14 shows the spectral absorbance curves of niobium, tantalum, tungsten, and titanium with absorption peaks at 400, 368, 475, and 480 mμ, respectively (365).

a. Procedure

The following general procedure can be used for the determination of niobium in the absence of interfering elements (227,247). The oxide (as ob-

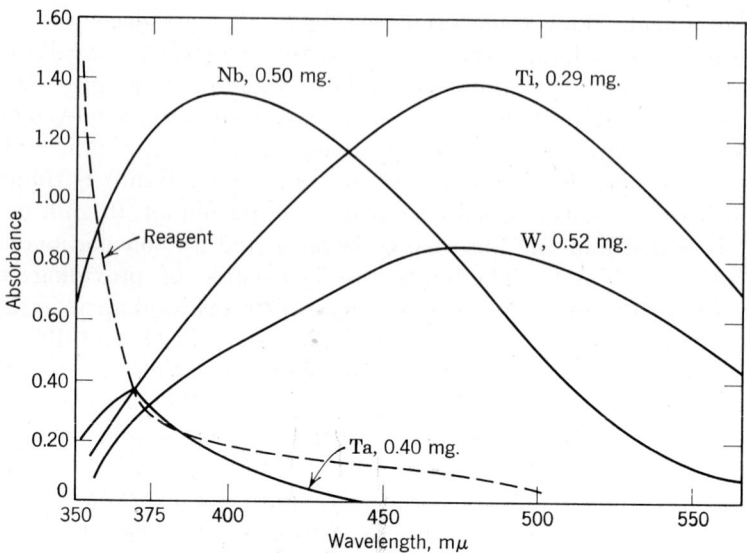

Fig. 14. Spectral absorbance curves for hydroquinone reaction products (365).

tained, for instance, by a cupferron precipitation using zirconium as collector after ion-exchange separations (see Section IX-A-2)) is fused in potassium bisulfate and the melt is dissolved in 4% ammonium oxalate solution. An aliquot containing about 200 γ of niobium and a total of 1 g. of $KHSO_4$ is treated with nitric and hydrochloric acids and evaporated to fumes with 10 ml. of $9M$ sulfuric acid. The solution is repeatedly heated to fumes of sulfuric acid with intermittent additions of water and a few drops of hydrogen peroxide until all organic matter is destroyed. To the cool solution, 1 drop of stannous chloride solution (10 g. in 100 ml. of $6M$ HCl) is added, followed by 25 ml. of hydroquinone solution (5 g. in 100 ml. of $18M$ H_2SO_4). The solution is transferred to a 50-ml. volumetric flask, using hydroquinone solution for washing purposes and to dilute the flask to volume. After mixing and standing for 10 minutes, the absorbance is measured in 5-cm. cells at 490 mμ, using a reagent blank as a reference.

b. Discussion

Reagent. It should be noted that, in the above procedure, the absorbance of the niobium–hydroquinone complex is measured at 490 mμ, although the maximum is at 395 mμ. This is due to the high blank at 395 mμ. It was shown, however (365), that the high blank values may be due to traces of nitrate or peroxide in the air, which tend to oxidize the reagent. When the hydroquinone solution is prepared by mixing the reagent and sulfuric acid in glass-stoppered, amber-colored bottles, the absorbance of the reagent blank at 395 mµ is reduced to an insignificant 0.02. It was also demonstrated that sunlight rapidly affects the stability of the reagent. Measurement of the absorbance of such solutions produced values ranging from 0.1 to 0.4 (against 95% sulfuric acid as the reference solvent), depending on the duration of sunlight exposure and intensity. It is, therefore, advisable to store the hydroquinone–sulfuric acid reagent in amber-colored bottles. The age of the reagent, prepared in accordance with the above instruction, also has some effect on the absorbance (365). The absorbance of the niobium hydroquinone complex is approximately 5% greater with reagent 24 hours old than with fresh reagent. After 24 hours, the efficiency of the reagent slowly diminishes, which is indicated by a drop in the absorbance of the niobium hydroquinone complex of about 2.5% for each 24-hour period.

Stability of the niobium hydroquinone complex. Maximum absorbance is obtained in approximately 10 minutes. After each additional half hour of standing, the absorbance decreases by approximately 2.5%. The hydroquinone photometric method has an appreciable temperature coefficient (about 0.6 to 0.7%/degree), so that all photometric readings must be carried out at a controlled temperature, or a separate set of calibration solutions must be carried along with each set of determinations (227).

c. Interferences

As is indicated in Fig. 14, tungsten, titanium, and, to a much lesser extent, tantalum interfere with the determination of niobium. The degree of interference is also demonstrated in Table XVII.

It can be seen that, at 400 mµ, niobium is the most sensitive element and that interference from 1% tantalum is equivalent to 0.08% niobium. In the presence of both titanium and tungsten, the method is useful only if

TABLE XVII

Slopes of Curves of Niobium, Tantalum, Titanium, and Tungsten at 400 and 500 mµ (365)

(Slopes as absorbance units per mg. of metal in 1 ml., added to 25 ml. of hydroquinone–sulfuric acid reagent

Element	400 mµ	500 mµ
Nb	3.07	1.16
Ta	0.250	0.00
Ti	2.60	4.71
W	1.00	1.66

titanium is determined separately and the tungsten is measured at a wavelength at which the absorbance due to niobium is comparatively weak. An inspection of Fig. 14 shows that, between 400 mμ (absorbance peak) and 500 mμ, the niobium absorbance decreases approximately 60%. At 500 mμ the absorbance due to tungsten is 97% of that of its absorption peak at 475 mμ, whereas at 400 mμ (absorption peak of niobium) it is reduced approximately 40%.

To determine both niobium and tungsten in a sample, the following equations may be used (365):

$$A_{400} = X(Nb) + Y(W) \tag{1}$$

$$A_{500} = X^1(Nb) + Y^1(W) \tag{2}$$

where

A_{400} and A_{500} = the total absorbance at 400 and 500 mμ, respectively, corrected for the hydroquinone and sample background colors

X and X^1 = slopes for niobium at 400 and 500 mμ, respectively (absorbance units/mg. of metal)

Y and Y^1 = slopes for tungsten at 400 and 500 mμ, respectively

Using the slopes listed in Table XVII we can reduce equations (1) and (2):

$$A_{400} = 3.07 \, Nb + 1.00 \, W \tag{3}$$

$$A_{500} = 1.16 \, Nb + 1.66 \, W \tag{4}$$

Insertion of the net absorbance values into equations (3) and (4) will give the niobium and tungsten values by simple calculation. The final results must be corrected for any tantalum and titanium in the sample. Each mg. of tantalum requires a 0.08-mg. correction to be applied to the niobium figure. Each mg. of titanium requires a correction of 0.85 mg. to be applied to the niobium figure and a correction of 2.8 mg. to be applied to the tungsten figure.

d. Application of Procedure

The procedure has been applied to the determination of both niobium and tungsten in stainless steel, after removal of the interfering titanium and other elements by hydrolysis the niobium and tungsten along the lines described in Section IV-G-1 (265). The impure oxides are fused in potassium bisulfate and the melt is leached in ammonium oxalate. In an aliquot the color is developed, after destruction of the oxalic acid with nitric, sulfuric, and phosphoric acids. Then absorbances of niobium and tungsten are measured at 460 and 525 mμ. The concentrations of niobium and tungsten are calculated by means of simultaneous equations.

The hydrolysis step is avoided in another procedure. It allows the direct determination of niobium and tungsten in high-temperature alloys after solution of the sample in nitric and hydrochloric acids, evaporation to fumes with sulfuric and phosphoric acids, and solution of the salts in 6M sulfuric acid (365). After reduction of iron and molybdenum with stannous chloride, an aliquot of the solution (1 ml.) is transferred into 25 ml. of hydroquinone (15 g. in 500 ml. of 95% sulfuric acid). A background solution is prepared by transferring 1 ml. of sample solution into 25 ml. of sulfuric acid. After 10 minutes of standing, the absorbance of the niobium and tungsten hydroquinone solution, of the background solution, and of the hydroquinone reagent are measured at 400 and 500 mμ against 95% sulfuric acid as a reference. The concentration of niobium and tantalum is calculated via simultaneous equations (see above).

The hydroquinone method also has been used to determine niobium in steels (227,287a) and in beryllium products (247), after removal of interfering elements by ion exchange, and in uranium or plutonium alloys, after extraction of the niobium into hexone from a 6.3M sulfuric–1.6M hydrofluoric acid medium (560).

4. 8-Quinolinol Method for Niobium

The determination of niobium, involving the spectrophotometric measurement of a chloroform solution of niobium oxinate that has been extracted from an ammoniacal citrate solution (pH 9.4), was introduced recently into the analytical repertoire (286). The absorbance of niobium oxinate shows a maximum at 385 mμ. Unreacted oxine absorbs strongly below 360 mμ; therefore, its concentration must be kept to a minimum. It is obvious that, for the determination of niobium in steels, the sample contains so many elements forming colored oxinates that it becomes necessary first to isolate the niobium by one or the other method outlined in Section IV. One method that has been successfully used (286) is based on decomposing the sample with hydrochloric and nitric acids, hydrolyzing the niobium by fuming with perchloric acid, and boiling with SO_2 water. The filtered residue is fused in potassium bisulfate and the melt is leached in citric acid. The niobium oxinate is subsequently formed and extracted in an aliquot. Of the more than 25 elements that may occur in steel, six elements, namely niobium, tantalum, tungsten, silicon, tin, and antimony, are precipitated by the perchloric acid hydrolysis. Of these, only niobium forms a colored oxinate, which can be extracted with chloroform from an ammoniacal citrate solution of pH 9.4. Other potentially interfering elements form soluble perchlorates due to the extended perchloric acid treatment (titanium, zirconium, and manganese), or their oxinates are not extracted from an ammoniacal citrate medium (molybdenum and vanadium).

More experimental data are required, however, before this procedure can be applied to steels of various composition.

The photometric oxinate procedure has also been extended to the determination of niobium and molybdenum in uranium-base alloys (394). In this procedure the sample is decomposed with nitric and hydrofluoric acids. After evaporation to fumes with sulfuric acid, the uranium is complexed with sodium fluoride solution (250 mg. of fluoride ion for each 1 g. of sample). An aliquot of the solution containing 5 to 100 γ of niobium is treated with 3 ml. of 8-quinolinol (1% w./vol. in 2.5% vol./vol. acetic acid). After adjustment of the pH to 5.1 with ammonium acetate and ammonium hydroxide, the niobium is extracted with 10 ml. of chloroform. The extract is washed with an ammonium carbonate–potassium cyanide solution (100 ml. of $2M$ $(NH_4)_2CO_3$ plus 20 ml. of $1M$ KCN in 1 liter) to remove residual uranium, and the niobium is determined by measuring the absorbance of the extract at 385 mμ. Even traces of iron interfere, but can be compensated by measuring first at 385 mμ (Nb + Fe), then at 470 mμ (Fe only). In the presence of molybdenum, the extract is divided into two fractions. One fraction is shaken with $0.05M$ oxalic acid solution, which removes niobium and iron. The absorbance of the other fraction is measured at 385 and 470 mμ, using the back-extracted solution as a reference (385 mμ = niobium + iron; 470 mμ = iron only). Molybdenum itself is determined by measuring the absorbance of the washed extract at 380 mμ. Up to 50 γ of aluminum, chromium, and manganese and 200 γ of tantalum, tungsten, and zirconium are not extracted under the above conditions and therefore do not interfere. Up to 200 γ of vanadium, nickel, and copper are removed by the carbonate–cyanide solution and therefore do not affect the accuracy of the niobium procedure.

The extraction of niobium, tantalum, and tungsten oxinates by the use of various solvents also has been recently mentioned in the Russian literature (13).

5. Molybdenum Blue Method for Niobium

Niobium forms a complex with sodium phosphate and ammonium molybdate. In addition to the yellow niobiophosphomolybdate, yellow phosphomolybdate is formed, which is destroyed by the addition of sulfuric acid prior to the reduction of the niobium complex with stannous chloride (130). A short outline of the procedure follows (413).

To a 10-ml. aliquot of the sample solution in a 50-ml. volumetric flask containing a maximum of 250 γ of niobium, 1 ml. of sulfuric acid, and 0.5 ml. of $0.5M$ hydrofluoric acid, the following reagents are added: 2 ml. of 0.06% disodium hydrogen phosphate solution, 5 ml. of 2% ammonium molybdate solution, and, after 15 minutes of standing, 10 ml. of $9M$ sulfuric acid to destroy the phosphomolybdate color. Exactly 30 seconds after the addition of the sulfuric acid, 3 ml. of 0.5% stannous chloride in $4M$ hydrochloric acid is added. The solution is immediately diluted to the mark with water and the absorbance is read within 5 minutes at 715 mμ against water as a reference.

(It should be noted that the decomposition of phosphomolybdate (also arsenomolybdate) requires about 20 seconds, whereas that of the niobiophosphomolybdate begins 35 seconds after the addition of the sulfuric acid (395).)

The molybdenum blue method described above has been applied to the determination of niobium in titanium (413) and in steel (130,395,440). Although it is very rapid, a great number of elements interfere.

Furthermore, the concentration of the reagents and the timing of their additions affect the stability of the various complexes to such an extent that the precision of results suffers.

The following elements must be absent because they form insoluble salts with one or the other reagent: Sb, Ba, Pb, Hg, Se, Ag, and Sr. W, Si, and V must be absent because they form heteropolymolybdenum blue complexes. As and Ce repress the niobium color. More than 2.5 mg. of Ti per 50 ml. of solution may cause high results if the niobium content is low, and low results if the niobium content is high. The following elements do not interfere in 50 ml. of solution: Al (50 mg.), Be (50 mg.), B (50 mg.), Cd (50 mg.), Ca (50 mg.), Co (50 mg.), Cr (50 mg.), Cu (25 mg.), Fe (15 mg.), K (50 mg.), Mg (50 mg.), Mn (50 mg.), Mo (5 mg.), Na (50 mg.), Ni (50 mg.), P (0.1 mg.), Si (50 mg.), Ta (5 mg.), Ti (2.5 mg.), Zn (50 mg.), Sn (50 mg.), Cl as NaCl (50 mg.), F as NaF (12 mg.), and NO_3^- (300 mg.).

The above procedure recently has been modified to the extent that the phosphate ion is omitted, so that the blue color formed is based on the reduction of molybdoniobic acid (223).

6. Pyrogallol Method for Niobium

Whereas pyrogallol has been widely used for the photometric determination of tantalum (see Section VII-B-8), it has found only limited application in the determination of niobium. In alkaline solutions pyrogallol reacts with niobium to form a yellow complex much more intensely colored than the corresponding tantalum complex, whereas in acid solutions the reverse is the case. For instance, the pyrogallol complex of 100 γ of Ta_2O_5 per ml., produced at pH 1 to 2, gives in 1-cm. cells an absorbance of about 1, but it decreases to about 0.04 if the complex is formed at pH 7 to 8. On the other hand, the absorbance of a solution containing 50 γ of Nb_2O_5 per ml. amounts to about 1 at pH 7 to 8 and about 0.005 at pH 1 to 2 (262). Therefore, whereas niobium has a negligible effect on the tantalum complex formed in an acid medium (the absorbance of 1 mg. of niobium is equivalent to that of 4 γ of tantalum (247)), tantalum interferes to a larger extent with the niobium complex formed in a slightly alkaline medium (the absorbance of 1 mg. of tantalum is equivalent to that of 26 γ of niobium (262)). It was shown recently that the reaction of niobium with pyrogallol is much more sensitive in the presence of a quaternary compound such as polyethoxy-15-stearyl methylammonium chloride (103a).

(The composition of the yellow niobium color formed in an alkaline solution containing ammonium oxalate has been investigated (336), and it can be assumed that the complex contains niobium and pyrogallol in a ratio of 1:1. It is not clear, however, whether the pyrogallol is added to the already existing oxotrioxalatoniobate (see Section II–E–4) or whether it replaces part of the oxalate.)

a. Procedure

The following general procedure has been suggested (262). Twenty-five mg. of the mixed oxides is fused in 0.5 g. of potassium bisulfate, the melt is leached in 1 g. of ammonium oxalate and 10 ml. of water, the solution is transferred to a 25-ml. volumetric flask, and is diluted to the mark. To a 1-ml. aliquot of this solution in a 50-ml. flask, 5 ml. of potassium bisulfate solution (0.1 g. per ml.) is added, followed by 20 ml. of 4% (w./vol.) ammonium oxalate solution and 20 ml. of alkaline pyrogallol solution (1 g. dissolved in 50 ml. of 20% sodium sulfite solution). After diluting to the mark with water, the absorbance is measured at 410 mμ against a reagent blank and then is calculated to Nb_2O_5. This value must be corrected for any tantalum present, which can be done simply by subtracting the apparent Nb_2O_5 weight from that of the original mixed oxides, thus providing an approximate Ta_2O_5 value. The absorbance reading for which this amount of tantalum is responsible is read from a tantalum correction curve (prepared under the conditions described above) and the weight of Nb_2O_5, equivalent to this absorbance, is subtracted from the original apparent figure for Nb_2O_5, thus giving the true value for Nb_2O_5.

b. Discussion

Unlike the tantalum complex, which shows an absorption peak in an acid medium at 398 mμ (247), the niobium complex has no useful absorption peak. Measurements at 410 mμ or thereabouts ensure minimum interference by other cations. All investigators use sodium sulfite to prepare a weakly alkaline solution (147,216,262,435,521,574), but the optimum suggested pH varies from 4.6 to 8.

c. Interferences

The interference of tantalum has already been discussed. Other metal oxides (1 mg.) cause positive errors equivalent to niobium oxide as follows (262): WO_3 = 0.11 mg.; MoO_3 = 0.84 mg.; ZrO_2 = 0.02 mg.; TiO_2 = 3.10 mg.; and SnO_2 < 0.01. Iron oxide and vanadium oxide must be absent since they form in the alkaline medium intense colored solutions. Fluorides prevent the formation of the niobium complex (209), but it has been stated that the following acids do not interfere (216): perchloric, nitric, hydrochloric, acetic, citric, succinic, and phthalic. On the other hand, it has been noted (262) that, because of the instability of pyrogallol toward oxidation agents, all reagent solutions must be freshly prepared.

7. Other Photometric Methods for Niobium

Both Nb^{+3} and Nb^{+5} form complexes with disodium-1,2-dihydroxybenzene-3,5-disulfonate (tiron) (178), but the complex with Nb^{+3} is not stable in air. The yellow complex of Nb^{+5} is stable in up to $5N$ hydrochloric acid and also in alkaline solution. Optimum color development, however, is obtained in a $1.6N$ hydrochloric acid solution. Tantalum does not interfere; the color interference of iron(III) and vanadium(V) can be prevented by reducing these substances with stannous tin or ascorbic acid. It is stated that the mean error in determining 2.5 to 8.5 mg. of niobium in 50 ml. of solution is 0.1 mg. To make the method more generally suitable, the combined oxides are fused in $K_2S_2O_7$, the melt is dissolved in a mixture of tartaric and hydrochloric acids, and the color is developed in an aliquot in a $10M$ HCl medium containing 2% tartaric acid (13a).

Niobium pentoxide is appreciably soluble in concentrated hydrochloric acid (565) (see Section II-E-4). The hydrochloric acid solution shows an absorption peak at 281 mμ (283). The sensitivity of the reaction is adequate (10 γ per ml. = absorbance of 1), but a great number of elements absorb in the region of the niobium peak, particularly V(V), Cr(III), Pb, Fe(III), Cu(II), Mo(VI), and Ti(IV). The interference of limited amounts of Fe and Cu can be avoided by reduction with stannous chloride.

Tantalum may be present in amounts 10 times that of the niobium. Small amounts of fluoride and sulfate do not interfere.

The yellow-brown color, due to Nb(III) through the reduction of Nb(V) with zinc in an ammonium oxalate–sulfuric acid solution, has been made the basis of another niobium procedure (339). More than traces of titanium interfere. The absorption peak lies in the 430 to 440 mμ region.

Pyrocatechol can be used as the photogenic agent if Nb(III) is produced by electroreduction (540).

It has also been noted that the solution obtained by fusing niobium pentoxide in potassium carbonate and extracting the melt with water shows a maximum absorption at 234.5 mμ and obeys the Beer-Lambert law within the concentration range of 5 to 25 γ per ml. (289).

Sulfosalicylic acid also has been suggested (128). Interfering elements (Fe,Ti) are removed by fusion with potassium hydroxide and extraction with water. Colored complexes due to small amounts of Mo, V, and U are removed by using a 1.1M hydrochloric acid medium. Sulfate, oxalate, fluoride, phosphate, tartrate, and arsenate interfere.

Of interest also is the fact that the yellow complex formed by niobium with ascorbic acid follows Beer's law over a wide range of concentration (272). The color is developed in a solution buffered on the slightly acid side with sodium acetate. The absorbance is measured at 345 mμ. Serious interference is caused by PO_4^{-3}, Br^-, F^-, UO_2^{+2}, V, Mo, and Ti. Tantalum can be similarly determined at 330 mμ, but the sensitivity is low.

Arsenazo, which has already been used as a reagent for the determination of tantalum (407), has now also been suggested for the photometric determination of niobium in titanium and aluminum alloys (408). No interference with the color due to 0.2 mg. of Nb in 100 ml. is caused by 0.8 mg. of Fe, 1.0 mg. of Cr, 4.0 mg. of Ni, 0.4 mg. of Cu, 1.0 mg. of Al, or 0.5 mg. of Mo. Quercetinsulfonic acid also has been mentioned as a potential niobium reagent (282b). It forms yellow chelates with niobium in neutral and slightly acid solutions.

8. Photometric Determination of Tantalum by Pyrogallol

Tantalum does not possess many characteristic reactions; therefore, analysts confronted with a mixture of earth acid oxides prefer to determine the niobium and calculate the tantalum "by difference." Pyrogallol is the only reagent that has found widespread use, and its basic features are discussed below in some detail.

The observation that tantalum forms with pyrogallol in acid solutions a yellow color, useful for analytical purposes, is due to several Russian investigators (7,313,434). Unfortunately, most of their publications are accessible to English-speaking chemists only in the form of abstracts. Available information frequently is incomplete and does not give a coherent picture of the effect of variations in the concentration of reagents and of the effect of interfering substances.

a. Procedure

The following procedure is presented to allow a discussion of all variables (247). The mixed oxides of tantalum and niobium, freed from titanium by tannin separations (see Section IV-G-3) or ion exchange (see Section IV-H-4), are fused in a Vycor or quartz crucible with 0.5 g. of potassium bisulfate, and the melt is dissolved in 15 ml. of hot ammonium citrate–ammonium oxalate solution (3 g. of ammonium citrate and 3 g. of ammonium oxalate dissolved in 100 ml. of $0.45M$ sulfuric acid). The solution, or an aliquot containing about 300 γ of tantalum, is transferred to a 25-ml. volumetric flask. (If an aliquot is taken, sufficient $KHSO_4$ is added to provide 0.5 g. of the salt.) 3.0 g. of pyrogallol is added, the solution is swirled until clear, and then is diluted to the mark with the ammonium citrate–ammonium oxalate solution. After 10 minutes of standing, the absorbance is measured at 398 to 420 mμ, using a reagent blank in the reference cell, carried through all steps of the procedure. The tantalum concentration is determined by reference to a calibration curve, prepared by adding pyrogallol to aliquots of a standard tantalum solution and diluting to 25 ml. with ammonium citrate–ammonium oxalate solution.

b. Discussion of Various Conditions

Acid Medium and Wavelength

Early work was carried out in a potassium bisulfate–ammonium oxalate solution (262,313,414,534) containing free sulfuric acid up to a concentration of $1N$. The optimum pH was found to be between 1 and 2 (262,534). In this medium the tantalum pyrogallol complex has an absorption peak at 398 mμ. If phosphoric acid is used to adjust the acidity to an optimum pH of 1.8 to 2.0 (265) (phosphoric acid reduces the interference of tungsten), this has little effect on the position of the absorption peak. On the other hand, if hydrochloric acid is used to adjust the acidity to an optimum $4N$ (this reduces the interference of the titanium complex to a minimum), the maximum absorbancy of the tantalum–pyrogallol complex is shifted all the way down to 315 mμ (137). If the solution contains, in addition, tartaric acid (to allow the determination of niobium in another aliquot by the thio-

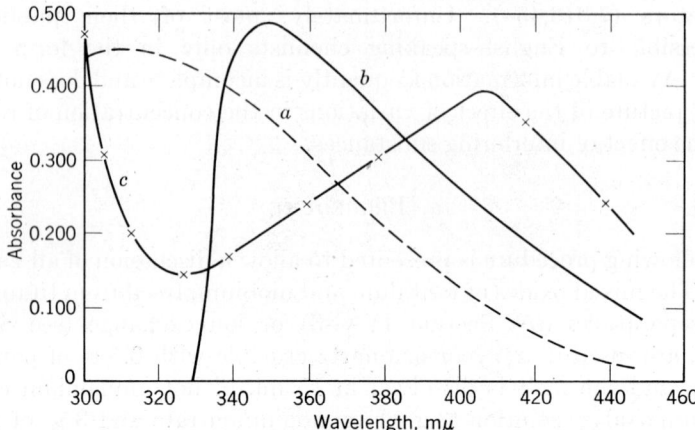

Fig. 15. Absorption spectra of the tantalum–pyrogallol complex in various media (362). Key: *a* is hydrochloric acid–ammonium oxalate; *b* is hydrochloric acid–tartaric acid–ammonium oxalate; and *c* is sulfuric acid–ammonium oxalate.

cyanate procedure), the absorption peak lies near 350 mμ (362). See Fig. 15.

Ammonium Oxalate

Several investigators (262,414) recommend a 4% ammonium oxalate concentration in conjunction with the sulfuric acid medium. Since, in a hydrochloric acid solution, ammonium oxalate has a small but pronounced effect on the absorbancy of the tantalum (this effect may be positive (362) or negative (137), depending on whether the solution contains also tartaric acid), investigators using the HCl system limit the ammonium oxalate concentration to 0.125 g. (137) or 0.375 g. (362) per 50 ml. Ammonium oxalate solutions, measured against water in the reference cell, were found to have an appreciable absorbance in hydrochloric acid solution (137). Therefore, a strict control of its concentration is advisable.

Potassium Bisulfate

Some analysts closely control the amount of bisulfate used (262,414), but others (137,247) report no, or only minor, effects on the absorbance of the tantalum pyrogallol complex.

Solution of Bisulfate Melt in Ammonium Oxalate

It has been noted in this author's laboratory and by others (137) that dissolution of the bisulfate melt in ammonium oxalate may yield clear solutions that give absorbances that cannot be duplicated on retrial. This may be due to colloidal particles causing surface phenomena acting at the interface of the melt and the extracting solution. Continuous stirring of the extracting solution for about one-half hour is, therefore, of critical importance in the preparation of a true solution of tantalum.

Pyrogallol

In any medium the absorbance of the tantalum–pyrogallol complex increases with the pyrogallol concentration (137,247,414). The use of greater or lesser amounts of pyrogallol causes a change in absorbance amounting to about 10% between and 8 and 16 g. per 100 ml. (247). Although an increase in the amount of pyrogallol used causes an increased absorbance, the solutions become turbid when the amount exceeds 16 g. per 100 ml. Because the aqueous solution of the reagent is unstable (265), some investigators prefer to add solid pyrogallol to the ammonium oxalate solution of the sample (247,414); others stabilize the reagent solution with stannous chloride (137,362). Dissolution of the reagent in dilute sulfuric acid (262, 385) or in an ammonium oxalate ammonium citrate mixture (175,227,247) is said also to increase the stability of the reagent.

Stability of the Tantalum–Pyrogallol Complex

In a hydrochloric acid medium the absorbance of the tantalum–pyrogallol complex, measured against a reference blank of similar age, decreases by only 1% in 7 hours (137). It has been stated (175) that in a sulfuric acid medium the absorbance increases 7 to 10% in a 3-hour period. The temperature coefficient for the complex is negligible in a hydrochloric acid medium (137). In a phosphoric acid medium, however, raising the temperature from 20 to 30°C. causes a 10% decrease in the absorbance (265). In a sulfuric acid–ammonium oxalate medium, the addition of ammonium citrate is said to eliminate an otherwise appreciable temperature coefficient (225).

c. Interfering Elements

To avoid the strong interference from platinum, the bisulfate fusion should be carried out in a Vycor crucible. The effect of various other elements depends to a large extent on the chosen medium and wavelength (Tables XVIII and XIX). For instance, in a hydrochloric acid–ammonium oxalate solution (where the pyrogallol complex is measured at 325 mμ), 5 mg. of TiO_2 causes only a 7% error in the determination of 1 mg. of Ta_2O_5.

TABLE XVIII
Interference of Various Cations in the Tantalum–Pyrogallol Method

1 mg. of metal oxide	Ta_2O_5 equivalent, in mg.	
	HCl–ammonium oxalate medium (325 mμ) (137)	H_2SO_4–ammonium oxalate medium (400 mμ) (262)
Nb_2O_5	0.024	0.007
Fe_2O_3	<0.002	0.19
WO_3	0.15	0.20
MoO_3	0.5	0.68
ZrO_2	0.01	0.00
TiO_2	0.01	4.00
V_2O_5	<0.005	0.44
SnO_2	<0.004	0.00
U_3O_8	0.05	
Sb_2O_3	0.1	
CuO	0.008	
Cr_2O_3, Al_2O_3, BaO, Bi_2O_3, NiO, PbO, BeO, Sc_2O_3, B_2O_3, Na_2O, ThO_2, SiO_2, K_2O, CaO, MgO, Y_2O_3, Ce_2O_3, Co_3O_4	Each less than 0.01	

TABLE XIX
Effect of Varying Amounts of Interfering Elements on the Determination of Tantalum by the Pyrogallol Method in a Hydrochloric Acid–Tartaric Acid–Ammonium Oxalate Medium (350 mμ) (362)[a]

Interference due to	Error (in %) when weight ratio of interference to Ta_2O_5 is:		
	1:1	10:1	100:1
TiO_2	+16	+185	+242
V_2O_5	+11	+ 11	+100
MoO_3	+ 6	+242	—
WO_3	+22	+157	—
U_3O_8	+11	+ 2	+ 96
ThO_2	− 3	+ 15	Precipitation
NiO	− 14	− 7	+ 22
Cr_2O_3	+30	+204	—
PtO_2	Precipitation	Precipitation	Precipitation
SnO_2	+ 2	+ 36	+ 38
MgO	+22	Precipitation	Precipitation
Bi_2O_3	− 2	+ 4	+ 58

[a] There is no significant interference from the following ions: Nb^{+5}, Fe^{+3}, Al^{+3}, Cu^{+2}, Sb^{+5}, Zr^{+4}, Mn^{+2}, Co^{+2}, PO_4^{-3}, F^-, alkaline earth, and alkali metal ions.

In a sulfuric acid–ammonium oxalate medium, however, where the color is measured at 400 mμ, 1 mg. of TiO_2 causes a 400% error in the determination of 1 mg. of Ta_2O_5. In contrast, niobium interference is greatest in the hydrochloric acid–ammonium oxalate solution measured at 325 mμ, and least in the same medium measured at 365 mμ, because of the presence of tartaric acid (362). Since the hydrochloric acid medium employs stannous chloride (137,362), Fe^{+3} is reduced to Fe^{+2} and does not interfere, even when present in substantial amounts. This allows the direct determination of the tantalum in ores after a fusion in potassium bisulfate and dissolution of the melt in tartaric acid (362). Since in the 400-mμ wavelength region titanium interference is strongest, its prior removal, preferably by ion exchange (see Section IV-H-4), is indicated. Alternatively, a hydrochloric acid medium should be used with a shift of the tantalum absorption peak into the ultraviolet and virtual elimination of the titanium–pyrogallol complex.

9. Other Photometric Tantalum Methods

The determination of tantalum by the hydrogen peroxide method has been discussed in Section VII-B-2. Gallic acid ($C_6H_2(OH)_3COOH$) is closely related to pyrogallol and also has been used for the determination of tantalum (182,373a). The tantalum–gallic complex absorbs over a broad region in the blue near-ultraviolet region. The adverse effect due to interfering ions can be minimized by a suitable choice of wavelength. A great number of other polyphenols have also been investigated (285).

Hydroquinone is best known as a reagent for niobium (see Section VII-B-3), but it has also been recommended for the determination of tantalum (144,559). Although the tantalum–hydroquinone complex shows maximum absorbance at 355 mμ, the color is measured at 375 mμ with a loss of 35% from the maximum absorbance value because of the extremely strong absorption of hydroquinone at the lower wavelength. Since the sensitivity of the tantalum hydroquinone reaction is poor (40 γ/ml. gives an absorbance of about 1 at 375 mμ), and since niobium interferes strongly, the method does not appear attractive for the determination of tantalum.

Trihydroxy-6-fluorone forms with tantalum in slightly acid solution a bright red precipitate or a colloidal solution which may be stabilized with gelatin (404). The minimum amount of tantalum detectable is 0.3 γ/ml. No interference is caused in a 10-ml. volume by 400 γ of Nb, 100 γ of Ti, 500 γ of Zr, 150 γ of W, 100 γ of Mo, 500 γ of Sb^{+5}, 3 γ of Sb^{+3}, 5 γ of Sn^{+4}, 3 γ of Ge, or 1000 γ of Fe. The red complex formed between tantalum and phenylfluorone (2,3,7-trihydroxy-9-phenyl-6 fluorone) also has been used to determine tantalum (343). Interference from other metals is avoided by a preliminary extraction of tantalum with methyl isobutyl ketone from HF–HCl solution. The organic solvent is evaporated to dryness with $HClO_4$ and HNO_3 and the residue is dissolved in dilute HF. EDTA is added to form complexes with coextracted metals, the solution is buffered to pH 4.5 with ammonium acetate, and the colored complex is then developed with phenylfluorone. Absorption measurements are made at 350 mμ.

The color reactions of tantalum and niobium with 3-flavone derivatives also have been examined (439). Tantalum forms colored complexes with quercetin and with 3-hydroxy-5,7,3',4-tetramethyloxyflavone. Niobium gives no color.

Tantalum in the range of 10 to 200 p.p.m. in oxalic–sulfuric–hydrochloric acid solution forms a deep yellow complex with catechol when the ratio of Ta_2O_5 to catechol is 1:1 (540). The absorption maximum lies at 395 mμ. Titanium interferes, as well as phosphate, fluoride, and tartrate ions. The interference of titanium can be minimized by using a 1.5 to 2N hydrochloric acid medium (463). The interference of many other elements can be avoided by the addition of tartrate or oxalate and EDTA (424,584). n-Butanol extracts tantalum and titanium from their complexes with catechol and oxalate, but niobium remains in the aqueous phase (583). The extraction is carried out at a pH of 3 and the titanium is subsequently removed by a washing of the extract with 5% sulfuric acid.

Methyl violet was recently introduced as a very sensitive reagent for tantalum (330,433). It was shown that the reaction product of the tantalum fluoride complex can be extracted with benzene (330) or toluene (433), and that under the optimum extraction conditions for tantalum (pH 1.9 to 2.3) the fluoride complex of niobium does not react with methyl violet. Optimum concentrations of F^- and methyl violet for forming the reaction product are 5 mg. per ml. and 0.02%, respectively. Toluene solutions of this compound have three absorption peaks, at 300 mμ (molar absorption coefficient 15,000), 550 mμ (22,700), and 600 mμ (30,900). The colored solution is stable for 2 hours and Beer's law is obeyed at concentrations up to 60 γ of tantalum per ml. Niobium does not interfere. Malachite green, which like methyl violet belongs to the triphenyl methane series, has also been suggested recently as an efficient tantalum reagent (278).

Other photometric methods for tantalum that have been suggested recently are based on the color it forms with arsenazo (o-(1,8-dihydroxy-3,6-disulfo-2-naphthylazo) benzene arsonic acid) (407) and with Rhodamine B (65a) and on the catalytic effect tantalum exerts on the oxidation of potassium iodide by hydrogen peroxide (576).

C. TITRIMETRIC DETERMINATION OF NIOBIUM

1. General Observations

Niobium(V) in various acid media can be reduced in a Jones Reductor to niobium(III). The solution containing the niobium(III) species is usually collected in ammonium ferric sulfate solution in which an amount of ferrous iron is produced equivalent to the amount of niobium present. The ferrous iron is determined by titration with standard dichromate or permanganate solution.

Although the titrimetric determination of niobium has been studied for almost 75 years by many investigators, no completely satisfactory procedure has been reported in the past. There is, however, reason to believe that the procedure most recently suggested (243) may obviate the difficulties inherent in previous procedures. A survey of earlier methods (490) indicates that the following media gave erratic results: solution of the bisulfate melt in sulfuric and succinic acids and reduction by amalgamated zinc in a Jones reductor (377); reduction of the fluoride solution with granulated zinc (337); and solution of a potassium carbonate melt in phosphoric acid and reduction by zinc dust (202). The difficulties were once believed to be associated with the formation of intermediate oxides of niobium, but the most likely explanation lies in the partial hydrolysis of the niobium with formation of a colloidal phase, only the portion in true molecular solution being reduced to niobium(III) (243,490). Observations by some investigators that the addition of known amounts of titanium (18,122,123,562,570) or iron (279) leads to almost stoichiometric values can probably be ascribed to the fact that solutions containing significant amounts of these elements are known to prevent or retard the hydrolysis of the niobium. In a recent publication it is claimed that niobium can be reduced quantitatively with cadmium in a hydrochloric acid–sulfuric acid medium (303a).

2. Procedure Based on Reduction in Sulfuric Acid Medium

a. Preparation of Solution

The following typical procedure is by chemists of the National Bureau of Standards (294), who emphasize the necessity of strict observance of all details. The niobium oxide (up to 300 mg.) is fused in potassium bisulfate (3 to 5 g.) in a covered quartz crucible and the melt is dissolved in 20 ml. of concentrated sulfuric acid. The cold salts are dissolved in 100 ml. of cold water with the addition of 1 to 2 ml. of hydrogen peroxide. After further dilution to 200 ml. with 20% (vol./vol.) of sulfuric acid, 2 g. of succinic acid is added and the solution is heated moderately until clear.

The Jones reductor has an I.D. of about 1 inch, and contains a column of 33 inches of amalgamated zinc, prepared as follows: 1000 g. of 20-mesh zinc metal is mixed for one minute with a 2% aqueous solution of mercuric chloride, the excess of which is discarded. The zinc is washed first with distilled water, then copiously with warm 1% (vol./vol.) of sulfuric acid, then again with water. Not more than 10 portions of sample solution or a total of 2.5 g. of Nb_2O_5 should be passed through the reductor. Shortly before use, the reductor is pretreated by pouring hot 1% (vol./vol.) sulfuric acid through it, followed by hot water (90°C.). The washings are discarded.

b. Reduction and Titration

A threefold excess of ferric iron is placed in the receiver, which is then attached to the reductor. 100 ml. of 20% (vol./vol.) sulfuric acid, heated to 60 ± 5°C., is poured into the reductor. With the aid of suction the acid, the solution of the niobium, and 150 ml. of 20% sulfuric acid containing 1% succinic acid, each at 65 ± 5°C., and 200 to 250 ml. of cold water are passed slowly and without interruption in succession through the reductor at such a rate that the total time of the passage is about 25 minutes. The solution of the reduced niobium is cooled during this operation by immersing the receiver in ice water. After removal of the receiver and washing of the outlet tube, 10 ml. of phosphoric acid (85%) is added and the ferrous iron [from the interaction of Nb(III) with Fe(III)] is titrated with a $0.1N$ solution of potassium permanganate, using orthophenanthroline as an indicator. The titration is corrected by deducting the amount of $KMnO_4$ consumed by a blank determination involving all reagents, obtained under identical conditions.

Of critical importance in the above procedure are the dimensions of the reductor, the mesh size and degree of amalgamation of the zinc, the acidity of the solution, and the rate at which the solution is passed through the reductor. This procedure is particularly suited as a supplement to Schoeller's tannin method, thus allowing the operator to carry out only a limited number of fractional precipitations, leading to a fraction that contains all the tantalum, most of the titanium, and some intentionally precipitated niobium (122,335) (see Section VII-A-2).

3. Procedure Based on Reduction in Hydrofluoric Acid Medium

Another procedure that may be somewhat superior to the above in regard to precision was recently introduced (243). Since the difficulties with the sulfuric acid system are commonly attributed to the formation of a colloidal phase and since the most stable complex of niobium in aqueous media is probably the fluoro complex, the new procedure involves a fluoride-containing medium. Because of the corrosive nature of hydrofluoric acid on glass ware, the apparatus used is constructed entirely of polyethylene and consists of heavy-gauge polyethylene tubing (length 60 cm., I.D. 1.80 cm.), narrowing at the base to a polyethylene stopcock. Polyethylene tubing is wrapped around the column and water is passed through it to maintain a cool, constant temperature within the column during the reduction. The collecting vessel is a 250-ml. polyethylene bottle fitted with a polyethylene screw cap having three holes in the top to accommodate the delivery tube from the column and the inlet and outlet tubes for nitrogen.

To carry out the procedure, up to 50 mg. of niobium oxide is dissolved in hydrofluoric acid. After evaporation to near dryness in a platinum crucible the residue is dissolved in 10 ml. of $2.5M$ hydrofluoric acid and the solution is transferred to a 2-ounce polyethylene bottle, using 10 ml. of water to rinse the crucible. After the addition of 26.5 ml. of $12M$ hydrochloric acid the solution is diluted to 50 ml. with water and is mixed.

The reductor is prepared by introducing previously amalgamated zinc into the column (500 g., 16 to 30 mesh, treated with 250 ml. of 2% $HgCl_2$ solution, washed with water, with 2% (vol./vol.) sulfuric acid, and again with water) and then by washing it with 150 ml. of $6M$ hydrochloric acid–$0.5M$ hydrofluoric acid. The flow of nitrogen is adjusted to 200 to 250 ml. per minute and the polyethylene collecting bottle is connected, containing 25 ml. of $0.04N$ ammonium ferric sulfate and 30 ml. of orthophosphoric acid.

The niobium solution is introduced into the column in 5- to 10-ml. portions, maintaining a constant depth of solution above the zinc. Subsequently the 2-ounce polyethylene bottle is washed three times with 5-ml. portions of $6M$ hydrochloric–$0.5M$ hydrofluoric acid wash solution and the rinsings are transferred to the column. The zinc is washed with an additional 15 ml. of the HCl–HF wash solution. The solution in the collecting bottle is diluted with oxygen-free water to 240 ml. and is then titrated with $0.1N$ potassium dichromate, using 5 drops of an aqueous 0.2% (w./vol.) solution of barium diphenylaminesulfonate as indicator. A blank value is established by passing 80-ml. portions of the HCl–HF wash solution through the column, adding 2.50, 5.00, and 10.00 ml. of $0.10N$ ammonium ferrous sulfate and titrating as described above. The difference between the titers for the solutions containing 5.00 and 10.00 ml. of the ferrous solution (or the difference between the 5.00- and 10-ml. solution), subtracted from the titer for that containing 5.00 ml. (or 10.00 ml.), gives the blank value.

An essential part of the above procedure is the design of the reductor, which contains a piece of tubing only long enough just to protrude into the collecting vessel. It is claimed that a longer delivery tube will cause oxidation of a small amount of niobium(III) by hydrogen ion.

D. POLAROGRAPHIC DETERMINATION OF NIOBIUM

The polarography of niobium is possible in a number of media and may be attractive in such cases in which interfering elements are either absent or can be eliminated easily.

Although tantalum does not yield a polarographic wave, it may interfere in some media through the formation of complexes containing both niobium and tantalum. These new complexes diffuse more slowly because of the heavy tantalum atom and thereby decrease the diffusion current due to the niobium reduction (288).

1. Nitrate System

Most early work was carried out in a nitrate–nitric acid medium (0.5 to $1.0M$ KNO_3 and $0.2M$ HNO_3) (519,586). The solution is prepared for polarographic measurements by fusing the niobium oxide in potassium carbonate and dissolving the melt in nitric acid. The limiting current is proportional to the niobium concentration, but the wave is actually due to the catalytic reduction of the nitrate ion in the presence of niobium (119). The reduction of the nitroniobate complex, $[NbO(NO_3)_5]^{-2}$, also has been conjectured (136). The wave height is independent of the nitrate concentration (between 0.2 and $2M$), but it increases with the nitric acid concentration. It is also increased by the addition of a very small amount of gelatin. The half-wave potential for the wave is -0.88 v. (in a $0.056M$ nitric acid solution) and -0.80 v. (in a $0.89M$ nitric acid solution) vs. S.C.E.

2. Sulfuric Acid System

The sulfuric acid system has also been investigated (208,209,315,316). No reduction occurs in 40% (vol./vol.) sulfuric acid (176). In 50% sulfuric acid the wave is poorly defined because of the small diffusion current plateau (155). Best results are obtained in a 70% (vol./vol.) solution. In this medium, the half-wave potential for $Nb(V) \rightarrow Nb(III)$ is -1.055 v. vs. S.C.E. and is due to a wave with two steps. The wave height is linearly related to the niobium concentration. Small amounts of titanium increase the height of the niobium wave, but larger amounts reduce it. In the presence of titanium, it is necessary to construct a graph showing the relation between diffusion current and the concentration of both niobium and titanium (208). To determine niobium in ores and similar products (315), the sample is decomposed in a platinum dish with sulfuric and hydrofluoric acids. The solution is evaporated to dryness, the residue is fused with 5 times its weight of $K_2S_2O_7$, and the melt is dissolved in 100 ml. of 10% (vol./vol.) sulfuric acid containing hydrogen peroxide. After filtration, 60 ml. of concentrated sulfuric acid is added and the solution is evaporated to fumes. After cooling, the solution is diluted with 30 ml. of water. Gelatine is added as maximum suppressor and hydrogen is passed through the solution prior to the polarographic measurement. Both niobium and titanium thus can be determined, with the $E_{1/2}$ of titanium being -0.574 v. vs. S.C.E. A similar but simplified method, suitable to mixtures of Nb_2O_5 and Ta_2O_5, also has been recommended (89a). It involves fusion of the oxides in pyrosulfate, followed by the steps described in the preceding procedure. A $6M$ acid medium has recently been suggested for the determination of niobium in zirconium-base alloys (518). The sample is decomposed in a mixture of sulfuric, hydrochloric, and fluoboric acids and is fumed with sulfuric acid. A cathode-ray polarograph is used to obtain the polarogram. The peak is at -0.950 v. vs. S.C.E. Tin does not but chromium does interfere.

A detailed study of the suitability of the $23N$ sulfuric acid system for the determination of niobium by oscillographic polarography was published recently (208a).

3. Hydrochloric Acid System

No reduction of niobium is possible in dilute hydrochloric acid solution (136,519) (probably because of the instability of the niobium solution). However, niobium gives a well-defined wave in very strong hydrochloric acid solution. The minimum hydrochloric acid concentration necessary to obtain the full wave height is about $11.6M$ (155). The reduction of the niobium is possible through the formation of such complexes as $Nb(OH)_2Cl_4^-$ (see Section II-E-6). The wave in concentrated hydrochloric acid is caused by the reduction of niobium(V) to niobium(IV) by a reversible 1-electron process ($E_{1/2} = -0.455$ v. vs. S.C.E.). A second wave appears but merges with the hydrogen discharge wave (118). [However, in a.-c. polarography the second peak appears immediately after the first one. Since its height is nearly the same as that of the first peak, it indicates that Nb(V) is first reduced to Nb(IV) and that the Nb(IV) is then reduced to Nb(III) (41).] To avoid the disproportionation of Nb(IV) into Nb(V) and Nb(III), some analysts add ethylene glycol (549,550).

To determine niobium in ores (155), the sample is fused in potassium carbonate and the melt is dissolved in dilute potassium hydroxide. After filtration (Fe, Ti, Zr, Ca, etc.), the solution is just acidified with dilute sulfuric acid. After the addition of 5 ml. of sulfuric acid and 5 ml. of hydrochloric acid (to reduce oxidized manganese compounds), the solution is evaporated to fumes, cooled, and diluted to 100 ml. with concentrated hydrochloric acid in a volumetric flask containing 1 ml. of 0.1% gelatin solution.

To determine niobium in steel (549), the sample is decomposed with perchloric acid (see Section IV-F-2) and the insoluble residue is fused with sodium carbonate. The melt is leached in water. Ethylene glycol is added and the solution is saturated with hydrogen chloride, and is then diluted with concentrated hydrochloric acid. Hydrogen is passed through the solution and the polarogram is recorded with $E_{1/2} = -0.395$ v. A method for the determination of niobium in columbite and tantalite is similar, but involves a preliminary fusion of the sample with sodium carbonate (550). Several hydrolysis steps (see Section IV-A-3) insure the removal of the following elements, which otherwise would interfere: Fe, Mo, Cu, Ti, Pb, Sn, As, Sb, V, W, and U. To determine niobium in titanium ores and pigments (41), prior removal of titanium by solvent extraction (381) has been suggested, using a hydrofluoric–sulfuric acid medium and isobutyl methyl ketone as extractant (see Section IV-J-3). The niobium is removed from the solvent by shaking with an aqueous solution of hydrogen peroxide. This solution is evaporated to fumes of sulfuric acid. After suitable dilution with concentrated hydrochloric acid and removal of oxygen with nitrogen or hydrogen, the polarogram is recorded, using a.-c. polarography, which produces an extremely well-defined peak at about -0.26 v. against the mercury pool, with a twofold increase in sensitivity over d.-c. polarography. Another useful peak appears at -0.40 v. The extraction process greatly reduces the number of other elements present in the final solution. However, the following elements may be partly extracted: V, As, Sb, Se, Te, W, Mo, Rh, Ge, and Ta. Owing to contamination, traces of Ti and Fe may also be present. It was found that 1 mg. of Ti, Fe, and Mo in 25 ml. of solution do not interfere. As and W do not interfere with the first peak, but do with the second. Sb interferes to some extent with the first, but not with the second peak. V interferes with both peaks (41).

4. Citrate System

A 10N hydrochloric acid–0.1% (w./vol.) potassium citrate solution allows the simultaneous polarographic determination of niobium and tungsten (396). Niobium gives in this medium a wave at -0.28 v. vs. S.C.E., and tungsten a wave at -0.42 v. The method is suited for the determination of both elements in steel. Sample preparation consists of a potassium carbonate fusion of the residue obtained by hydrolysis. The melt is leached in water and the polarogram is recorded in an aliquot of the filtered solution after adding the proper amounts of hydrochloric acid and potassium citrate. The citric acid medium has also been chosen by other investigators. It was noted that, of the various complexes that niobium forms, niobium citrate is quite stable to hydrolysis and is easily prepared (176). This system is useful to determine 0.1 to at least 250 γ of niobium per ml. The half-wave potential for the wave is -0.86 v. *vs.* S.C.E. from $2M$ citric acid solution and is well separated from the uranium wave. There is, however, interference from Sb, Ti, and Mo. Mn, Ni, Cr, Fe, and W do not interfere. The citrate medium was recently reinvestigated (288). Plots of E *vs.* log $(i/id-i)$ show that, in 0.4 to $1.1M$ citric acid, the niobium wave is reversible below pH 4, although below pH 2 measurements become difficult because of hydrogen reduction. Use of citric acid solutions at pH 3 yields reproducible results. The half-wave potential for this wave is -0.951 v. *vs.* S.C.E. in $1M$ citric acid. A number of other citrate complexes exist at different pH values (154,288), but are not attractive for polarographic measurements.

5. EDTA System

Preliminary work (177) indicated that in an EDTA medium at pH 3 niobium gives two waves. The first at -0.6 v. *vs.* S.C.E. represents reduction to Nb(IV), whereas the second represents reduction to Nb(III), but coincides with hydrogen reduction. The half-wave potential of the first wave depends on the pH: the complex is more easily reduced at a low pH. In the range pH 2 to 3, the wave is reversible. Outside this range the wave indicates irreversible reduction. Some analysts (288) feel that the EDTA medium cannot be considered suitable for the determination of niobium since the diffusion current does not vary linearly with the niobium concentration. It also was noted that wave heights for solutions containing more than $10^{-3}M$ niobium were not reproducible because of hydrolysis of the uncomplexed niobium. On the other hand, it was stated (76) that, if the niobium concentration is kept below $0.4 \times 10^{-3}M$, reliable results can be obtained. It must also be recognized that preliminary heating is required to obtain a maximum diffusion current (289b). Such a method has been recommended (76) for the determination of niobium in highly alloyed steels. It consists in fusing the acid-insoluble residue in potassium hydroxide in a gold crucible, leaching the melt in water, transferring an aliquot to a beaker, adding EDTA, and adjusting the pH to 1.9. In an aliquot of this solution, molybdenum (which otherwise would interfere with the polarographic determination of the niobium) is removed by an 8-hydroxyquinoline extraction, using benzene as a solvent. In the aqueous phase the polarogram is then recorded between -0.4 and -0.9 v., measuring the height of the niobium wave ($E_{1/2} = -0.65$ v.) vs. the mercury pool anode. A recently published procedure confirms the suitability of EDTA; its authors suggest a sample solution containing 2.5 to 25 mg. of dissolved niobium per 100 ml., $0.1M$ in EDTA, with the pH adjusted to 3 with dilute sulfuric acid (512a).

6. Other Systems

Polarographic steps have been found in tartaric, oxalic, malic, lactic, and gluconic acids (176). Phosphoric acid (287) and pyrophosphoric acid (317) have also been mentioned recently. Pyrophosphoric acid ($20M$) appears to be particularly attractive, since in this medium niobium, iron, and titanium can be determined by a single polarogram (317a).

E. SPECTROGRAPHIC DETERMINATION OF NIOBIUM AND TANTALUM

The literature contains a great number of spectrographic methods for the determination of small and large amounts of niobium and tantalum. A detailed discussion of these methods is not attempted, since the various investigators used such a variety of equipment and techniques as to make a comparison of results extremely difficult. It should be emphasized, however, that the sensitivity of the niobium and tantalum methods used by various investigators depends to a large extent on the mode of excitation. The degree of interference by other elements, however, varies with the dispersion of the spectrograph. Therefore, in the following, only examples of various techniques used in the determination of niobium and tantalum in typical metallurgical products can be presented.

1. High-Grade Ores

a. No Preliminary Separations

The concentration of niobium and tantalum may range from 5 to 80%. Although synthetic standards can be prepared for calibration purposes, samples previously analyzed by chemical methods are much to be preferred. One technique (555) consists in transferring 5 g. of 400-mesh sample to a porcelain evaporating dish. One ml. of a palladium solution (0.01 g./ml.) is added to serve as an internal standard. Care must be taken that the solution is transferred into the ore and does not touch the container. The liquid is evaporated under an infrared lamp. The ore is then oven-dried at 110°C. for 2 hours and, after cooling, is blended thoroughly in an agate mortar. The concentrate is packed into the 3/16-inch-deep cup of a graphite anode; a 1/4-in.-diameter tipped graphite rod serves as cathode. A high-voltage sparking of 60 seconds' duration will have a sensitivity for a minimum of about 5% of niobium and tantalum. Line pairs free of interference are:

Nb 3094.2 A.	Pd 3421.2 A.	(5–20% Nb_2O_5)
Nb 3194.9 A.	Pd 3421.2 A.	(20–80% Nb_2O_5)
Ta 2635.6 A.	Pd 3421.2 A.	(5–20% Ta_2O_5)
Ta 3311.2 A.	Pd 3421.2 A.	(20–80% Ta_2O_5)

Additional constituents of the sample may be determined simultaneously using the following wavelengths: Ti (3329.9 A.), Sn (2863.3 A.), Mn (2688.2 A.), and Fe (2621.7 A.).

A somewhat similar technique has been recommended for the determination of niobium and tantalum in tantalite–niobite (70). However, cobalt is used as the internal standard and excitation is by an intermittent arc. The lines chosen are Nb = 2590.94 A., Ta = 2603.57 A., and cobalt = 2582.24 A. Iron influences the intensity ratio of cobalt to niobium in concentrations up to 10% of Fe_2O_3, above which the effect is constant. Manganese does not interfere. Another method (203), which does not claim the highest degree of accuracy but has great speed, consists in mixing 1 part of sample with 10 to 100 parts of ZrO_2. BaF_2–graphite is used as a buffer. The 2891.8 and 2963.3 A. lines are used for the determination of tantalum in tantalite by another arc technique (1).

b. After Preliminary Separations

Standards for the above techniques consist of pulverized concentrates of identical or similar composition. Since such standards frequently are not available or the matrix of the sample to be analyzed is unknown, a more generally applicable technique consists of carrying out preliminary chemical separations, as described in Section IV. In one procedure (555), 5 g. of sample is decomposed in a polyethylene beaker with hydrofluoric acid. The hydrofluoric acid solution of the sample is contacted with methyl isobutyl ketone, which extracts the niobium and tantalum (see Section IV-J). In the organic phase, niobium and tantalum are precipitated with ammonia. The filtered precipitate is ignited and weighed. To 1 g. of the combined oxides, 1 ml. of standard palladium solution is added (0.01 g./ml.). After drying, the oxide samples are sparked for 30 seconds in 3/16-inch graphite electrodes; 150 v. and 360 μh. are suitable parameters. Standards are readily prepared by coprecipitation of known amounts of niobium and tantalum with ammonia.

In another procedure (452) (which is intended for the analysis of columbite but which can be modified to be suitable for tantalite as well), the combined oxides of niobium, tantalum, zirconium, tungsten, and most of the titanium, are obtained by a technique that involves p-hydroxy-phenyl-arsonic as precipitant (see Section IV-G-4). The oxides are treated with acetone and dried under a heat lamp. Ten mg. of dry sample is briquetted with 100 mg. of spectro-pure briquetting graphite into a 1/4-inch-diameter pellet with a briquetting press. A separate pellet is briquetted from 40 mg. of sample and 100 mg. of graphite for the determination of tungsten.

Synthetic standard samples are prepared, using the oxides of the five elements. A set of five synthetic standards, having the following composition, in percentages, is used:

Oxide	I	II	III	IV	V
Ta_2O_5	15	4	9	6	11.5
TiO_2	0.1	5	0.3	2	0.9
ZrO_2	0.05	0.9	2	0.4	0.15
WO_3	0.1	1.5	0.6	3	0.3
Nb_2O_5	84.75	88.6	88.1	88.6	87.15

This set of standards permits the determination of elements over the ranges:

	%
Ta_2O_5	4–15
TiO_2	0.1– 5
ZrO_2	0.05– 2
WO_3	0.1– 3
Nb_2O_5	75–98 (by difference)

The standard samples are briquetted into pellets exactly as described above for the analytical sample.

Pellets containing the samples and standards are transferred onto $3/8$-inch graphite electrodes shaped in such a way that the pellets fit snugly into the crater. The $3/8$-inch graphite counter electrode, with a hemispherical tip, is placed in the upper electrode holder. A 3-mm. analytical gap is maintained and the samples and standards are excited with a high-voltage air-interrupted condensed spark. 20000 v. and 300 μh. are suitable parameters. For the determination of tungsten, the exposure is 60 seconds; for all other elements, a 25-second exposure is sufficient. For tungsten, the first-order 3650 to 5050 A. region is used; for all other elements, the second-order

2750 to 3450 A. region is used. The following analytical line pairs are recommended:

Element	Analytical line, A.	Internal standard Nb line, A.	Conc. range, % oxide
Tantalum	2963.322	2961.628	4–15
	3311.162	3104.260	4–15
Titanium	3078.645	2961.628	0.1–1.0
	3105.084	3104.260	0.8–5.0
Zirconium	3306.278	3104.260	0.05–2.0
	3391.975	3104.260	0.05–0.3
Tungsten	4294.614	4303.881	0.12–3.0

A third procedure involves an arc technique after a preliminary fusion of the sample in $KHSO_4$ and precipitation of the earth acids with SO_2 water (244). The line pairs Nb 2716.69–Ta 2714.68 and Nb 2927.82–Ta 2933.67 are used. The method allows the determination of 1 to 99% of niobium or tantalum.

For the determination of tantalum (0.2 to 1%) in pyrochlore concentrate (50 to 60% Nb_2O_5), the following technique has been recommended (256). 0.300 g. of −320 mesh sample is mixed with an equal amount of pure lithium metaborate. To this mixture is added 1.00 g. of pure briquetting graphite. After mixing, the graphite-containing mixture is pressed in a briquetting press to give a rod of 6 mm. diameter. For excitation, a low-voltage d.-c. spark is used. The constants of the discharge circuit are capacitance, 58 $\mu f.$, and inductance, 250 $\mu h.$ The gap between the lower sample electrode (cathode) and the graphite 4-mm. counter electrode is 4 mm. The transmittances of the tantalum line at 2400.63 A., the background at 2400.45 A., and the niobium line at 2405.9 A. are measured. The analytical curve is established by using synthetic standard samples. The concentration of these standards is determined by the successive-addition-and-extrapolation technique. Since the niobium content (50 to 60%) is determined by an independent chemical or instrumental method, the analytical curve is constructed with the log intensity ratios on the ordinate and the log of the ratios of Ta_2O_5/Nb_2O_5 concentration on the abscissa.

2. Low-Grade Ores and Residues

a. No Preliminary Separations

If a purely spectrographic approach is desired, it must be remembered that both niobium and tantalum form carbides and, therefore, are very refractory. Of the elements with which they may be associated, some are very volatile (Na and K), others are less volatile (Si, Sn, Al, Ca, Fe, Mn, and Mg), and there are some elements (W, Ti, and Zr) that are approximately as refractory as niobium and tantalum. Anticipating this, the spectrographer often makes a preliminary moving-plate study of his samples. This procedure involves taking a series of exposures at intervals of about 10 to 30 seconds while the arc is running. From the spectrogram he prepares a curve of intensity *vs.* time. He then can decide what additions are required to control the volatilization of the niobium and tantalum. Quartz and carbon powders have been recommended for this purpose (1). Strontium carbonate (318), sodium carbonate (447), and lithium carbonate (266), together with carbon powder, have been used by others.

In one method (409), molybdenum (internal standard) is first introduced into graphite, which then is mixed in a 1:1 ratio with the mineral sample prior to excitation by an a.-c. arc. The analytical lines used are Nb 2950.9 A., Ta 2685.1 A., and Mo 2930.5 A. In still another niobium procedure (536), one part of sample is ground with 10 parts of a flux consisting of equal weights of titanium dioxide and ammonium sulfate. Small pellets, prepared from this mix., are arced on copper supporting electrodes and the region 3700 to 5000 A. is recorded. The line pair Ti 4145.0–Nb 4100.9 A. is used.

For the determination of small amounts of niobium and tantalum as found in wolframite, only graphite is added (203). A d.-c. arcing of 15 amp. and 15 seconds' duration permits the determination of about 0.10% each of niobium (4058.9 A.) and tantalum (3311.2 A.) (555). A similar procedure has been recommended for the determination of small amounts of niobium and tantalum in TiO_2 (526). The line pairs Nb 2950.88 | Ti 2958.28 A. and Ta 2714.67 | Ti 2713.76 A. were found usable.

b. After Preliminary Separations

The preliminary separations depend to a large extent on the matrix of the sample. Iron and calcium, for instance, can be removed by treating the finely pulverized ore with hydrochloric acid. The residue remaining after filtration can then be sparked by the technique described above, involving palladium as an internal standard (555). Since traces of niobium and tantalum are widely disseminated in granite and other silicate rocks, a treatment with hydrofluoric acid will aid in the concentration of the earth acids. One such procedure (525) employs, after a treatment of the sample with HF and H_2SO_4, pyrogallol as a precipitant and silica and gelatin as collectors. A similar procedure, for the determination of niobium, uses tin as collector and tannin as precipitant (405). The ignited oxides, mixed with powdered quartz containing ThO_2 as internal standard, are subjected to arc excitation, and the line pairs Nb 3163.40–Th 3154.73 A. or Nb 2927.81–Th 2942.86 A.

are measured. In a third procedure, after an $HF\text{-}HNO_3\text{-}H_2SO_4$ treatment, SO_2 water is employed to recover soluble earth acids (266). Portions of the ignited precipitate are mixed with twice their weight of lithium carbonate and are then subjected to arc excitation. The mixture is pre-arced for 30 seconds and the shutter is then opened for 10 seconds. The tantalum 3311.0 A. line and the 3310.5 A. niobium line are measured to determine the Ta:Nb ratio. Niobium is determined by a photometric procedure.

3. Tantalum in Niobium and Niobium in Tantalum Metals or Oxides

These methods are discussed in Sections VIII-B-1 and IX-D.

4. Alloys

a. Semiquantitative Analysis

Between 4200 and 6100 A. a number of niobium lines are found that are suitable for the visual estimation of the niobium concentration by employing a "steelscope" (64). If 1.5% niobium is present in a sample of stainless steel, the intensity of the niobium 5095.3 A. line is about equal to the iron 5098.7 A., and the intensity of the niobium 4573.1 A. line is equal to the chromium 4569.6 A. line.

b. No Preliminary Separations

Several point-to-plane methods have been recommended involving both spark excitation (133,210,237,322,480) and a.-c. arc discharge (436). In steels the spark technique allows the determination of 0.1% of niobium and 1% of tantalum, using the line pairs Nb 4058.93 | Fe 4118.6 A. or Nb 4100.92 | Fe 4118.6 A. and Ta 2387.1 | Fe 2387.46 A. (480). It is stated that the arc-discharge technique lowers the limit of detection of tantalum to 0.01% using the line pair Ta 3311.16 | Fe 3165.86 A., whereas that of the niobium is a rather poor 0.3% (Nb 3299.6 | Fe 3165.01 A.).

When solid pieces of metal and suitable standards of similar composition are not available, or when the homogeneity of the sample is in question, the spectrographer frequently prefers to dissolve the sample and synthetic standards in acids. Aliquots of the solution are transferred to graphite electrodes and are, after drying, subjected to various modes of excitation. If iron is used as an internal standard or if a definite amount of another internal standard such as palladium (555) is added to the solution, several "fills" may be transferred to the graphite cup with intermittent drying by an infrared lamp. If a spark technique is used, greater sensitivity may be obtained by transferring the sample solution to both electrodes

(204). With palladium as internal standard, the following lines are suitable for densitometer readings (555): Ta 3311.2 or 3317.9 A.; Nb 4058.9 or 3094.2 A.; and Pd 3027.9 or 3421.2 A. With iron as the internal standard (204) the following line pairs are useful: Nb 3130.8 | Fe 3116.6 A. and Nb 2883.3 | Fe 2880.8; Ta 2653.3 | Fe 2669.5 A. The method is sensitive enough to allow the determination of 0.1% of niobium and 0.3% of tantalum.

For the determination of niobium in bismuth alloys, the porous-cup method has been applied, using platinum as internal standard (132). The same procedure is also valid for the determination of niobium and tantalum in lithium metal (422).

Owing to the notably poor spectral sensitivity of tantalum, the above solution techniques cannot be applied to steel samples containing less than 0.25% of tantalum. In such cases, a much higher sensitivity can be obtained by using a "salt" technique which, in addition, permits the incorporation of a carrier or buffer (214). Such a procedure by necessity involves arc excitation (in this case a.-c. arc). Whereas the limits of detecting tantalum can thus be lowered to 0.05%, the precision is somewhat less ($\pm 10\%$) than in methods involving spark excitation. Sample preparation depends somewhat on the composition of the sample. For steels, an acceptable technique (214) consists in dissolving the sample and standards (synthetic if required) in aqua regia and fuming the salts to complete dryness with sulfuric acid. The dry salts are mixed with an equal volume of graphite powder. Triplicate electrodes are loaded for each sample, with approximately 50 mg. of the above-mentioned mixture being required to pack each electrode. Excitation is by a 2410-v., 4- to 8-amp. alternating current, using a 50-second exposure time. The spectral lines used are Nb 2950.88 / Fe 2964.63 A. and Ta 2714.67 / Fe 2886.32 A.

c. After Preliminary Separations

Because it is frequently difficult, if not impossible, to obtain standard samples for the preparation of spectrographic calibration curves, many analysts prefer first to isolate the earth acids and then to subject the mixture to spectrographic analysis. Prior chemical separations are time-consuming, but they allow the analysts to increase manyfold the lower limits of sensitivity. Many such procedures have been suggested and they are usually based on one or the other separation described in Section IV. Probably the most common approach is based on hydrolysis (214,578). If, subsequently, an a.-c. arc technique is employed, the lines Nb 2715.5 | Ta 2714.67 A. (214) or Nb 3310.5 | Ta 3311.0 A. (578) can be used to determine the $Nb_2O_5:Ta_2O_5$ ratio.

F. THE DETERMINATION OF NIOBIUM AND TANTALUM BY X-RAY FLUORESCENCE

1. Instrumental Considerations

Fluorescence x-ray spectroscopy, sometimes referred to as x-ray emission spectroscopy, represents a valuable tool for the determination of widely varying concentrations of niobium and tantalum in a great number of substances. Since instrumentation has been improved to the extent that mutual radiation interference of niobium and tantalum can be controlled or eliminated, the chemical separation of the two elements can in most instances be avoided.

In early work, the fact that niobium $K\alpha_1$ and $K\alpha_2$ lines have wavelengths slightly less than half that of the $L\alpha_1$ line of tantalum caused complication, in as much as the second-order K lines of niobium were difficult to resolve from the $L\alpha_1$ line of tantalum, which is the most sensitive line for analytical purposes. Early investigators who encountered this problem solved it by making measurements at several angles and by comparing the integrated intensity of the unresolved Ta-Nb lines with that of a single niobium line (59). Another approach consisted of operating the x-ray source at 18 kv., thus eliminating the niobium K lines (excitation potential 18.986 kv.), but at the price of low analytical-line intensity for tantalum. In spite of this resolving power-intensity problem, satisfactory methods were worked out to determine the Nb:Ta ratio in the mixed oxides (101). In 1952 spectrographs became available which had sufficient resolving power to allow the direct measurement of the $L\alpha_1$ line of tantalum (77). When determining a trace of tantalum in a niobium matrix, better results were obtained by the selection of pulse height (338).

The sensitivity obtainable by x-ray fluorescence techniques frequently matches or even exceeds that of optical spectroscopy. The lower limit of detection is about 0.001% of niobium and 0.03% of tantalum.

Further improvements in sensitivity and selectivity could be achieved by the introduction of high-voltage sources. At present most high-voltage sources for ordinary spectrographs are designed for a 60 kv. maximum, which is too low to excite the K lines of tantalum. Sources capable of exciting the Ta K lines have been built, but only at considerable increase in cost.

2. Effect of Other Elements

X-ray spectroscopy, like emission spectroscopy, when used for the quantitative determination of niobium and tantalum is dependent on the availability of standards of nearly identical chemical composition and identical physical characteristics. When considering the matrix effect, it must be recognized that most elements in one way or another decrease or increase the intensity of the niobium and tantalum radiation. For instance, in the four-element system, niobium, tantalum, iron, and titanium, the following observations have been made (387). Iron and titanium increase the intensity of the niobium because of the substitution of a light element for tantalum, which is a strong absorber of niobium radiation. The effect of titanium is greater than that of iron. In the presence of iron and titanium the intensity of the tantalum radiation is decreased because of the absorption of tantalum radiation by iron and titanium. The effect due to iron is greater than that due to titanium. Tests carried out in Ledoux & Company's laboratory indicate that other elements with which niobium and tantalum may be associated similarly affect the niobium and tantalum radiation. For instance, manganese enhances the niobium $K\alpha$ and decreases the tantalum $L\alpha$ radiation. The total effect is about the same as that of iron. Light elements such as silicon, aluminum, and calcium have severe effects on both the niobium and tantalum. The magnitude of these effects is complex and usually depends on the relative concentration of all constituents. The interelement effects are not critically dependent on the particular line studied, i.e., Nb $K\beta$ compared to Nb $K\alpha$, unless there are line interferences or the absorption edge of an element is located between the lines. Thus, the presence of zirconium, for example, results in the enhancement of Nb $K\alpha$ and the severe absorption of Nb $K\beta$ (387).

3. Effect of Physical Characteristics of the Sample

Since x-ray spectroscopy is essentially a surface analysis, the physical characteristics of chemically identical compounds must be closely controlled. Some investigators report that mechanically mixed standards (prepared from -325-mesh oxides, mixed for 5 to 10 minutes, passed through a 200-mesh screen, and ground under acetone for 30 minutes) are adequate (100,150), but other analysts have expressed doubt as to the reliability of results using such standards (2,392). In Ledoux & Company's laboratory, significant differences were observed, depending on the history of the oxides. Oxides obtained by precipitation with ammonia could not be used interchangeably with those obtained by precipitation with cupferron or tannin.

4. Elimination of Interferences

A number of methods have been suggested to obviate errors due to differences in the chemical composition or physical characteristics of the sample.

a. Use of Correction Factors

In one technique, the number of variables in the chemical composition is reduced to four (niobium, tantalum, iron, and titanium) by carrying out a preliminary cupferron separation, which also serves to provide oxides of identical physical characteristics (387). An elaborate system was developed that uses arithmetic correction factors based on empirical calibration curves to compensate for interelement effects. After chemical preparation of the oxides, materials such as ores, metals, and liquids fall within the scope of the method. The correction factors derived are independent of variations in instrumentation such as detectors, crystals, and x-ray tubes, and are applicable to any combination of the four elements listed above (see Table XX).

TABLE XX

Factor Addition for the Determination of Niobium and Tantalum in a Nb_2O_5–Ta_2O_5 Mixture Containing TiO_2 and Fe_2O_3 (387)

Element	Contaminant, %		Individual correction factors		Sum[a]	Actual factor
	Fe_2O_3	TiO_2	Fe_2O_3	TiO_2		
Nb_2O_5	10	10	1.17	1.27	1.44	1.44
	25	10	1.42	1.27	1.69	1.71
	50	10	1.85	1.27	2.12	2.10
	10	25	1.17	1.67	1.84	1.82
	25	25	1.42	1.67	2.09	2.04
Ta_2O_5	10	10	1.08	1.02	1.10	1.09
	25	10	1.21	1.02	1.23	1.23
	50	10	1.47	1.02	1.49	1.51
	10	25	1.09	1.07	1.16	1.16
	25	25	1.24	1.07	1.31	1.35

[a] Equals 1 plus the sum of the decimal fraction of the individual correction factors.

The above procedure is time-consuming, particularly in regard to the necessity of establishing a "family of curves" involving from zero to 100% of all four elements. Tests carried out in the author's laboratory indicate that comparable results can be achieved by adding to the samples and standards a large excess of the interfering element, thus minimizing the effect of smaller differences in the matrix (279).

b. Preliminary Fusion of Sample

When the sample represents a simple or reproducible chemical system, such as pyrochlore or titanium–niobium alloy, fusion of the sample and standards with borax (181,135a) or lithium metaborate (256) serves well to establish a common physical basis. In the case of pyrochlore (256), the method consists in fusing 0.5 g. of 200-mesh sample with 1.500 g. of lithium metaborate mixture (5 parts of $LiBO_2$ and 1 part of MoO_3; the latter serves as internal standard). The fusion is carried out at 1000°C. in a graphite crucible. The cool melt is crushed and passed through a 320-mesh screen prior to measuring the intensities of the Nb K and Mo K lines. The intensity ratio of these lines is a measure for the Nb_2O_5 concentration. It is stated that the standard deviation for a single determination is 0.21% Nb_2O_5 at a concentration level of 50% Nb_2O_5. A similar method has been used in the laboratory of the author for the determination of tantalum in microlite. However, potassium bisulfate is used as a flux and hafnium is used as the internal standard (279). In a method for the determination of niobium and titanium (181), the oxides are fused with borax in a platinum crucible at 900°C. The bubble-free homogeneous melt is poured onto a smooth-surfaced aluminum plate maintained at 400°C. After cooling to room temperature, the $K\alpha$ radiations of niobium ($\lambda = 0.748$) and titanium ($\lambda = 2.749$) are measured.

c. Other Miscellaneous Applications

For the determination of tantalum in commercial niobium oxide, standards of zirconium oxide containing varying amounts of tantalum were found useful (100). Zirconium and niobium have very similar x-ray absorption characteristics, and an investigation showed that working curves of tantalum oxide in niobium oxide are practically identical to those of analogous zirconium standards. The background is measured at the Ta $L\alpha$, peak position with a 100% zirconium oxide standard. (If niobium oxide free of tantalum oxide is available, standards can be prepared using only the Nb_2O_5–Ta_2O_5 system.) The results obtained by this procedure (molybdenum tube target, sodium chloride analyzing crystal, standard argon-filled Geiger tube) are in close agreement with those obtained by the use of the addition technique. It was noted that, of the common impurities, titanium, iron, and tin reduce the Ta $L\alpha$ line intensity, whereas silica and the alkali phosphates or sulfates increase this intensity. Since titanium is the only impurity other than tantalum usually found in any significant concentration in most commercial niobium oxide, a simple mathematical correction similar to that described above is possible.

For the determination of niobium in ores (preferably after a chemical decomposition to provide a common matrix), either molybdenum or zirconium is sometimes used as internal standards, depending on the composition of the ore (99). The spectral lines chosen for comparison are Nb $K\alpha$ with either Mo $K\alpha$ or Zr $K\alpha$. These lines have suitable intensities and can be completely resolved from Nb $K\alpha$. Suitable comparison lines used with Ta $L\alpha$ are Hf $L\alpha$, W $L\alpha$, Zn $K\alpha$, or Cu $K\alpha$ (99,392).

The niobium and tantalum content of tungsten ore (wolframite) can be determined by simply comparing the intensity of the $K\alpha$ radiation of niobium and the $L\alpha$ radiation of tantalum, using a set of chemically analyzed standards. The $L\alpha$ line of tantalum can be completely resolved from W $L\alpha$, but it is advisable to run a background correction. This procedure has been used successfully to determine as little as 0.03% of Ta_2O_5 and 0.005% of Nb_2O_5 in wolframite concentrate (279).

G. ELECTRON PROBE MEASUREMENTS OF NIOBIUM COMPOUNDS

Since niobium is of current interest as a container for liquid metal heat exchangers, because of its low nuclear cross section and relatively small mass transfer, and since the metal requires at elevated temperatures protective cladding, problems of intermetallic diffusion of niobium with the components of stainless steel (Cr, Fe, Ni, Mo) arise that require answers in predicting alloy reactions and designing better alloys for bonding to niobium.

The analytical tool that has made these rapid and extensive diffusion studies feasible is the electron probe microanalyzer. In this technique, a fine beam of electrons generates characteristic x-rays from the elements contained in a 1-μ diameter area on the specimen surface (60). By scanning the polished cross section of a diffusion zone and relating x-ray intensity to composition, the complete and continuous composition curve is obtained in a single operation. This replaces the lengthy wet-chemical or radioactive-tracer analysis of individual layers, physically removed from a diffusion zone. In actual operation, electrons of 25 K.e.v. energy are focused to a beam of 1 to 2 μ diameter with a two-lens electron optics system. Characteristic x-rays from the specimen are passed through the beryllium window of the vacuum electron probe instrument to the x-ray spectrometer in air. Curved lithium fluoride analyzing crystals of either 20 or 40 cm. radius of curvature are used with argon or krypton Geiger counters and standard amplifier, scaler, and recorder circuits. The method also has been used to determine the composition and extent of phases occurring in bimetal couples at 1000°C. (497). The five binary systems studied are Nb–Pt, Nb–Se, Nb–Zn, Nb–Co, and Ni–Ta.

H. RADIOCHEMICAL METHODS

1. Use of Radioactive Tracers

The recent development of analytical procedures for the determination of niobium and/or tantalum have been greatly aided by monitoring various fractions with radioactive niobium or tantalum tracers. Niobium-95 ($t_{1/2}$ = 35 days) and tantalum-182 ($t_{1/2}$ = 115 days) have been widely used to check various parameters of the photometric niobium thiocyanate procedure (384,547), of the paper chromatographic separation of niobium from tantalum (85), of the tannin procedure (57), and of the separation of niobium from zirconium-95 by cation exchange (462) or by liquid–liquid extraction procedures involving thenoyltrifluoroacetone (389), tributylphosphine oxide (548), and butyl phosphoric esters (466). Radioactive tracers have also been employed to monitor the extraction of niobium with diisopropyl ketone from HCl media (248), to check on the efficiency of various hydrolytic precipitation methods for niobium and tantalum (17, 71,96) and on the separation of niobium from tungsten involving magnesium salts (95).

2. Isotope Dilution Procedures

^{95}Nb and ^{182}Ta also have been used in conjunction with an isotopic dilution technique suitable for the determination of traces of niobium and tantalum in granite (464). The method is based on a preliminary concentration of niobium and tantalum following an HF–H_2SO_4 attack by three applications of the fluoride–pyrogallol method (see Section IV-G-5). The oxides are fused with $K_2S_2O_7$ (for Ta) or Na_2CO_3 (for Nb). The melts are extracted with dilute oxalic acid. In one aliquot, tantalum is determined photometrically with rhodamine-B (502) and, in another, radiometrically. Alternatively, the niobium is determined colorimetrically with KCNS (see Section VII-B-1) and radiometrically. Based on a knowledge of the amounts of radioactive isotopes added, the niobium and tantalum content of granite thus can be calculated. In another isotope dilution procedure (9), the preliminary separation of niobium and tantalum from each other and from titanium and zirconium is made by precipitation with benzeneselenous acid and by extraction with 8-hydroxyquinoline before the final determination by means of the radioactive isotopes Nb-95 and Ta-182.

3. Neutron-Activation Analysis of Niobium and Tantalum

This technique is based on the induction of radioactivity into the stable isotopes of niobium and tantalum. This induced activity decays with its own characteristic radiation and half life, and the radioactivity is then a measure of the amount of niobium or tantalum present. Briefly, the chief advantages of the technique lie in its freedom from blank difficulties and from contamination errors after irradiation. In addition, since usually an inactive isotopic carrier is added before the sample is processed chemically, chemical separations at the microgram level are avoided.

a. DETERMINATION OF NIOBIUM

Until recently it was believed that the method of neutron activation of niobium was too insensitive to natural niobium (34) to be of practical use, because the 6.6-minute half life of the activity induced on neutron bombardment (^{94m}Nb) was too short. The introduction of mechanical devices, however, allow the rapid and controlled transfer of samples to the neutron source and back to the counting equipment, and the 6.6 minutes of half life is sufficient to allow the determination of even small amounts of niobium. The decay scheme of ^{94m}Nb is presented in Fig. 16. On comparing the observed photopeak with the barium K x-ray (32 K.e.v.), it was found that the niobium peak does not have an energy of 42 K.e.v., as might be expected from its decay scheme, but has instead an energy of 16.8 ± 0.7 K.e.v. Since the conversion coefficient of ^{94m}Nb is large (520), it can be postulated that the 42-K.e.v. gamma from isomeric

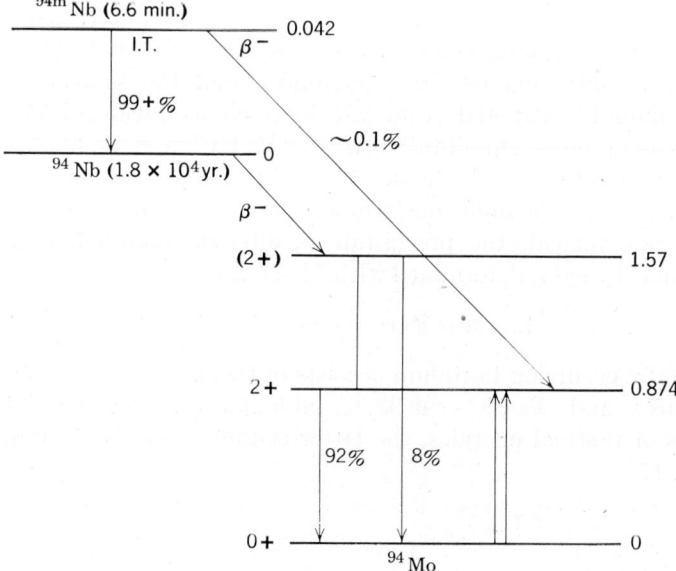

Fig. 16. Decay scheme of ^{94m}Nb (81).

transition is almost completely converted to an electron and an x-ray
The niobium x-ray can be observed by means of a 3- × 3-inch NaI
(Tl) crystal. A krypton-filled x-ray proportional counter whose output
is fed into a 100-channel analyzer proved less satisfactory (30). Because
of the short half life of the niobium, chemical separations should be avoided
wherever feasible. One method that has been suggested is based on the
collection of the niobium by absorption onto silica gel by boiling the
solution containing the niobium with an excess of nitric acid (453). It is
stated that, by means of ^{95}Nb tracer, a 30.35% chemical recovery has been
demonstrated. This technique has been applied to the determination of
niobium in rutile. The finely ground sample is activated in the pneumatic
tube for 15 minutes. The irradiated sample, enclosed in a gelatin capsule,
is then fused with 3 to 4 g. of sodium peroxide and the melt is cooled and
dissolved in dilute nitric acid. The solution is filtered to remove any
undissolved material. Concentrated nitric acid is added to the filtrate,
followed by 0.5 g. of silica gel, and is boiled for one minute. The solution
is filtered through a medium stainless steel filter funnel and the silica gel
is dried with acetone and analyzed with the 100-channel analyzer. Preliminary results indicate that niobium in amounts as low as 10^{-2} γ can be
determined following this 10-minute separation.

Another separation that requires about 12 minutes with a yield of
approximately 30% is suitable for the determination of small amounts of
niobium in rocks and minerals (82). The finely pulverized irradiated
sample is fused in potassium bisulfate, the melt is leached in saturated
oxalic acid containing carrier niobium(V), and the solution is filtered.
The niobium is extracted from $9M$ hydrochloric acid solution with 4-methyl-2-pentanone equilibrated with $9M$ HCl and is finally back-extracted with dilute oxalic acid. After the addition of calcium nitrate
(to eliminate the complexing oxalic acid) and rendering the solution
slightly ammoniacal, the precipitate is filtered, mounted, and counted.
The chemical yield is determined with ^{95}Nb tracer.

b. Determination of Tantalum

Naturally occurring tantalum consists of two isotopes, Ta-180 (0.0123%
abundance) and Ta-181 (99.987% abundance). On irradiation with
neutrons of thermal energies, the latter isotope gives rise to radionuclides
(see Fig. 17).

$$^{181}\text{Ta} \xrightarrow{n,\gamma} {}^{182m}\text{Ta}(t_{1/2} = 18.5 \text{ min.})$$
$$\downarrow$$
$$\longrightarrow {}^{182}\text{Ta}(t_{1/2} = 115.1 \text{ days}) \xrightarrow{\beta^-} {}^{182}\text{W(stable)}$$

Isotopic activation cross sections, σ, for the nuclear reactions are:

$$^{181}\text{Ta}(n,\gamma)^{182m}\text{Ta} \qquad \sigma = 0.03 \text{ barns}$$

$$^{181}\text{Ta}(n,\gamma)^{182}\text{Ta} \qquad \sigma = 21 \text{ barns}$$

Using a flux of 10^{12} neutrons cm.$^{-2}$ second^{-1} (Harwell) and measurement of the β^- activity of ^{182}Ta, it has been estimated that as little as 1×10^{-10} g. of tantalum can be determined under ideal conditions (275).

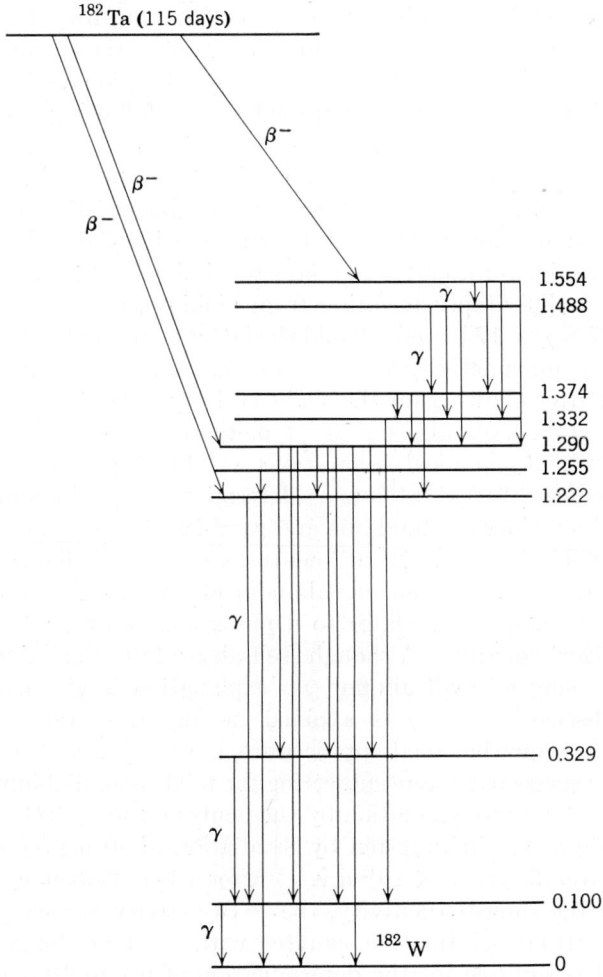

Fig. 17. Decay scheme of ^{182}Ta, with energies given in M.e.v. (390).

Although this high sensitivity can be achieved only through relatively long irradiations (about one week), a more rapid analysis is feasible through utilization of the 18.5-minute ^{182m}Ta, with the resulting advantage of considerably shorter irradiation times but with a sensitivity only of the order of 4×10^{-8} g. (556). With the shorter irradiation time it may be possible to do the analysis nondestructively, depending naturally upon the nature of other activities induced in the matrix, especially as the concentration of tantalum increases.

Because niobium and most other elements with which tantalum is associated in metals and minerals (such as titanium, iron, tin, zirconium, and silicon) have small neutron capture cross sections and short lives, the 115-day ^{182}Ta is ideally suited for neutron-activation analysis, either by nondestructive gamma-ray spectrometry or following radiochemical separations.

In an early method for the determination of tantalum in ferroniobium and niobium ores (297), the sample and standards consisting of iron powder mixed with known amounts of tantalum were irradiated in an "atomic pile" and set aside for about seven days to eliminate or reduce the activity of short-life radioisotopes resulting from niobium, and particularly from zirconium-97 ($t_{1/2} = 17$ hours), should the latter be present in large amounts (342). The gamma activity was then measured and the tantalum content determined, after allowing for the slight activity of the iron powder. Several investigators have also described methods for the determination of tantalum without chemical separations (57,148,231,342). In one such recent procedure (231), tantalum is determined in a niobium matrix by submitting the oxides of standards and samples (15 to 45 mg.) to irradiation at a flux density of 1.2×10^{12} neutrons /sec./cm.^{2}, allowing to "cool" for one month, and evaporating aliquots of the solution (KHSO$_4$ melt dissolved in ammonium oxalate) to dryness and counting the β activity with a Geiger counter. Although self-absorption during counting is avoided by using a small aliquot, no explanation is given of how self-shielding (discussed below) is avoided during irradiation of the solid sample. In a somewhat related procedure, a preliminary chemical separation has been suggested before subjecting the mixture of niobium, tantalum, and titanium oxides to irradiation by the neutron source (384). Tantalum has been determined in tungsten by irradiation of 10 mg. of sample at a thermal neutron flux of 1×10^{15} n.v.t. After a 1- to 2-week cooling period for decay of the tungsten activity, the ^{182}Ta activity is measured on a 2- $\times$ 2-inch NaI(Tl) scintillation counter with a single-channel analyzer. The method is suitable for the determination of 0.1 to 10% of tantalum (102a).

Self-Shielding Effects

Since tantalum has an appreciable neutron capture cross section in the thermal region of the neutron spectrum ($\Sigma = 22$ barns) and, in addition, shows a resonance peak of more than 10,000 barns in the intermediate or epithermal region of the spectrum (260), the possibility of self-shielding during irradiation, leading to a reduction of the effective neutron flux and causing unequal activation of samples and standards, represents a major problem (34,201). To investigate this phenomenon, varying weights of the pure oxide (34) or of oxide and metal sample (201) were irradiated simultaneously. The results show a significant decrease in specific activity in the oxide standards with an increase in sample weight (indicative of self-shielding), and demonstrate the unsuitability of pure tantalum compounds as standards in the determination.

Attempts have been made to eliminate these effects by mixing the samples and standards with powdered graphite or sucrose (201) or with other substances of low cross section, such as silica or alumina (34). Homogeneous solid solutions, however, are difficult to prepare and at best can only partially reduce the self-shielding effects, since self-shielding still occurs within the individual particles.

When liquid samples are used the effect of the absorption of resonance neutrons, which contributes so significantly to self-shielding, can be overcome only by either preparing standards of the same tantalum content as the sample or by dissolving the solid standards and the samples and diluting to the same concentration (201). If the samples and standards are both diluted to a concentration less than that giving 1% self-shielding error, the two concentrations need not be matched exactly, but if a preliminary analysis is made to enable the concentrations to be matched fairly closely, much higher concentrations and therefore shorter irradiation times can be used.

In one such procedure intended for the analysis of tantalum in high-tantalum ferroalloys (201), matched solutions (HF and HNO_3) of standards and samples are prepared in the range of 5 to 10 mg. of Ta/ml. If the solutions cannot be matched, they are diluted to less than 2 mg. of Ta/ml. for less than 1% self-shielding. Weighed 0.2-ml. portions of these solutions in silica ampules are irradiated in a neutron flux of 10^{12} neutrons/cm.^{2}/second for 30 minutes or proportionally longer for the more dilute solutions. After allowing the short-lived activities to decay for 24 hours, the tantalum activity is determined on a gamma-ray spectrometer, using a scaler connected into a fairly wide channel (about 5 v.). If a spectrometer is not available a normal gamma-ray counter, biased to just count the high-energy tantalum activity, can be used. The activities of standards and samples are compared and the results calculated in the usual way. The accuracy of the results obtained is impressive. For instance, on a sample containing 13.18% tantalum the above procedure gave results of 13.1 ±0.3%.

In a second procedure (34) (intended for the determination of tantalum and tungsten), silicate samples ground to -100 mesh (40 to 80 mg.), or fine drillings of metallic samples (50 to 100 mg.) are irradiated in silica ampules. To obviate difficulties with self-shielding, the standards are in the form of solution and contain less than 6 mg. of tantalum and less than 0.5 mg. of tungsten per gram of solution. After the decay of short-lived activities, silicate samples are fused in sodium peroxide and metallic samples are dissolved in acid. The melt or solution of metallic samples or standards are transferred to beakers containing 35 mg. of tantalum as the oxalate complex, 35 mg. of tungsten as sodium tungstate, and tartaric acid.

Tantalum and tungsten are subsequently collected with tannin plus cinchonine (486). The filtered and washed precipitate is fumed with nitric and sulfuric acids. After dilution with dilute hydrochloric acid, the insoluble oxides are collected by centrifugation and dissolved in $6M$ sulfuric–$0.4M$ hydrofluoric acid mixture. Subsequently, tantalum and tungsten are separated by extracting the former into diisopropyl ketone or hexone (see Section IV-J-3). Tantalum in the organic phase is then back-extracted with 5% hydrogen peroxide, precipitated with ammonia, and subjected to a second DIPK or hexone extraction. After transferring the tantalum again into the aqueous phase with hydrogen peroxide and precipitating it with ammonia, the hydrated oxide is dissolved in a minimum volume of hydrofluoric acid, diluted to 5 ml. with water, and precipitated by the addition of 10 drops of $0.1M$ tri-2,2'-dipyridyl ferrous sulfate. The centrifuged precipitate is transferred with alcohol to a counting tray, the solvent is removed by drying, and the chemical yield (40 to 60%) is determined. After counting, the radiochemical purity is determined by beta absorption curves and gamma spectra. The procedure also provides a detailed description for the recovery of the tungsten in the aqueous layer of the DIPK or hexone extraction step.

The initial steps of another procedure suitable for the determination of tantalum in rocks (390) resemble those of the previous method (34), in as much as powdered portions of the sample (0.5 g.) and standards in the form of tantalum oxalate (0.1 ml. of Ta solution, 50 mg./liter) are simultaneously irradiated in silica tubes. Following a "cooling" period of a few days, the following scheme is used.

The powdered samples are fused in a platinum crucible with potassium carbonate (4 g.) and potassium nitrate (0.1 g.). The cooled melt is dissolved in a polyethylene beaker in water and acidified with HF, and the tantalum is precipitated in a polyethylene centrifuge tube with ammonia, following the addition of tantalum carrier solution (20 mg. of Ta). After centrifuging and discarding the supernate, the precipitate is twice washed with concentrated nitric acid. The precipitate is dissolved in 10 ml. of a solution $6M$ in H_2SO_4 and $10M$ in HF. The tantalum is extracted with 10 ml. of tri-n-butyl phosphate and the organic phase is twice washed with 5-ml. portions of $0.5M$ HF. After the addition of petroleum ether to the TBP phase, the tantalum is precipitated with $6M$ NH_4OH and is collected by centrifugation. Following two washings of the precipitate with ammonium nitrate solution and two washings with hot nitric acid, the precipitate is finally washed into a porcelain crucible, dried, ignited, and weighed to determine its yield. (In this procedure the standards are not carried through all steps of the procedure, but are merely precipitated with ammonia.) The final Ta_2O_5 precipitates are counted through a standard Al–Pb sandwich absorber with a γ-scintillation counter.

It should be noted that in any determination of tantalum by neutron activation some consideration must be given to the possibility of formation of ^{182}Ta through nuclear reactions involving elements other than tantalum. For instance, ^{182}Ta could be produced by the following reactions:

$$^{185}Re\ (n,\alpha)\ ^{182}Ta$$

$$^{182}W\ (n,p)\ ^{182}Ta$$

$$^{183}W\ (\gamma,p)\ ^{182}Ta$$

However, the first two reactions may be brought about only by fast neutrons.

4. Radiochemical Separations Involving Niobium-95

a. Niobium Activity in Aqueous or Organic Solutions

The niobium isotope most commonly analyzed for is niobium-95 ($t_{1/2}$ = 35 days; 6% fission yield; 0.16 M.e.v. beta; and 0.75 M.e.v. gamma). Niobium-95 is the daughter of zirconium-95 ($t_{1/2}$ = 65 days; fission yield $\sim$6.4%; 0.4 and $\sim$0.8 M.e.v. beta; and 0.72 M.e.v. gamma (255)). The following procedure has been used extensively by the Oak Ridge National Laboratory (415) and is suitable for the determination of niobium activity in aqueous or organic solutions.

The niobium sample and carrier are complexed with oxalic acid in acidic solution to prevent precipitation of the carrier and to promote interchange between the carrier and niobium-95. A homogeneous precipitation of niobium pentoxide is carried out by adding potassium chlorate and warming the solution in a water bath to destroy the oxalate ion. Niobium precipitates as the pentoxide and is thus separated from the bulk of the other activities and elements.

The precipitate is dissolved in hydrofluoric acid and zirconium activity is removed as barium fluorozirconate, while niobium remains in solution as the soluble fluoride complex. Fluoride ions are removed by precipitating niobium with ammonium hydroxide and discarding the filtrate. The precipitate is dissolved in oxalic acid and a homogeneous precipitation of niobium pentoxide is again made with potassium chlorate. The niobium oxide is ignited, weighed as the pentoxide, and mounted, and the beta activity is counted.

The procedure must be varied slightly when certain interfering ions, such as fluoride, fluosilicate, orthophosphate, molybdenum, tellurium, or protactinium, are present. Fluoride ions will cause incomplete precipitation of niobium from the oxalate complex by potassium chlorate. Fluorides can be removed either by making an initial ammonium hydroxide precipitation of the niobium before adding the oxalic acid or by complexing the fluoride ions with either aluminum nitrate or boric acid. Orthophosphates coprecipitate with niobium pentoxide; if this occurs, the orthophosphate impurity will accompany the niobium through the entire procedure. However, if a high concentration of iodate ion is added, the coprecipitation of the phosphate can be eliminated (229). The presence of protactinium may require several barium fluorozirconate precipitations to effect its coprecipitation. Molybdenum and tellurium are removed by scavenging in oxalic acid solution with a heavy metal sulfide. A complete analysis requires an over-all time of approximately two hours and gives a chemical yield of 60 to 70%.

b. In Fission Products of Plutonium Fuels (369)

Because of the high activity of the plutonium (1 mg. ~75,000,000 counts/minute; 30% counter efficiency), a preliminary removal of Pu(IV) is carried out in $8M$ nitric acid, using Dowex-1 as the adsorbing resin (429). This separation is not required if operations are carried out in a dry box.

Ten mg. of niobium carrier (in oxalic acid) is added to the irradiated plutonium aliquot plus 10 mg. each of zirconium and tellurium hold-back carriers. After allowing the solution to stand for 15 minutes, an equal volume of $16M$ nitric acid is added, the mixture is heated, and 0.5 g. of sodium bromate is added to destroy the oxalate and allow niobic acid to precipitate. The latter is centrifuged and washed once with $8M$ nitric acid.

It is then dissolved in 1 ml. of saturated oxalic acid and 10 mg. of copper carrier is added. After adjusting the acidity to $1M$ with hydrochloric acid, copper is precipitated as sulfide, removed by centrifugation, and discarded. Next, 10 mg. of tellurium carrier is added, precipitated with SO_2 from a $3M$ hydrochloric acid solution, and, after centrifugation, is discarded. In the supernatant liquid the niobium is precipitated by the addition of $8M$ nitric acid and sodium bromate. The precipitate is removed by centrifugation and is dissolved in 2 ml. of $12M$ sulfuric acid plus two drops of $27M$ hydrofluoric acid. The solution is transferred to a polyethylene tube and, after the addition of 2 ml. of HF, is extracted with 2 ml. of tributylphosphate. The organic phase is transferred to another tube and the niobium is precipitated by the addition of ammonium hydroxide and petroleum ether. After centrifugation, the niobic acid is washed into a platinum crucible, dried, ignited, weighed to determine the yield, and mounted for counting.

c. In Uranyl Nitrate Solution from Irradiated Uranium (391)

The sample solution (1 to 10 ml.) is placed in a 50-ml. glass centrifuge tube. Ten mg. of zirconium carrier (10 mg. of Zr/ml. as nitrate) is added, followed by a volume of concentrated nitric acid equal to that of the solution in the tube. After the addition of 1 ml. of saturated oxalic acid and 2 ml. of niobium carrier (10 mg. as oxalate), the solution is heated to near boiling and the niobium is precipitated by the addition of 0.5 g. of potassium bromate. From this point on the method follows closely the scheme described in the previous procedure (369). The procedure has been tested for separation from possible interfering activities by using the following radioactive tracers: Cs-137, Ba-137 (equilibrium mixture), Ce-144, Pr-144 (equilibrium mixture), Ru-106, Rh-106 (equilibrium mixture), Mo-99, Tc-99m (equilibrium mixture), and Zr-95. It was shown that the separation from Cs, Ce, Pr, Ru, Rh, Mo, Tc, and Zr is adequate for the fission-product mixtures normally encountered. Aluminum absorption curve measurements and γ-scintillation spectrometer measurements have indicated the presence of a β-particle of ca. 0.15 M.e.v. and a γ-ray of 0.77 M.e.v. No other β or γ activities could be detected. However, a cooling time of greater than 30 days is required to eliminate any tellurium activity. It should be added that the aluminum absorption curve of pure 35-day Nb-95 cannot be obtained until about two weeks after the last step in any radiochemical procedure for the activity analysis of niobium in mixed fission products. This is due to the presence of conversion electrons from the decay of 90-hour ^{95m}Nb (514).

d. In Uranium Target (515)

This procedure achieves a decontamination factor of 10^6 for all other fission products. The uranium target is dissolved in concentrated hydrochloric acid containing niobium carrier in the oxalate form. A few drops of nitric and hydrofluoric acids are added to clear the solution. The solution is adjusted to $6M$ and is extracted three times with half volumes of diisopropyl ketone. The organic layer is discarded. The aqueous layer is made $6M$ in HCl and $9M$ in HF and is then extracted three times with half volumes of diisopropyl ketone. The combined ketone layers are washed three times with a solution $6M$ in H_2SO_4 and $9M$ in HF. Subsequently, the niobium is extracted by three extractions with water and is precipitated with ammonia. The precipitate is washed with slightly basic ammonium nitrate solution, then with concentrated nitric acid. To eliminate residual activities, the Nb_2O_5 is dissolved in $6M$ sulfuric acid and $9M$ hydrofluoric acid and is subjected to a second series of extractions with diisopropyl ketone. The combined ketone layers are again washed with $6M$ sulfuric–$9M$ hydrofluoric acid solution. After back-extraction with water the niobium is finely precipitated with ammonia and ignited in a platinum crucible to Nb_2O_5.

e. After Cupferron Extraction (207)

In this separation of niobium from other fission activities, zirconium is removed as barium fluorozirconate: any uranium(IV) present, as well as rare earth activities, are carried down as the fluorides at this state. After the addition of boric acid niobium is converted to its cupferrate, which is extracted into chloroform. This step gives an effective separation from uranium(VI). The cupferron complex is destroyed and the niobium is precipitated with ammonia. The oxide is dissolved in concentrated sulfuric acid and decontamination from tin and antimony is effected by means of a sulfide precipitation, using copper as carrier. Further decontamination is achieved by additional precipitations of Nb_2O_5 as the hydrous oxide, solution of the latter in tartaric acid, and reprecipitation with nitric acid. The hydrous oxide is dissolved in a solution containing hydrochloric and hydrofluoric acids and, after complexing with boric acid, the niobium is again subjected to a cupferron chloroform extraction. The chloroform layer is again destroyed, a second sulfide scavenger separation is carried out, and the niobium is twice more precipitated, once with ammonia and the second time with nitric acid, with an intermittent solution of the precipitate in tartaric acid. The final precipitate is dissolved in ammonium oxalate solution and the niobium is finally precipitated with cupferron and ignited to Nb_2O_5. In the above procedure, the beta activity is counted without the use of an absorber. The isotopes counted are Nb-97 (68 minutes), Nb-96 (23.3 hours), and Nb-95 (35 days).

f. After Ion-Exchange Separations

In an attempt to prepare a source of niobium containing no zirconium and as large a ratio as possible of Nb-95m to Nb-95, it is necessary to let as much zirconium-95 as is available grow for a while, strip off all the niobium it contains, leave it again for a while, and then strip the niobium again before the short-lived isomer can decay appreciably. This can be achieved by anion exchange using a $3N$ HCl–$0.05N$ HF medium (372). In another procedure intended for the analysis of radionuclides in rainout, an elegant ion-exchange scheme was devised that is based on the retention of Zr-95–Nb-95 in a $0.1M$ HF medium by Dowex-1, while the other activities, Ce-144–Pr-144, Cs-134, Co-60, Zn-65, and Sr-85, pass through the column unadsorbed (311). The HF medium was also used to isolate the Zr-95–Nb-95 activities from other fission products by adsorption on Dowex-1 resin and direct counting of the resin by a gamma-scintillation spectrometer (333). A method for the rapid separation and determination of Np, Pu, U, Zr, Nb, and Mo in mixed fission products has been developed (572). The fission-product sample in concentrated HCl is added to the column (Dowex-2) without tracers or carriers. The activities are eluted sequentially and are determined directly in a γ-ray scintillation well counter or a multichannel γ-ray spectrometer.

For the determination of Zr–Nb-95 in a fission-product mixture (194) the activity of Ru-103 is removed by precipitation with thioacetemide, using Ru as carrier. Sr-85 and rare earth activities are removed in a $0.1N$ H_2SO_4–$0.3N$ HF medium with the help of zirconium and cerium carriers. Zr–Nb-95 are first adsorbed on Dowex-2 resin to remove the unadsorbed Cs-137 activity, then are eluted with $6N$ HCl–$0.5N$ HF and are counted together.

5. Other Radiochemical Procedures

For the isolation of tantalum isotopes produced by deuterium bombardment (^{182}Ta, ^{183}Ta, ^{184}Ta) from W, Re, and niobium isotopes, after dissolution of the target, three approaches have been suggested: (1) precipitation with ammonia, using $Fe(OH)_3$ as collector; (2) ion exchange (HCl–oxalate medium); and (3) liquid-liquid extraction in an HF–HCl medium, using diisopropyl ketone (135).

The recently introduced instrumental analysis using the complement nuclide subtraction technique has been used to determine Nb-95 in the presence of other radionuclides (419). A number of other methods not mentioned above have been described in a publication in one of the series on radiochemistry sponsored by the National Research Council (514).

TABLE XXI
Tentative ASTM Specifications (1962) for Niobium

Element	Maximum, in p.p.m., except where otherwise stated		
	Type 1 (Reactor-grade niobium)	Type 2 (Commercial-grade niobium)	Type 3 (Reactor-grade niobium + 1 Zr)
Carbon	100	500	100
Nitrogen	300	400	300
Oxygen	300	800	300
Hydrogen	20	50	20
Zirconium	500	1000	0.8–1.2%
Iron	500	500	500
Tantalum	1000	2000	1000
Titanium	500	500	500
Silicon	300	500	300
Boron	2	—	2
Tungsten	500	500	500
Molybdenum	500	500	1000
Aluminum	100	500	—
Beryllium	100	—	—
Cadmium	5	—	5
Chromium	100	500	—
Cobalt	30	—	30
Copper	100	—	—
Lead	50	—	50
Lithium	10	—	—
Magnesium	50	—	—
Manganese	100	500	100
Nickel	200	500	200
Tin	200	500	—
Vanadium	200	500	200
Ytterbium	10	—	—
Zinc	50	—	—
Uranium	2	—	—
Hafnium	100	—	100
Calcium	100	—	—
Sodium	100	—	—
Chlorine	300	—	—
Niobium, by difference, minimum	99.6%	99.2%	98.5%

VIII. ANALYSIS OF NIOBIUM AND TANTALUM METALS, ALLOYS, AND COMPOUNDS

Because of the specific uses of niobium and tantalum in a number of areas, various impurities that may be present in the two metals must be closely controlled. This has recently led to the formation of tentative specifications covering the chemical composition of niobium and tantalum (see Tables XXI and XXII). Undoubtedly, these specifications will be modified in time and extended to include additional impurities. Nevertheless, these specifications are reproduced here to represent a convenient listing of those elements for which analytical procedures are required at present.

Because of the close relationship of the analytical chemistry of niobium and tantalum, many of the methods developed for the analysis of niobium can be applied to tantalum as well. As a matter of fact, the niobium task force of ASTM Committee E-3, Division R, also has been charged with the development of methods for the analysis of tantalum and has indicated that most niobium methods can be extended to tantalum.

TABLE XXII
Proposed ASTM Specifications (1960) for Unalloyed Tantalum in the Following Grades: Powder Metallurgy, Arc Cast, and Electron-Beam Cast

Element	Maximum limit, in %
Carbon	0.03
Oxygen	0.03
Nitrogen	0.015
Hydrogen	0.01
Niobium	0.10
Iron	0.02
Titanium	0.01
Tungsten	0.03
Silicon	0.02
Nickel	0.02

A. DETERMINATION OF INTERSTITIAL ELEMENTS IN NIOBIUM AND TANTALUM

A comprehensive review of suitable methods was recently published (355), on which part of the following report is based.

1. Oxygen

a. By Neutron Activation

Since oxygen has an almost zero cross section for thermal neutrons, this method is not suitable for oxygen in niobium and tantalum. Fast neutrons (14 M.e.v.) have already successfully been used to determine oxygen in beryllium (117), and undoubtedly will prove satisfactory for niobium and tantalum. The reaction produces protons and radioactive N-16, according to the equation $^{16}O(n,p)^{16}N$. The ^{16}N decays with a half life of 7.4 seconds by emission of β-rays with a maximum energy of 10 M.e.v. (used for the oxygen determination) along with high-energy γ-rays. Based upon a technique first proposed for determining oxygen in beryllium metal powders (421), as little as 25 p.p.m. of oxygen in tantalum has been determined as follows (49). The metal or alloy sample is surrounded by LiF during the activation. The neutron reaction $^{6}Li(n,\alpha)^{3}H$ produces a source of tritons, ^{3}H, which in turn activate the oxygen nuclei in the sample to produce ^{18}F by the secondary reaction $^{16}O(H_3,n)^{18}F$. ^{18}F decays by positron emission and with a half life of 112 minutes. The positrons (0.69 M.e.v.) are accompanied by annihilation gamma radiation (0.51 M.e.v.). These gamma radiations are used to complete the analysis by gamma-scintillation spectrometry. The comparator method of analysis is used and both nondestructive and separation methods of analysis have been employed. One of the complications limiting the use of the method is the oxygen content of the lithium compounds that are employed (68).

It has been postulated that, with linear accelerators having energies of 25 to 30 M.e.v. and intensities of 10 r. per minute, the reaction $^{16}O(\gamma,n)^{15}O$ may be used for the determination of oxygen with a sensitivity of $10^{-4}\%$ (74). An autoradiographic method following proton bombardment ($^{18}O(p,n)^{18}F$) has also been investigated (290), but was found to be too insensitive to determine oxygen in niobium.

b. By Halogenation

It was shown that bromine trifluoride reacts with a number of metal oxides according to the general formula (252):

$$3MO_2 + 4BrF_3 \rightarrow 3MF_4 + 2Ba_2 + 3O_2$$

Since the reactions with niobium and tantalum pentoxides are quantitative, it can be assumed that the method would be applicable to the determination of oxygen in the two metals. Other halogenation procedures that are based on the volatilization of metal by gaseous bromine, chlorine, or hydrogen chloride and weighing of the unreacted metal oxide, are not applicable to the determination of oxygen in niobium and tantalum because of the danger of volatilizing of niobium or tantalum oxyhalogen compounds. This is particularly true in the bromination–carbon reduction procedure (116), in which chemists in this author's laboratory succeeded in volatilizing 0.5-g. portions of Nb_2O_5 and Ta_2O_5 without leaving any unreacted oxide.

c. By Vacuum Fusion

The familiar vacuum-fusion method is the referee method for the determination of oxygen (and hydrogen) in most metals. It involves melting a metal sample in a graphite crucible in vacuum and collecting, measuring, and analyzing the evolved gases. The extraction is usually made at 1600°C. or higher and the gases evolved are principally carbon monoxide, hydrogen, and nitrogen.

Niobium and tantalum have melting points exceeding feasible operating temperatures for the vacuum-fusion furnace. To overcome this, specimens are sometimes dissolved in a degassed bath of another metal contained in the vacuum-fusion crucible. Iron or steel have been successfully used in the determination of oxygen in niobium and tantalum (378,406), as have tin (416b), iron plus tin (50) and platinum (234). Particularly attractive is the platinum flux technique (233), in which the specimen may be imbedded in the platinum flux by pressing or may be wrapped in the flux. This technique has been extensively tested by the Oxygen Subgroup, Niobium Task Force, Division R, Committee E-3, ASTM (25), and in all probability it will be recommended as a standard method. Most laboratories use extraction temperatures between 1790 and 1950°C. and extraction times from 10 to 50 minutes. It appears that 1900°C. for 15 minutes could be recommended. It is generally felt that the niobium content of the final bath should not exceed 20 weight %. The gas extracted in the vacuum-fusion methods can be analyzed in a number of ways (see Section VIII-A-2). Low-pressure analysis through absorption by dry chemicals or through fractional freezing are most commonly used. The gas also can be analyzed by a mass spectrograph or a gas chromatograph.

d. By Inert-Gas Fusion

In this procedure, purified argon is passed over an inductively heated graphite crucible containing the sample. Oxides are reduced to carbon monoxide, which is carried by the argon over iodine pentoxide. This leads, in turn, to the formation of carbon dioxide. The carbon dioxide thus formed is measured in a conductivity cell containing dilute $Ba(OH)_2$ or NaOH, and the change in the resistance is noted. This is translated to per cent of oxygen by means of a graph prepared from results on primary standards. The inert gas-fusion technique as applied to niobium and tantalum has been mainly used in connection with a platinum flux technique, maintaining a platinum to sample ratio of about 8:1. Since the measuring device (Wheatstone bridge arrangement) is not as sensitive and precise as a McLeod Gage of the vacuum fusion, the method suffers somewhat in the below-100-p.p.m. range when compared to vacuum fusion. However, the apparatus is much less expensive than vacuum fusion equipment, is easy to operate, and is particularly suitable for niobium and tantalum where the reduction of the oxides by the graphite is rapid.

e. By Diffusion Extraction

It was shown (162,232) that reliable oxygen results can be obtained by omitting any flux or bath and extracting the oxygen as carbon monoxide at 2000°C. in vacuum. Although the mechanism of the extraction is not fully understood, it appears that the oxygen diffuses to the sample surface, where it is desorbed as carbon monoxide. The source of the carbon is probably vapor from the graphite crucible because the vapor pressure of graphite is about 3.4×10^{-6} mm. of mercury (506). An additional mechanism may be based on the interaction of a volatile lower-valence niobium oxide (as in electron beam melting) with the graphite vapor. In the inert gas-fusion procedure, 90 to 100% recoveries of oxygen have been reported when the platinum flux or bath is omitted. This technique is now under further study by the ASTM oxygen group. The recovery may be based on a diffusion mechanism similar to that of the vacuum-fusion technique or may be based (if the temperature exceeds 2468°C., the melting point of niobium) on the well-known interaction of niobium oxide and niobium carbide: $Nb_2O_5 + 5NbC = 7Nb + 5CO$.

f. By Emission Spectrometry

Emission spectrometry is a recent approach now developed to a state in which it is well suited to the determination of oxygen in niobium and tantalum (161,168,169,170). A d.-c. carbon-arc discharge is used to extract the oxygen content of the sample into a static argon atmosphere. A special electrode assembly, which provides a molten Pt–Nb or Pt–Ta bath after the arc is initiated, is employed. The oxygen content is rapidly evolved from this alloy system as carbon monoxide into the argon atmosphere. In d.-c. carbon-arc discharges in argon, the carbon monoxide is dissociated and the emission spectrum of oxygen is excited. The intensity ratio of oxygen 7772 A./argon 7891 A. is related to the oxygen content of the samples. The concentration range from 0.002 to 0.5% can be covered. In terms of technique, the method is much similar but faster than vacuum fusion. Potentially, 70 to 80 specimens can be analyzed in a day using a direct-reader recording device.

g. By Mass Spectrometry

This is not an independent method but an adjunct to a gas-extraction method such as vacuum fusion. The sensitivity of the mass spectrometer is such that 10^9 molecules per second give unit deflection. Unfortunately, the H_2O, CO, CO_2, and N_2 are, as in all vacuum systems, present as residual gases and reduce the application range of the method. Even so, the precision of $\pm\ 0.0001\%$ on a 1-g. sample is 10 to 100 times the minimum sensitivity of the spectrometer.

h. By Isotopic Dilution (220)

This method is based on the equilibration of the sample's content of normal (99.757%) ^{16}O with a known amount of oxygen enriched in ^{18}O, as, for example, from a master alloy. Assuming that the extraction of oxygen-containing gas from the equilibrated complex is not affected by mass difference, only a fraction of the total gas need be removed to arrive at an assay of the total oxygen content. The extracted gas is analyzed on a mass spectrograph to determine the degree of dilution of the ^{18}O during the equilibration. The original oxygen content can then be calculated. The effect of ^{17}O enrichment does not seriously complicate the procedure. The method should be workable for niobium and tantalum at intermediate and high oxygen contents.

2. Hydrogen

a. Vacuum-Fusion Methods

Hydrogen is removed rapidly and quantitatively from niobium and tantalum when melted in vacuum and therefore is conveniently obtained as a by-product of the oxygen determination. In the conventional vacuum-fusion equipment (400), the extracted gas mixture, consisting of hydrogen, nitrogen, and carbon monoxide, is pumped by means of a mercury-diffusion pump into a vacuum chamber of known volume. The pressure of the collected gas is measured by a McLeod gage. The total pressure P_T of the gases present is equal to the sum of the pressures each gas would exert alone if it were present alone in the given volume.

$$P_T = P_{N_2} + P_{CO} + P_{H_2}$$

After the total pressure of the gases is determined they are circulated through a tube containing copper oxide and through a tube containing anhydrone. The hydrogen and carbon monoxide are converted to H_2O and CO_2. The anhydrone absorbs the water, the CO_2 is frozen out in a trap cooled to $-196\,°C.$, and the nitrogen is returned to the McLeod gage, where its pressure is determined.

Subsequently, the frozen CO_2 is vaporized by allowing the temperature to rise and its pressure, together with that of the nitrogen, is measured. The known pressures of the nitrogen plus carbon dioxide and of the total gases supply all the necessary data for a determination of the weight of each gas present.

Although hydrogen can thus be obtained as an adjunct to an oxygen determination, a more convenient apparatus (499) utilizes a specially designed palladium diffusion valve that allows the rapid determination of hydrogen alone by vacuum fusion. A sensitivity of 0.2 p.p.m. is indicated.

b. Diffusion Method

This technique, also referred to as vacuum extraction, was originally introduced for the determination of hydrogen in titanium (54,115), but was found to also yield rapidly (2 to 3 minutes) and quantitatively at temperatures above $1100\,°C.$, the hydrogen in niobium and tantalum. The amount of gas evolved is measured as a pressure (PV) product, which can be converted readily into mass units according to the general gas law $PV = (W/M)RT$, where W is the weight of gas having molecular weight M, T is absolute temperature, and R is the gas constant. Although in the above procedure it is assumed that the total gas content is due to hydrogen, interference may only be expected if the sample contains significant amounts of both oxygen and carbon. Since the hydrogen is evolved very rapidly and its release can be followed on a Pirani gage coupled to a strip-chart recorder, any interference due to carbon monoxide or other gas would be immediately detected by the slope of the recording. However, equipment is commercially available in the United States that either incorporates a palladium diffusion valve or measures the gas pressure before oxidation of the hydrogen to H_2O and after absorption of the latter by anhydrone.

c. Spectrographic Methods

Studies on the determination of hydrogen have revealed that the H line at 6563 A. is readily detectable when 200-mg. portions of the sample are arced in graphite electrodes at 30 amp. (167). The behavior of the hydrogen:argon intensity ratios during the arcing cycle indicates that the hydrogen is evolved rapidly during the first few seconds and then appears to be removed from the system as the arcing proceeds, probably through chemical absorption. A marked enhancement of the hydrogen line intensity occurs as the arcing cycle is conducted at lower pressures. Although maximum H line intensities are observed at pressures less than 100 mm. of Hg, it is desirable to compromise the advantages of maximum H line intensities with the disadvantages of increased metal volatilization at these low pressures. For quantitative determinations, an argon pressure of 125 mm. Hg, an arcing period of 30 amp., and exposures during the 10- to 40-second period are used.

A spectral isotopic method has recently received some attention (581, 582). In this procedure, the sample is placed in an exchanger tube of known volume (e.g., 150 ml.). The exchanger is evacuated, filled with deuterium, and heated until equilibrium is reached as a result of isotopic exchange between the sample and the surrounding gas. The determination of the isotopic composition of the gas is achieved by passing a portion through a high-frequency discharge tube of 1- or 2-mm. diameter at a rate of 1 or 2 ml. per minute and a pressure of 2 to 5 mm. (Hg), and by photoelectric measurement of the ratio of intensities of the Bahner lines of hydrogen and deuterium.

3. Nitrogen

a. Vacuum-Fusion Method

It is generally agreed that the vacuum-fusion technique, as at present operated, does not furnish very reliable nitrogen figures for those elements (particularly Group IVA) forming very stable nitrides (67). However, niobium and tantalum nitrides have a stability intermediate between the stable nitrides of Group IV and the unstable nitrides of Group VI (see Section II-E-7), and in metallurgical treatment of tantalum it has been observed that all the absorbed nitrogen is slowly liberated between 1900 and 2150°C. in a vacuum (257,293). During a recent study of the Kjeldahl procedure by members of an ASTM subgroup on nitrogen of the Columbium Task Force, several of the cooperating laboratories submitted results obtained by vacuum fusion that compared favorably with those obtained by the chemical procedure (24). One laboratory used a 0.3-g. sample, a platinum flux technique, and an outgassing time of the sample of 30 minutes at 1900°C. (360).

b. Inert Gas-Fusion Procedure

Not much is known about the stability of the nitrides at high temperatures (> 2000°C.) if no vacuum is applied. However, a recent announcement by a manufacturer claims that a new instrument is suited for the simultaneous determination of nitrogen and oxygen in metal samples (320). In this procedure, the sample is introduced into a high-frequency induction furnace, where it is fused in a graphite crucible. The nitrogen and oxygen (as CO) are swept into a specially designed trap (liquid nitrogen cooled) by the inert gas carrier. When released from the trap, gases are separated in a chromatographic column. The gases are then passed through a thermistor, which makes up one leg of a Wheatstone bridge. This unbalances the bridge and the imbalance is read on two programmed integrators, one for nitrogen and one for oxygen, and the results are plotted on a graph previously established with standard samples. Although this procedure appears attractive in every respect, no data are available yet on whether niobium and tantalum nitrides break down in the absence of a vacuum, when the higher temperatures of the inert gas apparatus are applied.

c. Kjeldahl Procedure

This procedure has been adopted by the Columbium Task Force of the ASTM Committee E-3, Division R, as a tentative standard and is based on solution of the sample at 90 to 95°C. in hydrofluoric and hydrochloric acids with the occasional addition of hydrogen peroxide. (Other solvents, such as hot concentrated sulfuric acid, also may be used.) All samples should be etched immediately prior to weighing with dilute HF and water-rinsed, followed by acetone rinsing and air drying. The solution of the sample is made basic with sodium hydroxide and nitrogen is separated as ammonia by steam distillation. Nessler reagent is added to the distillate to form the yellow ammonia complex. Photometric measurement is made at approximately 430 mμ. The method is suitable for the determination of nitrogen within the concentration range of 0.0020 to 0.12 mg. per 50 ml. of solution, using a cell depth of 2 cm. If preferred (538), the blue color formed when sodium phenate is added to the distillate, which has been treated with hypochlorous acid, may be used to determine the nitrogen content of the sample (479).

For low-nitrogen samples (N<0.01%) the above colorimetric techniques are preferable, but in case of a higher nitrogen content the ammonia may be absorbed by dilute boric acid and titrated with standard acid, using a mixed indicator (methyl red–bromcresol green).

d. Alkali-Fusion Method (234)

The niobium, in the form of a fine powder or millings, is fused in a nickel boat inside a Vycor combustion tube with sodium hydroxide at 600°C. for 1.5 hours in a stream of hydrogen. The ammonia formed during the fusion is absorbed in water or boric acid and is determined by the usual volumetric or colorimetric procedures.

4. Carbon

Since the carbon content of niobium and tantalum produced today rarely exceeds a few hundredths of one per cent, the gravimetric procedure based on the combustion of the metal in a globar furnace and absorption, after removal of impurities, of the CO_2 by ascarite is justified only in cases in which a high degree of accuracy is not required. The same applies for the gasometric procedure in which the CO_2 is collected and measured in a gas buret. The Carbon Subgroup of the Columbium Task Force, ASTM, is investigating a procedure based on ignition of the sample in an induction furnace, using an appropriate flux, and measuring the change in conductivity of a barium hydroxide solution in which the carbon dioxide is absorbed. In determining the carbon content of niobium and tantalum by the conductometric technique the main difficulty lies in obtaining a quantitative combustion. Because these metals do not properly load the induction coil, thus causing incomplete combustion of the sample, high-purity iron and tin are used as fluxes. Both niobium and tantalum burn easily and completely in oxygen; this author is therefore not convinced that induction heating, which limits the amount of sample and necessitates the use of fluxes (which always contain some carbon), is in this case superior over igniting of a suitable amount of sample in a conventional combustion tube and finishing the determination conductometrically. Alternatively, the method of combustion by high-frequency induced radiant heating (the sample in a porcelain crucible is subjected to the radiation from a surrounding platinum cage that is heated to about 1600 °C.) (505), in conjunction with a conductometric finish, deserves further attention. Carbon in tantalum metal also has been determined by the low-pressure combustion method (542) (the CO_2 is collected in a trap cooled by liquid nitrogen and its volume after evaporation is measured with a McLeod gage (564).

B. DETERMINATION OF METALLIC IMPURITIES IN NIOBIUM AND TANTALUM

1. Spectrographic Procedures (554)

a. General Considerations

Because of the complexity of the analytical chemistry of niobium and tantalum, the determination of metallic impurities in the two metals should be carried out by spectrographic techniques wherever routine determinations in a large number of production samples are required. However, even if the spectrographic approach is followed, it is not possible to completely avoid the complexity of the two elements, since their emission spectrum in the ultraviolet region consists of many lines: 3303 for niobium and 2164 for tantalum (236).

Only a few methods have been published for the analysis of impurities in niobium and tantalum. They usually employ the pentoxides, frequently along with a carrier or buffer plus an internal standard. Most procedures use the d.-c. arc as source unit in order to obtain the highest spectral sensitivities. The refractory nature of the oxides of niobium and tantalum (m.p. near 1500°C.) and of the metals themselves (m.p. near 2500 and 3000°C.), and their tendency to form even more refractory carbides, frequently suggest the use of buffers or carriers. Because of the low volatility of these refractories, fractional distillation of the more volatile impurities occurs during arcing with only traces of niobium and tantalum being carried into the arc. However, only a limited control can be exercised over this type of volatilization, and spectra, therefore, may be quite erratic. This, however, may sometimes be tolerated because of the higher sensitivity obtainable. A detailed procedure suitable for the determination of impurities in both niobium and tantalum is presented in Section IX-D.

b. Outline of Several Techniques

In one procedure (180), 14 elements (Cd, B, Si, Fe, Ta, Zr, Mn, Mo, Cr, Al, Sn, Ti, Co, and Ni) are determined in niobium by a technique using lithium carbonate as a buffer and lanthanum as the internal standard. The lithium carbonate is chosen because of its simple spectrum, its ability to suppress background, and its smooth-burning qualities. The lanthanum is used as internal standard because of the relative simplicity of its spectrum and its favorable volatilization characteristics. Since lithium carbonate suppresses the principal lines of boron and cadmium (level of interest = 1 p.p.m.) to the point of nondetectability, separate portions of the niobium oxide are used, omitting the buffer. Because of sensitivity and volatilization requirements, tantalum and zirconium also are determined from spectrograms produced by the excitation of the unbuffered oxides. Standards are prepared by the addition of the oxide of the metals sought to niobium oxide prepared from high-purity (electron-beam melted) niobium metal. The homogenizing of the standard mixtures is accomplished by placing the plastic vial in a dental amalgamator ("Wig-L-Bug") and mixing for several 3-minute periods. Synthetic standards of the concentration ranges of interest are prepared by appropriate dilutions with niobium oxide.

Another procedure involves equal amounts of Nb_2O_5 and a carrier consisting of one part of strontium chloride and one part of graphite. A one-minute excitation with a 15-amp. d.-c. arc has been recommended for the determination of the following elements in niobium: Mg, Mn, Cr, Co, Ni, Ti, Sn, Fe, Pb, Si, Cu, Nb, Mo, and Al (203). In a similar procedure, silver chloride is recommended as a carrier for the determination of the following elements in tantalum: Ni, Cr, Si, Mn, Sn, Pb, Co, Fe, Mg, Mo, Nb, Al, and Ti (400a).

Lanthanum carbonate has been proposed in another procedure covering 23 elements in niobium metal (368). Germanium oxide, serving as buffer and internal standard, is added to 325-mesh tantalum metal powder for the analysis of five elements (31). A carrier consisting of silver metal, silver chloride, and barium fluoride is added to the pentoxide of tantalum for the determination of 13 elements (321). However, neither of these two last reports makes provision for determining the niobium content in tantalum. In another procedure, pressed pellets of Ta_2O_5 are arced for the determination of six elements (341); tantalum metal powder is directly arced for determining higher levels of titanium and niobium (340).

Another laboratory divides the impurities in niobium into four separate groups, each group requiring a different procedure (173). Elements of Group I (Al, Be, Co, Cr, Fe, Mg, Mn, Mo, Ni, Pb, Si, Sn, V, and Yb) are analyzed by mixing 50 mg. of sample oxide and 50 mg. of SP-2 graphite powder in an agate mortar and pressing the mixture into graphite electrodes. Standards are prepared by dry grinding of the desired high-purity oxides with high-purity niobium oxide. These master standards are diluted with matrix oxide to obtain the standard level desired and are homogenized with the "Wig-L-Bug." Excitation is with a d.-c. arc (3.4-m. Wadsworth mount, 15,000 lines grating, 250 v., 15 amp., no preburn, and 40-second exposure). No internal standards are used and the following second-order spectral lines are suggested: Al = 3082.15; Be = 3130.42 and 3321.34; Co = 3453.51; Cr = 2835.63; Fe = 3047.15; Mg = 2795.53; Mn = 2794.82; Mo = 3170.35; Ni = 3414.76; Pb = 2833.07; Si = 2881.58; Sn = 2839.99; V = 3185.39; and Yb = 3694.20 A.

For the determination of Group II elements (Ta, Ti, and Zr), excitation is with a high-voltage spark (no preburn and a 60-second exposure), and the following analytical and internal standard lines are used:

Element	Wavelength, A.	Internal standard	Wavelength, A.
Ta	2685.11	Nb	2687.15
Ti	3234.52	Nb	3230.24
Zr	3391.97	None	—

For the determination of Group III elements (B and Cd), 80 mg. of sample oxide is mixed in an agate mortar with 20 mg. of CuOHF. Fifty mg. of the mixture is weighed into the sample electrode, which is covered with a boiler cap. Excitation is with the d.-c. arc (250 v., 10 amp., no preburn, and a 60-second exposure period). Lines used are B = 2496.78 and Cd = 2288.02 A. For the determination of tungsten, 100 mg. of sample and 10 mg. of sodium chloride is mixed in an agate mortar. The mixture is pressed into electrodes and excited with a 250-v., 6-amp. d.-c. arc, and the tungsten line at 4008.75 A. is finally measured.

For the lithium determination, the oxide is mixed with graphite and CuOHF and excited in a d.-c. arc. For the determination of zinc, the niobium oxide is weighed into a deep electrode and excited in a d.-c. arc.

A technique similar to the one described above has been used for the determination of impurities in tantalum. Ta_2O_5 mixed with half its weight of carbon powder is used in conjunction with an intermittent a.-c. arc for the determination of W, Ti, Mo, V, Cr, Fe, Nb, Ni, Si, Mn, and Mg (416).

Dry mixed oxides also are successfully used as standards by another installation employing a large Hilger spectrograph for the determination of 27 elements in niobium and 25 elements in tantalum (538). Another installation pelletizes 150 mg. of Ta_2O_5 with 450 mg. of briquetting graphite and preheats the pellets at 400°C. before excitation. A 10-second preburn and a 30-second exposure by a 3-amp. arc excitation are used for the determination of the following elements in tantalum: Ni, Cr, Si, Mn, Sn, Pb, Co, Fe, Mg, Mo, Nb, Al, Ti, and Zr (400a). This technique also has been used to determine W, V, and Mn (416a).

c. Other Procedures

Major alloying constituents, Mo (5 to 10%), Zr (1 to 8%), Ti (5 to 10%), and W (10 to 20%), have been successfully determined by a rotating-disc technique employing an a.-c. spark (366,366a). The sample and standards were prepared by solution in hydrofluoric acid with the addition of a little nitric acid.

A method for intermediate levels of tantalum in niobium has been reported in the Russian literature (527), as has a procedure for the spectrographic determination of titanium (0.006 to 0.1%) and tantalum (0.008 to 0.3%) in Nb_2O_5 and titanium (0.005 to 0.1%) and niobium (0.005 to 0.5%) in Ta_2O_5 (524). In both procedures, standards are prepared by the use of pure metal oxides. Titanium can be determined in a niobium matrix by a d.-c. arc technique in which the sample (10 mg.) is mixed with an equal amount of carbon powder. It is stated that after 2.5 minutes the whole of the titanium present has passed into the arc. The method is sensitive to 0.002% (110). Bi, Cd, Sn, Pb, and Sb have been determined in Ta_2O_5 prepared by low-temperature ignition, by volatilization of the impurities at 1400°C., condensation onto copper electrodes, and excitation by a condensed spark discharge (585). Small amounts of tin in Nb_2O_5 can be determined by arcing the sample in a carbon cathode and using lead as an internal standard (254). A determination of impurities in ferroniobium has been described, involving a technique in which the powdered sample, mixed with Fe_2O_3 and graphite in a 1:4:5 ratio, is pressed into a ring 0.5 mm. thick on a copper or nickel disc, which can be rotated at 1 r.p.m. and which serves as lower electrode. Spectra from which the content of Al, Ti, Zr, Sn, Cu, Mn, and Cr can be obtained are excited by a 4-amp. arc, with an upper carbon electrode (501).

2. X-Ray Fluorescence Procedures

Not much information is available on the application of x-ray fluorescence to the determination of impurities and alloying elements in niobium or tantalum. Small amounts of tantalum ($\sim$0.1%) in niobium can be determined by this technique, using the second-order Ta $L\alpha_1$ line (188). This procedure requires a long counting time and careful background corrections. Another laboratory covers the concentration range of 0.05 to 0.70% tantalum by using an oxide-cellulose pellet technique. The fluorescent x-radiation from the pellet is dispersed and the integrated intensity of the analytical line is compared to that of a nondispersed beam from an external standard (93). A technique described in Section VII-F-4, for the determination of tantalum in commercial niobium oxide, naturally can be applied also to the determination of tantalum in the oxides obtained after ignition of niobium metal. Because of the much greater sensitivity of the niobium $K\alpha$ radiation, the determination of niobium in tantalum can be extended to below 100 p.p.m.

X-ray fluorescence methods have been used for determining Mo, Nb, Th, W, Y, and Zr as impurities or alloying constituents in tantalum (230). Since the principles involved undoubtedly can be extended to the determination of other elements, it is given here in some detail.

The sample is dissolved in hydrofluoric acid with the addition of a little nitric acid. After evaporation to about 2 ml., the solution is diluted in a 5-ml. volumetric quartz flask with 35% HF.

For the determination of 25 to 2000 p.p.m. of molybdenum, niobium, and zirconium in 1-g. tantalum samples, scattered background radiation is used as an internal standard to reduce errors caused by daily fluctuations in line voltage, variations in x-ray tube voltage and current settings, fluctuations in noise of the electronic circuitry, and errors in sample preparation. A total of 128,000 counts are accumulated at the $K\alpha$ lines for molybdenum, niobium, and zirconium and at the background 24.00° 2θ angle (Philips Instrument, lithium fluoride crystal). The total of 128,000 counts is required to obtain a statistical counting error sufficiently small to yield a sensitivity approaching 25 p.p.m. To eliminate bubble formation on the Mylar window of the sample cell during the long counting times involved, the window is coated with a nonwetting agent such as Desicote.

For the determination of tungsten (0.5 to 10%) in tantalum, the W $L\alpha_1$ line is used because it is the most intense line available. To correct for errors caused by fluctuations in x-ray tube voltage and current, incoherent or Compton scattered radiation is selected as an internal standard. Because niobium may be present, the use of a pulse-height analyzer is advisable for eliminating interference of the second-order niobium $K\alpha$ line with the Compton line for tungsten. If no pulse-height analyzer is employed, each 0.1% of niobium causes a negative absolute error of 0.06% in tungsten values.

For the determination of 0.025 to 10% of yttrium, a sample containing 0.25 to 5 mg. of yttrium is dissolved in HF and HNO_3, thorium is added as a carrier (3 mg.), the insoluble fluorides are collected by centrifugation, and the residue is washed with dilute hydrofluoric acid. The fluorides are evaporated in a platinum dish with 4 drops of sulfuric acid, the residue is dissolved in water, and the solution is transferred to a 5-ml. volumetric flask and diluted to volume with water. The time to accumulate 64,000 counts is measured at the $K\alpha$ line for yttrium and the $L\alpha_1$ line for thorium, and the intensity ratio is compared to ratios obtained for known standards.

The procedure for determining thorium is similar to that described for yttrium, except that the sample size is adjusted to contain 0.25 to 7 mg. of thorium and that 1.00 mg. of yttrium is added as an internal standard.

A liquid sample technique has been used routinely to analyze major alloy constituents (W, Mo, Ti, and Zr) in General Electric's F-48 and F-50 niobium-base alloys (268,332).

3. Radioactivation Analysis for Trace Elements in Niobium

Niobium has a low neutron absorption cross section of 1.1 barn. Therefore, there are no self-shielding effects and the activation of the matrix is negligible. Hence, activation analysis represents a very useful tool for the analysis of trace impurities in niobium and especially for the determination of high-cross-section impurities in reactor-grade material. Because not much information has appeared yet in the published literature, the following is largely based on a report given at the 1961 Gatlinburg Conference (157).

A comparator method is usually used for the determination of an element by activation analysis. This technique involves irradiating simultaneously the sample as well as a known amount of the element to be determined. With these conditions, a simple comparison of the activities and weights of the sample and the comparator results in the concentration of the desired element.

After irradiation of known amounts of samples and comparators, the analysis may be completed by a nondestructive or by a radiochemical separation technique.

a. Nondestructive Technique

In the nondestructive technique, the sample and comparator are transferred from the irradiation container to suitable counting mounts, i.e., glass culture tubes or beta card mounts. Then the gamma-ray spectrum of the sample is compared with the gamma-ray spectrum of the comparator by a 3- × 3-inch diameter NaI (Tl-activated) crystal and a multichannel analyzer. The area of the desired photopeak is compared with the comparator and the concentration of the desired element is calculated. An example of this technique is the determination of chlorine in niobium metal. Other examples appear in Table XXIII.

For the determination of chlorine in niobium, 150 mg. of sample is weighed in triplicate and placed in polystyrene vials 1 cm. in diameter × 3 cm. long, which are then placed in a 6-ounce polyethylene bottle along with 10 mg. of NH_4Cl (also in triplicate). The polyethylene bottles are then irradiated for 30 minutes in the graphite reactor. After irradiation, the NH_4Cl comparators are dissolved and suitable dilutions are made. The niobium metal samples are transferred to glass culture tubes. The activity of the 1.63- or 2.15-M.e.v. gamma ray of the 37.5-minute ^{38}Cl is measured by the multichannel analyzer. The time at the start of the count is recorded. This procedure is repeated for each sample and comparator. A suitable time reference point is picked and all activities, both samples and comparators, are corrected to this time. Using the sample weight, the activity of the chlorine in the sample corrected for decay, and the weight of chloride in the standard, with its activity corrected for decay, a comparison may be made and the concentration of the chlorine in the niobium determined.

TABLE XXIII
Determination of Trace Impurities in Niobium Metal by Nondestructive Radioactivation Analysis Techniques (Oak Ridge National Laboratory)

Element	Irradiation time	Radio-nuclide measured	Half life	Decay mode	Particle	Gamma	Practical limit, γ^a
Al	1 min.	^{28}Al	2.3 min.	β^-	2.9	1.78	16
Cl	30 min.	^{38}Cl	37.5 min.	β^-, γ	4.8; 1.1; 2.8	2.15; 1.63	3
Mn	5 hr.	^{56}Mn	2.58 min.	β^-, γ	2.8; 1.0; 0.7	0.84; 1.81; 2.1	0.75
Th	1 min.	^{233}Pa	27.4 days^b	β^-, γ	0.26; 0.15	0.31; more	2
V	1 min.	^{52}V	3.76 min.	β^-, γ	2.5	1.44	1
W	16 hr.	^{187}W	24 hr.	β^-, γ	0.6; 1.3	0.69; 0.48; 0.13; more	0.001

^{a} Practical limit depends upon neutron flux, sample matrix, irradiation time, and reactor position.
^{b} Daughter radioactivities.

b. After Radiochemical Separations

An example of an analysis using the radiochemical separation technique is the determination of silicon in niobium. Other examples appear in Table XXIV. For the determination of silicon, 200 mg. of niobium is weighed in triplicate and wrapped in aluminum foil. The comparator standard used for this analysis is 100 mg. of $(NH_4)_2SiF_6$, also wrapped in aluminum foil. Both the samples and comparator standards are placed in a pneumatic rabbit and irradiated for 2.5 hours in the pneumatic tube. After irradiation, the samples and comparators are transferred to high-density polyethylene bottles. The niobium samples are dissolved in 2 ml. of conc. HF plus 5 ml. of conc. HNO_3. After solution of the sample, 20 mg. of silicon carrier and 20 ml. of conc. H_2SO_4 are added and the SiF_4 is distilled into approximately 10 ml. of water.

The polyethylene bottle is heated by placing it in a boiling-water bath. After allowing 10 minutes for the distillation, the distillate is removed from the system and 30 ml. of saturated $Al(NO_3)_3$ solution is added to complex the fluoride, followed by 30 ml. of conc. H_2SO_4. The solution is just heated to a boil, cooled, and centrifuged. The silica is washed twice with $6M$ HCl, filtered, and ignited at 1000°C. to SiO_2. After ignition, the SiO_2 is weighed to determine the yield of the carrier and then is mounted on a beta card mount.

The ^{31}Si activity is measured by a Geiger-Müller counter at 30-minute intervals to determine the radiochemical purity. Each sample and comparator is treated in the same manner.

c. By Differential-Count Method

Another technique frequently used in nondestructive analysis for short half-life isotopes is known as the "differential count." This technique is at its best for half lives of 30 seconds or less, but it has been used successfully for half lives of as long as 10 minutes. The technique is as follows.

The sample (20 to 50 mg.) is irradiated for 20 seconds and allowed to decay for 20 seconds. The gamma-ray spectrum is measured on the multichannel analyzer for one minute. Immediately after the initial analyzer counting interval, the sample is counted for one minute in the complemented or subtract mode of the analyzer. The resulting gamma-ray spectrum consists of only those activities that have decayed in one minute.

TABLE XXIV

Determination of Trace Impurities in Niobium Metal by Radiochemical Separation Techniques (Oak Ridge National Laboratory)

Element	Irradiation time	Radio-nuclide measured	Half life	Decay mode	Particle	Gamma	Practical limit, γ
Br	16 hr.	^{82}Br	35.9 hr.	β^-, γ	0.44	0.77; 0.55; 0.61; and more	1
Cd	1 week	^{115}Cd	43 days	β^-, γ	1.6	0.94; 1.30	10
Co	1 week	^{60}Co	5.27 yr.	β^-, γ	0.31	1.17; 1.33	0.01
Cr	1 week	^{51}Cr	27.8 days	EC, γ		0.32	0.1
Cu	16 hr.	^{64}Cu	12.8 hr.	EC, β^-, β^+, γ	0.6	1.34	0.002
Fe	1 week	^{59}Fe	45.1 days	β^-, γ	0.46; 0.27	1.10; 1.29; 0.19	5
Ni	5 hr.	^{58}Co^{a,b}	71.3 days	EC, β^+, γ	0.48	0.81; 1.6	1
P	1 week	^{32}P	14.3 days	β^-	1.7		0.05
S	1 week	^{32}P^a	14.3 days	β^-	1.7		5
Si	2.5 hr.	^{31}Si	2.65 hr.	β^-, γ	1.5	1.26	0.1
Sn	1 week	^{125}Sn	9.4 days	β^-, γ	2.3; 0.4	1.07, more	0.5
Ta	1 week	^{182}Ta	112 days	β^-, γ	0.5; 0.4	1.12; 1.22; 0.10; 0.22; and more	2
U	16 hr.	^{239}Np	2.33 days	β^-, γ	0.44; 0.31	0.11; 0.22; 0.28; and more	0.02
Zn	1 week	^{69}Zn	245 days	EC, $\gamma(\beta^+)$	0.3	1.11	0.3
Zr	1 week	^{95}Zr	63.3 days	β^-, γ	0.40; 0.36	0.72; 0.76	200

^{a} Produced by a n,p reaction.
^{b} Nickel is also determinable by the ^{64}Ni $(n,\gamma)^{65}$Ni reaction. ^{65}Ni decays with a 2.6-hour half life.

4. Determination of Impurities In Niobium and Tantalum by Chemical Methods

The determination of impurities in niobium and/or tantalum by instrumental procedures usually requires the availability of suitable standards. To provide such standards chemical procedures are required. Frequently, such chemical procedures are also preferred by the analyst who only determines impurities sporadically or is only interested in a limited number of elements.

a. Determination of Tantalum in Niobium

Pyrogallol After Ion Exchange

Niobium, titanium, and all other elements that would interfere with the pyrogallol procedure are removed by passing a solution of 4% (vol./vol.) HF and 14% (w./vol.) of ammonium chloride through a small column bed containing Dowex-1, 8 to 10% crosslinkage (26). The tantalum is subsequently eluted from the column by a 4% (w./vol.) ammonium fluoride–14% ammonium chloride (w./vol.) solution of pH 5.5. To the eluate boric acid is added, followed by HCl. The tantalum is subsequently precipitated with cupferron, using zirconium as collector, and is determined by the pyrogallol procedure described in Section VII-B-8.

Pyrogallol After Solvent Extraction

In another procedure (112), Nb_2O_5 is fused in $K_2S_2O_7$, the melt is dissolved in $0.4M$ HF–$2M$ H_2SO_4, and the tantalum is extracted with cyclohexanone (3 × 13 ml.), which is then washed with H_2O (3 × 3 ml.). Tantalum is re-extracted into a solution of ammonium oxalate containing boric acid. After destruction of organic matter the pyrogallol procedure is finally applied.

Other Tantalum Procedures

The separation and colorimetric determination of tantalum by extraction of the methyl violet complex (330,433) was already mentioned in Section VII-B-9.

Derivatives of 2,3,7-trihydroxy-6-fluorone also have been used for the determination of tantalum in niobium after removal of the bulk of the latter by extraction with acetone–isobutyl alcohol from an HF–HNO_3 medium (404). Tantalum forms a complex salt with malachite green in the presence of hydrofluoric acid, and this salt can be extracted with benzene or xylene (278). Although this procedure was only recently suggested for the determination of tantalum in niobium and therefore has not yet been critically examined by other workers, it appears to be simple and rapid.

b. Determination of Niobium in Tantalum

A modification of the above ion-exchange procedure can be used (279) (see Section IV-H-4). The column must be large enough to hold 1 g. of tantalum (1 inch diameter, 8 inches Dowex-1 resin). Iron, titanium, tungsten, and zirconium are removed by passing a 5:4:11 mixture of HCl–HF and H_2O through the resin bed. Niobium is subsequently eluted with 4% (vol./vol.) HF–14% (w./vol.) NH_4Cl. After complexing the HF with boric acid, niobium is precipitated with cupferron, using zirconium as collector. It is finally determined photometrically by the hydroquinone method (see Section VII-B-3). A similar ion-exchange procedure has been used to separate the niobium from the tantalum, followed by a spectrophotometric determination of the former by the thiocyanate procedure (55). In another procedure (538), tantalum is first removed by precipitation of K_2TaF_7 (see Section V-A-5) before the thiocyanate procedure is applied. A direct photometric thiocyanate procedure suitable for the determination of 0.01 to 0.30% of niobium also has been described (238).

c. Determination of Iron

The following procedure has been approved by an ASTM task group for the determination of iron in niobium (22): 0.5 g. of sample is decomposed in a platinum crucible in 2 ml. of HF with the addition of a minimum amount of HNO_3. After warming to expel oxides of nitrogen, the solution is washed into a 150-ml. beaker containing a mixture of 5 ml. of tartaric acid solution (335 g./liter) and 15 ml. of boric acid solution (50 g./liter). The solution is stirred and the color developed as follows: 5 ml. of hydroxylamine hydrochloric solution (100 g./liter), 10 ml. of ortho-phenanthroline solution (1 g./liter), 2 ml. of ammonia, and 10 ml. of ammonium acetate (335 g./liter) are added in succession, mixing the solution after each addition. The pH of the solution is adjusted to 5.5 with NH_4OH or HCl and then is transferred to a 100-ml. volumetric flask and diluted to volume with water. After mixing, the flask is placed on a water bath at 60 to 70°C. for 20 minutes. After cooling, the absorbance is measured in a 1-cm. cell at 508 mμ against a reagent blank. The above procedure covers the concentration range of 0.030 to 0.34 mg. of iron. By suitable changes of sample weight and cell depth, this range can be extended. An almost identical procedure has been recommended for the determination of iron in tantalum (239).

Bathophenanthroline (4,7-diphenyl-1,10-phenanthroline) has recently been suggested for the determination of iron in niobium and tantalum (191, 426). An extraction step involving either chloroform–ethanol or n-amyl alcohol increases the sensitivity of the method and also avoids possible turbidity due to partial hydrolysis of niobium and tantalum. Reduction of iron to the ferrous state is achieved either with ascorbic acid and hydroxylamine hydrochloric (426) or with sodium hydrosulfite (191).

Iron also can be determined in both niobium and tantalum by an 8-hydroxyquinoline photometric procedure. Following solution of the sample in HF and HNO_3, the iron oxinate complex is extracted in a polyethylene container with MIBK in an aliquot of the solution, using an ammoniacal tartrate medium (538).

d. Determination of Tungsten and Molybdenum

Dithiol Procedure

The following procedure has been approved by an ASTM subgroup for the determination of molybdenum and tungsten in niobium (23,251a): 0.5 g. of sample is heated with 10 ml. of sulfuric acid and 3 g. of ammonium bisulfate, until dissolved. The cooled solution is washed into a 50-ml. volumetric flask and is diluted to the mark using H_2SO_4 (1:2) to effect the transfer and dilution. A 15-ml. aliquot of this solution is evaporated to fumes of sulfuric acid; 15 ml. of HCl (1:2) is added, followed by 10 drops of HF and 12 drops of hydroxylamine hydrochloride solution (1 g./10 ml.). The solution is washed into a 125-ml. separatory funnel and cooled to 20°C., and the molybdenum is extracted as follows.

Ten ml. of dithiol solution is added (5.0 g. dissolved in 500 ml. of water (at 35°C.) containing 20 g. of sodium hydroxide; 10 ml. of thioglycolic acid is added and the solution is diluted to 1:1 with water. The reagent thus prepared is stable for at least a month if kept in a polyethylene bottle in a refrigerator). The solution is mixed thoroughly and allowed to stand for 15 minutes; 20 ml. of carbon tetrachloride is then added and the solution is shaken for 2 minutes. The carbon tetrachloride layer is transferred into a dry 50-ml. volumetric flask. The aqueous phase is washed by twice shaking with 10-ml. portions of carbon tetrachloride, adding the washings to the volumetric flask. The solution is diluted to volume with carbon tetrachloride, mixed, and its absorbance is measured in a 1-cm. cell at 680 mμ against a reagent blank carried through the above procedure.

The aqueous solution in the separatory funnel containing the tungsten is washed into a small flask and is heated to fumes of sulfuric acid. After the addition of a little nitric acid, the solution is refumed. To the cool solution 25 ml. of titanous sulfate solution (4 g. of titanium dissolved in water and 25 ml. of sulfuric acid and diluted to 1 liter) is added, followed by 20 ml. of hydrochloric acid and 10 drops of hydrofluoric acid. After heating on a water bath (80 to 90°C.) for 5 minutes, 10 ml. of dithiol solution and 10 more drops of hydrofluoric acid are added and the heating is continued for an additional 15 minutes. Ten more drops of HF are added and the solution is cooled and transferred to a 125-ml. separatory funnel, using 2 to 3 ml. of carbon tetrachloride to effect the transfer. Then 7 ml. of carbon tetrachloride is added and the solution is shaken for 2 minutes. The organic layer is transferred into a dry 25-ml. flask. The aqueous layer is washed with two 5-ml. portions of carbon tetrachloride and the washings are added to the volumetric flask. The solution is diluted to volume with carbon tetrachloride and mixed, and its absorbance is measured at 640 mμ against the reagent blank carried through all steps of the procedure.

The method, as described, is suited for the determination of 50 to 1000 p.p.m. of both molybdenum and tungsten, but the scope of the procedure can be extended by suitably varying sample weight, aliquot size, amounts of reagents used, and light path.

Thiocyanate Procedure

Another procedure has been described that is based on the determination of both molybdenum and tungsten as the thiocyanates (538). The bulk of the tantalum or niobium is removed as insoluble sodium niobate or tantalate by fusing the oxides, obtained by ignition of 2 g. of metal, in a nickel crucible with sodium hydroxides, leaching the melt in water, and filtering. To an aliquot of the solution, 10 ml. of ammonium thiocyanate (50% w./vol.) and 5 ml. of oxalic acid (2% w./vol. in 20% HCl) are added, and the solution is acidified with 8 ml. of sulfuric acid (1:2). After the addition of 1 ml. of ferric sulfate solution (1 ml. = 0.25 mg. of Fe), the cold solution is transferred to a separatory funnel, 25 ml. of MIBK saturated with 2% (w./vol.) of SnCl$_2$ solution and 2 ml. of hydrazine hydrochloride (5% w./vol.) are added, and the solution is shaken for 1 minute. The aqueous phase contains the tungsten and is transferred to another separatory funnel containing 1 g. of oxalic acid. After washing the organic phase in succession with 20% hydrochloric acid and 2% oxalic acid in 20% hydrochloric acid, the organic phase is transferred to a 25-ml.

volumetric flask, diluted to volume with MIBK, and its absorbance measured at 470 mµ (molybdenum).

The aqueous solution containing the tungsten is cooled to below 10°C. and 55 ml. of a saturated stannous chloride solution in 1% HCl is added. After 20 minutes of standing, 25 ml. of MIBK is added, the solution shaken, the aqueous layer discarded, and the organic layer washed in succession with 2% oxalic acid in 20% HCl, then with 3% oxalic acid in 35% HCl. The organic layer is transferred to a 25-ml. volumetric flask, diluted to volume with MIBK, and the absorbance of the solution is measured at 420 mµ (tungsten).

e. Determination of Silicon and Phosphorus (538)

The sample (1 g.) is ignited in a platinum crucible to the oxide, which is fused in a nickel crucible with NaOH (4 g.); the melt is then leached in NaCl solution (1% w./vol.) in polyethylene. The diluted solution is filtered (plastic funnel) into a 100-ml. polyethylene volumetric flask.

For the determination of silicon, a 10-ml. aliquot is transferred to a separatory funnel (polyethylene pipette) containing 60 ml. of a purified molybdate solution (mixture of 70 ml. of ammonium molybdate (10%), 350 ml. of H_2O, 35 ml. of HCl (1:1), and 1 drop of H_2O_2, extracted with 50 ml. of MIBK). After 5 minutes of standing, 10 ml. of oxalic acid (5% w./vol.)–HCl (3M) mixture is added, followed by extraction with MIBK (25 ml.). The aqueous phase is discarded and the molybdenum blue color is developed by the addition of hydrazine hydrochloride (10 ml. of 1% (w./vol.) solution in 10% HCl) and stannous chloride (10 ml. of 1% (w./vol.) in 10% HCl). The aqueous phase is discarded and the MIBK layer is transferred to a 25-ml. volumetric flask and is diluted to volume with the same solvent; the absorbance then is measured at 800 mµ in a 1-cm. cell. The range is 2 to 40 p.p.m. of Si.

For the determination of phosphorus, another aliquot of the alkaline solution is transferred to a separatory funnel containing 20 ml. of purified molybdate solution (mixture of 100 ml. of ammonium molybdate (10% w./vol.), 100 ml. of HCl, 8M, 1 drop of H_2O_2, extracted with 50 ml. of MIBK). After 5 minutes, the phosphomolybdate is extracted with 25 ml. of MIBK. The aqueous phase is discarded and in the organic phase the molybdenum blue color is developed with hydrazine hydrochloride (10 ml. of 1% solution in 10% HCl) and stannous chloride (10 ml. of 1% solution in 10% HCl). The aqueous phase is discarded and the MIBK layer is transferred to a volumetric flask and diluted to volume with the same solvent; the absorbance is measured at 800 mµ in a 1-cm. cell. The range is 10 p.p.m. of P and higher.

f. Determination of Zirconium

For amounts of zirconium within the concentration range of 250 to 2000 p.p.m., an ASTM subgroup of the Columbium Task Force is investigating a method based on the adsorption of niobium and simultaneous elution of the zirconium by Dowex-1 resin, using a 5:4:11 mixture of HCl–HF–H_2O (28). Subsequently, the zirconium is recovered by a cupferron precipitation. The ignited precipitate is dissolved in hydrofluoric acid and is evaporated to fumes with perchloric acid. The solution is transferred to a 100-ml. volumetric flask and diluted with water. The zirconium color is then developed by the addition of chloranilic acid, the absorbance being measured in 5-cm. cells at 540 mμ. The above procedure is inadequate for amounts of zirconium below 250 p.p.m.

The following procedure has been extensively used in the author's laboratory (279), and involves the identical ion-exchange procedure used in the chloranilic method and the collection of zirconium with cupferron, using a few milligrams of iron as a collector. The precipitate is fumed to dryness with nitric and perchloric acids, and zirconium is determined by the TOPO–pyrocatechol–pyridine procedure (579). For the determination of larger amounts of zirconium, as in the niobium–1% zirconium alloy, the oxides obtained by ignition of the metal can be fused in potassium carbonate. After filtration of the aqueous solution of the melt, the precipitate is ignited, fused in potassium bisulfate, the melt dissolved in dilute sulfuric acid containing hydrogen peroxide, and the zirconium precipitated as the phosphate.

g. Determination of Titanium

An ASTM subgroup of the Columbium Task Force is evaluating a procedure using sulfosalicylic acid as chromogenic reagent (27). In this procedure, the sample (0.5 g.) is decomposed with nitric and hydrofluoric acids and evaporated to fumes with 25 ml. of sulfuric acid. After refuming with the intermittent addition of water, the cool, clear solution is transferred to a dry 50-ml. volumetric flask, using H_2SO_4 to effect the transfer. Ten ml. of sulfosalicylic acid solution (1% w./vol. in H_2SO_4) is added, the solution is diluted to volume with H_2SO_4, and is mixed. After one hour of standing, the absorbance is read in 5-cm. cells at 450 mμ against a reference solution containing all the reagents. The above procedure covers the concentration range of 10 to 150 p.p.m. and is not subject to interference from elements in the amounts listed in the Specifications in Section VIII. Since niobium contributes slightly to the color with sulfosalicylic acid, for very accurate work a titanium-free niobium oxide (see Section IX-D-2-a for preparation) should be used to prepare the reference solution.

The determination of titanium by the hydrogen peroxide is useful only for larger amounts of titanium, since the niobium peroxide complex itself contributes slightly to the color. For amounts of titanium below 20 p.p.m. and as low as 2 p.p.m. the TOPO–thiocyanate procedure (580) has been successfully used in the author's laboratory after adsorbing niobium and tantalum on Dowex-1 resin in a 5:4:11 medium of HCl–HF–H_2O and recovering the titanium in the eluate with cupferron, using zirconium as a collector.

h. Determination of Cobalt

A simple method consists in solution of the sample in HF + HNO_3, extraction of the thiocyanate complex with MIBK in a medium buffered with ammonium acetate, and measuring the absorbance at 625 mμ (538). Iron does not interfere in this procedure since it remains complexed as the fluoride. The niobium thiocyanate complex is not formed at this low acidity. If a 1-g. sample is used, the sensitivity of the method is 10 p.p.m.

i. Determination of Nickel

After decomposition of the sample with nitric and hydrofluoric acids, the solution is neutralized with ammonia in the presence of tartaric acid and made alkaline with sodium bicarbonate. Dimethyl glyoxime solution is added and the nickel complex is extracted with chloroform, then is back-extracted with citric acid solution. Upon the addition of bromine water, ammonia, and dimethyl glyoxime, the nickel complex is formed, which is finally measured at 475 mμ (538). If a 1-g. sample is used the sensitivity of the method is 10 p.p.m.

j. Determination of Copper

After decomposition of the sample with nitric and hydrofluoric acids, dilution with tartaric acid, neutralization with ammonia, and reduction with ascorbic acid, the neocuproïne complex of copper is formed and extracted with chloroform. Its absorbance is measured at 457 mμ (538). If preferred, the sodium diethyl dithiocarbonate method can also be used for the photometric determination of the copper (239). Using a 1-g. sample the sensitivity of the method is 5 p.p.m.

k. Determination of Boron

The sample is decomposed in a quartz flask, which is equipped with a reflux condenser, by heating at a low temperature with sulfuric acid containing ammonium sulfate (538). After the decomposition is complete, the boron is distilled in the usual way with methyl alcohol and is determined photometrically with curcumin or carminic acid. Using a 2-g. sample the sensitivity of the method is 0.5 p.p.m. of boron.

l. Simultaneous Determination of Lead, Copper, Zinc, and Cadmium

The following polarographic procedure provides for the determination of Cu (2 p.p.m.), Pb (4 p.p.m.), Cd (2 p.p.m.), and zinc (2 p.p.m.), using a 5-g. sample (538). The sample is decomposed in a platinum dish with nitric and hydrofluoric acids and the solution is evaporated to a syrupy consistency. The solution is transferred to an Erlenmeyer flask containing 150 ml. of tartaric acid solution (20% w./vol.). After adjustment of the pH to 6.5 with ammonia, the solution is transferred to a separatory funnel and is extracted with 20 ml. of dithizone solution (0.1% w./vol. in chloroform). The organic layer is transferred to a beaker and is held. The pH of the aqueous solution is raised to 8.5 with ammonia and a second dithizone extraction is carried out. The organic layer is combined with the first one and is evaporated with nitric and sulfuric acids to destroy completely all organic matter. The dry salts are dissolved in 2 to 3 ml. of hydrochloric acid and 2 ml. of ammonium chloride solution (50% w./vol.). The solution is evaporated in a small dish until salts crystallize. The salts are washed with water into a 5-ml. flask and the solution is neutralized with ammonia and rendered acid with 2 drops of hydrochloric acid. After the addition of two drops of hydroxylamine hydrochloride (15% w./vol.), the solution is boiled for a minute and cooled; 0.5 ml. of ammonium acetate (30% w./vol.) and 0.5 ml. of sodium carboxy methyl cellulose solution (0.1% w./vol.) are added, and the solution is diluted to volume in a 5-ml. volumetric flask, mixed, and polarographed.

m. Determination of Tin and Antimony

The following polarographic method covers the determination of 2 p.p.m. of antimony and 3 p.p.m. of tin (538). The sample (5 g.) is decomposed with nitric and hydrofluoric acids and the solution is evaporated until incipient crystallization. The salts are washed into a solution containing 200 ml. of tartaric acid (2.5% w./vol.). After the addition of a few milligrams of arsenic, the sulfides are precipitated with H_2S filtered and dissolved in a few milliliters of hot sodium sulfide solution. The solution is filtered, oxidized with hydrogen peroxide, and evaporated until salts crystallize out. They are dissolved in 2 ml. of hydrochloric acid, the solution is transferred to a 5-ml. flask, and 0.5 ml. of sodium hypophosphite solution (50% w./vol.) is added, followed by one drop of a 5% (w./vol.) solution of mercuric chloride. After heating on the water bath to precipitate the arsenic, the solution is cooled, diluted to volume in a 5-ml. flask and, is polarographed after the addition of 0.5 ml. of sodium carboxy methyl cellulose solution,.

n. Determination of Lead and Cadmium (402)

For the determination of zero to 6 p.p.m. of lead and/or cadmium, the sample (1 g.) is decomposed with HF and HNO_3 and evaporated to incipient crystallization. The salts are dissolved in 1 to 2 ml. of HF, and washed into an alkaline potassium tartrate solution. After the addition of 2 ml. of potassium cyanide (2% w./vol.) and 5 ml. of sodium diethyl dithiocarbonate solutions (1% w./vol.), Bi, Cd, Pb, and Tl are extracted with chloroform. Pb and Cd are back-extracted with $0.2N$ HCl. To determine lead, one-half of the solution is treated with 2.5 ml. of $0.2N$ HCl, 2 ml. of $4.5N$ NH_4OH, 0.4 ml. of 5% (w./vol.) KCN, and 0.5 ml. of 20% (w./vol.) hydroxylamine, and then is shaken with 1 ml. of 0.002% (w./vol.) dithizone in chloroform. The color of the extract is compared with that of standards containing zero to 3γ of lead. To determine cadmium, the other half of the aqueous extract is mixed with 1.5 ml. of 10% (w./vol.) NaOH solution, 0.1 ml. of 0.5% (w./vol.) KCN solution, and 1 ml. of 0.002% (w./vol.) dithizone solution. The color of the extract is compared with standards containing 0 to 3γ of cadmium.

o. Determination of Sodium and Potassium (538)

The method is undoubtedly suitable also for the determination of lithium. The sample (1 g.) is decomposed in a platinum dish with HF and HNO_3 and evaporated to strong fumes with 15 ml. of H_2SO_4. In order to facilitate the complete removal of fluorides, the solution is taken up in water and refumed. To hydrolyze the earth acids completely, the cold salts are taken up in 50 ml. of water, washed into a 250-ml. volumetric flask, and boiled after the addition of 40 ml. of saturated SO_2 water. The cold solution is diluted to volume, mixed, and allowed to stand overnight. Portions of the supernatant clear solution are then examined by the usual flame spectroscopic procedures.

p. Determination of Arsenic

A procedure has been described that is based on the precipitation of magnesium ammonium arsenate (following decomposition of the sample in hot H_2SO_4 containing K_2SO_4 and solution of the salts in ammonium tartrate), using phosphate as collector (403). The precipitate is dissolved in sulfuric acid, the arsenic is liberated with zinc as arsine, and finally is determined by exploiting the arsenomolybdenum blue complex.

IX. SELECTED LABORATORY PROCEDURES

In the preceding sections an attempt was made to review various techniques and procedures in depth and to strengthen the presentation with up-to-date references. It should thus be possible for the practicing analytical chemist to find additional details in the original sources.

It was, therefore, felt that the scope of this section can be limited to the presentation of details of three basic procedures. One covers the analysis of niobium- and tantalum-bearing ores and concentrates; another deals with the analysis of steels and high-temperature alloys containing niobium and/or tantalum. The third describes the determination of impurities in niobium and tantalum metals by spectrographic techniques.

A. THE DETERMINATION OF NIOBIUM AND TANTALUM IN ORES AND CONCENTRATES INVOLVING ION EXCHANGE (282)

1. Apparatus and Reagents

The columns are made of polystyrene and are 15 inches long and 1 inch in inside diameter. If a number of determinations are carried out, it is desirable to arrange the columns so that a number can be operated with a minimum of attention. Plastic columns suitable for such an assembly were developed by Ledoux & Company and are equipped with "Dole-type" fittings of polystyrene. Inlet and outlet tubes are of polyethylene. Flexible connections are made of Tygon tubing. The flow of solutions can be controlled by hosecock clamps on these flexible connections. An assembly of this type is illustrated in Fig. 18.

Resin-Dowex-1, 100 to 200 mesh, 8 to 10% divinyl-benzene cross-linkage is used. The resin as received from the manufacturer is transferred to a polyethylene beaker and is washed several times with 200 ml. $3M$ nitric acid with intermittent decantations. Subsequently, the resin is washed with water and with hydrochloric acid and hydrochloric–hydrofluoric acid mixtures. It is finally transferred to the columns, containing a circle of Orlon or equivalent HF-resistant cloth, until a settled resin bed of 10 inches has been obtained. Before the sample is introduced into the columns, the resin is treated several times with the 5:4:11 HCl–HF–H_2O mixture.

Fig. 18. Multiple arrangement of ion-exchange columns in the laboratory of Ledoux & Company, Teaneck, New Jersey.

Reagents 5:4:11 mixture. Add 250 ml. of hydrochloric acid ($12M$) to 300 ml. of water, then add 200 ml. of hydrofluoric acid ($24M$) and dilute to 1 liter with water.

NH_4Cl–HF solution. Dissolve 140 g. of ammonium chloride in 500 ml. of water. Add 40 ml. of hydrofluoric acid and dilute to 1 liter with water.

NH_4Cl–NH_4F solution. Dissolve 140 g. of ammonium chloride in 500 ml. of water. Add 40 ml. of hydrofluoric acid. Adjust the acidity to a pH of 5.5 with ammonium hydroxide and dilute to 1 liter with water.

Cupferron solution. Dissolve 6 g. of cupferron in 80 ml. of cold water. Filter and dilute to 100 ml. Prepare fresh and keep cooled to 5°C.

2. Procedure

a. Decomposition of Sample

HF–HCl attack. Transfer 0.5 to 1 g. of high-grade material to a 200- to 400-ml. polyethylene beaker. Add 25 ml. of hydrochloric acid, 20 ml. of hydrofluoric acid, and cover with a thin sheet of polyethylene wrap which is held tightly to the sides of the beaker with a rubber band. Place on a steam bath and heat for several hours, with occasional mixing of the solution by swirling until the sample is completely decomposed. Disregard any undissolved cassiterite and/or zircon. Remove the wrap, add 55 ml. of water, and introduce the solution into the column.

Sodium bisulfate fusion of sample. This attack is as efficient as the HF–HCl decomposition. However, it is more time-consuming and requires more attention.

Transfer sodium bisulfate (fused) to a 100-ml. porcelain crucible. Heat until effervescence stops. Pour the molten salt into a quartz crucible. About 10 g. is ample for a 0.5- to 1-g. sample. Transfer the sample on top of the cold bisulfate. Heat the covered crucible, first with a small flame, finally for about half an hour with the full heat of a Bunsen burner. Allow the fusion to cool and solidify. Loosen the melt by a light tapping of the crucible and transfer to a polyethylene beaker. Rinse the quartz crucible and cover with the 5:4:11 mixture and add to the polyethylene beaker. Add sufficient 5:4:11 mixture to provide approximately 100 ml. of volume. Cover with a polyethylene wrap, as was indicated above, and place on a steam bath until the melt has completely dissolved. Then transfer the cooled solution onto the resin bed.

Sodium peroxide fusion of slags, residues, and other low-grade substances, as well as simpsonite. Fuse 0.5- to 2-g. sample in a nickel crucible in the usual way with 5 to 10 g. of sodium peroxide. Leach the cooled melt in a polyethylene beaker with 50 ml. of water. Rinse the crucible with hot water and add to the main solution. Carefully neutralize with $18N$ sulfuric acid to a pH of 5 to 7. Dilute with water to 110 ml., add 50 ml. of hydrochloric acid and 40 ml. of hydrofluoric acid, and heat on a steam bath until a clear solution results. If a black residue of nickel oxide persists, add a few drops of hydrogen peroxide and continue heating until the solution is clear. Transfer the cooled solution onto the resin bed.

If preferred, acidify the aqueous solution of the melt with hydrochloric acid, precipitate the impure rare earth acids with ammonia, and dissolve the precipitate in the 5:4:11 mixture.

b. Elution of Impurities

Wash the beaker with 5:4:11 mixture and transfer the washings to the column, allowing the solution to drain to the top of the resin each time. Finally wash down the sides of the column with the same solution and allow 300 ml. of eluate to pass through the resin bed. This may be done manually by adding from time to time new eluant, allowing the solution to drain to the top of the resin each time, or by automatic delivery (Fig. 18).

The flow rate of the column is about 100 to 125 ml. per hour. To reduce the evaporation of HCl and HF, the eluate may be received conveniently in polyethylene bottles (such as empty HF bottles). Discard the eluate unless determinations of other constituents of the sample are desired.

c. Elution of Niobium

Replace the beaker or flask with a 400- to 600-ml. polyethylene beaker and pass through the column 350 ml. of NH_4Cl–HF solution, maintaining the technique and flow rate described before. The eluate contains the niobium, contaminated possibly by traces of tin and/or antimony.

d. Elution of Tantalum

Again change beakers and pass 350 ml. of NH_4Cl–NH_4F eluant through the resin bed. The eluate contains the tantalum, possibly contaminated with bismuth.

e. Determination of Niobium

If the sample contains considerable cassiterite, a small amount of tin is rendered soluble by the HF–HCl attack or by the $NaHSO_4$ fusion. If a Na_2O_2 fusion was employed, all the tin is introduced into the column and significant amounts may accompany the niobium. Antimony need only be considered in the case of the rare stibiotantalite. It may partly dissociate in the column and accompany the niobium.

Unless the absence of tin and/or antimony is known or established, it is therefore always advisable to remove these two elements by a hydrogen sulfide separation. (If a photometric determination of the niobium is intended, tin need not be removed (188).) Add to the niobium eluate 15 g. of tartaric acid and 15 g. of boric acid. Stir until dissolved. Pass a brisk stream of hydrogen sulfide into the cold solution for 15 minutes. If necessary, filter and wash with cold 5% (vol./vol.) sulfuric acid containing a little tartaric acid and hydrogen sulfide. Discard the precipitate. Expel the hydrogen sulfide by gentle boiling. Cool the solution to below 10°C., add 60 ml. of hydrochloric acid followed by 60 ml. of cupferron and some filter pulp, and filter, preferably using mild vacuum. If the niobium content is known or suspected to be below 5 mg., use 10 mg. of zirconium in the form of zirconium sulfate and only 10 ml. of cupferron solution to collect the niobium. Wash the precipitate thoroughly with 5% (vol./vol.) hydrochloric acid containing a little cupferron. Ignite in a tared platinum dish or crucible, first at a low temperature, and finally to constant weight at 1100°C., and weigh as Nb_2O_5. If a photometric determination of the niobium is desired, the ignited precipitate can be treated as described in Section VII-B-2 or 3 (227).

f. Determination of Tantalum

Only in the rare case that the sample contains bismuth is it necessary to carry out an H_2S separation.

In the absence of bismuth, add to the eluate containing the tantalum 60 ml. of hydrochloric acid and 15 g. of boric acid. Stir until dissolved. Precipitate the tantalum with cupferron, filter, wash, and ignite as described for niobium. Weigh as Ta_2O_5. If a photometric determination of the tantalum is desired, the ignited precipitate is treated as described in Section VII-B-8 (227).

Determination of impurities. Since the first eluate contains various constituents of the original sample (Fe, Ti, W, Mo, U, and Mn, as well as Zr if a Na_2O_2 attack was used), it can be employed for the determination of one or the other elements. Some can be collected with cupferron after complexing the hydrofluoric acid with boric acid. Others are more conveniently determined after expulsion of the hydrochloric and hydrofluoric acids by evaporation with sulfuric acid.

g. Regeneration of Resin

After the elution of tantalum the resin can be readied for the next sample by passing two column volumes (about 75 ml.) of the 5:4:11 mixture through the column. To avoid contamination of the resin with undecomposed cassiterite and zircon, it is advisable to prevent the transfer of these insoluble substances onto the column. In extreme cases the solution can be introduced onto the resin bed by filtration. There is virtually no limit to the number of separations that can be carried out with the same resin. The columns illustrated in Fig. 18 have been in uninterrupted use since 1954, and each columnh as accommodated more than 600 samples.

B. THE DETERMINATION OF NIOBIUM AND TANTALUM IN STEELS AND HIGH-TEMPERATURE ALLOYS

The principles underlying the determination of niobium and tantalum in various alloys containing numerous elements in virtually unlimited combinations and proportions are similar to those described in the preceding procedure. The aim is to obtain a solution containing HCl, HF, and H_2O

in a 5:4:11 ratio. When this solution is transferred onto a resin bed of Dowex-1, the following elements among many others are not adsorbed by the resin and are therefore separated from niobium and tantalum: Fe, Ni, Co, Cr, W, Mo, Ti, Zr, Hf, Al, Cu, V, and U. The separation of niobium from tantalum is carried out as described in the ore procedure.

Decomposition of the sample. Decompose steels and high-temperature alloys in polyethylene beakers with a 10:1 mixture of hydrochloric and nitric acids (sample size depends on the concentration of niobium and/or tantalum and on whether the final measurement is by a photometric or gravimetric procedure). After decomposition is complete, add sufficient hydrofluoric acid (1 to 5 ml.) to prevent the hydrolysis of niobium, tantalum, tungsten, or molybdenum. Evaporate the solution to dryness on a steam bath and remove residual nitric acid by adding 5 ml. of hydrochloric acid and 3 ml. of hydrofluoric acid, and by repeating the evaporation.

Dissolve the salts in a 5:4:11 mixture of HCl, HF, and H_2O and continue as outlined in detail in the ore procedure.

C. DETERMINATION OF NIOBIUM AND TANTALUM IN NIOBIUM-, TANTALUM-, TITANIUM-, TUNGSTEN-, OR MOLYBDENUM-BASE ALLOYS

Decompose these alloys in a platinum or polyethylene container with HF and HNO_3, remove the latter by repeated evaporations with hydrofluoric acid, and dissolve the residue in the 5:4:11 mixture of HCl, HF, and H_2O.

D. SPECTROGRAPHIC DETERMINATION OF IMPURITIES IN NIOBIUM AND TANTALUM

In the following, one procedure is described in detail. Although it largely represents the efforts of one organization (555), it has been routinely applied to a great number of samples. Naturally, it can be modified to suit varying requirements.

The procedure employs the d.-c. arc source, samples in the form of the pentoxides, ammonium chloride as a carrier, and cobalt as an internal standard. The oxide form is ideal for two reasons. First, homogeneous synthetic standards can be prepared easily by doping the pentoxide with either solutions or dry oxides of the impurities. Second, niobium or tantalum, practically in any form, may be converted readily to the pentoxide. Samples of metal (powders, turnings, drillings) may be directly ignited in air. Carbides, hydrides, and nitrides also can be converted in this way. Other compounds such as the pentahalides, the complex fluotantalates, and the oxyfluoniobates may be precipitated from solutions with ammonium hydroxide; the resulting precipitate can then be ignited to the oxide.

The use of ammonium chloride as carrier is suggested by the fact that some of the more important impurities in niobium and tantalum compounds are also refractory. The use of a buffer that would suppress, for instance, the lines of tantalum would also suppress niobium, zirconium, and titanium. The use of ammonium chloride, on the other hand, results in little or no suppression of the line spectra. At the same time, it carries a uniform amount of sample into the arc gap and the intensity of the spectra produced is quite uniform and reproducible. Thus ammonium chloride provides a medium for mechanically carrying pentoxide into the arc gap as well as a chloride medium that may lead to the formation of the volatile pentachlorides of niobium and tantalum.

1. Tantalum

a. Preparation of High-Purity Tantalum Oxide

The ideal matrix material is Ta_2O_5 with no detectable amounts of any impurities. Solvent-extraction methods are best suited for the preparation of such ultrapure oxide. The following procedure, a variation of techniques described in Section IV-J-3, is based on the extraction of tantalum from a concentrated fluoride solution of the metal with methyl isobutyl ketone. Trace impurities accompanying the tantalum are then removed from the ketone with a series of sulfuric acid washes.

One hundred g. of tantalum metal powder is dissolved in a platinum or plastic container by small additions of the metal to 100 ml. of concentrated HF. To the filtered solution (polyethylene funnel) in a polyethylene bottle, 200 ml. of methyl isobutyl ketone is added and the solution is vigorously shaken for about one minute. The organic phase is decanted off and the aqueous phase containing the impurities is discarded. The organic phase is now washed with 10 ml. of $6M$ H_2SO_4 and the aqueous phase is discarded. After dilution of the organic phase to 200 ml. with MIBK, three more washings with 10-ml. portions of $6M$ H_2SO_4 serve to remove the last traces of impurities. The last traces of niobium are removed by adding 5 ml. of HF and by three further washings with $3M$ H_2SO_4.

The tantalum is precipitated in the organic phase in a platinum or polyethlene container with ammonia. The precipitate, after washing with deionized water, is finally ignited to Ta_2O_5.

b. Preparation of Synthetic Standards

A master blend of primary standards now can be prepared from the pure tantalum oxide by the addition of the oxides or solutions of each metallic impurity to be determined. Then, by suitable dilution of the blend with pure Ta_2O_5, a series of standards can be prepared covering a broad range of concentrations.

The elements added as standardized solutions are Mn, Fe, Al, Ca, Ni, Cu, and Sn. It is more convenient to add the oxides of the following elements: Nb, W, Zr, Ti, Mo, and Si. Separate standards should be prepared to cover the higher percentages of any of the elements, another set for the boron, and an additional set for sodium and potassium. The following technique was found advantageous.

1. Appropriate weights of the dry oxides are added to a convenient weight of Ta_2O_5 (30 g.) and are thoroughly mixed in an agate mortar or mechanically in large plastic vials.

2. To the mixed oxides now transferred to a platinum evaporating dish are added, one at a time, the appropriate volumes of standard solutions of the rest of the elements. They are added in such a way that the liquid does not come in contact with the sides of the dish. All solutions should be in a hydrochloric acid medium. Nitric acid cannot be tolerated because of the formation of aqua regia and the subsequent attack on the platinum. The mixture is dried after each addition under an infrared lamp and, after the final addition, is dried for one hour.

3. After an ignition at low heat, the blend is returned to the plastic vial and is blended for 15 minutes to ensure homogeneous distribution of all elements.

4. Separate weighed portions of the master blend are diluted with pure Ta_2O_5 to serve as final standards. A 0.20% Nb_2O_5 in the master blend, for instance, will provide standards of 0.18%, 0.16%, ..., 0.02%, 0.01% by dilutions of 9:1, 8:2, ..., 1:9, 0.5:9.5. In this method, possible weighing errors in one of the individual standards would be evident on the analytical curve.

c. Preparation of Carrier—Internal Standard

Each standard, and later the sample, is blended with an equal weight of NH_4Cl–$CoSO_4$ carrier, which is prepared as follows: 2 g. of pure cobalt metal is dissolved in 60 ml. of $6M$ sulfuric acid, any precipitable impurities are removed with ammonia, and the filtrate, after acidification with sulfuric acid, is diluted to 1000 ml. in a volumetric flask. A 20-ml. aliquot is evaporated to dryness in a glazed porcelain casserole and is oven-dried. The dry salts are removed with a platinum spatula and are pulverized with 19 g. of pulverized, reagent-grade NH_4Cl.

d. Preparation of Samples

The following compounds and various metallic forms of tantalum are readily converted to Ta_2O_5 by ignition: tantalic acid, hydrides, carbides, nitrides, tantalum metal powder, foil, and fine wire. Platinum crucibles should always be used, since residual fluorides are present in some tantalum materials; these would etch porcelain or silica crucibles. Samples of millings, turnings, and bar croppings require a longer ignition period. A flow of oxygen over the crucibles will facilitate oxidation. It may be necessary in the case of more massive samples to cool the crucible, pulverize off the oxide already formed, and reignite in order to insure complete oxidation.

The most common tantalum salt, K_2TaF_7, cannot be ignited directly to the oxide. This compound will, however, easily dissolve in a platinum or plastic dish in hot, dilute hydrofluoric acid. From this solution hydrated Ta_2O_5 and certain impurities present in the salt can be precipitated by the addition of an excess of ammonia. The precipitate is washed by decantation, with intermittent additions of 2% NH_4Cl solution. The precipitate is next centrifuged in a polyethylene tube and then is transferred to the platinum dish for the ignition to Ta_2O_5.

Anhydrous tantalum pentachloride ($TaCl_5$) or the oxychloride ($TaOCl_3$) react vigorously with concentrated HCl. The suspension of the oxides is diluted with water and precipitated with ammonia, followed by ignition to the oxide.

Solution samples of TaF_5, such as those encountered in organic extractions, are treated with excess ammonium hydroxide and then handled in the same manner.

The Ta_2O_5 obtained from any of these various types of samples is then blended with an equal weight of carrier–internal standard. Moving camera studies have shown that boron and silicon, however, can be detected at lower levels when the oxide is arced directly without carrier. A longer arcing period is required in this case.

The tantalum pentoxide obtained from the direct oxidation of metallic tantalum differs physically from the pentoxide obtained by ignition of the chemically precipitated tantalic acid. The metal yields an oxide that is coarser and denser. Even after long pulverizing it still exhibits arcing characteristics different from those of the "chemical" oxide. The "metallic" oxide produces a more intense spectrum when arced either directly or with a carrier. Burn-off patterns obtained on direct-reading spectrographs illustrate this difference even more clearly.

The body-centered cubic structure of tantalum metal more closely resembles the pseudohexagonal form of Ta_2O_5 than does the tantalic acid. Oxygen atoms can be added without much disruption of the metal structure, forming the dense oxide. When tantalic acid is ignited, however, bonds with H_2O, NH_4F, etc., must be broken before the pentoxide forms. Such severe action causes this pentoxide to be a fine, impalpable powder.

These conditions necessitate the preparation of separate sets of standards. Metal powder standards may be prepared by adding the various impurity elements to an oxide obtained by ignition of an especially pure tantalum powder. Alternatively, less-pure powder may serve as secondary standards. Such powders are first converted to chemically precipitated Ta_2O_5 by dissolving the powder in hydrofluoric acid, and subsequently by precipitating with NH_4OH and thioacetamide and igniting in platinum. Values of the impurity levels in powders are then assigned by analyzing them spectrographically against the chemically precipitated primary standards. In practice, the directly calcined powders, with a range of values, are used in preparing the analytical curves for all powder samples. In a similar manner, standards for sintered or arc-cast metal should be the oxides obtained by the direct calcining of metal thus prepared. Values for such standards are again obtained by an analysis of the metal dissolved, precipitated, and compared with the primary standards.

Sodium and potassium lines are very difficult to measure to low levels in the complex tantalum spectrum. High concentrations of these lines, furthermore, seriously suppress the spectra both of the matrix and of the other impurities. A chemical separation of sodium and potassium from the tantalum matrix can be effected by the volatilization of tantalum as the pentachloride (see Section IV-D-3). Equipment for accomplishing this need not be elaborate. A one-foot length of Vycor tube, containing 5 g. of sample in a porcelain boat, forms a convenient combustion tube. Chlorine is fed from a cylinder into an H_2SO_4 scrubbing bottle and then into the combustion tube, maintained just below 700°C. Chlorine is suitable for metallic tantalum, but volatilization of the Ta_2O_5 requires sulfur monochloride (S_2Cl_2), formed through the reaction of chlorine with solid sulfur (see Section IV-D-3).

Sodium and potassium present in the original sample remain as chlorides in the boat (as do $MgCl_2$, $CaCl_2$, and rare earth chlorides). After the residue has cooled, 5 ml. of warm $6N$ HCl is added to dissolve the residue. One ml. of the standard $CoSO_4$ is added and the solution is transferred to a microbeaker and diluted to 10 ml. with water. Aliquots of the solution are added to spectrographic electrodes, dried under an infrared lamp, and then are arced directly.

e. Spectrographic Conditions

Spectrographs that cover the wavelength region of 2400 to 4100 A. in the first order and that have a dispersion of at least 5 A./mm. must be employed to record all the line pairs listed below. Some spectrographs can be conveniently used only up to 3700 A., which makes the most sensitive lines of niobium, tungsten, and potassium unavailable for measurements. Suitable electrodes are high-purity graphite rod, $1/4$ inch I.D. A cathode with a conical tip arcs well; the anode has a $3/16$-inch deep cup; a $1/32$-inch hole drilled horizontally through the cup well into the cavity serves to vent some of the pressure built up by the NH_4Cl during arcing. Electrodes

TABLE XXV
Analytical Line Pairs and Impurity Limits in Tantelum (First-Order Spectrum)

Line pairs, A.	Concentration, range, %
Arced 10 seconds with carrier	
Nb 4058.9/Ta 4058.5	0.01 – 0.20
Nb 3225.5/Ta 3245.3	0.10 – 1.00
Nb 2883.2/Ta 2890.0	1.00 – 5.00
Nb 4100.4/Ta 4105.0	4.00 –20.00
Mn 4031.5/Co 3997.9	0.001– 0.05
W 4008.9/Co 3997.9	0.01 – 0.50
Ca 3968.5/Co 3997.9	0.001– 0.05
Al 3961.5/Co 3997.9	0.001– 0.05
Fe 3734.9/Co 3997.9	0.003– 0.30
Ni 3392.9/Co 3442.9	0.001– 0.05
Zr 3391.9/Ta 3403.0	0.01 – 0.20
Ti 3372.8/Ta 3309.8	0.01 – 0.50
Ti 3322.9/Ta 3325.7	0.10 – 1.00
Cu 3247.5/Co 3442.9	0.005– 0.10
Mo 3194.0/Ta 3189.7	0.003– 0.30
Sn 3175.0/Ta 3189.7	0.01 – 0.20
Si 2516.1/Ta 2546.8	0.005– 0.20
Si 2514.3/Ta 2532.1	0.10 – 1.50
Si 2435.2/Ta 2429.7	0.40 – 5.00
Arced 25 seconds without carrier	
B 2497.7/Ta 2488.7	0.0001–0.05
Arced 10 seconds as dried solution of nonvolatile chlorides without carrier	
K 4044.1/Co 3409.2	0.01 – 0.50
Na 3302.3/Co 3409.2	0.01 – 0.50

must be prearced for about 20 seconds; this purifies the graphite and also results in a more stable arcing of the sample. Electrodes can be conveniently filled by gently tamping them into the oxide–carrier blend. A small hole pinched vertically through the packed charge with a fine tantalum or platinum wire causes a more even burning of the arc.

A direct-current arc (220 v. adjusted to deliver 10 to 20 amp.) will furnish the necessary sensitivity. Exposure times of 10 to 15 seconds, along with suitable light-intensity controls, will result in spectra of proper line density. A slit width of 40 μ or less will satisfactorily resolve most of the sensitive impurity lines. Emulsions sensitive in the ultraviolet wavelength region will cover this entire tantalum range. Normal procedures are used for processing, measuring, and calibrating the film. The analytical line pairs are recorded in Table XXV.

2. Niobium

The techniques for analyzing niobium materials parallel closely those employed for tantalum. The pentoxide is again ideally suited as the common starting point for both standards and samples. Ammonium chloride serves as carrier for the d.-c. arc procedure. Because of certain spectral interferences, it has been found necessary to use the second-order ultraviolet spectra. Thus it is impossible to obtain adequate resolution in the first-order spectrum to determine zirconium, tungsten, titanium, and other impurities. Particularly, it is impossible to measure tantalum in the first-order niobium spectrum lower than about 0.15%. In the second order the sensitivity limit of tantalum is approximately 0.10%. Theoretically, the lower limit that can be observed in the second order should be less than that obtained in the first order. However, the limiting amount of an element that can be observed is reached when its most sensitive line cannot be distinguished above the continuous background. In the second order, resolution of lines is twice that in the first order. Increasing the dispersion reduces the amount of continuous background in the wavelength region near the line, and thus increases the line-to-background ratio.

a. Preparation of Synthetic Standards

The solvent-extraction technique employed in the preparation of ultrapure Ta_2O_5 can be modified to produce a high-purity Nb_2O_5. This in turn serves as a base for the standards.

At high fluoride concentrations both niobium and tantalum readily extract into the organic phase as NbF_7^{-2} and TaF_7^{-2} (see Section IV-J-3).

When the organic phase is treated by an aqueous phase, niobium, probably as $(NbOF_5)^{-2}$, can be back-extracted and thus separated from the tantalum, which does not readily form a $(TaOF_5)^{-2}$. Thus, when 100 g. of niobium metal is dissolved in 200 ml. of hydrofluoric acid and is contacted with 200 ml. of MIBK, the aqueous phase can be discarded since

TABLE XXVI
Analytical Line Pairs and Impurity Limits in Niobium

Line pairs, A.	Concentration range, %
Second-order spectrum	
Arced 30 seconds with carrier	
W 4008.8/Co 3997.9	0.01 –0.20
Arced 15 seconds with carrier	
Al 3961.5/Co 3997.9	0.0001–0.005
Yb 3694.2/Co 3997.9	0.01 –0.20
Pb 3683.5/Co 3997.9	0.001 –0.05
Ti 3642.7/Nb 3645.4	0.01 –0.50
Cr 3605.3/Co 3431.6	0.001 –0.05
Zr 3601.2/Nb 3600.0	0.002 –0.15
Zr 3601.2/Nb 3598.3	0.10 –0.50
Zr 3614.7/Nb 3624.4	0.25 –1.50
Ni 3414.8/Co 3431.6	0.0001–0.01
Cu 3273.9/Co 3431.6	0.0001–0.005
Mo 3194.0/Nb 3192.8	0.001 –0.05
V 3183.9/Nb 3182.6	0.001 –0.05
Sn 3175.0/Nb 3175.5	0.005 –0.05
Ta 2714.7/Nb 2714.1	0.10 –1.00
First-order spectrum	
Arced 10 seconds with carrier	
Mg 2802.7/Nb 2801.5	0.0001–0.005
Fe 2719.0/Co 3431.6	0.001 –0.10
Mn 2576.1/Nb 2576.6	0.005 –0.05
Si 2514.3/Nb 2707.8	0.01 –0.50
Be 2348.6/Nb 2707.8	0.0005–0.005
Ta 2714.7/Nb 2786.4	0.15 –1.50
Arced 25 seconds without carrier	
Zn 3345.0/Nb 3339.5	0.0005–0.005
Pb 2833.1/Nb 2829.9	0.001 –0.05
B 2497.7/Nb 2510.5	0.0001–0.005
Cd 2288.0/Nb 2296.0	0.0001–0.005
Arced 10 seconds as dried solution of nonvolatile chlorides without carrier	
Li 3232.7/Co 3409.2	0.005 –0.50

it contains only impurities. After three washings with 10-ml. portions of 6M sulfuric acid the niobium is back-extracted into the aqueous phase with three 100-ml. portions of water. The organic phase containing the tantalum is discarded. Three washings of the aqueous phase with 50-ml. portions of MIBK removes the last traces of tantalum. Pure Nb_2O_5 now can be precipitated in the aqueous solution with NH_4OH.

A series of standards containing the impurities can be prepared by successive dilutions of a master blend containing impurities added as either oxides or standard solutions. Procedures identical to those employed for tantalum are used here.

b. Preparation of Carrier and of Samples

The same NH_4Cl–$CoSO_4$ used for tantalum is added to Nb_2O_5 samples, with one part of carrier to one part of sample. Although cobalt is not ordinarily detected in niobium samples, the specifications (see Table XXI) may require its determination. In such cases, only NH_4Cl is employed as the carrier.

Niobium occurs in compounds similar in composition to those of tantalum. Consequently, they can be handled by the same techniques described earlier: direct ignition of metal samples to Nb_2O_5, precipitation of soluble salts, such as K_2NbOF_5, $NbCl_5$, and NbF_5, with NH_4OH, and subsequent ignition to Nb_2O_5. The sulfur monochloride volatilization technique employed in determining sodium and potassium in tantalum affords a convenient method for determining these same alkalies, as well as lithium, in niobium.

c. Spectrographic Conditions

The factors considered in the tantalum procedure are just as applicable to niobium. Generally, the photographing of second-order spectra requires longer arcing periods than does that of the first order. Spectrum Analysis #1 film is suitable for photographing the second-order ultraviolet. The trace impurities that may be detected in niobium are listed in Table XXVI.

REFERENCES

1. Addink, N. W. H., *Spectrochim. Acta*, **5**, 495 (1953); **7**, 45 (1955).
2. Adler, I., and J. M. Axelrod, *Spectrochim. Acta*, **7**, 91 (1955).
3. Adler, N., and C. F. Hiskey, *J. Am. Chem. Soc.*, **79**, 1834 (1957).
4. Albrecht, W. M., W. D. Klopp, B. G. Koehl, and R. I. Jaffee, paper presented to the Fall Meeting, American Institute of Mining, Metallurgical, and Petroleum Engineers, Chicago, 1959.
5. Albright, C. H., Kawecki Chemical Co. The assistance of Mr. Albright in the preparation of this Section is gratefully acknowledged.

6. Aleksandrova, L. S., and K. V. Chmutov, *Izv. Akad. Nauk S.S.S.R.*, **1960**, 80, through *Anal. Abstr.*, **8**, 516 (1961).
7. Aleksewskaya, N. W., and M. S. Platonov, *Zh. Prikl. Khim.*, **10**, 2139 (1937).
8. Alexander, K. M., and F. Fairbrother, *J. Chem. Soc.*, **1949**, 223, 2472.
9. Alimarin, I. P., and G. N. Bilimovich, *Collection Czech. Chem. Commun.*, **26**, 255 (1961).
10. Alimarin, I. P., and T. A. Burova, *Zh. Prikl. Khim.*, **181**, 289 (1945); through *Chem. Abstr.*, **40**, 3362 (1946).
11. Alimarin, I. P., and B. I. Frid, *Zavodsk. Lab.*, **7**, 913 (1938).
12. Alimarin, I. P., and B. I. Frid, *Zavodsk. Lab.*, **7**, 110 (1938).
13. Alimarin, I. P., and I. M. Gibalo, *Fiz. Khim.*, **1956**, 185; *Ref. Zh. Khim.*, **1958**, Abstr. No. 7540; through *Anal. Abstr.*, **5**, 3695 (1958).
13a. Alimarin, I. P., I. M. Gibalo, and Kuang Jung Ch'in, *Zh. Analit. Khim.*, **17**, 60 (1962).
13b. Alimarin, I. P., O. M. Petruklin, and Yun-Hsiang Tse, *Dokl. Akad. Nauk SSSR*, **136**, 1073 (1961); *Ref. Zh. Khim.*, **1961**, No. 20 D36.
14. Alimarin, I. P., and R. L. Podvalnaja, *Zh. Analit. Khim.*, **1**, 30 (1946).
15. Alimarin, I. P., E. S. Pozhevalskii, I. V. Puzdrenkova, and A. P. Golovina, *Trudy Kommiss. Anal. Khim., Akad. Nauk S.S.S.R.*, **8**, 152 (1958); *Ref. Zh. Khim.*, **1958**, Abstr. No. 77181; through *Anal. Abstr.*, **6**, 2027 (1959).
16. Alimarin, I. P., and E. I. Stepanynk, *Zavodsk. Lab.*, **22**, 1149 (1956).
17. Amano, H., *J. Japan Inst. Metals*, **22**, 625 (1958).
18. American Society for Testing and Materials, *ASTM Methods for Chemical Analysis for Metals*, Philadelphia, 1950.
19. Am. Soc. Testing Materials, *Ibid.*, 1956.
20. Am. Soc. Testing Materials, Committee B-2, *Tentative Specification for Primary Columbium Metal*, March, 1961.
21. Am. Soc. Testing Materials, Committee E 3, Division F, *Tentative Method for the Determination of Columbium in Ferrocolumbium*, 1961.
22. Am. Soc. Testing Materials, Committee E-3, Division R, *Report by Iron Subgroup of Columbium Task Force*, 1961.
23. Am. Soc. Testing Materials, Committee E-3, Division R., *Report by Molybdenum–Tungsten Subgroup of Columbium Task Force*, 1961.
24. Am. Soc. Testing Materials, Committee E-3, Division R, *Report of Nitrogen Subgroup, Columbium Task Force*, 1961.
25. Am. Soc. Testing Materials, Committee E-3, Division R, *Report of Oxygen Subgroup, Columbium Task Force*, 1959.
26. Am. Soc. Testing Materials, Committee E-3, Division R, *Report by Tantalum Subgroup of Columbium Task Force*, 1961.
27. Am. Soc. Testing Materials, Committee E-3, Division R, *Method Under Investigation by Titanium Subgroup of Columbium Task Force*, 1960.
28. Am. Soc. Testing Materials, Committee E-3, Division R, *Method Under Investigation by Zirconium Subgroup of Columbium Task Force*, 1962.
29. Anders, F. J., *Proc. Met. Soc. A.I.M.E., Conf. on Refractory Metals and Alloys* (*Detroit*), 219 (1961).
30. Anders, O. U., *U. S. At. Energy Comm. Rept.*, **AECU-3513** (1957).
31. Andrychuk, D., and J. Massengale, paper presented at the 10th Chicago Symposium on Spectroscopy, 1959.
32. Articolo, O. J., *U. S. At. Energy Comm. Rept.*, **KAPL-M-OJA-1** (1959).

33. Athavale, V. T., V. P. Menon, and Ch. Venkateswarlu, *Analyst*, **85,** 208 (1960).
34. Atkins, D. H. F., and A. A. Smales, *Anal. Chim. Acta*, **22,** 462 (1960).
35. Atkinson, R. H., J. Steigman, and C. F. Hiskey, *Anal. Chem.*, **24,** 477 (1952).
36. Atkinson, R. H., J. Steigman, and C. F. Hiskey, *Anal. Chem.*, **24,** 481 (1952).
37. Backer, R. O., V. R. Wiederkehr, and G. W. Goward, *U. S. At. Energy Comm. Rept.*, **WAPD-CTA-GLA-204** (1958).
38. Bacon, A., and G. W. C. Milner, *Anal. Chim. Acta*, **15,** 129 (1956).
39. Bagshawe, I. B., and W. T. Elwell, *J. Soc. Chem. Ind.*, **66,** 398 (1947).
39a. Bakes, J. M., G. R. E. C. Gregory, and P. G. Jeffery, *Anal. Chim. Acta*, **27,** 540 (1962).
40. Bakish, R., *J. Electrochem. Soc.*, **105,** 574 (1958).
41. Balchin, L. A., and D. I. Williams, *Analyst*, **85,** 503 (1960).
42. Balke, C. W., "Columbium, Metal and Alloys" in *Encyclopedia of Chemical Technology*, R. E. Kirk and D. F. Othmer, Eds., Interscience, New York, 1954, Vol. 4, p. 314.
43. Balke, C. W., in *Corrosion Handbook*, H. H. Uhlig, Ed., Wiley, New York, 1948.
44. Balke, C. W., and C. E. Balke, United States Patent 2,205,386 (1940).
45. Bamberger, E., *Ber.*, **52,** 1116 (1919).
46. Bandi, W. R., E. G. Buyok, L. L. Lewis, and L. M. Melnick, *Anal. Chem.*, **33,** 1275 (1961).
47. Banks, C. V., K. E. Burke, J. W. O'Laughlin, and J. A. Thompson, *Anal. Chem.*, **29,** 995 (1957).
48. Banks, C. V., and E. M. Fornfeld, Report CC2938, 1945, in C. J. Rodden, Ed., *Analytical Chemistry of the Manhattan Project*, McGraw-Hill, New York, 1950, p. 466.
49. Bate, L. C., and G. W. Leddicotte, Paper #40, Pittsburgh Conference on Analytical Chemistry and Applied Spectroscopy, 1958.
50. Beach, A. L., and W. G. Guldner, *ASTM Spec. Tech. Publ. No.* **222** (1958).
51. Bedford, M. H., *J. Am. Chem. Soc.*, **27,** 1216 (1905).
52. Belekar, G. K., and V. T. Athavale, *Analyst*, **82,** 630 (1957).
53. Benedict, M., and T. H. Pigford, *Nuclear Chemical Engineering*, McGraw-Hill, New York, 1957.
54. Benner, F. C., "Analytical Techniques for Vacuum Processed Metals and Alloys," Part IX, Analytical Techniques, in *Vacuum Metallurgy*, R. F. Bunshah, Ed., Reinhold, New York, 1958.
55. Bergstresser, K. S., *Anal. Chem.*, **31,** 1812 (1959).
56. Berry, B. E., G. L. Miller, and S. V. Williams, *B. I. O. S. Final Report No.* **803** (1946); through G. L. Miller, *Tantalum and Niobium*, Academic Press, New York, 1959.
57. Beydon, J., and C. Fisher, *Anal. Chim. Acta*, **8,** 538 (1953).
58. Bhattacharya, H., *J. Indian Chem. Soc.*, **29,** 871, 891 (1952).
59. Birks, L. S., and E. J. Brooks, *Anal. Chem.*, **22,** 1017 (1950).
60. Birks, L. S., and R. E. Seebold, *J. Nucl. Materials*, **3,** 249 (1961).
61. Bishop, C. R., and M. Stern, *J. Metals*, **13** (2), 144 (1961).
62. Bisichi Tin Company of Nigeria, private communication (1960).
63. Blasius E., and A. Czekay, *Z. Anal. Chem.*, **156,** 81 (1957).
64. Blinov, V. I., *Zavodsk. Lab.*, **14,** 1494 (1948); through *Chem. Abstr.*, **46,** 7931g (1952).

65. Blocher, J. M., in *High Temperature Technology*, I. E. Campbell, Ed., Wiley, New York, 1956.
65a. Blymu, I. A., and G. N. Shebalkova, *Tr. Kazakhsk. Nauchn.-Issled. Inst. Mineral'n. Syr'ya*, **1961**, 265; *Ref. Zh. Khim.*, **1962**, No. 15 D62.
66. Bogdanova, V. I., *Akad. Nauk S.S.S.R.*, **1959**, 224; *Ref. Zh. Khim.*, **1960**, Abstr. No. 38,376.
67. Booth, E., F. J. Bryant, and T. Parker, *Analyst*, **82**, 50 (1957).
68. Born, H. J., and P. Wilkniss, *Intern. J. Appl. Radiation Isotopes*, **10**, 133 (1961).
69. Borneman, I. D., *Symposium-Materialy Sovesh-Chaniya po Primeneniyu Iomogo Obname v Tsvetnoi Metallurgi M.*, **80** (1957); *Ref. Zh. Khim.*, Abstr. No. 60580; through *Anal. Abstr.*, **6**, 506 (1959).
70. Boudergues, R., *Chim. Anal.*, **42**, 421 (1960).
71. Boyd, T. F., and M. Galan, *Anal. Chem.*, **25**, 1568 (1953).
72. Brauer, G., and H. Müller, in *Plansee Proceedings 1958*, F. Benesovsky, Ed., Springer, Vienna, 1959.
73. Braun, H., R. Kieffer, and K. Sedlatschek, in *Plansee Proceedings 1958*, F. Benesovsky, Ed., Springer, Vienna, 1959.
74. Breger, A. Kh., B. F. Ormont, B. I. Viting, V. M. Grizhko, V. A. Kozlov, V. S. Kutseva, B. A. Chapyzhnikov, and L. V. Chepel, *Trudy Kommiss. Anal. Khim., Akad. Nauk S.S.S.R.*, **10**, 137 (1960); *Ref. Zh. Khim.*, **1961**, Abstr. No. 2D92.
75. Brewer, L., and H. Haraldsen, *J. Electrochem. Soc.*, **102** (7), 399 (1955).
76. Brindley, D. J., *Analyst*, **85**, 877 (1960).
77. Brissey, R. M., *Anal. Chem.*, **24**, 1034 (1952).
78. British Iron and Steel Research Association, Methods of Analysis Committee, *J. Iron St. Inst.*, **187**, 341 (1957).
79. Brode, W. R., *Chemical Spectroscopy*, Wiley, New York, 1943.
80. Browning, E., *Toxicity of Industrial Metals*, Butterworths, London, 1961.
81. Brownlee, J., in R. S. Maddock and W. W. Meincke, Eds., *U.S. At. Energy Comm. Rept.*, **AECU-4438, TID-11009**, Progress Report #8 (1959).
82. Brownlee, J., in R. S. Maddock and W. W. Meincke, Ed., *U.S. At. Energy Comm. Rept.*, **AECU-4438, TID-11009**, Progress Report #9 (1960).
83. Brüning, K., K. Meier, and H. Wirtz, *Metall und Erz*, **36**, 551 (1939).
84. Bruninx, E., J. Eeckhout, and J. Gillis, *Anal. Chim. Acta*, **14**, 74 (1956).
85. Bruninx, E., J. Eeckhout, and J. Gillis, *Mikrochim. Acta*, **4–6**, 689 (1956).
86. Bruninx, E., and J. W. Irvine, Jr., in "Radioisotopes in Scientific Research," *Proc. First UNESCO Conf.*, Paris, 1957, Vol. 2, p. 232.
87. Bruninx, E., and J. W. Irvine, Jr., *Angew. Chem.*, **70**, 77 (1958).
88. Bryson, T. C., G. W. Goward, and M. D. Hartmann, *U.S. At. Energy Comm. Rept.*, **WAPD-CTA-GLA-146** (1956).
89. Bukhsh, M. N., and D. N. Hume, *Anal. Chem.*, **27**, 116 (1955).
89a. Buncak, P., *Chem. Prumsyl*, **11**, 634 (1961); through *Anal. Abstr.*, **9**, 4157 (1962)
90. Bunney, L. R., N. E. Ballou, J. Pascual, and S. Foti, *Anal. Chem.*, **31**, 324 (1959).
91. Burstall, F. H., P. Swain, A. F. Williams, and G. A. Wood, *J. Chem. Soc.*, **1952**, 1497.
92. Burstall, F. H., and A. F. Williams, *Analyst*, **77**, 983 (1952).
93. Burton, R., R. M. Jacobs, and E. R. Valecko, *U. S. At. Energy Comm. Rept.* **WAPD-CTA(GLA)-617-1** (1958).
94. Bykova, V. S., *Acad. Sci. U.R.S.S., Compt. Rend.*, **18**, 665 (1938); through *Chem. Abstr.*, **32**, 6172^8 (1938).

95. Bȳkovskaya, Yu. I., *Trudȳ Kommiss. Anal. Khim.*, **9,** 323 (1958); *Ref. Zh. Khim.,* **1959,** Abstr. No. 27,115.
96. Bȳkovskaya, Yu. I., *Trudȳ Kommiss. Anal. Khim.*, **9,** 329 (1958); *Ref. Zh. Khim.,* **1959,** Abstr. No. 27,047.
97. Cabell, M. J., and I. Milner, *Anal. Chim. Acta*, **13,** 258 (1955).
98. Campbell, I. E., *High Temperature Technology*, Wiley, New York, 1956.
99. Campbell, W. J., and H. F. Carl, *Anal. Chem.*, **26,** 800 (1954).
100. Campbell, W. J., and H. F. Carl, *Anal. Chem.*, **28,** 960 (1956).
101. Carl, H. F., and W. J. Campbell, *ASTM Spec. Tech. Publ. No.* **157** (1953).
102. Carmichael, R. L., in *Technology of Columbium*, B. W. Gonser and E. M. Sherwood, Eds., Wiley, New York, 1958.
102a. Carth, R., *Anal. Chem.*, **34,** 1607 (1962).
103. Casey, A. T., and A. G. Maddock, *J. Inorg & Nucl. Chem.*, **10,** 289 (1959).
103a. Catoggio, J. A., and L. B. Rogers, *Talanta*, **9,** 377 (1962).
104. Chaigneau, M., *Compt. Rend.*, **244,** 900 (1957).
105. Chan, F. L., *Talanta*, **7,** 253 (1958).
106. Charlot, G., and J. Saulnier, *Chim. Anal.*, **35,** 51 (1953).
107. *Chem. Eng. News*, **37** (42), 52 (1959).
108. *Chem. Eng. News*, **37** (43), 82 (1959).
109. *Chemical Week*, "Special Report," May 6, 1961, p. 89.
110. Chernikhov, Yu. A., Sh. G. Melamed, and B. M. Dobkina, *Zavodsk. Lab.*, **24,** 677 (1958).
111. Chernikhov, Yu. A., R. S. Tramm, and K. S. Pevzner, *Zavodsk. Lab.*, **22,** 637 (1956).
112. Chernikhov, Yu. A., R. S. Tramm, and K. S. Pevzner, *Zavodsk. Lab.*, **25,** 398 (1959).
113. Chesneau, G., *Compt. Rend.*, **149,** 1132 (1909).
114. Clauss, F., and C. Barrett, in *Technology of Columbium*, B. W. Gonser and E. M. Sherwood, Eds., Wiley, New York, 1958.
115. Codell, M., *Analytical Chemistry of Titanium Metals and Compounds*, Interscience, New York, 1959.
116. Codell, M., S. Kallmann, and G. Norwitz, *ASTM Spec. Tech. Publ. No.* **222** (1958).
117. Coleman, R. F., and J. L. Perkin, *Analyst*, **84,** 233 (1959).
118. Cozzi, D., and S. Vivarelli, *Ricerca Sci.*, **23,** 2244 (1953); *Z. Elektrochem.*, **58,** 177 (1954).
119. Cozzi, D., and S. Vivarelli, *Z. Elektrochem.*, **57,** 406 (1953).
120. Craven, S. W., *J. Soc. Chem. Ind.*, **66,** 400 (1947).
121. Crouthamel, C. E., B. E. Hjelte, and C. E. Johnson, *Anal. Chem.*, **27,** 507 (1955).
122. Cunningham, T. R., *Ind. Eng. Chem., Anal. Ed.*, **10,** 233 (1938).
123. Cunningham, T. R., *Ind. Eng. Chem., Anal. Ed.*, **5,** 305 (1933); **6,** 287 (1934).
124. Dams, R., and J. Hoste, *Talanta*, **9,** 86 (1962).
125. Dana, J. D., and S. D. Dana, *The System of Mineralogy*, 7th ed., C. Palache, H. Berman, and C. Frondel, Eds., Wiley, New York, 1944, Vol. I.
126. Darnell, J. R., and L. F. Yntema, in *Technology of Columbium*, B. W. Gonser and E. M. Sherwood, Eds., Wiley, New York, 1958.
127. Das, M. S., Ch. Venkateswarlu, and V. T. Athavale, *Analyst*, **81,** 239 (1956).
128. Das Gupta, A. K., and S. K. Dhar, *J. Sci. Ind. Research*, **12-B,** 396 (1953).
129. Davis, M. V., and D. T. Hauser, *Nucleonics*, **16,** 87 (1958).

130. Davydov, A. L., Z. M. Vaisberg, and L. E. Burkser, *Zavodsk. Lab.*, **13**, 1038 (1947).
131. Dayton, S. H., *Mining World*, July, 1958, p. 40.
132. Delaney, J. D., and L. E. Owen, *Anal. Chem.*, **23**, 577 (1951).
133. Delespaul, I., and F. Lievens, *Spectrochim. Acta*, **16**, 1486 (1960).
134. DeMent, J., and H. C. Dake, *Rarer Metals*, Chemical Pubishing Co., Brooklyn, 1946.
135. Demildt, A., and J. Hoste, *Bull. Soc. Chim. Belg.*, **70**, 145 (1961).
135a. De Wappner, B. G., *Anales Asoc. Quim. Arg.*, **49**, 285 (1961).
136. Dhar, S. K., *Anal. Chim. Acta*, **11**, 289 (1954).
137. Dinnin, J. I., *Anal. Chem.*, **25**, 1803 (1953).
138. Doan, U. M., and C. Duval, *Anal. Chim. Acta*, **6**, 83 (1952).
139. Dobina, A. A., and M. S. Platonow, *Zh. Prikl. Khim.*, **14**, 421 (1941).
140. Dobkina, B. M., and T. M. Malyntina, *Zavodsk. Lab.*, **24**, 1336 (1958).
141. Dobkina, B. M., and E. I. Petrova, *Zavodsk. Lab.*, **23**, 421 (1957).
142. Dobkina, B. M., and E. I. Petrova, *Zavodsk. Lab.*, **25**, 1064 (1959).
143. Driggs, F. H., U. S. Patent 1,815,054 (1931).
144. Dufour, R. F., *U.S. At. Energy Comm. Rept.*, **KAPL**, **MHME**-1 (1960).
145. Duke, J. F., and W. Stawpert, *Analyst*, **85**, 671 (1960).
146. Duwez, P., and F. Odell, *J. Electrochem. Soc.*, **97** (10), 299 (1950).
147. Eder, A., *Arch. Eisenhüttenw*, **26**, 431 (1955).
148. Eicholz, G. G., *Nucleonics*, **10** (12), 58 (1952).
149. Ekeberg, A. G., *Ann. Chim.*, **43**, 276 (1802).
150. Electro Metallurgical Company, Union Carbide Corporation, *Method No. 22-41, 73-257* (1950).
151. Ellenberg, J. Y., G. W. Leddicotte, and F. D. Moore, *Anal. Chem.*, **26**, 1045 (1954).
152. Elinson, S. V., K. I. Petrov, and A. T. Rezova, *Zh. Analit. Khim.*, **13**, 576 (1958).
153. Elliot, R. P., Preprint No. 143, *Metal. Progr.*, **76**, 242 (1959).
154. Elson, R. E., *J. Am. Chem. Soc.*, **75**, 4193 (1953).
155. Elving, P. J., and E. C. Olson, *Anal. Chem.*, **28**, 338 (1956).
156. Elwell, W. T., and D. F. Wood, *Anal. Chim. Acta*, **26**, 1 (1962).
157. Emery, J. F., W. T. Mullins, L. C. Bate, and G. W. Leddicotte, "Trace Element Determination in Niobium and Zirconium by Radioactivation Analysis," paper presented at the 5th Conference on Analytical Chemistry in Nuclear Reactor Technology, Gatlinburg, Tennessee, 1961.
158. Engels, J., Gesellschaft für Elektrometallurgie mbH., Weisweiler, Germany, private communication (1961).
159. Epremian, E., in *Technology of Columbium*, B. W. Gonser and E. M. Sherwood, Eds., Wiley, New York, 1958.
160. Epstein, R. Ya., and G. P. Ginberg, *Trudy Nauch.-Issled. Inst. Geol. Arktiki*, **119**, 84 (1961); *Ref. Zh. Khim.*, **1961**, Abstr. No. 19D 61.
161. Evens, F. M., and V. A. Fassel, *Anal. Chem.*, **33**, 1056 (1961).
162. Fagel, J. E., R. F. Witbeck, and H. A. Smith, *Anal. Chem.*, **31**, 1115 (1959).
163. Fansteel Metallurgical Corp., *Columbium*, 1946.
164. Fansteel Metallurgical Corp., *The Metal Tantalum*, 1953.
165. Faris, J. P., *Anal. Chem.*, **32**, 520 (1960).
166. Faris, J. P., and J. Brody, paper #34, given before the Pittsburgh Conference on Analytical Chemistry and Applied Spectroscopy, 1961.
167. Fassel, V. A., in *Encyclopedia of Spectroscopy*, G. L. Clark, Ed., Reinhold, New York, 1960.

168. Fassel, V. A., and M. Evens, Second Conference, Gatlinburg, Tennessee, *U.S. At. Energy Comm. Repts.*, **TID-7568** (Part 1) (1958).
169. Fassel, V. A., W. A. Gordon, and R. J. Jasinski, *Proc. 2nd Intern. Conf. Peaceful Uses At. Energy, Geneva, 1958*, **A/Conf. 15/P/917** (1958).
170. Fassel, V. A., W. A. Gordon, and R. W. Tabeling, *ASTM Spec. Tech. Publ. No.* **222** (1958).
171. Faye, G. H., *Chemistry in Canada*, **10**(4), 90 (1958).
172. Faye, G. H., and W. R. Inman, *Can. Dept. of Mines and Tech. Surveys, Res. Rept. No.* **MD210** (1957).
173. Featheringham, J. A., C. F. Lentz, and R. M. Jacobs, *U.S. At. Energy Comm. Rept.*, **WAPD-CTA(GLA)-631-1-5** (1958).
174. Feigel, F., and A. Caldas, *Mikrochim. Acta*, **7–8**, 1310 (1956).
175. Ferraro, T. A., Jr., *Watertown Arsenal Report No.* **WALTR 823/2** (1961).
176. Ferrett, D. J., and G. W. C. Milner, *J. Chem. Soc.*, **1956**, 1186.
177. Ferrett, D. J., and G. W. C. Milner, *Nature*, **175**, 477 (1955).
178. Flaschka, H., and E. Lassner, *Mikrochim. Acta*, **7–8**, 778 (1956).
179. Fletcher, J. M., D. F. C. Morris, and A. G. Wain, *Trans. Inst. Min. Metall.*, **65**, 487 (1955).
180. Fornwalt, D. E., and M. K. Healy, *Appl. Spectroscopy*, **13**, 38 (1957).
181. Fornwalt, D. E., and J. Komisarek, Second Conference, Gatlinburg, Tennessee, *U.S. At. Energy Comm. Rept.*, **TID-7568**, Part 1 (1958).
182. Freund, H., K. H. Hammil, and F. C. Brissownette, Jr., *U. S. Bur. Mines, Rept Invest. No.* **5242** (1956).
183. Freund, H., and A. E. Levitt, *Anal. Chem.*, **23**, 1813 (1951).
184. Fricioni, R., Allegheny Ludlum Steel Corp. Brackenridge, Pa., private communication (1961).
185. Fridman, I. D., and I. N. Yudina, *Zh. Prikl. Khim.*, **32**, 1914 (1959).
186. Friederich, E., and L. Sittig, *Z. Anorg. Chem.*, **144**, 169 (1925).
187. Friedrich, H. J. Thesis, Technische Hochschule, Hanover, 1955.
188. Friedrich, H. J., Hermann C. Starck Goslar, Germany, private communication (1962).
189. Fucke, H., and J. Daubländer, *Forschungsber. Krupp*, **2**, 174 (1939).
190. Furey, J., Union Carbide Metals Co., Niagara Falls, New York, private communication (1951).
191. Gahler, A. R., R. M. Hammer, and R. C. Shubert, *Anal. Chem.*, **33**, 1937 (1961).
192. Gebhardt, E., and H. Preisendanz, *Z. Metallkunde*, **46**, 560 (1955).
193. Gebhardt, E., H. D. Seghezzi, and Durrschnabel, in *Plansee Proceedings 1958*, F. Benesovsky, Ed., Springer, Vienna, 1958.
194. Geiger, E. L., *Anal. Chem.*, **31**, 806 (1959).
195. Geld, I., and J. Carroll, *Anal. Chem.*, **21**, 1098 (1949).
196. Gentry, C. R., and L. G. Sherrington, *Analyst*, **73**, 65 (1948).
197. Gerlach, W., and E. Riedl, *Chemische Emissionsspektralanalyse*, Leopold Voss, Leipzig, 1936, Vol. III.
198. Gesellschaft Deutscher Metallhütten-und Bergleute, *Analyse der Metalle*, Springer, Berlin, Vol. 2, 1953.
199. Gesellschaft Deutscher Metallhütten-und Bergleute e.V., Unterausschuss Stahlveredler, in *Z. Erzbergbau und Metallhüttenwesen* **11** (10), 465 (1958).
200. Gesellschaft Deutscher Metallhütten-und Bergleute e.V., Unterausschuss Stahlveredler, in *Z. Erzbergban und Metallhüttenwesen*, **12** (12), 612 (1959).

201. Gibbons, D., Isotope Division, Atomic Energy Research Establishment, Harwell, private communication (1959).
202. Giles, W. E., *Chem. News*, **95**, 37 (1907).
203. Gillespie, J., Wah Chang Corporation, New York, private Communication (1961).
204. Gillis, J., and J. Eeckhout, *Anal. Chim. Acta*, **1**, 377 (1947).
205. Gillis, J., Z. Eeckhaut, P. Cornoud and A. Speecke, *Meded. Kon. VI Akad. Weten. Belg.*, **15**, 13 (1953).
206. Gillis, J., J. Hoste, P. Cornoud, and A. Speecke, *Meded. Vlaam. Chem. Ver.*, **15**, 63 (1953).
207. Gilmore, J. S., *U.S. At. Energy Comm., Rept.*, **La-1721** (1958).
208. Gokhshtein, Ya. P., *Zavodsk. Lab.*, **22**, 38 (1956).
208a. Gokhshtein, Ya. P., L. A. Genkina, and A. M. Demkin, *Akad. Nauk Moldavsk. SSR, Materialy Pervogo Vses. Soveshch.*, **1962**, 34; *Chem. Abstr.*, **59**, 2149 (1963).
209. Gokhshtein, Ya. P., L. A. Genkina, and A. M. Demkin, *Zavodsk. Lab.*, **25**, 1042 (1959).
210. Goleb, J. A., *Anal. Chem.*, **28**, 965 (1956).
211. Golubstsova, R. B., *J. Anal. Chem. U.S.S.R.*, **6**, 34 (1951).
212. Golubstsova, R. B., *J. Anal. Chem. U.S.S.R.*, **6**, 357 (1951); through *Brit. Abstr. C*, **140**, 1062 (1952).
213. Gonser, B. W., in *Technology of Columbium*, B. W. Gonser and E. M. Sherwood, Eds., Wiley, New York, 1958.
214. Gordon, N. E., Jr., R. M. Jacobs, and M. C. Rickel, *Anal. Chem.*, **25**, 1031 (1953).
215. Gorshchenko, Ya. G., M. I. Andreeva, and A. G. Babkin, *Zh. Prikl. Khim.*, **32**, 1904 (1959); through *Anal. Abstr.*, **7**, 2699 (1960).
216. Goto, H., and Y. Kakita, *Sci. Rept. Res. Inst. Tohoku Univ.*, **2**, 249 (1950); through *Brit. Abstr. C*, **140**, 1903 (1951).
217. Goward, G. W., T. M. Reinhold, and V. R. Wiederkehr, *U. S. At. Energy Comm., Rept.*, **WAPD-CTA(GLA)-543** (1958).
218. Grimaldi, F. S., *Anal. Chem.*, **32**, 119 (1960).
219. Grimaldi, F. S., and M. M. Schnepfe, *Anal. Chem.*, **30**, 2046 (1958).
220. Grosse, A. V., A. D. Kirschenbaum, and R. D. Mossman, *Trans. Am. Soc. Metals*, **46**, 525 (1954).
221. Gulbransen, E. A., and K. F. Andrew, *J. Electrochem. Soc.*, **96**, 364 (1949).
222. Gulbranson, E. A., and K. F. Andrew, *J. Metals*, **180**, 586 (1950).
223. Guyon, J. C., G. W. Wallace, Jr., and M. G. Mellon, *Anal. Chem.*, **34**, 640 (1962).
224. Hagel, W. C., *ORNL Metallurgy Div. Rept.*, **CF-51-8-256** (1951).
225. Hague, J., National Bureau of Standards, private communication (1954).
226. Hague, J. L., E. D. Brown, and H. A. Bright, *J. Res. Natl. Bur. Std.*, **53**, 261 (1954).
227. Hague, J. L., and L. A. Machlan, *J. Res. Natl. Bur. Std.*, **62**, 11 (1959).
228. Hague, J. L., and L. A. Machlan, *J. Res., Natl. Bur. Std.*, **62**, 53 (1959).
229. Hahn, R. B., and C. L. Burros, *Anal. Chem.*, **23**, 1713 (1951).
230. Hakhila, E. A., and G. R. Waterbury, *Talanta*, **6**, 46 (1960).
231. Halverson, G., and A. Shtasel, *Anal. Chem.*, **33**, 1627 (1961).
232. Hansen, W. R., and M. W. Mallett, *Anal. Chem.*, **29**, 1868 (1957).
233. Hansen, W. R., M. W. Mallett, and M. J. Trzeciak, *Anal. Chem.*, **31**, 1115 (1959).
234. Harris, W. F., in *Technology of Columbium*, B. W. Gonser and E. M. Sherwood, Eds., Wiley, New York, 1958.
235. Harrison, G. R., *Wavelength Tables*, Wiley, New York, 1939.
236. Harvey, C. E., *Spectrochemical Procedures*, Applied Research Laboratories Press, 1950.

237. Hasler, M. F., C. E. Hervey, and H. W. Dietert, *Ind. Eng. Chem., Anal. Ed.*, **15**, 102 (1943).
238. Hastings, J., and T. A. McClarity, *Anal. Chem.*, **26**, 683 (1954).
239. Hastings, J., T. A. McClarity, and E. J. Broderick, *Anal. Chem.*, **26**, 379 (1954).
240. Hatchett, C., *Phil. Trans.*, **92**, 49 (1802).
241. Hauser, O., and A. Lewit, *Angew. Chem.*, **25**, 100 (1912).
242. Hayashi, S., *J. Chem. Soc. Japan, Pure Chem. Sect.*, **70**, 376 (1949), **71**, 17 (1950); through *Chem. Abstr.*, **45**, 2818,a,b 5063c (1951).
242a. Headridge, J. B., and E. J. Dixon, *Analyst*, **87**, 32 (1962).
243. Headridge, J. B., and M. S. Taylor, *Analyst*, **87**, 43 (1962).
244. Herman, P., *Spectrochim. Acta.*, **3**, 389 (1948).
245. Herrmann, M., *Ind. Chim. Belge*, **23**, 123 (1958).
246. Hermann, R., *J. Prakt. Chem.*, **38** (1846).
247. Hibbits, J. O., H. Oberthin, R. Liu, and S. Kallmann, *Talanta*, **8**, 209 (1961).
248. Hicks, H. G., and R. S. Gilbert, *Anal. Chem.*, **26**, 1205 (1954).
249. Higbie, K. B., and J. R. Werning, *U. S. Bur. Mines Rept. No.* **5239** (1956).
250. Hillebrand, W. F., and G. E. F. Lundell, *Applied Inorganic Analysis*, Wiley, New York, 1953.
251. Hiskey, C. F., L. Newman, and R. H. Atkinson, *Anal. Chem.*, **24**, 1988 (1952).
251a. Hobart, E. W., and E. P. Hurley, *Anal. Chim. Acta*, **27**, 144 (1962).
252. Hoekstra, H. R., and J. J. Katz, *Anal. Chem.*, **25**, 1608 (1953).
253. Hogarth, D. D., *Can. Mineralogist*, **6**, 610 (1961).
254. Holdt, G., and H. Schäfer, *Z. Anal. Chem.*, **146**, 4 (1955).
255. Hollander, J. M., I. Perlman, and G. T. Seaborg, *Revs. Mod. Phys.*, **25**, 469 (1953).
256. Hollandsche Metallurgical, Bedrijven, Arnhem, private communication (1960).
257. Horn, F., and W. Zweigler, *J. Am. Chem. Soc.*, **69**, 2762 (1947).
258. Huffman, E. H., and G. M. Iddings, *J. Am. Chem. Soc.*, **74**, 4714 (1952).
259. Huffman, E. H., G. M. Iddings, and R. C. Lilly, *J. Am. Chem. Soc.*, **73**, 4474 (1951).
260. Hughes, D. J., and J. A. Harvey, *U. S. At. Energy Comm. Rept.*, **BNL 325** (1955).
261. Hunt, E. C., A. A. North, and R. A. Wells, *Analyst*, **80**, 172 (1955).
262. Hunt, E. C., and R. A. Wells, *Analyst*, **79**, 345 (1954).
263. Hunt, E. C., and R. A. Wells, *Analyst*, **79**, 351 (1954).
264. Hurd, D., *Chemistry of the Hydrides*, Wiley, New York, 1952.
265. Ikenberry, L., J. L. Martin, and W. G. Boyer, *Anal. Chem.*, **25**, 1340 (1953).
266. Ingles, J. C., "Manual of Analytical Methods for the Uranium Concentration Plant," *Mines Branch Monograph* **866**, Mines Branch Dept. of Mines & Techn. Serv., Ottawa, Canada, 1959.
267. *Intern. Sci. and Tech.*, Prototype Issue 18–27 (1961).
268. Isaacs, O. L., *General Electric Rept.*, **R61 FPD33** (1961).
269. Jaboulay, B. E., *Rev. Métallurgie*, **45**, 343 (1948).
270. Jaboulay, B. E., *Chim. Analytique*, **37**, 198 (1955).
271. Jaeger, F. M., and W. A. Veenstra, *Rec. Trav. Chim. Pays-Bas*, **53**, 677 (1934).
271a. Jaillet, J. B., St. Lawrence Columbium and Metals Corp., Montreal, private communication (1963).
272. Janauer, G. E., and J. Korkisch, *Anal. Chim. Acta*, **24**, 270 (1961).
273. Jantsch, G., in *Handbuch der Analytischen Chemie*, W. Fresenius and G. Jander, Eds., Springer, Berlin, 1956, Part II, Vol. IVb, Va/b.
274. Jefferey, P. G., *Analyst*, **82**, 66 (1957).

275. Jenkins, E. N., and A. A. Smales, *Quart. Rev. Chem. Soc.*, **10**, 83 (1956).
276. Johnson, C. M., *Iron Age*, **157**, 66 (1946).
277. Jones, R. J., *Columbium (Niobium) and Tantalum*, Dept. of Mines & Technical Surveys, Canada, Memorandum Series No. 135 (1957).
278. Kakita, Y., and H. Goto, *Anal. Chem.*, **34**, 618 (1962).
279. Kallmann, S., Ledoux & Co., Teaneck, New Jersey, unpublished laboratory notes (1961).
280. Kallmann, S., manuscript in preparation for publication in *Analytical Chemistry*.
281. Kallmann, S., R. Liu, and H. Oberthin, *Wright Air Development Center Tech. Rept.*, **57**, 229 (1957).
282. Kallmann, S., H. Oberthin, and R. Liu, *Anal. Chem.*, **34**, 609 (1962).
282a. Kallmann, S., "Analysis of Ten Representative Samples of Brazilian Pyrochlore," Ledoux & Co., 1963.
282b. Kanno, T., *Japan Analyst*, **10**, 38 (1961).
283. Kanzelmeyer, J. H., and H. Freund, *Anal. Chem.*, **25**, 1807 (1953).
284. Kanzelmeyer, J. H., J. Ryan, and H. Freund, *J. Am. Chem. Soc.*, **78**, 3020 (1956).
285. Karyakin, Yu. V., and P. M. Telezhnikova, *Zh. Prikl. Khim.*, **19**, 435 (1946).
286. Kassner, J. L., A. Garcia-Porrata, and E. L. Grove, *Anal. Chem.*, **27**, 492 (1955).
287. Kawahara, M., H. Mochizuki, and R. Kajiyama, *Japan Analyst*, **8**, 25 (1959).
287a. Kawahata, M., H. Mochizuki, and T. Misaki, *Bunseki Kagaku*, **10**, 1016 (1961); through *Chem. Abstr.*, **59**, 4537h (1963).
288. Kennedy, J. H., *Anal. Chem.*, **33**, 943 (1961).
289. Kharlamov, I. P., P. Ya. Yakovler, and M. I. Lȳkova, *Zavodsk. Lab.*, **24**, 928 (1958); through *Anal. Abstr.*, **4**, 1697 (1959).
289a. Kidman, L., and G. White, *Metallurgia*, **64**, 153 (1963).
289b. Kirby, R., and H. Freiser, *Anal. Chem.*, **35**, 122 (1963).
290. Kirtchik, H., *General Electric Rept.*, **R61FPD33** (1961).
291. Kling, H. P., in *Technology of Columbium*, B. W. Gonser and E. M. Sherwood, Eds., Wiley, New York, 1958.
292. Klinger, P., and W. Koch, *Arch. Eisenhüttenw.*, **13**, 127 (1939).
293. Klopp, W. D., C. T. Sims, and R. I. Jaffee, in *Technology of Columbium*, B. W. Gonser and E. M. Sherwood, Eds., Wiley, New York, 1958.
294. Knowles, H. B., and G. E. F. Lundell, *J. Res. Natl. Bur. Std.*, **42**, 405 (1949).
295. Koerner, E. L., and M. Smutz, *U. S. At. Energy Comm. Rept.*, **ISC-793** (1956).
296. Kofstad, P., *J. Inst. Metals*, **90**(7), 253 (1962).
297. Kohn, A., *Compt. Rend. Sci. Paris*, **236**, 1419 (1953); *Chim. & Ind. (Paris)*, **71**, 69 (1954).
297a. Kokubu, N., and K. Matsudea, *Rika Gaku Kenkyusho Hokoku*, **38**, 21 (1962); through *Chem. Abstr.*, **58**, 5035e (1963).
298. Kolk, A. J., M. E. Sibert, and M. A. Steinberg, in *Technology of Columbium*, B. W. Gonser and E. M. Sherwood, Eds., Wiley, New York, 1958, p. 44.
299. Kolmeschate, G. J. van, in *Handbuch der Analytischen Chemie*, W. Fresenius and G. Jander, Eds., Springer, Berlin, 1957, Part III, Vol. Vb.
300. Körbl, J., and R. P̆ibil, *Chem. Listy*, **51**, 1061 (1957).
301. Korchynsky, M., and A. R. Gahler, paper presented before the Refractory Metals Symposium, April 12–13, 1962, Chicago.
302. Kosta, L., and M. Dular, *Talanta*, **8**, 265 (1961).
303. Kosta, L., and L. Ravnik, *Proc. 2nd Intern. Conf. Peaceful Uses At. Energy, Geneva, 1958*, **Rep. A/CONF. 14/P/482** (1958).

303a. Kotlyar, E. E., and T. N. Nazarchuk, *Zh. Analit. Khim.*, **18** (4), 474 (1963).
304. Kraus, K. A., and G. E. Moore, *J. Am. Chem. Soc.*, **71**, 3855 (1949).
305. Kraus, K. A., and G. E. Moore, *J. Am. Chem. Soc.*, **73**, 9 (1951).
306. Kraus, K. A., and G. E. Moore, *J. Am. Chem. Soc.*, **73**, 13 (1951).
307. Kraus, K. A., and G. E. Moore, *J. Am. Chem. Soc.*, **73**, 2900 (1951).
308. Kraus, K. A., and F. Nelson, *ASTM Spec. Tech. Publ. No.* **195** (1958).
309. Kriege, O. H., *U. S. At. Energy Comm. Rept.*, **LA-2049** (1956).
310. Kriege, O. H., and R. D. Gardner, *U. S. At. Energy Comm. Rept.*, **LA-2032** (1956).
311. Krieger, H. L., J. E. Gilchrist, and S. Gold, *Talanta*, **6**, 254 (1960).
312. Krishna, Rao, B. S., D. V. N. Sarma, and B. S. V. Raghava Rao., *Z. Anal. Chem.* **160**, 351 (1958).
313. Krivoshlykov, N. F., and M. S. Platonov, *Zh. Prikl. Khim.*, **10**, 184 (1937).
314. Kroll, W. J., and F. E. Bacon, U. S. Patent 2,427,360, Sept. 16, 1947.
315. Krylov, E. I., and V. S. Kolevatova, *Zavodsk. Lab.*, **21**, 911 (1955).
316. Krylov, E. I., V. S. Kolevatova, and V. A. Samarina, *Dokl. Akad. Nauk S.S.S.R.*, **98**, 593 (1954).
317. Kurbatov, D. I., *Zh. Analit. Khim.*, **14**, 743 (1959).
317a. Kurbatov, D. I., and I. S. Skorynina, *Akad. Nauk Moldavsk. SSR, Materialȳ Pervogo Vses. Soveshch.*, **1962**, 248, 258. through *Chem. Abstr.*, **59**, 2156 (1963).
318. Kvalheim, A., through W. Fresenius and G. Jander, Eds., *Handbuch der Analytischen Chemie*, Springer, Berlin, 1957, Part III, Vol. Vb.
319. Kvalheim, A., S. Rutlin, and K. A. Arnseth, *Tidsskr. Kjemi, Bergvesen Met.*, **12**, 93 (1952).
320. Laboratory Equipment Corporation, St. Joseph, Mich., *Leco Nitrox-6*, 1962.
321. Laib, R. D., paper preesnted at Eleventh Chicago Symposium on Spectroscopy, 1960.
322. Landis, F. P., and L. P. Pepkowitz, *Anal. Chem.*, **27**, 141 (1955).
323. Langmyhr, F. J., and T. Hongslo, *Anal. Chim. Acta*, **22**, 301 (1960).
324. Larsen, E. M., in *Encyclopedia of Science and Technology*, McGraw-Hill, New York, 1960.
325. Lassner, E., *Chemist Analyst*, **51** (1), 14 (1962).
326. Lassner, E., and R. Scharf, *Talanta*, **7**, 12 (1960).
327. Lassner, E., and R. Scharf, *Chemist Analyst*, **50**, 69 (1961).
328. Lassner, E., and H. Weisser, *Z. Anal. Chem.*, **157**, 343 (1957).
329. Latimer, W. M., *The Oxidation States of the Elements and Their Potentials in Aqueous Solution*, 2nd ed., Prentice-Hall, New York, 1952.
330. Lauer, R. S., and N. S. Polvektov, *Zavodsk. Lab.*, **25**, 903 (1959).
331. Lauw-Zecha, A. B. H., S. S. Lord, and D. N. Hume, *Anal. Chem.*, **24**, 1169 (1952).
332. Laux, P. G., and O. L. Isaacs, *General Electric Rept.*, **R59FPD934** (1959).
333. Leaf, A. C., *Talanta*, **6**, 265 (1960).
334. Leddicotte, G. W., and F. L. Moore, *J. Am. Chem. Soc.*, **74**, 1618 (1952).
335. Ledoux & Co., Teaneck, New Jersey, *Laboratory Manual*, 1959.
336. Lee, K. S., E. O. Price, and J. E. Land, *J. Am. Chem. Soc.*, **78**, 1325 (1956).
337. Levy, A. G., *Analyst*, **40**, 204 (1915).
338. Liebhafsky, H. A., H. G. Pfeiffer, E. H. Winslow, and P. D. Zemany, *X-Ray Absorption and Emission in Analytical Chemistry*, Wiley, New York, 1960.
339. Lokka, L., *Comm. Géol. Finlande, Bull. No.* **149** (1950).
340. Lomonosova, L. S., *Izvest. Akad. Nauk S.S.R.*, **14**, 692 (1950).
341. Lomonosova, L. S., *Zavodsk. Lab.*, **21**, 1080 (1955).

342. Long, J. V. P., *Analyst*, **76,** 644 (1951).
343. Luke, C. L., *Anal. Chem.*, **31,** 904 (1959).
344. Majumdar, A. K., and A. K. Mukherjee, *Anal. Chim. Acta*, **19,** 23 (1958).
345. Majumdar, A. K., and A. K. Mujherjee, *Anal. Chim. Acta*, **21,** 245 (1959).
346. Majumdar, A. K., and A. K. Mukherjee, *Anal. Chim. Acta*, **21,** 330 (1959).
347. Majumdar, A. K., and A. K. Mukherjee, *Anal. Chim. Acta*, **22,** 25 (1960).
348. Majumdar, A. K., and A. K. Mukherjee, *Anal. Chim. Acta*, **22,** 514 (1960).
349. Majumdar, A. K., and A. K. Mukherjee, *Anal. Chim. Acta*, **23,** 246 (1960).
350. Majumdar, A. K., and A. K. Mukherjee, *Naturwissenschaften*, **44,** 491 (1957).
351. Majumdar, A. K., and A. K. Mukherjee, *Naturwissenschaften*, **45,** 239 (1958).
351a. Majumdar, A. K., and A. K. Mukherjee, *Z. Anal. Chem.*, **189,** 339 (1962).
351b. Majumdar, A. K., and B. K. Pal, *Anal. Chim. Acta*, **24,** 497 (1961).
351c. Majumdar, A. K., and B. K. Pal, *Anal. Chim. Acta*, **27,** 356 (1962).
351d. Majumdar, A. K., and B. K. Pal, *Z. Anal. Chem.*, **184,** 115 (1962).
352. Majumdar, A. K., and J. B. Ray Chowdhury, *Anal. Chim. Acta*, **19,** 18 (1958).
353. Majumdar, A. K., and J. B. Ray Chowdhury, *Naturwissenschaften*, **44,** 420 (1957).
354. Malissa, A., *Mikrochim. Acta*, **6,** 726 (1958).
355. Mallett, M. W., *Talanta*, **9,** 133 (1962).
356. Mambetov, A. A., and N. A. Pzaeva, *Azerbaidzh. Sel-Khoz. Inst.*, **11,** 221 (1960); *Ref. Zh. Khim.*, **1960,** Abstr. No. 88299; through *Anal. Abstr.*, **8,** 3683 (1961).
357. Marignac, J., *Am. Chim. Phys.*, **8,** 4 (1866).
358. Marignac, J., *Am. Chim. Phys.*, **8,** 5 (1866); **9,** 249 (1866).
359. Marignac, J., *Arch. Sci. Phys. Nat.*, **29,** 265 (1867).
360. Martin, J. P., National Research Corporation, Cambridge, Mass., private communication (1959).
361. Marzys, A. E. O., *Analyst*, **79,** 327 (1954).
362. Marzys, A. E. O., *Analyst*, **80,** 194 (1955).
363. May, S. L., A. W. Henderson, and H. A. Johansen, *Ind. Eng. Chem.*, **46,** 2495 (1954).
364. McDuffie, B., W. R. Bandi, and L. M. Melnick, *Anal. Chem.*, **31,** 1311 (1959).
365. McKaveney, J. P., *Anal. Chem.*, **33,** 744 (1961).
366. McKaveney, J. P., Crucible Steel Company, private communication (1961).
366a. McKaveney, J. P., and G. L. Vassilaros, *Anal. Chem.*, **34,** 384 (1962).
367. McKinley, T. D., E. I. du Pont de Nemours & Co., private communication (1959).
368. McKinley, T. D., L. S. Brooks, and L. R. Hoidal, paper presented at Pittsburgh Conference on Analytical Chemistry and Applied Spectroscopy, 1959.
369. Meadows, J. W. T., G. M. Matlack, and G. B. Nelson, *Talanta*, **6,** 246 (1960).
370. Meggers, W. F., C. H. Corliss, and B. F. Scribner, *Natl. Bur. St., Monograph #32* (1961).
371. Melikow, P., and L. Pissarjewsky, *Z. Anorg. Chem.*, **20,** 340 (1899).
372. Mellish, C. E., *U. K. At. Energy Auth. Rept.*, **A.E.R.E. 1/M 39** (1955).
373. Mellor, J. W., *A Treatise in Quantitative Inorganic Analysis*, Charles Griffin and Co., London, 1913.
373a. Menon, V. P., N. Mahadavan, K. Srinivasula, and C. Venkateswalu, *J. Sci. Ind. Res. (India)*, **21B,** 20 (1962); through *Anal. Abstr.*, **9,** 3166 (1962).
374. Mercer, R. A., and R. A. Wells, *Analyst*, **79,** 339 (1954).
375. Mercer, R. E., and A. F. Williams, *J. Chem. Soc.*, **1952,** 3399.
376. Merill, H. B., *J. Am. Chem. Soc.*, **43,** 2378 (1921).
377. Metzger, F. D., and C. E. Taylor, *Chem. News*, **100,** 154, 257, 270 (1909).

378. Mikhailov, G. V., Z. V. Turvotseva, and R. Sh. Khalitov., *Zh. Analit. Khim.*, **12,** 338 (1957).
379. Miller, G. L., *Tantalum and Niobium*, Academic Press, New York, 1959.
380. Milner, G. W. C., *U. K. At. Energy Auth. Rept.*, **C/R 852** (1956).
381. Milner, G. W. C., G. A. Barnett, and A. A. Smales, *Analyst*, **80,** 380 (1955).
382. Milner, G. W. C., and J. W. Edwards, *Anal. Chim. Acta*, **13,** 230 (1955).
383. Milner, G. W. C., and A. A. Smales, *Analyst*, **79,** 315 (1954).
384. Milner, G. W. C., and A. A. Smales, *Analyst*, **79,** 425 (1954).
385. Milner, G. W. C., and A. J. Wood, *U. K. At. Energy Auth. Rept.*, **C/R 895** (1952).
386. *Mining World*, January, 1958, p. 38.
387. Mitchell, B. J., *Anal. Chem.*, **30,** 1894 (1958).
388. Moore, F. L., *Anal. Chem.*, **27,** 70 (1955).
389. Moore, F. L., *Anal. Chem.*, **28,** 997 (1956).
390. Morris, D. F. C., and A. Olya, *Talanta*, **4,** 194 (1960).
391. Morris, D. F. C., and D. Scargill, *Anal. Chim. Acta*, **14,** 57 (1956).
392. Mortimore, D. M., P. A. Romans, and J. L. Tews, *Appl. Spectroscopy*, **8,** 24 (1954).
393. Moshier, R. W., and J. E. Schwarberg, *Anal. Chem.*, **29,** 947 (1957).
394. Motojima, K., and H. Hashitani, *Anal. Chem.*, **33,** 48 (1961).
395. Mukaewaki, K., *Japan Analyst*, **8,** 219 (1959).
396. Mukhina, Z. S., and A. A. Tikhonova, *Zavodsk. Lab.*, **22,** 1154 (1956).
397. Müller, H., University of Breisgau, Thesis, 1958.
398. Münchow, P., *Chemiker Z.*, **84,** 490 (1960).
399. Mundy, R. J., *Anal. Chem.*, **27,** 1408 (1955).
400. National Research Corporation, Cambridge, Mass., *Operating Instructions for Vacuum Fusion Apparatus*, 1960.
400a. National Research Corporation, Cambridge, Mass., private communication (1961).
401. Nazarenko, I. I., *Akad. Nauk S.S.S.R.*, **188** (1957); *Ref. Zh. Khim.*, **1958,** Abstr. No. 46395.
402. Nazarenko, V. A., and E. A. Birynk, *Zavodsk. Lab.*, **25,** 28 (1959).
403. Nazarenko, V. A., G. V. Flyantikova, and N. V. Lebedeva, *Zavodsk. Lab.*, **23,** 891 (1957).
404. Nazarenko, V. A., and M. B. Shustova, *Zavodsk. Lab.*, **23,** 1283 (1957); through *Anal. Abstr.*, **5,** 2600 (1957).
405. Nedler, V. V., *Zavodsk. Lab.*, **23,** 1136 (1957); through *Anal. Abstr.*, **5,** 2600 (1957).
406. Niebuhr, J., in *Plansee Proceedings, 1958*, F. Benesovsky, Ed., Springer, Vienna, 1959.
407. Nikitina, E. I., *Zh. Analit. Khim.*, **13,** 72 (1958).
408. Nikitina, E. I., *Zavodsk. Lab.*, **27,** 663 (1961).
409. Nomokonova, N. A., and I. F. Morozova, *Inst. Min. Tsvetn. Metallurg. S.S.S.R.*, **1957,** [4 (19)], 14; *Ref. Zh. Khim.*, **1958,** Abstr. No. 70505; through *Anal. Abstr.*, **6,** 1621 (1959).
410. Norton, J. T., and A. M. Lowry, *J. Metals*, 1–2, 137 (1949).
411. Norwitz, G., M. Codell, and F. Venderame, *Frankford Arsenal Report* #**MR-545,** *Project No. TB4-15C-3* (1953).
412. Norwitz, G., and M. Codell, *Frankford Arsenal Report* #**MR-552,** *Project No. TB4-15C-3* (1953).
413. Norwitz, G., and M. Codell, *Anal. Chem.*, **26,** 1230 (1954).

414. Norwitz, G., M. Codell, and J. J. Mikula, *Anal. Chim. Acta*, **11**, 173 (1954).
415. Oak Ridge National Laboratory, *Master Analytical Manual, TID-7015* (Section 2), 1957.
416. Oda, N., and M. Idohara, *J. Japan Inst. Metals (Sendai)*, **25**, 52 (1961).
416a. Oda, N., and M. Idohara, *J. Japan Inst. Metals (Sendai)*, **25**, 52 (1961).
416b. Oda, N., N. Katayana, and K. Endo, *Trans. Japan Inst. Metals*, **3**, 142 (1962); *J. Japan Inst. Metals (Sendai)*, **25**, 693 (1961).
417. Oka, Y., and M. Miyamoto, *J. Electrochem. Soc. Japan*, **17**, 114, 183 (1949); through *Chem. Abstr.*, **48**, 14136, 14137 (1954).
418. Oka, Y., and M. Miyamoto, *Sci. Rept. Res. Inst. Tohoku Univ.*, **1949**, 1, Ser. A, 115; through *Brit. Abstr.*, **C-485**, 3892 (1950).
419. Olson, D. G., *Talanta*, **6**, 201 (1960).
420. Oshman, V. A., *Zavodsk. Lab.*, **12**, 154 (1946).
421. Osmond, R. G., and A. A. Smales, *U. K. At. Energy Auth. Repts.*, **AERE C/R, 1233** (1953).
422. Owen, L. B., and J. Y. Ellenburg, *Anal. Chem.*, **23**, 1823 (1951).
423. Palilla, F. C., N. Adler, and C. F. Hiskey, *Anal. Chem.*, **25**, 926 (1953).
424. Patrovsky, V., *Chem. Listy*, **51**, 968 (1957); **52**, 255 (1958).
424a. Patrovsky, V., *Coll. Czech. Chem. Commun.*, **27**, 1824 (1962).
425. Pemberton, R., *Analyst*, **77**, 287 (1952).
426. Penner, E. M., and W. R. Inman, Department of Mines and Techn. Surveys, Mines Branch, Ottawa, *Internal Report MS-62-46* (1962).
427. Pennington, U. E., *J. Am. Chem. Soc.*, **18**, 52 (1896).
428. Peterson, R. C., W. M. Fassel, and M. E. Wadsworth, *Trans. Am. Inst. Min. Met. Eng.*, **200**, 1038 (1954).
429. Phillips, G., and E. N. Jenkins, *J. Inorg. & Nucl. Chem.*, **4**, 220 (1957).
430. Pickup, R., *Colonial Geol. Mineral Resources*, **5**, 174 (1955).
431. Pickup, R., *Colonial Geol. Mineral Resources*, **3**, 358 (1953).
432. Pied, H., *Compt. Rend.*, **179**, 897 (1924).
433. Pilipenko, A. T., and V. A. Obolonchik, *Akad. Nauk Ukr. S.S.R.*, **8**, 131 (1960); *Ref. Zh. Khim.*, **1961**, Abstr. No. 10D90; *Anal. Abstr.*, **8**, 4597 (1961).
434. Platonov, M. S., N. F. Krivoshlykov, and A. A. Marakaev, *Zh. Obschei. Khim.*, **6**, 1815 (1936).
435. Platonov, M. S., and N. F. Krivoshlykov, *Trudy Vsesoynz. Konferentsii, Anal. Khim.*, **2**, 359 (1943); through *Chem Abstr.*, **39**, 3492 (1945).
436. Poehlman, W. J., and R. E. Sarnowski, *J. Opt. Soc. Am.*, **42**, 489 (1952).
437. Ponomarev, A. I., and A. Ya. Sheskol'skaya, *Zh. Analit. Khim.*, **12**, 355 (1957); **14**, 67 (1959).
438. Ponomarev, A. I., and A. Ya. Sheskol'skaya, *Akad. Nauk S. S. S. R.*, **1960** (4), 240; *Ref. Zh. Khim.*, **1961**, Abstr. No. 1D98.
439. Popa, G., D. Negoiu, and G. Bainlesco, *Z. Anal. Chem.*, **165**, 16 (1959).
440. Popel, A. A., and L. P. Maksimova, *Zh. Khim.*, **1957**, Abstr. No. 19572.
441. Popova, N. M., and A. F. Platonova, *Zavodsk. Lab.*, **16**, 1182 (1950).
442. Powell, A. R., and W. R. Schoeller, *Analyst*, **50**, 485 (1925).
443. Powell, A. R., W. R. Schoeller, and C. Jahn, *Analyst*, **60**, 506 (1935).
444. Powers, M. C., *X-Ray Fluorescent Spectrometer Conversion Tables*, Philips Electronics, Mount Vernon, New York, 1957.
445. Rankama, K., *Compt. Rend. Soc. Géol. Finlande*, **14** (1939) and **15** (1941).
446. Rankama, K., *Comm. Géol. Finlande, Bull. No.* **133**, 1 (1944).

447. Rankama, K., and O. Joensun, *Comm. Géol. Finlande, Bull. No.* **138**; *Compt. Rend. Soc. Géol. Finlande, No.* **19**, 8 (1946).
448. Rankama, K., and Th. G. Sahama, *Geochemistry*, The University of Chicago Press, 1949.
449. Remy, H., *Treatise on Inorganic Chemistry*, translated by J. S. Anderson, Elsevier Publishing Co., New York, 1956, Vol. II.
450. Rienäcker, G., and W. Schiff, *Z. Anal. Chem.*, **94**, 415 (1933).
451. Risi, L. R., Shieldalloy, Newfield, New Jersey, private communication (1958).
452. Rizak, F., Vanadium Corporation of America, private communication (1961).
453. Roake, W. E., *U. S. At. Energy Comm. Repts.*, **AEC-D**, **3201** (1951).
454. Robins, D. A., through G. L. Miller, *Tantalum and Niobium*, Academic Press, New York, 1959.
455. Rodden, C. J., Ed., *The Analytical Chemistry of the Manhattan Project*, McGraw-Hill, New York, 1950.
456. Roos, C. E., and G. Kneip, paper presented before the American Physical Society Meeting, Washington, 1961.
457. Ruff, O., and F. Thomas, *Z. Anorg. Allgem. Chem.*, **148**, 1 (1925).
458. Ryabchikov, D. I., and M. P. Volynets, *Zh. Analit. Khim.*, **4**, 700 (1959).
459. Saint-James, R., and T. Lecomte, *Anal. Chim. Acta*, **24**, 155 (1961).
459a. Saitu, K., and T. Takenchi, *Bunseki Kagaku*, **10**, 1013 (1961); through *Anal. Abstr.*, **58**, 6183 (1963).
460. Sandell, E. B., *Colorimetric Determination of Traces of Metals*, 2nd ed., Interscience Publishers, New York, 1950.
461. Sandell, E. B., *Colorimetric Determination of Traces of Metals*, 3rd ed., Interscience Publishers, New York, 1959.
462. Sano, H., and R. Shiomi, *J. Inorg. & Nucl. Chem.*, **5**, 251 (1958).
463. Sarma, B., and J. Gupta, *J. Indian Chem. Soc.*, **32**, 285 (1955).
464. Savostin, A. D., and I. P. Alimarin, *Vestn. Moskov. Univ.* **45**, 1960; through *Anal. Abstr.*, **7**, 4227 (1960).
465. Sax, N. I., *Handbook of Dangerous Materials*, Reinhold, New York, 1951.
466. Scadden, E. M., and W. E. Ballou, *Anal. Chem.*, **25**, 1602 (1953).
467. Schäfer, H., *Angew. Chem.*, **71**, 153 (1959).
468. Schäfer, H., *Z. Für Naturforschung*, **3b**, 376 (1948).
469. Schäfer, H., L. Bayer, and C. Pietruck, *Z. Anorg. Allgem. Chem.*, **265**, 258 (1951)
470. Schäfer, H., L. Bayer, and C. Pietruck, *Z. Anorg. Allgem. Chem.*, **266**, 140 (1951)
471. Schäfer, H., and G. Breil, *Z. Anorg. Allgem. Chem.*, **267**, 265 (1952).
472. Schäfer, H., and K. D. Dohmann, *Z. Anorg. Allgem. Chem.*, **300** (1–2), 1 (1959).
473. Schäfer, H., C. Goser, and L. Bayer, *Z. Anorg. Allgem. Chem.*, **265**, 258 (1951).
474. Schäfer, H., and C. Pietruck, *Z. Anorg. Allgem. Chem.*, **264**, 2 (1951).
475. Schäfer, H., and C. Pietruck, *Z. Anorg. Allgem. Chem.*, **266**, 152 (1951).
476. Schäfer, H., C. Pietruck, and U. Grözinger, *Z. Anal. Chem.*, **141**, 24 (1954).
477. Schäfer, H., and F. Schulte, *Z. Anal. Chem.*, **149**, 73 (1956).
478. Scherff, H. L., and G. Herrmann, *Z. Elektrochem.*, **64**, 1022 (1960).
479. Scheurer, P. G., and F. Smith, *Anal. Chem.*, **27**, 1616 (1955).
480. Schliessmann, O., *Archiv. Eisenhüttenw.*, **15**, 167 (1941).
481. Schoeller, W. R., *Analyst*, **54**, 453 (1929).
482. Schoeller, W. R., *Analyst*, **57**, 750 (1932).
483. Schoeller, W. R., *Z. Anal. Chem.*, **96**, 252 (1934).
484. Schoeller, W. R., *The Analytical Chemistry of Tantalum and Niobium*, Chapman & Hall, Ltd., London, 1937.

485. Schoeller, W. R., and E. C. Deering, *Analyst*, **52**, 633 (1927).
486. Schoeller, W. R., and C. Jahn, *Analyst*, **52**, 506 (1927).
487. Schoeller, W. R., and C. Jahn, *Analyst*, **54**, 320 (1929).
488. Schoeller, W. R., and A. R. Powell, *J. Chem. Soc.*, **120**, 1931 (1921).
489. Schoeller, W. R., and A. R. Powell, *Analysis of Minerals and Ores of the Rarer Elements*, 3rd ed., Hafner Publishing Co., New York, 1955.
490. Schoeller, W. R., and E. F. Waterhouse, *Analyst*, **49**, 215 (1924).
491. Schofield, T. H., *J. Inst. Metals*, **85**, 372 (1957).
492. Schönberg, N., *Acta Chem. Scand.*, **8**, 240 (1954).
493. Schwarzkopf, P., and R. Kieffer, *Refractory Hard Metals*, The Macmillan Co., New York, 1953.
494. Scott, I. A. P., and R. J. Magee, *Talanta*, **1**, 329 (1958).
495. Scribner, B. F., Chief, Spectrochemistry Section Analytical and Inorganic Chemistry Division, National Bureau of Standards, Washington, D. C., private communication (1961).
496. Sears, G. W., and L. Quill, *J. Am. Chem. Soc.*, **47**, 922 (1925).
497. Seebold, R. E., and L. S. Birks, *J. Nucl. Materials*, **3**, 260 (1961).
498. Semmel, J. W., Preprint No. 161, *Metal Progr.*, **76**, 282 (1959).
499. *Serfass Gas Analyzer*, Fisher Scientific Company, Pittsburgh, 1957.
500. Shaeppi, I. J., and W. D. Treadwell, *Helv. Chim. Acta*, **31**, 577 (1948).
501. Shaevich, A. B., Ya. M. Kalinski, N. I. Chabanenko, and M. A. Pereplkina, *Zavodsk. Lab.*, **24**, 1478 (1958).
502. Shapiro, M. Ya., *Zh. Prikl. Khim.*, **11**, 1028 (1938), through *Chem. Abstr.*, **33**, 1624 (1938).
503. Sharp, H. T., *Chem. Eng.*, **63** (8), 103 (1956).
504. Shcherbakov, V. G., and Z. K. Stegendo, *Zavodsk. Lab.*, **26**, 139 (1960).
505. Simons, E. L., J. E. Fagel, Jr., and E. W. Balis, *Anal. Chem.*, **27**, 1123 (1955).
506. Simpson, O. C., R. J. Thorn, and G. H. Winslow, *U. S. At. Energy Comm. Rept.*, **AECD-2680** (1949).
507. Sims, C. T., *J. Metals*, **10**, 340 (1958).
508. Sims, C. T., *J. Metals*, **13**, 316 (1961).
509. Sims, C. T., W. Klopp, and R. Jaffee, *Trans. Am. Soc. Metals*, **51**, 263 (1959).
510. Smith, H. R., C. D'A. Hunt, and C. W. Hanks, *J. Metals*, **11**, 112 (1959).
511. Smithels, C. J., *Metals*, Reference, Butterworths, London, 1955, Vol. I.
512. Sociètè Gènèrale Mètallurgique de Hoboken, Belgian Patent 470,892 (February, 947).
512a. Spauszus, S., and J. Hupfer, *Chem. Tech. (Berlin)*, **13**, 750 (1961).
513. Speecke, A., and J. Hoste, *Talanta*, **2**, 332 (1959).
514. Steinberg, E. P., *U. S. At. Energy Comm. Rept.*, **ANL-6356** (1961).
515. Stevenson, P. C., and H. G. Hicks, *Anal. Chem.*, **25**, 1517 (1953).
516. Stockhausen, C. J., and D. M. Zall, paper given at the Pittsburgh Conference on Analytical Chemistry and Applied Spectroscopy, 1954.
517. Stonhill, L. G., *Analyst*, **83**, 642 (1958).
518. Stricos, D. P., *U. S. At. Energy Comm. Rept.*, **KAPL-M-DPS-3** (1960).
519. Stromberg, A. G., and L. M. Reinus, *Zh. Fiz. Khim.*, **20**, 693 (1946).
520. Strominger, D., J. M. Hollander, and G. T. Seaborg, *Rev. Mod. Phys.*, **30-2**, 585 (1958).
521. Swoboda, K., and A. Eder, *Mikrochem.*, **36**, **37**, 813 (1951).
522. Szymanski, H. A., and J. H. Archibald, *J. Am. Chem. Soc.*, **80**, 1811 (1958).

523. Tanenbaum, M., and W. V. Wright, *J. Metals*, **14**, 367 (1962).
524. Tarasevich, N. I., and G. V. Kozireva, *Vestn. Moskov. Univ.*, **3**, 185 (1959).
525. Tarasevich, N. I., and K. A. Semenenko, *Zh. Anal. Khim.*, **14**, 705 (1959).
526. Tarasevich, N. I., K. A. Semenenko, and N. F. Melekhina, *Vestn. Moskov. Univ., Ser. Khim.*, **1960 II**, No. 2, 64; through *Anal. Abstr.*, **8**, 75 (1961).
527. Tarasevich, N. I., and A. A. Zheloznova, *Vestn. Moskov. Univ.*, **12 (5)**, 156 (1957).
528. Taylor, D. F., "Tantalum and Tantalum Compounds," in *Encyclopedia of Chemical Technology*, R. E. Kirk and D. F. Othmer, Eds., Interscience, New York, 1954, Vol. 13, p. 600.
529. Taylor, D., Fansteel Corp., North Chicago, Ill., private communication (1962).
530. Telep, G., and D. F. Boltz, *Anal. Chem.*, **24**, 163 (1952).
531. Tews, J. L., and S. L. May, *U. S. Bur. Mines Rept. Invest. No.* **USBM-U-252** (1957).
532. Tews, J. L., and S. L. May, in *Technology of Columbium*, B. W. Gonser and E. M. Sherwood, Eds., Wiley, New York, 1958.
533. Thanheiser, G., *Mitt. Kaiser-Wilhelm Inst. Eisenforsch.*, Düsseldorf, **22**, 255 (1940).
534. Thanheiser, G., *Arch. Eisenhüttenw.*, **14**, 371 (1940).
535. Theodore, M. L., *Anal. Chem.*, **30**, 465 (1958).
536. Thorne, R. P., and B. M. Childs, *U. K. At. Energy Auth. Res. Establ.* **C/R, 1232** (1955).
537. Tikhomiroff, N., *Compt. Rend. Acad. Sci., Paris*, **236**, 12, 1263 (1953).
538. Tombu, C., Société Générale Métallurgique de Hoboken, *Tantale, Niobium, Analyses des Produits Pure Osxyde et Metal*, Laboratory Manual, 1961, and private communication.
539. Tomicek, O., and V. Holocek, *Chem. Listy*, **46**, 11 (1952).
540. Tomicek, O., and L. Jerman, *Chem. Listy*, **46**, 144 (1952).
541. Tompkins, J., X. Kyhm, and W. E. Cohn, *J. Am. Chem. Soc.*, **69**, 2769 (1947).
542. Torrisi, A. F., J. L. Kernahan, and R. E. Fryxell, *Anal. Chem.*, **26**, 733 (1954).
543. Tottle, C. R., *Nucl. Eng.*, **3**, 212 (1958).
544. Tramm, R. S., and K. S. Pevzner, *Zavodsk. Lab.*, **22**, 1025 (1956).
545. Traub, K. W., *Ind. Eng. Chem., Anal. Ed.*, **18**, 122 (1946).
546. Treadwell, W. D., H. Guyer, R. Hauser, and G. Bischofsberger, *Helv. Chim. Acta.*, **35**, 2248 (1952).
547. Troitskii, K. V., *Zh. Analit. Khim.*, **12**, 349 (1957).
548. Umezawa, H., and R. Hara, *Anal. Chim. Acta*, **23**, 267 (1960).
549. Vivarelli, S., *Chim. Ind.*, **37**, 1026 (1955).
550. Vivarelli, S., and D. Cozzi, *Chim. Ind.*, **35**, 637 (1953).
551. Von der Porten, K., Ledoux & Co., Teaneck, New Jersey, private communication (1961).
552. von Kobell, F., *Ak. München, Ber.*, **150** (1860); **210** (1861).
553. Wacker, R. E., and W. H. Baldwin, *U. S. At. Energy Comm. Repts.*, **ORNL-637** (1950).
554. Waehner, K. A., Fansteel Metallurgical Corp. The assistance of Mr. Waehner in the preparation of this section is gratefully acknowledged.
555. Waehner, K. A., Fansteel Metallurgical Corp., private communication (1961).
556. Wahl, W. H., Union Carbide Nuclear Company, Tuxedo, New York, private communication (1962).
556a. Wakamatsu, S., *Japan Analyst*, **9**, 507 (1960).

557. Ward, F. N., and A. P. Marranzino, *Anal. Chem.*, **27**, 1325 (1955); *Science*, **119**, 655 (1954).
558. Wasilewski, R. J., *J. Am. Chem. Soc.*, **75**, 1001 (1953).
559. Waterbury, G. R., and C. E. Bricker, *Anal. Chem.*, **29**, 1474 (1957).
560. Waterbury, G. R., and C. E. Bricker, *Anal. Chem.*, **30**, 1007 (1958).
561. Waterbury, G. R., and C. F. Metz, *Talanta*, **6**, 237 (1960).
562. Weihrich, R., *Die Chemische Analyse in der Stahlindustrie*, 2nd ed. by J. Kassler, Enke, Stuttgart, 1939.
563. Weiss, L., and M. Landecker, *Z. Anorg. Allgem. Chem.*, **64**, 65 (1909).
564. Wells, J. E., *J. Iron Steel Inst.*, **166**, 113 (1950).
565. Wernet, J., *Z. Anorg. Allgem. Chem.*, **267**, 213 (1952).
566. Werning, J. R., K. B. Higbie, J. T. Grace, B. F. Speece, and H. L. Gilbert, *Ind. Eng. Chem.*, **46**, 644 (1954).
567. Wilkins, D. H., *Talanta*, **2**, 355 (1959).
568. Williams, A. F., *J. Chem. Soc.*, **3155** (1952).
569. Williams, L. R., and T. J. Heal, *Plansee Proceedings, 1958*, F. Benesovsky, ed., Springer, Vienna, 1959.
570. Wirtz, H., *Z. Anal. Chem.*, **117**, 6 (1939); **122**, 88 (1941).
571. Wirtz, H., and H. Rothmann, *Z. Erzbergw. Metalhüttenw.*, **11**, 465 (1958).
572. Wish, L., *Anal. Chem.*, **31**, 327 (1959).
573. Wood, G. A., *D.S.I.R. Teddington, Rept.*, No. **CRL/AE 62** (1950).
574. Wood, D. F., and I. R. Scholes, *Anal. Chim. Acta*, **21**, 121 (1959).
575. Yajima, S., E. Shikata, and C. Yamaguchi, *Japan Analyst*, **7**, 721 (1958).
576. Yatsimirskii, K. B., O. M. Drobysheva, and V. O. Rigni, *Zh. Analit. Khim.*, **14**, 60 (1959).
577. Yntema, L. F., and A. Percy, in *Rare Metals Handbook*, C. A. Hempel, Ed., Reinhold, New York, 1954.
578. Young, J. F., *Iron Age*, **168**, 91 (1951).
579. Young, J. P., and J. C. White, *Talanta*, **1**, 263 (1958).
580. Young, J. P., and J. C. White, *Anal. Chem.*, **31**, 393 (1959).
581. Zaidel, A. N., A. A. Petrov, and K. I. Petrov, *Fiz. Sb. L'vovsk. Univ.*, **4**, 206 (1958); *Ref. Zh. Khim.*, **1959**; Abstr. No. 67682.
582. Zaidel, A. N., *Spectrochim. Acta*, **10**, 369 (1958).
583. Zaikovskii, F. V., *Zh. Analit. Khim.*, **11**, 269 (1956).
584. Zaikovskii, F. V., *Zh. Analit. Khim.*, **11**, 553 (1956).
585. Zakharov, E. I., L. V. Lipis, and K. I. Petrov, *Zh. Analit. Khim.*, **14**, 135 (1959).
586. Zeltzer, S., *Collection Czech. Chem. Commun.*, **4**, 319 (1932).

Part II
Section A

TECHNETIUM

By James W. Cobble, *Department of Chemistry, Purdue University, Lafayette, Indiana*

Contents

I. Introduction	408
II. Properties	409
A. Physical Properties	409
B. Nuclear Properties	411
C. Optical Properties	412
D. Chemical Properties	413
E. Electrochemical Properties	417
III. Separation and Isolation	418
A. General Procedures	418
B. Separation Methods	418
1. Precipitation	418
2. Electroplating	419
3. Solvent-Extraction Behavior	420
4. Ion Exchange	421
5. Selective Volatility	421
IV. Detection and Identification	421
A. Radioactivity	423
B. Emission Spectroscopy	423
C. Spot Tests	423
V. Toxicology and Industrial Hygiene	423
VI. Determination of Technetium	424
A. Gravimetric Determinations	424
1. As Metallic Technetium	424
2. As Technetium Dioxide	426
3. As Tetraphenylarsonium Pertechnetate	426
4. As Nitron Pertechnetate	426
5. As Technetium Sulfides	427
6. As Ammonium Pertechnetate	427
B. Titrimetric Determinations	427
1. Neutralization Reactions	427
2. Redox Reactions	427
C. Polarographic Methods	428
D. Spectrophotometric Methods	428
E. Radiochemical Methods	430
VII. Recommended Laboratory Procedures	431
A. Gravimetric Methods	431
1. Nitron Pertechnetate ($C_{20}H_{16}N_3 \cdot HTcO_4$)	431
2. Tetraphenylarsonium Pertechnetate ((C_6H_5)$_4$-AsTcO$_4$)	431

Contents (*continued*)

    B. Spectrophotometric Methods.................... 431
        1. Ascorbic Acid Reduction and Tc(V) Thiocyanate
           Complex................................... 431
        2. Thioglycolic Acid–Tc(VII) Complex............. 432
  References... 432

I. INTRODUCTION

The element of atomic number 43, now called technetium, was first reported by the Noddacks (47) simultaneously with their discovery of rhenium in 1925. On a chemical basis, this discovery was certainly anticipated and to be expected. However, in subsequent years, milligram and, eventually, gram quantities of rhenium became available, whereas no weighable amounts of element 43 were forthcoming. Consequently, credit for the discovery of the element is now almost universally given to Perrier and Segré (49), who, in 1937, 1938, and 1939, produced isotopic quantities of element 43 by transmutation reactions of deuterons and protons on molybdenum. Their name, technetium (symbol Tc), has been accepted officially (32).

Soon after the discovery of nuclear fission, it was realized that isotopes of technetium should be found among the fission products, and in 1940 Segré and Wu (56) reported a six-hour activity (now known to be Tc^{99m}) separated from such a source. In 1946 and 1947 a long-lived isotope of technetium (Tc^{99}) was found and isolated from both irradiated uranium reactor fuel and molybdenum, and several workers separated small but weighable amounts of the element at that time (44,48). The isolation of the first gram of fission-product technetium metal was reported by Cobble *et al.* (16) in 1952; at the present time (1963) gram quantities are being offered for sale by the Isotopes Division of the U. S. Atomic Energy Commission for about $90 per gram.

Concurrently with the chemical investigations, studies in nuclear spectroscopy have proposed some 21 isotopes of technetium, 10 of which are isomeric states. To date the isotopes of technetium having the longest known half lives are Tc^{97} ($\sim 10^5$ years), Tc^{98} (2.6×10^6 years), and Tc^{99} (2.2×10^5 years). Although these species have half lives too short to have existed for the long cosmological times involved since the creation of the elements, there is the possibility that the ground state of Tc^{98} might be long-lived enough to have survived since that time. However, the latest searches, taking advantage of the best radiochemical procedures available, have not found any evidence of primordial terrestrial technetium (7).

Technetium has been observed in certain stars (38,39), and this discovery may prove highly significant with respect to cosmological theory. An analysis of the nuclear reactions proposed for the synthesis of technetium in stellar matter is beyond the scope of the present discussion, but the facts strongly suggest that a continuous production of elements is taking place in the universe (34).

Undoubtedly, fission-product Tc^{99} exists on the earth as a result of nuclear detonations and from the spontaneous and natural neutron-induced fission of uranium and thorium. Recently, Kenna and Kuroda (35) have reported the isolation of a millimicrogram of Tc^{99} from 5.3 kg. of African pitchblende.

The element technetium occupies an interesting and key position in the middle of the transition elements, and a number of chemical studies have been carried out to define the properties of the element and its compounds. Recent articles have summarized these in detail (3,8,60). Radiochemical procedures applicable to trace quantities also have been reviewed (4). As expected, the chemistry of the element lies between that of manganese and rhenium, but it more closely resembles that of rhenium. Analytically, many of the procedures described in this Treatise for rhenium can be used equally well for technetium.

At present there are few, if any, practical uses for the element, although this is largely due to its radioactivity and high cost. Cartledge has reported that the TcO_4^- ion is quite superior to the CrO_4^{-2} ion in inhibiting corrosion of mild carbon steels (13); as little as 5 p.p.m. of technetium protects these steels from attack in aerated distilled water up to at least 250°C.

Because of the short-ranged radioactivity associated with weighable amounts of Tc^{99}, special precautions should be used in handling technetium. The specific activity is about 20 $\mu c.$ per milligram, and the low energy of the beta radiation is absorbed by ordinary glassware. However, ingestion of technetium must be prevented. Although preliminary reports indicate that the body eliminates technetium rather rapidly (28), more recent evidence suggests that it is concentrated in the thyroid tissue (6).

II. PROPERTIES

A. PHYSICAL PROPERTIES

The atomic weight of $_{43}Tc^{99}$ has not been determined chemically or fixed by international agreement. However, fission-product technetium is essentially monoisotopic (30), and the nuclear mass (55) on the chemical scale, 98.91, is recommended for use. The half life of $_{43}Tc^{99}$ is 2.12×10^5 years, and it decays by weak ($\beta^-_{max} = 0.290$ M.e.v.) beta emission (59).

TABLE I
Properties of Technetium Compounds

Compound	Color	Solubility	Additional comments
Tc_2O_7	Yellow	V.s., dioxane	Deliquescent
$TcO_3(?)$	Red	S.	Disprop. to TcO_4^- and TcO_2
TcO_2	Black	Insol.	D., 6.9; s. H_2O_2
$Tc_2O_3(?)$	Black	Insol.	S. H_2O_2
NH_4TcO_4	White	V.s.	D., 2.73; decomp. to $TcO_2 \sim 550°C$.
$KTcO_4$	White	S. (0.63M)	M.p., 540°C.; sublimes at 1000°C.
K_2TcCl_6	Gold	S. (decomp.)	Hydrolyzes in water to TcO_2
$TcCl_4$	Red	S. (decomp.)	
$TcCl_6$	Green	S. (decomp.)	Decomp. to $TcCl_4$ on standing
TcF_6	Gold-yellow	Decomp.	M.p., 33°C.
$HTcO_4$	Yellow[a]	S.	Deliquescent
Tc_2S_7	Black	Insol.	Decomp. to ReS_2 on heating; s. $9M$ HCl
TcS_2	Black	Insol.	S. H_2O_2
$TcO_3F(?)$			
TcO_3Cl	Red-violet in conc. $HCl-H_2SO_4$	S.	
$TcOCl_3(?)$	Light-brown	S.	
$(C_6H_5)_4AsTcO_4$	White	Insol.	More soluble than Re salt
$AgTcO_4$	White	S.s.	
$CsTcO_4$	White	S.	
Nitron (TcO_4)	White	Insol.	
$Tc_2(CO)_{10}$	Colorless	S. ether, toluene; d. water	M.p. 159–160°C.

[a] The red color often reported for this species and its concentrated solutions is believed to be due to small amounts of TcO_3 formed from reduction of Tc(VII) by traces of dust or other organic matter.

The silvery gray metal is hexagonal close-packed and isomorphous with rhenium; the lattice constants are $a = 2.375$ and $c = 4.391$ A., with a density of 11.48. The melting point of the pure metal has been reported to be 2140°C., and the first, second, and third ionization potentials are 7.28, 15.26, and 31.9 v., respectively. The ground-state electronic structure of the gaseous Tc atom is $4d^5 5s^2$, $^6S_{5/2}$. The magnetic susceptibility of the metal indicates that it is weakly paramagnetic (45), with $\chi = 270 \times 10^{-6}$ (c.g.s. units).

Technetium forms a number of soluble compounds. The potassium salt, $KTcO_4$, is more soluble than the corresponding $KReO_4$. Tc_2O_7 dissolves readily to form the strong acid, $HTcO_4$, and K_2TcCl_6 is soluble in acids (in water it rapidly hydrolyzes to precipitate TcO_2). NH_4TcO_4, like NH_4ReO_4, is soluble, and can be dried without decomposition at temperatures slightly above 100°C.

Tc_2S_7, TcS_2, and TcO_2 are all insoluble in water and dilute acids, but can be dissolved in H_2O_2 or, more rapidly, in ammoniacal H_2O_2. The red halide, $TcCl_4$, dissolves but presumably hydrolyzes except in strong acid solutions. Table I summarizes the known properties of some of the compounds.

B. NUCLEAR PROPERTIES

A summary of some of the most important isotopes and their nuclear properties is given in Table II.

Of these isotopes, Tc^{99} is the one that is encountered in weighable amounts, and it can be separated from the fission products or made from

TABLE II
Some Isotopes of Technetium

Mass	$T_{1/2}$	Principal radiations (in M.e.v.)	Method of preparation and additional comments
Tc^{95}	20.0 hr.	β^+ E.C. $\gamma = 0.762, 0.932, 1.071$	$Mo^{95}(p,n)$
Tc^{95m}	60 days	E.C., $\beta^+ = 0.68, 0.46$ $\gamma = 0.03896, 0.2042, 0.584$	$Mo^{95}(p,n)$
Tc^{96}	4.20 days	E.C., $\gamma = 0.771, 0.806, 0.842, 1.119$	$Mo^{96}(p,n)$
Tc^{96m}	51.5 min.	$\gamma = 0.0344$	$Mo^{96}(p,n)$
Tc^{97}	2.6×10^6 yr.	E.C.	Daughter Ru^{97}
Tc^{97}	92 days	$\gamma = 0.097$	$Mo^{97}(p,n)$
Tc^{98}	1.5×10^6 yr.	$\beta^- \approx 0.3$ $\gamma = .740, .65$	$Mo^{98}(p,n)$; $Tc^{98}(n,\gamma)Tc^{99m}$, $\sigma^a \approx$ 2 b.
Tc^{99}	2.2×10^5 yr.	$\beta^- = 0.290$ No γ	Fission; $Tc^{99}(n,n')Tc^{99m}$, $\sigma \approx 0.05$ b. $Tc^{99}(n,\gamma)Tc^{100}$, $\sigma = 22$ b.
Tc^{99m}	6.0 hr.	$\gamma = 0.14$	Daughter Mo^{99}
Tc^{100}	15.8 sec.	$\beta^- = 2.9, 3.4$ $\gamma = 0.54, 0.60$	$Mo^{100}(p,n)$; $Tc^{99}(n,\gamma)$
Tc^{101}	13.2 min.	$\beta^- = 1.32, 1.07$ $\gamma = 0.31, 0.54$	$Mo^{100}(d,n)$

[a] Cross section values, σ, are given in barns (10^{-24} cm.^{2}) for thermal neutrons.

neutron bombardment of molybdenum: $Mo^{98}(n,\gamma)Mo^{39} \xrightarrow{\beta^-} Tc^{99}$. Tc^{100} is of particular interest, since its short half life and high cross section for formation from Tc^{99} by neutron irradiation provide a sensitive ($\sim 2 \times 10^{-11}$ g.) analytical method for the detection of technetium by activation analysis.

Occasionally it is of interest to "tag" trace technetium solutions and compounds by a more active isotope of higher specific activity. Tc^{95} (60 days) and Tc^{96} (4.20 days) are useful for this purpose. Actually, in practice, proton bombardment of natural molybdenum gives rise to a number of isotopes, of which Tc^{95m}, Tc^{95}, Tc^{96}, and Tc^{97m} have half lives greater than 20 hours. By use of separated molybdenum isotopes, however, it is possible to obtain selected technetium nuclides of higher purity for tracer purposes.

C. OPTICAL PROPERTIES

The arc and spark spectra of technetium have been determined (37), and the most sensitive lines are summarized in Table III.

TABLE III
Principal Optical Emission Spectral Lines

Wavelength, A.	Relative intensity	Interferences[a]
4297.06	500	None
4262.26	400	None
4238.19	300	None
4031.63	300	None[b]
3636.10	400	Cu, 3636.916 A.
3195.21	200	Ru, 3195.157 A.
3212.01	300	Re, 3211.756 A.
3237.02	400	None
2647.02	600	Ru, 2647.315 A.
2610.00	800	Ru, 2610.078 A.
2543.34	1000	None

[a] Interferences refer to neighboring elements in the periodic table, rhenium and copper.
[b] This line appears to have the least interference of all those examined.

The principal x-ray emission spectrum of technetium has been determined (51), and is as follows: K_{α_1}, 673.57 (X.U.); K_{α_2}, 677.90; $K_{\beta_{1,3}}$ 600.20; and $K_{\beta_{2,4}}$, 588.99.

The absorption spectra for a number of technetium species in solution have been determined. Fig. 1 is a composite diagram of the particularly

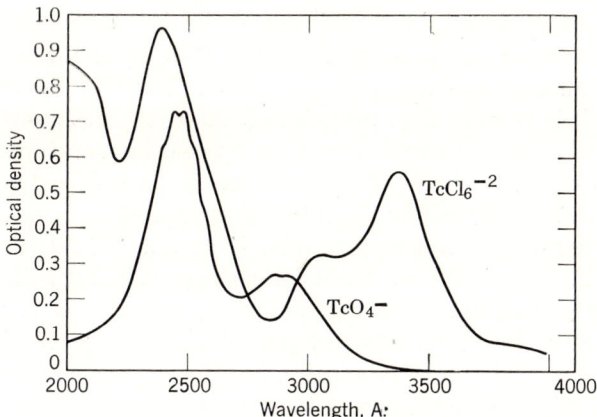

Fig. 1. The ultraviolet absorption spectra of the TcO_4^- (18) and $TcCl_6^{-2}$ (in $1M$ HCl) (11) ions.

important ultraviolet absorption spectra for two species, TcO_4^- and $TcCl_6^{-2}$ (11). Table IV gives a summary of the significant features of the spectra.

The high molar extinction coefficients listed in Table IV make spectrophotometric methods attractive for microquantities of the element. As little as 0.1 γ/ml. can be detected using the Tc(V) thiocyanate complex.

TABLE IV
Absorption Spectra of Technetium Solutions

System	λ_{max}, A.	$\epsilon_{\lambda_{max}}$[a]
TcO_4^- in water	2440 (doublet)	6220
	2875	2360
$TcCl_6^{-2}$ in $1M$ HCl	2400	18,900
	3380	11,000
Tc(VII)–1,5-diphenyl carbohydrazide complex in CCl_4	5200	50,000
Tc(V)–thiocyanate complex in organic solvents	5130	52,200
Tc(VII)–thioglycolate complex in ether	6550	~1800
Tc(IV)–ferrocyanide complex in water	6800	10,800

[a] Molar extinction coefficient.

D. CHEMICAL PROPERTIES

The aqueous chemistry of technetium is characterized by multiple oxidation states, numerous complex ions, and few, if any, simple ionic species. The chemistry of technetium and rhenium are so similar that the differences between the two elements becomes of major interest. Fortunately, a number of these are now known, so that identification and separation of the two elements in their mutual presence is possible.

Technetium metal burns in oxygen at elevated temperatures to form the volatile and deliquescent heptoxide Tc_2O_7. It is oxidized by hot nitric acid and bromine water but, in contrast to rhenium, is only slowly attacked by alkaline hydrogen peroxide, if at all. The hydrohalic acids have little effect upon the metal, but aqua regia and hot sulfuric acid will oxidize it into solution. The free halogens, except iodine, should react with the metal; fluorine reacts to form TcF_6 (57), and chlorine reacts to give $TcCl_6$ (19). Thermodynamically, in solution, the end result of treatment by an oxidizing agent of a potential greater than about 0.5 v. is the pertechnetate ion, TcO_4^-. The corresponding pertechnetic acid, $HTcO_4$, is strong, and a pK of -1.5 may be estimated from the corresponding values of -1.25 and -2.25 listed for $HReO_4$ and $HMnO_4$, respectively (15). Acidic solutions of TcO_4^- may be dehydrated to give successively concentrated solutions of $HTcO_4$ (aq), $HTcO_4$ (c), and Tc_2O_7. Because of the difficulty in maintaining equilibrium water-vapor conditions in such evaporations under ordinary conditions, however, volatilization loss of technetium almost always occurs. Table V lists the vapor pressures of Tc_2O_7 and saturated solutions of $HTcO_4$ as a function of temperature, calculated from data in the literature (58).

TABLE V
Vapor Pressure of Technetium Heptoxide and Saturated Aqueous Pertechnetic Acid

Temperature, °C.	Tc_2O_7 v.p., in mm. Hg	H_2O over $HTcO_4$ (satd.)[a]
0	1.9×10^{-8}	0.32
25	2.6×10^{-6}	1.7
100	0.10	69
120	1.0 (m.p.)	—
150	3.0	—
177	—	760 (b.p.)
200	34	—
310	760 (b.p.)	—

[a] Calculated from vapor pressure equations given in Reference 58.

The heptoxide of technetium is considerably more volatile than Re_2O_7, and this fact has been used as the basis for a method of separation of the two compounds. Solutions or solids containing rhenium and technetium can be treated with fairly concentrated sulfuric or perchloric acids and distilled from a glass distilling flask. The addition of sodium peroxydisulfate is advantageous if sulfuric acid is used. After the initial excess water has been distilled over, Tc_2O_7 will distill from the system as the boiling point of the mixture begins to rise. Appreciable amounts of halide ions will interfere with the distillation until they have volatilized as the corresponding acid. Appreciable amounts of Re_2O_7 will not distill from the mixture until the boiling point approaches that of the constant-boiling acids.

Solutions of pertechnetates may be reduced by many common reducing agents but, except in complexing media or in concentrated sulfuric acid solutions, the result of this reduction is solid TcO_2 (hydrate) or the hydrotechnetate(VI) ion. This latter species has not been positively identified in solution; there is evidence (23) that it is analogous to the hydrorhenate-(VI) species, ReH_8^{-2}, solid compounds of which have the formula K_2ReH_8 (25). Evidence has been obtained for green Tc(III) and pink Tc(IV) in phosphate buffer solutions by controlled-potential electrolysis (42). These species are undoubtedly complex in nature. Concentrated hydrochloric acid reduces TcO_4^- (but not ReO_4^-) to Tc(V), possibly to $TcOCl_4^-$, which disproportionates in $1M$ HCl to $TcCl_6^{-2}$ and TcO_4^- (11).

The action of hydrogen sulfide or aqueous sulfides on soluble pertechnetates results in precipitation of the heptasulfide, Tc_2S_7, usually with excess sulfur. This excess sulfur can be removed from the dried sulfide by a suitable organic solvent such as carbon bisulfide. The sulfide is soluble in concentrated hydrochloric acid, is oxidized by hot nitric acid, and is readily dissolved by alkaline hydrogen peroxide. Tc_2S_7 cannot be precipitated from $9N$ HCl (because of reduction of TcO_4^-), even though Re_2S_7 can, and this fact is one basis for chemical separation of the two elements. The disulfide can be prepared by heating Tc_2S_7 to 300 to 500°C. in nitrogen or in a vacuum. It is a dark, inert substance, soluble only in strong oxidizing agents. Both sulfides can be oxidized to the heptoxide or tetrachloride by the action of oxygen or chlorine at elevated temperatures. In hydrogen, both sulfides are reduced to the metal at higher temperatures (600°C.).

Technetium dioxide is formed by pyrolysis of ammonium pertechnetate at 400 to 500°C.:

$$NH_4TcO_4 \rightarrow TcO_2 + 2H_2O + 1/2 N_2 \qquad (1)$$

The hydrated dioxide also can be made by action of metallic zinc on an acid solution of TcO_4^-, and is deposited by electrolysis of neutral or alkaline pertechnetates. Hydrated ReO_2 cannot be prepared by electrolysis in this manner, and this electrochemical method can be used to separate Tc from

Re. Electrolysis from dilute sulfuric acid solutions leads to the formation of bright metallic deposits of both metals.

Anhydrous TcO_2 reacts only slowly with ceric ion and other oxidizing agents, but the hydrous material dissolves more rapidly. Both substances react rapidly with alkaline H_2O_2. The dioxide is extremely stable, and can be volatilized without decomposition at temperatures >1000°C.

The formation of what may have been Tc_2O_3 has recently been reported (53,54). The material appeared as a black suspension resulting from a 4-electron coulometric reduction of pertechnetate ion at a massive mercury electrode, in either dilute acidic or alkaline media in the absence of complex anions. In this respect it is interesting to note that Nelson (46) reported a similar precipitate, not behaving like TcO_2, from the hydrolysis of a technetium halide now known to have contained $TcCl_4$ in acid solutions. This species may have disproportionated to some extent under these conditions, and would tend to verify the work of Salaria, Rulfs, and Elving (53,54) referred to above.

The reaction of Tc_2O_7 with CCl_4 in a sealed bomb leads to the formation of blood-red crystals of $TcCl_4$ (36); the corresponding $ReCl_4$ has not been reported. The complex halide K_2TcCl_6 has been prepared by reduction of TcO_4^- with iodide in concentrated hydrochloric acid solutions. It is quite unstable toward hydrolysis into TcO_2 (hydrate) in water. Recently the preparation of K_2TcBr_6 and K_2TcI_6 have been reported and the spectra of the $TcBr_6^{-2}$ and TcI_6^{-2} ions in solution have been measured (12). $TcCl_6$ has been prepared by direct chlorination of the metal (19); it dissolves in water to give TcO_2 and TcO_4^-.

In the preparation of the yellow Tc_2O_7 by high-temperature oxidation of the metal or other low oxidation states, traces of red deposits are sometimes noted in the combustion tube. Indeed the first attempts in preparing fractional milligram quantities of the oxide led to reports of a red volatile substance, and a preliminary formula was determined to be TcO_3. This latter report was shown to be incorrect when larger quantities of technetium became available, but the nature of the red deposit is interesting. ReO_3 is a well-known compound and is red, and it seems logical to assume that the red material noted above comes from incomplete oxidation of technetium to give TcO_3.

No successful preparation of pure TcO_3 in sufficient quantities to definitely establish its formula has been reported; however, trace quantities of the red substance when placed in water dissolve to give a pink solution, which rapidly precipitates the black TcO_2 and forms TcO_4^-. These observations lend weight to the assignment of the formula TcO_3 to the red substance. Thermodynamic considerations of the known bond energies

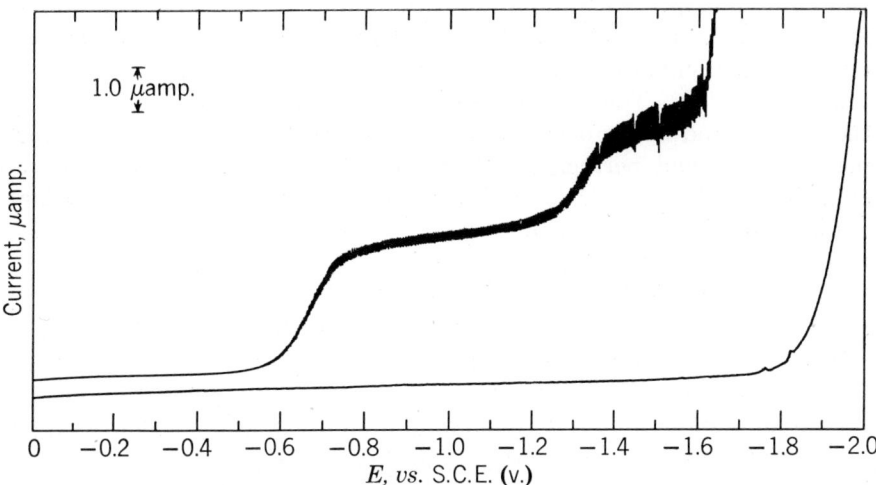

Fig. 2. Polarogram showing the reduction of a $10^{-4}M$ TcO_4^- solution in pH 7 phosphate buffer (42). The lower line represents the residual current curve of the buffer. Reproduced by permission of Pergamon Press from *Advances in Polarography*.

in TcO_2 and Tc_2O_7, and in ReO_2, ReO_3, and Re_2O_7, actually predict that TcO_3 should undergo disproportionation in aqueous solutions:

$$H_2O + 3TcO_3 \rightarrow 2TcO_4^- + TcO_2 + 2H^+ \qquad (2)$$

E. ELECTROCHEMICAL PROPERTIES

The limiting conductance of TcO_4^- has been reported to be 58.1 (42), somewhat higher than ReO_4^-, as expected. The TcO_2/TcO_4^- electrode appears to be reversible, and by a combination of thermochemical and electrochemical measurements an oxidation–reduction potential diagram has been constructed (17):

$$Tc \xrightarrow{-0.240} TcO_2 \xrightarrow{-0.8} TcO_3 \xrightarrow{-0.7} TcO_4^-$$

with -0.782 spanning TcO_2–TcO_4^- and -0.472 spanning Tc–TcO_3.

Cartledge and Smith report -0.738 v. for the TcO_2/TcO_4^- couple (14).

Polarographic studies on technetium were reported in 1949 (9), and very extensive analytical work has been done by Miller, Kelley, and Thomason (42). Polarographic waves have been obtained in both acid and basic solutions, and in $0.1M$ NH_4Cl, $0.1M$ KCl, borate buffer (pH 10), and phosphate buffer (pH 7) solutions. Suitable polarographic analyses can be carried out in the $0.1M$ KCl and pH 7 and 10 solutions (see Fig. 2). Salaria, Rulfs, and Elving, in recent polarographic studies (54), have obtained good results for small amounts of technetium in pH 2 hydrochloric acid at -0.14 v. The waves at -0.91 and -1.12 v. also are useful, and these results and similar ones in pH 14 media make possible the estimation of pertechnetate and perrhenate when present together. Evidence for an aqueous technetium hydride species recently has been reported from polarographic work in $2M$ KCl (20).

III. SEPARATION AND ISOLATION

A. GENERAL PROCEDURES

The largest amounts of technetium are now produced from the fission of uranium and are found in the so-called waste products (i.e., solvent-extraction raffinates). It is separated from most other fission products and actinide elements by certain solvent-extraction, ion-exchange, precipitation, and volatilization procedures, some of which will be described. Another important separation problem is the one of separation of rhenium and technetium. The chemistry of these two elements is so similar that special techniques have had to have been devised to effect their separation and identification in the presence of each other.

One aspect of the chemistry of technetium must always be considered carefully in any assay or separation procedure. Boiling concentrated acid solutions of technetium can lead to vaporization loss of Tc_2O_7, which is even more volatile than the corresponding rhenium compound. Consequently, procedures involving the initial dissolution of technetium compounds either should be carried out in dilute acid solutions or the system should be designed so as to collect the vapors in a suitable trap. A better procedure would be the use of alkaline fusion methods, using sodium peroxide or sodium carbonate–sodium nitrate mixtures.

For pure compounds of technetium, alkaline peroxide usually acts rapidly to oxidize and dissolve the sulfides and TcO_2; the halides dissolve and decompose in pure water, usually resulting in a TcO_2 precipitate. Therefore, alkaline H_2O_2 is a good solvent for the halides also. The metal cannot be dissolved in this reagent, but it will dissolve in nitric acid or, slowly, in ceric solutions in dilute sulfuric acid.

B. SEPARATION METHODS

1. Precipitation

Technetium sulfide, like rhenium, follows the acid-insoluble elements, and Tc_2S_7 is quite insoluble. As little as 3 mg./liter (4) can be precipitated in dilute acids (up to $4N$ HCl or $8N$ H$_2$SO$_4$). Bromine water is usually added first to ensure that all of the technetium is present as Tc(VII); the process of precipitation is more rapid and complete at elevated temperatures. Digestion at 90°C. is usually effective in coagulating the precipitate and helping to ensure complete precipitation. Under these conditions both rhenium and ruthenium sulfides also precipitate. Ruthenium can be removed previously by precipitation of RuO$_2$ from dilute acid solutions, using ethyl alcohol as the reducing agent. Molybdenum can be removed by prior precipitation, using 8-hydroxyl quinoline or α-benzoin oxime. Rhenium and technetium can be separated by selective sulfide precipitation from $10N$ HCl. In this case, the solution containing rhenium and technetium is heated for about 1 hour in $10N$ HCl and H$_2$S is added. Re$_2$S$_7$ will precipitate, but apparently the TcO$_4^-$ is reduced to TcOCl$_4^-$ and possibly to TcCl$_6^{-2}$ in the concentrated acid, and is not carried down with the rhenium.

Tetraphenylarsonium chloride, (C$_6$H$_5$)$_4$AsCl, forms the insoluble tetraphenylarsonium pertechnetate from solutions containing as little as 5 mg./liter (4). In basic solutions molybdenum does not interfere, but high concentrations of ClO$_4^-$ will precipitate the tetraphenylarsonium perchlorate, and traces of Ru tend to be carried. Rhenium, of course, also is quantitatively precipitated. The tetraphenyl precipitates can be dissolved in an alcohol solution, and the rhenium and technetium subsequently separated by ion exchange (8). Nitron, C$_{20}$H$_{16}$N$_3$, will precipitate both TcO$_4^-$ and ReO$_4^-$ quantitatively, although the pertechnetate is somewhat more soluble.

The pertechnetates are usually more soluble than the corresponding perrhenates, and thallous, silver, and potassium pertechnetates are not suitable for quantitative work.

2. Electroplating

Technetium can be electroplated from solution under a number of conditions. Technetium dioxide results from almost all of these procedures, but there has been one report of apparently metallic technetium being plated from $2N$ H$_2$SO$_4$ containing trace fluoride (8). Quantitative yields (22) of the dioxide have been reported for tracer technetium on a platinum cathode in dilute sulfuric acid (pH 2.36) at -0.8 v. vs. S.C.E. However, no reports of yields this high have been made for weighable quantities of technetium. Unfortunately, under many of the procedures molybdenum and rhenium will also partially be deposited.

3. Solvent-Extraction Behavior

Many of the ionic species and compounds of technetium readily extract into a number of organic solvents. Extensive studies have been reported by Gerlit, using 21 organic solvents (24), and by Boyd and Larson, using 34 solvents (10). The TcO_4^- ion extracts into hexone and butex, as does ReO_4^-. In acid solutions TcO_4^- extracts well into alcohols, ketones, and tributyl phosphate; in basic solutions, ketones and cyclic amines such as pyridine (26) and 2,4-lutidine (50) are quite effective. In many cases added electrolytes, such as sodium carbonate and sodium nitrate, aid the extraction. In $5N$ NaOH, hexone extracts TcO_4^-, ReO_4^-, MoO_4^{-2}, and Ru(IV) (tracer), with partition coefficients of 17, 8.2, <0.001, and <0.001, respectively, and provides a feasible separation of technetium from fission products. Tertiary amines and quaternary ammonium salts appear to give the largest partition coefficients in organic solvents (10). Tri-n-octyl phosphine oxide (TOPO) also is an extractant.

Recently, Rimshaw and Malling (50) developed a method of preparing pure ammonium pertechnetate from fission-product solutions, involving extraction of TcO_4^- from $4.0N$ $NaNO_3$–$0.5N$ NaOH solutions, using pyridine and its methyl-substituted derivatives. The method is equally useful in analytical or preparative work. By proper choice of conditions, rhenium can be separated as well. The solvents can be removed by steam distillation, and the addition of $(NH_4)_2CO_3$ results in precipitation of the NH_4TcO_4.

Perhaps the most effective extraction system yet reported involves the tetraphenylarsonium salt into chloroform (61) to form neutral salt solutions:

$$(C_6H_5)_4As^+(aq.) + TcO_4^-(aq.) \rightleftharpoons (C_6H_5)_4AsTcO_4 \text{ (chloroform)} \quad (3)$$

$$K = 3 \times 10^6$$

The technetium can be recovered from this very stable organic precipitate by passing an alcoholic solution of the tetraphenylarsonium pertechnetate through a bed of the chloride form of a strong-base anion exchanger. The pertechnetate ion is strongly absorbed and the soluble tetraphenylarsonium chloride is washed through the bed. TcO_4^- can then be removed by elution, using dilute perchloric acid (8).

Only limited information has been reported on extraction of other oxidation states. The Tc(V) thiocyanate complex extracts into alcohols, ethers, ketones, esters, and into solutions of trioctylphosphine oxide or trioctylamine hydrochloride in cyclohexane or 1,2-dichloroethane. The $TcCl_6^{-2}$ ion may be extracted quantitatively from acid solutions into alcohols and by trioctylamine hydrochloride dissolved in cyclohexane. $TcCl_6^{-2}$ also extracts into chloroform in the presence of high concentrations of the tetraphenylarsonium ion (2).

4. Ion Exchange

The pertechnetate ion, TcO_4^-, is strongly adsorbed by strong-base anion exchangers, as are all XO_4^- ions. The process has been used in Re/Tc separations (5) using ammonium sulfate–ammonium thiocyanate solutions at pH 8.3 to 8.5 on Dowex-2 resin. Tc/Mo separations have been reported using thiocyanate on Amberlite IRA-400 resins (27). The best eluant, however, for Re/Tc separations appears to be ClO_4^- (10). Hydrochloric and nitric acids have been used in similar separations. Most of these procedures involve long time periods and are not directly suited to quantitative analytical work. The reader is referred to the monograph by Anders (4) for further references and discussion.

5. Selective Volatility

Since Tc_2O_7 is more volatile than Re_2O_7, partial separations have been reported, based upon selective distillation. The main problem is that procedures devised for obtaining quantitative distillation of technetium almost always also involve partial distillation of rhenium. If good Tc/Re separations are required, quantitative yields of technetium cannot be obtained. Furthermore, other elements forming volatile compounds, such as RuO_4 and various forms of molybdenum, also may codistil.

Distillation from perchloric acid gives good yields but only partial separations from rhenium and ruthenium. Molybdenum distillation can be reduced by the addition of phosphoric acid (43). Better separations of Tc/Ru can be carried out by sulfuric acid distillations, but quantitative yields are difficult to obtain. The presence of oxidizing agents, such as Ce(IV) or sodium peroxydisulfate, will increase the Tc yield, but Ru will then codistil as RuO_4.

Re/Tc separations have been reported using HCl and concentrated H_2SO_4 (43), wherein Tc does *not* volatilize since it is reduced by high-activity chloride to the $TcCl_6^{-2}$ state.

IV. DETECTION AND IDENTIFICATION

The chemistry of technetium is similar to that of rhenium in many ways, and consequently many of the methods of identifying rhenium can be applied equally well to technetium (see Part II, Volume, 7, of this Treatise).

TABLE VI
Spot-Plate Tests for Technetium[a]

Reagent	Medium	Initial oxidation state	Colors	Sensitivity limit	Interferences[b]
Potassium thiocyanate	6N HCl	IV	Purple to pink will not extract into CCl_4	—	Re, Fe(III)
Potassium thiocyanate + $SnCl_2$	6N HCl	IV, VII	Yellow	—	Re, possibly Fe(III)
Dimethylglyoxine + $SnCl_2$	10N HCl	IV, VII	Green	0.04 γ/ml.	Very specific for Tc
Nitron	Slightly acidic	VII	White ppt'n.	—	Re, Mo
Tetraphenylarsonium chloride	Acidic	VII	White ppt'n.	—	Re, Mo
Thiourea	2N HNO_3	VII	Orange	0.04 γ/ml.	Re, Mn, Ru do not interfere
Potassium xanthate	In HCl or HNO_3; extract into CCl_4	IV, VII	Purple	0.02 γ/ml.	Mo
Sulfosalicylic acid + $SnCl_2$	Acidic	VII	Red	—	Mn, Re, Mo, Th, Fe do not interfere
Potassium ferrocyanide	Conc. HCl	VII	Blue	—	Re does not interfere

[a] For details of the procedure the reader is directed to References 1 and 33.
[b] Some 20 elements were tested for interference, including Re, Mn, Ru, Mo, and Fe.

A. RADIOACTIVITY

Because of the high specific activity of Tc^{99}, which is the only available isotope involving chemically detectable amounts of technetium, the *absence* of radioactivity in a sample suspected of containing technetium can be definitive. However, the beta radiation is very weak (β^-_{max} = 300 k.e.v.), and it is absorbed by ordinary glassware and many thick-walled Geiger counters. Consequently, thin-window counters, windowless flow counters, or air-ionization chambers must be used in testing for its presence. The specific activity of 7×10^5 disintegrations/mg. will allow the detection of few-microgram quantities in the absence of other radioactive contaminates.

Unfortunately, however, since all technetium isotopes have been synthesized by nuclear means, extraneous radioactivity due to other nuclides is very probable in samples in which the element is being sought. Consequently, its radioactivity is usually used to infer its presence and the contamination of equipment in laboratories where weighable quantities are being used in chemical research.

B. EMISSION SPECTROSCOPY

The most intense and easily identified emission spectrographic lines of technetium, at wavelengths of 2543.34 and 4031.63, 4262.26, and 4297.06 A., are frequently used for qualitative identification; the 4031.63 A. line is particularly free from interference, and as little as 10^{-7} g. can be inferred by the copper-spark method.

C. SPOT TESTS

Recently, some specific and sensitive reactions suitable for spot-plate analyses have been proposed (1,33), and are summarized in Table VI.

A pertechnetate solution in hydrochloric acid may be reduced with ascorbic acid in the presence of potassium thiocyanate to give red and yellow colors, which can be extracted into a small volume of ether or ester solvents such as butyl acetate. Molybdenum interferes, but rhenium develops a color only very slowly (in hours), compared to the few minutes required for technetium. If uranium is present the addition of sufficient fluoride ion will eliminate its interference (29).

V. TOXICOLOGY AND INDUSTRIAL HYGIENE

All of the known technetium isotopes are radioactive, and the species available in weighable amounts, Tc^{99}, has a specific activity of about 20 μc./mg. Fortunately its long half life, 200,000 years, and its low beta energy ($\beta^-_{max.}$ = 290 k.e.v.) somewhat reduce its radiological hazard. However, not much is known about the physiological consequences of ingestion of technetium, except perhaps that it tends to concentrate in thyroid tissue (6). The kidneys and lungs also are listed as critical organs (62). The International Atomic Energy Agency (31) has classified Tc^{99} as a Class 3 or moderately dangerous isotope according to relative toxicity. Federal regulations (62) have set the limits for the maximum permissible concentrations of Tc^{99} that can be present in laboratory air or discharged into domestic sewer water as 1×10^{-2} $\mu c./ml.$ of water and 2×10^{-6} $\mu c./ml.$ of air, respectively, for soluble forms, and 5×10^{-3} $\mu c./ml.$ of water and 6×10^{-8} $\mu c./ml.$ of air, respectively, for insoluble forms of technetium. Normally, these limits can easily be met in well-ventilated and frequently monitored hoods. The use of rubber gloves is necessary to prevent ingestion. The volatility of Tc(VII) from hot concentrated acid solutions increases its hazard somewhat. All high-temperature operations (e.g., reduction, pyrolysis, and evaporation) must be carefully controlled and observed to prevent the spread of technetium-bearing fumes. Most glassware that has been used with milligram quantities of technetium compounds will become contaminated and can be cleaned only with difficulty. No radioactive solutions containing sizable quantities of appreciable specific activity should ever be stored in sealed containers because of the possibility of leakage produced by the gas pressure resulting from radiation decomposition of the solvent.

VI. DETERMINATION OF TECHNETIUM

Some of the important methods that can be used for the determination of technetium are summarized in Table VII.

A. GRAVIMETRIC DETERMINATIONS

1. As Metallic Technetium

Under certain conditions, solutions of pertechnetates and selected pure compounds of technetium are easily reduced to metallic technetium in a stream of hydrogen at elevated temperatures (900 to 1000°C.), and this method probably is the most accurate yet known for technetium analysis. This method excludes any compounds containing elements or combinations thereof that will not decompose and/or volatilize under the conditions of the reduction. Consequently, solutions of NH_4TcO_4 and other ammonium salts (except perchlorates and nitrates) can be evaporated and reduced directly. Pure oxides, sulfides, and halides can be treated similarly.

TABLE VII
Methods for the Determination of Technetium

Method	Conditions	Sensitivity	Interferences and comments
Gravimetric			
Metallic technetium	H_2 reduction at 1000°C.	Limited only by microbalance	Limited to certain pure compounds; ClO_4^- and NO_3^- interfere
Technetium dioxide	Pptn. in HCl, using Zn	Limited only by microbalance	Accuracy limited to $\geq 0.5\%$
Tetraphenylarsonium pertechnetate	Acid to basic solution	Conc. must be ≥ 5 mg./liter of Tc	Mo, Re, ClO_4^-
Nitron Pertechnetate	Neutral pptn.	Conc. must be ≥ 5 mg./liter of Tc	Mo, Re
Spectrophotometric			
Technetium(V) thiocyanate	585 mμ	10^{-6} g.	NO_3^- interferes if $>6\%$
Pertechnetate ion	247 mμ	10^{-6} g.	Re, Mo
	289 mμ	10^{-6} g.	Mo
Polarographic			
Pertechnetate in pH 7 phosphate buffer	-0.68 v. *vs.* S.C.E.	$\times 10^{-8}$ g.	ϕ_4As^+, Pb(II), Tl(I), Cr(III), Cd(II); common fission products do not interfere
	-1.35 v. *vs.* S.C.E.	5×10^{-8} g.	Fe(III), Re(VII), Cr(III), Mn(II,VII), Co(II), Cr(VI)
Activation analysis			
$Tc^{99}(n,\gamma)Tc^{100}$	Slow neutron activation in reactor; flux $> 5 \times 10^{11}$ cm.$^{-2}$ sec.$^{-1}$	2×10^{-11} g.	Can be used in presence of Re

For other compounds and solutions, TcO_2 sometimes can be quantitatively precipitated away from extraneous ions and then be washed and redissolved in ammoniacal hydrogen peroxide. Evaporation of this solution results in pure ammonium pertechnetate for reduction. When pure the metal does not tarnish or otherwise react rapidly with even moist laboratory air, and it is an excellent gravimetric standard.

2. As Technetium Dioxide

Soluble compounds of technetium usually can be reduced to the TcO_2 (hydrate) by treatment with granulated zinc in dilute (3 to $6N$) hydrochloric acid (45). The initially fine precipitate can be coagulated by carefully neutralizing the solution with aqueous ammonia. Although the resulting hydrous oxide can be filtered, dried, weighed, and reacted with an excess of standardized ceric sulfate, followed by back-titration with iodide, the reaction is slow and the best procedure is to dehydrate the dioxide *in vacuo* at 500°C. The product sublimes at temperatures >1000°C.

Technetium dioxide also results from the decomposition of NH_4TcO_4 *in vacuo* at 400 to 500°C. It can therefore be used to determine the technetium content of sulfides and halides by treatment with ammoniacal hydrogen peroxide, evaporation to dryness, and careful ignition. The ammonium salts are sublimed away and/or decomposed. Nitrates and perchlorates are to be avoided since they frequently result in excessive sputtering of the sample out of the combustion boat.

3. As Tetraphenylarsonium Pertechnetate

Like rhenium, technetium(VII) forms an insoluble tetraphenylarsonium pertechnetate, although it is more soluble than the rhenium salt and is not suitable for samples $\leq$ 5 mg. The precipitation is apparently quantitative and can be carried out in dilute acids or bases (it dissolves in concentrated H_2SO_4 and is oxidized by strong and hot $HClO_4$ and HNO_3); for quantitative work the precipitation should be carried out near 0°C. Thermal stability studies have not been reported, but presumably the material can be dried satisfactorily at 110°C.

4. As Nitron Pertechnetate

The same conditions as those given for nitron perrhenate are valid for technetium, although no extensive literature is available on the subject. Because it is more soluble than the corresponding rhenium salt, the method is not suitable for samples $\leq$5 mg.

5. As Technetium Sulfides

The precipitation of technetium heptasulfide or the use of TcS_2 as either a gravimetric procedure or standard is not recommended.

6. As Ammonium Pertechnetate

Under certain limited conditions the use of ammonium pertechnetate may prove advantageous as a basis of gravimetric estimation. The first manner in which this substance can result is from the treatment of various oxides of technetium with ammoniacal hydrogen peroxide. The resulting solution, when evaporated, yields pure ammonium pertechnetate, which once was thought to be unstable near 100°C. (52). However, it is now known to be stable at this temperature for many hours (18). In fact, it can be sublimed (preferably *in vacuo*) without decomposition if the temperature is kept below 350 to 400°C. (18,63).

If other ammonium compounds are present along with the ammonium pertechnetate (such as $(NH_4)_2SO_4$ and NH_4Cl), these must be removed by volatilization and/or decomposition at higher temperatures (400 to 500°C.). Under these conditions anhydrous TcO_2 results.

B. TITRIMETRIC DETERMINATIONS

1. Neutralization Reactions

Pure Tc_2O_7 or pertechnetic acid, $HTcO_4$, are strong acids and may be titrated with standard base, using a pH meter or any suitable indicator. The method is limited to these two compounds and their solutions.

TcO_2 (and presumably TcO_3) will react slowly with excess hydrogen peroxide. After some heating to destroy the remaining peroxide the resulting $HTcO_4$ solutions can be titrated in the normal manner.

2. Redox Reactions

Very little has been reported on the redox titrations of technetium compounds. Most of the known stable compounds and their solutions react too slowly for direct titration. However, technetium in TcO_2 and its hydrate were determined (18) by treating the compounds with excess standardized acid ceric sulfate solutions, and back-titrating the resulting solution with standard iodide (or ferrous). No data on other compounds using this method have been reported.

Crouthamel (21) has obtained potentiometric evidence for soluble Tc-(VI), Tc(V), and Tc(IV) species in the titration of Tc(VII), using Ti(III) in strong (12M) H_2SO_4. The titration to Tc(V) might be useful as a method of estimation of technetium content. In the opinion of this author, a number of the lower oxidation states, prepared by titrimetric or controlled-potential reduction, could be easily adapted to convenient redox titration methods. However, the exact state of reduction will have to be well defined. It should be noted that neighboring ruthenium forms a number of noninteger stable oxidation states.

C. POLAROGRAPHIC METHODS

The most extensive studies on the polarography of technetium that have been published are from the Oak Ridge (42) and University of Michigan laboratories (54). Technetium(VII) was studied in various media, some of which provide the basis for sensitive and accurate methods of analysis. Table VIII summarizes the polarographic behavior.

Of all the systems, the -0.68-v. (vs. S.C.E.) wave in the phosphate (pH 7) buffer appears very good for analytical purposes, and the diffusion current has been shown to be a linear function of the concentration over the 0.1 to 1.0 γ/ml. concentration range. Neither ruthenium, rhenium, nor other common fission products interfere, but the tetraphenylarsonium ion is reduced at more positive potentials and must therefore be removed before analysis. Analyses at pH 2 in chloride media at -0.14 v. (vs. S.C.E.) also are satisfactory.

D. SPECTROPHOTOMETRIC METHODS

When acid technetium(VII) solutions are treated with concentrated ($\sim 4M$) ammonium thiocyanate, reduction takes place to a stabilized, intense red Tc(V) thiocyanate complex (see Table IV). The intensity of the color makes it useful for quantitative work (21). The reduction takes about 1 to 3 hours, and although it can be carried out faster using auxiliary reducing agents, such as Ti(III) or Sn(II), the reduction then generally proceeds below the $+5$ oxidation state. The thiocyanate complex is soluble in ethers, ketones, and esters, and is conveniently separated and concentrated in these solvents. Unfortunately, rhenium interferes, although reduction of the Re(VII) is said to be much slower.

A better procedure involving the Tc(V) thiocyanate complex uses ascorbic acid as a reductant, with Fe(III) to prevent further reduction to lower oxidation states (29). Under these conditions, Tc(VII) is reduced in a few minutes. Rhenium is reduced only very slowly under the same conditions. Molybdenum does not interfere with the Tc(V) complex, since it absorbs at 450 mμ, whereas the technetium complex absorbs at 510 mμ (see Fig. 3). The exact locations of the peaks vary somewhat with solvent.

TABLE VIII
Polarographic Behavior of Tc(VII)

Media	$E_{1/2}$(v. vs. S.C.E.)	Comments
2M, 10M H_2SO_4	No well-defined waves	Reduction takes place from −0.5 v. to hydrogen evolution, but no limiting current noted
4M $HClO_4$	No well-defined waves	Reduction takes place from −0.5 v. to hydrogen evolution, but no limiting current noted
0.1M NaOH	−0.85	Tc(VII) → Tc(V)
	−1.15	Ill-defined; Tc(V) → Tc(IV)
	−1.3	Large catalytic wave
0.1M NH_4Cl, 0.01% gelatin; pH = 8.5 (NH_4OH)	−0.8	
	−1.2	Catalytic ?
Borate buffer (pH 10)	−0.79	Well-defined Tc(VII) → Tc(V) ?
Phosphate buffer (pH 7)	−0.68	Well-defined Tc(VII) → Tc(IV) ?
	−1.35	Well-defined; requires addition of potassium pyrophosphate to buffer
2M KCl	−1.15	Well-defined if Tc(VII) ≤ 0.1mM; Tc(VII) → Tc(−I) (TcH_8^{-2} ?)
0.1M KCN	−0.81	Well-defined Tc(VII) → Tc(IV)
0.5M KCl (pH 2, HCl)	−0.14	Tc(VII) → Tc(III)
	−0.91	Tc(VII) → TcH_3 (?)
	−1.12	Catalytic
0.25M K_2SO_4 (pH 13, KOH)	−0.81	Tc(VII) → Tc(IV)
	−1.02	Tc(IV) → Tc(III)
	−1.60	Catalytic

A Tc(VII) complex with thioglycolic acid has been noted (40). This species has an absorption peak at 655 mμ; Re(VII), halides, SO_4^{-2}, and PO_4^{-2} do not interfere, but Mo(VI), Ru(VII), and $Cr_2O_7^{-2}$ do. This method, as well as other methods depending upon an initial Tc(VII) oxidation state, can be used for analysis of lower oxidation states by prior treatment with alkaline peroxide, followed by boiling to destroy the excess H_2O_2. Careful neutralization and cooling will prevent volatilization loss of technetium.

A quantitative procedure for analysis of solutions containing a few p.p.m.

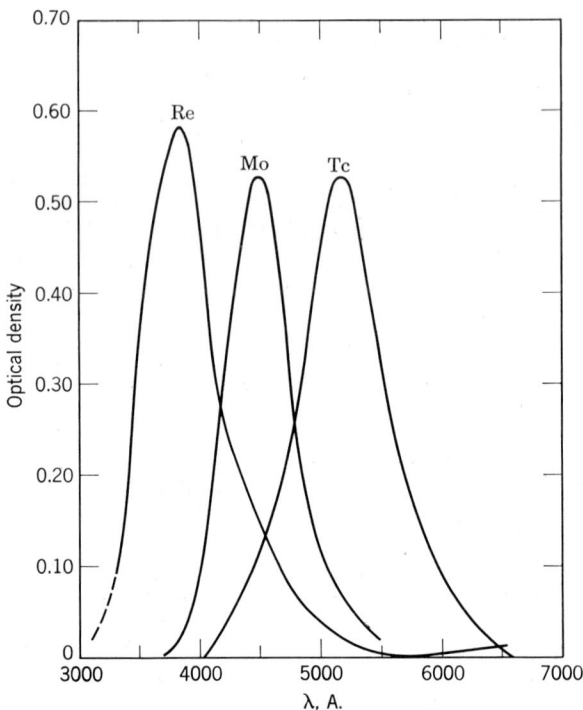

Fig. 3. Comparative absorption spectra of the Re(V), Mo(V), and Tc(V) thiocyanate complexes in n-butyl acetate (the concentrations are all approximately equal).

of technetium, as TcO_4^- in the presence of much larger quantities of rhenium, has been reported, using the Tc(IV)–ferrocyanide complex (1). The complex is prepared by reduction of TcO_4^- in acid solutions, using bismuth amalgam. It has a maximum absorption at 680 mμ, with a molar extinction coefficient of approximately 10,800.

The Tc(VII)–1,5-diphenyl carbohydrazide complex extracted into CCl_4 has been noted as a method for determining technetium (41).

The pertechnetate ion itself strongly absorbs, with maxima at 244 and 288 mμ. It has been used to analyze Tc(VII) solutions in the absence of Ru and Mo. Re also interferes with the 244-mμ peak. The strong peaks due to $TcCl_6^{-2}$ at 240 and 339 mμ also could be used effectively.

E. RADIOCHEMICAL METHODS

The radioactivity associated with Tc^{99} ($\beta^-_{max.} = 290$ k.e.v.) makes it possible to determine the technetium content of samples directly, but to only modest accuracy ($\pm 5\%$). The difficulty lies in the fact that the low-energy radiations are strongly absorbed by the sample itself and a number of corrections must be made to the raw counting data. Such a discussion is beyond the scope of the present work; the reader is referred to the chapters in Part I of this Treatise on radioactivity measurement and instrumentation.

The activation analysis of technetium in the presence of rhenium has been studied (7). It is extremely sensitive.

VII. RECOMMENDED LABORATORY PROCEDURES

A. GRAVIMETRIC METHODS

1. Nitron Pertechnetate ($C_{20}H_{16}N_3 \cdot HTcO_4$)

Use the procedure recommended for rhenium in Part II, Volume 7, p. 527, of this Treatise for samples ≥ 5 mg of Tc. The technetium factor is 0.2137.

2. Tetraphenylarsonium Pertechnetate ((C_6H_5)$_4$AsTcO$_4$)

Use the procedure recommended for technetium in Part II, Volume 7, p. 527 of this Treatise for samples ≥ 5 mg of Tc. The technetium factor is 0.1811.

B. SPECTROPHOTOMETRIC METHODS

1. Ascorbic Acid Reduction and Tc(V) Thiocyanate Complex (29)

The sample, dissolved in approximately 50 ml. of water, and contained in a separatory funnel, is adjusted to an acidity between 2.5 and 3.5N, using HCl or NH$_4$OH as necessary (Note 1). Add 1 ml. of a 0.037M FeCl$_3$ solution, which is also 0.1M in HCl, and 5 ml. of 6.6M ammonium thiocyanate. Mix the contents well and add 1 ml. of a solution of ascorbic acid, made by dissolving 5 g. of the acid in 100 ml. of water (Note 2). Mix, and add 25 ml. of n-butyl acetate, accurately delivered with a pipette. Shake vigorously for about a minute and allow the phases to separate. Discard the lower aqueous phase and wash the butyl acetate extract by adding 1 ml. of the thiocyanate solution and 20 ml. of a 2.5M NaH$_2$PO$_4$ solution. Again shake for 30 seconds, discarding the aqueous phase, and repeat the thiocyanate–phosphate wash.

Using a small plug of cotton in the stem of the funnel as a moisture filter, drain out a few milliliters of the n-butyl acetate extract and discard. Fill a 1- or 5-cm. spectrophotometer cell and measure the absorbance at 585 mμ ($\epsilon \approx 50{,}000$) against n-butyl acetate. Calibrate the procedure by the use of known technetium samples. If molybdenum is present (Note 3), measure the absorbance at some other wavelength, such as 585 mμ ($\epsilon \approx 16{,}500$). Note Fig. 3.

Note 1. It is important that the acidity falls between these limits to prevent subsequent reduction and/or disproportionation of the Tc(V) complex.
  2. This solution should not be kept longer than a week.
  3. Re does not interfere with this method since it reduces much too slowly; no interference was noted with moderate amounts of Cr, Fe, Mn, Mo, Ni, Ru, V, and W. Nitrate will not interfere if <6%. The method can be used in the presence of large excesses of uranium if sufficient HF is present to complex the uranium (for details, see Reference 29).

2. Thioglycolic Acid–Tc(VII) Complex (40)

An aliquot containing from 10 to 100 γ of Tc(VII) (Note 1) is pipetted into a 5-ml. volumetric flask. To each aliquot add 1 ml. each of $1M$ sodium acetate and 10% thioglycolic acid (pH 8.0 ± 0.2) (Note 2). Heat the flasks in boiling water for 15 minutes. Cool and add water to exactly fill the volumetric flask (Note 3). Fill a spectrophotometer cell with this solution and read the absorbance against a blank of the reagents at 655 mμ ($\epsilon \approx 1800$). The amount of TcO_4^- is determined by comparing to a calibration curve determined from known amounts of technetium treated in the same manner (Note 4).

Note 1. If the sample contains technetium in some other oxidation state, it should first be treated with ammoniacal hydrogen peroxide, followed by boiling to destroy the excess H_2O_2. The absorbance–concentration plot is linear over the 2 to 40 γ/ml. range.
  2. The pH 8.0 thioglycolic acid is prepared by neutralizing the 10% acid to this pH with dilute ammonium hydroxide.
  3. It is desirable to check and readjust, if necessary, the pH at 8.0 again at this step in the procedure.
  4. Cl^-, Br^-, I^-, ReO_4^-, PO_4^{-3}, and SO_4^{-2} do not interfere; some interference is noted with MoO_4^{-2}, RuO_4^-, and $Cr_2O_6^{-2}$.

REFERENCES

1. Al-Kayssi, M., R. J. Magee, and C. L. Wilson, *Talanta*, **9**, 125 (1962).
2. Alperovitch, E., Doctoral thesis, Columbia Univ., New York, 1954.
3. Anders, E., *Ann. Rev. Nucl. Sci.*, **9**, 203 (1959).
4. Anders, E., "The Radiochemistry of Technetium," *U. S. At. Energy Comm.*, **NAS-NS 3021** (1960).
5. Atteberry, R. W., and G. E. Boyd, *J. Am. Chem. Soc.*, **72**, 4805 (1950).

6. Bauman, E. J., *Am. J. Physiol.*, **185**, 71 (1956).
7. Boyd, G. E., and Q. V. Larson, *J. Phys. Chem.*, **60**, 707 (1956).
8. Boyd, G. E., *J. Chem. Ed.*, **36**, 3 (1959).
9. Boyd, G. E., and L. B. Rogers, as reported by L. B. Rogers, *J. Am. Chem. Soc.*, **71**, 1507 (1949).
10. Boyd, G. E., and Q. V. Larson, *J. Phys. Chem.*, **64**, 988 (1960).
11. Busey, R. H., *U. S. At. Energy Comm.*, **ORNL-2782** (1959).
12. Busey, R. H., and Q. V. Larson, *U. S. At. Energy Comm.*, **ORNL-2584**, 6 (1958).
13. Cartledge, G. H., *J. Am. Chem. Soc.*, **77**, 2658 (1955).
14. Cartledge, G. H., and W. T. Smith, Jr., *J. Phys. Chem.*, **59**, 1111 (1955).
15. Clifford, A. F., *Inorganic Chemistry of Qualitative Analysis*, Prentice-Hall, Englewood Cliffs, N. J., 1961, p. 445.
16. Cobble, J. W., C. M. Nelson, G. W. Parker, W. T. Smith, Jr., and G. E. Boyd, *J. Am. Chem. Soc.*, **74**, 1852 (1952).
17. Cobble, J. W., W. T. Smith, Jr., and G. E. Boyd, *J. Am. Chem. Soc.*, **75**, 5777 (1953).
18. Cobble, J. W., Doctoral thesis, Univ. of Tennessee, 1952.
19. Colton, R., *Nature*, **193**, 872 (1962).
20. Cotton, R., J. Dalziel, W. P. Griffith, and G. Wilkinson, *J. Chem. Soc.*, **1960**, 71.
21. Crouthamel, C. E., *Anal. Chem.*, **29**, 1756 (1957).
22. Flagg, J. F., and W. E. Bleidner, *J. Chem. Phys.*, **13**, 269 (1945).
23. Floss, J. G., and A. V. Grosse, *J. Inorg. Nucl. Chem.*, **16**, 44 (1960).
24. Gerlit, J. B., *Proc. Intern. Conf. Peaceful Uses At. Energy, Geneva, 1955*, **7**, 145 (1956).
25. Ginsberg, A. P., J. M. Miller, and E. Koubek, *J. Am. Chem. Soc.*, **83**, 4909 (1961).
26. Goishi, W., and W. F. Libby, *J. Am. Chem. Soc.*, **74**, 6109 (1952).
27. Hall, N. F., and D. H. Johns, *J. Am. Chem. Soc.*, **75**, 5787 (1953).
28. Hamilton, J. G., *U. S. At. Energy Comm.*, **UCRL-98** (1948).
29. Howard, O. H., and C. W. Weber, *Anal. Chem.*, **34**, 530 (1962).
30. Inghram, M. G., D. C. Hess, Jr., and R. J. Hayden, *Phys. Rev.*, **72**, 1269 (1947).
31. International Atomic Energy Agency, *Safe Handling of Radioisotopes*, Vienna, Austria, 1960.
32. International Union of Pure and Applied Chemistry, Committee on New Elements, *Chem. Engr. News*, **27**, 2996 (1946).
33. Jasim, F., R. J. Magee, and C. L. Wilson, *Talanta*, **2**, 93 (1959).
34. Kenna, B. T., *J. Chem. Ed.*, **39**, 436 (1962).
35. Kenna, B. T., and P. K. Kuroda, *J. Inorg. Nucl. Chem.*, **23**, 142 (1961).
36. Knox, K., et al., *J. Am. Chem. Soc.*, **79**, 3358 (1957).
37. Meggers, W. F., *Spectrochim. Acta*, **4**, 317 (1951).
38. Merrill, P., *Astrophys. J.*, **116**, 21 (1952).
39. Merrill, P., *Publs. Astron. Soc. Pacific*, **68**, 70 (1956).
40. Miller, F. J., and P. F. Thomason, *Anal. Chem.*, **32**, 1429 (1960).
41. Miller, J. F., and H. E. Zittel, Abstracts of Papers, Sixth Conference on Analytical Chemistry in Nuclear Reactor Technology, Oak Ridge National Laboratory, Oak Ridge, Tenn., Oct., 1962, paper 17.
42. Miller, H. H., M. T. Kelley, and P. F. Thomason, in I. S. Longmuir, Ed., *Advances in Polarography*, Pergamon, London, 1959, Volume 2, p. 716.
43. Morgan, F., and M. L. Sizeland, *U. K. At. Energy Authority Repts.*, **AERE C/M96** (1950).

44. Motta, E. E., G. E. Boyd, and Q. V. Larson, *Phys. Rev.*, **72**, 1270 (1947).
45. Nelson, C. M., G. E. Boyd, and W. T. Smith, Jr., *J. Am. Chem. Soc.*, **76**, 348 (1954).
46. Nelson, C. M., Doctoral thesis, Univ. of Tennessee, 1952.
47. Noddack, W., I. Tacke, and O. Berg, *Naturwiss.*, **13**, 567 (1925).
48. Parker, G. W., J. Reed, and J. W. Ruch, *U. S. At. Energy Comm.*, **AECD-2043** (1948).
49. Perrier, C., and E. Segré, *J. Chem. Phys.*, **5**, 712 (1937); **7**, 155 (1939).
50. Rimshaw, S. J., and G. F. Malling, *Anal. Chem.*, **33**, 751 (1961).
51. Rogosa, G. L., and W. F. Peed, *Phys. Rev.*, **100**, 1763 (1953).
52. Rulfs, C. L., and W. W. Meinke, *J. Am. Chem. Soc.*, **74**, 235 (1952).
53. Salaria, G. B. S., C. L. Rulfs, and P. J. Elving, *J. Chem. Soc.*, **1963**, 2479.
54. Salaria, G. B. S., C. L. Rulfs, and P. J. Elving, *Anal. Chem.*, **35**, 979 (1963).
55. Seeger, P. A., *Nucl. Phys.*, **25**, 1 (1961).
56. Segré, E., and C. S. Wu, *Phys. Rev.*, **57**, 552 (1940).
57. Selig, H., C. L. Chernick, and J. G. Malm, *J. Inorg. Nucl. Chem.*, **19**, 377 (1961).
58. Smith, W. T., Jr., J. W. Cobble, and G. E. Boyd, *J. Am. Chem. Soc.*, **75**, 5773 (1953).
59. Strominger, D., J. M. Hollander, and G. T. Seaborg, *Rev. Mod. Phys.*, **30**, 585 (1958).
60. Tribalat, S., *Rhenium and Technetium*, Gauthier-Villars, Paris, 1957.
61. Tribalat, S., and J. Beydon, *Anal. Chim. Acta*, **8**, 22 (1953).
62. U. S. National Bureau of Standards, *Maximum Permissible Amounts of Radioisotopes in the Human Body and Maximum Permissible Concentrations in Air and Water*, Handbooks **52** and **69**, Government Printing Office, Washington, D. C., 1953, 1959.
63. Wu, C. S., personal communication (1950); as reported in Reference 4.

Part II
Section A

ACTINIUM, ASTATINE, FRANCIUM, POLONIUM, AND PROTACTINIUM

By Jacob Sedlet, *Argonne National Laboratory, Argonne, Illinois*

Contents

```
I. Introduction.........................................  439
   A. Radioactivity of the Elements...................  439
   B. Occurrence and Sources..........................  441
      1. Occurrence in Nature.........................  441
      2. Sources of the Elements......................  445
   C. Industrial and Research Uses....................  447
   D. Toxicology and Industrial Hygiene...............  447
      1. Radioactivity Hazards........................  447
      2. Maximum Permissible Radiation Exposures......  448
      3. Safe Handling of Radioactive Materials.......  449
II. Actinium (Atomic Number 89)......................  450
   A. Introduction....................................  450
   B. Properties......................................  451
      1. Physical Properties..........................  451
         a. Nuclear Properties........................  451
         b. Other Physical Properties.................  451
      2. Electrochemical Properties...................  451
      3. Optical Properties...........................  452
      4. Chemical Properties..........................  453
         a. Tracer Chemistry..........................  454
   C. Separation and Isolation........................  456
      1. Dissolution of Samples.......................  456
         a. Uranium Ores and Ore-Processing Residues..  456
         b. Biological Samples........................  456
         c. Radium, Uranium, and Thorium..............  456
      2. Separation...................................  457
         a. Introduction..............................  457
         b. Coprecipitation and Precipitation.........  457
            (1) Fluorides.............................  457
            (2) Oxalates..............................  458
            (3) Hydroxides............................  458
            (4) Sulfates..............................  459
            (5) Precipitation of Other Elements from Ac-
                tinium................................  459
         c. Solvent Extraction........................  459
            (1) 2-Thenoyltrifluoracetone (TTA)........  459
            (2) Organic Phosphates....................  461
            (3) Alcohol...............................  462
```

Contents (*continued*)

    d. Ion-Exchange Methods 463
     (1) Cation-Exchange Systems 463
     (2) Anion-Exchange Systems 465
    e. Paper-Chromatographic Separations 465
   D. Detection and Identification 466
    1. Emission Spectroscopy 466
    2. Radiochemical Methods 466
     a. Growth of Decay Products into Purified Actinium-227 467
     b. Separation and Counting of Decay Products ... 468
     c. Gamma-Ray Counting 470
   E. Determination 472
    1. Introduction 472
    2. Determination by Separating Actinium 472
     a. Separations 472
     b. Counting 474
    3. Determination by Separating Actinium Decay Products 477
     a. Thorium-227 478
     b. Active Deposit 479
     c. Francium-223 480
    4. Other Methods 480
   F. Determination in Specific Materials 481
    1. Radioactive Ores and Residues 481
    2. Irradiated Radium 482
    3. Biological Samples 482
   G. Recommended Procedures 483
    1. Extraction with 2-Thenoyltrifluoroacetone 483
    2. Separation and Counting of Francium-223 483
    3. Extraction with Di(2-Ethylhexyl) Phosphoric Acid 485
III. Astatine (Atomic Number 85) 487
  A. Introduction 487
  B. Properties 487
   1. Physical Properties 487
   2. Chemical Properties 489
    a. Oxidation States 489
    b. Volatility 491
    c. Solvent Extraction 492
  C. Separation and Isolation 492
   1. Separation from Irradiated Targets 492
    a. Distillation 492
    b. Chemical Separations 493
   2. Coprecipitation 495
   3. Solvent Extraction 496
   4. Separation from Biological Materials 497
  D. Detection 497
  E. Determination 498
  F. Recommended Procedures 500
   1. Determination in Solution or in Precipitates 500
   2. Determination by Coprecipitation with Silver or Palladium 501
   3. Determination in Biological Materials by Coprecipitation with Tellurium 501

ACTINIUM, ASTATINE, FRANCIUM, POLONIUM, PROTACTINIUM

Contents (*continued*)

IV. Francium (Atomic Number 87)	501
A. Introduction	501
B. Properties	503
1. Physical Properties	503
2. Chemical Properties	503
a. Coprecipitation	504
(1) Perchlorates	504
(2) Picrates	504
(3) Tartrates	504
(4) Chloroplatinates	504
(5) Other Chloroanions	505
(6) Cesium Cobaltinitrite and Iodate	505
(7) Cesium Silicotungstate	505
(8) Heteropolyacids	505
b. Volatility	506
C. Separation and Isolation	507
1. Coprecipitation Methods	507
2. Chromatographic Methods	509
3. Ion-Exchange Resin Separations	510
4. Solvent Extraction	511
5. Other Separations	511
D. Detection	512
E. Determination	513
F. Separation and Determination in Specific Materials	516
1. Actinium	516
2. Irradiated Thorium and Uranium	517
G. Recommended Procedures	517
1. Isolation of Francium from Thorium Targets with Silicotungstic Acid	517
2. Separation of Francium-221 with Silicotungstic Acid	519
3. Separation of Francium-223 from Actinium-227 by Precipitation of Other Elements	520
V. Polonium (Atomic Number 84)	520
A. Introduction	520
B. Properties	522
1. Physical Properties	522
a. Nuclear Properties	522
b. Other Physical Properties	524
2. Electrochemical Properties	525
3. Optical Properties	526
4. Chemical Properties	527
a. Polonium Metal	527
b. Properties of the Oxidation States	527
c. Tracer Chemistry	530
C. Separation and Isolation	531
1. Dissolution of Samples	531
2. Separation	532
a. Coprecipitation	532
(1) Tellurium	532
(2) Lead Tellurate	533
(3) Selenium	533
(4) Maganese Dioxide	533

Contents (*continued*)

     (5) Hydroxides............................ 533
     (6) Other Carriers........................ 533
    b. Spontaneous Deposition and Electrodeposition. 534
     (1) Spontaneous Deposition................ 534
     (2) Electrodeposition..................... 536
    c. Solvent Extraction........................ 536
     (1) Alcohols.............................. 536
     (2) Chelating Agents...................... 536
     (3) Ethers and Ketones................... 538
     (4) Tributylphosphate..................... 538
     (5) Tri-*n*-Benzylamine.................... 539
    d. Ion-Exchange Resins...................... 539
    e. Volatility Separations..................... 541
    f. Chromatographic Separations............... 542
  D. Detection and Identification.................... 543
    1. Coprecipitation............................ 544
    2. Electrodeposition......................... 544
    3. Alpha-Particle Counting................... 544
    4. Gamma- and X-Ray Counting.............. 545
    5. Detection of Milligram Amounts of Polonium.... 547
  E. Determination................................ 547
    1. Preparation of Sample for Measurement......... 547
    2. Counting Techniques...................... 548
    3. Miscellaneous Methods.................... 549
  F. Determination in Specific Materials............... 549
    1. Biological Materials....................... 549
    2. Irradiated Bismuth and Other Heavy-Metal Targets.. 550
    3. Ores, Water, and Air...................... 551
    4. Analysis of Curie Amounts of Polonium Compounds..................................... 552
  G. Recommended Procedures....................... 553
    1. Deposition on Silver Foil................... 553
    2. Determination of Polonium-210 in Uranium Ores.. 553
    3. Determination of Polonium in Irradiated Bismuth by Solvent Extraction...................... 554
    4. Separation of Pure Polonium from Radium-DEF Mixtures................................... 554
    5. Paper-Chromatographic Separation of Polonium, Bismuth, Tellurium, and Selenium............ 554
    6. Determination of Polonium in Urine and Tissue... 555
VI. Protactinium (Atomic Number 91).................... 555
  A. Introduction................................... 555
  B. Properties.................................... 557
    1. Physical Properties........................ 557
     a. Nuclear Properties..................... 557
     b. Other Physical Properties............... 557
    2. Electrochemical Properties.................. 557
     a. Spontaneous Deposition................ 557
     b. Electrodeposition...................... 559
     c. Electrode Potential.................... 559
     d. Polarographic Behavior................. 559
    3. Optical Properties......................... 559

Contents (*continued*)

a. Emission Spectrum	559
b. Absorption Spectrum	560
4. Chemical Properties	561
a. Protactinium(V)	561
b. Protactinium(IV)	563
C. Separation and Isolation	564
1. Dissolution of Samples	564
2. Separation	566
a. Coprecipitation	566
b. Solvent Extraction	571
c. Ion-Exchange Separations	578
(1) Anion-Exchange Resins	578
(2) Cation-Exchange Resins	580
d. Chromatographic Separations	580
D. Detection and Identification	581
E. Determination	585
F. Determination in Specific Materials	587
1. Ores and Ore-Processing Residues	587
2. Irradiated Thorium	591
3. Geological Samples	593
G. Recommended Procedures	594
1. Determination of Protactinium in Uranium Ore Residues by Alpha Pulse-Height Analysis	594
a. Alpha Pulse-Height Analysis Following Chemical Separation	594
b. Alpha Pulse-Height Analysis Without Chemical Separation	595
2. Determination of Protactinium-231 in Ore Residues by Gamma-Ray Spectrometry	596
3. Determination of Protactinium-231 by Differential Gamma-Ray Spectrometry	597
4. Determination of Protactinium-233 in Neutron-Irradiated Thorium	598
5. Preparation of Carrier-Free Protactinium-233	598
References	599
General References	608

I. INTRODUCTION

A. RADIOACTIVITY OF THE ELEMENTS

All isotopes of the elements discussed in this chapter are radioactive, and this nuclear instability greatly affects the choice of techniques and procedures that are used in their determination. In most cases it is preferable or necessary for the final measurement in an analysis to consist of a quantitative measurement of the amount of radioactivity associated with the element and/or its decay products. This measurement is extremely sensitive, and amounts usually undetectable by any other means may be detected in this way. Several methods are available for this measurement, but those usually used respond to and record individual nuclear events, i.e., the nuclear particle emitted by the decaying atom, or other consequences of nuclear decay, i.e., the emission of an x-ray or gamma-ray. This type of radioactivity measurement is termed "counting." Radioactivity detection methods record only a fraction of the rays or particles emitted by the radioactive source. The determination of the absolute amount of radioactive material in the source, in terms of nuclear disintegrations per unit time, requires a knowledge of this fraction and of the nuclear decay scheme. In many cases only the relative amounts of radioactive materials are needed, and this requires only that the factors affecting the fraction recorded remain constant among the samples to be compared. A discussion of counting methods and other techniques for radioactivity measurements cannot be given in this chapter. Detailed discussions of these techniques may be found elsewhere in this Treatise and in other publications. A selection of references on this subject is given at the end of the chapter.

In considering the chemistry of radioactive elements it is useful to distinguish between two broad ranges of concentration, "trace" and "ordinary," or macroscopic, concentrations. In trace concentrations the element usually can be detected only through its radioactivity, or in special cases, by fluorometric or mass-spectrographic methods. At macroscopic concentrations the element can be detected by the methods used for nonradioactive elements. At trace concentrations precipitation reactions are not observed because the solubility product of the compound cannot be exceeded. However, other methods of separation, such as solvent extraction, volatilization, separation on ion-exchange resins, and coprecipitation with carriers, can be carried out. In trace concentrations an element may exhibit anomalous behavior, compared with ordinary concentrations. Adsorption on impurities or on the walls of the containers, hydrolysis, and colloid formation may occur, and the tracer can partially or wholly disappear from solution or may no longer be in true solution. This type of behavior must always be considered when dealing with trace concentrations, even under conditions in which it is not important at ordinary concentrations. For example, whereas the loss of a microgram of protactinium by adsorption on the walls of the container will cause no appreciable error in the analysis of several milligrams, a tracer solution may contain no more than a few micrograms. Differences between chemical behavior at trace and ordinary concentrations also may occur due only to the effect of concentration on reaction rates and equilibria conditions. A thorough discussion of the behavior of tracer solutions is given in Reference 298.

At very high concentrations of radioactive elements, the nuclear particles emitted during decay can cause troublesome nuclear and chemical reactions. Nuclear particles have sufficient energy to decompose water, discolor or craze glassware, cause oxidation–reduction reactions, raise the temperature of the radioactive material, and excite light emission in the medium through which they move. Many elements, particularly those of low atomic number, emit neutrons when bombarded by alpha particles, and strong sources of alpha-emitting elements may cause the emission of appreciable numbers of neutrons. Thus the neutron hazard must be considered when handling large amounts of polonium or actinium fluorides.

Of the elements discussed in this chapter, astatine and francium are too unstable to be obtainable in macroscopic amounts, and can only be detected by radioactivity methods. Weighable amounts of actinium, polonium, and protactinium are available in limited quantities, but most workers encounter them only in tracer concentrations. When macroscopic amounts are used the simplicity of radioactivity measurements, and the experimental difficulties and health hazards associated with analyzing large samples, often make the use of minimal quantities preferable, and final measurements are usually made by counting.

B. OCCURRENCE AND SOURCES

1. Occurrence in Nature

Isotopes of the elements discussed in this chapter occur in nature only as members of the radioactive decay series shown in Figs. 1 to 4. The figures give the half lives, in the usual time units, and the symbols for the historical names of the nuclides. Each series may be characterized by the formula for the mass numbers of the nuclides, $4n + a$, where n and a are integers. The value of a is constant for all the nuclides in a given series. Several collateral series that decay into one of these four principal series have been produced synthetically. The thorium, uranium–radium, and uranium–actinium series occur naturally because one member of each series, the parent, has a sufficiently long half life to have existed through geological times. The $(4n + 1)$, or neptunium, series was discovered after the long-lived members of this series, neptunium-237 and uranium-233, were produced synthetically. Very small amounts of these nuclides are produced in uranium and thorium ores as a result of neutron reactions:

$$^{238}_{92}(n,2n) \rightarrow {}^{237}_{92}U \xrightarrow[6.7 \text{ days}]{\beta} {}^{237}_{93}Np$$

$$^{232}_{90}Th\;(n,\gamma)\;{}^{233}Th \xrightarrow[23.3 \text{ minutes}]{\beta} {}^{233}_{91}Pa \xrightarrow[27.4 \text{ days}]{\beta} {}^{233}_{92}U$$

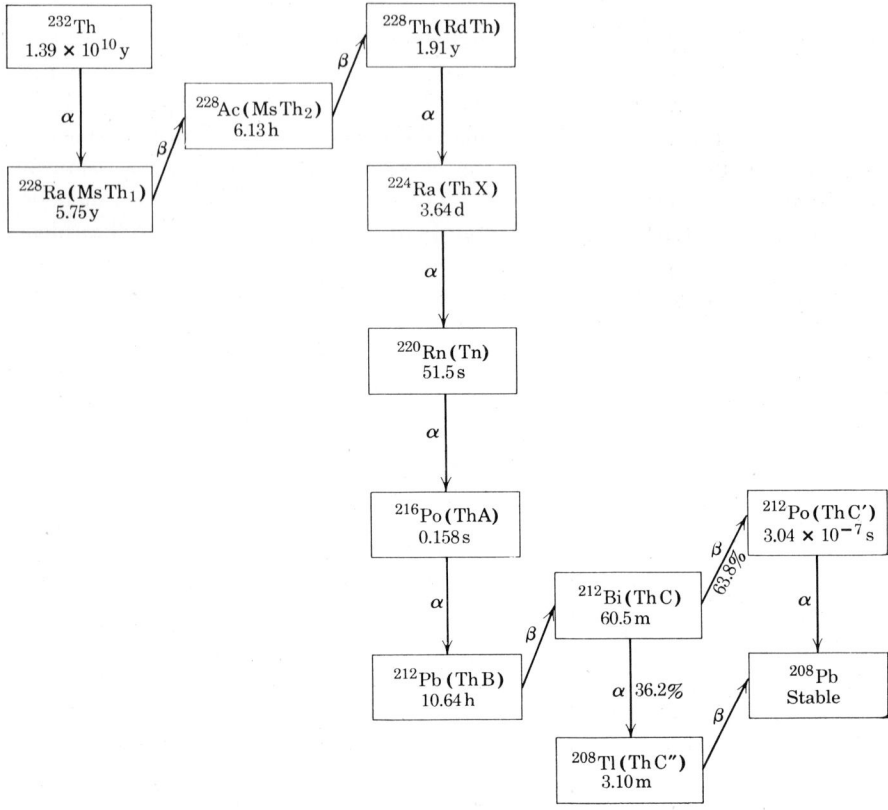

Fig. 1. The thorium (4n) series. Key to abbreviations and symbols: y = years; d = days; h = hours; m = minutes; and s = seconds.

Uranium contains about one part by weight of the neptunium series in 10^{13} parts of uranium-238 (229). Significant amounts of the members of this series are available only from the production of neptunium-237 and uranium-233 in nuclear reactors. Francium and astatine isotopes occur in the main decay sequence of the neptunium series, but are extremely rare in nature since they are produced only by infrequent branch decays in the natural series.

The principal isotope of polonium, mass number 210, occurs naturally as the last radioactive member of the uranium series. One ton of uranium as the ore contains about 0.1 mg. of polonium-210. Since uranium is widely distributed in nature in low concentration, its descendents are present in very small traces in most natural materials. Small amounts

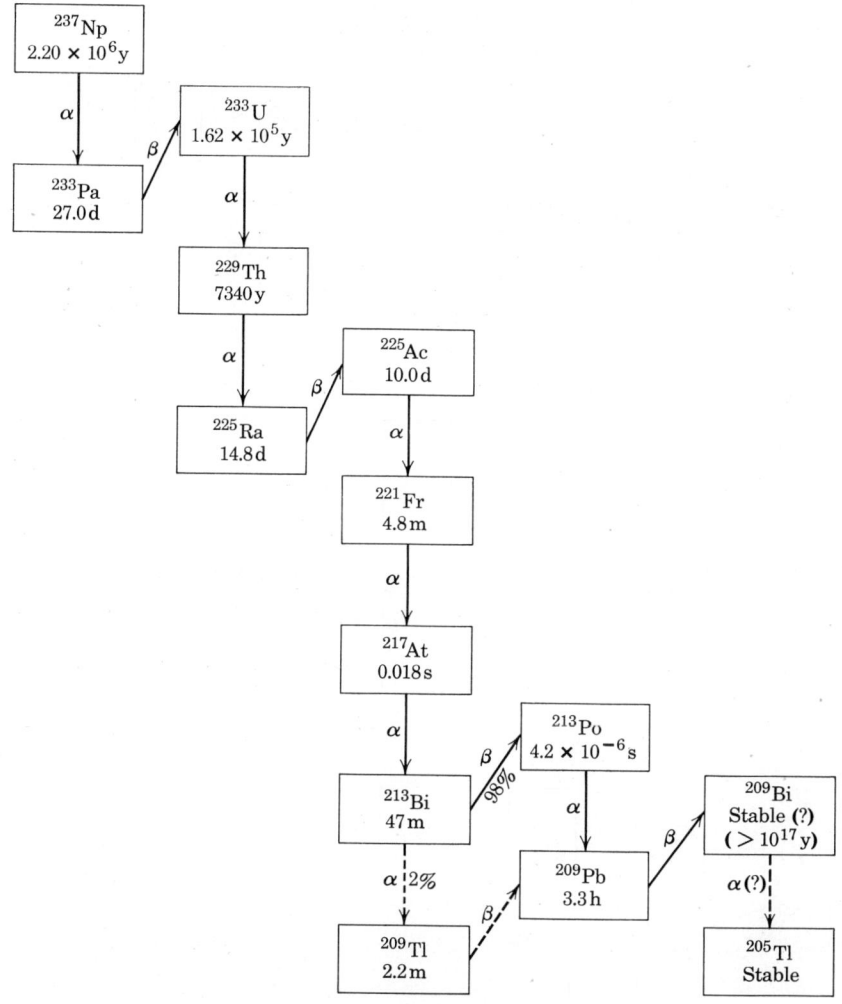

Fig. 2. The neptunium $(4n + 1)$ series.

of radon, the rare gas daughter in the uranium series, emanate continuously from solids containing radium, and polonium-210 is therefore normally present in air from the decay of radon (see Fig. 3). The concentration in the air is of the order of 10^{-6} disintegrations per minute per liter (44). The principal isotopes of protactinium and actinium, mass numbers 231 and 227, respectively, occur as descendents of uranium-235 in the $(4n + 3)$ series. The uranium-235 isotope, in turn, occurs in natural

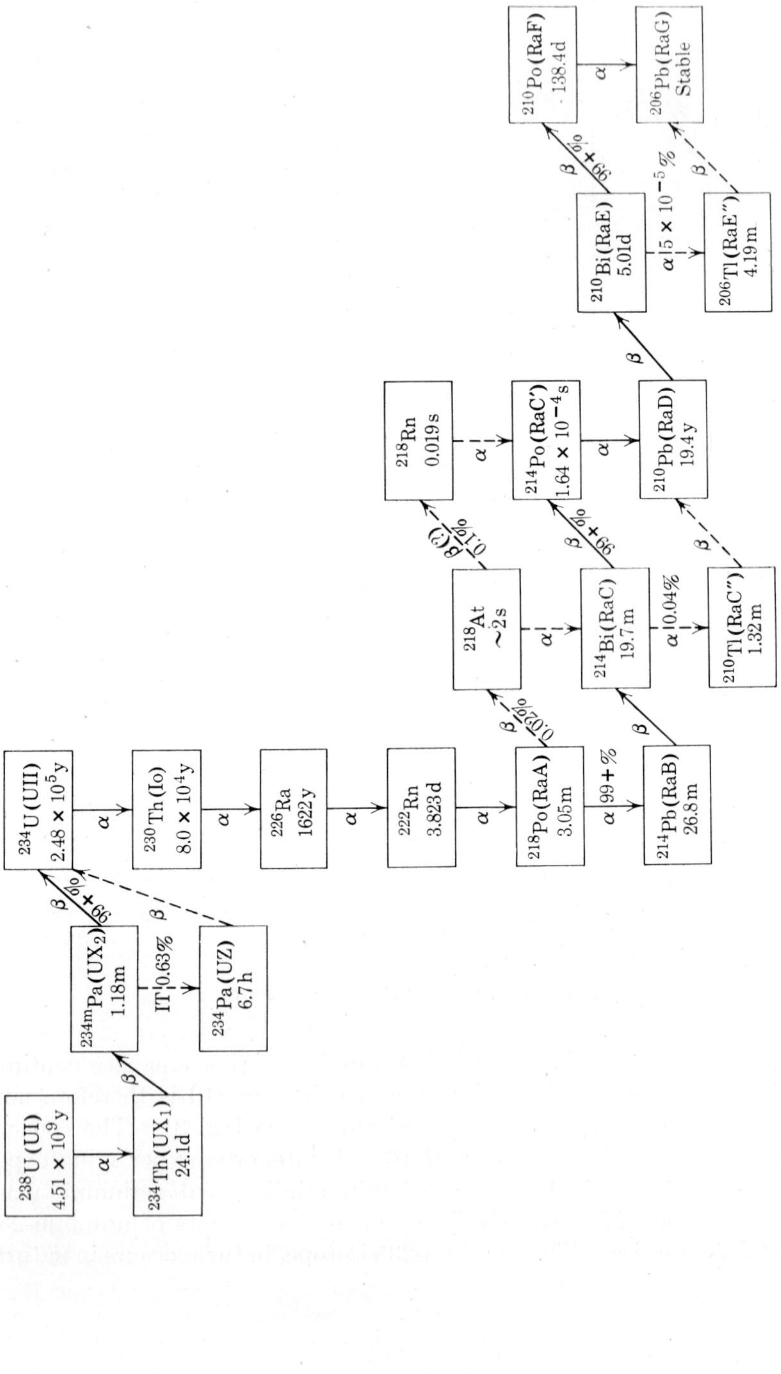

Fig. 3. The uranium–radium ($4n + 2$) series.

ACTINIUM, ASTATINE, FRANCIUM, POLONIUM, PROTACTINIUM

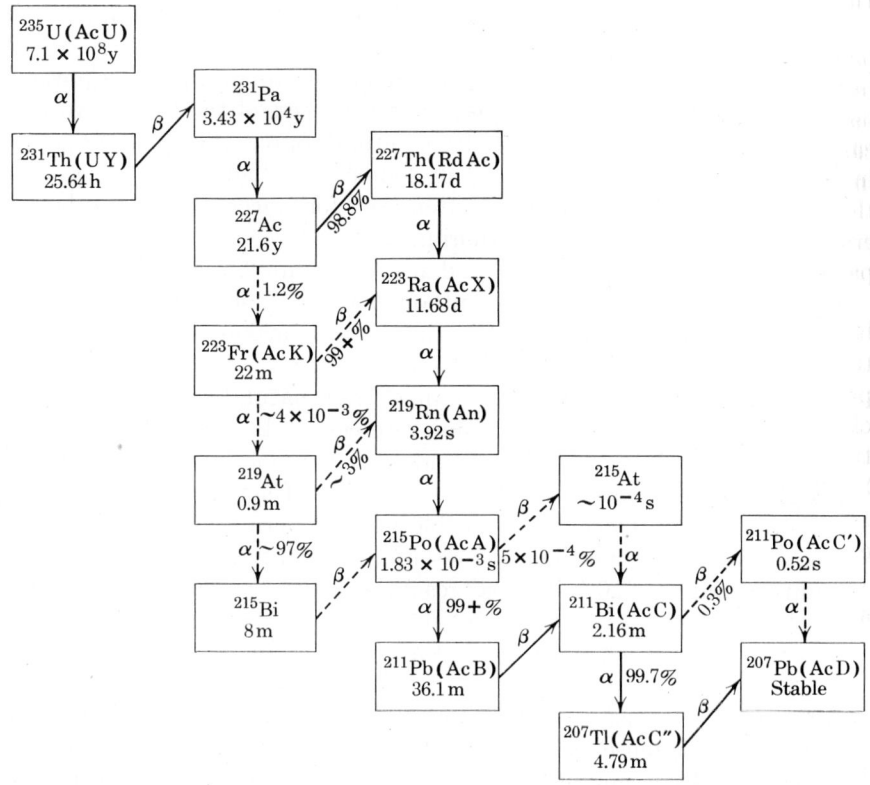

Fig. 4. The uranium–actinium ($4n + 3$) series.

uranium to the extent of 0.7%. One ton of uranium as the ore contains about 0.15 mg. of actinium and about 250 mg. of protactinium.

2. Sources of the Elements

Protactinium-231 is obtained almost exclusively from uranium ores. The large-scale processing of uranium has produced a large supply of residues from which protactinium has been extracted in spite of its initial low abundance in the ore. Milligram and gram amounts have been separated at several American laboratories under the auspices of the Atomic Energy Commission (151,162). The largest supply, more than 100 g., has been separated by British workers from the sludge remaining after the ether extraction of uranium from ore (52,142,143). As will be seen, this is a remarkable achievement in view of the difficult chemical behavior of protactinium. A synthetic route to protactinium-231 is by neutron irradiation of ionium (thorium-230), but only small quantities have been made in this way. Gram amounts of ionium for this purpose can be obtained from uranium ores, but the success of the British workers in obtaining protactinium directly from natural sources probably eliminates the ionium route as an important source. Protactinium-233, a beta emitter with a 27-day half life, is often used for tracer studies and is prepared by neutron irradiation of thorium-232.

In spite of its low natural abundance, actinium-227 can, in principle, be obtained from uranium ores, although the chemical similarity of actinium to the rare earth impurities in the ores makes the separation of pure actinium from this source extremely difficult. Weighable amounts of pure actinium have not been obtained from natural sources. Concentrated sources of actinium containing rare earths, known as "actiniferous lanthanum," have been prepared and used for tracer studies and spectroscopic measurements. Although separation of actinium from the rare earths with which it occurs is possible with modern techniques, the best source of milligram quantities at present is by neutron irradiation of gram amounts of radium-226 in nuclear reactors (100):

$$^{226}_{88}\text{Ra}\ (n,\gamma)\ ^{227}_{88}\text{Ra}\ \xrightarrow[41.2\ \text{min.}]{\beta}\ ^{227}_{89}\text{Ac}$$

Although the separation from radium and its descendants is difficult because of the intense radioactivity associated with radium, purification from rare earths is not necessary. Since gram amounts of pure protactinium have been prepared, a potential natural source of actinium-227 is from the alpha decay of protactinium-231. One gram of protactinium-231 will produce about 0.02 mg. of actinium-227 per year by radioactive decay.

The principal source of milligram amounts of polonium-210 is transmutation of bismuth in a nuclear reactor:

$$^{209}_{83}\text{Bi}\ (n,\gamma)\ ^{210}_{83}\text{Bi}\ \xrightarrow[5.0\ \text{days}]{\beta}\ ^{210}_{84}\text{Po}$$

Small amounts are obtainable from old or spent radon tubes used for radon therapy, and from aged radium salts. Milligram quantities have been separated from lead residues resulting from uranium processing, since lead-210 serves as a long-lived parent of polonium-210.

The principal source of francium is actinium-227, which decays by alpha emission to francium-223 in 1.2% of its disintegrations. Astatine isotopes are prepared by helium ion bombardment of bismuth or other heavy metals in a cyclotron. Other isotopes of actinium, polonium, protactinium, and francium also have been prepared by charged-particle bombardment of heavy metals.

C. INDUSTRIAL AND RESEARCH USES

All the uses of these elements are based on their radioactivity. Polonium, because of its high specific activity, is used as a source of alpha particles to study chemical and nuclear reactions induced by alpha particles, and for activation analysis by alpha-particle irradiation. Its use in the study of nuclear reactions has been largely supplanted by the cyclotron, but it was used as the source of alpha particles in the discovery of the neutron. It is used mixed with light elements, such as beryllium, to produce sources of fast neutrons by the (α,n) nuclear reaction, and as a heat source for self-contained portable thermoelectric generators. As a source of ionizing radiation, it is used to study the ionizing effects of alpha particles in matter, to eliminate undesirable electrostatic charges, and it has been used on electrodes of spark plugs for internal-combustion engines to facilitate sparking. Actinium and protactinium have no significant industrial uses, due in part to their unavailability until recently. They are of scientific interest as members of the actinide series, and the chemistry of protactinium is of special interest because it is an intermediate in the production of uranium-233 from thorium-232 by neutron irradiation. Actinium, if available in sufficient quantity, would probably find uses similar to those of polonium, since it has several alpha-emitting descendants.

Francium concentrates in some types of cancerous tissues in experimental animals, and this property may find use in the diagnosis of certain types of cancer (246). Astatine concentrates in the thyroid, as does iodine, and may find application in the radiation therapy of certain thyroid disorders (113).

D. TOXICOLOGY AND INDUSTRIAL HYGIENE

1. Radioactivity Hazards

The elements discussed in this chapter are dangerous because of their radioactivity. Two general types of hazards are involved in their use: (*1*) irradiation of the body or parts of the body by the rays and particles emitted by external radioactive sources, and (*2*) irradiation of body organs by radioactive materials taken into the body. External hazards are limited to beta particles and x- and gamma-radiation, since alpha particles have a very short range in matter and will be absorbed by material thicker than 5 to 10 mg./cm.^{2} (e.g., a sheet of paper or a rubber glove). All types of ionizing radiation are hazardous if emitted by materials deposited in the body because of the close contact between the tissue and source. Contamination of the intact skin by radioactive materials is generally an external hazard, although short-range particles can produce local skin damage if the material is not removed. Absorption through unbroken skin is very slight, but since the possibility of cuts or abrasion always exists, skin contamination should be removed promptly to eliminate this hazard. Internal deposition of radioactive materials is potentially more dangerous than external exposure to radiation, since very small amounts can be toxic and internal deposition is more difficult to detect and measure than external exposure.

Radioactive materials may be taken into the body through the broken skin by ingestion, and by inhalation. Ingestion is the least hazardous since absorption into the blood stream by this route is small. Absorption through the broken skin is more hazardous since the material is directly available to the blood stream and, if soluble, is quickly carried to one or more organs where it can remain for extended periods. This type of intake, however, can be avoided with reasonable care. The most common route of intake is by inhalation. In this case the lung is irradiated, and an appreciable fraction can be deposited in other organs. This fraction varies with the chemical and physical properties of the material involved. If taken into the body in a form available to the circulatory system, actinium and protactinium tend to concentrate in the bone, astatine in the thyroid, francium in the muscle, and polonium in the liver and kidneys.

2. Maximum Permissible Radiation Exposures

These values are determined by the International Commission on Radiological Protection (ICRP) and the National Committee on Radiation Protection (137,138,223), and their recommendations are followed in all work with radioactivity. For those working with radioactive materials the recommended maximum radiation dose to most organs from external radiation is 3 rems in any 13 consecutive weeks, and the maximum total accumulated dose is $5(N - 18)$ rems, where N is the age in years and is greater than 18; if the exposure is before age 18 the yearly dose should not exceed 5 rems. The unit of dose, the rem, is that dose of ionizing radiation that will produce the same biological effect as that produced by one roentgen of high-voltage x-radiation. It is very difficult to calculate the dose from a given quantity of radioactive material, and in practice the dose is measured with appropriate instruments. Empirical equations are available that relate the external dose with the amount of a radioactive nuclide expressed in disintegrations per unit time or similar units, and these may be used as guides (160).

The maximum permissible amounts that may be fixed in the body, and the air concentrations that will result in these amounts being fixed if contaminated air is breathed continuously, are given in Table I for the important isotopes of the elements discussed in this chapter, except francium. The values for francium have not been given by the ICRP. It should be noted that internally deposited radionuclides as well as external radioactive sources will result in a dose to the body, and the total internal and external dose should not exceed the maximum dose given above for external exposure. The exposure of any worker should, of course, be as small as possible regardless of the magnitude of the maximum permissible amounts.

TABLE I
Maximum Permissible Amounts for Internal Radiation Exposure (138)

Nuclide	q^a	$(MPC)a^b$
^{227}Ac	0.03	2×10^{-12}
^{228}Ac	0.04	2×10^{-8}
^{211}At	0.02	7×10^{-9}
^{210}Po	0.03	2×10^{-10}
^{231}Pa	0.02	10^{-12}
^{233}Pa	40	2×10^{-7}

[a] The maximum permissible amount that may be tolerated in the body for 50 years, expressed in microcuries (1 $\mu c.$ = 2.22×10^6 disintegrations per minute).

[b] The maximum permissible air concentration, in $\mu c./cm.^3$, for an exposure of 40 hours per week for 50 years.

Chemical poisoning is not an important consideration in the use of these elements, even when macro amounts are involved, since the radioactive toxicity is greater than the chemical toxicity. This means that the precautions taken to prevent radiation overexposure will be sufficient to prevent any conceivable chemical poisoning. One important difference between chemical and radioactive toxicity is that radioactive materials can be dangerous even when outside the body, whereas this is not the case for chemical poisons.

3. Safe Handling of Radioactive Materials

Radioactive materials can be handled safely, although it frequently requires more time to perform the same operations with radioactive than with stable materials. The precautions necessary for safe operation depend upon the amounts and upon the chemical and radioactive properties of the materials involved. The operations should be carried out in a manner designed to prevent irradiation of the body by sources external to it and to prevent the intake of radioactive substances. In the former case, shielding, usually of high-atomic-number materials (lead, with zinc bromide windows) is interposed between the worker and source. Equipment is available for performing most laboratory operations remotely behind thick shields. This type of protection is necessary for gamma-ray emitters. To prevent body intake the material must be contained in such a way as to prevent its appearance in the air breathed by the worker, and to prevent skin contamination. For these reasons work with alpha emitters is carried out in hoods with good ventilation, or better, in sealed glove box systems. Volatile materials, such as astatine and certain polonium compounds, are particularly dangerous. If both penetrating radiation (gamma-rays and high-energy beta particles) and short-range radiation (alpha and low-energy beta particles) are present in large amounts, both massive shielding and sealed systems are required.

The effectiveness of the safety precautions should be monitored regularly by measuring of the external radiation dose with the proper instruments and by collecting and analyzing air samples. The instrumentation for this is available commercially. Hazards may be reduced by using the least amount of material necessary for successful results. Excellent microchemical techniques have been developed for use with radioactive materials, partly for safety reasons and partly because many active elements are available only in small amounts. Work with tracer levels does not require elaborate protection; however, care must be exercised continuously, and this requirement is one of the best assurances that the work will be done safely. There is an extensive literature on the safe handling of radioactive materials (10,24,33,136–138,151,160,224). These works, and the references given in them, should be consulted before using radioactive materials.

II. ACTINIUM (ATOMIC NUMBER 89)

A. INTRODUCTION

Actinium was discovered in 1899 by A. Debierne and independently by F. Giesel in 1902 in the rare earth fraction of pitchblende. The isotope involved in both cases was actinium-227. The similarity of the chemical properties of actinium to those of the rare earths was soon recognized. An account of the early history of actinium is given in Reference 10. Actinium is now generally considered to be the first member of the actinide series of transition elements, involving valence electrons in the $5f$ and $6d$ orbitals (151), and it occupies a position in the periodic table similar to lanthanum.

B. PROPERTIES

1. Physical Properties

a. Nuclear Properties

The nuclear properties of the known actinium isotopes are summarized in Table II. Additional details on the decay energies and gamma-ray abundances may be found in References 56 and 283. This information is often useful in identifying the isotopes. The two naturally occurring actinium isotopes (mass numbers 227 and 228) are those most often encountered in analytical problems. Actinium-227, the longest-lived isotope, occurs as the daughter of protactinium-231 in the uranium–actinium decay series. It is the only actinium isotope with a sufficiently long half life to permit preparation in milligram amounts. Actinium-228 occurs in the thorium series and is important analytically in the determination of its parent, radium-228. Radium-228 is a beta emitter with a half life of 6.7 years, and it is important in radium toxicology. Since its radiations are too weak to be easily detected, it is often determined by analyzing for its actinium daughter.

b. Other Physical Properties

Actinium metal is highly reactive and oxidizes rapidly in moist air to become coated with the oxide. It glows weakly in the dark with a blue color as a result of its radioactivity. Its intense radioactivity probably contributes to its chemical reactivity (151). The melting point of the metal is approximately 1050°C. (281) and the boiling point is approximately 3300°C. (77).

The crystal structure of the actinium compounds thus far prepared are structurally similar to the corresponding lanthanum compounds (80). From crystal structure studies, the Ac(III) ion has a radius of 1.10 A. (304). It is, therefore, the largest and most basic tripositive ion. La (III), with an ionic radius of 1.04 A., is the second largest tripositive ion. Since both ions have an inert gas electronic configuration, their chemical similarity is not surprising (102).

2. Electrochemical Properties

The standard electrode potential of actinium has not been determined. The formal electrode potential for the reaction $Ac \rightarrow Ac^{+3}$ (aq.) is ~ 2.6 v.

TABLE II
Nuclear Properties of Actinium Isotopes (56,116,283,299)[a]

Mass number	Type of decay	Half life	Energy of principal radiations (m.e.v.)	Source
221(?)	α	< 0.1 s	α: 7.6	Dtr ^{225}Pa(?)
222	α	5.5 s	α: 6.96	Dtr ^{226}Pa
223	α, 99% EC, 1%	2.2 m	α: 6.64	Dtr ^{227}Pa
224	EC, ~90% α, ~10%	2.9 h	α: 6.17 γ: with EC, 0.133, 0.217	Dtr ^{228}Pa
225	α	10.0 d	α: 5.82, 5.78, 5.72, others γ: 0.0366, 0.0384, 0.0628, 0.0873, 0.0994, 0.150, 0.187	Dtr ^{225}Ra in ^{237}Np series; dtr ^{229}Pa, ^{225}Th; Th—spall
226	β, ~80% EC, ~2%	29 h	β: ~1.2 γ: with β, 0.0721, 0.158, 0.230, with EC, 0.0676, 0.185, 0.253	Dtr ^{230}Pa; Th, U—spall
227	β, ~99% α, 1.2%	21.6 y	α: 4.94, others γ: 0.037(?)	Natural source: dtr^{231}Pa in U–Ac series; ^{226}Ra (n,γ); dtr ^{227}Ra
228 (MsTh$_2$)	β	6.13 h	β: 0.45, 0.64, 1.1, 1.7, 1.85, 2.18 γ: 0.0568, 0.0781, 0.0978, 0.113, 0.128, 0.179, 0.184, 0.232, 0.336, 0.410, 0.458, 0.907, 1.035, others	Natural source: dtr ^{228}Ra in Th series
229	β	66 m		Dtr ^{229}Ra
230(?)	β	< 1 m	β: ~2.2	Dtr ^{230}Ra

[a] Key to abbreviations and symbols: α—alpha particle; β—negative beta particle; γ—gamma-ray; EC—orbital electron capture; n—neutron; s—second; m—minute; h—hour; d—day; y—year; dtr—daughter; spall—spallation reaction; and (?)—identification uncertain.

(151). The metal is too electropositive to be cathodically deposited as the metal from aqueous solution, although adherent deposits of unknown composition can be obtained by electrolysis of tracer solutions (52,134, 206,289).

3. Optical Properties

The most intense lines in the emission spectrum of actinium are given in Table III. The principal lines useful in the spectrographic identification of actinium are the second and third lines given for singly ionized actinium (AcII).

TABLE III
Emission Spectrum Lines of Actinium (203)

Class	Wavelength, A.			
AcI	4179.98	4183.12	6359.87	6691.26
AcII	3863.11	4088.43	4168.40	5918.87
AcIII	3392.77	3487.60	4569.87	

The absorption spectrum of Ac(III) in hydrochloric acid solution has been studied (102). Since the ion is colorless in solution, no absorption bands occur in the visible region. From 300 to 400 mμ a slight amount of absorption takes place, and a pronounced peak is observed at 250 mμ.

4. Chemical Properties

Actinium has one known oxidation state, (III), and its chemical behavior is generally very similar to that of the rare earths. The structure of nine actinium compounds has been determined by x-ray diffraction, and all are isostructural with the corresponding lanthanum compounds (80). Its separation from the rare earths, particularly lanthanum, is the most difficult problem in actinium chemistry. Separation from its neighbors in the actinide group is considerably easier. Like the rare earths, actinium forms an insoluble hydroxide, fluoride, oxalate, carbonate, phosphate, and fluosilicate (102). Actinium is more basic than lanthanum, in agreement with its higher ionic radius.

Actinium trifluoride can be precipitated from an actinium solution by the addition of hydrofluoric acid. The other halides are water-soluble. Actinium forms two series of halides: the normal trihalides and the oxyhalides, produced by partial hydrolysis of the solid trihalide at elevated temperatures. All the halides except the two iodides have been well characterized (102).

Actinium hydroxide precipitates as a white, gelatinous solid on addition of sodium or ammonium hydroxide to an actinium solution (80). On the basis of tracer studies early workers have reported that the hydroxide is more soluble than the corresponding lanthanum compound (10).

Actinium oxalate is precipitated from dilute (0.1N) hydrochloric acid by the addition of ammonium oxalate. Oxalic acid under similar conditions gives no precipitate (80). The oxalate is soluble in mineral acids, insoluble in neutral or alkaline solution. The solubility in 0.5N oxalic acid–0.1N nitric acid solution is approximately 0.025 mg./ml. (267). White actinium phosphate is precipitated from 0.1N hydrochloric acid by sodium dihydrogen phosphate. When dried at 70°C., it corresponds to the formula $AcPO_4 \cdot 1/2 H_2O$. A large excess of saturated potassium sulfate produces a white crystalline precipitate in $0.1 N$ sulfuric acid, probably a double salt of actinium and potassium sulfates (102).

a. Tracer Chemistry

Much of the coprecipitation behavior of tracer concentrations of actinium may be predicted from the solubility of its compounds, as given above. In general, tracer actinium will be coprecipitated or "carried" out of solution by forming an insoluble compound in the solution with a reagent whose anion also forms an insoluble compound with actinium. Thus, one would predict that insoluble phosphates, fluorides, etc., especially of lanthanum, will carry actinium. The fraction of tracer actinium carried, however, depends on a variety of conditions: the concentration of the carrier (coprecipitating compound), the acid concentration, the temperature, the presence or absence of interferring ions, the rate of addition of the precipitating agent, and other factors. A summary of important coprecipitation reactions for actinium is given in Table IV.

Many separation procedures from other radioactive elements may be devised from the information in Table IV. A notable feature of many of the coprecipitation reactions is the detrimental effect of ammonium ion on the fraction carried, and care must be taken to avoid a large excess. Lanthanum is the preferred carrier for actinium if a lanthanum–actinium separation is not required later in the separation procedure. Actinium-227 from natural sources always contains some rare earths, so additional carrier is generally not required in this case. The effect of lanthanum ion on the carrying by barium sulfate should be noted. The lanthanum apparently displaces the actinium from the precipitate. This effect may be used to advantage in the separation of radium and actinium by carrying the radium on barium sulfate. In this case lanthanum ion is added as a "holdback-carrier." The influence of lanthanum on coprecipitation by other carriers has not been reported, but the same effect may occur in other cases of coprecipitation by adsorption. For an excellent discussion of the coprecipitation of tracers, see References 103 and 298. Since actinium hydroxide is more soluble than lanthanum hydroxide, complete precipitation of the lanthanum is required to obtain quantitative carrying on lanthanum hydroxide. The same problem arises in carrying by lanthanum phosphate, but in the case of lanthanum oxalate carrying is essentially complete when only 75% of the lanthanum has precipitated (200).

Actinium forms a number of organic complexes that are useful in the separation and purification of actinium by solvent extraction. Thenoyltrifluoroacetone, tributyl phosphate, and various alkylphosphoric acids have been used. The complexes of these compounds with actinium are much more soluble in nonpolar solvents than in aqueous solution, and this property is used in solvent-extraction separations.

TABLE IV
Coprecipitation Behavior of Actinium (102,200,298)

Carrier	Conditions	% Carried	Remarks
Hydroxides of Al(III), Fe(III), Y(III), Zr(IV), rare earths	Complete precipitation of carrier with NH_4OH or $NaOH$.	~100	High NH_4^+ conc. reduces carrying.
$Th(OH)_4$	Precipitated with pyridine from alcoholic solution.	0	
$Th_2O_7 \cdot 4H_2O$	Precipitated with H_2O_2.	0	
$BaCl_2 \cdot 2H_2O$	Precipitated from concentrated HCl.	0	
LaF_3	Precipitated with HF from HNO_3 concentrations up to $5N$	~100	NH_4^+ decreases carrying to ~85%.
$ZrO(IO_3)_2$	Precipitated with $0.25M$ KIO_3 from $0.2N$ HNO_3.	97	
$ZrO(IO_3)_2$	Precipitated with $0.1M$ KIO_3 from $0.2N$ HNO_3.	~70	High NH_4^+ conc. decreases carrying even further.
$BiPO_4$	Precipitated with $0.5M$ H_3PO_4 from hot $0.1N$ HNO_3.	95	High NH_4^+ conc. decreases carrying.
$BiPO_4$	Same as above, except $1N$ HNO_3.	4	High NH_4^+ conc. decreases carrying even further.
$LaPO_4$	Complete precipitation of La from hot H_3PO_4 solution by neutralization with NH_4OH.	~100	
$La_2(C_2O_4)_3$	Precipitated from hot ammoniacal or slightly acid solution.	~100	Poor carrying at room temperature.
$La_2(C_2O_4)_3$	Precipitated from $0.5N$ HNO_3.	7	
$La_2(CO_3)_3$	Precipitated from alkaline solution.	~100	
$ZrO(H_2PO_4)_2$	Precipitated from $1M$ HNO_3 with H_3PO_4.	<1	
$BaSO_4$	Precipitated from dilute HNO_3 with H_2SO_4.	11–96	La(III) decreases carrying.
$PbSO_4$	Precipitated from $6M$ H_2SO_4.	~90	
Acid-insoluble sulfides (PbS, Bi_2S_3)	Precipitated from acid solution.	~0	

C. SEPARATION AND ISOLATION

1. Dissolution of Samples

a. Uranium Ores and Ore-Processing Residues

Since actinium is always present in trace concentrations in these materials, a significant fraction of the actinium may be adsorbed on a small amount of insoluble residue remaining after a dissolution step. Such residues should be completely dissolved and analyzed for actinium. If the residues prove to be free from actinium, they usually can be disregarded when additional similar samples are dissolved.

Samples may be decomposed by treatment with mineral acids, singly or in mixtures, or by fusion with sodium peroxide, sodium carbonate, or potassium pyrosulfate. Any insoluble residue may be further dissolved by additional treatments after the residue has been separated from the soluble portion. Residues containing silica may be dissolved by conventional treatment with hydrofluoric acid or by fusion with potassium fluoride, but the fluoride must be completely removed by fuming with the high-boiling acids or by fusion with pyrosulfate (274), since actinium is carried by some insoluble fluorides. Barium or lead sulfate residues from ore processing may be dissolved by metathesis with carbonate, either in solution or by fusion, or by reduction to the sulfide by sintering with zinc dust.

b. Biological Samples

Tissues and organs from animal studies of actinium metabolism are first wet ashed with nitric acid (or other oxidizing agents) or ignited at elevated temperatures. The inorganic ash will usually dissolve readily in nitric or hydrochloric acid. If a small amount of acid-insoluble material remains, it may be dissolved by repeated treatment with mineral acids, including hydrofluoric acid if silica is present, or by fusion.

c. Radium, Uranium, and Thorium

Radium salts chosen for neutron irradiation to produce actinium should be water-soluble. However, it is best to dissolve the salt in dilute mineral acid, approximately $0.1M$, to prevent adsorption of actinium and its daughters on the container walls. The choice of acid will depend upon the chemical separations to be used later. Uranium metal and oxide cyclotron targets will dissolve in nitric acid, but thorium oxide and metal require in addition a small amount of hydrofluoric or fluosilicic acid (about $0.05M$) for rapid and complete dissolution (270).

2. Separation

a. Introduction

The actinium separations usually required are: *(1)* separation from macroscopic amounts of nonradioactive material to avoid self-absorption difficulties in counting; *(2)* separation from other naturally occurring radioactive elements to permit accurate counting; and *(3)* separation from fission products produced along with actinium in the bombardment of heavy elements in accelerators. Because the chemical properties of actinium closely resemble those of the rare earths, separation from the latter elements is extremely difficult. This problem is always present in separating actinium from natural sources. The magnitude of this problem is indicated by the fact that pure actinium has not been obtained from natural sources. Fractional crystallization of a number of salts has been used to obtain rare earth fractions enriched in actinium, but this method is chiefly of historical interest and will not be discussed here.

b. Coprecipitation and Precipitation

(1) Fluorides

Coprecipitation of tracer actinium with a rare earth fluoride is often used to separate actinium from metals whose fluorides are soluble, including uranium, francium, and, under proper conditions, protactinium. Using 0.3 mg./ml. of lanthanum carrier, McLane and Peterson (200) found that more than 98% of the actinium precipitated with lanthanum fluoride from acid concentrations up to $5M$ nitric acid when the solutions are made $3M$ in hydrofluoric acid. However, from $2M$ nitric acid–$5M$ ammonium nitrate solution, only 87% of the actinium was carried. Fluosilicic acid also reduced the amount carried. Since lanthanum is very difficult to separate from actinium, cerium(III) or probably yttrium may be used as the carrier in place of lanthanum if an actinium–rare earth separation is desired later. Cerium may be oxidized to the tetravalent state and precipitated as the iodate with iodic acid, leaving most of the actinium in solution; yttrium may be separated by a solvent-extraction method to be discussed later. Thorium fluoride also carries actinium from acid solutions (123).

In precipitating an ionic carrier compound, the ion whose chemistry resembles that of the tracer ion is generally added to the solution first and, after being thoroughly stirred into the solution, is completely precipitated by adding an excess of the other ion (or ions) of the compound. The most rapid and efficient carrying is ordinarily obtained when the reagents are added in this order since precipitation occurs throughout the solution and the carrying can take place during the actual precipitation process. If the reagents are added in the reverse order the ion added last would be completely precipitated in a small fraction of the solution and much of the tracer would have to be carried on a "preformed" precipitate. In some cases the fraction of the tracer carried is strongly dependent on the order of addition of the reagents. When lanthanum fluoride is precipitated by adding an excess of hydrofluoric acid to a solution containing lanthanum ion carrier and trace protactinium, at least 70% of the protactinium is carried by adsorption in spite of the fact that protactinium forms a stable fluoride complex that is soluble under these conditions. Since actinium also is carried completely, no separation is obtained. However, when the lanthanum is added to a solution containing an excess of hydrofluoric acid, little, if any, protactinium is carried while the actinium is still carried efficiently (53). This behavior indicates that protactinium does not adsorb on a preformed precipitate, that actinium is carried on lanthanum fluoride by isomorphous replacement, and that this replacement can occur before or after the lanthanum fluoride has been formed.

(2) Oxalates

Coprecipitation with a rare earth oxalate (200) also is useful as a concentration and separation step. Thorium oxalate is insoluble in acid solution, so separation from thorium isotopes is not obtained. Separation is obtained from iron, aluminum, and a number of other nonactive elements, as well as from uranium. Carrying of tracer actinium by lanthanum oxalate is complete at a pH of 4.5 when the precipitate is digested at 75 to 80°C. for 30 minutes. Lanthanum ion concentrations of 0.1 mg./ml. are sufficient. Yttrium oxalate also has been used (250), and presumably other rare earth oxalates are also effective. Purification of macro amounts of actinium from iron, aluminum, and other impurities has been accomplished by homogeneous precipitation of the oxalate from very dilute nitric acid solutions (267).

(3) Hydroxides

Coprecipitation with an insoluble hydroxide (aluminum, iron, thorium, or rare earth hydroxides may be used) will separate actinium from radium and francium. Carrier concentrations greater than 0.1 mg./ml. are necessary and the carrier should be completely precipitated. The solution must be free from carbonate ion to avoid carrying radium. Some of the radium may be adsorbed on the hydroxide, and several reprecipitations are usually necessary to obtain complete separation. Precipitation will be incomplete if a large excess of ammonium ion is present, so highly acid solutions should be nearly neutralized with sodium hydroxide, the solution boiled to remove carbon dioxide, and the carrier precipitated with carbonate-free ammonia. Weighable amounts of actinium behave in a similar fashion.

(4) Sulfates

Variable amounts of trace actinium will carry on lead or barium sulfate precipitated from dilute nitric acid with sulfuric acid (200,298). Nearly complete carrying is obtained when the precipitation is carried out from a hot solution and the precipitate is digested for several hours at an elevated temperature. The presence of macroscopic amounts of lanthanum will give low results, presumably by displacing the actinium from the precipitate (200). Separation of actinium from daughters cannot be accomplished in this manner, but separation from large amounts of inactive solids may be obtained. Coprecipitation with lead sulfate has been used to separate actinium and its daughters from urine salts (258). The lead sulfate was dissolved in hydrochloric acid and the lead removed by sulfide precipitation, leaving the actinium in solution. Carrier concentrations of 0.5 mg./ml. of Ba and 1 mg./ml. of Pb are sufficient.

(5) Precipitation of Other Elements from Actinium

A good separation from thorium may be obtained by precipitating thorium peroxyhydroxide with hydrogen peroxide, leaving the actinium in solution (118). The precipitation is made from a slightly acid solution and the precipitate is digested near boiling for a few minutes.

The lead, bismuth, thallium, and polonium daughters of actinium will coprecipitate with lead or bismuth sulfide from slightly acid solution while the actinium remains in solution. Macroscopic amounts of inactive metals that form acid-insoluble sulfides also may be separated in this way. Trace amounts of radium can be removed by carrying on barium chromate from acetate-buffered solutions (96). Actinium remains in solution if the pH is less than 5.7 (102). Macroscopic amounts of radium can be precipitated as the nitrate from concentrated nitric acid solutions, leaving actinium in solution. Tracer radium can be removed in a similar fashion, using barium as the carrier.

c. Solvent Extraction

(1) 2-Thenoyltrifluoroacetone (TTA)

Liquid–liquid extraction with TTA was used in the first isolation of a pure actinium compound from neutron-irradiated radium (100). TTA is a fluorinated β-diketone that can react in its enol form with metal ions to form chelate compounds that are soluble in nonpolar solvents. In use the TTA is dissolved in a solvent such as benzene or xylene and is contacted with the aqueous solution containing the ions to be separated. The reaction may be represented as:

$$M^{+n} \text{ (aq.)} + n\text{HTTA (org.)} \rightarrow M(\text{TTA})_n \text{ (org.)} + H^+ \text{ (aq.)}$$

where HTTA is the enol from of TTA and (aq.) and (org.) refer to the aqueous and organic phases, respectively. The equilibrium constant, K, is given by:

$$K = \frac{[M(\text{TTA}_n)] \text{ (org.)}}{[M^{+n}] \text{ (aq.)}} \times \frac{[H^+]^n \text{ (aq.)}}{[\text{HTTA}]^n \text{ (org.)}}$$

The ratio [M](org.)/[M](aq.) for a given metal ion thus varies inversely as the nth power of the hydrogen ion activity and directly as the nth power of the TTA activity. By varying the concentration of TTA in the organic phase and the acid concentration in the aqueous phase, it is possible to vary the fraction extracted over wide limits for many metal ions and, in this way, effect numerous separations. In practice, it is common to keep the TTA concentration constant and vary the acidity.

The pH-dependence for a particular ion is such that essentially no extraction into the organic phase occurs below a given pH. As the acidity is reduced the ion begins to extract, and the fraction extracted increases from essentially zero to essentially one over a short pH range. The fraction extracted remains high as the acidity is further decreased, until a pH is reached at which the ion precipitates as the hydroxide, hydrolyzes, or otherwise no longer remains in solution. The ion can be re-extracted into an aqueous solution sufficiently acid to destroy the chelate complex by reversing the reaction. This technique is particularly adaptable to the extraction of tracer concentrations of ions, since the pH and TTA concentration do not change significantly during the extraction and the problem of solubility of the chelate in the organic phase does not arise. However, macroscopic concentrations of ions can be extracted if proper attention is given to these factors.

Table V gives the pH conditions for extraction into $0.25M$ TTA in benzene for actinium and other elements commonly encountered in actinium separations. Examination of the table shows that actinium can be extracted into a TTA solution at pH 6 while the radium remains unextracted, and that at pH 1 actinium remains in the aqueous phase while thorium is completely extracted and polonium is partially extracted. These two extractions separate actinium from radium and from all radium and

TABLE V
Extraction of Actinium, Radium, Daughter Elements, and Lanthanides from Aqueous Solution by an Equal Volume of 0.25M Thenoyltrifluoroacetone in Benzene (100,213)

Metal	pH(0)[a]	pH(1)[a]
Ac	3	6
Th	0	1
Po	0	2
Bi	1	3
Tl(III)	1	4
Pb	2	4.7
Lanthanides	2–3.2	3.6–4.7
Tl(I)	3.5	8
Ra	(No extraction at pH $\leq$ 6)	

[a] Key: (0) = pH at which extraction begins; (1) = pH at which extraction is essentially complete. See text.

actinium daughters except lead and bismuth isotopes (thallium daughters are short-lived and disappear by radioactive decay). The lead and bismuth isotopes, as well as any remaining polonium, may be removed by adding lead carrier to the actinium solution and precipitating lead sulfide. This method was used by Hageman (100) to separate 1.27 mg. of pure actinium-227 from 1 g. of neutron-irradiated radium, and serves as a useful analytical procedure to separate actinium from its daughters (see Section II-G-1). Large amounts of elements that extract into TTA solution at a pH less than 6 are better separated by other methods, although small amounts can be removed from actinium by TTA extraction (123). Among these elements are Zr(IV), Fe(III), Pu(IV), and Np(IV), in addition to those given in the table. Quantitative separation of actinium from the lanthanide elements is difficult, as indicated by the data in Table V, but may be possible by multiple extraction or careful pH control (111), using high concentrations of TTA to extract weighable amounts of the rare earths (127).

(2) Organic Phosphates

Actinium can be extracted into tributyl phosphate from aqueous solutions containing high concentrations of nitrate salts (229). Ammonium nitrate or aluminum nitrate are effective. From concentrated nitric or hydrochloric acid solutions, extraction is too low to be useful as a separation method (96). The rare earths, particularly the higher members of the group, are extracted to a much greater degree, so this solvent may find application in certain lanthanide–actinium separations.

Some of the mono- and diesters of phosphoric acid are very powerful extractants for actinium. They may be used to extract actinium quantitatively from dilute acid solution. The presence of nitrate salts in the aqueous phase is not required as it is for tributyl phosphate. Dibutyl phosphoric acid (25% in carbon tetrachloride) will extract actinium from $0.1N$ nitric acid solution (173). Peppard and his co-workers (230) have made a detailed study of the extraction of actinium, americium, and some of the lathanides into di- and mono-(2-ethylhexyl) phosphoric acids and the corresponding p-octyl phenyl esters. The extraction of these elements was directly proportional to the third power of the ester concentration in the organic phase and inversely proportional to the third power of the acid concentration in the aqueous phase. Quantitative extraction of actinium can be obtained from acid concentrations in the range 0.01 to $0.1N$ with ester concentrations of the order of several molar. The ester may be dissolved in inert nonpolar solvents such as carbon tetrachloride or benzene. The di(2-ethylhexyl) ester has been used in the determination of actinium in solutions resulting from the processing of uranium ores (249). As is the case with TTA, these extractants are not specific, since a number of other elements also extract; among these are the lanthanides, thorium, uranium, and polonium. However, many separations are possible because of the large differences in extractability between some elements. For example, whereas thorium and actinium are both extracted into dibutyl phosphoric acid from $0.1N$ nitric acid, actinium may be back-extracted into $4N$ nitric acid while the thorium remains in the organic phase (173). Commercial sources of the phosphate esters are generally mixtures, and must be purified before use to obtain reproducible results. Purification is readily accomplished by solvent extraction and by washing with alkaline solutions. The original literature should be consulted for details.

(3) **Alcohol**

Actinium can be separated from its radium, thorium, and bismuth daughters by a dry extraction method developed by Haïssinsky (105) and modified by McLane and Peterson (202,248). A mixture of the dry nitrate salts of actinium, thorium, and radium is extracted with absolute ethyl or isopropyl alcohol. The actinium and thorium dissolve, leaving radium in the residue. If only tracer amounts of the elements are present, lanthanum (or cerium), thorium, and barium are added as carriers. The thorium is removed from the alcohol solution by precipitation as the pyridine complex salt, leaving actinium and most of the bismuth and lead decay products in solution. The decay products may be removed by adding lead carrier and precipitating lead sulfide. Carrier-free actinium-228 can be prepared from its parent, radium-228, by using cerium in place of lanthanum as the carrier for the actinium. After the lead sulfide precipitation the cerium is oxidized with silver oxide and cerium and silver iodates are precipitated with iodic acid.

d. Ion-Exchange Methods

The most difficult separation involving actinium, its separation from lanthanum, has been accomplished on cation resin columns. Like solvent extraction and other chromatographic methods, ion-exchange separations have the advantage of not requiring the use of carriers when dealing with tracer concentrations. Compared to solvent extraction, ion exchange has the disadvantage that the eluted material often is contained in a large volume of solution. The time required to make a column separation is also a disadvantage, since daughter growth during the separation will complicate counting the actinium and additional separations may then be required. However, its ability to perform many actinium separations that are extremely difficult to obtain by other means make ion-exchange separations a very useful method.

(1) Cation-Exchange Systems

On cation-exchange resin columns, actinium elutes after the rare earths because of its larger ionic radius. Elution with citrate has been used to separate tracer concentrations of lanthanum-140 and actinium-228. McLane and Peterson (201) used a column of Amberlite IR-1 resin (ammonium form), 35 cm. long, and an eluting agent of $0.25M$ citrate solution (a mixture of diammonium citrate and citric acid) at pH 3.09. The tracers were added to the column in a small volume of the citrate solution and the column is immediately eluted with additional solution. After most of the lanthanum had been removed the eluting solution was changed to $0.5M$ monoammonium citrate (pH 3.76) to remove the actinium faster. Separation was almost complete, but there was a slight overlapping of the peaks that indicated a longer column would be necessary for complete separation. The separation of microgram and trace quantities of actinium from milligram amounts of lanthanum can be done on an Amberlite IR-100 column (ammonium form), using a 0.5% solution of ammonium citrate adjusted to pH 5.5 with ammonium hydroxide (292,293). In this work the lanthanum was removed from the column in 210 minutes, followed by the actinium. Separation was not complete, but again a longer column probably would have been effective.

Hageman (102) gives the distribution coefficients of lanthanum and actinium between the ammonium form of Dowex-50 and 0.25M citrate solution as a function of pH between 2 and 4.6. The ratio of the distribution coefficients is quite large, greater than 30 between pH 3.3 and 3.8, so separation should be readily achieved with this system. Elution with hydrochloric acid also separates actinium from the rare earths and alkaline earths. Diamond, Street, and Seaborg (63) studied the behavior of a number of tripositive actinides and lanthanides, as well as radium, strontium, and barium in the Dowex-50–hydrochloric acid system. Separations of actinium were obtained using acid concentrations between 3 and 12M. Actinium always eluted last. The best separation of actinium from lanthanum was obtained with 6M hydrochloric acid.

Elution from Dowex-50 resin with EDTA has been used to separate radium, bismuth, lead, and actinium (66). The specific application was the preparation of radium-228 from its parent, thorium-232. Thorium nitrate was dissolved in water and the radium-228 was coprecipitated with barium chloride by adding concentrated hydrochloric acid. To separate the radium-228 from its daughters the barium chloride precipitate was dissolved in water and added to the column. The column was eluted with 0.01M EDTA solution adjusted to pH 9 with ammonium hydroxide. The order of elution was actinium, bismuth, lead, and radium.

The separation of actinium from radium and decay products on a Dowex-50 column does not require elution with a complexing agent, since the ions to be separated vary greatly in their affinity for the resin (3,99). Actinium can be adsorbed completely on the resin from a 2M nitric acid solution. The thorium daughter is also completely adsorbed, although the bivalent cation daughters are only partially adsorbed. Continued washing of the column removes the remainder of the bivalent metals. The eluting solution is then changed to 4M nitric acid to elute the actinium while the thorium remains on the column. This technique has been used to separate milligram amounts of actinium from irradiated radium. A similar method was used to separate neutron-irradiated actinium-227 from its daughters (45). The Dowex-50 column was operated at 60°C. and the solutions were preheated before being added to the column. The actinium was added to the column in 2M hydrochloric acid. The column was eluted with 2M hydrochloric acid to remove the bismuth, lead, and francium daughters, then with 3M nitric acid to remove radium. The actinium was then removed with 6M nitric acid while the thorium-227 remained on the column.

Trace thorium (as the UX$_1$ isotope) was separated from actinium-228 (74) by adsorption on Dorex-50, followed by elution with oxalic acid.

The tracers were added to the column in 0.1N hydrochloric acid, and the thorium was eluted with 7% oxalic acid at 81°C. After the bulk of the thorium had been removed, the column was washed with water and the actinium was eluted at 81°C. with 5% citric acid at pH 3. The purity of each fraction was about 95%.

(2) **Anion-Exchange Systems**

Although actinium does not form anionic complexes to any appreciable extent in acid solution, Danon (59) obtained almost complete separation of actinium and lanthanum by anion exchange in lithium nitrate solution. Actinium and lanthanum in 8.5M lithium nitrate solution were added to a Dowex-1 column pretreated with lithium nitrate solution, and the column was eluted with 4.3M lithium nitrate. Actinium eluted before lanthanum, and the author points out that the same anomolous behavior is observed in the separation of actinium from the lanthanides by fractional crystallization of the lanthanide–magnesium double nitrates. In the latter separation actinium concentrates in the middle fraction instead of the least-soluble fraction, as expected from its greater basicity. This behavior in the two separations is correlated by assuming that the adsorbability of the nitrates of actinium and rare earths on the resin is determined by the solubility of the nitrates. The more stable is the complex the less soluble is the salt.

c. Paper-Chromatographic Separations

Adloff and Perey (1,243) have developed procedures for the separation of francium-223 (AcX) from actinium-227 and other actinium daughters by migration on filter paper. The separation was done in ammonium carbonate solution since actinium and its daughters form insoluble carbonates or hydroxides and move very little, whereas francium, as an alkali metal, is soluble in the solution. In another method for this separation (28) the actinium is adsorbed on a column of cellulose and zirconium oxide, and the francium is removed with phenol (liquefied with 2N hydrochloric acid). These separations are useful in the determination of actinium-227 by separating and counting AcX, the daughter fo med by the rare alpha decay of actinium-227 (Sections II-E and II-G).

Good separation of many of the rare earths, including lanthanum, from actinium has been obtained by paper chromatography and electrochromatography (181). In the electrochromatographic separations 1% citric acid was used as the electrolyte. In one separation several milligrams of rare earths in a citrate solution containing actinium were added to the center of a sheet of Whatman No. 1 paper, and the separation was run at 300 v. for 45 minutes. Actinium traveled farthest to the cathode. Continuous electrochromatographic separations also were performed by inserting the top of a sheet of paper into a trough of the electrolyte, with a small section of the top inserted into the solution to be separated. The voltage was applied at the bottom of the sheet. The paper-chromatographic separations were obtained using a solvent of butanol, acetylacetone, acetic acid, and water. The separation of trace amounts of actinium from lanthanum has been accomplished by a descending technique, using a developing solution of butanol saturated with a solution of $7M$ lithium nitrate–$2M$ nitric acid (61).

D. DETECTION AND IDENTIFICATION

1. Emission Spectroscopy

Actinium may be detected and identified by its characteristic emission spectrum if it is present in sufficient quantity (Section II-B-3). A detection limit of 0.02 γ of actinium-227 in 0.05 ml. of solution was obtained by exciting the spectrum in a copper spark (31). This method was used to analyze for actinium in the processing of irradiated radium. Since actinium-227 is normally not encountered in such high concentrations, and since the other actinium isotopes are too short-lived to be accumulated in such large amounts, emission spectroscopy has limited application. The method is considerably simpler than methods based on radioactivity measurements and should be strongly considered for actinium-227 when sufficient amounts are available.

2. Radiochemical Methods

Since actinium isotopes are usually encountered only in tracer concentrations, the most useful qualitative as well as quantitative methods will depend upon some type of chemical separation followed by the counting of the characteristic radiations emitted by actinium and its daughters. The methods given below for qualitative identification can, with varying degrees of difficulty and accuracy, be made quantitative. In the radioactivity measurements, use is made of the types and energies of the radiations and the radioactive half life. These properties, coupled with some knowledge of the chemical behavior of the radioactive material, will characterize the isotope. That is, if the unknown activity behaves as actinium in one or more chemical separations (enough to eliminate other possible active elements in the sample), emits the characteristic radiations, and exhibits the proper rate of change of activity with time, the identity is established. Alternately, one may show the presence of a known decay product supported by the actinium parent. This is done by separating the decay product from the sample, allowing it to accumulate ("grow in") for a known length of time, and then characterizing the daughter nuclide by its radioactive or chemical properties or both.

a. Growth of Decay Products into Purified Actinium-227

The presence of actinium-227 and its radiochemical purity may be shown by separating the actinium from other radioactive elements, including its daughters, and comparing the rate of growth of the daughter activity with the rate calculated from the half lives. The growth rate for the actinium decay series may be calculated from the Bateman equations for radioactive chains, and such calculations for the natural decay series

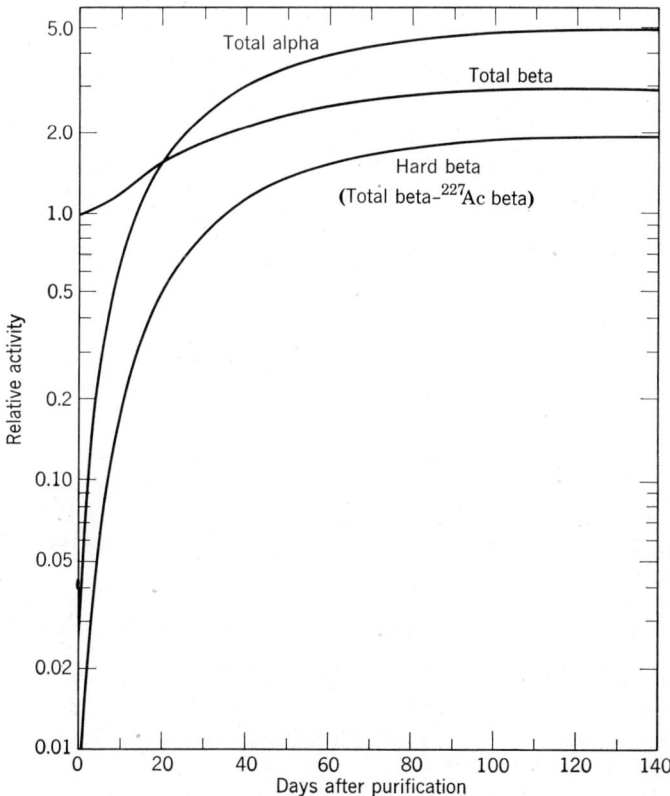

Fig. 5. Growth of alpha and beta activity in initially pure actinium-227.

have been given in a very useful form by Kirby (161). The calculated rate of growth of activity in purified actinium is given in Fig. 5. The activity is plotted relative to the initial actinium-227 activity. The maximum activity is reached at 175 days, after which the decay of actinium-227 becomes apparent. It is best to follow the growth of the alpha or hard beta activity at least during the first few weeks, when the rate of growth is greatest. The radioactive chain is given in Fig. 4. The beta particle emitted by actinium-227 is relatively weak, 0.046 M.e.v., and difficult to detect unless the sample is very thin ("weightless") and is placed inside the counting chamber (a windowless or internal counter). Thus, if solids are present in the sample or if the sample is counted in the usual windowed counter, the growth will follow the "hard" beta curve in Fig. 5 that begins at an activity of zero, rather than the total beta curve that begins at an activity of 0.988. Because of its low energy the actinium beta particle was not detected directly until 1935 (118), long after actinium was discovered. The curves in Fig. 5 were obtained by assuming equal counting efficiencies for all countable alpha and beta particles. In the case of samples that are thick compared with the least energetic beta particle, the theoretical beta growth will not be observed since self-absorption varies with beta energy. An empirical growth curve obtained by counting actinium-227 in samples of similar thickness must be used in this case.

The behavior of a rare gas member of a decay chain must be considered in work of this type. When counting actinium-227 or its thorium or radium daughters, the theoretical growth will not be obtained if the emanation daughter, radon-219, leaves out of the sample before it decays. The half-life of radon-219 is short, so the problem is not serious unless the sample is very porous. In practice the radioactive sample is evaporated to dryness on a small disk or dish and the dried sample is counted. If the sample is ignited for a few seconds prior to counting the loss of radon-219 is very small. Most inorganic compounds emanate poorly, especially after ignition. Even from samples containing very little residue the radon-219 loss is only about 2% (45).

b. Separation and Counting of Decay Products

More definite identification may be made by separating and counting the actinium daughters after they have been allowed to accumulate in a sample for some time. A period of a few days or weeks is sufficient unless the amount of actinium is very small. Separated thorium-227 (RdAc) exhibits the characteristic growth and decay shown in Fig. 6. This curve also was calculated from the Bateman equations. If other thorium

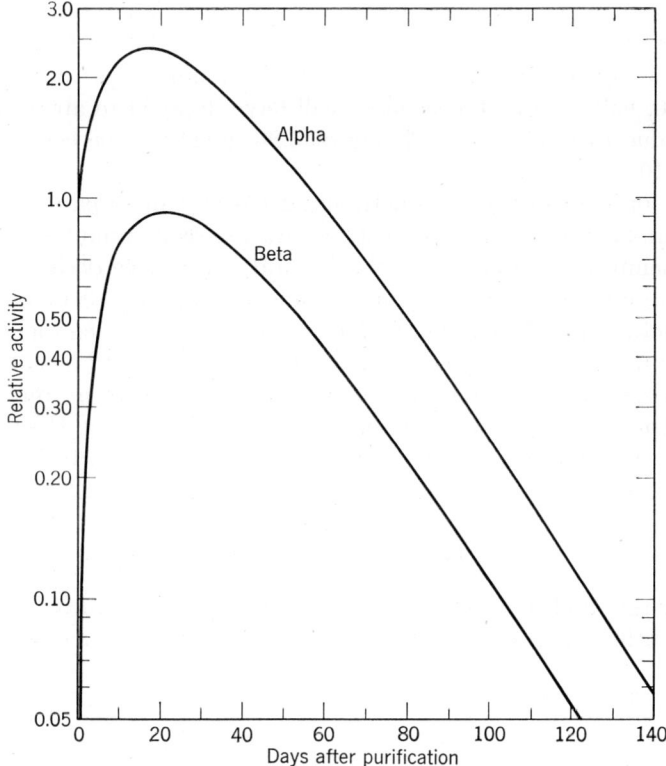

Fig. 6. Growth and decay of alpha and beta activity in initially pure thorium-227.

isotopes might be present the activity should be followed at least until after the maximum activity is reached (about 20 days after separation) and the activity begins to decrease. The radium-223 and lead-211 daughters also may be separated and counted to identify actinium-227. Radium-223 decays with a half life of 11.2 days after a rapid growth of daughters that is essentially complete in a few hours; lead-211 is a beta emitter with a 36.1-minute half life. These daughters may be separated by the methods given earlier. For example, thorium-227 may be separated by TTA extraction from $0.1N$ nitric acid, radium-223 may be coprecipitated with barium chloride from concentrated hydrochloric acid solution, and lead-211 may be coprecipitated with lead sulfide from acid solution.

Radon-219 (actinon) may be separated and detected by passing a gas through a solution containing the sample and into a counting chamber. The gas to be used depends on the type of counter, e.g., air for an air-

470   A. SYSTEMATIC ANALYTICAL CHEMISTRY OF ELEMENTS

filled ionization chamber, or argon for a proportional counter. After the air flow is stopped the activity due to actinon (half life of 3.92 seconds) will have decayed completely at the end of a minute, that due to thoron (radon-220, half life of 54.5 seconds) will have decayed to about one-half, and that due to radon-222 (half life of 3.83 days) will be essentially unchanged (102).

In 1.2% of its disintegrations actinium decays by alpha-particle emission to francium-223, a beta emitter with a half life of 22 minutes. Since no other francium isotope occurs in the natural radioactive decay series, the separation and characterization of this nuclide is an excellent qualitative test for actinium-227. Because of its 22-minute half life the francium separation must be done rapidly, in an hour or less. Methods of separating and counting francium for the qualitative as well as quantitative analysis of actinium-227 are described in Sections II-E-3-c and IV-F-1. This method is 100 to 500 times less sensitive than those described earlier because of the low abundance and short half life of francium-223.

c. Gamma-Ray Counting

The energy and relative abundances of the gamma-rays associated with the actinium series, as well as their rate of growth or disappearance

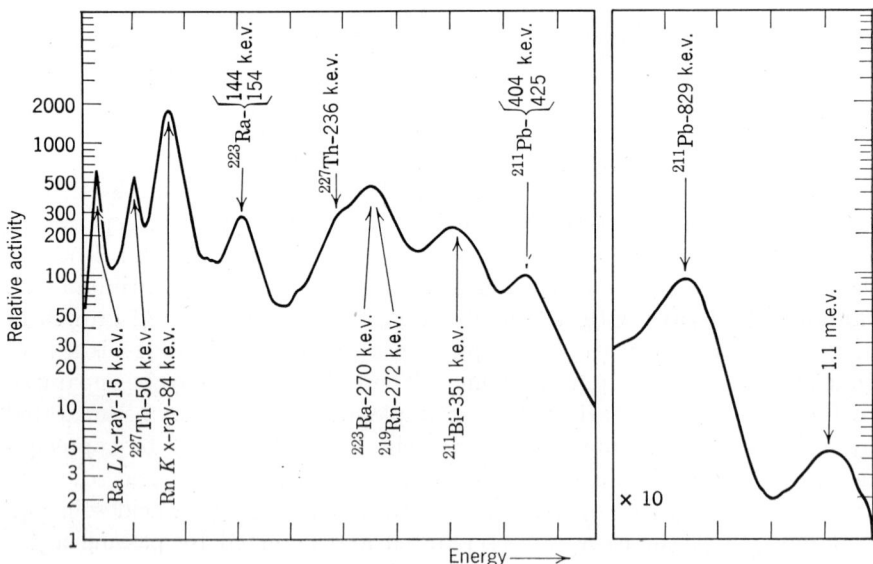

Fig. 7. Gamma-ray spectrum of actinium-227 and decay products.

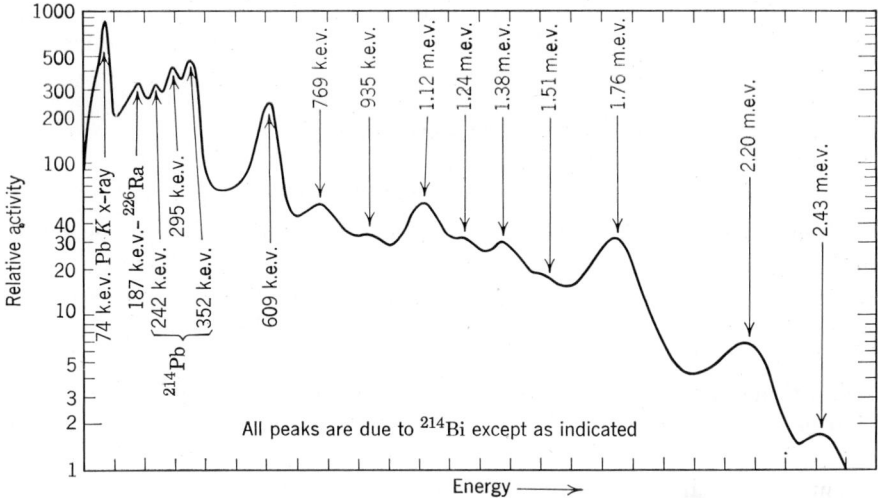

Fig. 8. Gamma-ray spectrum of radium-226 and decay products.

in purified actinium-227 or its daughters also may be used for identification. The gamma-ray energies are best characterized by scintillation counting with a sodium iodide–thallium activated crystal and multiplier phototube combination, coupled to a linear amplifier and multichannel analyzer.

A gamma-ray spectrum of actinium-227 and its decay products, taken with a cylindrical crystal 4 inches in diameter and 4 inches high and a 256-channel pulse-height analyzer, is shown in Fig. 7. All the peaks in this spectrum are due to actinium decay products. In addition to the gamma-rays, two x-rays appear in the spectrum. The x-rays observed during some decay processes are emitted by the daughter nucleus, and are the characteristic x-rays of decay products and not of the decaying nucleus. A 37 k.e.v. gamma-ray from actinium-227 has been reported (283), but its existence is questionable. Therefore the rate of change of the gamma-ray peaks with time must be known before the presence of actinium-227 can be established from the gamma-rays of its decay products. That is, it is necessary to know that the daughter products detected are, in fact, supported by the actinium-227 parent. The usefulness of gamma-ray spectrometry will depend on the identity of the other radioactive elements in the sample. For comparison purposes, gamma-ray spectra of the other long-lived members of the uranium–radium and uranium–actinium series, obtained with the same instrument as the actinium spectrum, are given in Figs. 8, 14, 15, 16, and 17. The detection of actinium in radium (Fig. 8) is of particular interest since actinium is generally obtained by neutron irradiation of radium-226. In this case the actinium daughter gamma-rays at 50, 150, and 829 k.e.v. and the K x-rays appear more useful.

E. DETERMINATION

1. Introduction

Except for emission spectroscopy, a quantitative determination of actinium consists of separating actinium or its daughters free from other radioactive elements, and from excessive amounts of inert material, and measuring the radioactivity of the separated product. The separation should preferably, but not necessarily, be quantitative. Quantitative separations will have the greatest reliability and are preferred. The separation need not be quantitative if the recovery is known and reproducible, as established by prior determinations on similar samples to which known amounts of actinium or daughters have been added or, better, by using a tracer to measure the chemical yield of the actual analysis.

Actinium-228 may be used as a tracer for actinium-227. A known amount of actinium-228 is added to the sample to be analyzed for actinium-227. The sample should be put into complete solution before beginning the separations to assure that complete isotopic exchange occurs between the two actinium isotopes. After separation the actinium-227 is determined by alpha counting and recovery is determined by counting the strong beta particle of actinium-228 under conditions in which the alpha particles are not counted. The analysis should be completed within a period of a few half lives of actinium-228 (6.13 hours) to reduce the additional error due to a large correction for decay. Many of the separation procedures described in the literature were designed to give a pure product and were not meant to be quantitative. See, for example, the compilations given in References 204, 205, and 280. These procedures may be used with appropriate tracers to obtain chemical yields.

2. Determination by Separating Actinium

a. Separations

Extraction into TTA is often used to separate actinium from its daughters and other activities except the rare earths (Section II-C-2-c.). The presence of large amounts of certain elements will interfere in the actinium extraction at pH 5.5 to 6 since they may not be completely removed by the first extraction at pH 1 (123,127). Among these are thorium, zirconium, plutonium(IV), neptunium(IV), bismuth(III), lead(II), and

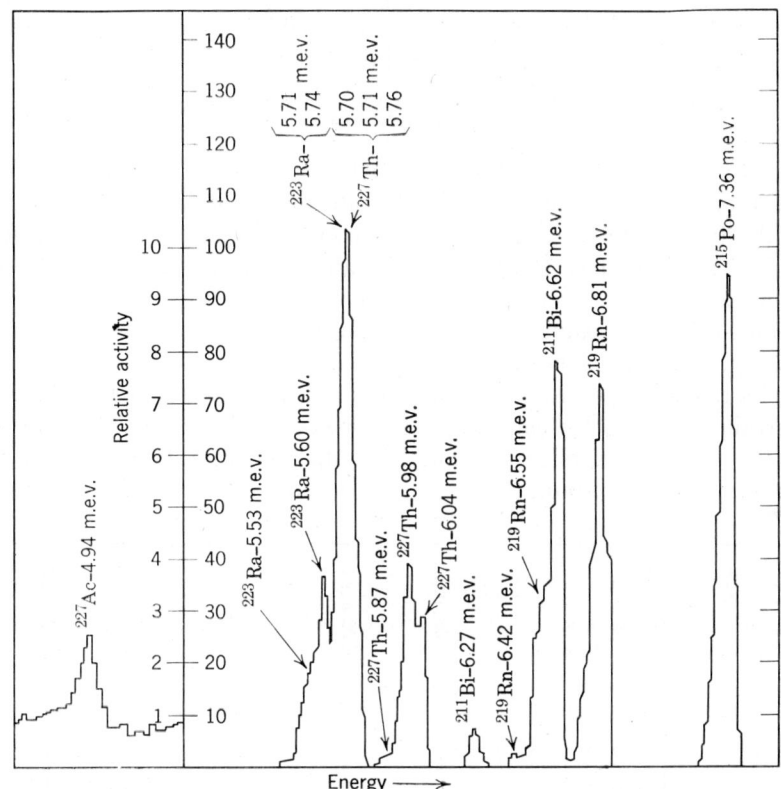

Fig. 9. Alpha-particle spectrum of actinium-227 and decay products.

iron(III). Methods used for the preliminary removal of large amounts of thorium include precipitation of thorium peroxyhydroxide (118), extraction into pentaether (dibutoxytetraethylene glycol (122,205)), extraction into dibutylphosphoric acid from $4M$ nitric acid (173), and extraction by 1-(3,4-dichlorophenyl)-4,4,5,5,6,6,6-heptafluoro-1,3-hexanedione (127,133). The latter compound is a fluorinated beta diketone, similar to TTA, but whose thorium complex is much more soluble in nonpolar solvents than the corresponding TTA complex. Coprecipitation with a rare earth fluoride will separate actinium from zirconium (in the absence of barium), from iron(III), and from plutonium and neptunium after their oxidation to the hexavalent state. Lead and bismuth can be removed by precipitation of lead sulfide from acid solution, and iron also can be removed by adsorption of iron(III) on an anion-exchange resin from concentrated hydrochloric acid.

474   A. SYSTEMATIC ANALYTICAL CHEMISTRY OF ELEMENTS

Many other methods can be devised by combining individual separation steps given in Section II-C.   Examples of these are given later.

b. COUNTING

The low energy of the actinium-227 beta particle and the low abundance of its alpha particle limit the direct counting of actinium to very thin (essentially weightless) samples to avoid counting losses due to self-absorption. In such samples the beta or the alpha particle can be counted in "windowless," or internal, counters. It should be possible to count the actinium-227 beta particle in a liquid scintillation counter with good efficiency, but this has not been reported as yet.

Alpha-particle spectrometry can be used to determine actinium-227 in certain mixtures by counting its 4.94-m.e.v. alpha particle, and this method has been applied to samples obtained from the processing of neutron-irradiated radium (87).   In addition to actinium-227, radium-226,

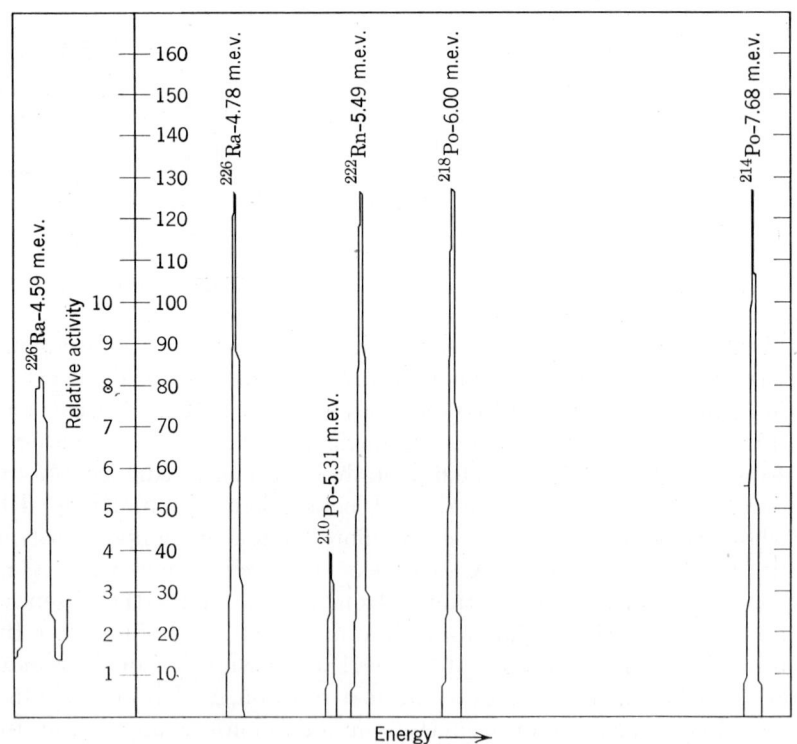

Fig. 10. Alpha-particle spectrum of radium-226 and decay products.

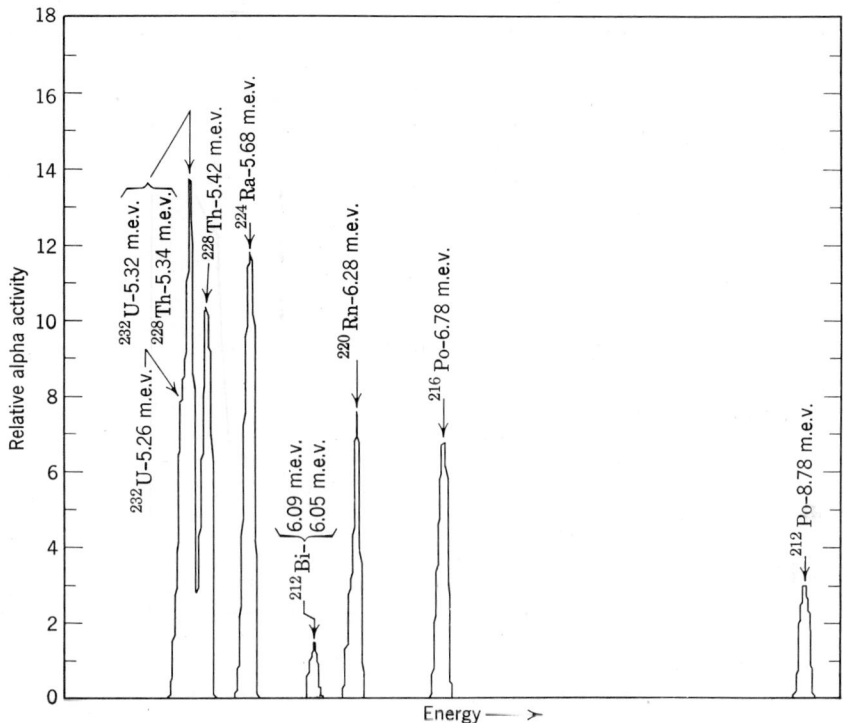

Fig. 11. Alpha-particle spectrum of thorium-228 and decay products (uranium-232 was used as the source of the thorium-228 chain).

and their decay products, these samples may also contain thorium-228 and its decay products if the radium contained radium-228 in addition to radium-226. The actinium-227 alpha particle can be distinguished from the other alpha particles in this mixture on the basis of its energy. Alpha-particle spectra of these decay chains are shown in Figs. 9, 10, and 11. These spectra were obtained with a silicon (surface barrier) solid state charged-particle detector coupled to a linear amplifier and 400-channel pulse-height analyzer. A spectrum of a radium–actinium mixture obtained from neutron-irradiated radium is given in Fig. 12 (87). This spectrum was obtained with a gridded ionization chamber and 30-channel pulse-height analyzer. Very thin uniform mounts are required for sharp alpha spectra, and to obtain them the sample solutions were evaporated on the counting plate using tetraethylene glycol as a spreading agent, or were sublimed onto the plate. The sensitivity of the analysis was only 255 disintegrations of actinium-227 on the counting plate, since actinium-

476   A. SYSTEMATIC ANALYTICAL CHEMISTRY OF ELEMENTS

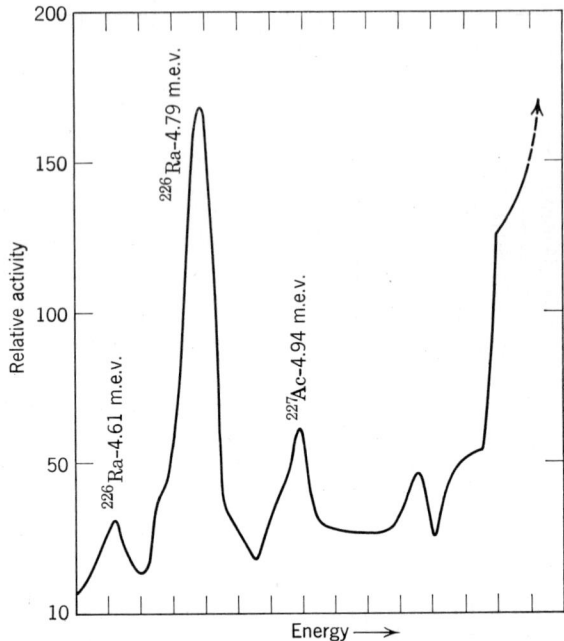

Fig. 12. Alpha-particle spectrum of a radium–actinium mixture (87). Reproduced by permission from *United Kingdom Atomic Energy Authority Reports*.

227 emits alpha particles in only 1.2% of its disintegrations. The gridded ionization chamber was formerly the standard detector for alpha spectrometry, but solid state detectors, because of their superior resolution, are replacing the ionization chamber for this purpose.

Usually the amount of actinium is calculated from the rate of growth of the daughter alpha activity in the first few weeks or days, or the daughters are allowed to come to equilibrium with the parent and the alpha, beta, or gamma activity is measured (102,180). Alpha counting is preferred because of its greater accuracy. The actinium activity is obtained by dividing the observed total activity by the relative activity at the time of counting (Fig. 5).

Any counting procedure requires a knowledge of the efficiency of the detector for the radiations being detected. This is determined by counting known amounts of activity under conditions identical to those used in the final determination, or by making appropriate corrections if the conditions of calibration and sample counting are not identical (see Section I-A).

Other actinium isotopes with half lives greater than a few hours may be counted without serious difficulty, except that the separation and counting of the short-lived isotopes must be done rapidly to minimize decay. The final calculations must include corrections for actinium decay or daughter growth if it occurs. Mixtures of actinium isotopes may be analyzed by observing the growth and/or decay for an appropriate length of time and calculating the composition of the mixture from the known half lives (161).

Perey (240) proposed a rapid and simple method for determing actinium-227 in the presence of moderate amounts of inactive material. The method consists of counting the energetic beta activity of the decay products that follow the rare gas daughter, actinon, in the decay chain (the "active deposit"). A known volume of the solution containing actinium in radioactive equilibrium is placed in a plastic dish and weighed. The weight is adjusted to 2.5 g. either by the addition of water or by evaporation, and the dish is sealed with a cover glass spread with a thin film of grease on the underside. Since some actinon is lost in this process, the sample is allowed to stand for about four hours to allow the active deposit to reach equilibrium. This growth is governed by the half life of lead-211, 36.1 minutes. The beta activity in the sample is then counted in an end-window beta counter. The presence of radioactive impurities can be detected by comparing the shape of the beta absorption curve with that obtained with pure actinium in equilibrium. This method requires that the actinium be in equilibrium with its daughters, and that other energetic beta emitters be absent. Calibration of the counting arrangement is done with a known amount of actinium-227 counted under the same conditions. With the counting arrangement used the method was sensitive to 6700 disintegrations per minute of actinium in 2.5 g. of solution.

3. Determination by Separating Actinium Decay Products

For these methods actinium-227 need not be in equilibrium with its daughters, but the time of the last separation from its daughters must be known in order to calculate the actinium activity from the growth equations. When other radioactive impurities or large amounts of rare earths are present, these methods are advantageous. The general procedure is as follows. The daughter to be counted is separated from the actinium sample and the time of separation is noted. The sample is allowed to remain undisturbed for a known interval to allow the daughter to accumulate, and the daughter is again separated and its activity determined by counting. The activity of the daughter at the time of the final separation is calculated from the appropriate growth or decay equation for the daughter. From this value and the time interval of daughter growth the actinium content is calculated, using the Bateman equation that describes the growth rate of the daughter into purified actinium. If the actinium is known to be in radioactive equilibrium the first separation and the growth period are unnecessary.

When thorium-227 is separated and counted the results are calculated as follows. The activity of thorium-227 is an initially pure sample of actinium-227 as given by

$$A_1/A_0 = \lambda_2/(\lambda_2 - \lambda_1)[\exp(-\lambda_1 t) - \exp(-\lambda_2 t)]$$

where A_1 is the activity of thorium-227 at the time of the second thorium separation, A_0 is the activity of actinium-227 at the time of the first thorium separation, λ_1 and λ_2 are the decay constants (ln 2/half life) of actinium-227 and thorium-227, respectively, and t is the time interval between the thorium separations. After the second thorium separation the total alpha activity of the separated thorium-227 is given by:

$$A/A_1 = 11.058 \exp(-\lambda_2 t_1) - 10.060 \exp(-\lambda_3 t_1) + 10^{-6} \exp(-\lambda_4 t_1)$$
$$+ 0.00239 \exp(-\lambda_5 t_1) - 10^{-1} \exp(-\lambda_6 t_1)$$

where A is the total alpha activity of thorium-227 and its decay products at any time, t_1, after the second thorium separation, and λ_2, λ_3, λ_4, λ_5, and λ_6 are the decay constants of thorium-227, radium-227, radon-219, lead-211, and bismuth-211, respectively. The constants before each exponential are functions of these decay constants (161). The alpha activity of the separated thorium-227 is determined at several times after separation, and an average value for A_1 is calculated by use of the second equation. This value of A_1 is then used in the first equation to obtain the actinium-227 activity, A_0.

a. Thorium-227

This nuclide may be separated by appropriate methods (e.g., TTA extraction), and its concentration and purity may be determined by alpha or beta counting, as given in Section II-D-2. Small amounts of other thorium isotopes may be tolerated since they can be distinguished by the rate of radioactive growth and decay. Rosholt (260) has described a procedure for the analysis of uranium, protactinium, thorium, radium, and lead isotopes in minerals to determine the state of radioactive equilibrium. Although the thorium-227 activity was used in this case to determine the amount of protactinium-231, the parent of actinium-227, the method is equally applicable to actinium determinations. The sample (1 g. or less) was fused with 4 to 5 g. of sodium peroxide and the melt was leached with water and acidified with hydrochloric acid. Bismuth and polonium isotopes were removed by coprecipitation with bismuth sulfide carrier. After boiling the solution to remove the hydrogen sulfide, thorium isotopes were coprecipitated with zirconium phosphate in the presence of lanthanum hold-back carrier to prevent adsorption of actinium isotopes. The time of this separation was noted. The zirconium phosphate precipitate was dissolved in oxalic acid and praseodymium oxalate was precipitated to carry the thorium. Fluorescent zinc sulfide was added to the mixture and the combined solids were separated and mounted for scintillation alpha counting with a multiplier phototube. Less than an hour should elapse between the zirconium phosphate separation and the oxalate separation. The amount of actinium can be calculated from the thorium-227 disintegration rate at the time of separation.

b. Active Deposit

Methods based on the separation and beta counting of the active deposit lead-211 and bismuth-211 also have been applied to actinium-227 in equilibrium with its daughters. In one method (239), lead sulfide is precipitated from an acid solution of the sample, separated, washed with cold $0.25N$ hydrochloric acid saturated with hydrogen sulfide to depress the solubility of lead sulfide, and the precipitate is counted. This method, sensitive to 22,000 disintegrations per minute of actinium-227, is not applicable to samples containing much thorium or radium since the active deposits of their decay chains interfere. In an interesting modification (241) to eliminate this interference, radium-223 is separated first by coprecipitation with barium sulfate carrier. After conversion of the barium sulfate to the carbonate and dissolution of the carbonate in hydrochloric acid, the solution is boiled under reduced pressure to remove radon isotopes and lead sulfide is precipitated to remove the active deposits. The active deposit of actinium grows in more rapidly than that of thorium and radium. Therefore a second lead sulfide precipitation is made 30 minutes after the first, and this precipitate, which contains very little of the thorium and radium daughters, is counted. By this method 22,000 disintegrations per minute of actinium-227 in the presence of one hundred times as much radium was analyzed with an accuracy of 5%. This procedure is applicable to radioactive minerals and ores.

Actinium-227, in the presence of thorium and radium, also has been determined by separating and counting the active deposit in an alpha pulse-height analyzer (138a). From the alpha-particle spectrum it is possible to distinguish bismuth-211 (actinium-C) in the presence of the alpha emitters of the radium and thorium active deposit.

c. Francium-223

The separation and counting of francium-223, mentioned in Section II-D-2, also may be used for quantitative determinations. In the original method given by Perey (234) for actiniferous rare earth mixtures, actinium-227 and all its daughters except francium-223 and thallium-207 are removed by coprecipitation with the rare earth carbonates. Thallium-207 and the remaining traces of radium-223 are removed by coprecipitation with barium chromate and lanthanum hydroxide, leaving the francium in solution. Hyde (124) separated francium-223 from actinium-227 and other daughters by coprecipitation with free silicotungstic acid from concentrated hydrochloric acid. The precipitate was dissolved in water and passed through a cation-exchange column. Francium is adsorbed, whereas the nonionic silicotungstic acid passes through. Elution of the column with concentrated hydrochloric acid produces a carrier-free solution of francium-223.

The francium-223 must be in equilibrium with its actinium parent before separation, but this requires only a few hours since the half life of the francium isotope is only 22 minutes. The separation must be completed in about an hour because of this short half life. The sensitivity of this method is much less than methods based on counting other daughters because francium-223 is produced in only 1.2% of the actinium disintegrations and because the francium decays rapidly. The procedure, however, is very specific since francium-223 is the only francium isotope of any consequence in nature.

4. Other Methods

Quantitative analysis by emission spectroscopy has been used to follow actinium in the processing of neutron-irradiated radium (31). Millicurie amounts of actinium have been determined by measuring the total gamma activity of the decay chain (in radioactive equilibrium) with an ionization chamber (180). The sensitivity has been increased to 0.007 μc. by using a high-pressure, argon-filled ion chamber (4). Improvements in this type of measurement, and its application to the analysis of many mixtures of radioactive nuclides, can probably be accomplished by counting with sodium iodide gamma-ray scintillation spectrometers.

Millicurie amounts of pure actinium-227 samples in equilibrium have been determined calorimetrically to an accuracy of a few per cent by measuring the amount of heat produced by the radioactive decay (177, 180, 268).

Delayed coincidence counting of radon-219 and its very short-lived daughter, polonium-215, provides a fairly specific method for actinium-227 (138a). The radon-219 is swept out of a solution of the sample and into a counting chamber by a current of air. The counter is arranged to count only pairs of alpha particles emitted within 5 μseconds of each other. Radon-219 emits an alpha particle to form polonium-215, which in turn decays by alpha-particle emission with a half life of 1.83 milliseconds. In the 5 μseconds allowed for coincidence counting, some of the polonium-214 produced in the chamber will decay, and very few consecutive alpha emissions will occur from the radon-220 and -222 decay chains (see Figs. 1, 3, and 4).

F. DETERMINATION IN SPECIFIC MATERIALS

1. Radioactive Ores and Residues

Determination in ores and minerals by separation of radium-223, followed by separation and counting of the active deposit and by separation of thorium-227, have been discussed in Section II-E-3. Another method uses TTA extraction to separate actinium, followed by separation and counting of francium-223 (177). Once the actinium is in the organic phase the francium can be readily separated by extraction with water (1).

Two methods have been described for determination in waste solutions from uranium processing plants ("mill effluents"). In one method (73), actinium is coprecipitated from the mill effluent with lanthanum fluoride carrier, and the precipitate is washed with a mixture of nitric and hydrofluoric acids containing a little dichromate ion to prevent reduction of U(VI) to U(IV). Tetravalent uranium coprecipitates with lanthanum fluoride. After treatment with sodium hydroxide to convert the fluoride to hydroxide, the precipitate is dissolved in $1.8M$ hydrochloric acid and is passed through an anion-exchange column. Lead, bismuth, and polonium are retained on the column; the lanthanum and actinium pass through and are precipitated with ammonium hydroxide. The hydroxide is dissolved in aluminum nitrate–dilute nitric acid solution, and the solution is extracted with TTA to remove thorium. The aqueous phase is absorbed on a cation-exchange column, and the lanthanum and actinium are eluted with citric acid at pH 5.0, leaving radium on the column. The lanthanum and actinium are then precipitated as the oxalate, and the precipitate is dried and counted. Actinium-228 is counted in an end-window beta counter to eliminate the low-energy beta particles emitted by actinium-227. This is done soon after the chemical separations, before the active deposit has grown in appreciably. Actinium-227 is determined by alpha counting of its decay products, or by beta counting in a windowless counter if actinium-228 is absent. Another method, based on extraction with di(2-ethylhexyl) phosphoric acid (249,250), is given under Recommended Procedures (Section II-G-3).

2. Irradiated Radium

Measurement by emission spectroscopy (31) and by alpha-particle pulse-height analysis (87) has been discussed (Sections II-D-1 and II-E-2-6). In the latter method a very high ratio of radium to actinium would probably introduce errors since the radium-226 alpha has an energy of 4.79 m.e.v., fairly close to that of actinium-227 (4.94 m.e.v.). Alpha pulse-height analysis is applicable, without preliminary separation, to process solutions from which much of the radium already has been removed.

The TTA extraction method given under Recommended Procedures also may be used, but if the radium-to-actinium ratio is very high the phases must be separated carefully and completely to prevent mechanical carry-over of small amounts of radium.

3. Biological Samples

Radium-228 has been determined in acid solutions of bone by separating and counting its immediate daughter, actinium-228 (279). The beta particle of radium-228 has an energy of only 0.012 m.e.v., too weak to be conveniently counted by ordinary methods. The actinium-228 was separated from its parent by nitric acid extraction from $0.1N$ nitric acid solution into a solution of 25% dibutyl phosphoric acid in carbon tetrachloride. Thorium, lead, bismuth, and polonium isotopes, present as decay products, are either partially or completely extracted, but actinium may be separated from thorium by re-extracting the organic phase with $4N$ nitric acid. Thorium remains in the organic phase, whereas actinium passes into the acid solution. Any lead, bismuth, or polonium isotopes in the nitric acid solution are removed by coprecipitation with lead sulfide. The supernatant from the sulfide precipitate contains the purified actinium.

Determinations of actinium-227 in rat urine by coprecipitation with lead sulfate (258) and in the urine of laboratory workers by coprecipitation with ceric phosphate (163) have been described. The distribution of actinium-227 and its thorium and radium daughters in rats has been studied by gamma-ray counting of tissues and organs with a high-pressure ionization chamber (4). By counting at various times the amount of each nuclide was calculated from the relative rate of growth and/or decay of the activity. Since the gamma-ray activity was measured, thin samples and, therefore, chemical separations were not required. The samples were counted without prior treatment.

G. RECOMMENDED PROCEDURES

1. Extraction with 2-Thenoyltrifluoroacetone (102)

This procedure is applicable to the separation of actinium from its daughters and to the determination of actinium in samples containing moderate amounts of other natural activities, including radium. The method is not applicable in the presence of large amounts of rare earths.

Adjust the actinium solution to $0.1N$ with mineral acid and shake or stir thoroughly for 10 to 20 minutes with twice its volume of $0.25M$ 2-thenoyltrifluoroacetone (TTA) in benzene. Allow the phases to separate, draw off the benzene layer (the aqueous layer may be washed with a small amount of benzene to complete the removal of the TTA), and add a new portion of TTA solution. Adjust the pH of the aqueous solution with sodium hydroxide solution to 5.5 to 6.0 as measured with a glass electrode (high concentrations of ammonium ion should be absent), and stir or shake again for 10 to 20 minutes. After the phases have separated, test the pH of the aqueous solution again. If the pH is not between 5.5 and 6.0, readjust the acidity with sodium hydroxide and extract again. Repeat until the pH at the end of the extraction is between 5.5 and 6.0. Remove the benzene layer to a clean container (a benzene wash may be used to assure complete transfer) and re-extract the actinium from the benzene by mixing with half its volume of $0.1N$ mineral acid. Allow the phases to separate and discard the benzene layer. Add 0.5 to 1.0 mg. of lead carrier per milliliter of solution and precipitate lead sulfide with hydrogen sulfide. Filter or centrifuge the sulfide precipitate. The filtrate contains actinium free from its daughters. Evaporate all or a known fraction of the solution to dryness in a suitable counting dish or disk and flame the sample in a gas burner. To determine actinium-227, count the alpha activity in the sample periodically for at least one week and calculate the actinium-227 content for each count from the actinium growth curve (Fig. 5), using the first TTA extraction as the zero time. The result from the individual counts should agree within the counting error if only actinium-227 and its decay products are present. Alternately, the comparison with the theoretical growth may be done graphically. For other actinium isotopes, count the sample in an appropriate counter and follow the decay to establish the half life of the isotope.

2. Separation and Counting of Francium-223

One method (234,237,238) is applicable to lanthanide–actinium-227 mixtures and to samples containing many other radioactive impurities. The sensitivity is about 1 μc. of actinium-227, using ordinary beta counters.

The actinium and francium must be in radioactive equilibrium, but this requires only about three hours after the last separation of francium-223.

Obtain the sample in a solution containing a slight excess of nitric or hydrochloric acid. Heat the solution and add a slight excess of sodium carbonate. Filter rapidly and wash the precipitate with water. This should require 2 to 3 minutes. The end of the filtration is the zero time for the francium separation. The carbonate precipitate contains the actinium and its daughters, except francium-223 and thallium-207, as well as any rare earths in the sample. (If the sample does not contain any rare earths, it is advisable to add several milligrams of lanthanum carrier.) Acidify the filtrate with hydrochloric acid and boil to destroy the excess carbonate ion. Add one drop of saturated potassium chromate solution, two drops of saturated barium chloride solution, and a few milligrams of lanthanum carrier. Add ammonium hydroxide until the solution is just alkaline. Filter and wash the precipitate, which contains any remaining traces of actinium-227, radium-223, and thorium-227, as well as the thallium-207. Combine the filtrate and washings, and evaporate the solution (or a suitable fraction if the activity is high) in a counting dish and count in a beta counter. Follow the decay of the activity to establish the 22-minute half life. Correct for the decay of francium-223 by extrapolating the counting rate to the zero time and calculate the actinium-227 activity from the branching ratio. Calibrate the counting system by performing a similar analysis using a known amount of actinium-227.

A second method (124) is based on the coprecipitation of francium with free silicotungstic acid. Some difficulty has been reported (1) in applying this method to samples containing a high proportion of rare earths. Otherwise the applicability and sensitivity are similar to that of the previous method.

Starting with a solution of the sample in concentrated hydrochloric acid, add a few drops of $0.4M$ silicotungstic acid, stir, centrifuge the precipitate of silicotungstic acid, and wash it twice with hydrochloric acid that has been saturated with hydrogen chloride gas. After decanting the final wash, remove the remaining supernatant carefully with a pipet. A column of Dowex-50 resin (250 to 500 mesh, ammonium form) about 1 cm. long and 4 mm. in diameter is prepared prior to analysis and is washed with distilled water. Dissolve the silicotungstic acid precipitate in distilled water and transfer it to the top of the column. Draw the solution through the column by suction at the rate of about 0.5 ml. per minute. Rinse the column with 1 to 2 ml. of distilled water at the same rate. A 15-ml. centrifuge tube with a side arm serves as a convenient receiving tube for the column. The side arm is connected to an aspirator to provide suction, and the resin column is inserted into the mouth of the tube through a rubber stopper. The francium is completely adsorbed on the resin if the acid concentration is less than $0.5M$, whereas the silicotungstic acid passes through the column. A white precipitate with solution properties distinct from the normal form of silicotungstic acid may form in the solution that passes through the column. Replace the receiving tube and remove the francium from the column by pulling 0.5 ml. of clean concentrated hydrochloric acid through the column in a period of 2 to 3 minutes. Evaporate this solution to dryness in a counting dish and complete the analysis, as given in the previous procedure. The zero time for the decay correction is the centrifugation time of the silicotungstic acid precipitate. Since the francium is obtained carrier-free, it is possible to calibrate the beta counter for francium-223 by counting the alpha particles emitted by its decay products after the francium has completely decayed. This requires a sample containing at least 10^7 disintegrations per minute of actinium-227. However, a standard actinium solution is not required because alpha counting of weightless samples can be done with good accuracy in ordinary alpha counters.

3. Extraction with Di(2-Ethylhexyl) Phosphoric Acid

This method (249,250) is applicable to uranium ores and ore-processing residues that do not contain significant amounts of the rare earth elements. It will not separate actinium from the rare earths. If the amount of rare earths is large enough to interfere with the counting, an additional separation from them is required.

Transfer the sample to a separatory funnel and add 2 ml. of 0.04% meta-cresol purple indicator and adjust the acidity to the orange-yellow color of the indicator, using $6M$ ammonium acetate or $6M$ nitric acid. Add 50 ml. of purified $1.5M$ di(2-ethylhexyl) phosphoric acid (EHPA) in heptane (Note 1) and shake for 2 minutes. Transfer the aqueous layer to a second separatory funnel, readjust the acidity, and repeat the extraction as before with a fresh portion of EHPA. Combine the EHPA fractions and wash them by shaking for two minutes with 25 ml. of $0.1M$ nitric acid–3% hydrogen peroxide solution (the hydrogen peroxide serves to remove vanadium). Discard the wash solution and repeat the washing step.

Remove the actinium from the organic layer by extracting with two 15-ml. portions of $2M$ nitric acid. Combine the nitric acid solutions in a separatory funnel and wash them with 50 ml. of benzene. Remove the aqueous layer to a beaker and add 2 ml. of $1.1M$ yttrium nitrate solution and 1 ml. of 0.04% meta-cresol purple indicator. Add 30 ml. of saturated oxalic acid solution whose pH has been adjusted to 2.3 with concentrated ammonium hydroxide. Heat the solution just to boiling and cool in an ice bath. Centrifuge the yttrium oxalate precipitate, discard the supernatant, and wash the precipitate with 20 ml. of 1% oxalic acid solution. Centrifuge and discard the wash solution. Add 1 g. of potassium chlorate and 4 ml. of concentrated nitric acid and heat in a boiling-water bath for 10 minutes. Dilute the solution to 20 ml. with water, cool, add 20 ml. of concentrated ammonium hydroxide, and stir well. Centrifuge the yttrium hydroxide precipitate, dissolve it in 10 ml. of $2M$ nitric acid, add 10 ml. of concentrated ammonium hydroxide, stir, and centrifuge. This step serves to remove calcium. Add 1 ml. of 0.04% meta-cresol purple indicator, dissolve the precipitate with enough $2M$ nitric acid so the indicator is just yellow. Extract for 3 minutes in a separatory funnel with 50 ml. of $1.5M$ EHPA. After 1 minute adjust the acidity with 1% ammonium hydroxide solution so the indicator is just yellow, and continue the extraction. Discard the aqueous phase. Wash the EHPA layer twice with 15-ml. portions of $0.1M$ nitric acid and discard the washings. Remove the actinium from the EHPA layer by contacting it for 2 minutes each with two 20-ml. portions of $0.8M$ hydrobromic acid (note the time). Combine the hydrobromic acid portions in a separatory funnel and add 50 ml. of 30% Aliquat 336 in benzene (Note 2). Transfer the aqueous phase to a beaker, wash the Aliquat solution with 5 ml. of $0.8M$ hydrobromic acid, and add the wash to the beaker. The extraction with Aliquat 336 separates lead, bismuth, and polonium from actinium. Combine and evaporate the hydrobromic acid solutions to dryness in a beaker and ignite at 600°C. for 10 minutes. Cool, add 10 ml. of concentrated nitric acid, and evaporate to 1 ml. Transfer the nitric acid quantitatively to a suitable counting dish, using $1M$ nitric acid to rinse the beaker; evaporate to dryness and bake the dish at 600°C. Cool and count as described in Section II-G-1. The chemical yield of actinium is about 84%.

Note 1. The EHPA reagent must be free from traces of metals and from the corresponding mono-ester. The mono-ester has an extremely high distribution coefficient for actinium and, if this compound is present, back-extraction of actinium into aqueous solution is very inefficient. To purify the reagent make a $1.5M$ EHPA solution in heptane (equal volumes of EHPA and heptane) and wash the solution, in order, with equal volumes of $1M$ nitric acid, saturated ammonium carbonate, $3M$ nitric acid, and twice with water.

2. Aliquat 336 is a mixed trioctyl and tridecyl ammonium nitrate available from General Mills Company. Pre-equilibrate the Aliquat solution by shaking it twice with equal volumes of $0.8M$ hydrobromic acid before use.

III. ASTATINE (ATOMIC NUMBER 85)

A. INTRODUCTION

Astatine was discovered in 1940 by Corson, MacKenzie, and Segré (54) in the reaction products obtained by bombarding bismuth with 32-m.e.v. alpha particles. They separated an activity with a 7.5-hour half life and showed it to be an isotope of element 85 of mass 211, produced by an (α,2n) reaction. Earlier claims of the discovery of element 85 in nature have been disproven or were inconclusive. For a review of the early searches for element 85 see Reference 298. Very small amounts of astatine do occur in nature (Figs. 1, 3, and 4). Francium-223 decays by alpha emission in 0.004% of its disintegration to form astatine-219. Branch-beta decay occurs to a small extent in the "A" decay products of the two uranium series, polonium-218 (RaA) and polonium-215, to form astatine-218 and astatine-215, respectively. The beta decay of polonium-216 in the thorium series to form astatine-216 has been reported, but appears doubtful. The principal decay sequence of the naturally occurring radioactive series does not include element 85. Minute amounts of neptunium-237 and uranium-233 have been found in radioactive ores produced by neutron reactions, and astatine-217 occurs in the main decay sequence of these nuclides (Fig. 2). A review of the natural occurrence of astatine is given in References 125 and 128.

Astatine is known only in tracer concentrations. The most stable isotope, astatine-210, has a half life of only 8.3 hours, and consequently it is necessary to work with concentrations of about 10^{-10} to $10^{-15}M$. It is unlikely that longer-lived isotopes can be produced (125,128). Since astatine can only be studied at very low concentrations, many of its chemical and physical properties are unknown. Astatine, as the heaviest halogen, is expected to be a very reactive element, so at tracer concentrations reactions with impurities become very important. In addition, adsorption and radiocolloid formation may obscure the desired reactions. As a result, astatine often exhibits erratic and nonreproducible behavior and it is not surprising that relatively little is known of its chemistry. Somewhat similar behavior has been observed with iodine at trace levels, so these difficulties with astatine are probably due to its low concentration.

B. PROPERTIES

1. Physical Properties

The nuclear properties of astatine isotopes are summarized in Table VI. Additional details on the nuclear properties, including gamma-ray abundances, in so far as they are known, and the energies of the x-rays emitted in some decay processes may be found in References 56 and 283. The most important isotopes are those of mass numbers 210 and 211.

TABLE VI
Nuclear Properties of Astatine Isotopes (56,116,283)[a]

Mass number	Type of decay	Half life	Energy of principal radiations (m.e.v.)	Source
<202	α, EC	43 s	α: 6.50	Bi(α,spall)
<203	α, EC	1.7 m	α: 6.35	Bi(α,spall)
203	α, EC	7 m	α: 6.10	Bi(α,10n), Au(C,6n)
204	EC	~25 m		Bi(α,9n)
205	α, EC	25 m	α: 5.90	Bi(α,8n), Au(C,4n)
206	EC	2.9 h		Bi(α,7n)
207	EC, ~90% α, ~10%	1.8 h	α: 5.75	Bi(α,6n)
208	EC	6.3 h		Bi(α,5n)
208m	EC, 99+% α, 0.5%	1.6 h	α: 5.65 γ: with EC, 0.18, 0.25, 0.66	Dtr ^{212}Fr
209	EC, ~95% α, ~5%	5.5 h	α: 5.64 γ: 0.195, 0.545, 0.780	Bi(α,4n)
210	EC, 99+% α, 0.17%	8.3 h	α: 5.519, 5.437, 5.355 γ: with EC, 0.245, 1.185, 1.604, others	Bi(α,3n)
211	EC, 59.1% α, 40.9%	7.2 h	α: 5.862 γ: 0.67	Bi(α,2n) Th, U-spal
212(?)	α	0.22 s		Bi(α,n)
213(?)	α		α: 9.2	Descendant ^{225}Pa(?)
214	α	~2 × 10^{-6} s	α: 8.78	Descendant ^{226}Pa (dtr ^{218}Fr)
215	α	~10^{-4} s	α: 8.0	Natural source: dtr ^{215}Po in U–Ac series, descendant ^{227}Pa (dtr ^{219}Fr)
216	α	~3 × 10^{-4}s	α: 7.79	Descendant ^{228}Pa (dtr ^{221}Fr)
217	α	0.018 s	α: 7.05	Dtr ^{221}Fr in ^{237}Np series
218	α, 99+% β, 0.1%	1.5–2.0 s	α: 6.63	Natural source: dtr ^{218}Po in U–Ra series
219	α, ~97% β, ~3%	0.9 m	α: 6.27	Natural source: dtr ^{223}Fr in U–Ac series

[a] Key to abbreviations and symbols: α—alpha particle; β—negative beta particle, γ—gamma-ray; EC—orbital electron capture; n—neutron; s—second; m—minute; h—hour; d—day; y—year; dtr—daughter; spall—spallation reaction; and (?)—identification uncertain.

Most of the chemical and biological studies of astatine have been made with the latter isotope.

Other physical properties of astatine or its compounds are, of course, not known. Theoretical calculations of the ionization potential of At(VII) have given results of 10.4, 9.65, and 9.5 electron volts. The calculated ionization potential of the At(I) ion is 18.2 e.v., and the ionic radius of the At(VII) ion has been calculated to be 0.62 A. References to the original literature are given in the review by Bagnall (10).

2. Chemical Properties

a. Oxidation States

At least four oxidation states of astatine have been established (6,7,147), with the oxidation numbers (-1), (0), $(+1)$, or $(+3)$, or a mixture of these states, and $(+5)$. The $(+7)$ state has not been identified. Redox reactions reported for astatine are given in Table VII. By analogy with io-

TABLE VII
Oxidation-Reduction Reactions of Astatine (6,7,147)

Reaction	Reagents
At(0 or +1) → At^{-1}	SO$_2$ in acid solution
	Zn in acid solution
	As(III), pH > 4
	Fe(CN)$_6{}^{-4}$, pH > 3, ionic strength < 1
	SnO$_2{}^{-2}$ in basic solution
At^{-1} → At(0)	I$_2$ in acid solution
	1:1 Fe(CN)$_6{}^{-4}$–Fe(CN)$_6{}^{-3}$, pH < 3
	As(III), pH < 4
At(0) → At($+x$)^a	Br$_2$ in acid solution
	VO$_2{}^+$ in light, >10^{-5} M I$_2$
	Fe(III) in light, Fe(III)/Fe(II) > 100
	Cl$_2$–Cl$^-$ mixture
At($+x$) → At(0)	VO^{+2} in the dark
	Fe(II) in acid solution
At($+x$) → AtO$_3{}^{-1}$	Cl$_2$ in acid solution, no added Cl$^-$
	Periodate in acid solution (reaction slow at room temp.)
AtO$_3{}^{-1}$ → At($+x$)	Cl$^-$ in acid solution
At(0), At^{-1}, At($+x$) → AtO$_3{}^-$	S$_2$O$_8{}^{-2}$, hot (reaction slow at room temp.)
	HClO
	Ce(IV) in acid solution
	I$_2$–I$^-$ mixture, Ag$^+$ added

^{a} The $+x$ state refers to the positive oxidation state, intermediate between the zero and +5 states.

dine, these states have been defined by their coprecipitation and solvent-extraction behavior.

The (−1) state, the astatide ion, is characterized by nearly complete coprecipitation with silver, lead, or thallium iodides. It is formed by the reduction of higher states, as given in Table VII. There is some disagreement as to whether sulfur dioxide reduces astatine to the (−1) or (0) state. Tellurium carrier, formed by reduction with sulfur dioxide in acid solution, carries astatine, presumably in the zero state, whereas precipitation of tellurium with stannite in alkaline solution does not carry astatine (54,184). This would imply that stannite reduction produces the (−1) state, although sulfur dioxide does not. Reported discrepancies in the coprecipitation of astatine with thallium and silver iodides following reduction with sulfur dioxide also can be explained by assuming that astatine is not reduced to the (−1) state. However, differences in experimental conditions and incomplete and slow reduction at low sulfur dioxide concentrations also could explain the results (6). This is an example of the difficulties encountered in interpreting results of the tracer behavior of astatine. Astatine is not reduced to the (−1) state by ferrous ion.

The zero oxidation state is usually designated as At(0) since the species present in solution is unknown. In the absence of other halogens the astatine may exist, at least partially, in the form of organic compounds formed by reaction of astatine with the organic impurities unavoidably present in the solution (6,7). In the presence of other halogens, At(0) forms interhalogen compounds (8,224). At(0) is the principal oxidation state of astatine in nitric acid solution in the absence of other oxidizing or reducing agents, and is the form obtained when astatine is distilled from irradiated bismuth at elevated temperatures (147). The zero oxidation state is characterized by high volatility, a tendency to be adsorbed on some surfaces, and extractability into organic solvents from acid solutions. These properties will be discussed in more detail below.

The intermediate positive oxidation state or states [At(+x)] is assigned to a form of astatine that is produced by the action of certain oxidizing agents, is not extracted into benzene or carbon tetrachloride, and is not coprecipitated with insoluble iodides or iodates. It may be the +1 or +3 state, or a mixture of the two. As indicated in Table VII, it is produced by two remarkable photochemical reactions (6,7), and the possibility of light-induced reactions should be considered when working with astatine.

The +5 state, astatate ion, is characterized by complete coprecipitation with silver, lead, or barium iodate and by nonextractability into organic solvents.

By determining the oxidation state of astatine resulting from reaction with various redox couples of known potential, Appelman (7) proposed the following potential scheme in $0.1N$ acid:

$$\text{At}^- \xrightarrow{-0.3 \text{ v.}} \text{At}(0) \xrightarrow{-1.0 \text{ v.}} \text{At}(+x) \xrightarrow{-1.5 \text{ v.}} \text{AtO}_3^- \xrightarrow{<-1.6 \text{ v.}} \text{H}_5\text{AtO}_6(?)$$

The perastate ion was not observed but, based on the experimental conditions used, its potential is less than -1.6 v.

b. Volatility

One of the outstanding properties of the zero oxidation state is its high volatility and its variable affinity for various surfaces. Use is made of this property in separating astatine from irradiated bismuth by heating the bismuth to its melting point and collecting the evolved astatine on a cold glass or metal surface (54). The astatine can be further purified by distilling it at room temperature from a glass surface or at an elevated temperature (approximately 400°C.) from a metal surface such as platinum. Astatine vapor is strongly adsorbed on platinum, silver, and gold and poorly adsorbed on aluminum, nickel, copper, and glass at room temperature. At 325°C. it is still strongly adsorbed on silver, but poorly adsorbed on the other materials mentioned (147). In evaporating weightless samples of At(0) on counting supports, the possibility of volatility losses must be considered (147). Platinum and silver appear to be the best materials for this purpose. Astatine is in general less volatile than iodine. This is shown by the distillation experiments of Johnson et al. (147). In distilling an astatine solution in $5M$ nitric acid containing 3 mg. of iodide, the first 10% of the distillate contained all of the iodine and only 2% of the astatine. It would be interesting to compare the distillation of trace concentrations of iodine under these conditions. The results of the distillation experiments are summarized in Table VIII. The results indicate

TABLE VIII
Distillation of Astatine Solutions (147)

Initial solution	Per cent left in still	Material balance, %
$16M$ HNO_3	86	87
$5M$ HNO_3 + 3 mg. I^-	98	—
$12M$ HCl	70	84
$1:1$ H_2SO_4	91	92
60% $HClO_4$	101	102
$0.5M$ H_2SO_4 + $0.05M$ Fe(II)	7.4	—
CCl_4	64	78
Benzene	62	—

that astatine in aqueous solution can be handled without loss in the presence of oxidizing agents, but not in organic solutions or in aqueous solutions containing certain reducing agents.

c. Solvent Extraction

The solubility of At(0) and of interhalogen compounds of astatine in organic solvents is very useful in separating and purifying astatine and in studying its chemistry. The extractability of At(0) into organic solvents is high but unpredictable and variable (147). High distribution coefficients are obtained with solvents such as carbon tetrachloride, benzene, ethers, and other oxygenated solvents from aqueous solutions that do not contain other halide ions. In the presence of high concentrations of other halide ions the extractability into the nonoxygenated solvents decreases markedly, although the extraction into the oxygenated solvents remains high. This is believed due to the formation of species such as $AtCl_2^-$ that would be insoluble in carbon tetrachloride but soluble in ethers as HAt_2S (224).

C. SEPARATION AND ISOLATION

The separation and isolation of astatine is generally limited to these situations: (1) separation from irradiated heavy-metal targets for the purpose of determining the nuclear properties of astatine isotopes or to obtain a stock solution of astatine for chemical and biological studies; (2) separation from solutions for analytical purposes in experimental work on the chemistry of astatine (in this case other radioactive isotopes except astatine daughters are ordinarily absent); and (3) separation from biological samples in the study of the distribution of astatine in liver systems and in the study of the biological effects of astatine (here also other radioactive elements, except daughters, are usually absent, although some studies have been made using radioiodine and astatine simultaneously).

1. Separation from Irradiated Targets

a. Distillation

A number of different procedures have been described for this separation (6,7,26,54,147,224,228,282). Bismuth is the usual target, and after irradiation in a cyclotron it is heated to its melting point or above (750 to 800°C.) either in a vacuum or in air, and the evolved astatine is collected on a water-cooled metal surface, such as platinum or aluminum, or in a glass U-tube cooled to liquid nitrogen temperatures. Further purification may be obtained by redistilling the astatine at a lower temperature. For this purpose the trap or metal plate is usually heated in a glass vacuum system and the volatilized astatine is condensed in a cooled U-tube. A redistillation is necessary if the bismuth is heated much above its melting point, since under these conditions some polonium and bismuth will also distill. Irradiated gold also can be treated in this way by dissolving it in molten bismuth. A stream of carrier gas, nitrogen or helium, is sometimes used in a closed system to assist in the distillation. The astatine is recovered from the cold U-tube trap by dissolving it in an appropriate solution, depending on the requirements of the experiment.

In a typical procedure (147), bismuth targets up to 40 mg./cm.^{2} thick were prepared by evaporation on one-mil thick aluminum disks in a high vacuum. After bombardment the astatine was distilled from the bismuth at its melting point in an all-glass system and collected in a 4-mm. diameter U-tube cooled by liquid nitrogen. Yields up to 95% were obtained with less than 0.001% of any polonium present in the original target distilling. When approximately equal numbers of polonium and astatine atoms were produced in the bombardment, a second distillation was performed by sealing off the tube containing the bismuth and then distilling the astatine at room temperature into a second U-tube cooled by liquid nitrogen. The yield was 83% of the astatine originally present in the target, with less than 10^{-6}% of the polonium coming over. Appleman (6), however, reports difficulty in distilling the astatine at the melting point of bismuth with thick targets, and in redistilling the astatine from glass at room temperature. More extreme temperatures were necessary for high yields. It should be pointed out that not all of these separations were designed to be quantitative. Many were intended only to give a pure product.

b. Chemical Separations

Two methods have been reported for the separation of astatine from bismuth by dissolving the bismuth in mineral acid and extracting the astatine with diisopropyl ether. In one method (26) the bismuth target was dissolved in concentrated hydrochloric acid, ferrous sulfate was added to insure reduction of astatine to At(0), and the astatine was extracted into isopropyl ether. The ether solution was purified by washing it with dilute sulfuric or hydrochloric acid. No bismuth or polonium alpha activity could be detected in the purified astatine. The polonium, bismuth, and lead daughters that grew into the astatine were removed periodically by adding one-tenth volume of 20% tributyl phosphate in isobutyl ether to the ether solution containing astatine, and extracting the daughters into $4M$ nitric acid–$4M$ hydrochloric acid solution.

In the second procedure (224) the bismuth target was dissolved in nitric acid, the solution was concentrated by heating, and concentrated hydrochloric acid was added to bring its concentration to $9M$. The astatine was extracted into isopropyl ether, the ether was washed with $8M$ hydrochloric acid, and the astatine was back-extracted with $0.5M$ sodium hydroxide solution. Small amounts of nitric acid do not interfere with the extraction. The ether was washed three times with saturated ferrous sulfate solution and three times with water before use. Isopropyl ether previously equilibrated with $8M$ hydrochloric acid is better than pure ether for the extraction, since its use minimizes volume changes. The yields in a single extraction are greater than 90%, and the presence of Bi(III) salts increases the extractability. Washing the ether phase with $8M$ hydrochloric acid remover about 10% of the astatine. Less than 0.01% of the bismuth is retained. Appelman (5) reports that the hydrochloric acid solution used to wash the ether phase should be about $1M$ in nitric acid, otherwise up to one-half of the astatine may be back-extracted in this wash, and that the back-extraction with sodium hydroxide is variable and often incomplete.

In another procedure (9), bismuth oxide targets are dissolved in perchloric acid and the bismuth is removed by precipitation with phosphate. The filtrate contains the astatine. A procedure for the separation of astatine from irradiated thorium is as follows (184). The thorium is dissolved in hydrochloric acid containing a little hydrofluoric acid to catalyze the reaction. The acidity is adjusted to $3M$ in hydrochloric acid and tellurium carrier, added as telluric acid, is precipitated as the metal with stannous chloride. The tellurium carries astatine, polonium, and certain fission products. The tellurium is separated and dissolved in a minimum amount of concentrated nitric acid, and the solution is made up to 3 ml. with concentrated hydrochloric acid. The solution is heated to 100°C. in a water bath, and a hydrochloric acid solution of the same concentration, saturated with hydrazine hydrochloride and sulfur dioxide, is added. The tellurium precipitates, carrying with it the astatine, while the polonium remains in solution. To separate the tellurium from the astatine, the precipitate is dissolved in a drop of concentrated nitric acid, the solution is made alkaline with $6N$ sodium hydroxide, and 2 ml. of 5% sodium stannite solution is added. The tellurium precipitates and carries a number of fission products, while the astatine remains in solution. Sodium fluoride is added to the filtrate to prevent the formation of colloids, since astatine is strongly adsorbed on colloidal tellurium. The filtrate is acidified to obtain a solution 0.3 to $0.5N$ in nitric acid, and the astatine is deposited from solution on a clean silver foil. The solution containing the foil is stirred for one hour to obtain the deposit. Some radioactive iodine is carried through this procedure, but it can be removed by dissolving the silver foil in nitric acid and precipitating silver chloride. If the precipitation is carried out in the absence of light and of reducing agents, metallic silver is not formed and the astatine is not carried. The astatine is presumably present in the zero oxidation state since it is not carried by silver chloride under these conditions, but At(0) is strongly adsorbed by metallic silver. The over-all yield of this procedure is 40 to 50%.

2. Coprecipitation

In the coprecipitation behavior of astatine, as in other aspects of astatine chemistry, quantitatively different results are reported by various workers and variable results are often found under presumably identical conditions. Again, the effect of unknown impurities and uncertainty of the oxidation state probably account for many of the discrepancies. Coprecipitation behavior is discussed in References 6, 54, and 147. Reference 6 contains considerable detailed information on various aspects of astatine chemistry and should be consulted by anyone working with astatine.

As mentioned previously, astatide ion carries nearly quantitatively on silver, lead, and thallium iodide from reduced solutions, acidic or basic. From solutions in which the astatine is only partially reduced by sulfur dioxide, it is only partially carried by thallium iodide, but completely carried by silver iodide. In the latter case, any At(0) is probably adsorbed on metallic silver. Complete coprecipitation from partially reduced solutions can also be obtained by adding iodine to the solution and precipitating an insoluble iodide. In this case, some insoluble triiodide precipitates, carrying any At(0). The At(0) and the iodine can be washed out of the precipitate with acetone, whereas astatide ion is not removed.

Aluminum, bismuth, ferric, and lanthanum hydroxides carry from 7 to 60% of the astatine from nitric acid solution, presumably as At(0). From solutions oxidized with hot persulfate, carrying on aluminum and bismuth hydroxides is unchanged, but carrying on ferric and lanthanum hydroxides increases to 85 to 99%. Variable amounts of At(0) are reported to precipitate with insoluble iodates. However, nearly complete coprecipitation is obtained with silver, lead, or barium iodates when the astatine is oxidized to the astatate ion by strong oxidizing agents such as hot persulfate. Nearly complete coprecipitation of astatine with insoluble sulfides from hydrochloric acid solution has been reported. If a nitric acid solution of astatine is made basic, astatine is completely coprecipitated with thallium or silver iodide.

Coprecipitation of astatine with tellurium is a useful method for separating astatine (Sections III-C-1-b and III-F-3). This reaction was described in the original paper on astatine (54). Elementary tellurium, precipitated by sulfur dioxide from $3M$ hydrochloric acid solution, carries astatine and gives an excellent separation from polonium. Tellurium precipitated with sodium stannite does not carry astatine, and this reaction affords a good method for separating it from bismuth, tellurium, selenium, and mercury. In coprecipitation with tellurium, astatine resembles polonium rather than iodine.

3. Solvent Extraction

In this case also, results are variable and unpredictable. When an aqueous acid solution at At(0) is extracted with carbon tetrachloride or benzene, the distribution coefficients (astatine concentration in organic phase/astatine concentration in aqueous phase) between one and ten have been obtained. If the organic phase is separated and repeatedly washed with portions of fresh aqueous solution, the distribution coefficients increase with each successive wash until the astatine left in the organic phase is almost completely unextractable into the aqueous phase. Likewise, if the initial aqueous phase is separated and washed with successive portions of the organic solvent, the distribution coefficients decrease rapidly until the remaining astatine is nearly unextractable into the organic phase (6,147). Once the astatine is extracted into carbon tetrachloride or benzene, it can be back-extracted into sodium hydroxide solution, nitric acid solutions containing Fe(II), potassium iodide, sulfur dioxide, Fe(III), Hg(II), or bromine ($< 0.1M$); potassium persulfate solution at 50°C.; hydrochloric acid; and sodium chloride or bromide solutions (6,147). With regard to back-extraction into sodium hydroxide, however, note the earlier discussion regarding isopropyl ether in Section III-C-1-b. It would appear that extraction with carbon tetrachloride or benzene is not a reliable analytical method. Extraction of astatine solutions containing halide ions into oxygenated solvents such as isopropyl ether and tributyl phosphate appear to be more efficient and reliable (6,8,30a,224). The extraction of astatine into diisopropyl ether from nitric acid and sulfuric acid, following oxidation with Ce(IV), also has been studied (299a).

An interesting separation is that used by Hyde and Ghiorso in the separation of astatine from a natural source (130). Astatine-219 is produced by the rare alpha branching of francium-223, which is in turn produced by the rare alpha branching of actinium-227. Starting with 20 mc. of actinium-227, the francium-223 daughter was separated by silicotungstic acid coprecipitation (Sections II-E-3 and IV-C-1), and the astatine-219 was separated from the francium solution by extraction from $3M$ hydrochloric acid into an equal volume of 10% tributyl phosphate in butyl ether. Astatine extraction is nearly complete, although radium, lead, and bismuth are only slightly extracted. The organic phase was washed with $3M$ hydrochloric acid, evaporated to dryness, and counted. A more rapid separation was achieved by evaporating the francium solution on a platinum filament and heating the platinum carefully to volatilize the astatine and retain the francium, lead, and bismuth on the platinum. The volatilized astatine was collected on a platinum plate positioned 1 to 2 mm. above the filament.

4. Separation from Biological Materials

Astatine has been separated from biological materials by oxidation of the organic matter with nitric perchloric acid mixtures, followed by coprecipitation of the astatine with elementary tellurium formed by the reduction of tellurous acid with sulfur dioxide (84).

The simultaneous separation of iodine and astatine from thyroid glands has been accomplished by a distillation technique (65). In these samples it was necessary to determine stable iodine, iodine-131, and astatine-211. The samples were first placed in a sulfuric acid–chromium trioxide–water mixture and distilled into carbon tetrachloride to remove chlorine. The chlorine fraction was discarded. After this distillation was complete, oxalic acid was added and the mixture was distilled into carbon tetrachloride to collect iodine together with some of the astatine-211. Ferrous sulfate was then added and the distillation was continued to collect the remainder of the astatine. The second distillate and the carbon tetrachloride were treated with sodium sulfite to reduce the astatine and iodine and extract them into the aqueous layer. Silver oxide and iodide were precipitated from aliquots of the aqueous layer and, without removing the supernatant, the slurry was evaporated to dryness in a tinned bottle cap. Astatine was retained in this process presumably because a layer of metallic silver was formed on the precipitate by reduction in contact with the tinned surface of the bottle cap. The silver layer served to retain the astatine during the evaporation. Astatine-211 was determined by alpha counting and iodine-131 by gamma counting. Stable iodine was determined in an aliquot of the aqueous phase by a spectrophotometric method. Astatine and iodine recoveries were about 90 and 98%, respectively.

D. DETECTION

Astatine isotopes can be detected by their characteristic nuclear properties, as given in Table VI and Section III-E. Chemical separation from other activities and from inert material if alpha counting is used may be required before the nuclear properties can be determined. Astatine's characteristic volatility and a knowledge of the method of production or the source of the activity also are useful in detecting this element. With regard to chemical separation, it is pointed out in the publication describing its discovery (54) that astatine can be separated from all other elements by one or more of the reactions given in the article.

The solvent-extraction procedure using diisopropyl ether and tributyl phosphate (TBP), described in Section III-C-1, has been used to identify unknown astatine isotopes produced in the bombardment of bismuth with helium ions (26). In the bombardment many other activities are formed, including polonium, bismuth, lead, and thallium isotopes, and fission products, as well as spallation products further removed from bismuth. The characterization of known daughters was used for the most part to identify new astatine isotopes. The energy of the bombarding particles also sets the order of the mass numbers produced. The astatine was first extracted into diisopropyl ether. The polonium, bismuth, and lead daughters that grow into the astatine were removed from the ether phase by adding TBP and extracting them into a nitric–hydrochloric acid mixture, while the astatine remained in the ether phase. Identification of the daughter nuclides then served to establish the presence of their astatine parents in the ether phase.

The decay scheme of astatine-211 permits identification of this isotope by counting the alpha particles emitted by astatine-211 and its polonium-211 daughter in an alpha pulse-height analyzer (128). The detection of the 5.86-m.e.v. alpha particle of astatine-211 and the 7.43-m.e.v. alpha particle of polonium-211 in the ratio of 2:3, and a half life of 7.2 hours for both alphas, are conclusive proof of the identity and purity of astatine-211.

E. DETERMINATION

Astatine must, of course, be determined by quantitative counting of the radiations emitted during its decay or emitted by its daughters. Astatine-211 decays 59% by orbital electron capture to polonium-211, which decays with a 0.5-second half life by alpha emission to stable lead-207. It also decays 41% by alpha emission to bismuth-207, which decays by electron capture with a long half life (8 years or more) to stable lead-207. Thus, astatine-211 may be determined by alpha counting, in which case one alpha particle is observed for each atom that decays, or by counting the 80-k.e.v. x-ray of polonium-211 that accompanies the decay of astatine-211 by electron capture.

Alpha counting may be done in conventional alpha counters: gas-flow proportional or ionization chambers or zinc sulfide scintillation counters. In this case it is necessary to prepare a dried sample on a counting support for introduction into the counter. Solutions may be evaporated to dryness on the support, and precipitates may be transferred to the support as a slurry and dried by evaporation. In the case of precipitates and samples containing solids, corrections for self-absorption of alpha particles in the dried sample mount must be made to obtain the absolute amount of activity. This requires a knowledge of the thickness of the sample mount, which can be obtained by weighing. The self-absorption due to a given thickness can be measured by counting a known amount of astatine (or other alpha emitters of similar energies) dispersed in the same thickness of material of the same or similar atomic number as the unknown sample. Essentially weightless samples eliminate the need for self-absorption corrections, but in the case of astatine the counting of such samples is complicated by the volatility of At(0). Glass cannot be used for a sample support because astatine is rapidly volatilized from the dried sample. Appelman (6) found that in preparing weightless mounts reproducible and adherent deposits could be obtained by evaporating the astatine on platinum or silver from hydrochloric acid solutions greater than $1M$. Under the other conditions tested the retention varied from 10% to 90% of that obtained with hydrochloric acid solutions. Even in the latter case it was not established that retention was complete. Johnson et al. (147) found that samples prepared by evaporating nitric acid solutions on platinum at 80°C. showed no loss of astatine from the dried sample in 24 hours, but they found that appreciable loss occurred during the evaporation. The loss, about 27%, was reproducible, and this method was used for routine assay.

Self-absorption and volatility problems can be avoided by alpha liquid-scintillation counting of astatine-211 (27). In this method the astatine solution is dissolved in a liquid scintillator and is counted by placing the solution in optical contact with a multiplier phototube. The phototube is connected to a suitable electronic system to amplify and record the alpha pulses. A two-thirds saturated solution of terphenyl in toluene can be used to count astatine-211 in organic solvents. For aqueous solutions, a scintillating solution containing 50 g. of 2,5-diphenyloxazole in p-dioxane can be used. Up to 25% water can be added to the dioxane solution. Alpha counting with 100% efficiency is claimed for this method. Liquid-scintillation counting is subject to "quenching" (reduction in counting rate) by a number of substances, including water, and this effect must be tested for all materials added to the scintillation solution with the sample.

The problems associated with alpha counting of astatine can be largely eliminated by counting the x-rays emitted by the polonium daughter following astatine decay by electron capture (6,7,84,112). This is best done by scintillation counting with a thallium-activated sodium iodide crystal and multiplier phototube assembly coupled to a single or multichannel pulse-height analyzer. By counting only the x-ray peak the highest sample-to-background ratio is obtained. Liquid samples may be counted directly without evaporation, and both liquid and solid samples require only that they be placed on or near the crystal in a reproducible geometry. Appelman (6) used a small well crystal, 1 inch thick and 1.75 inches in diameter, covered with a 0.01-inch thick aluminum foil. The samples were counted in a glass vial inserted in the well. The geometry was 70%. By counting only the x-ray peak, 40% of the x-rays entering the crystal were counted with a background of 7 counts per minute. Precipitates were slurried into the vial with propanol. Since the precipitates settled to the bottom, they were counted with a higher efficiency than solutions and corrections for this were made. Small corrections for x-ray absorption by the precipitates themselves also were required. The long-lived bismuth-207 daughter of astatine-211 decays by electron capture, and x-rays are emitted in this decay. Samples assayed by x-ray counting three or more days after purification from bismuth should be recounted after the astatine has entirely decayed and the long-lived bismuth-207 activity has been subtracted from the original count of the sample. For small amounts of astatine, alpha counting is more sensitive than x-ray counting.

Other astatine isotopes can be determined by alpha or x-ray counting, depending upon their mode of decay. If gamma-rays are emitted during the decay, as in the case of isotopes of mass numbers 208m, 209, and 210, these may be conveniently counted with a sodium iodide crystal.

All counters must, of course, be calibrated if absolute rather than relative determinations are required, and corrections for decay must be made if the activity at some previous time is needed.

F. RECOMMENDED PROCEDURES

1. Determination in Solution or in Precipitates (6)

This method is applicable to solutions of sufficient concentration such that a few milliliters or less can be taken for analysis, or to precipitates weighing less than about 50 mg. The astatine must be free of radioactive impurities. If other activities are present, one or more of the separations previously discussed must be made first. Place a suitable aliquot of the solution in a glass vial and dilute to a predetermined volume that will be used for all determinations to be compared. Dilute all aqueous solutions to the desired volume with 2N sulfuric or perchloric acids and dilute all organic solutions with butyl carbitol. Slurry precipitates into a glass vial with propanol. Count under the polonium x-ray peak (approximately 80 k.e.v.) with a sodium iodide crystal and apply the necessary corrections. In the counting arrangement used (10) the correction for settling of the precipitate amount to 4% relative to 0.5-ml. solutions and 10% relative to 1-ml. solutions. The correction for x-ray absorption by the precipitates amounted to about 4% per 10 mg. of precipitate.

2. Determination by Coprecipitation with Silver or Palladium (9)

The method is useful in separating astatine from solutions containing inactive iodine carrier and has been applied to nitric and perchloric acid solutions. Add an excess of silver or palladium ion to precipitate all the iodine. Add sufficient sulfite to precipitate the excess silver or palladium. Any astatine not carried by the iodide precipitate is carried by the precipitated metal. (With a nitric acid solution, sufficient sulfite must be added to neutralize all or almost all of the acid.) Separate the precipitate, mount, and count in an alpha counter. Less than 0.5% of the astatine remains in solution.

3. Determination in Biological Materials by Coprecipitation with Tellurium (84)

Place a sample of the tissue (less than 10 g.) in a borosilicate beaker and digest with a minimum volume of 9N perchloric acid containing 30% by volume of 16N nitric acid. (The reaction is smooth if 10 to 30% nitric acid is present.) After the organic matter is oxidized, evaporate to 10 to 15 ml. of concentrated perchloric acid, cool, and add 5 mg. of tellurium as tellurous acid and 1 ml. of 12N hydrochloric acid. Reduction of tellurium in perchloric acid is slow unless dilute hydrochloric acid is present. Pass a stream of sulfur dioxide through the solution to precipitate tellurium metal. Centrifuge, wash the precipitate three times with water, transfer it to a counting dish, and count in an alpha counter.

IV. FRANCIUM (ATOMIC NUMBER 87)

A. INTRODUCTION

Francium, the heaviest alkali metal, was discovered in 1939 by Mlle. M. Perey as the product of the rare alpha-branching decay of actinium-227 (232,233,235). She found that actinium emits an alpha particle in 1.2% of its decays (see Fig. 4), and she separated the product of this decay, a beta emitter of 22-minute half life (francium-223) that had the properties of an alkali metal. The alpha activity of actinium-227 had been observed earlier by other workers, but was ascribed to protactinium impurity. Earlier claims to the discovery of element 87 have been shown to be erroneous. For an account of these claims see Reference 298.

The only other francium isotope that occurs in nature is francium-221, a member of the neptunium decay series (see Fig. 2). It is present in radioactive minerals because of the steady production of small amounts of neptunium-237 and uranium-233 by neutron reactions on uranium and thorium. A number of other isotopes have been artificially produced, but francium-223 remains the most stable francium isotope, and the chance of discovery of longer-lived isotopes is remote (128). Consequently, as is the case for astatine, francium can be prepared only in tracer concentrations. The minute amounts that can be obtained, and its great instability, make studies of francium chemistry very difficult, and many of its chemical

TABLE IX
Nuclear Properties of Francium Isotopes (56,116,283)[a]

Mass number	Type of decay	Half life	Energy of principal radiations (m.e.v.)	Source
212	EC, 56% α, 44%	19.3 m	α: 6.34, 6.39, 6.41	Th, U(p, spall)
217(?)	α			Descendant ^{225}Pa (dtr ^{221}Ac)
218	α	~0.005 s	α: 7.85	Descendant ^{226}Pa (dtr ^{222}Ac)
219	α	0.02 s	α: 7.30	Descendant ^{227}Pa (dtr ^{223}Ac)
220	α	27.5 s	α: 6.69	Descendant ^{228}Pa (dtr ^{224}Ac)
221	α	4.8 m	α: 6.33, 6.22 γ: 0.216	Dtr ^{225}Ac, ^{221}Rn in ^{237}Np series
222	β, 99+% α, ~0.1%	14.8 m		Th(p, spall)
223	β, 99+% α, ~0.004%	22 m	α: 5.34 β: 1.15 γ: 0.0498, 0.080, 0.215, 0.31	Natural source: dtr ^{227}Ac in U–Ac series

[a] Key to abbreviations and symbols: α—alpha particle; β—negative beta particle; γ—gamma-ray; EC—orbital electron capture; n—neutron; p—proton; s—second; m—minute; h—hour; d—day; y—year; dtr—daughter; spall—spallation reaction; and (?)—identification uncertain.

and physical properties cannot be determined experimentally. Even tracer studies must be completed in a very short time. Francium is the most unstable of the first 101 elements.

B. PROPERTIES

1. Physical Properties

The nuclear properties of the francium isotopes are given in Table IX. The most important isotopes are those of mass numbers 212 and 223, and all of the chemical studies of francium have been made with these. Francium-223 is most commonly used since it can be readily isolated free of other activities from actinium-227 and emits an energetic beta particle that is readily detected by conventional counters. Francium-212 has been used by Russian workers in the study of francium chemistry (178). Since it decays in part by alpha emission, francium-212 can be counted in the presence of beta emitters and can, therefore, be used in combined francium–radiocesium studies. In fact, during its production of high-energy spallation reactions on thorium and uranium, the cesium isotopes formed by fission need not be separated for chemical studies of francium for this reason.

Other physical properties have, of course, not been determined directly. Calculations of the ionization potential of francium have given results of 3.83, 4, and 4.24 e. v. The calculated ionization potential of the Fr(I) ion is 21.5 e. v. and the ionic radius of the Fr(I) ion has been calculated to be 1.8 A. References to the original literature are given in the review by Bagnall (10).

2. Chemical Properties

On the basis of its behavior at tracer concentrations, francium has the properties expected for the heaviest alkali metal. It presumably exists in aqueous solution as a singly charged positive ion. From coprecipitation data, its hydroxide, carbonate, chloride, sulfide, sulfate, nitrate, fluoride, and chromate are soluble. Thus, the ordinary precipitation methods used to separate other elements from the alkali metals also are applicable to francium. It coprecipitates partially or completely with many insoluble cesium compounds, and the methods used to separate cesium from potassium and rubidium should also separate francium from potassium and rubidium. As would be expected, cesium is the best carrier for francium and cesium–francium separations are difficult. Like the heavier alkali metals, francium appears to form a tetraphenylborate. On cation-exchange resin columns it elutes after cesium, as would be expected. It shows little tendency to form complex ions. The difficulties that appear in attempting to establish the true chemical properties of astatine from its tracer chemistry do not arise in the case of francium.

a. Coprecipitation

Unless indicated otherwise the coprecipitation behavior described below was determined by Perey (236,242). Francium remains in solution when other elements are precipitated with hydrogen sulfide, ammonium sulfide, ammonium or sodium hydroxides, sodium carbonate, hydrogen peroxide, fluoride, or chromate. It is not carried by manganese dioxide, tellurium metal, or silver chloride (178). Coprecipitation occurs with the following compounds.

(1) Perchlorates

Sixty per cent of the francium in solution is carried when cesium perchlorate is precipitated under these conditions: (1) several milligrams of cesium chloride is added to a small volume of a cold solution of francium, a concentrated solution of sodium perchlorate in alcohol is added, and the perchlorate precipitate thus formed is washed with alcohol. Thallium isotopes also coprecipitate; and (2) the perchlorate is precipitated at room temperature with perchloric acid from an alcohol solution containing $0.01M$ cesium chloride and francium (178). Potassium perchlorate precipitated at room temperature with perchloric acid from an alcohol solution containing $0.03M$ potassium chloride carries 60% of the francium in solution (178).

(2) Picrates

Cesium picrate carries 50% of the francium when it is precipitated by the addition of a picric acid solution in alcohol to a cesium carrier solution containing francium. Thallium isotopes also coprecipitate.

Cesium dipicrylaminate, precipitated by the addition of an excess of ammonium dipicrylaminate to a $0.01M$ cesium chloride solution containing francium, also carries 50% of the francium.

(3) Tartrates

Precipitation of cesium tartrate from acetic acid solution carries a part of the francium. Potassium tartrate precipitated from dilute acetic acid solution does not carry francium or cesium, but does carry Tl(I) isotopes.

(4) Chloroplatinates

Francium coprecipitates nearly quantitatively with the chloroplatinates of cesium, rubidium, and potassium. Thallium isotopes also coprecipitate.

It is interesting that complete carrying of francium on cesium chloroplatinate was obtained with excess cesium carrier as well as with excess chloroplatinic acid (178). This implies that carrying is by isomorphous displacement and that francium chloroplatinate is less soluble than the cesium chloroplatinate (298).

(5) Other Chloroanions

Cesium chlorobismuthate and chloroantimonate carry at least 90% of the francium. Cesium hexachlorostannate, precipitated with an excess of stannic chloride, precipitates francium almost completely. Thallium isotopes also coprecipitate. From dilute solution, rubidium is not precipitated.

(6) Cesium Cobaltinitrite and Iodate

Cesium–sodium cobaltinitrite, precipitated from acetic acid solution, carries francium almost completely. Partial carrying is obtained on cesium iodate.

(7) Cesium Silicotungstate

Francium carries quantitatively with this compound formed by the addition of a silicotungstic acid solution to a hydrochloric acid solution ($6M$ or less) of cesium carrier and francium. Similar heteropolyacid salts such as silicomolybdate and phosphotungstate also are good carriers for francium (124,129).

(8) Heteropolyacids

Free silicotungstic acid, precipitated from concentrated or saturated hydrochloric acid solution, carries francium quantitatively. This carrier is very useful since it can be dissolved instantly in a little water and reprecipitated by the addition of excess hydrochloric acid, and since the francium can be separated from the silicotungstic acid carrier by dissolving it in a little water and passing it through a short column of Dowex-50 cation-exchange resin. The francium is adsorbed while the nonionized silicotungstic acid passes through. The francium can be eluted quickly from the resin with concentrated hydrochloric acid. Cesium also is carried, but rubidium, radium, and barium are not. Sodium and ammonium ion interfere in the coprecipitation and in their presence francium is incompletely carried. This carrier was introduced by Hyde (124), who explains this coprecipitation on the basis that many salts of heteropolyacids are isomorphous with the corresponding free acid and that this isomorphism may extend to the francium and cesium compounds. Other heteropolyacids may be substituted for silocotungstic acid; phosphotungstic acid, for example, also carries cesium from concentrated hydrochloric acid.

The coprecipitation of francium and cesium with a number of di-, tri-, and tetraheteropolyacids has been studied as a function of mineral acid concentration, heteropolyacid concentration, and the absence or presence of alkali metal carrier (164a,178). Some of the results are summarized in Table X. In general the solubility of the heteropolyacid determines the efficiency of coprecipitation, and the best carrying is obtained with the least soluble acids. Some acids carry francium to a greater extent than cesium, but complete separation could not be obtained either by repeated precipitations or on columns of the ammonium salts of the heteropolyacids (164a). Of particular importance is the variable effect of cesium carrier on the coprecipitation of francium with heteropolyacids. The presence of cesium can increase or decrease the fraction carried, depending on the relative solubilities of the free acid and its cesium salt.

TABLE X

Coprecipitation of Carrier-Free Francium and Cesium from $12M$ HCl and HNO_3 with Heteropolyacids (164a)

Heteropolyacid	Per cent precipitated			
	Francium		Cesium	
	HCl	HNO_3	HCl	HNO_3
$H_3PW_{12}O_{40} \cdot nH_2O$	99	96	98	97
$H_4SiW_{12}O_{40}$	98	a	97	a
$H_3PMo_{12}O_{40}$	98	93	95	93
$H_4SiMo_{12}O_{40}$	97	80	94	75
$H_3PMo_{10}W_2O_{40}$	97	86	96	86
$H_3PW_{10}V_2O_{39}$	82	a	79	a
$H_3PMo_{10}V_2O_{39}$	87	46	83	42
$H_3AsMo_4W_8O_{40}$	92	75	87	40
$H_3AsMo_{10}V_2O_{39}$	82	57	71	51
$H_5BW_{10}V_2O_{39}$	77	39	64	30
$H_4GeW_{10}V_2O_{39}$	68	a	32	a
$H_3PMo_8W_2V_2O_{39}$	92	74	87	59

^{a} The heteropolyacid did not precipitate from $12M$ HNO_3.

b. Volatility

The volatility of francium is an important analytical property. Francium (and cesium) will distill from a platinum backing heated to dull red heat (101). Use has been made of this property in preparing francium mounts for counting (124), and in the calibration of counters for francium-223 (126). In the actinium decay chain it appears that actinium-227, radium-223, and thorium-227 will not volatilize at temperatures that will distill francium. However, lead and particularly thallium isotopes are volatile at these temperatures (1). With regard to the temperature and rate of volatility, according to one report (1), francium begins to volatilize from a heated platinum filament at $225 \pm 50°C$. in air and at $110 \pm 15°C$. in a vacuum. At temperatures well above these threshold values francium is volatilized rapidly and quantitatively. The volatilization takes less than one minute and the volatilized activity appears to be quantitatively condensed on a water-cooled metal foil mounted a few millimeters above the heated platinum filament. According to another report (178), no volatilization occurs in 4 minutes below 400°C. from a platinum disk on which a neutral francium chloride solution is evaporated, and complete volatilization requires about 20 minutes at 800°C. and about 10 minutes at 1000°C. It appears that the quantitative aspects of francium volatility must be determined for the particular experimental conditions used. Cesium and francium cannot be separated by volatilization of their chlorides. In a study of the volatility of francium halides (165), it was found that the greatest difference in volatility between cesium and francium halides occurs with the iodides, and these two elements might be separated by distillation of their iodides.

C. SEPARATION AND ISOLATION

Because of the great instability of francium isotopes, all practical separation methods must be rapid, i.e., they should not require more than a few half lives. Thus, in the case of the longest-lived isotope (22 minutes), separation times are limited to one or possibly two hours. If more time is spent in the preliminary separations the counting determination is complicated by large decay corrections, and the absence of longer-lived active impurities becomes very important. This requirement limits considerably the chemistry that may be used. For example, solids that precipitate slowly and time-consuming ion-exchange elution methods cannot be used.

1. Coprecipitation Methods

Francium can be separated quantitatively from all other elements, except cesium, that occur with it in uranium minerals or are produced in the irradiation of heavy metal targets, by precipitation of the other elements while francium remains in solution. For those elements present only in trace concentrations, appropriate carriers may be added to obtain their separation. The number and type of precipitations depend on the composition of the sample. This type of separation is well illustrated by the original method devised by Perey (234,235) for the determination of actinium-227 by separating and counting francium-223. Metal ions that are insoluble in alkaline solution are precipitated with sodium carbonate. This removes the rare earths and most of the actinium and its decay products except thallium. A mixed barium–lanthanum chromate precipitation from alkaline solution removes thallium and the traces of actinium and radium remaining after the carbonate precipitation. The francium remains in solution free from other metal ions. The details of this separation are given in Section II-G-2.

Another procedure, not necessarily quantitative, based on the same principle has been described for the separation of francium from proton-irradiated uranium (178). The uranium is dissolved in nitric acid, cesium ion is added as a carrier for francium, and bismuth sulfide is precipitated from the solution. This precipitate carries with it some of the fission products and heavy metals produced during the irradiation. The filtrate is boiled to remove hydrogen sulfide, and the uranium is precipitated with ammonium hydroxide. Several "scavenging" precipitations of ferric hydroxide are made from the filtrate to remove traces of other active metals. The final filtrate contains the francium and the cesium and rubidium fission products, plus any elements not carried by bismuth sulfide, ammonium diuranate, and ferric hydroxide (e.g., the alkaline earth fission products). The ferric hydroxide filtrate is evaporated to a small volume and the francium, cesium, and rubidium activities are removed by adding alcohol and perchloric acid to precipitate cesium perchlorate. This isolation requires 35 to 40 minutes. Other methods may be devised from the information given previously on coprecipitation reactions of francium.

Procedures for the separation and isolation by direct coprecipitation of francium, without preliminary separation of other elements, are given by Hyde (124). One method used for the isolation of francium from irradiated thorium targets uses successive coprecipitations with cesium silicotungstate and cesium perchlorate. This method is based on one originally devised for the radiochemical determination of fission-product cesium (86). Since cesium perchlorate does not carry francium completely, this method cannot be regarded as quantitative. A better procedure is the interesting coprecipitation with free silicotungstic acid described earlier. The procedure has been applied to the separation of francium isotopes from thorium targets, to the separation of 4.8-minute francium-221 from its 10-day actinium-225 parent in the uranium-233 series, and to the separation of francium-223 from actinium-227. The details are given in Section IV-G. Some difficulty has been reported in the application of this separation to actinium-227 (1). With actinium samples containing a high proportion of rare earths the silicotungstic acid precipitate is reported to contain some rare earth material, and the mixed precipitate is contaminated with some radium-223. In most cases this contamination could be removed by dissolving and reprecipitating the silicotungstic acid. Also, if francium-223 is repeatedly separated from the same actinium solution by addition of more silicotungstic acid after allowing the francium to grow back, the amount of francium removed in subsequent separations may be 25% or more below the initial amount. The exact cause of this declining yield was not determined, but at least the first separation is quantitative (124). Additional study of this method for the determination of actinium-227 is warranted because of its great simplicity.

2. Chromatographic Methods

Perey and Adloff (1,234,244,245) have devised a rapid method for the separation of francium-223 from actinium-227 by filter-paper chromatography. A small volume (about 0.1 ml.) of the solution is deposited about 2.5 cm. from the end of a strip of Schleicher and Schull filter paper 15 cm. long and 1 cm. wide. The paper is placed in a closed container with the upper 2 cm. inserted into the eluant, 10% aqueous ammonium carbonate solution, and the remainder of the paper allowed to hang vertically downward. An atmosphere of saturated water vapor at 60°C. is maintained in the container. The eluant is absorbed by the paper and moves past the actinium and down the paper. Since actinium and its thorium, radium, and lead daughters are insoluble in carbonate solution, they move very little (radium-223 stays entirely in the first 5 cm. of the paper strip). The francium and the thallium-207 move about 10 to 12 cm. in about 15 minutes. The francium can be leached from the paper with water. The position of the francium can be determined by calibration runs made with francium or radioactive cesium tracer. Thallium-207 is not separated by this method; it can be allowed to decay since its half life is 4.8 minutes, or it can be separated before or after the chromatographic step. Since prior separation of most of the rare earths is necessary, the thallium separation can be incorporated in this step. In the original procedure the rare earths were precipitated as the carbonates, and the thallium coprecipitated with lead sulfide. Other methods may be used. The chromatographic separation can be done in as little as 6 minutes and the preliminary separations in as little as 25 minutes.

The separation of francium from the other members of the actinium-227 chain has also been accomplished by electromigration on paper (1a). The electrolyte is an aqueous solution of the ammonium salt of nitrilotriacetic, citric, oxalic, or ethylenediaminetetraacetic acid. These ligands complex all the elements in the mixture except francium. By using a strong electric field, 80 v. per cm., the separation can be accomplished in less than 1 minute.

A chromatographic method has been described (78) for repeated separations of francium-223. The actinium-227 sample, containing some hydrogen peroxide, is adsorbed on a column composed of a 10:1 mixture of cellulose and zirconium oxide. The francium is eluted with phenol that has been liquefied by equilibration with $2N$ hydrochloric acid. The zirconium oxide and hydrogen peroxide are used to depress the elution of thorium-227. Some of the thallium-207 elutes with the francium and is removed by coprecipitation.

3. Ion-Exchange Resin Separations

The difficulty of conducting ion-exchange studies with the short-lived francium isotopes has severely limited work in this field. Only one study of the behavior of francium on ion-exchange resins appears to have been made (178,179). The cation-exchange resins used were KU-2, a sulfonated polystyrene resin, and KU-1, a bifunctional resin that contains, in addition to the strongly acidic sulfonic acid group, the weakly acidic phenolic group, —OH. The second group undergoes exchange only in alkaline solutions. Both resins are made in the U.S.S.R. The distribution coefficients of francium, cesium, and rubidium between the resin and hydrochloric acid solutions were measured as a function of acid concentration in batch-equilibration experiments. For cesium and francium the greatest separation factors (ratio of the distribution coefficients of francium to cesium) were obtained at 5 to $6M$ hydrochloric acid with both KU-1 and KU-2 (separation factors of 2 and 3, respectively) and at $0.7M$ hydrochloric acid with KU-1 (separation factor of 3). Using columns of these resins the investigators could not separate francium, cesium, and rubidium by elution from KU-2 resin with $6M$ hydrochloric acid. However, nearly complete separation was obtained with KU-1 resin by elution with $5.5M$ or $0.7M$ hydrochloric acid. A column 13 cm. long and 0.3 cm. in diameter was used. The column was washed with hydrochloric acid of the same concentration used for the elution ($0.7M$ or $5.5M$), and the francium, cesium, and rubidium were added to the column in a small volume (0.05 to 0.1 ml.) of acid of the same concentration as that used for the elution. With $0.7M$ acid, elution required 4 to 5 hours, too long to be practical in the case of francium. With $5.5M$ acid the elution required only 20 to 40 minutes, a reasonable time in view of the half life of francium-212 and -223.

Hyde (129) suggests that certain inorganic ion exchangers may be useful in the separation of francium, but apparently this has not been tested.

4. Solvent Extraction

In common with the other alkali metals, francium has quite limited solubility in organic solvents. One report (221) states that francium-223 can be extracted into nitrobenzene from an aqueous solution $0.05M$ in sodium tetraphenyl borate and buffered at pH 9 with sodium borate. Using equal volumes of the two phases, 99% of the francium is extracted. Radium-223 is 90% extracted. However, if 1% EDTA is added to the aqueous solution the radium extraction is completely inhibited while the francium extraction remains quantitative. The francium may be extracted from the nitrobenzene with $1M$ hydrochloric acid. This method separates francium from all members of the actinium series except thallium-207.

5. Other Separations

Because of the actinium branching ratio, the francium activity in an equilibrium mixture of actinium-227 and its decay products is only about one one-hundredth of that due to each of the other daughters. Under these conditions the separations must be quite good to obtain radiochemically pure francium. The separation of francium-223 is simplified and shortened if it need only be separated from actinium. From the actinium-decay chain (Fig. 4) it is seen that the rate of growth of daughters (except francium-223) is determined by the half lives of thorium-227 (18.2 days) and radium-223 (11.7 days). The decay products following radium-223 have considerably shorter half lives and come to equilibrium with radium-223 rapidly, compared with the rate of growth of thorium-227 and radium-223 into pure actinium. Thus, if thorium-227 and radium-223 are separated from actinium, the radium-223 decay products will decay rapidly and considerable time will elapse before significant amounts of actinium daughters, except francium-223, will be present. In the meantime, francium-223 grows to its equilibrium value within three hours, and the mixture will consist essentially of only actinium and francium-223 (129).

Methods of separating thorium and radium are discussed in Section II, and these may be used to advantage in purifying actinium-227 from its descendants prior to the francium separation. Some of the methods for separating thorium from actinium are precipitation of thorium peroxyhydroxide; coprecipitation of thorium with ceric iodate or hydroxide (by dropwise addition of ammonia, ceric hydroxide can be precipitated before most of the more-soluble actinium hydroxide precipitates); extraction of thorium into organic phosphates (e.g., tributylphosphate, dibutylphosphoric acid, or di-(ethylhexyl) phosphoric acid) or thenoyltrifluoroacetone (TTA) solutions; or adsorption of thorium on ion-exchange resins from nitric acid solution. Separation of radium from actinium may be accomplished by coprecipitation with a variety of insoluble barium compounds, by coprecipitation of actinium with rare earth or other hydroxides (some radium-223 will also carry, so several reprecipitations are usually necessary to obtain a clean separation), by extraction of actinium into an organic phosphate or TTA solution, or by elution from cation-exchange columns.

A simple method for separating francium-223 from actinium after the actinium has been purified from its descendants is to extract the actinium into a TTA–benzene solution, allow the francium to grow into equilibrium, and extract the francium from the organic layer into distilled water. The removal of francium is essentially instantaneous. If the TTA concentration in benzene is increased and the ratio of benzene to water is high (perhaps 10:1), the extraction of actinium into the water is kept to a minimum (1). This same principle could presumably be applied to other organic solvents containing actinium.

D. DETECTION

As with other trace radioactive elements, qualitative detection consists of obtaining sufficient knowledge of the chemical behavior and the radioactive decay properties of the unknown activity to establish its identity. A knowledge of the method of production or the source of the activity also is helpful. The chemical properties of a francium isotope are evaluated by using the coprecipitation, volatility, and solvent-extraction properties discussed earlier. For isotopes of known nuclear properties the half life and, if necessary, the type, abundance, and energy of the emitted radiations and particles establish the identity. For new isotopes of unknown decay properties the source or method of production, chemical behavior, and identification of known daughters can be used. Thus, if the alpha decay of an actinium isotope is established, and an unknown activity that follows the chemical behavior expected of francium is isolated, the identity of a francium isotope is established. In this connection, the half life and type of decay theoretically expected for an unknown isotope of a given mass number is useful (125,128). The method used by Perey (232,233,235) in the discovery of francium is, of course, an excellent qualitative method.

An example of the qualitative detection of francium is the discovery of the isotope of mass 212 by Hyde, Ghiorso, and Seaborg (131). An unknown alpha activity produced in irradiated thorium was identified as francium by its coprecipitation with cesium silicotungstate and cesium perchlorate, and its failure to carry with ferric hydroxide, thorium fluoride, and barium carbonate. It was possible to separate it from cesium by elution from a short column of colloidal Dowex-50 with $1M$ nitric acid. The alpha activity eluted last, as expected. The volatility of the separated francium fraction was that expected for radon and astatine (daughters of francium-212) and francium-223 followed the alpha activity exactly through the chemical separations.

E. DETERMINATION

The amount of francium is determined, following its separation from interfering activities, by counting in appropriate instruments. The efficiency of the counting system for the particular sample used must be known to determine the absolute amount of francium. For alpha or beta counting the solution, or a fraction, is usually evaporated on a suitable backing material. Gamma- and beta-ray counting of francium-223 may be done directly in solution if desired. Any conventional beta counter may be used; gamma-rays are counted most efficiently with a thallium-activated sodium iodide crystal. If a sufficient quantity of radiochemically pure and carrier-free francium-223 is available, the efficiency of a beta counter for this isotope may be accurately standardized by counting the alpha particles emitted by its daughters (126). A sample of the activity is evaporated in the center of a platinum disk. An asbestos washer is placed on the disk and a clean platinum collector disk is placed a few millimeters away. The bottom disk is heated with a gas torch for one second to heat the platinum to redness. The francium is volatilized to the collector foil, leaving behind any traces of actinium-227, thorium-227, and radium-223 present at the time. This instant is taken as zero time. The beta-counting rate of the volatilized francium is followed and extrapolated to zero time. Several hours later, after the francium has completely decayed, the alpha activity is measured in an internal alpha counter (52% counting efficiency for weightless samples on platinum). Four alpha particles are emitted in the decay chain following francium-223 (see Fig. 4). The alpha activity decays with the half life of radium-223 (11.7 days), and one-fourth of the alpha activity is due to this nuclide. The radium-223 activity is extrapolated to zero time and the initial beta-disintegration rate of francium-223 is calculated from the fact that 1000 disintegrations of francium-223 produces 1.31 disintegrations per minute of radium-223. This method of standardization is desirable because alpha counting is ordinarily more accurate than beta counting.

In counting francium isotopes, corrections for decay and for the growth of daughters always must be made. An initially pure sample of francium-223 will decay with its half life of 22 minutes for about three half lives, but eventually a long-lived component will appear due to daughter activities. The alpha particles of radium-223, radon-219, and polonium-215 can be blocked from the beta counter by placing an aluminum absorber (5 to 10 mg./cm.^{2} thick) between the sample and the counter, but the energetic beta particles of lead-211 and thallium-207 cannot be removed. These activities grow to a counting rate a few tenths of a per cent of the original counting rate. When the counting rate of the daughter activities becomes an appreciable fraction of the francium-223 counting rate, the results lose accuracy (129). If the sample is counted at this point in its decay, the long-lived activity must be subtracted from the total activity to obtain an accurate francium-223 counting rate. This may be done by following the rate of decay until the long-lived component ("tail") is observed. The results are plotted on semilog paper to obtain a curve of the logarithm of the counting rate as a function of time and the "tail" is extrapolated back in time and its value subtracted from the total counting rate at any time in the conventional manner (see the general references on radioactivity measurements in the bibliography). Alternately, the sample can be counted after essentially all the francium-223 has decayed (about 10 half lives), and the counting rate at this time subtracted from the earlier counting rates. Since the "tail" decays with the half life of radium-223 (11.7 days), decay corrections for this activity can readily be made, if necessary, before the subtraction.

Francium-212 decays 44% by alpha emission and 56% by electron capture, according to the following scheme:

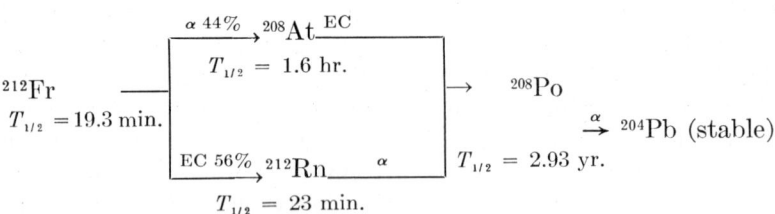

where $T_{1/2}$ is the half life and α and EC indicate decay by alpha-particle emission and orbital electron capture, respectively. This isotope may be determined by counting its alpha activity. The calculated alpha activity of initially pure francium-212 as a function of time after separation from

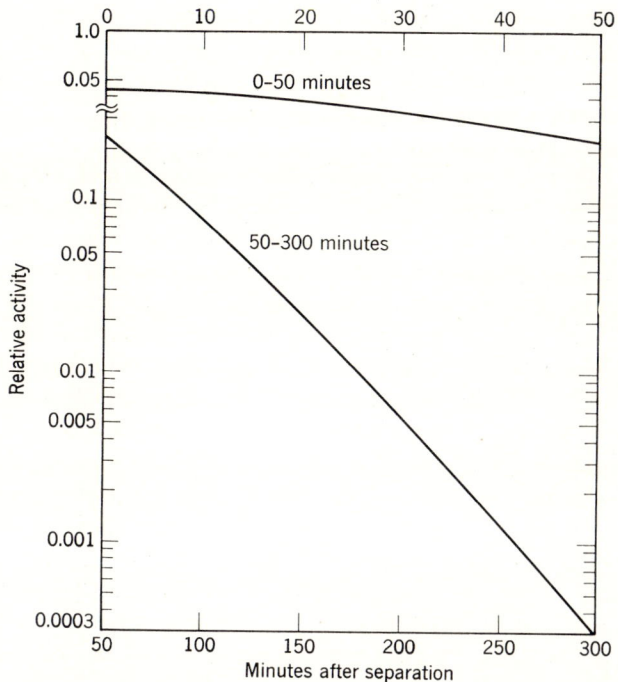

Fig. 13. Decay of alpha activity in initially pure francium-212.

its daughters is plotted in Fig. 13. The decay curve does not show the 19.3-minute half life of francium-212 because of the growth of radon-212. The alpha activity decays with a gradually decreasing apparent half life of about 45 minutes one-half hour after separation to about 25 minutes 4 to 5 hours after separation. At about 400 minutes a "tail" in the decay curve due to 2.3-year polonium-208 becomes noticeable and, eventually, after the francium and radon have decayed, the sample decays with the half life of polonium-208. If the time of separation of francium-212 from its daughters is known the francium activity can be obtained from the counting rate at any subsequent time by dividing the counting rate by the relative activity for that time obtained from the decay curve. If the sample is counted at a time when the polonium-208 contributes a significant fraction of the total activity (after 400 minutes), a correction for this contribution must be made in a manner similar to that described for francium-223.

The decay curve was calculated assuming complete retention of radon-212. If some of the radon diffuses out of the sample before it decays the sample will appear to decay faster than calculated from the half lives. Emanation of radon isotopes from solid samples is not very significant for the short-lived radon isotopes such as radon-219 (half life of 3.92 seconds), but may be significant for radon-212 in very thin samples. Most inorganic solids emanate very little, so that samples containing some solids, such as carrier precipitates, can be evaporated to dryness and counted without a significant radon loss. However, very thin or weightless samples may present a problem in this respect, and the radon loss should be determined. This can be done by placing a sample in a sealed counting chamber or one with very little gas flow for a short period (about 10 minutes), removing the sample and quickly resealing the chamber, and counting. The presence of radon activity in the counter can then be detected. Another method is to follow the decay of a purified francium-212 sample and to compare the observed decay with the rate calculated by assuming complete radon retention. The fractional loss can then be determined. Unsealed liquid samples and samples on filter paper also emanate. Such samples, as well as thin evaporated samples, may be sprayed with a thin layer of paraffin, lacquer, or similar material to prevent radon loss.

Gamma-ray counting may be used to determine francium-221 and francium-223 (126). Francium-212 also may be determined by counting the x-rays emitted in the electron-capture decay of francium-212 and its astatine-208 daughter.

F. SEPARATION AND DETERMINATION IN SPECIFIC MATERIALS

1. Actinium

The determination of francium-223 in actinium preparations as a means of measuring the amount of actinium has been discussed previously. However, it might be useful to summarize here some pertinent facts regarding this determination. Francium is not determined in actinium for its own sake, since its activity in a known amount of actinium is known. The amount of francium is measured to determine the actinium content of an unknown sample. However, if the amount of actinium is determined by another method, the sample may be used as a standard source of francium-223 to standardize the entire determination or to calibrate a counter for francium-223.

Because of the short half life of francium-223 and the low branching ratio (1.2%) leading to its formation by decay of actinium-227, the determination of actinium by this method is at least one-hundred times less sensitive than determinations based on counting actinium-227 itself or its other daughters. However, when sufficient actinium is available this method is extremely useful, because actinium-227 determinations are difficult by other means.

Detailed procedures for this determination are given in Section II.

2. Irradiated Thorium and Uranium

The preferred method for the separation of francium from thorium targets is by coprecipitation with free silicotungstic acid (124). The method is rapid, simple, specific (except for cesium), essentially quantitative, and yields a carrier-free product. Its application to uranium targets apparently has not been reported, but if uranium does not interfere it is certainly superior to the more complex method described in Section IV-C-1 (178).

The determination of beta-emitting francium isotopes is complicated by the presence of cesium fission products in the target, since these fission products are also beta emitters. All methods reported to date have not separated cesium from francium. The amounts of many of the cesium isotopes can be calculated from the decay curve of the separated francium and can then be subtracted from the total activity. This procedure is relatively accurate for cesium isotopes with half lives considerably longer than the francium isotopes, since their activity can easily be determined after the francium decay is complete. For the few cesium isotopes whose lifetimes are of the same order as the francium isotopes of interest, this procedure is less accurate since the complex decay curve of the mixed activities is difficult (but not impossible) to resolve. In the latter case, measurement of the gamma-ray energy spectrum will be useful in identifying the nuclides present. Francium-212 can be readily determined in the presence of beta emitters by alpha counting.

It should be pointed out that quantitative determination in targets is rarely required. Frequently the francium is separated only to establish its identity and approximate activity. This would require only qualitative knowledge of the radioactive cesium contamination in order to avoid mistaking cesium isotopes for francium.

G. RECOMMENDED PROCEDURES

1. Isolation of Francium from Thorium Targets with Silicotungstic Acid (124)

Dissolve the thorium target (in the form of metal foil about $0.5 \times 1.5 \times 0.005$ inches) in 5 ml. of hot concentrated hydrochloric acid containing a few drops of $0.2M$ ammonium fluosilicate to catalyze the dissolution of the thorium. Dilute the resulting solution with 15 ml. of ice-cold concentrated hydrochloric acid that has previously been saturated with hydrochloric acid gas. Centrifuge and remove the small amount of insoluble material. To the clear supernatant, stirring vigorously, add three drops of a $0.4M$ silicotungstic acid solution. Centrifuge the precipitate of free silicotungstic acid and wash it twice with 15 ml. of concentrated hydrochloric acid presaturated with hydrogen chloride gas. After the last decantation remove the remaining drops of excess hydrochloric acid carefully with a pipet. Dissolve the precipitate in 1 ml. of distilled water.

Before the bombardment prepare a column of Dowex-50 cation-exchange resin with the approximate dimensions 1 cm. × 4 mm. The column is prepared from 250 to 500 mesh resin in the ammonium form and is carefully rinsed with redistilled water. The column is inserted through a rubber stopper into a 15-ml. centrifuge tube containing a side arm connected to an aspirator. This arrangement permits solutions to be drawn rapidly through the column. Pipet the solution of francium and silicotungstic acid onto the top of the column and draw it through the column by suction at the rate of about 0.5 ml. per minute. Rinse the column with 1 to 2 ml. of redistilled water at the same rate. The silicotungstic acid passes through the column. A white precipitate with solution properties different from ordinary silicotungstic acid may form in the solution that passes through the column. The francium adsorbs strongly on the column if the acid concentration is less than $0.5M$. The success of this step depends on careful removal of the excess hydrochloric acid from the silicotungstic acid precipitate. Insert the column in a second 15-ml. side-arm centrifuge tube and desorb the francium by pulling 0.5 ml. of clean concentrated hydrochloric acid through the column in a period of 2 to 3 minutes. The resulting 0.5 ml. of solution is a pure, carrier-free solution of francium uncontaminated with any radioactivity except cesium fission products, and this solution may be counted as indicated in Section IV-E. The separation may be completed in 30 minutes or less. To avoid the appearance of a small amount of solid matter in the final solution, use very clean glassware, rinse the resin carefully to remove the excess ammonium ion, use very pure distilled water, and prepare the hydrochloric acid from this water and hydrogen chloride gas.

Occasionally small amounts of radioactive contaminants are encountered—due to the presence of a small amount of insoluble matter formed in the dissolution step—which are sometimes hard to remove by centrifugation. When 10 to 15 minutes more can be spent on purification, one of the following steps may be introduced between the silicotungstic acid precipitation and the ion-exchange step to remove any possible contaminants (notably protactinium) by coprecipitation. To the 1.0-ml.

solution of silicotungstic acid in water add 1 mg. of thorium ion and precipitate thorium hydroxide by adding 100 to 200 mg. of solid calcium hydroxide. Centrifuge, decant, and saturate the supernatant with hydrogen chloride gas. Centrifuge the silicotungstic acid precipitate, remove the supernatant completely, dissolve the precipitate in 1.5 ml. of redistilled water, and continue with the ion-exchange step. Manganese dioxide may be used instead of thorium hydroxide to "scavenge" radioactive impurities. For this precipitation, dissolve the silicotungstic acid precipitate in 1.5 ml. of water. Add one drop of a 56% solution of manganous nitrate and one drop of $1M$ potassium permanganate to precipitate manganese dioxide. After two minutes, centrifuge the precipitate and transfer the supernatant to a 15-ml. centrifuge cone. Add 8 ml. of cold concentrated hydrochloric acid and pass gaseous hydrochloric acid through this solution to precipitate the silicotungstic acid. Centrifuge, discard the supernatant, and dissolve the precipitate in 4 drops of pure water. Add 5 ml. of cold concentrated hydrochloric acid and reprecipitate silicotungstic acid as before with gaseous hydrochloric acid. Centrifuge, remove all the excess hydrochloric acid and proceed with the ion-exchange step. The francium recovery by this procedure, including the "scavenging" precipitations, is greater than 96%, as determined by using cesium tracer as a stand-in for francium.

2. Separation of Francium-221 with Silicotungstic Acid (124)

Francium-221 (half life of 4.8 minutes) is isolated from its 10-day parent, actinium-225, which has been previously isolated from the decay products of uranium-233 or from cyclotron-bombarded thorium. The shortness of the half life requires that the separation be as brief as possible. If the actinium is first purified carefully the number of possible contaminating radioactivities is quite limited, and a single precipitation of silicotungstic acid and a single acid wash give a nearly quantitative separation of francium-221 from its parent. If a small amount of silicotungstic acid can be tolerated in the francium fraction, the separation can be reduced to 10 minutes by the following procedure. Obtain a concentrated hydrochloric acid solution of actinium-225 about 5 ml. or less in volume and free from other radioactive elements (see Section II for actinium separations). Allow the solution to stand for about 30 minutes after the last francium separation. In this period (about six half lives) francium-221 reaches 98.4% of its equilibrium value. Precipitate silicotungstic acid as given in the previous procedure and wash the precipitate once with 15 ml. of concentrated hydrochloric acid. Dissolve the precipitate in 0.5 ml. of distilled water, add 1 to 2 ml. of ethyl ether, and stir. A three-phase system is produced in which the bottom layer consists of a liquid compound of silicotungstic acid and ether, the upper phase is ethyl ether, and the intermediate phase is a water solution of francium.

3. Separation of Francium-223 from Actinium-227 by Precipitation of Other Elements

Details are given in Section II-G-2-a. The method is applicable to pure actinium or actinium mixed with rare earths.

V. POLONIUM (ATOMIC NUMBER 84)

A. INTRODUCTION

Polonium was discovered in 1898 by Pierre and Marie Curie. Mme. Curie noted that pitchblende was considerably more radioactive than could be accounted for by its uranium content, and suspected the presence of another more radioactive element. In the processing of large amounts of pitchblende the Curies found a new activity that concentrated in the bismuth fraction, and they eventually identified the activity as due to a new element. The isotope discovered was polonium-210, the last radioactive member of the uranium–radium decay series (Fig. 3). It is the longest-lived naturally occurring polonium isotope and it is the one that has been used almost exclusively in studying the macrochemistry of polonium.

The early studies on polonium chemistry were made with submicrogram amounts of polonium-210 obtained from uranium ores. Since it is the second descendant of the long-lived lead-210, the lead residues from uranium ore refining were used as a source of polonium. Trace amounts also were separated from radon tubes used for therapeutic purposes. After the decay of the radon and its short-lived descendants, the tubes contain only the radium DEF chain (a mixture of lead-210, bismuth-210, and polonium-210 that decays with the half life of lead-210, 19.4 years), and fairly pure polonium can be obtained from this source. Much of the early separation studies of polonium were concerned with separating polonium from lead and bismuth in the radium DEF chain. Prior to 1944 only one sample of a weighable amount of polonium (100 γ) was prepared. However, since the advent of strong sources of thermal neutrons from nuclear reactors, milligram amounts of polonium-210 can be prepared by the irradiation of stable bismuth-209 with neutrons. The production of polonium by this method was developed by Mound Laboratory in the

United States, beginning in 1944. The extensive work on polonium done at this laboratory is detailed in Reference 218.

The high specific activity of polonium-210 (4.5 curies per milligram) makes the use of large amounts of this isotope very difficult. There is a considerable health hazard associated with the use of millicurie amounts of polonium. Since this hazard is due almost entirely to the highly ionizing, but short-range, alpha particles, protection from external radiation is not required. However, the polonium must be handled in such a way as to prevent its intake into the body by inhalation, ingestion, or through openings in the skin. Many polonium compounds are volatile, and this requires extra care to avoid spreading polonium contamination over the work area. This usually means that the polonium must be contained in closed glove boxes, and vessels containing the polonium must not come in contact with the bare skin. Under such conditions even ordinary chemical manipulations cannot be performed with the usual ease and speed. Problems associated with the safe handling of polonium are discussed by Bagnall (10,12,13), and its use in macroscopic amounts should not be undertaken unless the proper health practices are observed. In addition to the health hazard the intense radioactivity of polonium causes decomposition of its solutions (accompanied by gas evolution) and of air, with the formation of nitrogen dioxide. Polonium chars organic compounds rapidly, and work with such materials must be completed quickly. The alpha particles also can oxidize lower-valence states of polonium and discolor, crystallize, and craze glassware. Solid polonium compounds, even at the 0.1-mg. level, emit a blue glow, and this color, together with the fluorescence induced in the containers due to alpha-particle bombardment, often make it difficult to assign a color to a new compound. Solid preparations also are scattered and decomposed by the energetic alpha-particle emission. The radioactivity also produces an appreciable heating effect, and polonium preparations are always appreciably above the ambient temperature. The temperature decreases as the sample decays. Because of the relatively short half life of polonium-210, stable lead-206 grows into the polonium at the rate of about 0.5% per day, rapidly decreasing the purity of polonium preparations. Helium, from the alpha particles, increases the pressure in closed systems.

In spite of these difficulties, considerable progress has been made in the chemistry of polonium using macroscopic amounts of the element. Utilizing modern methods and equipment and exercising care, large amounts of radioactive materials can be handled safely. Considering the elaborate equipment available today, one can only marvel at the excellent work done by the early workers with highly radioactive materials.

B. PROPERTIES

1. Physical Properties

a. Nuclear Properties

The pertinent nuclear properties of polonium isotopes are given in Table XI. Additional data on decay schemes and energies, and references to the original literature, are given in Reference 283. Gamma and x-ray energies are given in Reference 56.

TABLE XI
Nuclear Properties of Polonium Isotopes (56,116,283)[a]

Mass number	Type of decay	Half life	Energy of principal radiations (m.e.v.)	Source
196(?)	α	1.9 m	α: 6.14	W(Ne,xn)
197(?)	α	~4 m	α: 6.04	Bi($p,13n$); W(Ne,xn)
198(?)	α	~6 m	α: 5.94	Bi($p,12n$)
199(?)	α	~11 m	α: 5.85	Bi($p,11n$); W(Ne,xn)
200(?)	α, EC	~11 m	α: 5.77	Bi($p,10n$)
201	α, EC	18 m	α: 5.67	Bi($p,9n$)
202	EC, 98% α, 2%	51 m	α: 5.78	Bi($p,8n$); W(Ne,xn) dtr ^{206}Rn
203	EC	42 m		Bi($p,7n$); Pb(α,spall)
204	EC, ~99% α, ~1%	3.8 h	α: 5.37	Bi($p,6n$); Pb(α,spall) dtr ^{208}Rn, ^{204}At
205	EC, 99+% α, 0.074%	1.8 h	α: 5.2	Pb(α,3n); dtr ^{205}At
206	EC, 95% α, 5%	8.8 d	α: 5.22 γ: 0.286, 0.338, 0.807, 1.031, others	^{204}Pb(α,2n); dtr ^{210}Rn, ^{205}At
207	EC, 99% α, 0.01% β$^+$, ~0.2%	5.7 h	α: 5.10 γ: 0.406, 0.743, 0.992, 1.149, others	^{206}Pb(α,3n); dtr ^{211}Rn, ^{207}At
208	α EC (weak)	2.93 y	α: 5.11 γ: with EC, 0.285, 0.60, 0.885	Pb + α; Bi + p or d; dtr ^{208}At
209	α, 99+% EC, ~0.5%	103 y	α: 4.88 γ: with α, 0.260; with EC, 0.91	Bi($p,2n$); dtr ^{209}At

TABLE XI (continued)

Mass number	Type of decay	Half life	Energy of principal radiations (m.e.v.)	Source
210 (RaF)	α	138.4 d	α: 5.305 γ: 0.804	Natural source, dtr ^{210}Bi in U–Ra series; ^{209}Bi(n,γ); Bi + p or d ^{208}Pb(α,n)
211m	α	25 s	α: 7.14, 7.85, 8.70 γ: 0.56, 1.06	
211 (AcC′)	α	0.52 s	α: 7.44 γ: 0.562, 0.88	Natural source, dtr ^{211}Bi in U–Ac series; dtr ^{211}At, ^{215}At
212 (ThC′)	α	3.04 × 10⁻⁷ s	α: 8.78	Natural source, dtr ^{212}Bi in Th series
213	α	4.2 × 10⁻⁶ s	α: 8.35	Dtr ^{213}Bi, ^{217}Rn in ^{237}Np decay series
214 (RaC′)	α	1.64 × 10⁻⁴ s	α: 7.68	Natural source, dtr ^{214}Bi in U–Ra series
215 (AcA)	α, 99+% β⁻, 5 × 10⁻⁴%	1.83 × 10⁻³ s	α: 7.36	Natural source, dtr ^{219}Rn, ^{215}Bi in U-Ac series
216 (ThA)	α	0.158 s	α: 6.78	Natural source, dtr ^{220}Rn in Th series
217	α	<10 s	α: 6.54	Dtr ^{221}Rn
218 (RaA)	α, 99+% β⁻, 0.02%	3.05 m	α: 6.00	Natural source; dtr ^{222}Rn in U–Ra series

ᵃ Key to abbreviations and symbols: α—alpha particle; β⁻—negative beta particle; β⁺—positive beta particle (positron); γ—gamma-ray; EC—orbital electron capture; n—neutron; p—proton; d—deuteron; s—second; m—minute; h—hour; d—day; y—year; dtr—daughter; spall—spallation reaction; and (?)—identification uncertain.

The most important and common isotope is 138.4-day polonium-210 (RaF). It decays by alpha-particle emission predominantly to the ground state of lead-206. A gamma-ray of 0.80 m.e.v. is emitted in only about $1.2 \times 10^{-3}\%$ of its decays. In these instances polonium-210 decays to an excited state of lead-206, which in turn reaches its ground state by gamma-ray emission. The rare alpha particle corresponding to this energy difference, about 4.5 m.e.v. or 0.80 m.e.v. less than the main alpha energy, has been observed. Two groups of lead x-rays (about 10 and 77 k.e.v.) of low intensity also are emitted following the alpha decay (218).

Short-lived polonium isotopes appear in the naturally occurring radioactive series as the "A" and "C-prime" decay products (Figs. 1, 3, and 4). These isotopes are ordinarily detected when measuring their longer-lived parents and are not themselves separated and measured in analytical work. The two longest-lived isotopes, masses 208 and 209, can be prepared by bombardment of lead or bismuth with charged particles. Polonium-209, with a half life of 103 years, would be particularly useful for macrochemical studies, but it is as yet impractical to produce milligram amounts.

The existence of a very long-lived polonium isotope in certain gold and tellurium minerals has been claimed, but the isotope has not been characterized (257). If it does exist, it probably has a low mass number.

b. Other Physical Properties

Polonium metal is silvery-gray in color and soft enough to be readily scratched with a dissecting needle (197). As a result of its intense radioactivity, large samples of polonium-210 excite light emission in quartz or glass containers and in the surrounding gas. As little as 0.2 γ is visible if spread over less than 1 mm. length in a 0.5-mm. capillary (218). The radioactivity also produces a heating effect (27.4 calories/hour/curie), and large samples are always at a temperature above that of the surroundings.

Polonium is a fairly volatile metal, and it distills readily in a vacuum and deposits as a bright mirror if it is pure (218). The vapor pressure between 438 and 745°C. is represented by the equation (41):

$$\log p = \frac{-5377.8 + 6.7}{T} + 7.2345 \ (\pm 0.0068)$$

TABLE XII
Physical Properties of Polonium Metal

Crystal structure	
α-Po	Simple cubic
β-Po	Simple rhombohedral
Density[a]	
α-Po	9.4 g./cm.^{3}
β-Po	9.2 g./cm.^{3}
Melting point	252°C.
Boiling point	962°C.
Heat of vaporization	24,597 cal./mole
Electrical resistivity	
α-Po	42 ± 10 μohm-cm. (0°C.)
β-Po	44 ± 10 μohm-cm. (0°C.)
α-Po	140 ± 10 μohm-cm. (20°C.)

[a] Calculated from x-ray data.

where p is the vapor pressure in millimeters of mercury and T is the absolute temperature. From this study the boiling point and heat of vaporization given in Table XII were calculated. Other physical properties are summarized in Table XII.

TABLE XIII
Reduction Potentials of Polonium Couples

Couple	Electrode potential, v.	Reference
Po(II) + 2e → Po(s)	+0.6[a]	(298)
	ca. +0.38 (1N HCl)	(18)
Po(IV) + 4e → Po(s)	+0.8[a]	(298)
	+0.76 (1N HNO$_3$)	(18)
	+0.55 (1N HCl)	(18)
Po(IV) + 2e → Po(II)	+1.0[a]	(298)
	+0.72 (1N HCl)	(18)
"PoO$_3$(s)" + 2e → Po(IV)	+1.1[a]	(298)
Po + 2e → H$_2$Po	> −1.1[a]	(298)

[a] Derived from published electrodeposition studies.

The two allotropic modifications of the metal given in Table XII can coexist between 18 and 54°C. Below 18°C. the alpha form is stable; above 54°C. the beta form is produced. At room temperature freshly prepared samples of the metal are in the high-temperature, or beta, form due to the heating effect of the alpha particles. As the sample decays and cools the low-temperature form is produced (29,30,196). The physical properties of polonium are closer to those of its neighbors in the periodic table, bismuth, lead, and thallium, than to its lighter homologue, tellurium. However, in chemical behavior polonium resembles tellurium rather than its neighbors (10). Pure polonium metal can be prepared by vacuum sublimation of polonium metal deposited from solution (11) or, better, by thermal decomposition of polonium sulfide in a vacuum (11,15,21).

2. Electrochemical Properties

A number of studies of the electrochemical behavior of polonium at tracer concentrations have been made in order to study the applicability of the electrochemical laws at very low concentrations. This work is discussed by Wahl and Bonner (298).

As might be expected from the difficulties encountered in working with polonium, the values of the oxidation-reduction potentials are not on firm ground and additional work remains to be done. Whether this work can be done with polonium-210 is doubtful. The difficulties encountered in using weighable amounts of polonium-210 are pointed out by Bagnall and Freeman in their work on the electrode potentials of polonium (18). However, some of the potentials are known in certain solutions with sufficient accuracy to permit their use for analytical purposes. A few measurements have been made using polonium electrodes and solutions of polonium, but most have been made by the "critical deposition potential" method (106,115,148,254). Some of the reported values of the reduction potentials for polonium are given in Table XIII. Other values are given in Reference 218.

In the table no attempt has been made to identify the ionic species involved in the electrode reactions, since the compositions of the ions are not well established. In nitric acid solution polonium is at least partially complexed; in hydrochloric acid polonium is highly complexed, probably as $PoCl_6^{-2}$ and $PoCl_4^{-2}$.

Wahl and Bonner (298) also have estimated potentials for polonium solutions at $10^{-9}M$, a concentration at which much experimental work has been done. Their results are as follows:

$Po(II) + 2e \rightarrow Po(s)$	$+0.3$ v.
$Po(IV) + 2e \rightarrow Po(s)$	$+0.65$ v.
$Po(IV) + 2e \rightarrow Po(II)$	$+1.0$ v.
"$PoO_3(s)$" $+ 2e \rightarrow Po(IV)$	$+1.4$ v.
$Po(s) + 2e \rightarrow H_2Po$	> -0.7 v.

From the potential of the Po/Po(IV) electrode, it appears that polonium lies between tellurium and silver in the electrochemical series, in agreement with its behavior in solution toward reducing agents. The reduction potentials of polonium in a number of acids and in sodium hydroxide also are given in Reference 218.

Electrolytic deposition and spontaneous deposition of polonium on a variety of metals is often used in separation, isolation, and analytical procedures. Spontaneous deposition of polonium on more active metals is probably the most common method of preparing tracer amounts of polonium for alpha counting, and this topic will be discussed later. The electrolysis of polonium solutions is discussed in Reference 218.

3. Optical Properties

The emission spectrum of polonium has been studied recently with the pure element (120,288). Using a high-frequency electrodeless discharge, 139 lines were observed. Thirty-five of these lines and 13 additional lines were observed in a spark discharge. The strongest lines are at 2450.11, 2558.01, 3003.21, and 4170.45 A.

The molecular spectrum of diatomic polonium has been studied for the molecules (^{210}Po–^{208}Po) and (^{208}Po–^{209}Po) (288). About 450 red-shaded band heads were observed with the 210–208 mixture and more than 500 with the 208–209 mixture. The isotope shifts of the band heads for the same transitions were determined, as well as the relative band-head intensities used for the isotopic analysis of 208–209 mixtures (120).

The absorption spectrum of polonium in hydrochloric acid solution has been studied from 220 to 795 mμ (119). In solutions less than $0.5M$ in hydrochloric acid an absorption peak occurs at 344 mμ. Above $0.5M$ hydrochloric acid a peak occurs at 418 mμ. In strong acid a peak is observed at 247 mμ and a shoulder at 260 mμ. The peak at 418 mμ can be used for the determination of polonium in solutions above $1M$ in hydrochloric acid. This peak obeys Beer's law over the range of polonium concentrations studied, 3×10^{-5} to $6 \times 10^{-5}M$. Over this range the absorbance index (the ratio of absorbance to polonium molarity) is 1.06×10^4.

4. Chemical Properties

a. Polonium Metal

Polonium metal reacts with oxygen to form a dioxide, PoO_2. The reaction is slow in dry oxygen at room temperature, but is rapid at 300°C. (218). Milligram amounts of polonium dissolve slowly in $2N$ hydrochloric acid, giving a solution of bivalent polonium that is rapidly oxidized to the quadrivalent state by the alpha-particle bombardment to form a yellow complex, presumably $PoCl_6^{-2}$. The presence of chlorine water increases the dissolution rate. The metal dissolves rapidly in concentrated hydrochloric acid (10). The solubility of polonium in nitric acid at 25°C. varies from 4 mg./liter in $0.1N$ acid to 777 mg./liter in $8N$ acid (227). The solubility increases sharply above $1N$ acid, and migration experiments indicate that polonium is present as an anionic nitrate complex at high nitric acid concentrations.

The solubility of polonium in hydrochloric and hydrofluoric acids is quite high compared to other acids. The solubility of polonium in sulfuric, phosphoric, and perchloric acids is low, although citric and oxalic acids are reported to increase the solubility in phosphoric acid (218).

b. Properties of the Oxidation States

By analogy with its lighter homologues in Group 6A of the periodic table, one would expect polonium to have stable oxidation states of -2,

+2, +4, and +6. A +3 state has been reported, but its existence remains questionable. The formation of polonides was first suggested by the observation that, when tracer polonium was melted with silver or copper, only a small proportion of the polonium distills from the melt, whereas electroplated or evaporated polonium distilled readily from these metals. Milligram amounts of a number of metal polonides have been prepared, but their behavior in solution apparently has not been studied. The crystallographic data for these compounds are given in References 10 and 218.

The +2 oxidation state is unstable in solution in the absence of reducing agents. Unless stabilized in this way, it is rapidly oxidized to the +4 state under the influence of its own alpha particles. It is probable that the bivalent state would be the stable oxidation state in the absence of ionizing radiation. The +2 state is obtained in concentrated hydrochloric acid solution by reduction of Po(IV) with sulfur dioxide or hydrazine in the cold or by arsenious oxide on warming. Polonium dichloride is a dark, ruby red solid that dissolves in dilute hydrochloric acid to give a pink solution (15). Polonium monosulfide is obtained as a black precipitate when hydrogen sulfide is passed through a solution of Po(II) or Po(IV) in dilute hydrochloric acid (22). The first step in the sulfide precipitation from solutions of the tetrachloride is evidently reduction to Po(II). The sulfide is soluble in concentrated hydrochloric acid and insoluble in ammonium sulfide, ethyl alcohol, acetone, or toluene. It is decomposed by strong oxidizing agents. The sulfide is important as the starting material for the preparation of the pure metal. To prepare polonium metal the sulfide is decomposed into the elements by heating at about 275°C. under 5 μ pressure. At this temperature the sulfur sublimes from the residue of lead sulfide and polonium. The polonium is distilled at 450 to 500°C. while the lead sulfide remains behind.

The +4 oxidation state is the most stable state in solution. A number of compounds of Po(IV) have been prepared. The tetrachloride is a bright yellow solid, soluble in water and hydrochloric acid. It is relatively volatile (boiling point, 390°C.) and, therefore, hydrochloric acid solutions of polonium must be evaporated carefully to prevent loss of polonium. When such samples are evaporated to dryness for counting, losses occur if the temperature rises above 150°C. The tetrabromide is a bright red solid, soluble in hydrobromic acid. The tetraiodide is a volatile black solid, slightly soluble in hydroiodic acid, and less stable than the corresponding bromide and chloride. These halides probably exist in solution as the PoX_6^{-2} complex, since insoluble compounds of the form M_2PoX_6 (where M is a singly charged cation such as an alkali metal ion, or tetramethylammonium) are precipitated from a solution of the polonium halide by the addition of the alkali halide or quaternary ammonium halide. In so far as they have been studied, these salts are isomorphous with the corresponding compounds of tellurium and other tetravalent metals (15,16,17,278). Quadrivalent polonium is reduced to the metal by stannous chloride under a variety of conditions. In dilute hydrochloric or hydrofluoric acid solution it is reduced to the metal by sulfur dioxide, sodium sulfite, or hydrazine. Tellurium and selenium precipitate as the metal under these conditions. However, in concentrated hydrochloric acid these reducing agents precipitate tellurium, leaving polonium in solution (10). Sodium or ammonium hydroxide precipitate polonium from solution as the hydroxide. In the case of Po(IV) the precipitate probably has the formula $PoO(OH)_2$. The bivalent hydroxide is rapidly oxidized, and its composition is unknown. The solubility of polonium hydroxide is about 0.27 mg./liter in water and 0.90 mg./liter in concentrated ammonium hydroxide. In potassium hydroxide the solubility increases with increasing alkali concentration, reaching 1.08 g./liter in $1.73N$ hydroxide (19). Thus the hydroxide is slightly amphoteric, and ammonia should be used as the precipitating agent.

Polonium forms complexes with a number of anions. Complexes have been reported with oxalic acid, phosphoric acid, acetic acid, halides, cyanide, tartaric acid, tributylphosphate, acetylacetone, and EDTA (10,75).

Experimental evidence for the existence of the +6 oxidation state is not conclusive. No compounds of hexavalent polonium have been characterized on the macroscopic scale. Some evidence for hexavalent polonium has been obtained from tracer experiments, the best of which is the work of Matsuura and Haïssinsky (195). These authors studied the distribution of polonium between mineral acids and methyl isobutyl ketone. The concentration of polonium, presumably in the +4 state, in the organic phase decreased when ceric ion or dichromate ion was added to the system. This indicates an oxidation to a less extractable higher state, presumably +6. Destruction of these oxidizing agents with peroxide established the original distribution in the case of Ce(IV) and an intermediate distribution with Cr(VI). The latter situation was believed due to incomplete reduction. The partition of polonium between the phases depended on the Ce(III)/Ce(IV) and Cr(III)/Cr(VI) ratios. From these reactions the authors estimated the potential of the Po(IV)/Po(VI) couple to be about +1.5 v. At the milligram level, Bagnall and his co-workers (11) observed reactions of polonium with chromic trioxide, potassium permanganate, and potassium chlorate–hydroxide mixtures that were consistent with the production of an unstable oxidation state greater than +4. The formation of a volatile polonium compound with fluorine (presumable PoF_6) has been reported (301), but the compound was too unstable to be isolated and characterized.

c. Tracer Chemistry

In the handling and analysis of tracer concentrations of polonium (as well as of many other elements) it is very important to recognize and consider the possibility of loss of polonium by hydrolysis and/or radiocolloid formation. In neutral or nearly neutral solutions tracer polonium has been observed to behave more like a colloid than an ionic species. In such solutions the results of dialysis, filtration, diffusion, sedimentation, and electrophoretic experiments indicate that polonium is present in colloidal form. Since it is difficult to reconcile the fact that the solubility products of the compounds in question have probably not been exceeded with the colloidal behavior of the tracer, the term radiocolloid has been applied to such materials. The exact nature of these radiocolloids is not certain, but since impurities have a marked influence on radiocolloidal behavior the radioactive element may be adsorbed on colloidal impurities. Radiocolloids are likely to form in solutions in which the radioactive element would be insoluble if present in macroscopic concentrations.

As an example of the radiocolloidal behavior of polonium, some recent filtration experiments may be cited (217). Trace polonium solutions in the pH range from 1 through 12 were filtered through membrane filters. Buffered and unbuffered solutions at several ionic strengths were used. Typical results showed that the polonium was completely, or nearly completely, filterable at pH 1 and pH 12. Between these extremes the filterability decreased as the solutions approached neutrality, until only 0 to 20% was filterable between pH 6 and 8. Therefore, in neutral, weakly acid, or weakly basic solutions, it should be noted that polonium may be removed from solution by adsorption on the glass walls of containing vessels or on insoluble impurities in the solution and that sedimentation will occur upon centrifugation. To avoid such losses polonium solutions should be made at least $1N$ in acid if possible, and insoluble or hydrolyzable impurities should be kept to a minimum. It has been reported that radiocolloid formation in very dilute acid solutions can be markedly reduced if paraffin-coated containers are used instead of bare glass containers (261). Additional information on radiocolloidal behavior may be found in References 10, 49, and 298.

In trace concentrations polonium can be reduced by stannous chloride, sulfur dioxide, and hydrazine to a state (probably the metal) in which it coprecipitates with elementary selenium, tellurium, antimony, and gold (10,298). Tellurium is the most common carrier. The conditions under which the reduction is carried out are important. For example, polonium is carried by tellurium or selenium precipitated from dilute hydrochloric acid with sulfur dioxide, but it is not carried by tellurium precipitated from concentrated hydrochloric acid. Polonium does not coprecipitate with tellurium when reduced with hydrazine in concentrated hydrochloric acid, although tellurium is reduced to the elementary state. Polonium is carried by selenium formed by reduction with hydrazine in hydrofluoric acid solution. Reduction of polonium to the metal (or other insoluble state) is also accomplished by treatment with Ti(III), hypophosphorous acid, sodium dithionate in cold acid solution, and formaldehyde in alkaline solution. Ferrous sulfate, formic acid and formaldehyde in acid solution, and hydroxylamine do not reduce polonium to the metal (10).

C. SEPARATION AND ISOLATION

1. Dissolution of Samples

Solid materials normally encountered in polonium analyses are radioactive minerals, neutron-irradiated heavy metals (particularly bismuth), biological materials, radium–DEF mixtures from spent radon tubes, and particulate material in air (dust, soil, etc.). Since polonium is soluble in the common mineral acids, these acids may be used if the sample itself is acid-soluble. However, any acid-insoluble residues should be checked for polonium before discarding them. The insoluble residues in test samples may be dissolved by fusion or hydrofluoric acid treatment. Silica-containing minerals can be attacked with hydrofluoric acid and the fluoride removed by fuming with sulfuric or perchloric acids. The usual fusion techniques also may be used to dissolve minerals.

Neutron-irradiated bismuth may be dissolved in nitric–hydrochloric acid mixtures. Molar ratios of hydrochloric acid to nitric acid between 4:1 and 4:3 are suitable (218). In the production of polonium from bismuth, it is desirable to remove or destroy the excess nitric acid since polonium recovery is more readily accomplished from chloride solutions. The nitric acid may be destroyed by adding formaldehyde or formic acid and heating the solution (218). Removal of nitric acid also may be accomplished by evaporating the solution to dryness and heating the residue, or by repeatedly adding hydrochloric acid and evaporating to a small volume (75,218). The evaporation is difficult to perform satisfactorily when large amounts of bismuth are involved because spattering occurs when the solution becomes concentrated. However, for the small amounts of bismuth normally used for analysis this method is suitable. Bismuth may be dissolved electrolytically in hydrochloric acid alone if a positive potential is applied to the bismuth (218). In its present state of development this method does not appear to be useful in analytical work. Simple treatment with hydrochloric acid alone does not dissolve bismuth readily.

In studies of polonium metabolism and the effect of polonium on living systems it is necessary to analyze tissue, vegetation, and other samples consisting largely of organic material. The general method used for such samples consists of wet oxidation of the organic matter with oxidizing acids. The resulting acid solution contains the polonium and the inorganic portion of the sample, and it is analyzed for polonium. High-temperature ignition of the organic samples is usually avoided because of the volatility of many polonium compounds. Typical oxidation procedures call for heating with nitric–perchloric acid mixtures (209), nitric–perchloric–sulfuric acid mixtures (262), and fuming nitric–sulfuric acid mixtures (135).

Radium–DEF mixtures from spent radon tubes can be dissolved with mineral acids. Dust and soil samples require treatment similar to that used for minerals to place them in solution.

2. Separation

a. Coprecipitation

(1) Tellurium

The coprecipitation of polonium with tellurium metal was discovered in early polonium studies (192,193), and tellurium remains a useful and frequently used carrier for polonium. It is particularly useful as a preliminary step in concentrating polonium from a large sample. Polonium is quantitatively carried on elementary tellurium precipitated by reduction of tellurate ion with stannous chloride (187,192,193,218) or sodium hypophosphite (135,262) from hydrochloric acid solution. The solution is usually heated to coagulate the precipitate. The tellurium carrier is ordinarily added as telluric acid or potassium tellurate; carrier concentrations of a few tenths of a milligram per milliliter are sufficient. Less than 2% of the polonium remains in solution at hydrochloric acid concentrations up to $3N$. In $6N$ acid about 5% of the polonium is not carried (187). The polonium can be separated from the tellurium carrier by dissolving the precipitate in a minimum of concentrated nitric acid, adding hydrochloric acid, and precipitating elementary tellurium with sulfur dioxide (198,218). If the hydrochloric acid concentration is greater than $2N$ more than 98% of the polonium remains in solution (187). At lower acid concentrations polonium coprecipitation becomes appreciable. Large amounts of nitric acid prevent complete precipitation of tellurium and must be avoided. Reduction with hydrazine also can be used to precipitate tellurium without coprecipitation of polonium. This procedure has been used to separate polonium from irradiated bismuth (218).

(2) Lead Tellurate

This compound, precipitated by heating a very dilute nitric acid solution of lead and telluric acid, carries polonium quantitatively (149). The lead can be removed by digesting the precipitate with sulfuric acid and separating the insoluble lead sulfate. To separate the tellurium the filtrate is evaporated to dryness, the residue is dissolved in concentrated hydrochloric acid, and the solution is boiled to reduce tellurium to the quadrivalent state. The tellurium can then be precipitated with sulfur dioxide, as described above.

(3) Selenium

Polonium is carried almost quantitatively on elementary selenium precipitated from hydrofluoric or hydrochloric acid solution with sulfur dioxide or hydrazine (10,298). In this respect selenium differs from tellurium, and the coprecipitation with selenium may be due to the formation of polonium selenide (10). The polonium may be almost completely leached from the selenium precipitate with concentrated hydrochloric acid.

(4) Manganese Dioxide

Manganese dioxide carries polonium almost quantitatively from dilute nitric acid solution (298). The precipitate can be formed conveniently by the addition of manganous and permanganate ions to the solution and warming, if necessary, to coagulate the precipitate.

(5) Hydroxides

The hydroxides of iron, bismuth, aluminum, and lanthanum have been used to carry polonium (10,298). Precipitation may be carried out with ammonium or sodium hydroxide. Concentrations of sodium hydroxide greater than $2N$ result in incomplete carrying, but ammonium hydroxide concentrations up to $7N$ have been used successfully.

(6) Other Carriers

Polonium coprecipitates with the acid-insoluble sulfides, such as those of copper, bismuth, and lead (298). Antimony and bismuth precipitated with pyrogallol from dilute nitric acid solution carry polonium quantitatively. Tracer bismuth also coprecipitates with antimony (10). Polonium is carried completely by metallic gold formed by reduction with stannous chloride in hydrochloric acid solution, but it is incompletely carried when the reduction is carried out with hydrogen sulfide. Lanthanum fluoride has been reported to carry 90% of trace polonium from $1N$ nitric acid (214).

b. Spontaneous Deposition and Electrodeposition

Deposition of polonium from solution on more active metals by chemical displacement and on inert electrodes by electrodeposition is a very useful analytical property. The deposition is readily done and it is a convenient way to prepare a uniform, thin sample for alpha counting. For these reasons most analytical procedures, particularly for trace amounts, use this method as the final step in the separation. When the polonium is available in the proper solution, deposition is also used as a concentration step earlier in the separation procedure.

(1) Spontaneous Deposition

Polonium in solution deposits readily on a number of metals. The metal is simply inserted in a solution of the proper acid concentration and the reaction is allowed to proceed to completion. If the deposited polonium is to be counted, a metal disk or foil is used. Experimental arrangements for this purpose are discussed below. If additional separations are to be performed, finely divided metal may be used, although this is usually not necessary.

Silver is the metal most often used for this purpose. Polonium deposition is complete under the proper conditions, and few other metals will deposit. The best conditions for quantitative deposition on silver are $0.5N$ hydrochloric acid at 70 to 97°C. for about 1.5 hours (69,72). Only a small fraction of any radium-D and radium-E present will deposit, and large amounts of bismuth do not interfere. Deposition also occurs from dilute nitric, sulfuric, or acetic acid. Deposition from acetic acid is improved by the presence of small amounts of hydrochloric acid, and dilute nitric–hydrochloric acid mixtures have been used successfully to deposit large amounts of polonium (10). During the deposition a black film, possible silver oxide or peroxide, may appear on the silver and interfere with the electrolysis. In such cases the deposition is improved and the silver is much cleaner if the solution is boiled with sulfur dioxide prior to the deposition or if hydrazine is present in the solution during the deposition. It is well to remove any silver sulfide film from the silver before use. The deposition of trace polonium from hydrochloric acid appears to be independent of the acid concentration above $0.1N$. The deposition is inhibited by gold, mercury, platinum, or tellurium, and these elements must first be removed from solution by reduction with hydrazine in 20% hydrochloric or acetic acid solution. Traces of Fe(III) also interfere; in the presence of $0.002M$ Fe(III) only 50% of the polonium is deposited (297). This interference may be eliminated by reduction of the Fe(III) to Fe(II) with ascorbic acid or sulfur dioxide or by complexing the iron with fluoride ion.

The deposition of milligram amounts of polonium on silver has been studied by Bagnall and his co-workers (11). In this case the deposition rate decreases with increasing hydrochloric acid concentration. Thus, the deposition from 0.5 or $1N$ acid is more than 99% complete in 30 minutes, whereas it is only 90% complete after 50 minutes from acid concentrations greater than $4N$. The best depositions are obtained from hot solutions in the presence of reducing agents (hydrazine or sulfur dioxide) and hydrocyanic acid. The cyanide dissolves the silver chloride formed in the displacement.

A study of the mechanism of the displacement process (291) has shown that two silver atoms replace one polonium ion from dilute nitric acid, although quadrivalent polonium would be expected to be present in solution. The ion involved may be PoO^{+2}.

Polonium can be recovered from deposits on silver foil by vacuum sublimation or dissolution of the foil in nitric acid and precipitation of the silver as silver chloride (11).

Copper, nickel, stainless steel, and presumably other metals more electropositive than polonium also can be used to deposit polonium from dilute hydrochloric acid solution. As with silver the deposition is more efficient at elevated temperatures. These metals are not to be preferred to silver since they react to some extent with the dilute hydrochloric acid solution, producing an inferior deposit, and since other elements often associated with polonium will also deposit. Bismuth-210, for example, will deposit on copper. However, from some complex solutions, such as biological fluids, deposition on silver is incomplete. Nickel in the form of powder or foil has proved to be very effective in such solutions. If other ions, active or inactive, that deposit and interfere with the radioactivity measurement are absent, the plated nickel foil can be counted directly. If additional separations are necessary the nickel can be dissolved. Polonium deposits on platinum or palladium saturated with hydrogen from $0.1N$ hydrochloric acid (10). The metal apparently acts as a hydrogen electrode under these conditions. Oxidizing agents and nitric acid must be absent. Deposition on platinum also occurs in the presence of reducing agents such as quinone and hydroquinone. Under these conditions radium-E also deposits. Milligram or trace amounts of polonium deposit readily on nickel from dilute hydrochloric acid solution (10,14,34), and several analytical methods using nickel have been reported. Acid concentrations between $0.1N$ and $1N$ are best. As would be expected, radium-E also deposits. Polonium also deposits on bismuth and gold (10), but these metals have apparently not had any analytical application. For analytical purposes the deposition of radium-E, a beta emitter, is not important since the polonium alpha particles can be counted without interference from beta particles.

(2) Electrodeposition

There are some advantages to electrodeposition over spontaneous deposition, particularly in the preparation of milligram amounts of polonium metal (75). In spontaneous deposition the electrode potential is fixed by the metal and the composition of the solution. Since the deposition potential of bismuth is close to that of polonium, bismuth also is deposited on many metals. Electrodeposition permits control of the plating potential and separation from bismuth can be assured. However, unless the potential is closely controlled, other metals will deposit after the polonium is plated. Electrodeposition often is used in the final purification of milligram amounts of the element. At this level, electrodeposition is more efficient than spontaneous deposition. In analytical procedures, spontaneous deposition on silver is generally preferred because of its simplicity. Polonium can be plated on a platinum cathode from acetic acid solution, and anodically or cathodically from nitric acid (10,75,298).

c. Solvent Extraction

(1) Alcohols

Diisopropyl and diisobutyl carbinol extract polonium efficiently from $6N$ hydrochloric acid solution, provided the nitric acid concentration is less than $1N$ (214). This extraction has been used to separate polonium from bismuth targets. Some other heavy-element alpha emitters, e.g., protactinium, also are extracted.

(2) Chelating Agents

There has been considerable interest in the polonium–dithiazone complex since the extraction of this complex was first reported (35,140).

Polonium can be extracted from aqueous solutions of the proper composition into tetrachloride (140) or chloroform (35) solutions of dithizone. The extraction is usually carried out from dilute hydrochloric or nitric acid. The extraction is about 95% complete between pH 0.2 and pH 5; at higher or lower acidities the amount extracted decreases. The polonium can be back-extracted into $4N$ hydrochloric acid. This extraction system can be used to separate radium–DEF mixtures since radium-D (lead) does not extract at pH less than 3, and radium-E (bismuth) does not extract appreciably (from hydrochloric acid) at pH less than about 0.5. A carbon tetrachloride solution of dithizone is also reported to extract polonium from ammoniacal cyanide–citrate solution, ammoniacal cyanide–tartrate solution and, acidic cyanide–acetone solution (189). The results given above were obtained with trace polonium. The extraction of millicurie amounts of polonium also has been studied (10,20). The results are not very reproducible due to radiation decomposition of the dithizone and due to hydrolysis of the polonium at pH 2. The pH dependence is somewhat different, but extraction still remains nearly complete at pH 0.2 to 1. Polonium dithizonate probably has the formula $PoO(Dz)_2$ (20).

Several other organic complexing agents have been studied as extractants for polonium. Polonium can be extracted with chloroform, carbon tetrachloride, or amyl alcohol from acetate solution containing sodium diethyldithiocarbamate. The extraction is 93% complete at pH 1, and less efficient at lower acidities (158). The thionalide salt of polonium can be extracted into chloroform from acetic acid solution. The extraction is 99% complete at pH 4 (158). Extraction of polonium from saturated ammonium acetate solution with a chloroform solution of 8-hydroxyquinoline (70 mg./100 ml.) was found to be 74% complete at pH 3.4 (158). In this case the volume ratio of the organic to aqueous phases was one to two. The extraction of polonium by thenoyltrifluoroacetone (TTA) as a function of pH has been studied in connection with the purification of actinium by TTA extraction (100). Extraction of polonium by a $0.25M$ TTA benzene solution begins at about pH 0 and is complete at about pH 2. This system will give good separation of polonium from lead, actinium, and thallium(I) since these elements are extracted only at lower acidities. Bismuth and thorium will extract with the polonium. In the reported experiments some difficulty was experienced in obtaining a material balance, presumably due to loss of polonium by hydrolysis. Acetylacetone also extracts polonium completely from halogen acid solutions (11,117). Cupferron in amyl acetate extracts polonium from hydrochloric acid solutions containing sulfur dioxide (190).

Since the compounds formed by polonium with these organic complexing agents are volatile at temperatures above 100°C., care must be exercised in evaporating solutions of these compounds for counting (20,158,189). It is best to evaporate the solution at room temperature or slightly higher to avoid loss of polonium. Some, but not all, of these compounds may be decomposed by adding a strong acid, thus rendering the polonium less volatile. For example, TTA complexes may be destroyed by adding a few crystals of trichloroacetic acid to the aliquot of the TTA–benzene solution being evaporated.

(3) Ethers and Ketones

Polonium can be extracted into a number of ethers and ketones under appropriate conditions. With diethyl ether (62,158) or isopropyl ether (158), polonium extraction from hydrochloric acid or hydrobromic acid is negligible. From $8M$ nitric acid containing hydrogen peroxide, however, diethyl ether extracts 60% of the polonium originally present in the aqueous phase. Isopropyl ether, methyl isobutyl ketone, and diisopropyl ketone extract polonium from hydrochloric acid–potassium iodide solutions. One report (158) states that 80% of the polonium can be separated from a radium–DEF mixture in $3N$ hydrochloric acid–$0.25M$ potassium iodide solution by shaking with four times its volume of isopropyl ether. Another report (46) states that 97% of the polonium can be extracted by diisopropyl ketone from a solution containing $3N$ hydrochloric acid and $0.5M$ potassium iodide. The same investigator (46) reports that this solvent also extracts polonium completely from $12N$ sulfuric acid, and no other naturally radioactive elements are extracted. Under a variety of other conditions polonium extraction was significant but not quantitative. These results were obtained using equal volumes of organic and aqueous phases. Other workers (175,195) report good extraction of polonium from hydrochloric acid by methylisobutyl ketone in the absence of potassium iodide. Polonium can also be extracted by mesityl oxide from nitric acid saturated with ammonium nitrate (194).

Bagnall (11) reports that methyl isobutyl ketone extracts polonium completely from halogen acid solutions over a wide range of acid concentrations, but that diisopropyl ketone, diphenyl ketone, and dibenzyl ketone (in chloroform) extract polonium only slightly. This work was presumably done with millicuric amounts of polonium.

(4) Tributylphosphate

Tributylphosphate (TBP) is an efficient extractant for polonium. Using a 20% TBP solution in dibutyl ether and an aqueous phase of $6N$

hydrochloric acid, an extraction coefficient of 110 in favor of the organic phase has been obtained with trace polonium (150). This extraction was used as part of a purification scheme to separate polonium from irradiated bismuth. The lead and bismuth daughters produced by decay of the low-mass polonium daughters being studied were removed from the TBP layer by washing it with $6N$ hydrochloric acid. The polonium can be back-extracted into nitric acid. Bagnall and Robertson (20) have studied the extraction of millicuric amounts of polonium into 10% TBP, using dekalin and kerosene as the organic diluent. A maximum partition coefficient of about 60 is obtained from aqueous solutions between 7 and $9M$ in hydrochloric acid. The extraction is strongly acid-dependent, and only about 0.2% is extracted from $1M$ hydrochloric acid. These workers recommend dekalin as the organic diluent for extraction of large amounts of polonium in place of an ether, since dibutyl ether is subject to radiation decomposition. The results were difficult to reproduce with millicuric amounts of polonium, probably because of decomposition induced by the alpha particles. Back-extraction from TBP into dilute hydrochloric acid should be readily accomplished. Because the density of TBP is very close to that of water, TBP is usually used mixed with a water-insoluble solvent with a density different from that of water to permit easy separation of the phases.

(5) **Tri-*n*-Benzylamine**

Polonium can be extracted from hydrochloric acid solution into a solution of tri-*n*-benzylamine (212). Using a 5% solution of the amine in chloroform as the solvent and an organic/aqueous ratio of 1.67:1, 99.4% of the polonium was extracted from $6N$ hydrochloric acid. This type of extraction depends on the formation of an anionic complex of polonium. In hydrochloric acid polonium exists as a chloro complex. The complex, being acidic, presumably forms an amine salt insoluble in aqueous solution but soluble in organic solvents because of the high molecular weight of the amine. If the complex formation is suppressed the amount extracted would be expected to decrease. Thus, from $5M$ hydrochloric acid–$1M$ nitric acid, only about 93% of the polonium is extracted, and from $6M$ nitric acid less than 10% is extracted. This extractant is not selective since other metal ions that form anionic chloro complexes will also extract.

d. Ion-Exchange Resins

The behavior of polonium with the cation-exchange resins Dowex-50, Amberlite IR-120 and IR-1, and Zeo-Karb 225 has been studied. Polonium has been separated from tellurium and radium-E (bismuth) on a column of Dowex-50 by washing the column first with 2N nitric acid to remove all the tellurium and bismuth and then with 2N hydrochloric acid to remove the polonium (187,255). In one case the reported polonium yield was 70% (187). The mixture was adsorbed on the resin from 0.1 to 0.3N hydrochloric acid solution. However, unusual behavior in the adsorption step was reported by Lindner (187). If the polonium was first carried on tellurium and if the tellurium was dissolved in concentrated nitric acid, evaporated to dryness, and redissolved in 0.2N hydrochloric acid, the polonium was adsorbed on the column. If a polonium solution in 2N hydrochloric acid was diluted to 0.2N hydrochloric acid, the polonium could not be made to adsorb on Dowex-50. When the same solution was boiled to near-dryness and repeatedly fumed with concentrated nitric acid to remove all the chloride, and the residue was dissolved in 0.2N hydrochloric acid, the polonium was adsorbed on the resin. The reason for this behavior is not clear. Bismuth and polonium have also been separated on Amberlite IR-1 under slightly different conditions (39).

The adsorption of millicurie amounts of polonium on Amberlite IR-120 and Zeo-Karb 225 has been studied in batch (equilibrium) experiments. Polonium is strongly adsorbed from nitric acid solution. Distribution coefficients (amount of polonium per gram of resin/amount of polonium per milliliter of solution), ranging from about 1000 at 0.1N nitric acid to about 3 at 5N nitric acid, were reported using Amberlite IR-120 (62). Another paper reports somewhat lower, but still appreciable, adsorption with both Amberlite IR-120 and Zeo-Karb 225 (22). Due to complex formation, oxalic acid lowers the distribution coefficient in the Amberlite IR-120-nitric acid system (62). At 1M nitric acid the distribution coefficient was 26.7 in the absence of oxalic acid and less than 1 at oxalic acid concentrations greater than 0.05M. No separations were attempted in this work, but the application of these results to column separations may be inferred. From hydrochloric acid solutions the adsorption is considerably less, as would be expected on the basis of the known chloro complex of polonium. With Amberlite IR-120, adsorption coefficients ranging from 150 at 0.05M hydrochloric acid to 1.3 at 0.2M hydrochloric acid and less than 1 at higher acidities were obtained (62).

Batch experiments have shown that polonium is very strongly adsorbed by the strong-base anion-exchange resins from hydrochloric acid solution (189). Distribution coefficients, using Dowex-1, varied from about 1.5×10^5 at 0.05N hydrochloric acid to about 2×10^4 in 12N hydrochloric acid, although the values at low acidities were not very reproducible. Distribution coefficients ranging from 123 at 0.8N nitric acid to 90 at 5N nitric acid were obtained with Dowex-1 in the nitrate form, and in some cases up to a week was required to reach equilibrium (189). Bagnall and his co-workers (22) report poor reproducibility in experiments with millicurie amounts of polonium due to the effect of the alpha radiation on the resin and solution; therefore, the separation of milligram amounts would probably be difficult. The separation of radium–D, –E, and –F on a column of Amberlite XE-98 (IR-411), a strong-base anion-exchange resin, has been reported (139). The mixture, in nitric or hydrochloric acid, is adsorbed on the top of a column of the resin in the chloride form. The lead-210 (RaD) is eluted first with 2 or $3N$ hydrochloric acid. Concentrated hydrochloric acid is then used to elute the bismuth-210 (RaE), and finally the polonium (RaF) is eluted with $7N$ nitric acid. The radiochemical purity of each fraction is said to be 99.9%. An anion-exchange separation of trace polonium, selenium, and tellurium also has been reported (269). The mixture, in concentrated hydrochloric acid, was adsorbed on a column of Dowex-1. Selenium was eluted with $6N$ hydrochloric acid, tellurium by $2N$ hydrochloric acid, and polonium by $1N$ nitric acid. The polonium could also be eluted with $2N$ perchloric acid, but if the polonium was allowed to stay on the column for 50 hours, only 10% of it could be eluted with perchloric acid, although $3N$ nitric acid was still effective. However, a recent report (187) states that while bismuth could be eluted from Dowex-1 (chloride form) with $3N$ sulfuric acid or $1N$ nitric acid, polonium could not be completely eluted by sulfuric acid in concentrations up to $3N$ or nitric acid in concentrations up to $16N$. In view of the differences in the ion-exchange behavior of polonium reported by various workers, the use of published procedures should be approached with caution until the reasons for these discrepancies are known.

Batch equilibrium experiments were also performed by Tompkins in a study of the state of polonium in aqueous solution (290). The distribution coefficients between Dowex-50 and Dowex-2 and various solutions were determined. Dowex-50 adsorbed polonium from perchloric acid solution up to the highest concentration used, $8M$. Distribution coefficients (K_D) of the order of 10^5 were obtained. Hydrogen peroxide lowered the K_D values by about a factor of two, and hydrochloric acid ($0.01M$) reduced the K_D to about 25. Distribution coefficients were also determined for lead-210 and bismuth-210. The separation factors (ratio of K_D's) indicated that column separations of these three nuclides could be best done with Dowex-2. This separation was later performed by others, as given above.

e. Volatility Separations

The volatility of polonium and many of its compounds has been mentioned earlier. Bagnall and Robertson (21) have used a distillation method for the preparation of milligram amounts of pure polonium metal from mixed polonium–lead sulfide. The mixed sulfide was decomposed by heating it at 275°C. under 5 μ pressure. The polonium was then distilled at the same pressure by heating at 450 to 500°C. The distillation efficiency is about 99.7% and the over-all recovery, including the sulfide precipitation, is 96 to 99%. The separation of bismuth and polonium by distillation also has been investigated. Polonium begins to sublime in air at 700°C., and is completely vaporized at 900°C., whereas very little bismuth sublimes below 1100°C. (75,218). The product of sublimation in air is probably the dioxide, but in vacuum the product is known to be the metal. Foils of various metals have been used to condense the polonium. Platinum and paladium are suitable for both vacuum and atmospheric-pressure distillations, but gold, because of its low melting point, is suitable only for vacuum distillations.

The volatility of a number of polonium compounds with organic complexing agents has been discussed in connection with the evaporation of solutions of these compounds for counting. One of these compounds, the diphenylcarbizide, has been used for the separation of radium–DEF mixtures by distillation. When a dilute nitric acid solution containing a radium–DEF mixture and diphenylcarbizide was distilled to one-half its original volume, essentially all of the radium–D and –E remained behind, although approximately one-half of the polonium was found in the distillate and one-third was found in the condenser (158). In general, volatility separations have not been successfully applied to the quantitative separation and analysis of trace polonium.

f. Chromatographic Separations

A simple and efficient separation of selenium, tellurium, polonium, and bismuth by paper chromatography has been demonstrated (57). The separation was done on Whatman 3 MM paper by ascending chromatography, using a developing solution of 60 g. of 49% hydrofluoric acid in 100 ml. of methylethyl ketone. All four elements were completely separated in 2 to 3 hours and the polonium was obtained in 100% yield. Hydrochloric acid in several organic solvents also was used, but polonium and tellurium could not be separated. The polonium–lead–bismuth separation has been studied (82), using Whatman No. 1 paper and two different solutions to develop the chromatogram: (1) 50 parts butanol, 15 parts pyridine, 5 parts concentrated hydrochloric acid, 10 parts acetic acid, and 10 parts water; and (2) 60 parts butanol, 12 parts concentrated hydrochloric acid, and 1 part concentrated sulfuric acid. Using the first solvent, and with lead and bismuth carriers present, polonium was completely separated from lead and bismuth. With the second solvent complete separation from bismuth was obtained only when bismuth and lead carriers were both present, whereas separation from lead was accomplished with or without carriers. Polonium has been separated from radium–D and –E on Whatman No. 1 paper, using butanol–hydrochloric acid mixtures as solvents (64,186).

The separation of lead-210 and bismuth-210 from polonium-210, gold, and mercury on a column of Whatman No. 1 cellulose paper has been reported (76). In this case the purpose was to obtain carrier-free lead-210 and bismuth-210 from spent radon seeds in the form of gold needles. In addition to the radium–DEF and gold, the sample also contained mercury from diffusion pumps used to pump radon. The sample was added to the column in $3N$ hydrochloric acid and eluted with butanol saturated with $3N$ hydrochloric acid. Gold, mercury, and polonium eluted first, followed by the lead and bismuth. Although polonium was not separated, it is likely that the separation could be modified to obtain the polonium in a pure state. The same authors report a faster method for the same separation if all the elements are present only in trace concentration (300). The technique of ascending paper chromatography is used with a developing solution of 20% butylphosphate in acetone.

Electromigration in paper also has been used for the separation of radium–DEF mixtures (23). Radium–D was separated from radium–E and polonium by this technique in oxalic, citric, tartaric, lactic, and acetic acids. Both bismuth and polonium tend to be adsorbed by the paper and trail. Polonium could be separated from radium–D and –E in EDTA solution.

The paper chromatographic separation of polonium from lead-210, bismuth-210, tellurium, and selenium in nitrate solution also has been studied (61). In this case the developing solution was 50% butanol–50% propanol, previously equilibrated with a $7M$ lithium nitrate–$2M$ nitric acid solution. Whatman No. 1 paper was used and the chromatogram developed for 18 hours at room temperature. Polonium traveled the greatest distance along the paper, so its nitrate complex is apparently the strongest of this mixture of ions (185).

D. DETECTION AND IDENTIFICATION

The qualitative detection of trace amounts of a radioactive element consists of establishing the chemical and radioactive decay properties. The decay properties useful in qualitative detection are the types, energies, and relative amounts of the emitted radiations and the half lives. The identification and rate of growth of known daughters also can be used in many cases, and this is often useful or necessary in characterizing new isotopes produced by nuclear reactions. The use of chemical separations combined with radioactivity measurements will be illustrated by the examples given below. The identification of macroscopic amounts of polonium will be treated separately.

1. Coprecipitation

Polonium coprecipitates with bismuth, copper, or lead sulfide from acid solution (298). No other naturally occurring alpha emitters coprecipitate, and polonium-210 can be readily distinguished from the naturally radioactive bismuth or lead isotopes by alpha counting. Lead has no alpha-active isotopes and the natural alpha-emitting bismuth isotopes all have half lives of 60 minutes or less. Thus, an alpha activity that carries on bismuth sulfide from acid solution and decays with the half life of polonium-210 (138.4 days) constitutes an identification of polonium in a natural source.

Coprecipitation with elementary tellurium under the conditions described earlier, followed by radioactivity counting to establish the type of radiation and half life, also serves to identify polonium.

Other carriers also can be used, but since they are not as specific as bismuth sulfide and tellurium the subsequent radioactivity measurements assume greater importance. However, if the known radioactive decay properties can be established for the unknown activity the identification may be considered as complete.

2. Electrodeposition

The spontaneous deposition of polonium on silver from dilute hydrochloric acid, followed by counting to establish the radioactive properties, is a good qualitative test for polonium. Traces of bismuth may deposit, but bismuth isotopes can be distinguished from polonium as indicated above.

3. Alpha-Particle Counting

The use of alpha counting for the identification of polonium has been mentioned. Since alpha particles are readily counted in the presence of beta particles, alpha-emitting polonium isotopes are readily identified in the presence of beta emitters. Also, polonium-210, the most important polonium isotope, has a half life sufficiently short (138.4 days) to be useful in its characterization. The measurement of alpha-particle energies is also useful. This measurement requires a very thin deposit of the activity ("weightless source") to avoid significant loss of the alpha-particle energy in the source. Weightless sources are best prepared by deposition (spontaneous or electrolytic) on a metal foil, by volatilization to a metal foil, or by evaporation of the active solution after purification by solvent extraction, ion-exchange separation, or paper-chromatographic separation. To obtain an alpha-particle energy spectrum the source is counted in a gridded ionization chamber or solid state alpha-particle detector, and the pulses from these detectors are linearly amplified and recorded by a multichannel pulse-height analyzer. These instruments are commercially available.

4. Gamma- and X-Ray Counting

When gamma-rays are emitted during the decay process (Table XI) the identification of the gamma-ray energies constitutes a good qualitative test. The absence of other nuclides that emit gamma-rays of similar energy must be known, or preliminary separations should be performed. Polonium-210 emits an 0.8-m.e.v. gamma-ray in $1.2 \times 10^{-3}\%$ of its disintegrations. The specificity of this gamma-ray for polonium detection can be seen from the spectra of the long-lived members of the uranium–radium and uranium–actinium series given in Figs. 7, 8, and 14 to 17

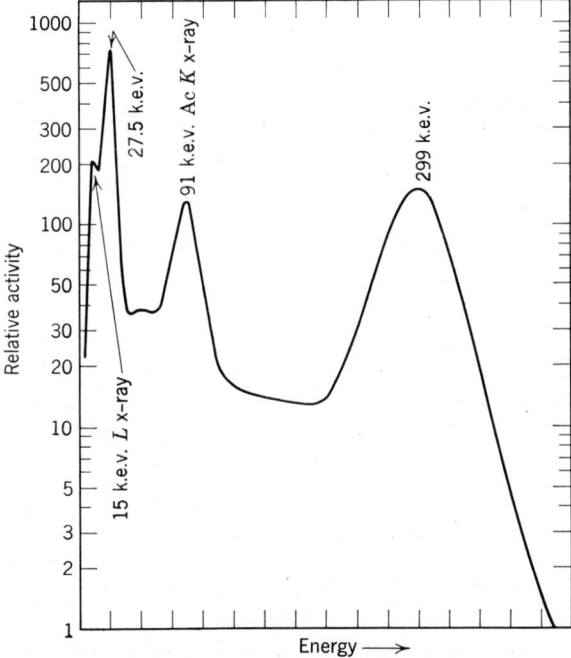

Fig. 14. Gamma-ray spectrum of protactinium-231.

Because of the rarity of this radiation, gamma-ray detection is sensitive to about one microcurie, whereas alpha counting is sensitive to several micro-microcuries. However, gamma-ray spectrometry is relatively simple with modern counting equipment. This counting is done with a sodium

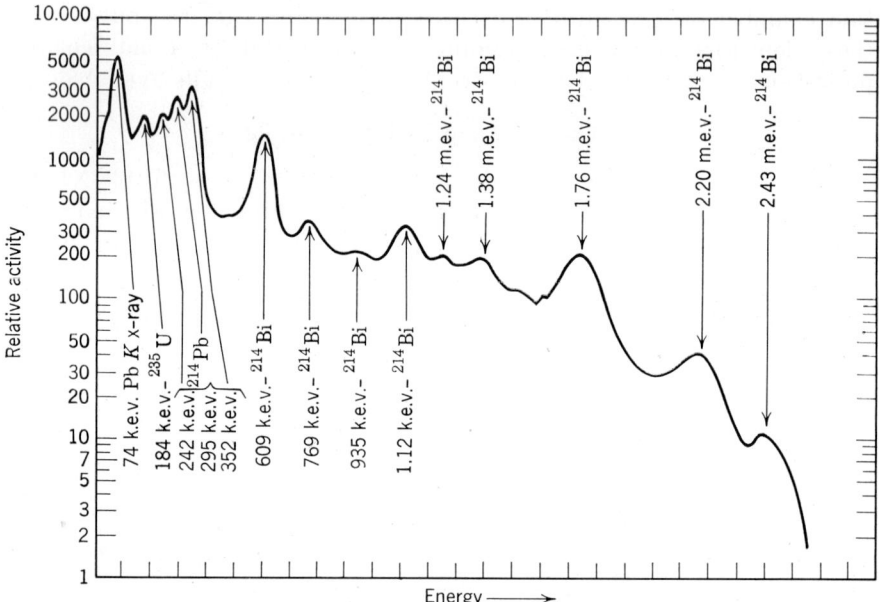

Fig. 15. Gamma-ray spectrum of pitchblende ore.

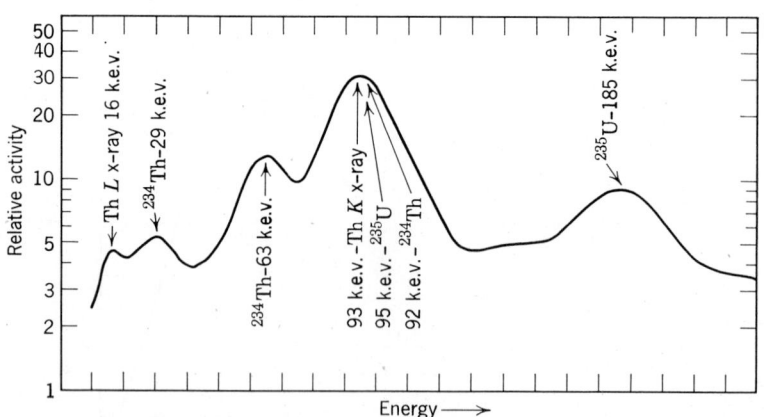

Fig. 16. Gamma-ray spectrum of uranium (natural isotopic abundance) containing equilibrium amounts of thorium-234 and protactinium-234.

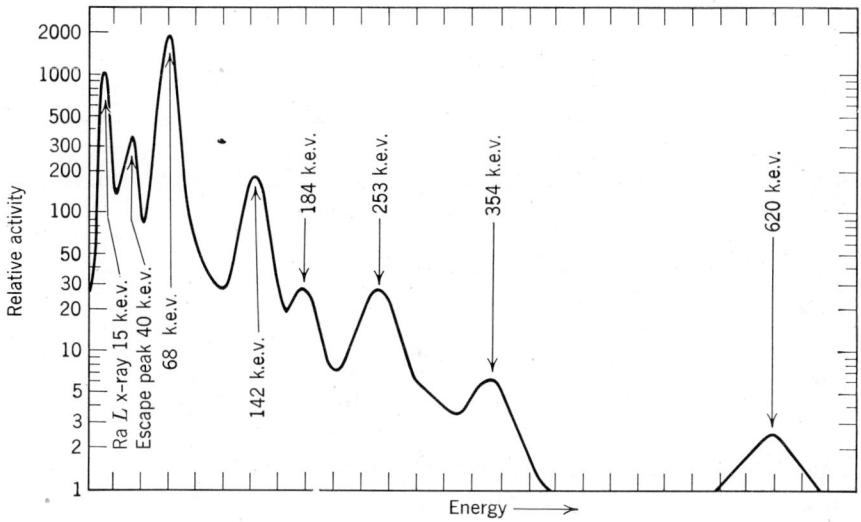

Fig. 17. Gamma-ray spectrum of thorium-230.

iodide crystal as the detector and a multichannel pulse-height analyzer to record the gamma-ray energies (56). Polonium nuclides that decay by electron-capture are best detected by characterizing the x-ray accompanying this type of decay with the same equipment used for gamma-ray counting.

5. Detection of Milligram Amounts of Polonium

Milligram amounts are encountered only in the production of polonium or in studies of polonium chemistry. In such cases other alpha emitters are not involved and the problem reduces to the detection of the alpha or gamma activity in, for example, irradiated bismuth, preparations of polonium compounds, and residues and wastes. Although any method suitable for trace polonium is applicable, gamma or alpha measurements with simple instruments, such as portable survey instruments, are ordinarily adequate. Emission spectroscopy is also applicable, and the molecular and atomic spectra of polonium have been discussed in Section IV-B-3.

E. DETERMINATION

1. Preparation of Sample for Measurement

The determination of polonium by radioactivity measurement consists of separating the polonium from interfering active and inactive materials, followed by a quantitative measurement of the activity. The separation may be done by one or more of the methods given earlier. The final pure polonium solution, or an aliquot, may be evaporated on a metal foil and counted for alpha activity, or the solution may be counted directly for gamma or x-rays if these measurements can be used. For alpha counting the polonium is usually plated spontaneously on a suitable metal foil, generally silver, and methods for this procedure will be discussed. The plating is best done from dilute hydrochloric acid solution, somewhat above $0.1N$ and at an elevated temperature (about 90°C.). The solution should be stirred during the plating period of about 1.5 hours.

A simple method of inserting the metal foil in the solution is to drill a small hole in the foil and suspend it in the solution with a glass hook (207,297). Both sides of the foil must be counted. To avoid this, some workers (135) have coated one side of the foil with a water-insoluble insulator such as vacuum wax. An improved method is to use a cylindrical container with the metal foil as the bottom. A number of such plating cells have been designed. One cell consists of a plastic cylinder (a polyfluorinated plastic will withstand the elevated plating temperature) threaded at the bottom. The foil is placed against the bottom end and secured in place by a threaded cap (94). Another cell is made from an infant nursing bottle. The bottom is cut off, and the foil kept in place with the original plastic screw cap supplied with the bottle (28). Other methods for securing the foil liquid-tight to the bottom of the cylinder have been described (297).

2. Counting Techniques

Alpha counting of plated polonium samples may be done in conventional alpha counters without interference from beta particles. A number of such counters are available commercially. For accurate alpha counting the samples must be essentially weightless to eliminate errors due to self-absorption of the alpha particles in the sample. With such samples the disintegration rate is obtained from the counting rate by applying corrections for the counter geometry and back-scattering of alpha particles from the metal foil. Thick samples may be counted, but experimentally determined corrections for self-absorption must be made to obtain the disintegration rates, and the counting accuracy suffers.

Solutions of alpha emitters also can be counted by liquid scintillation techniques. The principal problem is that of obtaining the active material in a solution that is soluble in the organic scintillating solution. Aqueous solutions can be dissolved in some of the solvents used for liquid scintillation counting (dioxane, toluene–ethyl alcohol mixtures, and others). The volume of aqueous solution used must be small compared with the volume of the scintillating solution to maintain a high counting efficiency. Gamma and x-ray counting may be used for the determination of some polonium nuclides, as discussed under qualitative detection. The same advantages and limitations apply. Quantitative measurements by these types of counting are discussed in Reference 56.

A detector for the direct measurement of the alpha activity due to polonium in solution has been described (85). The detector consists of an air-ionization chamber with a mica window through which the alpha particles can enter the sensitive volume of the ion chamber. The probe is immersed directly in the solution to be analyzed. Polonium concentrations down to about 10^7 disintegrations per minute per milliliter can be measured. This type of measurement is, of course, less sensitive than evaporation of the solution prior to alpha counting, but is useful for some applications.

3. Miscellaneous Methods

A number of other methods have been used for polonium-210 determinations. These include calorimetric and alpha- and gamma-ionization measurements. These and other methods are described in Reference 218. They are not commonly used, but find application under specialized conditions.

F. DETERMINATION IN SPECIFIC MATERIALS

1. Biological Materials

Because of its high specific alpha activity, polonium is quite toxic when deposited in the body. As a result, considerable research has been conducted on the behavior of polonium in experimental animals, and techniques have been developed for the determination of polonium in animal tissues and in human excreta. A simple method for urine consists of acidifying the samples to $0.5N$ in hydrochloric acid and placing a clean silver disk in the solution (297). The solution is stirred for about one hour at an elevated temperature. The disk is then removed, washed with water, dried, and counted. At Mound Laboratory copper was substituted for silver, and the deposition was carried out at room temperature (207). Other workers reported variable results with copper, but found that deposition on nickel for one hour gave polonium recoveries of $98 \pm 10\%$ (70).

Urine volumes much greater than 100 ml. cannot be analyzed conveniently in this way, and some workers have found that the plating efficiency and speed decrease with increasing volume. Therefore, for low polonium concentrations requiring sample volumes up to 1 liter, a concentration step is desirable. Such procedures use nitric, sulfuric, and perchloric acids to oxidize the organic matter (135,262). Typically, the sample is evaporated in the presence of nitric acid and sulfuric acid to about 100 ml. Additional nitric acid is added and the mixture is heated until charred. After additional heating with fuming nitric acid the last trace of organic matter is oxidized by fuming with perchloric acid. The usual precautions required in the use of perchloric acid must be observed. After the oxidation is complete the solution is diluted with water and 10 mg. of tellurium carrier is added as potassium tellurate. The carrier is reduced to the metal with sodium hypophosphite at near-boiling temperatures. The precipitated tellurium, containing the coprecipitated polonium, is filtered or centrifuged and is dissolved in a saturated solution of bromine in concentrated hydrochloric acid. The tellurium is precipitated from this solution with hydrazine and is separated. Under these conditions the polonium is not carried. The solution is adjusted to about $0.5N$ in hydrochloric acid and the polonium is deposited on silver from this solution. Blood and solid biological samples are treated similarly. Most of the organic matter is oxidized with fuming nitric acid and sulfuric acid, the last traces are then oxidized with perchloric acid, and the analysis is completed as for urine. Some authors have used perchloric–nitric acid mixtures for the entire group of oxidations (209), but extreme care is necessary because of the well-known explosive nature of perchloric acid–organic mixtures. It is desirable to oxidize most of the organic matter before using perchloric acid. After oxidation the polonium may be plated directly, following proper adjustment of the acidity, or the polonium may be carried on tellurium as above. Other procedures that have been used for the dissolution of biological samples are oxidation with perchloric acid and hydrogen peroxide (219), repeated treatments with nitric acid alone or with nitric and sulfuric acids if the sample contains much fat (271,276), and oxidation with nitric acid and hydrogen peroxide (50). A comparative study of several of these procedures is given by Black (32).

Ashing at elevated temperatures in a furnace cannot be used since polonium will be volatilized at the temperatures required to oxidize the sample.

2. Irradiated Bismuth and Other Heavy-Metal Targets

The polonium content of large pieces of irradiated bismuth has been determined by measuring the gamma radiation with an electroscope (218). Gamma-rays are emitted by polonium and by neutron-activation products of certain impurities in the bismuth and its jacket. However, for any given lot of bismuth and jacket material the gamma radiation due to the impurities is a constant fraction of the total gamma radiation. Thus, for a given lot of bismuth the polonium content can be estimated from the total gamma radiation. The amount of heat generated by the decay of polonium also has been used (218). This calorimetric measurement must be made about 30 days after irradiation to allow the bismuth-210 to decay. An alpha-particle ionization chamber also has been used to determine the polonium content of unjacketed irradiated bismuth (218). Another method used at Mound Laboratory consists simply of dissolving the bismuth, evaporating a suitable aliquot to dryness, and alpha counting. For lower-activity samples the solvent-extraction method of Moore, using diisopropyl carbinol (214), or the paper-chromatographic method of Crouthamel (57) can be used to separate the polonium prior to counting. Both methods have been described earlier. Karraker and Templeton separated polonium from irradiated lead or bismuth by coprecipitation with tellurium metal in the presence of thallium and lead holdback carriers (150). Additional purification was accomplished by extraction into tributylphosphate–dibutylether solution. These separations may be used to obtain a pure polonium fraction for counting.

Polonium has been separated from tungsten targets by dissolving the tungsten in a hydrofluoric–nitric acid mixture and electrolyzing the polonium on a platinum cathode in the presence of 1% tartaric acid (225). Polonium has been separated from heavy-metal targets by an anion-exchange method (188). The sample, in about 5 ml. of 7 to $8N$ hydrochloric acid, is passed through a column of Dowex-1 resin in the hydrogen form. The column is then washed in succession with hydrochloric acid solutions of these concentrations: $12N$, $6N$, $0.1N$, $12N$, and $0.1N$. The polonium is then eluted with concentrated nitric acid. This solution is fumed with perchloric acid and diluted with 5% citrate–hydrochloric acid solution at pH 1, and the polonium is plated on a silver disk for about two hours. The disk is then washed with water and acetone, dried, and counted. The reported yield is about 60%.

3. Ores, Water, and Air

Natural concentrations of radium–D, –E, and –F have been determined in air by analyzing rainwater and air-filter samples (94, 159). The papers were oxidized in a nitric–sulfuric acid mixture and the nitric acid was removed by fuming. After dilution of the acid solution with water, lead sulfate was precipitated to carry radium–D, the precipitate was dissolved, and the lead was extracted with dithizone. Tellurium was precipitated from the lead sulfate filtrate to carry radium–E and polonium. The tellurium was dissolved, the polonium was removed by plating on silver, and the bismuth was removed by dithiazone extraction. The same chemical separations were used for rainwater.

Higher concentrations of polonium-210 in air have been measured by bubbling the air through hydrochloric acid to absorb the polonium, followed by coprecipitation with tellurium metal by reduction of telluric acid with stannous chloride, and alpha counting of the tellurium precipitate (38). Polonium in waste water has been determined by coprecipitation of the polonium with calcium tannate at pH 8, oxidation of the tannate with aqua regia, and plating of the polonium on a silver foil from $1N$ hydrochloric acid (275). Other carriers should serve equally well for the initial concentration step.

Polonium in uranium ores has been determined as follows. The sample is fused with sodium carbonate, dissolved in 1 to $2N$ hydrochloric acid, and the solution is filtered. The polonium is plated from the solution on a nickel disk at 75°C. for 200 minutes with stirring (2). To standardize the method, ores of known uranium content and known to be in radioactive equilibrium were analyzed by the same procedure. Polonium-210 in ores also has been determined by coprecipitation with bismuth sulfide (259,260). This method is given as a recommended procedure.

4. Analysis of Curie Amounts of Polonium Compounds

In the preparative chemistry of polonium on the milligram and submilligram scale one is faced with the problem of analyzing an extremely radioactive compound for small amounts of stable elements and ions in the presence of larger amounts of these elements and ions in the starting materials. The polonium analysis consists of obtaining a solution of the compound from which a known fraction can be evaporated and alpha-counted. Separations from other alpha emitters is, of course, not necessary, and the specific activity of the preparation is sufficiently great to make chemical separation of the polonium unnecessary. In the analysis of the elements of radicals combined with polonium, interference is often encountered because of chemical reactions induced by the polonium alpha particles. Prior removal of the polonium eliminates this problem. Methods of handling the analyses of prepared polonium compounds can be illustrated by examples from the work of Bagnall and his colleagues (10–22). One method used to analyze polonium chromate consisted of determining chromate by titrating the iodine liberated by its reaction with iodide ion (11). Since iodine was also liberated by the action of the polonium alpha particles on iodide ion, the polonium chromate was decomposed with $0.1N$ potassium hydroxide and the insoluble polonium hydroxide was filtered. The chromate was determined in the filtrate, and the polonium was determined by dissolving the hydroxide in acid and alpha counting an aliquot of the acid solution. Since only milligram or smaller amounts of the polonium compounds were prepared, it was necessary to remove the excess reactants carefully and completely. In the case of precipitation of insoluble compounds, the excess of precipitant was removed by washing.

G. RECOMMENDED PROCEDURES

1. Deposition on Silver Foil (72)

Obtain the sample in about 200 ml. of $0.5N$ hydrochloric acid. Gold, mercury, platinum, tellurium, and iron(III) must be absent. If trivalent iron is present, add 200 mg. of ascorbic acid. (The other elements can be removed by reduction with hydrazine in 20% hydrochloric acid.) Suspend a silver foil, 0.003 inch thick by 1.5 inch in diameter, in the solution by means of a glass rod hooked through a 1-mm. hole in the foil. Alternately, a plating cell designed to expose only one side of the silver disk can be used (see Section V-E-1). Insert the beaker or cell in a water bath at 97°C. Stir the solution for 1.5 hours, and wash down the sides of the beaker with distilled water after one hour. Remove the silver foil, wash with a gentle stream of water, allow to dry, and count in an alpha counter. If both sides of the foil were exposed to the solution, each side must be counted. The recovery is greater than 99%.

2. Determination of Polonium-210 in Uranium Ores (259,260)

Fuse the sample, not to exceed one gram, at almost red heat with 4 to 5 g. of sodium peroxide in a nickel crucible. Allow the melt to disintegrate in water and neutralize with hydrochloric acid. Dilute the solution to 100 ml. while adjusting the acidity to $1.2N$ hydrochloric acid. The sample should be completely in solution. If all of the silica is not taken into solution, a smaller sample should be used. Add 6 mg. of bismuth carrier (1 ml. of $0.028M$ bismuth chloride in 5% hydrochloric acid), heat the solution to 45°C., and saturate with hydrogen sulfide. Filter the precipitate of bismuth sulfide, which contains the polonium-210, on a Millipore or Whatman No. 42 filter (in a small Buchner or Hirsch funnel). Wash the precipitate with water to collect it in a predetermined area on the paper (0.5 to 1.5 inches in diameter). Continue filtering for several minutes to remove as much water as possible. Remove the paper, glue or otherwise attach the paper to a counting dish, and dry with moderate heat. Count the paper in an alpha scintillation counter after about one day. This delay is necessary to allow for the decay of short-lived bismuth, polonium, and lead isotopes.

3. Determination of Polonium in Irradiated Bismuth by Solvent Extraction (214)

Dissolve the bismuth target in 6 to $8M$ nitric acid and dilute to a suitable volume in a volumetric flask. Remove a suitable aliquot to a separatory funnel and add hydrochloric acid until a concentration of $6M$ is obtained. The nitric acid concentration should be maintained at less than $1M$. Add a double volume of diisopropyl or diisobutyl carbinol, previously shaken for 5 minutes with an equal portion of $6M$ hydrochloric acid, to the funnel and mix the phases for 5 minutes. Allow the layers to separate, transfer the aqueous phase to a second funnel, and repeat the extraction. Combine the organic phases and centrifuge for 2 minutes. Evaporate a suitable aliquot to dryness on a platinum counting plate. Do not flame. Count the dried counting plate in an alpha counter.

4. Separation of Pure Polonium from Radium–DEF Mixtures (35)

Obtain the radium–DEF in a solution of 0.3 to $0.5N$ hydrochloric acid. In a separatory funnel, contact this solution for several minutes with an equal volume of a solution of dithizone in chloroform (0.1 g./liter). Discard the aqueous phase. Wash the chloroform solution with $0.5N$ hydrochloric acid to remove traces of radium–E that extract with the polonium. Discard the wash solution. Back-extract the polonium from the dithizone–chloroform solution into $4N$ hydrochloric acid.

5. Paper-Chromatographic Separation of Polonium, Bismuth, Tellurium, and Selenium (57)

Use Whatman 3 MM strips 2.0 cm. wide and sufficiently long to permit the development of a chromatogram 25 cm. in length by an ascending technique. Add the sample mixture (traces of selenium and polonium, and up to about 0.3 mg. of bismuth and tellurium) near one end of the paper and allow to air-dry for one hour. Insert the paper into the developing solution (contained in a polyethylene cylinder) of 60 g. of 49% hydrofluoric acid per 100 ml. of methylethyl ketone, and develop the chromatogram for approximately 2 to 3 hours. The position of each element can be determined by scanning the paper strip with a thin-window

Geiger counter. The order of the elements is bismuth, polonium, tellurium, and selenium, with the bismuth closest to the starting point.

6. Determination of Polonium in Urine and Tissue (121)

Place solid samples (preferably 500 g. or less) in a glass vessel (beaker or Erlenmeyer flask) of such capacity that the sample occupies 10 to 20% of the volume. Urine samples may constitute 30% of the vessel volume. A smaller vessel may be used and the sample added as the evaporation progresses. Add 2 ml. of concentrated nitric acid for each gram of solid material up to 50 g., and 1 ml. for each gram over 50. Allow to stand several hours or overnight at room temperature. For urine samples add about 0.5 ml. of nitric acid for each milliliter of sample. Standing overnight is not necessary in this case.

Place vessel on hot plate and gradually heat to boiling. Use glass beads if bumping occurs and a silicone antifoaming agent as needed to prevent excessive foaming. Continue to heat at a smooth boil until the mixture is a clear yellow color. Evaporate to a small volume and add 25 ml. of water. If the solution is not clear the presence of fat is indicated, and the boiling with nitric acid should be continued until a clear solution is obtained.

Neutralize the solution with sodium hydroxide, then make it $0.5N$ in hydrochloric acid. Add a 100-mg. tablet of ascorbic acid if the presence of iron is suspected, and complete the analysis according to the procedure given in section V-G-1.

A faster and more efficient digestion may be performed by using a combination of perchloric and nitric acids. See References 28, 32, 276, and 297 for details. However, this method on occasion has led to explosive mixtures, and extreme care is required when oxidizing organic matter with perchloric acid.

VI. PROTACTINIUM (ATOMIC NUMBER 91)

A. INTRODUCTION

Protactinium is the third member of the actinide series, occupying the position between thorium and uranium. In its highest valence state, (V). it resembles the Group Vb elements rather than exhibiting characteristics similar to those of the rare earths. The first protactinium isotope discovered was UX_2 (mass number 234), an early member of the uranium–radium decay series (Fig. 3). This isotope was discovered by Fajans and Gohring in 1913. Because of its short half life, 1.2 minutes, little work could be done on its chemical properties. The long-lived isotope, protactinium-231, was discovered in 1917 by Hahn and Meitner and, independently, by Soddy and Cranston. This isotope occurs in the uranium–actinium decay series and is the parent of actinium-227 (Fig. 4). A. V. Grosse was the first to isolate weighable amounts of protactinium-231 and study its chemistry on a milligram scale. For an account of the early history of protactinium see References 110 and 298. One gram of naturally occurring uranium, in radioactive equilibrium with its decay products, contains 3.1×10^{-7} g. of protactinium-231, about the same as the radium-226 concentration.

Knowledge of the chemistry of protactinium has been slow to accumulate, primarily because its nonreproducible and complicated behavior in solution makes it difficult to obtain in the pure state and because, until recently, there has been no industrial need for such knowledge. However, the chemistry of protactinium has recently become important since protactinium-233 is an intermediate in the production of uranium-233 from thorium-232 in breeder reactors:

$$^{232}\text{Th}(n,\gamma) \rightarrow {}^{233}\text{Th} \xrightarrow[\beta]{23.3\text{m}} {}^{233}\text{Pa} \xrightarrow[\beta]{27.4\text{ d}} {}^{233}\text{U}$$

Gram quantities of protactinium have been separated from uranium refinery residues in recent years. These residues are available in quantity because large amounts of uranium are now being refined. The largest of such efforts was carried out by British workers, who obtained 125 g. of 99.9% pure protactinium-231 (51,53,143,222). The methods used for the isolation of smaller amounts of protactinium by American workers at Argonne National Laboratory (68,176,287) and at Mound Laboratory (162,265), by French workers (109), and by Russian workers (273) also have been described. Protactinium-231 can also be produced by neutron bombardment of thorium-230 (ionium) in a nuclear reactor. The chemical separation of protactinium from thorium is considerably easier than its separation from natural sources, but the ionium must first be separated from uranium ores and it is doubtful if this process will become a major source of protactinium-231. Because of its availability one can expect considerable clarification in the chemistry of protactinium in the future. This information will be of considerable interest not only in the work on breeder reactors, but also because of the position of protactinium in the periodic system and its erratic behavior. The unreproducible and frustrating behavior sometimes observed in protactinium solutions is probably due to its insolubility under numerous conditions, and its resulting tendency to hydrolyze and form colloids. It is hoped the studies now being done with gram quantities of protactinium will clarify the chemistry of this element.

B. PROPERTIES

1. Physical Properties

a. Nuclear Properties

The nuclear properties of protactinium are summarized in Table XIV. Additional information on decay energies and genetic relationships may be found in References 56, 116, and 283. The isotope of mass 231 is the only one with a sufficiently long half life to be obtainable in weighable amounts. The isotope of mass 233 is useful in tracer studies and is readily made by irradiating natural thorium in a nuclear reactor. The isotopes of masses 226, 227, 228, and 230 are parents of radioactive decay chains that decay into the natural radioactive series.

b. Other Physical Properties

Protactinium metal is grey in color, malleable, and approximately as hard as uranium. In air it becomes covered with a thin layer of the oxide. It has a body-centered, tetragonal crystal structure and a calculated density of 15.37 g./cm.^{3} (109). Its estimated melting point is about 1600°C. (58) and its estimated vapor pressure at 1927°C. is 5.1×10^{-5} atm. (220).

The electronic structure of the element has not been definitely established. Katz and Seaborg (151) suggest that the outer electronic structure of the gaseous atom is either $5f^26d^17s^2$ or $5f^16d^27s^2$, in keeping with an actinide structure. Haïssinsky and Bouissières (109) claim that the similarity of protactinium to tantalum and niobium and the ionic radii suggest that there are three $6d$ electrons and no $5f$ electrons. However, Katz and Seaborg point out that the energy differences between the $6d$ and $5f$ orbitals are very small, within chemical binding energies, so electrons from both orbitals can be involved in chemical reactions. Protactinium is generally considered to be an actinide element, although the number of its $5f$ electrons is uncertain and may be different in different compounds and complexes.

2. Electrochemical Properties

a. Spontaneous Deposition

Trace quantities of protactinium will deposit spontaneously on lead, iron, or zinc from dilute sulfuric acid (109). Deposition on lead was used to separate UX$_2$ (protactinium-234) from UX$_1$ (thorium-234) in the discovery of protactinium-234 (71). Higher deposition efficiencies are obtained from

558   A. SYSTEMATIC ANALYTICAL CHEMISTRY OF ELEMENTS

TABLE XIV
Nuclear Properties of Protactinium Isotopes (56,116,283)[a]

Mass number	Type of decay	Half life	Energy of principal radiations (m.e.v.)	Source
225(?)	α	2.0 s		Th–spall
226	α	1.8 m	α: 6.81	Th–spall
227	α, 85% EC, 15%	38.3 m		Th–spall
228	EC, 98% α, 2%	22 h	α: 6.09, 5.85 γ: with EC, 0.185, 0.224, 0.329, 0.445, 0.968, 1.572, others	Th–spall; dtr ^{228}U
229	EC, 99+% α, 0.25%	1.5 d	α: 5.69	^{230}Th$(d,3n)$; dtr ^{229}U
230	EC, 85% β$^-$, 15% β$^+$,(?) 0.03%	17.7 d	β$^-$: 0.41 γ: with EC, 0.255, 0.445, 0.953, 1.013, others; with β$^-$, 0.0517	Th, Pa + d or α; ^{233}U + d
231	α	3.43 × 10^4 y	α: 5.05, 5.02, 5.00, 4.94, others γ: 0.0275, 0.057, 0.0967, 0.101, 0.299, others	Natural source: dtr ^{235}U in U–Ac series; ^{230}Th$(n, γ)$; ^{232}Th $(n,2n)$
232	β$^-$	1.31 d	β$^-$: 0.26, 0.37, others γ: 0.047, 0.109, 0.389, 0.455, others	Th + d or α; Pa + d or n
233	β$^-$	27.0 d	β$^-$: 0.15, 0.26, 0.57 γ: 0.301, 0.313, 0.341, others	Th + n, d, or α; dtr^{233} Th
234m (UX$_2$)	β$^-$, 99+% IT, 0.63%	1.18 m	β$^-$: 0.058, 1.50, 2.31 γ: 0.76, 1.00, 1.84, 1.49, 1.7, others	Natural source: dtr ^{234}Th in U–Ra series
234 (UZ)	β$^-$	6.66 h	β$^-$: 0.16, 0.32, 0.53, 1.13 1.4 γ: 0.043, 0.099, 0.153, 0.255, 0.368, others	Natural source: IT of UX$_2$ in U–Ra series
235	β$^-$	23.7 m	β$^-$: 14.	U + p or d; dtr ^{235}Th
237	β$^-$	11 m		U + d

[a] Key to abbreviations and symbols. α—alpha particle; β$^-$—negative beta particle; β$^+$—positive beta particle (positron); γ—gamma-ray; EC—orbital electron capture; IT—isomeric transition; n—neutron; p—proton; d—deuteron; s—second; m—minute; h—hour; d—day; y—year; dtr—daughter; spall—spallation reaction; and (?)—identification uncertain.

hydrofluoric acid solution. Camarcat and his co-workers (47) found that from fluoride solutions appreciable deposition occurred on elements more electropositive than tantalum and on lead, which has a sufficiently negative reduction potential in fluoride solution. At pH 2.1, using a 20-minute deposition period, the percentages deposited varied from about 90% for beryllium to 27% for zinc. The poor reproducibility found in some cases was attributed primarily to attack of the metal by the hydrofluoric acid solution and to the presence of zirconium used as a carrier in the protactinium purification. From $1N$ hydrochloric acid solution the deposition was quite low.

b. Electrodeposition

Protactinium can be electrolyzed from dilute hydrofluoric acid solution on platinum, silver, copper, nickel, or lead cathodes. The metal deposit is adherent, and up to 0.1 mg./cm.^{2} can be plated. Trace protactinium has been deposited on platinum cathodes by electrolyzing slightly acid hydrochloric acid–ammonium chloride solutions (210). Approximately 95% was electroplated in 15 minutes.

c. Electrode Potential

The reversible electrode potential has not been measured. From critical deposition potentials the normal reduction potential of the Pa–PaF$_7^{-2}$ couple is estimated to be about -1.03 v. (60,74).

d. Polarographic Behavior

A study has been made of the polarographic behavior of 99% protactinium in oxalate solution in attempts to establish valence states less than +5 (67). Two waves were obtained; each was irreversible and involved a one-electron change. The first half-wave potential was -1.74 v. vs. the saturated calomel electrode. This wave was attributed to the reduction of Pa(V) to Pa(IV) and was hydrogen-ion independent. The second wave, attributed to the reduction of Pa(IV) to Pa(III), varied from -1.92 v. at pH 2.5 to -2.55 v. at pH 11.6.

3. Optical Properties

a. Emission Spectrum

The emission spectrum of protactinium has been examined by copper-spark excitation (79). The most sensitive and distinctive lines are at 3957.8, 3054.6, 3053.5, and 2743.9 A. The limit of sensitivity by this method of excitation is about 0.5 γ of protactinium-231.

b. Absorption Spectrum

The absorption maxima of protactinium solutions found by various investigators are summarized in Table XV. The absorption peak in the neighborhood of 260 mµ in hydrochloric acid solution is apparently due to hydrolyzed protactinium, whereas protactinium(V) in true solution does not absorb strongly down to about 210 mµ. Several investigators (42, 45,

TABLE XV
Absorption Maxima of Protactinium in Aqueous Solutions

Experimental conditions[a]	Absorption maxima (mµ)	Reference
$HClO_4$, low acidity	208	(294)
$HClO_4$, intermediate acidity	235	(294)
$HClO_4$, high acidity	255–260	(294)
H_2SO_4, low acid	215, 280–300	(294)
H_2SO_4, 6M	255, 298	(294)
H_2SO_4, 15M–16M	323	(294)
HCl, 0.5M–2M	213, 258	(294)
HCl, 1M H$^+$, low Cl$^-$	228, 236, 275	(294)
HCl, 2M H$^+$, high Cl$^-$	236, 267	(294)
HCl, 5M–8M	228, 269, 247	(294)
HCl, >4N	<210	(48)
HCl, Pa hydrolyzed	260	(48)
HCl, >5M	208	(302)
HCl, <5M	260	(302)
Pa(IV), 1M HCl	224, 255, 276	(8)
Pa(V), 1M HCl	<220	(8)
HCl, >6M	213	(42)
HCl, 1.7M	262, 300	(42)
Pa(IV), HCl (1.7M H$^+$, 7M Cl$^-$)	229, 260, 282	(42)
Pyrogallol in oxalate solution	355	(48)
Gallic acid in oxalate solution	343	(48)
Catechol in oxalate solution	323	(48)

[a] All solutions contained pentavalent protactinium unless otherwise indicated.

302) have noted that freshly prepared protactinium solution, in a solvent-extractable state, does not show the 260-mµ peak, but this peak can appear on standing or dilution and is accompanied by the formation of colloidal or hydrolyzed protactinium.

Pyrogallol forms a colored complex with protactinium, and this reaction has been suggested as the basis for a spectrophotometric determination (48).

The yellow color of the protactinium–thenoyltrifluoroacetone (TTA) complex in benzene has been used as the basis for a spectrophotometric determination at 430 to 440 mμ (210a). The system obeys Beer's law in the concentration range from 1.3 to 8.7 γ of protactinium-231 per milliliter. The protactinium was extracted into a solution of TTA in benzene from 6M hydrochloric acid. At this acidity few other metal ions extract, although Fe(III) will extract to some extent and form a colored TTA complex. The TTA extraction of protactinium is discussed in Section VI-C-2-b.

4. Chemical Properties

a. Protactinium(V)

The most stable oxidation state of protactinium in solution is +5, and this state is always encountered in solution unless a specific attempt is made to obtain a lower state. Unless otherwise indicated, all discussions of oxidized protactinium will refer to the pentavalent state.

One of the great difficulties in working with protactinium in solution is its insolubility in a large variety of aqueous solutions. Unless complexed, for example, by fluoride, sulfate, or oxalate, it shows a marked tendency to hydrolyze into insoluble compounds or polymerize as colloids. In this state it no longer has the properties of protactinium in true solution, and may deposit on the walls of glass vessels. This property probably accounts for the earlier difficulties encountered in separating protactinium from natural sources and for the variable and nonreproducible behavior often reported by investigators. Some observations by Casey and Maddock (48) illustrate this problem very well. A protactinium solution (several micrograms per milliliter) in 6M hydrochloric acid was stored in a polyethylene vessel for six weeks. During this period the extractability of the protactinium in diisopropyl carbinol was 99%, as measured in aliquots of this solution. After six weeks, and in the course of one day, the extractability dropped to 2%. The resulting solution appeared to contain a colloidal protactinium compound. The reaction could not be reversed by heating the solution or increasing the acid concentration to 11M. This decrease in extractability was always accompanied by the appearance of an absorption band at 260 mμ. Thus, the possibility of hydrolysis should always be considered when working with protactinium. Colloidal protactinium can be brought into true solution by treatment with a complexing agent such as hydrofluoric acid or sulfuric acid. In the latter case it is best to fume the solution. If protactinium-233 is used as a tracer to follow protactinium-231 through a series of chemical reactions, steps should be taken to insure that both isotopes exist as the same chemical species, i.e., that exchange has taken place. This can be done by forming the fluoride complex. In fluoride solution protactinium appears to be perfectly stable and does not undergo hydrolysis. Peppard et al. (230) report no hydrolysis in such a solution after it was left standing for five years.

In the pentavalent state, protactinium resembles tantalum and niobium, although it is more basic. Thus, niobium and tantalum are soluble in strongly basic solution while protactinium is insoluble or very slightly soluble.

Soluble protactinium exists in weakly acid solution in the form of oxygenated or hydrolyzed cations (e.g., PaO_2^+ and $Pa(OH)_n^{+(5-n)}$) and in strong acids in the form of anionic complexes (e.g., $PaOCl_6^{-3}$), although there is disagreement as to the exact composition of the ionic species. Anionic complexes are formed with anions such as fluoride, sulfate, chloride, bromide, thiocyanate, oxalate, citrate, and tartrate. Kirby (162) states that only hydrofluoric and sulfuric acid permanently dissolve macroscopic concentrations of protactinium and that with all other mineral acids precipitation or colloid formation eventually occurs. The best solvent for protactinium is hydrofluoric acid, in which the stable complex, PaF_7^{-2}, is formed. From fluoride solutions protactinium can be precipitated with potassium and barium as K_2PaF_7 (67) and $BaPaF_7$ (93), respectively.

The only study of the solubility of protactinium in the common mineral acids is the preliminary work of Thompson (285). This work was done with only 0.5-mg. portions of protactinium and about 0.05-ml. portions of solvent, and anomalous behavior was sometimes observed, so high accuracy cannot be expected. The solubilities in perchloric acid were low, about 0.03 g./liter in 1.7 to $11M$ solutions. In hydrochloric acid the solubility was less than 0.01 g./liter below $5N$ but increased to 0.3 g./liter in $9.6N$ acid. The solubility in dilute nitric acid was also small, 0.04 to 0.06 g./liter below about $5M$, but it increased to 6.6 g./liter at $13.8M$ and then decreased to 4.2 g./liter at $15.3M$. A maximum solubility at an intermediate concentration was also observed with sulfuric acid. The solubility varied from 0.8 g./liter in $0.9N$ acid to 6.8 g./liter in $9N$ acid to 0.09 g./liter at $32.5N$. A single determination in hydrofluoric acid gave a solubility of 3.9 g./liter in $0.05N$ acid. The results indicate that the solubilities are low in the absence of complexing anions. The relatively high solubility in nitric acid compared to hydrochloric and sulfuric acids is surprising. Hydrofluoric acid is the preferred solvent and should be used when possible.

Protactinium is insoluble in basic solution and is precipitated from solution by the addition of ammonium hydroxide, sodium hydroxide, or soluble carbonates. The phosphate of protactinium is insoluble in moderately strong acid solution. This behavior is similar to that of zirconium.

Sodium iodate, added to a protactinium solution, produces a precipitate that is insoluble in 5N sulfuric acid and tartaric acid but is readily soluble in 0.2N hydrofluoric acid (109). Iodate precipitation from acid solution provides excellent separation from phosphate ion (162). Other insoluble protactinium salts are the phenylarsonate, the double sulfate formed with potassium sulfate, the tannate, the pyrogallate, and the peroxide, and the double fluorides formed with barium and potassium fluorides.

Kirby reports that the phosphate can be dissolved by digestion with 18N sulfuric acid alone, followed if necessary by dilution with hydrochloric acid containing some hydrogen peroxide (162). In the absence of potassium the sulfate produced by dilution of a sulfuric acid solution, which is concentrated in protactinium, redissolves on addition of sufficient ammonium sulfate, indicating complex formation. Freshly precipitated protactinium hydroxide is soluble in oxalic, tartaric, or citric acid, undoubtedly due to complex formation (36,109,162).

Protactinium does not normally show amphoteric properties in aqueous solutions, but recent experimental work (144,145) indicates that protactinium forms an anion when fused with sodium or potassium hydroxides. After fusion with the alkali hydroxide and dissolution of the melt in water, trace protactinium was found to migrate when chromatographed on filter paper (and was, therefore, in solution under alkaline conditions) and moved as an anion under paper electrophoresis. Adsorption studies indicated that the protactinate ion was polyvalent. After fusion, as described above, the protactinium did not hydrolyze in solutions as dilute as 0.1N potassium hydroxide and when heated in solution to 100°C. The investigators also report that trace protactinium was partially soluble even when treated with 6N potassium hydroxide if the starting solution, protactinium-233 in 6N hydrochloric acid, was evaporated to dryness and moistened with hydrochloric acid three times.

b. Protactinium(IV)

The first evidence for the existence of tetravalent protactinium in solution was obtained by Haïssinsky and Bouissières (107,108,109). These investigators reduced protactinium(V) in solution to protactinium(IV) by treatment with zinc amalgam, chromium(II), or titanium(III). The reduction by zinc occurs in hydrochloric, hydrobromic, or sulfuric acids (42,42a). In the tetravalent state protactinium forms an insoluble fluoride, and can be coprecipitated with lanthanum fluoride. In this respect it is similar to thorium and the lanthanides and differs from protactinium(V). Protactinium(IV) does not form a soluble fluoride complex. In addition, the phosphate, hypophosphate, iodate, phenylarsonate, and the potassium double sulfate are slightly soluble, whereas soluble complexes are formed with carbonate, citrate, and tartrate. The polarographic evidence for oxidation states less than five was mentioned previously. Fried and Hindman (81) prepared protactinium(IV) in solution in the absence of other metal ions by reducing protactinium pentachloride to the tetrachloride with hydrogen at 600–800°C., and dissolving the tetrachloride in oxygen-free hydrochloric acid. The presence of the tetravalent state was confirmed by comparing the absorption spectrum of the reduced state with that of Ce(III) and with Pa(V), and by titration of the Pa(IV) with Np(IV). The potential for the reaction: $Pa(V) + e \rightarrow Pa(IV)$ was estimated to be -0.1 v.

The tetravalent state appears to be fairly stable in solution, although not as stable as uranium(IV). Thus, it was found that, in the absence of oxygen, about 5 to 10% was oxidized in 3 days and appreciable oxidation occurred 75 minutes after passing oxygen through the solution. Haïssinsky and Bouissières found that 64% of the Pa(IV) was oxidized by air in 42 minutes.

C. SEPARATION AND ISOLATION

1. Dissolution of Samples

From the previous discussion of the solubility of protactinium and of the tendency of protactinium to hydrolyze under a variety of conditions, the problems of dissolution are apparent. The bulk of the protactinium can remain insoluble under conditions that will dissolve the bulk of the macroscopic constituents of the sample. The presence of foreign ions may change the expected behavior of protactinium. As a result, chemical separations that work well with the pure tracer may fail when applied to a complex sample. Changes in protactinium concentration can also produce changes in chemical behavior, and it is dangerous to extrapolate chemical behavior at trace concentrations to macroscopic concentrations without experimentally determining the validity of the extrapolation. The nature of protactinium in solution is such that many pitfalls await those inexperienced with the chemistry of this element. Careful observation is always necessary.

Irradiated thorium metal or oxide can be readily dissolved by nitric acid (8 to 16N) containing a small amount (about 0.01N) of hydrofluoric or fluosilicic acid as a catalyst (270). The presence of fluoride ion insures the solubility of the protactinium.

A variety of methods have been used to dissolve ores and ore-processing residues, but no general method can be given. Mixtures of mineral acids, fusion with the common fluxes, or a combination of these, have been used. Hydrofluoric acid is often used since it complexes protactinium strongly and dissolves silica, on which protactinium is often found strongly adsorbed. In some reported methods it was not known if the protactinium dissolved quantitatively.

Uranium ores (1 g. or less) have been dissolved by fusion with sodium peroxide and treatment of the melt with hydrochloric acid (260), but since the protactinium content was determined by analyzing for one of its radioactive descendants, thorium-227, it was not known if the protactinium was in true solution. Golden and Maddock (93) used two methods for dissolving a silicious residue consisting mainly of lead, barium, and calcium sulfates. In one method, the sample was treated with 60% oleum to dissolve the sulfates, and the silicious residue, containing the protactinium, was separated and dissolved in 40% hydrofluoric acid. In the second method the sample was treated directly with hydrofluoric acid, dissolving the protactinium and leaving the sulfates as a residue. Sulfate residues have also been dissolved by fusion with sodium hydroxide, leaching the cooled melt with hydrochloric acid, fusion of the remaining solids with sodium carbonate, leaching with water, and dissolution of the carbonates in acid (153). The protactinium in the carbonate residue from uranium ore processing has been dissolved as follows. The residue was treated with an excess of nitric acid. The protactinium remained in the undissolved portion. When this portion was separated and heated with hydrofluoric acid containing sulfuric or perchloric acid, most of the protactinium dissolved (93,153,166,172). The acid-insoluble residue may be dissolved by fusion with alkali carbonate and bisulfate.

To recover protactinium from the sludge remaining from the ether extraction of uranium from ores, Collins and his co-workers (53) leached with $4N$ nitric acid (containing $0.1N$ hydrofluoric acid) for 8 hours at room temperature. Over 95% of the protactinium dissolved. It is surprising that such a simple treatment could be effective, but the authors point out that most batches of sludge contained some fluoride ion, and that some hydrofluoric acid was added to the nitric acid in treating possible fluoride-deficient samples.

Hahn and Meitner (104) mixed a silicious residue with tantalum pentoxide and fused the mixture with bisulfate. After washing the melt with water the protactinium and tantalum were dissolved in hydrofluoric acid. Kirby (162) recommends the following treatments, in the order given:

1. Digest the material with concentrated hydrochloric acid to remove iron.

2. Digest the residue from step *1* with hot, concentrated sulfuric acid.

3. Digest the residue from step 2 with 25 to 48% hydrofluoric acid or a hydrofluoric acid–sulfuric acid mixture.

4. Digest the residue from step 3 with a hot nitric–sulfuric acid mixture.

5. Digest the residue from step 4 with hot 40 to 50% sodium hydroxide.

6. Fuse the remaining residue with sodium carbonate, sodium hydroxide, potassium hydroxide, or potassium bisulfate.

Follow the protactinium-231 during the treatment by counting the residues with a gamma-ray spectrometer. Protactinium-231 emits a 27.5-k.e.v. gamma-ray that is unique among the natural radioisotopes.

2. Separation

a. Coprecipitation

Most analytical procedures used for samples containing a complex mixture of other ions depends on a coprecipitation step for the initial separation or concentration. Protactinium is carried from acid solution by many insoluble compounds, provided complexing agents are absent, and practically any precipitate formed from alkaline solution carries protactinium. The coprecipitation behavior of protactinium is summarized in Tables

TABLE XVI
Coprecipitation of Protactinium (67,286): Macroscopic Concentrations in the Presence of Gross Amounts of Impurities

Precipitate	Precipitant	Conditions	% Carried
$ZrO(H_2PO_4)_2$	H_3PO_4	20% HCl, H_2O_2	20–95[a]
$Th(C_2O_4)_2$	$H_2C_2O_4$	Dilute HCl	90–95
K_2TaF_7	KF		90–95
SiO_2			90–95
MnO_2	Mn(II) + $KMnO_4$	1M HNO_3	15–95[a]
$ZrO(IO_3)_2$	HIO_3	1M HNO_3	35–95[a]
$ZrO(IO_3)_2$	HIO_3	0.1M HNO_3	45–95
Ta salicylate	Salicylic acid	0.1M HNO_3	89–95
K_2CeF_6	KF	0.1M HNO_3	60–95
Ta, Ti, Nb oxyhydrates	Heat	0.05–0.1M HNO_3	70–90
Ta and Tb tannates	Tannic acid	pH 5–7	95–99
Bi, Hg, Sb sulfides	H_2S	1.1% HCl	25–63
Re_2S_5	$Na_2S_2O_4$	15% HCl	40–60
Hydroxides and other precipitates from alkaline solution in absence of HF			95–100

[a] The lower and intermediate values were obtained with solutions containing considerable amounts of foreign ions.

XVI and XVII. The coprecipitation of protactinium is affected by many variables, so the efficiency of a given carrier should be determined in advance in the particular solution of interest, using known amounts of protactinium. Some workers have found that in the presence of silica, titanium, or tantalum, protactinium often seems to form colloidal solutions from which it is not coprecipitated by the carriers usually effective in acid solution. From such solutions coprecipitation is apt to be more complete if the protactinium solution is freshly prepared by precipitating and dissolving a hydroxide. Upon standing, and particularly on warming, the protactinium reverts to the noncoprecipitable state (286). Several authors have expressed the opinion that in most cases protactinium is carried by adsorption processes rather than by isomorphous replacement (93,286), so different behavior under varying conditions is not surprising. The adverse effect of zirconium(IV) on the carrying of protactinium by some precipitates indicates coprecipitation by adsorption.

Some of the more important carriers that have been studied in detail recently are discussed below.

Manganese Dioxide. Manganese dioxide, precipitated from 1 to $5N$ nitric acid solution, carries more than 95% of the protactinium. The precipitate is formed by the addition of potassium permanganate to the nitric acid solution containing manganeous nitrate. The precipitation of about 1 mg./ml. of manganese is sufficient. Katzin and Stoughton have made a detailed study of the separation of protactinium from irradiated thorium (in solution as the nitrate) using this carrier (152). After the addition of the permanganate, digestion with heating is necessary for good carrying. Thorium concentrations up to $0.65N$ can be tolerated. Thorium is carried almost completely at trace concentrations, and small but detectable amounts are carried at macroscopic concentrations. Members of the thorium decay chain, principally ThB (lead-212) and ThC (bismuth-212) also are carried. The precipitate may be dissolved in hydrochloric acid or hydroxylamine and reprecipitated. The carrying is poor in the presence of fluoride or citrate ion, and the protactinium can be leached out of the precipitate by $10N$ hydrofluoric acid (93,152). Goodall and Moore (95) report that heating was not necessary to obtain good carrying and that low concentrations ($0.01M$) of zirconium(IV) or phosphate ions completely prevented the formation of the precipitate.

Lead Dioxide. This compound, formed by the oxidation of Pb(II) with sodium bismuthate at 100°C., carries protactinium well from $1M$ nitric acid containing no more than $1M$ Th(IV). Precipitation is quantitative with 0.05 to $0.1M$ lead dioxide (95). Zirconium(IV) or phosphate ions interfere.

TABLE XVII

Coprecipitation of Protactinium: Tracer Concentrations Under Controlled Conditions (67,93,95,123,151,277,286,298)

Precipitate	Amount of carrier, mg./ml.	Precipitant	Acidity	Comments	% Carried
$Th(C_2O_4)_2$	10–20	$H_2C_2O_4$	1.0–1.5N HCl		95–100
ThF_4	2–14	HF	1N HF	1N HF	0.7
$Th(IO_3)_4$	5–10	HIO_3	2–6N HNO_3		90–100
Th_2O_7	10	H_2O_2	0.2N HCl		92
$K_2Th(SO_4)_3$	10	K_2SO_4			82
K_1ThF_6	2	KF	1N HNO_3	Carrying increases with increasing HF conc.	30–90
K_2UF_6					
$ZrO(IO_3)_2$	2–5	HIO_3	0.5–2N HNO_3		100
$ZrO(H_2PO_4)_2$	0.5	H_3PO_4	1–9N HNO_3		98
$ZrO(H_2PO_4)_2$	2.5	H_3PO_4	1.9M H_2SO_4	H_2O_2 present	100
LaF_3	0.1	HF	0.5–10N HCl	Zr(IV) greatly decreases carrying	70–98
LaF_3	0.1	La(III)	1N HF	La added last	1
$BiPO_4$	2.5	H_3PO_4	1N HNO_3	0.6M H_3PO_4 digested 1 hr. at 75°C.	99

Carrier	Amount	Reagent	Conditions	Notes	Yield %
BiPO$_4$	2.0	H$_3$PO$_4$	0.6N HF 0.5N HNO$_3$	0.7M H$_3$PO$_4$	18
MnO$_2$	1.0	Mn(II) + KMnO$_4$	0.5–8N HNO$_3$		95
MnO$_2$	1.2	Mn(II) + KMnO$_4$	1N HNO$_3$	0.2N KF	5
PbO$_2$	10–20	NaBiO$_3$	1M HNO$_3$	Carrier added as Pb(II)	99
SnO$_2$	10		1M HNO$_3$	SnO oxidized to SnO$_2$ by heating	99
Ce(IO$_3$)$_4$	1.0	HIO$_3$	0.5N–6N HNO$_3$		100
K$_2$CeF$_6$	1.0	KF	2N KF, 1N HF	KF added last	99
BaF$_2$, CaF$_2$	10.0	HF	0.1N HNO$_3$	HF added last	85–100
CaF$_2$	1.0	Ca(II)	2N H$_2$SO$_4$ 1N HF	Ca added last	10
BaSO$_4$	1.0	Ba(II)	2N H$_2$SO$_4$	Ba added last	2
			0.08N KOH acidified with H$_2$SO$_4$		85
Ta$_2$O$_5$	Zr-2	2.5N HF			95
Ba$_2$ZrF$_6$	Ba-10				
Zr mandelate	0.1	Mandelic acid	6M HCl		90
Zr phenylarsonate		NH$_4$ phenylarsonate	5M HCl		100

Stannic Oxide. Stannic oxide carries protactinium essentially completely from $1M$ nitric acid–$1M$ Th(IV) solutions (95). Hydrofluoric acid at a concentration of $0.015M$ does not interfere. The oxide is formed by adding stannous oxide to the solution and digesting for 2 hours at 100°C. Stannic oxide added directly is not as efficient as that produced by the oxidation of the stannous oxide. Zirconium(IV) or phosphate ions interfere.

Zirconium Phosphate. This carrier was first used by Grosse in 1928 in the separation of protactinium from natural sources, and has been used successfully by other workers (151,208). In $3N$ nitric acid, 96% of the protactinium is coprecipitated when only 70% of the zirconium carrier is precipitated with phosphoric acid. Carrying is efficient up to $9N$ nitric acid. Thorium is also carried to a small degree. The precipitate can be dissolved in hydrofluoric acid or oxalic acid. The precipitate may be washed repeatedly with $6N$ hydrochloric acid to remove impurities without removing protactinium. The precipitation also may be carried out in strong hydrochloric or dilute sulfuric acid solution.

Zirconium Iodate. This precipitate carries protactinium quantitatively from $3N$ nitric acid solution (190). The carrying is 99% complete when 95% of the zirconium has been precipitated. Since thorium iodate is insoluble in acid solution, thorium must be absent.

Lanthanum Fluoride. The essential features of carrying by this precipitate are given in Table XVII. Carrying is apparently by adsorption, since Zr(IV) greatly inhibits carrying, and the order of addition of reagents is very important.

Barium Fluozirconate. This precipitate carries about 95% of the protactinium from $2.5N$ hydrofluoric acid using carrier concentrations of 2 mg./ml. of Zr(IV) and 10 mg./ml. of Ba(II) (190). The barium is added last, at least 30 minutes after the addition of zirconium.

Barium Fluoride. From $2.5N$ hydrofluoric acid more than 99% of the protactinium is carried using a barium carrier concentration of 19 mg./ml. The barium fluoride can be preferentially dissolved in cold water. For example, dissolution of 50% of the precipitate dissolves less than 0.01% of the protactinium (190).

Zirconium Mandelate. This precipitate has been reported to carry more than 90% of the protactinium from hydrochloric acid solution under the following conditions: zirconium concentration greater than 0.1 mg./ml., acid concentration $6N$, and precipitate digested for 1 hour and allowed to stand for 24 hours after precipitation (277). Under these conditions less than 0.2% of polonium, less than 1% of thorium-234, and less than 0.03% of radium are carried.

b. Solvent Extraction

The solvent extraction of protactinium was first studied by Hyde and Wolf (132). Protactinium can be extracted from the common mineral acids, except hydrofluric acid, into a wide variety of organic solvents. The most efficient and commonly used solvents are some of the higher molecular weight alcohols, ketones, and organic phosphates. Low molecular weight ethers are not useful. The extractability into a number of solvents is summarized in Table XVIII. Other compounds also have been tested (67, 93). Diisopropyl ketone, diisobutyl carbinol, and diisopropyl carbinol are preferred by most workers, and the choice between them is often made on the basis of availability and purity of commercial sources. Golden and Maddock (93) found diisopropyl ketone the most satisfactory solvent for both macroscopic and trace amounts. Other workers (98,162,216) recommend the use of diisobutyl carbinol because of the high extraction capacity, and decontamination obtainable with this solvent, and because it is readily available in pure form. Diisopropyl carbinol has been used to extract macroscopic amounts of protactinium.

Fluoride ion inhibits the extraction in all cases in which its effect has been studied. Concentrations of greater than $0.005M$ fluoride ion seriously reduce the extraction by diisopropyl ketone (93). Hydrofluoric acid is, therefore, frequently used to back-extract protactinium from the organic to the aqueous phase. The protactinium–fluoride complex can be destroyed and the protactinium returned to an extractable state by the addition of aluminum(III), boric acid, or thorium(IV) (167). The protactinium can be cycled between the organic and aqueous phases in this way. This procedure is superior to the use of dilute acid solutions for back-extraction since it insures that the protactinium will not hydrolyze in the aqueous phase.

As shown in the table, the high distribution coefficients with alcohols and ketones are obtained from solutions of high acidity. Hydrochloric acid is preferred, especially for macroscopic amounts, since protactinium has less tendency to hydrolyze in this acid than in nitric acid. It is neither desirable nor necessary to use high concentrations of salts. In dilute nitric acid protactinium extraction is low, even in the presence of high salt concentrations. Uranium and thorium, on the other hand, can be extracted at very low acidities if the salt concentration is high. This behavior permits a separation of protactinium from uranium and thorium (123,127).

A study of the diisopropyl ketone–hydrochloric acid system for analytical separations has been made by Golden and Maddock (93). Using a protactinium concentration of 0.02 mg./ml., an extraction coefficient of 100 was

TABLE XVIII
Extraction of Protactinium into Organic Solvents

Organic phase	Aqueous phase	% Extracted by equal volume	Reference
Diisopropyl carbinol	6N HCl	95	(90, 93)
Diisopropyl carbinol (saturated with 6M HCl)	6N HCl	99.6	(93)
Diisopropyl carbinol (saturated with 6M HCl)	6N HNO₃	89.5	(211)
Diisopropyl carbinol	1.8M Th(NO₃)₄ 2.0M HNO₃	55	(67)
Diisopropyl carbinol	2.0M Th(NO₃)₄ 3.0M HNO₃	92	(67)
Diisobutyl carbinol	6M HCl	99.9	(211)
Diisopropyl ketone	6M HCl	99	(93)
	2.6M NH₄NO₃ (or Mn(II), Th(IV), Al(III) nitrates) 1.0M HNO₃	75	(67)
Methyl isobutyl ketone	2.5M Ca(NO₃)₂ 1M HNO₃	69	(67)
	4M AlCl₃	79	(67)
Phorone	6M HCl	90	(93)
Acetophenone	5M HCl	95^a	(90)
Mesityl oxide	4.7M HCl	94	(93)
Diethyl ketone	6N HCl	87	(93)
Amyl alcohol	6N HCl	79	(93)
	4M AlCl₃	90	(67)
Octanol	3M NH₄NO₃ 1M HNO₃	75	(67)
2,2-Dichlorodiethyl ether	4M AlCl₃	89	(67)
	6M CaCl₂, 6M HCl	83	(67)
	8M HCl	98	(90)
Benzonitrile	8M HCl	98^a	(90)
5% Methyldioctyl amine in xylene	6M HCl	95	(211)
5% Methyldioctyl amine in chloroform	6M HCl	88.5	(211)
5% Tributyl amine in benzene	7.2M HCl	92^a	(90)
0.5M Thenoyltrifluoro-acetone (TTA) in xylene	6M HCl	88.5	(211)
	2M HCl	95.8	(211)
0.2M TTA in benzene	1M HCl	97^a	(252)
	2M HCl	87^a	(252)
	9M HCl	14	(252)

TABLE XVIII (continued)

Organic phase	Aqueous phase	% Extracted by equal volume	Reference
Tributylphosphate	6M HCl	99^a	(90)
	12M HCl	100	(231)
30% Tributylphosphate in kerosene	12M HNO$_3$	90	(114)
Di(2-ethylhexyl)phosphoric acid (5 × 10^{-3} M in kerosene)	2M HNO$_3$	99	(114)
0.1M Dibutyl phosphoric acid in isoamyl acetate	2M HNO$_3$	99.9	(273a)
0.1M Di(isoamyl) phosphoric acid in isoamyl acetate	2M HNO$_3$	99.9	(273a)
Nitrobenzene	8.5M HCl	92^a	(90)

^{a} Estimated from graphs.

obtained from about 4M hydrochloric acid; at 0.002 mg./ml. an acid concentration of 6M hydrochloric acid was required to obtain a similar extraction coefficient. These workers proposed a separation procedure as follows: (1) make the protactinium solution 7M in hydrochloric acid and 0.5M in hydrofluoric acid and extract with diisopropyl ketone; (2) retain the aqueous phase, add aluminum chloride to complex the fluoride ion, and extract the protactinium with diisorpropyl ketone; (3) wash the ketone phase with 8M hydrochloric acid; and (4) re-extract the protactinium into 8M hydrochloric acid–0.5M hydrofluoric acid. This cycle may be repeated if necessary. The yield of a single cycle is about 98%. This procedure was found to give good separation from iron, polonium, uranium, titanium, vanadium(V), manganese(II), and thorium. Fair separation was obtained from zirconium, tin, and cadmium. Extraction into diisopropyl ketone is less efficient from nitric acid and hydrobromic acid solutions than from hydrochloric acid, presumably because the protactinium complexes with nitrate and bromide ions are weaker than the chloro complex (89). The extraction of protactinium-233 from nitrate solutions of neutron-irradiated thorium with diisobutyl ketone was studied by Oliver and his co-workers (226). Although good distribution coefficients in favor of the organic phase can be obtained, extraction from hydrochloric acid is to be preferred for analytical purposes.

Diisopropyl carbinol has been used very successfully to extract protactinium (211,216). From 6M hydrochloric acid, one extraction with an equal volume of solvent extracts more than 99% of trace protactinium. Thorium is not appreciably extracted. The extraction of trace niobium, zirconium, uranium, and ruthenium can be greatly inhibited by extracting from 4% oxalic acid–6M hydrochloric acid solution. The oxalate complex of protactinium under these conditions is not strong enough to affect its extraction. The presence of macroscopic amounts of niobium carrier apparently inhibits the extraction of protactinium from hydrochloric acid–oxalic acid mixtures, but this interference can be avoided if the solution also contains sulfuric acid at a concentration of about 4M (216). However, from sulfuric acid niobium also extracts to an appreciable extent. Niobium can be separated quantitatively from protactinium by extracting it from 6M hydrofluoric acid–6M sulfuric acid; the protactinium remains in the aqueous phase. Kirby (162) finds that the presence of sulfuric acid is necessary to obtain complete extraction of protactinium by diisopropyl carbinol even in the absence of niobium, and reports that an aqueous phase containing 9N sulfuric acid and 6N hydrochloric acid has been useful for all protactinium concentrations and degrees of purity. The presence of sulfuric acid has not been reported as necessary by other workers.

Protactinium can be extracted from acid solution into organic phosphates. Tributylphosphate is more efficient from chloride solutions (90, 231) than from nitrate solutions (114). As is the case for ketones and alcohols the extraction is highly acid-dependent, and efficient extraction requires high acid concentrations. Peppard, Mason, and Gergel (231) separated thorium, protactinium, and uranium from each other by extraction with tributylphosphate. A multiple extraction technique, the "countercurrent distribution" method of Craig (55), was used to obtain pure fractions in high yield. Uranium and protactinium were separated from thorium by extraction from 5M hydrochloric acid. Uranium was separated from the protactinium by extraction from 5M hydrochloric acid–0.5M hydrofluoric acid. The uranium was removed from the organic phase by contacting it with a solution of 0.5M hydrochloric acid–1M hydrofluoric acid. Equilibrium was obtained rapidly, in less than three minutes of contact. The distribution coefficients of each of these elements as a function of hydrochloric acid concentration is given in this report (231).

The acidic organic phosphates are powerful extractants for protactinium at much lower acidities than used with tributylphosphate. Thus, a distribution coefficient (protactinium concentration in organic phase/protactinium concentration in aqueous phase) of 10 was obtained between 30% tributylphosphate in kerosene and 12M nitric acid; with di(2-ethylhexyl)-phosphoric acid the distribution coefficient between $5 \times 10^{-3}M$ solvent and 2M nitric acid was 270 (114).

The extraction of protactinium and several other metal ions with di(2-ethylhexyl)phosphoric acid from nitric acid oxalic acid mixtures has been studied (157). As the oxalic acid concentration is increased the distribution coefficients for protactinium and thorium decrease more rapidly than for uranium, due to the formation of oxalate complexes. Separations based on the relative stability of the oxalate complexes are possible. The organic phosphates are good extractants for a number of metal ions and are, therefore, not as selective for protactinium as the propyl and butyl ketones and carbinols.

The separation of protactinium-233 and uranium-233 from irradiated thorium has been accomplished by extracting the protactinium and uranium with 1% tributylphosphine oxide in toluene from $6N$ hydrochloric acid. The protactinium was separated from the uranium by scrubbing the organic layer with 4 to $6N$ hydrochloric acid saturated with sodium fluoride to remove protactinium (141).

Butex, or penta ether (the dibutylether of diethylene glycol), also has been used to extract protactinium. From $8M$ nitric acid a distribution coefficient of about 2 has been obtained (114). The extractability of a number of metal ions including protactinium into TBP and butex has been compared (114).

The extraction of protactinium by thenoyltrifluoroacetone (TTA) is strongly acid-dependent, as are all TTA extractions of metal ions. However, as seen from the data in Table XVIII, the extraction of protactinium is quite high even from acid solutions as high as $6M$. At such high acidities only a few ions, such as zirconium, hafnium, and iron, are extractable, whereas most other elements likely to be present as impurities can only be extracted at lower acidities. Thus, solutions of protactinium in TTA can be purified from many elements by washing with strong mineral acid solutions (127).

The addition of 3% TTA to diisopropyl carbinol was found to give better extraction of protactinium from thorium nitrate solutions than the pure carbinol, and less thorium was extracted (247). Synergistic extraction of protactinium by mixtures of solvents has also been observed by Goble and Maddock (91). These workers found maxima in the distribution coefficients using solvent mixtures of diisopropyl ketone and carbinol, diisopropyl carbinol and acetophenone, and many others.

High molecular weight amines are also good extractants for protactinium. The extraction from nitric acid solution is not as complete as from hydrochloric acid solution (114,154). This type of extraction depends on the formation of an amine salt with an anionic metal ion complex, and therefore, many metal ion forming a negatively charged complex ion will extract to some extent. Methyldioctylamine will extract protactinium quantitatively from 1.8M phosphoric acid (216). This is a useful procedure since protactinium is difficult to extract from phosphoric acid solution with the solvents ordinarily used. Solvent extraction by organic amines has been reviewed by Moore (215).

The cupferron complex of protactinium has been extracted efficiently into solvents such as benzene, ether, chloroform, and amyl acetate (190). Distribution coefficients greater than 10 were obtained from solutions between 0.1 and 4N in mineral acid.

The problems that can arise in obtaining reproducible results with protactinium are typified by the difficulties that have been encountered in solvent extraction. Protactinium is frequently found in a "nonextractable" state. The sudden reversion to this state in 6M hydrochloric acid has been described in Section VI-B-4. Some of the distribution coefficients reported in the literature were determined by back-extraction from the organic solvents and, therefore, any nonextractable protactinium was not included in this distribution. Hardy and his co-workers (114) found about 5% of a $10^{-5}M$ protactinium solution to be nonextractable by tributylphosphate. Using the same solvent, Peppard and his co-workers (231) found about 1% nonextractable in their work; the nonextractable fraction converted to an extractable form in 4 to 13.3M hydrochloric acid. The rate of conversion increased with increasing temperature. However, Casey and Maddock (48) could not convert unextractable protactinium to a form soluble in diisopropyl carbinol by increasing the acidity or temperature. Golden and Maddock (93) also encountered solutions from which only part of the protactinium could be extracted. These solutions were sometimes obtained after dissolution of hydroxide precipitates. Colloidal silica was very effective in preventing extraction, indicating that protactinium is readily adsorbed from solution. However, adsorption on glass walls is not the only means by which protactinium is rendered nonextractable (although this can occur) since the same effect has been observed in polyethylene containers (48). The effect of the container material on the state of protactinium is not clear. Hardy (114) used polyethylene vessels to study the extraction of protactinium from nitric acid solutions by several solvents and found that the distribution coefficients increased with time of extraction less the polyethylene had previously been treated several times with the solvent. The increase with time was believed due to the leaching of some impurity in the polyethylene by the solvent. In glass the distribution coefficients was independent of time after one minute. Golden and Maddock (93) used polyethylene for most of their extractions. They report that glass vessels adsorbed protactinium even from hydrochloric acid solution and gave lower distribution coefficients than polyethylene. Peppard and his co-workers (231) found that the distribution of protactinium between diisopropyl ketone or tributylphosphate and hydrochloric acid solutions was the same in glass or polyethylene. Peppard suggests that ketone-soluble phosphate esters might have been present as plasticizers in the polyethylene equipment used by Golden and Maddock. Since such compounds are powerful extractants, they might be responsible for the higher extractabilities in polyethylene compared to glass. The distribution coefficients obtained by Peppard for the ketone agree very well with those found by Golden and Maddock in glass. Goble and Maddock, however, present evidence that any impurities in their polyethylene equipment did not affect the extraction, and that protactinium in hydrochloric acid solution, even $8M$, becomes colloidal and unextractable with time. These factors must apparently be determined by each investigator for his own equipment and protactinium. If the nature of the vessel surface has a large effect on the extraction of protactinium, widely different results are possible for trace and macroscopic amounts. The extrapolation from one extreme of concentration to another must be done cautiously. The use of data obtained by others under superficially similar conditions must also be used with caution since many seemingly minor variables affect the behavior of protactinium.

The purity of the solvent also must be considered if reproducible results are to be obtained, although this requirement is not limited to protactinium extraction. The organic phosphoric acids from commercial sources are usually mixtures of the mono-, di-, and possibly triderivatives of phosphoric acid. Since these compounds have widely different distribution coefficients for a metal ion, the solvent must be purified. Purification based on differences in solubility and acidity is not difficult. Tributylphosphate is subject to hydrolysis to mono- and dibutyl phosphoric acid. The latter compounds are much more efficient extractants than tributylphosphate, so purification before use by washing with alkaline solutions is required. Tributylphosphate also can hydrolyze during an extraction from acidic solutions, so contact time should be kept to a minimum. The original literature should be consulted for details on purification. The increased extraction of mixtures relative to the pure solvents also requires high-purity solvents to obtain reproducible results. Commercial sources of the ketones and carbinols may contain the other compound as an impurity, and such solvents must be purified before use to avoid anomalous results.

Bouissières and Vernois (37) report that tracer Pa(IV) is not extracted by tributylphosphate in benzene or by methyl isobutyl ketone from $6M$ hydrochloric acid. The behavior toward the ketone is similar to that of thorium(IV) and zirconium(IV), whereas the behavior toward tributylphosphate is similar to that of thorium(IV) but opposite to that of zirconium(IV). They also found that protactinium(IV) is extracted somewhat more efficiently by TTA from $6M$ hydrochloric acid than is protactinium(V).

c. Ion-Exchange Separations

(1) Anion-Exchange Resins

The anion-exchange behavior of protactinium can be used to obtain many useful separations. Kraus and Moore (169) found that protactinium-233 is strongly adsorbed on Dowex-1 from hydrochloric acid solutions greater than $4M$ and could be eluted with less-concentrated acid solutions. Although equilibrium was reached only after several days of contact in batch experiments, efficient column separations of protactinium from thorium, uranium, tantalum, niobium, and zirconium were devised. The sample is adsorbed on the column from strong hydrochloric acid and eluted with hydrochloric acid–hydrofluoric acid mixtures or dilute hydrochloric acid alone. Thorium is not adsorbed on the resin and passes through unchanged (171), although the other elements are adsorbed along with protactinium as anionic chloro complexes. Elution with $9M$ hydrochloric acid–$0.004M$ hydrofluoric acid removes zirconium first, followed by the protactinium. Niobium is eluted with $9M$ hydrochloric acid–$0.18M$ hydrofluoric acid and tantalum with $1M$ hydrofluoric acid–$4M$ ammonium chloride (168). Ferric iron (170) and hexavalent uranium (171) were eluted last with dilute hydrochloric acid.

Other workers have also used hydrochloric–hydrofluoric acid mixtures. Barnett (25) purified protactinium by recycling through a Dowex-1 anion resin column. The protactinium was adsorbed on the column from $9M$ hydrochloric acid and eluted with $7M$ hydrochloric acid–$0.05M$ hydrofluoric acid. To recycle the protactinium through the resin the fluoride in the eluate was complexed with $0.1M$ boric acid or $0.1M$ aluminum chloride. The protactinium could be readsorbed on the resin from this mixture, and eluted as before. Protactinium has been separated from its daughters (radium, thorium, bismuth, and polonium isotopes) by adsorbing the sample on Dowex-1 from concentrated hydrochloric acid, and washing the column with $9M$ hydrochloric acid to remove the daughters. The protactinium was then eluted with $9M$ hydrochloric acid–$0.04M$ hydrofluoric acid (174). A similar procedure has been used to purify 125 g. of protactinium obtained from uranium processing residues. For the final purification the protactinium solution was adsorbed on a column of an anion-exchange resin, Deacidite F.F., from $8M$ hydrochloric acid and eluted with $8M$ hydrochloric acid–$0.5M$ hydrofluoric acid. The decay products of protactinium and many other impurities were not adsorbed, although the iron impurity remained adsorbed during the elution with the hydrofluoric–hydrochloric acid mixture. The original solution contained fluoride ion, which was removed from the protactinium by complexing with aluminum ion (143).

Protactinium and tantalum have been separated on a column of Amberlite 1R-4B by elution with $6.5N$ hydrofluoric acid adjusted to a pH 3 with ammonium hydroxide. The tantalum elutes first, and about 90% of the protactinium elutes free of tantalum (146). Maddock and Pugh (191) separated zirconium and protactinium on a column of Amberlite IRA-400. The mixture was adsorbed on the column from strong hydrochloric acid ($ca.$ $9M$), and the zirconium was eluted with 6 to $7M$ hydrochloric acid and the protactinium with more dilute acid ($3M$ or less). At least 95% of the zirconium and not more than 0.1% of the protactinium was found in the zirconium fraction and more than 95% of the protactinium was found in the dilute acid fraction.

The behavior of a number of metal ions, including protactinium, on the anion-exchange resin Dowex-2 has been studied (43). The equilibrium distribution coefficients (the ratio of the amount of metal ion adsorbed per gram of dry resin to the amount per milliliter of solution) were determined as a function of acid concentration for hydrochloric, nitric, and sulfuric acids. From these results the authors suggest several possible separations, although these separations were not actually carried out on resin columns. The following separations are among those that appear possible with Dowex-2. Thorium, uranium, and protactinium could be separated in hydrochloric acid solution. As is the situation in the Dowex-1–hydrochloric acid system, thorium is not adsorbed at any acid concentration and uranium and protactinium are strongly adsorbed at high acid concentrations. At intermediate acidities (4 to $5M$), protactinium is less strongly adsorbed than uranium and could be eluted, leaving uranium on the resin. Uranium could then be eluted with $0.1N$ acid. Zirconium and protactinium could be separated using the Dowex-2–nitric acid system. At about $8M$ nitric acid the distribution coefficient for zirconium is about 10 and for protactinium about 1.

Milligram amounts of protactinium have been separated from neutron-irradiated thorium-230 (ionium) on Dowex-1 (284). In this work it was difficult to elute protactinium completely from Dowex-1 with hydrochloric acid, the degree of difficulty increasing with increasing protactinium concentration.

The behavior of protactinium(IV) on the anion-exchange resins Dowex-1 and Amberlite IR-401 also has been investigated (251). Protactinium-(IV), like thorium(IV), is not adsorbed on these resins from strong hydrochloric acid solution.

(2) Cation-Exchange Resins

There are only a few reports of the behavior of protactinium on cation-exchange resins. Hyde, in his review (127), quotes some unpublished work by Sullivan, Studier, and Elson at Argonne National Laboratory reporting that thorium and protactinium could be completely adsorbed from dilute nitric acid solution ($0.1M$ to $2.0M$) and that the thorium eluted with $0.2M$ ammonium sulfate solution at pH 3.4, whereas the protactinium could be eluted with oxalic acid solutions at pH 3 to 5. Kirby, in his review (162), gives the results of his otherwise unpublished work on the separation of protactinium-231 from its decay products, actinium-227, thorium-227, and radium-223. The protactinium ($10^{-5}M$), in dilute nitric–hydrofluoric acid solution, is passed through a Dowex-50 column. The decay products are completely adsorbed and 99.5% of the protactinium passes through. Other workers report somewhat different behavior on ZeoKarb 225 cation resin (114) with a $10^{-5}M$ protactinium solution in $0.01M$ hydrofluoric acid–$6M$ nitric acid. From this solution 16% of the protactinium was adsorbed. In the absence of hydrofluoric acid, 73% was adsorbed. Another thorium–protactinium separation on Dowex-50 resin has been reported. The mixture is adsorbed on the resin column from nitric acid solution and the protactinium eluted with 0.8 to $2M$ nitric acid or 0.05 to 0.7% oxalic acid solution (155,156). Completely different behavior in nitric acid is reported by Katz and Seaborg (151), who state that protactinium is not eluted from strong cation-exchange resins by $2M$ nitric acid.

d. Chromatographic Separations

A few paper-chromatographic separations of protactinium have been described. These are summarized in Table XIX.

The migration on paper of protactinium and a number of other metal ions as a function of hydrochloric and hydrofluoric acid concentration in butanol–water solutions (295) and as a function of hydrochloric acid concentration in acetone–water solutions (182) has been studied.

The paper-electrophoretic behavior of protactinium in hydrochloric–hydrofluoric acid mixtures has been studied (296). Protactinium was

TABLE XIX
Paper-Chromatographic Separations of Protactinium

Developing solvent[a]	Elements separated	Reference
Butanol (50), 12N HCl (25), 20N HF (1), H$_2$O (24)	Pa, Ta, Nb	(183,295)
2-Methyltetrahydrofuran containing 2.5% HNO$_3$	Pa, U	(303)
1M HF in ethylmethyl ketone	Pa, Nb	(83)
Butanol (50), 12N HCl (25), 20N HF (5), H$_2$O (20)	Pa, Ti	(295)
Butanol (50), 12N HCl (25), 20N HF (5), H$_2$O (20)	Pa, Bi	(295)
Butanol (50), 12N HCl (33), 20N HF (1), H$_2$O (16)	Pa, Fe	(295)

[a] The numbers in parentheses indicate the relative amounts of each component in the mixture.

separated from zirconium, uranium, thorium, and iron with an electrolyte of 0.6M hydrochloric acid–2N hydrofluoric acid. The separation from iron and zirconium was obtained with hydrofluoric acid alone as the electrolyte.

D. DETECTION AND IDENTIFICATION

Protactinium is generally identified through its radioactive decay properties, although chemical separations may be required to obtain the protactinium free from interfering radioactive nuclides. Protactinium-231 can be detected in uranium residues by its distinctive gamma-ray spectrum (Fig. 14). The 27.5-k.e.v. gamma-ray is particularly useful for this purpose since the only interference in the two uranium decay series is the 29-k.e.v. gamma-ray emitted by thorium-234, the first daughter of uranium-238 (see Figs. 7, 8, and 15 to 18). This gamma-ray is of low abundance and can be distinguished from the 27.5-k.e.v. protactinium gamma-ray by decay measurements, since the half life of thorium-234 is only 24.1 days. If the thorium-234 in the sample is supported by its uranium parent the 185-k.e.v. uranium-235 gamma-ray also will be detected, and the distinction can still be made. A gamma-ray spectrum of a uranium ore residue showing the protactinium gamma-ray is given in Fig. 18 (162).

The alpha-particle energy spectrum is useful if the sample contains sufficient activity (several hundred disintegrations per minute per milligram), so that a sample weighing a milligram or less can be counted. Thin samples are necessary to avoid obscuring the energy spectrum by self-absorption of alpha particles in the sample. The applicability of alpha-particle spectrometry can be seen by comparing the spectra given in

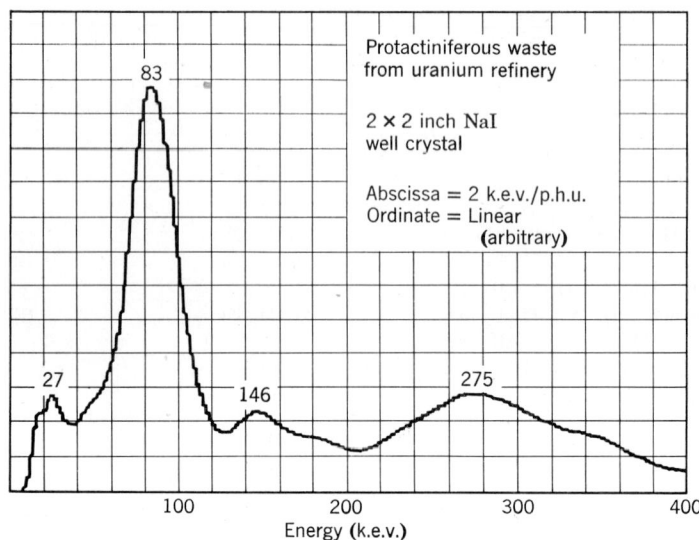

Fig. 18. Gamma-ray spectrum of uranium ore residue (courtesy of Harold W. Kirby, Mound Laboratory, Monsanto Chemical Company).

Figs. 9, 10, 11, 19, and 20. The 50-m.e.v. alpha particle of protactinium can be readily distinguished in the presence of other natural activities in uranium residues (88). An alpha spectrum of such a residue is shown in Fig. 21. Protactinium-231 can be characterized by its emission spectrum (Section VI-B-3), although this has apparently not yet been applied to protactinium concentrates from uranium refining.

Protactinium-233, a beta emitter, can be characterized by its half life of 27.4 days and its gamma-ray spectrum. Other protactinium isotopes can be identified by one or more of the following: chemical properties, type of decay, half life, energies and relative abundances of emitted particles and radiations, identification of known daughters, and method of production. The quantitative methods given in Sections VI-E and -F can, of course, be used for qualitative detection if the half-lives of the protactinium isotopes are sufficiently long to permit carrying out the chemical manipulations.

An example of the use of known daughters is the discovery of the beta emitter, protactinium-232, in deuteron-irradiated thorium through identification of its daughter, uranium-232 (92). The use of coprecipitation methods to distinguish protactinium-231 from other long-lived, naturally occurring alpha emitters is illustrated by the following series of reactions: (1) zirconium phosphate will carry protactinium and thorium from acid

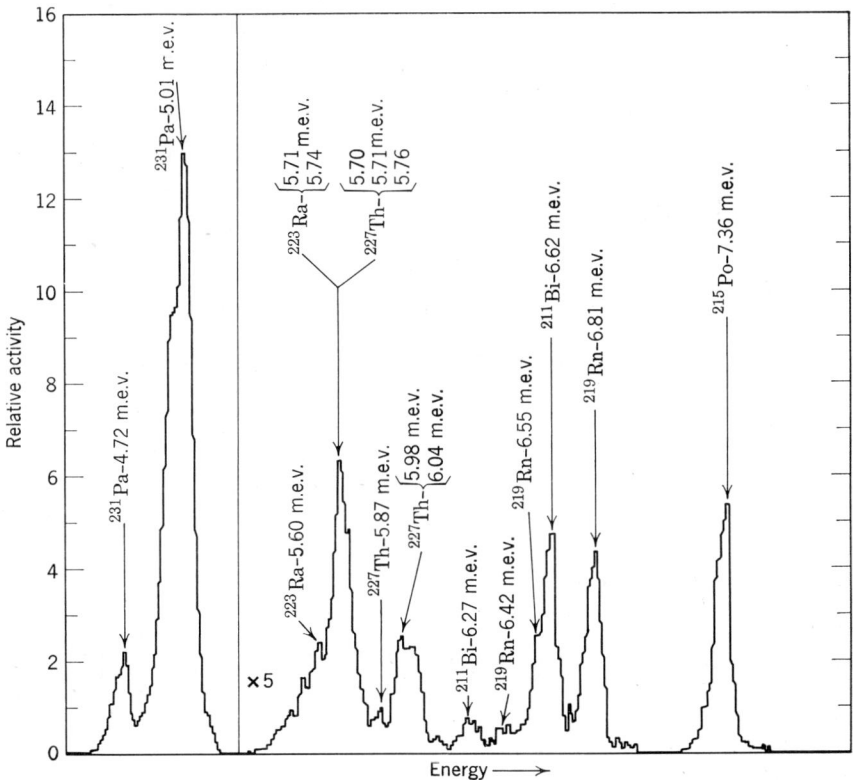

Fig. 19. Alpha-particle spectrum of protactinium-231 and decay products.

solution but not uranium or radium; (*2*) thorium fluoride will carry thorium isotopes but not protactinium or polonium; (*3*) lanthanum fluoride will carry protactinium when the fluoride is added last, but not uranium; (*4*) barium sulfate will carry radium but not protactinium; thorium isotopes are partially carried; (*5*) manganese dioxide will carry polonium in the presence or absence of fluoride ion, but will carry protactinium well only in the absence of hydrofluoric acid; (*6*) silver metal will remove polonium from hydrochloric acid solutions; and (*7*) when thorium-232 is added to a detectable concentration, manganese dioxide will not carry ionium (thorium-230) significantly (153).

Casey and Maddock (48) have studied the reactions of protactinium with several organic reagents with a view toward developing colorimetric methods for the element. Colored solutions or precipitates were formed with alizarin, quinalizarin, 8-hydroxyquinoline, tannic acid, pyrogallol,

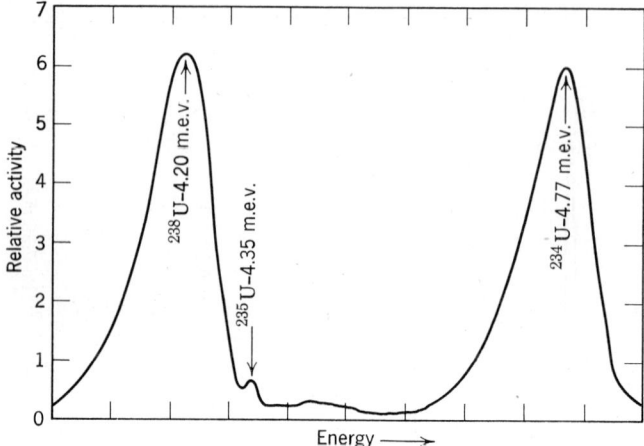

Fig. 20. Alpha-particle spectrum of uranium (natural isotopic abundance).

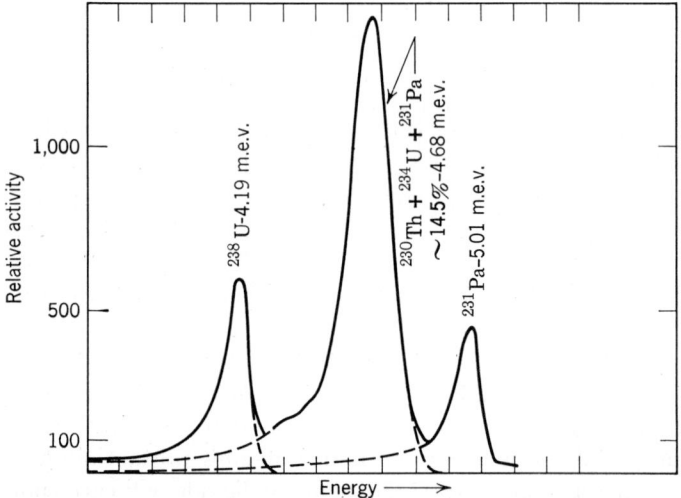

Fig. 21. Alpha-particle spectrum of uranium ore residue (88). Reproduced by permission from *United Kingdom Atomic Energy Authority Reports*.

and gallic acid. A number of reagents that form colored products with chemically similar elements gave no reaction with protactinium. All the tests taken together comprise a unique test for the detection of titanium, zirconium, hafnium, vanadium, niobium, tantalum, molybdenum, tungsten, and protactinium in mixtures of these elements. The original paper should be consulted for details.

E. DETERMINATION

Methods for measuring the amount of protactinium will be discussed in this section. Separations that can be used to obtain the protactinium in a form suitable for this measurement are discussed in Section VI-C-2, and separations that have been used for the more common types of samples are described in Section VI-F.

After separation and purification macroscopic amounts of protactinium can be determined gravimetrically by precipitation with ammonium hydroxide and ignition to the oxide, Pa_2O_5 (272). This is rarely done, and the common procedure involves a radioactivity measurement.

Protactinium-231 is generally determined by alpha counting, which is applicable over a wide range of concentrations. It is the most sensitive method, but can also be used for large amounts by counting a suitable fraction of the sample. Except as described below, the sample must be free from other alpha emitters. The sample may be counted in a gas-flow proportional or ionization counter, zinc sulfide scintillation counter, or a solid state alpha-particle detector by evaporating the pure protactinium solution, or a known fraction, on a support (counting planchet) that is not attacked by the solution. For internal gas-flow counters a conducting support, i.e., a metal, is necessary. Since alpha particles have a very short range, the dried sample must be free from extraneous material to avoid losses due to self-absorption of alpha particles in the sample mount. This requires careful purification to eliminate solid impurities. Thick samples, such as those obtained when carriers are used to separate protactinium, can be counted, but with less accuracy than "weightless" samples. Self-absorption corrections are difficult to make accurately. Kirby (162) points out that the alpha-particle energy (about 5 m.e.v) and the specific activity of protactinium-231 are sufficiently low to make alpha counting especially susceptible to self-absorption. Relative counting requires that the samples to be compared must be similar to thickness and back-scattering, and otherwise must be counted under identical conditions. Absolute counting requires in addition that the relation between the counting rate and the disintegration rate be known. This is best done by standardizing a strong protactinium-231 source or solution in a low-geometry alpha counter, and using this standard to calibrate a counter for routine use. However, reasonable accuracy can be obtained without such standardization in an internal gas-flow proportional or pulse-ionization counter. Weightless, small-area sources in such counters have counting efficiencies between 50.5% and 51% for samples mounted on steel and between 51.5% and 52% for samples mounted on platinum. The "geometry" is 50%, and the additional counting efficiency is due to back-scattering of alpha particles. The counting rates of protactinium preparations several years old must be corrected for the growth of alpha-emitting daughters or the preparations must be purified from these alpha emitters.

It is not always necessary to purify protactinium samples from all other alpha emitters before counting. For some mixtures of alpha emitters the fraction due to protactinium-231 may be determined in an alpha pulse-height analyzer. The sample, however, must be extremely thin to obtain accurate results. This method has been used very successfully in the determination of protactinium in uranium refining residues without chemical separation (88).

Protactinium-231 emits gamma-rays in its decay process, and consequently it can also be determined by gamma-ray counting. This method has the advantage that source thickness is not critical for the more energetic gamma-rays, and with modern gamma-ray scintillation spectrometers, measurements can be made in the presence of some other gamma emitters. This method has been used at Mound Laboratory to determine protactinium-231 in uranium ore residues and to follow the chemical separation of protactinium from such residues (266). The counting was done with a thallium-activated sodium iodide crystal optically coupled to a multiplier phototube. The pulses from the phototube were linearly amplified and analyzed with a single-channel pulse-height analyzer to obtain the gamma-ray spectrum. The modern multichannel pulse-height analyzers would be more efficient for this purpose. The gamma-ray spectrum is given in Fig. 14. Quantitative measurements were made by determining the number of 0.299-m.e.v. gamma-rays, since the lower-energy gamma-rays are more sensitive to self-absorption. To eliminate gamma-rays of this energy due to other naturally occurring nuclides, thorium fluoride was precipitated from the sample to carry these nuclides, while the protactinium remained in solution. Absolute measurements were made by counting pure protactinium-231 under the same conditions used in counting samples. In this determination care must be exercised to avoid counting gamma-rays in the vicinity of 300 k.e.v. emitted by radium-223 and its decay product, bismuth-211 (Fig. 7). Even if the protactinium has been separated from radium at some stage of the chemical separation, radium-223 will grow back into the protactinium and interfere. The unknown and standard protactinium must, therefore, be counted within one week after separation from radium or the counting rate must be corrected for the radium-223 contribution. A procedure for this correction, using a sample of actinium-227 to obtain the daughter gamma-ray spectrum, has been described by Kirby (162).

Microgram quantities of protactinium-231 can be determined spectrophotometrically, as described in Section VI-B-3.

Protactinium-233 may be determined by beta or gamma counting. Its beta particles are relatively weak (95% have energies less than 0.3 m.e.v.), and to obtain reproducible result self-absorption and window and air absorption (if an end-window counter is used) must be carefully controlled. As in the case of protactinium-231, absorption problems are considerably reduced if gamma-ray counting is used. One method of calibrating a beta counter for absolute counting of protactinium-233 is as follows (199). A sample of pure neptunium-237 (a long-lived alpha emitter that decays to protactinium-233) is electrodeposited on a platinum disk from basic carbonate solution at 8 v. and 2 amp. for 45 minutes. The neptunium disintegration rate is determined by alpha counting, and the growth of protactinium-233 beta activity in the deposited sample is followed until equilibrium is reached. The protactinium disintegration rate and counting efficiency is calculated from the theoretical growth equation:

$$C_p/E = D_p = D_n (1 - \exp[-\lambda t])$$

where C_p is the counting rate of protactinium-233 in the beta counter, E is the efficiency of the beta counter, D_p is the disintegration rate of protactinium-233 at time t, after the electrodeposition, D_n is the disintegration rate of neptuniun, and γ is the decay constant of protactinium-233. The best results are obtained at equilibrium ($t = \sim 270$ days) when D_n equals D_p. This procedure is relatively simple since the alpha disintegration rate can be determined with reasonable accuracy in an internal proportional counter. However, the counting efficiency obtained is valid only for samples mounted as described, that is, samples with negligible self-absorption, mounted on a backing of the same atomic number as the standard, and covering the same area if a windowed beta counter is used. A more generally applicable procedure has been described by Fudge and Woodhead (83). The disintegration rate of a pure protactinium-233 solution is standardized in a 4 π beta counter, and this solution is used to calibrate a gamma scintillation counter. Subsequent samples were counted in the scintillation counter. McCormack and his co-workers (198) also used 4 π beta counting and alpha counting of neptunium-237 in equilibrium with protactinium-233 to standardize counters for protactinium-233.

F. DETERMINATION IN SPECIFIC MATERIALS

1. Ores and Ore-Processing Residues

If the specific activity of the sample is sufficiently high (several hundred disintegrations per minute per milligram) the protactinium-231 content may be determined directly by counting in a gridded ionization chamber or with a solid state alpha-particle detector. Very thin samples are required. The pulses from either type of detector are linearly amplified and sorted by pulse height in a multichannel pulse-height analyzer. Using the resulting alpha-particle energy spectrum the protactinium concentration is calculated from the activity of its 5-m.e.v. alpha particle. The system must be calibrated with alpha emitters of known energy and with alpha sources of known disintegration rate. This method was used by the British workers in analyzing uranium residues in the separation of gram quantities of protactinium (88,142). The samples, either liquid or solid, were spread on the counting planchet in a uniform layer by evaporation with tetraethylene glycol. The total residues on the planchets varied from 25 γ to 1 mg. The protactinium content varied from about 3 to 50 γ/g. and accounted for approximately 20% of the total alpha activity in the samples. The remaining alpha activity was due primarily to uranium-234 and -238 and thorium-230. Since the alpha peaks due to thorium-230, uranium-234, and the lower-energy alpha particles of protactinium overlapped the 5.01-m.e.v. protactinium peak slightly, a correction in the counting rate under this peak was required (Fig. 21). The sensitivity of this method was about 0.1 p.p.m. Because no chemical separations are required, this method is to be preferred when it is applicable. The same workers also used a chemical method (142). A known amount of protactinium-233 was added as a tracer to determine the chemical yield, and ferric hydroxide was precipitated to carry the protactinium. The precipitate was dissolved in nitric acid containing a little hydrofluoric acid and the protactinium was coprecipitated with either barium fluozirconate or manganese dioxide. The precipitate was dissolved in $8M$ hydrochloric acid–aluminum chloride solution and the protactinium was extracted with diisobutyl ketone. The protactinium was back-extracted with dilute hydrofluoric acid, and aliquots of this solution were counted for alpha activity (to determine protactinium-231) and beta activity (to determine protactinium-233). The authors found that some of the protactinium was lost on the walls of polyethylene containers before complete exchange of the protactinium isotopes occurred. Storage of the solutions in hydrofluoric acid before analysis would probably eliminate this difficulty. A similar chemical method for the same types of samples was used by another group of workers (53). Golden and Maddock (93) also used protactinium-233 as a tracer in analyzing uranyl solution and the heavy-metal sulfate precipitate from uranium ore processing. The sulfate precipitates were dissolved in hydrofluoric acid and the protactinium was separated by coprecipitation with barium fluoride and manganese dioxide, followed by extraction with diisopropyl ketone from hydrochloric acid–aluminum chloride solution. An appreciable fraction of the polonium follows this procedure, as evidenced by the fact that ignition of the extract reduces its alpha activity. However, most of the polonium could be separated by re-extraction from the solvent into 7 to $8M$ hydrochloric acid–$0.05M$ hydrofluoric acid. The recovery of protactinium-233 was 80 to 85%. Uranyl solutions were made $8M$ in hydrochloric acid and $0.5N$ in hydrofluoric acid and protactinium-233 tracer was added. Extraction of this solution with diisopropyl ketone removed many impurities (e.g., iron and polonium). Extraction after adding aluminum chloride to complex the fluoride ion brought the protactinium into the solvent layer. Washing this layer with $8N$ hydrochloric acid removed those elements whose extraction was enchanced by aluminum chloride (e.g., uranium and zirconium). The protactinium was then back-extracted with 6 to $8N$ hydrochloric acid–$0.5N$ hydrofluoric acid. The next cycle avoided the use of aluminum ion by using repeated precipitation with ammonia and dissolution in hydrochloric acid, or continued evaporation with hydrochloric acid, to remove fluoride. In either case the efficiency of a single cycle was at least 98%. Two cycles were found to be sufficient to obtain good radiochemical purity.

Another procedure for the analysis of protactinium-231 in ores that uses protactinium-233 tracer to measure the chemical yield is as follows (166). The sample is heated with nitric, hydrofluoric, and sulfuric acids, and the tracer is added. Hydrofluoric acid is used to leach the protactinium from the sample and provide a medium for exchange of the protactinium isotopes. The mixture is fumed and the residue is treated with nitric acid. At this point the solution usually contains more than 75% of the protactinium. The residue is discarded, and zirconium iodate is precipitated from the solution to carry the protactinium. The precipitate is dissolved in concentrated hydrochloric acid and a silver strip is inserted in the solution to remove polonium by spontaneous deposition on the silver. The solution is then contacted with diisopropyl carbinol to extract the protactinium. The carbinol solution is counted for alpha and beta activity, and the alpha spectrum is determined in a pulse-height analyzer. The protactinium-231 is determined from the alpha spectrum and total alpha activity. The extraction step, in addition to providing separation from thorium isotopes, produces a solution that gives an essentially weightless mount for the alpha-particle spectrometry. The protactinium yield is determined from the protactinium-233 recovery. More than 90% of the extracted alpha activity was due to protactinium-231.

Another group of workers used a coprecipitation method for the analysis of ore residues (153). The principal long-lived alpha emitters in these residues were radium, polonium, ionium (thorium-230), and protactinium.

The series of reactions that were used is given in Section VI-D because it will also serve as a qualitative test for protactinium in these samples. In performing the analyses, solid samples were dissolved by treatment with hydrofluoric and hydrochloric acids and by fusion with the common fluxes. The chemical yield, as determined with protactinium-233 tracer, was 50 to 60%.

The use of protactinium-233 as a tracer for protactinium-231 is an attractive procedure because of the unpredictable behavior of protactinium. However, it is necessary to establish that exchange has actually occurred. The best medium for this exchange is hydrofluoric acid because of the strong fluoride complex formed by protactinium. When solids are treated with hydrofluoric acid or otherwise dissolved, complete exchange is certain only if all of the protactinium has actually been dissolved. If sufficient protactium is present, this can be determined by gamma-ray spectrometry. In the absence of evidence that all, or a known fraction, of the protactinium-231 in the sample has dissolved, the most that can be said with certainty is that the fraction of protactinium-231 is complexed with fluoride in the same chemical form as the tracer. Adding the tracer to the solid before dissolution does not, of course, mean that complete exchange will occur.

Protactinium-231 also can be determined from the activity of its decay products if the state of equilibrium of the decay chain is known. A procedure for its determination in ores based on measurement of thorium-227 has been described (260). The method assumes that the following chain is in radioactive equilibrium.

$$^{231}Pa \xrightarrow[3.43 \times 10^4 \text{ years}]{\alpha} {}^{227}Ac \xrightarrow[22 \text{ years}]{\beta} {}^{227}Th \xrightarrow[18.6 \text{ days}]{\alpha}$$

The protactinium-231 concentration is calculated from the amount of thorium-227 in the sample. This equilibrium is attained in about 10 actinium-227 half lives, so the time required is very short with respect to geological periods. The chemical separations used are as follows. The sample is fused with sodium peroxide and the melt is dissolved in water; it is acidified with hydrochloric acid and bismuth sulfide is precipitated from the solution to carry polonium and bismuth isotopes. The filtrate is boiled to remove hydrogen sulfide, and zirconium phosphate is precipitated. This precipitate carries thorium and protactinium isotopes. The zirconium phosphate precipitate is dissolved in oxalic acid and parescodymium oxalate is precipitated. This precipitate, which contains protactinium and thorium isotope, is counted for alpha activity at daily intervals for six to ten days. For uranium ores containing little thorium, the thorium-232 and thorium-228 activity is negligible, and the thorium-227 concentration is determined from the growth of thorium-227 daughters (see Fig. 6). The method is applicable to ores with a protactinium-231 to thorium-232 ratio of 50 to 1 or greater. The chemical yield and counting efficiency are determined by carrying a sample of known protactinium-231 content through the procedure.

A complex procedure for the determination of protactinium in ore residues that is designed to insure complete dissolution of the protactinium without the use of hydrofluoric acid or fusions is given by Kirby (162). The sample is leached with hydrochloric acid and the residue is separated. The residue is repeatedly treated with hydrochloric and sulfuric acids containing hydrogen peroxide, and the solution is extracted with diisobutyl carbinol. This treatment is continued until the residue is free from protactinium, as shown by gamma-ray counting of the 27.5-k.e.v. peak. Titanium(III) phosphate is precipitated from the original hydrochloric acid solution to carry protactinium, the precipitate is dissolved in sulfuric and hydrochloric acid, and the protactinium is extracted with diisobutyl carbinol. The organic extracts are finally combined, and are contacted with $1N$ nitric acid–$0.05N$ hydrofluoric acid to back-extract the protactinium; the protactinium-231 content is determined by counting the 0.3-m.e.v. gamma-ray.

2. Irradiated Thorium

The determination of protactinium-233 in neutron-irradiated thorium is important in the production of uranium-233 from thorium-232. The sample may contain any or all of the following: thorium, uranium, and fission products from the neutron irradiation. The most difficult fission products to separate are niobium and zirconium because of their chemical similarity to protactinium. One method is based on solvent extraction with diisobutyl carbinol (216) from $6M$ hydrochloric acid–4% oxalic acid. The oxalic acid is added to inhibit the extraction of niobium, which is produced in trace amounts as a fission product in the neutron bombardment. In the absence of oxalic acid, about 90% of the niobium-95 is extracted. Oxalic acid also inhibits the otherwise small extraction of trace zirconium and ruthenium (present as fission products) and uranium-233 when present in trace or macroscopic amounts. If thorium is present in the sample, oxalic acid is omitted in the original extraction since thorium oxalate would precipitate. In this case the extraction is made from hydrochloric acid alone, but since the organic phase is washed with hydrochloric acid–oxalic acid solution, the zirconium, niobium, ruthenium, and uranium are separated at this stage. Iron(III), if present, is extracted to the extent of 80 to 90%, but the addition of stannous chloride to reduce the iron results in the extraction of only 0.1% of the iron. Under these conditions about 1% of the tin is extracted. The fission product, antimony-125, may be present is some solution, and is also largely extracted. The organic phase containing the protactinium-233 is contacted with $6M$ sulfuric acid–$6M$ hydrofluoric acid to return the protactinium to the aqueous phase. In this step only a few per cent of the antimony in the organic phase is back-extracted, and a repetition of the extraction cycle may be done to reduce the antimony comtamination to negligible proportions. Free iodine (also a fission product) follows protactinium into the organic phase, but less than 0.5% is back-extracted. Other procedures based on the extraction of protactinium by diisobutyl or diisopropyl carbonol from nitric or hydrochloric acid solutions has also been reported (97,117,156).

A procedure based on coprecipitation with niobium has been used (83). The sample is first treated with hydrofluoric acid to remove any thorium as the insoluble fluoride. The excess fluoride is removed by fuming with perchloric acid, and the residue is taken up with oxalic acid solution. Niobium carrier and zirconium, cerium, and strontium holdback carriers are added, the solution is acidified with nitric acid, and the oxalic acid is destroyed with potassium perchlorate. The precipitate of niobium oxide that forms, carrying the protactinium, is separated, washed, and dissolved in oxalic acid. Zirconium holdback carrier and ethylenediaminetetraacetic acid (to complex the zirconium) are added, the acidity is adjusted to pH 5, and the niobium (and protactinium) are precipitated with tannic acid. The tannate is ignited to the oxide and dissolved in nitric–hydrofluoric acid, and the protactinium is separated from niobium by coprecipitation with barium fluozirconate. After several reprecipitations (the precipitate is soluble in boric acid) the barium fluozirconate is counted. When the protactinium activity in the sample exceeds 10^6 disintegrations per minute, the niobium–protactinium separation can be done by paper chromatography. The tannate precipitate is dissolved in hydrofluoric acid. The separation is performed on one one-hundreth of the solution, using Whatman No. 1 paper and $1M$ hydrofluoric acid in ethylmethyl ketone to develop the chromatogram. Another procedure consists simply of extracting the protactinium from nitric solution into $0.5M$ thenoyltrifluoroacetone in 1-hexanol (263). The organic layer is counted with a gamma-ray scintillation spectrometer. Since separation from zirconium and niobium is incomplete the gamma activity due to these fission products must be subtracted from the protactinium gamma activity to obtain an accurate result. This can be done by counting the zirconium–niobium-95 gamma-ray peak at about 0.76 m.e.v., and correcting the protactinium-233 gamma-ray activity at

0.32 m.e.v. for the zirconium–niobium contribution. Another procedure (198) uses extraction with diisopropyl carbinol from 6M nitric acid, followed by back-extraction with 6N hydrochloric–0.5N hydrofluoric acid, adsorption on the anion-exchange resin Amberlite C.G. 400 from hydrochloric acid solution, and elution with 11M hydrochloric acid–0.5N hydrofluoric acid. In this case protactinium-231 was added as a tracer to determine the chemical yield. The yields were 70 to 80%.

3. Geological Samples

The determination of protactinium-231 in ocean water and ocean sediments is of importance in geologic oceanographic studies. The age of deep-sea sediments and cores has been calculated from the activity ratios of protactinium-231 to thorium-230 and protactinium-231 to uranium-235 in such materials. One method that has been used is a modification of a procedure designed for the determination of protactinium-233 in the presence of fission products. In the original method (253), protactinium-231 is added as a tracer for the chemical yield determination, and lanthanum fluoride is precipitated from acid solution to remove fluoride-insoluble fission products. The protactinium is then coprecipitated from the supernatant with barium fluozirconate, the precipitate is dissolved in a boric acid–nitric acid mixture, and the protactinium is carried on the iodate precipitate formed by the addition of iodic acid. The iodate is of unknown composition, and may contain both barium and zirconium. The iodate precipitate is dissolved in a small volume of hydrochloric acid, the solution is diluted with 6M hydrochloric acid, and the protactinium is extracted with diisopropyl carbinol. The organic layer is washed twice with 6M hydrochloric acid and the protactinium is then back-extracted into 2M nitric acid–0.5M hydrofluoric acid. The above separations are repeated twice, beginning with the lanthanum fluoride precipitation, except that the final back-extraction is made into a small volume of 5 to 6N hydrofluoric acid. The latter solution is evaporated and counted. In the procedure as modified for geological samples (264), protactinium-231 is determined with the aid of protactinium-233 as a tracer. Iron(III) is added as a carrier to the acidified ocean water sample, and the protactinium is coprecipitated with ferric hydroxide by the addition of ammonium hydroxide. The protactinium is separated from the iron carrier by two coprecipitations with zirconyl iodate from 4M nitric acid. After dissolution of the iodate, protactinium is extracted from 6M hydrochloric acid with diisobutyl carbinol, and is back-extracted into dilute hydrofluoric acid. The latter solution is evaporated and counted. Sediments are digested with sulfuric and hydrofluoric acids, the hydrofluoric acid is removed by fuming, and the above procedure is followed. Tests made with a solid uranium standard containing a known amount of protactinium-231 indicated that the chemical yield was correctly measured by the protactinium-233 recovery.

Two other methods for the determination of protactinium in geological samples have been reported (40). A chemical procedure consisted of the following steps. The sample solution, in $10M$ hydrochloric acid, was adsorbed on a column of Dowex-1 anion-exchange resin, and the column was washed with $10M$ hydrochloric acid to remove thorium, radium, and lead isotopes. The protactinium was eluted with $9M$ hydrochloric acid–$0.1M$ hydrofluoric acid, the eluate was evaporated to dryness, and the residue was taken up in $6M$ hydrochloric acid. The protactinium was extracted from this solution with a solution of $0.25M$ thenoyltrifluoroacetone in benzene, and the organic phase was evaporated and counted. The second procedure consisted of counting the radon-219 and polonium-215 daughters of protactinium-231. Methane was swept through a solution of the sample and into an alpha proportional counting chamber, carrying with it the radon isotopes in the sample. The counter was operated so it accepted only paired pulses occurring within 10 milliseconds, and thus is sensitive only to the decay of radon-219 and polonium-215 since polonium-215 has a 1.83-millisecond half life. The polonium daughters of the other naturally occurring radon isotopes are sufficiently long so they do not decay significantly within 10 milliseconds after their production from their radon parents. The method is very attractive because a counter operated as specified has essentially no background, and no protactinium separations are required. However, careful calibration and standardization of the entire system is necessary because of the short half life of radon-219 (3.9 seconds). The protactinium-231 must be in equilibrium with its daughters. The thorium and radium precursors of radon-219 are sufficiently short-lived so that any deficiency in these nuclides may be eliminated by allowing time for their growth. However, if the original sample is deficient in actinium-227 (half life of 22 years), low results will be obtained.

G. RECOMMENDED PROCEDURES

1. Determination of Protactinium in Uranium Ore Residues by Alpha Pulse-Height Analysis

a. Alpha Pulse-Height Analysis Following Chemical Separation (166)

Weigh an amount of sample undergoing about 2000 alpha disintegrations per minute into a platinum dish. Moisten the sample with water and add $3M$ nitric acid until gas evolution, due to carbonates, stops. To the suspension add sufficient protactinium tracer to "drown out" the natural beta activity. Add concentrated hydrofluoric acid and one drop of concentrated sulfuric acid and heat under an infrared lamp for about 30 minutes. Allow to cool, centrifuge, and wash the residue with hydrofluoric acid. Combine the supernatant and washings, and evaporate the solution until strong sulfur trioxide fumes are evolved. Add a small amount of $5M$ nitric acid to the residue and allow to stand for about 30 minutes with occasional stirring. Centrifuge and discard any residue. Add sufficient zirconium carrier to the solution to give a concentration of 0.1 mg./ml. and precipitate zirconium iodate by the addition of iodic acid. Centrifuge the iodate precipitate and dissolve it in a minimum quantity of concentrated hydrochloric acid. Suspend a clean silver strip into the solution overnight to remove polonium. Adjust the acidity of the hydrochloric acid solution to 2 to $4M$ and extract with an equal volume of diisopropyl carbinol. Evaporate the carbinol layer on a platinum counting plate, and count in a beta counter and in an alpha pulse-height analyzer. The protactinium yield is calculated from the protactinium-233 recovery and the alpha activity due to protactinium-231 is determined from the alpha-particle spectrum.

b. Alpha Pulse-Height Analysis Without Chemical Separation (88)

This method is applicable to residues containing at least 0.1 p.p.m. of protactinium-231. It was originally applied to the residue remaining after the ether extraction of uranium from solutions of uranium ores. These samples contained thorium-230 and uranium isotopes in addition to protactinium-231. Although the details of calculating the results apply only to such mixtures, this procedure illustrates a technique that can be used to resolve other complex alpha spectra.

For solid samples, weigh out between 0.25 and 1 mg. of the finely ground material on a stainless steel counting disk. Wet the sample with a drop of dilute nitric acid to prevent loss of the dry powder. Add a drop of tetraethylene glycol to the disk and evaporate to dryness on a hot plate. Rotate the disk during the evaporation to aid in obtaining a uniform sample distribution. Ignite the dried sample and count in an alpha pulse-height analyzer. Evaporate nitric acid solutions with tetraethylene glycol in the same manner, using a sample size such that the ignited disk will contain less than 1 mg. of solids, and count as for solid samples. Use the smallest sample volume that will give satisfactory counting rates.

The high-energy tail of the 4.7-m.e.v. protactinium alpha peak overlaps on the low-energy tail of the peak due to the 5.01-m.e.v. protactinium alpha particle (Fig. 21). To correct for this, extrapolate the low-energy side of the 5.01-m.e.v. peak, assuming it to have the same shape as the low-energy side of the 4.7-m.e.v. peak. Similarly, extrapolate the high-energy side of the 4.7-m.e.v. peak, assuming it to have the same shape as the high-energy side of the 5.01-m.e.v. peak. These extrapolations permit the isolation of the complete 5.01-m.e.v. peak. Integrate the number of counts under both extrapolated peaks. The total protactinium activity can now be calculated from the counting efficiency, since it is known that 85.5% of the protactinium disintegrations produce alpha particles under the 5.01-m.e.v. peak, while the remainder appear under the 4.7-m.e.v. peak. The peak at about 4.7 m.e.v. is due to thorium-230, uranium-234, and protactinium-231. The peak at 5.01 m.e.v., after extrapolation, is due only to protactinium-231.

2. Determination of Protactinium-231 in Ore Residues by Gamma-Ray Spectrometry (226)

Weigh about 5 g. of residue into a small vial suitable for gamma-ray counting with a sodium iodide crystal. Determine the gamma-ray counting rate under the 300-k.e.v. peak. A solid or well-type sodium iodide crystal connected to a multiplier phototube, linear amplifier, and multi- or single-channel pulse-height analyzer may be used. The most convenient counting system uses a well-type crystal and multichannel analyzer. Transfer the sample to a beaker, add 100 ml. of $9N$ hydrochloric acid and 1 to 2 ml. of 48% hydrofluoric acid. Heat until the sample dissolves. Add more hydrofluoric acid if necessary. Cool the solution to room temperature. Add slowly, with stirring, 10 ml. of a thorium solution (1 mg. of thorium per ml. of $1N$ hydrochloric acid). Allow the mixture to stand for 5 minutes. Add an additional milliliter of 48% hydrofluoric acid, and allow to stand for 5 minutes. Add a second 10-mg. portion of thorium carrier, stir, and allow to stand for 5 minutes. Filter the thorium fluoride. Transfer the fluoride and paper to the bottle used to count the original sample. Determine the gamma counting rate under the 300-k.e.v. peak. Make a blank determination to correct for the radioactivity of the thorium carrier. Obtain the protactinium counting rate by subtracting the gamma counting rate in the fluoride precipitate from that in the original sample. Count a standard protactinium-231 sample at 300 k.e.v.; from this standardization the protactinium-231 content in the sample can be calculated.

3. Determination of Protactinium-231 by Differential Gamma-Ray Spectrometry (164)

This procedure is limited to the analysis of samples containing only protactinium and its descendants, the actinium-227 chain. Obtain a sample of actinium-227 in equilibrium with its decay products (AEM). The actinium should have been prepared by neutron irradiation of radium-226 to avoid possible protactinium-231 contamination. Prepare a radiochemically pure protactinium-231 standard by one of the following two methods. If the protactinium is in a dry state, dissolve it in hot concentrated sulfuric acid. Dilute the solution to $18N$, cool, and add an equal volume of $12N$ hydrochloric acid plus one or two drops of 30% hydrogen peroxide. If the protactinium is in aqueous solution, adjust the concentration to $9N$ sulfuric acid–$6N$ hydrochloric acid. If fluoride is present it must be removed by fuming with sulfuric acid or complexing with aluminum ion or boric acid. Extract the solution with 2 ml. of diisobutyl carbinol (DIBC) diluted to 50% with benzene or other inert diluent. Wash the organic phase with fresh $9N$ sulfuric acid–$6N$ hydrochloric acid. The organic phase will contain radiochemically pure protactinium-231. If the protactinium is originally in DIBC or other organic solution, wash the organic solution with two or three volumes of $9N$ sulfuric acid–$6N$ hydrochloric acid solution to remove decay products. The protactinium should be repurified at least once a week.

Determine the gamma-ray activity of the unknown, the protactinium standard, and the AEM standard at 90 and 300 k.e.v., using a gamma-ray spectrometer (sodium iodide crystal, linear amplifier, and pulse-height analyzer) as in the previous procedure. Determine the alpha disintegration rate of the AEM and protactinium samples. The alpha disintegration rate of protactinium-231 in the unknown is calculated by determining the fractional contribution of protactinium and AEM to the gamma-ray counting rate in the 300-k.e.v. region.

Calculations

Solve the simultaneous equations below for (Pa):

$$C(90) = R(1)(Ac) + R(2)(Pa)$$

$$C(300) = (Ac) + (Pa)$$

where $R(1)$ = ratio of gamma-ray activity at 90 k.e.v. to that at 300 k.e.v. for the AEM sample; $R(2)$ = same ratio for protactinium standard; $C(90)$ and $C(300)$ = gamma-ray activity of the unknown sample at 90 and

300 k.e.v., respectively; (Ac) = gamma counting rate at 300 k.e.v. of actinium and its descendants in the unknown sample; and (Pa) = gamma counting rate of protactinium at 300 k.e.v. in the unknown sample.

Disintegrations per minute of protactinium-231 (d.p.m.) are calculated from

$$\text{d.p.m.} = (\text{Pa}) \times \text{R}(3)$$

where $R(3)$ is the ratio of alpha disintegrations per minute to the gamma counting rate at 300 k.e.v. for the protactinium standard.

4. Determination of Protactinium-233 in Neutron-Irradiated Thorium
(256)

Dissolve the sample in hydrochloric acid (containing $0.001M$ hydrofluoric acid in the case of thorium metal or difficultly soluble thorium compounds).

Adjust the acidity of a suitable aliquot to $6M$ hydrochloric acid. Extract for 5 minutes with an equal volume of diisobutyl carbinol (previously treated for 5 minutes with an equal volume of $6M$ hydrochloric acid). Allow the phases to separate and discard the aqueous phase. Wash the carbinol phase for 1 to 2 minutes with an equal volume of $6M$ hydrochloric acid. Scrub the organic phase three times for 5 minutes each with equal volumes of $6M$ hydrochloric acid–4% oxalic acid. Discard the scrub solutions. Strip the organic phase of protactinium by extracting for 3 minutes with an equal volume of $6M$ sulfuric acid–$6M$ hydrofluoric acid solution. Discard the organic phase. Wash the aqueous phase for 3 minutes with an equal volume of diisobutyl carbinol and discard the organic phase. Count a suitable aliquot of the aqueous phase for protactinium-233.

5. Preparation of Carrier-Free Protactinium-233 (93)

Thorium carbonate is irradiated with neutrons in a nuclear reactor to produce the protactinium-233. After irradiation, dissolve the carbonate in $8N$ hydrochloric acid and extract with diisopropyl ketone. Strip the ketone layer with $2N$ hydrochloric acid to return the protactinium to the aqueous layer. Adjust the acidity of the aqueous layer to $8N$ hydrochloric acid and repeat the extraction and stripping. Alternately, dissolve the carbonate in $8N$ hydrofluoric acid and saturate the solution with aluminum chloride before the first extraction. The separation is completed as before. This modification gives improved yields.

ACKNOWLEDGMENT

The author is grateful for the assistance provided by William Fairman, Fred Iwami, and Dale Henderson, all of Argonne National Laboratory, in obtaining the alpha and gamma spectra shown in Figs. 7 to 11, 14 to 17, 19, and 20.

REFERENCES

1. Adloff, J. P., cited in E. K. Hyde, *The Radiochemistry of Francium*, U. S. At. Energy Comm., **NAS-NS-3003**, 23, 27, 91 (1960).
1a. Adloff, J. P., and R. Bertrand, *Compt. Rend.*, **254**, 2575 (1962).
2. Ancaraini, L., and L. Riva, *Studi Ric. Div. Geomineraria, Comit. Nazl. Ric. Nucl.*, **2**, 155 (1959); through *Comm. Energie At. (France) Rapport*, **CEA-tr-I-137** (1960).
3. Andrews, H. C., and F. Hageman, U. S. At. Energy Comm., **ANL-4176** (1948).
4. Anthony, D. S., S. E. Campbell, G. R. Hagee, and E. S. Robajdek, *Radiation Res.*, **4**, 286 (1956).
5. Appelman, E. H., *The Radiochemistry of Astatine*, U. S. At. Energy Comm., **NAS-NS-3012** (1960).
6. Appelman, E. H., U. S. At. Energy Comm., **UCRL-9025** (1960).
7. Appelman, E. H., *J. Am. Chem. Soc.*, **83**, 805 (1961).
8. Appelman, E. H., *J. Phys. Chem.*, **65**, 325 (1961).
9. Aten, A. H. W., T. Doorgeest, V. Hollstein, and H. P. Moeken, *Analyst*, **77**, 774 (1952).
10. Bagnall, K. W., *Chemistry of the Rare Radioelements*, Academic Press, New York, 1957.
11. Bagnall, K. W., J. H. Freeman, D. S. Robertson, P. S. Robinson, and M. A. A. Stewart, U. K. At. Energy Authority Repts., **AERE C/R 2566** (1958).
12. Bagnall, K. W., "Some Aspects of Polonium Chemistry," in *Proc. Intern. Conf. Peaceful Uses At. Energy, Geneva, 1955*, **7**, 386 (1956).
13. Bagnall, K. W., *Quart. Rev. (London)*, **11**, 30 (1957).
14. Bagnall, K. W., and R. W. M. D'Eye, *J. Chem. Soc.*, **1954**, 4295.
15. Bagnall, K. W., R. W. M. D'Eye, and J. H. Freeman, *J. Chem. Soc.*, **1955**, 2320.
16. Bagnall, K. W., R. W. M. D'Eye, and J. H. Freeman, *J. Chem. Soc.*, **1955** 3959.
17. Bagnall, K. W., R. W. M. D'Eye, and J. H. Freeman, *J. Chem. Soc.*, **1956**, 3385.
18. Bagnall, K. W., and J. H. Freeman, *J. Chem. Soc.*, **1956**, 2770.
19. Bagnall, K. W., and J. H. Freeman, *J. Chem. Soc.*, **1957**, 2161.
20. Bagnall, K. W., and D. S. Robertson, *J. Chem. Soc.*, **1957**, 509.
21. Bagnall, K. W., and D. S. Robertson, *J. Chem. Soc.*, **1957**, 1044.
22. Bagnall, K. W., D. S. Robertson, and M. A. A. Stewart, *J. Chem. Soc.*, **1958**, 3633.
23. Bailey, R. A., and L. Yaffe, *Can. J. Chem.*, **38**, 1871 (1960).
24. Barnes, D. E., and D. Taylor, *Radiation Hazards and Protection*, Pitman Publ. Corp., New York, 1958.
25. Barnett, M. K., *J. Inorg. Nucl. Chem.*, **4**, 358 (1957).
26. Barton, G., A. Ghiorso, and I. Perlman, *Phys. Rev.*, **82**, 13 (1951).
27. Basson, J. K., *Anal. Chem.*, **28**, 1472 (1956).
28. Baxter, R., and D. Wood, U. S. At. Energy Comm., **UR-269** (1953).
29. Beamer, W. H., and C. R. Maxwell, *J. Chem. Phys.*, **14**, 569 (1946).

30. Beamer, W. H., and C. R. Maxwell, *J. Chem. Phys.*, **17**, 1293 (1949).
30a. Belyaev, B. N., V. Yun-Yui, E. N. Sinotova, L. Nemet, and V. A. Khalkin, *Radiokhimiya*, **2**, 603 (1960).
31. Birks, F. T., *U. K. At. Energy Authority Rept.*, **AERE C/R 2474** (1959).
32. Black, S. C., *U. S. At. Energy Comm.*, **UR-463** (1956).
33. Blatz, H., Ed., *Radiation Hygiene Handbook*, McGraw-Hill, New York, 1959.
34. Bonet-Maury, P., *Ann. Phys.*, **11**, 253 (1929).
35. Bouissières, G., and C. Ferradini, *Anal. Chim. Acta*, **4**, 610 (1950).
36. Bouissières, G., and S. Odiot, *Bull. Soc. Chim. France*, **1951**, 918.
37. Bouissières, G., and J. Vernois, *Compt. Rend.*, **244**, 2508 (1957).
38. Bouteiller, E., *Comm. Energie At. (France) Rapport*, **CEA 1697** (1960).
39. Broido, A., J. D. Teresi, and P. Tompkins, *U. S. At. Energy Comm.*, **MDDC-598** (1946).
40. Broecker, W. S., *U. S. At. Energy Comm.*, **TID-6095** (1960).
41. Brooks, L. S., *J. Am. Chem. Soc.*, **77**, 3211 (1955).
42. Brown, D., A. J. Smith, and R. G. Wilkins, *J. Chem. Soc.*, **1959**, 1463.
42a. Brown, D., and R. G. Wilkins, *J. Chem. Soc.*, **1961**, 3804.
43. Bunney, L. R., H. E. Ballou, J. Pascual, and S. Foti, *Anal. Chem.*, **31**, 324 (1959).
44. Burton, W. M., and N. G. Stewart, *U. K. At. Energy Authority Rept.*, **AERE HP/R 2084** (1960).
45. Cabell, M. J., *Can. J. Chem.*, **37**, 1094 (1959).
46. Cairo, A., "Separation of Polonium with Diisopropylketone," in *Proc. 2nd Intern. Conf. Peaceful Uses At. Energy, Geneva, 1958*, **7**, 331 (1958).
47. Camarcat, M., M. G. Bouissières, and M. Haïssinsky, *J. Chim. Phys.*, **46**, 153 (1949).
48. Casey, A. T., and A. G. Maddock, *J. Inorg. Nucl. Chem.*, **10**, 58 (1959).
49. Charles G. W., D. J. Hunt, G. Pish, and D. L. Timma, *U.S. At. Energy Comm.*, **MLM-941** (1954); *J. Opt. Soc. Am.*, **45**, 869 (1955).
50. Cember, H., J. A. Watson, and T. B. Grucci, *Nucleonics*, **12**, No. 8, 40 (1954).
51. *Chem. Eng. News*, **39**, 48 (Aug. 7, 1961).
52. Cotelle, S., and M. Haïssinsky, *Compt. Rend.*, **206**, 1644 (1938).
53. Collins, D. A., J. J. Hillary, and G. M. Phillips, *U. K. At. Energy Authority Rept.*, **DEGR-18W** (1959).
54. Corson, D. R., K. MacKenzie, and E. Segré, *Phys. Rev.*, **58**, 672 (1940).
55. Craig, L. C., *Anal. Chem.*, **21**, 85 (1949).
56. Crouthamel, C., Ed., *Applied Gamma-Ray Spectrometry*, Pergamon Press, New York, 1960.
57. Crouthamel, C., and C. Gatrousis, *Talanta*, **1**, 39 (1958).
58. Cunningham, B. B., "Thermodynamics of the Heavy Elements," in *Proc. Intern. Conf. Peaceful Uses At. Energy, Geneva, 1955*, **7**, 225 (1956).
59. Danon, J., *J. Inorg. Nucl. Chem.*, **7**, 422 (1958).
60. Danon, J., and C. Ferradini, *Compt. Rend.*, **234**, 1361 (1952).
61. Danon, J., and M. C. Levi, *J. Chromatog.*, **3**, 193 (1960).
62. Danon, J., and A. A. L. Zamith, *J. Phys. Chem.*, **61**, 431 (1957); *Nature*, **177**, 746 (1956).
63. Diamond, R. M., K. Street, Jr., and G. T. Seaborg, *J. Am. Chem. Soc.*, **76**, 1461 (1954).
64. Dickey, E. E., *J. Chem. Ed.*, **30**, 525 (1953).
65. Durbin, P. W., J. G. Hamilton, and M. W. Parrott, *U. S. At. Energy Comm.*, **UCRL-2792** (1954).

66. Duyckaerts, G., and R. Lejeune, *J. Chromatog.*, **3**, 58 (1960).
67. Elson, R. E., "The Chemistry of Protactinium," in G. T. Seaborg and J. J. Katz, Eds., *The Actinide Elements*, McGraw-Hill, New York, 1954, Chap. 5.
68. Elson, R. E., G. W. Mason, D. F. Peppard, P. A. Sellers, and M. H. Studier, *J. Am. Chem. Soc.*, **73**, 4974 (1951).
69. Erbacher, O., and K. Philipp, *Z. Physik.*, **51**, 309 (1928).
70. Eutsler, B. C., M. F. Milligan, and M. C. Robbins, *U. S. At. Energy Comm.*, **LA-1904** (1955).
71. Fajans, K., and O. Gohring, *Phyzik. Z.*, **14**, 877 (1913).
72. Feldman, I., and M. Frisch, *Anal. Chem.*, **28**, 2024 (1956).
73. Feldman, M. H., *U. S. At. Energy Comm.*, **WIN-121** (1961).
74. Ferradini, C., *J. Chim. Phys.*, **53**, 714 (1956).
75. Figgins, P. E., *The Radiochemistry of Polonium*, U. S. At. Energy Comm., **NAS-NS-3037** (1961).
76. Fink, R. W., G. W. Warren, R. R. Edwards, and P. E. Damon, *Phys. Rev.*, **103**, 651 (1956).
77. Foster, K. W., *U. S. At. Energy Comm.*, **MLM-901** (1953).
78. Gouagre, J., and W. W. Meinke, *U. S. At. Energy Comm.*, **AECU-3887**, 48 (1958), "A Source Reservoir for Francium-223," in *Radioisotopes in the Physical Sciences and Industry*, International Atomic Energy Authority, Vienna, 1962, p. 511.
79. Fred, M., and F. S. Tompkins, *J. Opt. Soc. Am.*, **39**, 357 (1949).
80. Fried, S., F. Hageman, and W. H. Zachariasen, *J. Am. Chem. Soc.*, **72**, 771 (1950).
81. Fried, S., and J. C. Hindman, *J. Am. Chem. Soc.*, **76**, 4863 (1954).
82. Frierson, W. J., and J. W. Jones, *Anal. Chem.*, **23**, 1447 (1951).
83. Fudge, A. J., and J. L. Woodhead, *Analyst*, **81**, 417 (1956).
84. Garrison, W., J. Gile, R. D. Maxwell, and J. G. Hamilton, *Anal. Chem.* **23**, 204 (1951).
85. Giorgi, A. L., M. G. Bowman, G. W. Rupert, and D. T. Vier, *U. S. At. Energy Comm.*, **LA-1551** (1953).
86. Glendenin, L. E., and C. M. Nelson, "Perchlorate Method for the Determination of Cesium Activity in Fission," in C. D. Coryell and N. Sugarman Eds., *Radiochemical Studies: The Fission Products*, McGraw-Hill, New York, 1951, Book 3, p. 1642.
87. Glover, K. M., A. B. Beadle, and F. J. G. Rogers, *U. K. At. Energy Authority Rept.*, **AERE C/R 2359** (1957).
88. Glover, K. M., and F. J. G. Rogers, *U. K. At. Energy Authority Rept.*, **AERE/2971** (1959).
89. Goble, A., J. Golden, and A. J. Maddock, *Can. J. Chem.*, **34**, 284 (1956).
90. Goble, A. G., and A. G. Maddock, *J. Inorg. Nucl. Chem.*, **7**, 94 (1958).
91. Goble, A. G., and A. G. Maddock, *Trans. Faraday Soc.*, **55**, 591 (1959).
92. Gofman, J. W., and G. T. Seaborg, "New Isotopes of Protactinium and Uranium," in L. I. Katzin, Ed., *Production and Separation of U^{233}*, U. S. At. Energy Comm., **TID-5222** (Pt. 1), Paper 2.4 (1960).
93. Golden, J., and A. G. Maddock, *J. Inorg. Nucl. Chem.*, **2**, 46 (1956).
94. Gomm, P. J., *U. K. At. Energy Authority Rept.* **AERE MED/M 22** (1958).
95. Goodoll, C. A., and R. L. Moore, *J. Inorg. Nucl. Chem.*, **11**, 290 (1959).
96. Gray, P. R., and S. G. Thompson, in *U. S. At. Energy Comm.*, **URCL-2069**, 29 (1952).

97. Gresky, A. T., *U. S. At. Energy Comm.*, **CF-56-2-157** (1956).
98. Gresky, A. T., et al., *U. S. At. Energy Comm.*, **ORNL-1367,** 71 (1952).
99. Hageman, F., *U. S. At. Energy Comm.*, **ANL-4215** (1948).
100. Hageman, F., *J. Am. Chem. Soc.*, **72,** 768 (1950).
101. Hageman, F., L. I. Katzin, M. H. Studier, G. T. Seaborg, and A. Ghiorso, *Phys. Rev.*, **79,** 435 (1950).
102. Hageman, F., "The Chemistry of Actinium," in G. T. Seaborg and J. J. Katz, Eds., *The Actinide Elements*, McGraw-Hill, New York, 1954, Chap. 2.
103. Hahn, O., *Applied Radiochemistry*, Cornell Univ. Press, Ithaca, N. Y., 1936.
104. Hahn, O., and L. Meitner, *Ber.*, **523,** 1812 (1919); *Physik. Z.*, **20,** 127 (1919).
105. Haïssinsky, M., *Compt. Rend.*, **196,** 1788 (1933).
106. Haïssinsky, M., *Experientia*, **8,** 125 (1952).
107. Haïssinsky, M., and G. Bouissières, *Compt. Rend.*, **226,** 573 (1948).
108. Haïssinsky, M., and G. Bouissières, *J. Chem. Soc.*, **1949,** 5256.
109. Haïssinsky, M., and G. Bouissières, *Bull. Soc. Chim. France*, **1951,** 146 (available in English translation as *U. S. At. Energy Comm.*, **AEC-Tr-1878** (1951)).
110. Haïssinsky, M., and G. Bouissières, "Protactinium," in P. Pascal, Ed., *Nouveau Traite de Chemie Minerale*, Masson et Cie, Paris, 1958, Tome XII, pp. 617–680 (available in English translation by H. W. Kirby as *U. S. At. Energy Comm.*, **MLM-1101-Tr**).
111. Hall, K. L., and D. H. Templeton, *U. S. At. Energy Comm.*, **URCL-957** (1950).
112. Hamilton, J., K. G. Scott, C. W. Asling, P. C. Wallace, and G. Thile, *U. S. At. Energy Domm.*, **AECD-3200,** 4 (1951).
113. Hamilton, J. G., C. Asling, W. Garrison, and K. Scott, *The Accumulation, Metabolism, and Biological Effects of Astatine in Rats and Monkeys*, Univ. of Calif. Press, Berkeley, 1953.
114. Hardy, C. J., D. Scargill, and J. M. Fletcher, *J. Inorg. Nucl. Chem.*, **7,** 257 (1958).
115. Hevesy, G., and F. Paneth, *Monatsh.*, **36,** 45 (1915); *Physik. Z.*, **15,** 797 (1914).
116. Hollander, J. N., I. Perlman, and G. T. Seaborg, *Rev. Mod. Phys.*, **25,** 469 (1953).
117. Hudgens, J. E., B. Warren, and F. L. Moore, *U. S. At. Energy Comm.*, **Mon-N-234** (1947).
118. Hull, D. E., W. F. Libby, and W. M. Latimer, *J. Am. Chem. Soc.*, **57,** 1649 (1935).
119. Hunt, D. J., *U. S. At. Energy Comm.*, **MLM-979** (1954).
120. Hunt, D. J., and G. Pish, *J. Opt. Soc. Am.*, **46,** 87 (1956).
121. Hursh, J. B., Ed., *U. S. At. Energy Comm.*, **AECU-4024,** 13 (1958).
122. Hyde, E. K., in W. W. Meinke, Ed., *U. S. At. Energy Comm.*, **AECD-2750,** 3 (1949).
123. Hyde, E. K., "Radiochemical Separations of the Actidine Elements," in G. T. Seaborg and J. J. Katz, Eds., *The Actinide Elements*, McGraw-Hill, New York, 1949, Chap. 15.
124. Hyde, E. K., *J. Am. Chem. Soc.*, **74,** 4181 (1952).
125. Hyde, E. K., *J. Phys. Chem.*, **58,** 21 (1954).
126. Hyde, E. K., *Phys. Rev.*, **94,** 1221 (1954).
127. Hyde, E. K., "Radiochemical Separation Methods for the Actinide Elements," in *Proc. Intern. Conf. Peaceful Uses At. Energy, Geneva, 1955*, **7,** 281 (1956).
128. Hyde, E. K., *J. Chem. Ed.*, **36,** 15 (1959).
129. Hyde, E. K., *The Radiochemistry of Francium*, *U. S. At. Energy Comm.*, **NAS-NS-3003** (1960).
130. Hyde, E. K., and A. Ghiorso, *Phys. Rev.*, **90,** 267 (1953).

131. Hyde, E. K., A. Ghiorso, and G. T. Seaborg, *Phys. Rev.*, **77**, 765 (1950).
132. Hyde, E. K., and M. J. Wolf, "A Survey of Solvents for the Extraction of U^{233} and Pa^{233} from Neutron-Irradiated Thorium," in L. I. Katzin, Ed., *Production and Separation of U^{233}, U. S. At. Energy Comm.*, **TID-5223** (Pt. 1), Paper 3.12, (1960).
133. Iddings, G. M., R. L. Tellefson, and R. N. Osborne, in M. Lindner, Ed., *U. S. At. Energy Comm.*, **UCRL-4377**, 40 (1954).
134. Imre, L., *Z. Physik. Chem.*, **127** (Abt. A.) 262 (1931).
135. Industrial Group, Risley, England, *U. K. At. Energy Authority Rept.*, **IGO-AM/W-167** (1958).
136. International Atomic Energy Agency, *Safe Handling of Radio-isotopes*, Vienna, 1958.
137. International Committee on Radiological Protection, *Brit. J. Radiol.*, **Suppl. 6** (1954).
138. International Committee on Radiation Protection, *Health Phys.*, **3**, 1 (1960); National Bureau of Standards, *Handbook No. 69*, Government Printing Office, Washington, D. C., 1959.
138a. Isabaev, E. A., U. K. Asylbaev, and V. V. Cherdyntsev, *Radiokhimiya*, **2**, 98 (1960).
139. Ishimori, T., *Bull. Chem. Soc. Japan*, **28**, 432 (1955).
140. Ishimori, T., and H. Sakaguchi, *J. Chem. Soc. Japan*, **78**, 327 (1950).
141. Ishimori, T., K. Watanabe, and K. Kimura, *J. At. Energy Soc. Japan*, **2**, 750 (1960).
142. Jackson, N., F. J. G. Rogers, and J. F. Short, *U. K. At. Energy Authority Rept.*, **AERE-R-3311** (1960).
143. Jackson, N., F. J. G. Rogers, and J. F. Short, *U. K. At. Energy Authority Rept.*, **AERE-R-3377** (1960).
144. Jakovac, Z., and M. Lederer, *J. Chromatog.*, **1**, 289 (1958).
145. Jakovac, Z., and M. Lederer, *J. Chromatog.*, **2**, 411 (1959).
146. Jeng-Tsong, Y., *Compt. Rend.*, **231**, 1059 (1950).
147. Johnson, G. L., R. F. Leininger, and E. Segré, *J. Chem. Phys.*, **17**, 1 (1949).
148. Joliot, F., *J. Chim. Phys.*, **27**, 119 (1930) (an English translation is available as *U. S. At. Energy Comm.*, **UCRL-Trans-70** (1953)).
149. Karl, A., *Sitzber. Akad. Wiss. Wien Abt. IIa*, **140**, 199 (1931); through *Chem. Abstr.*, **25**, 5618 (1931).
150. Karraker, D. G., and D. H. Templeton, *Phys. Rev.*, **81**, 510 (1951).
151. Katz, J. J., and G. T. Seaborg, *The Chemistry of the Actinide Elements*, Methuen and Co., London, 1957.
152. Katzin, L. I., and R. W. Stoughton, *J. Inorg. Nucl. Chem.*, **3**, 229 (1956).
153. Katzin, L. I., Q. Van Winkle, and J. Sedlet, *J. Am. Chem. Soc.*, **72**, 4815 (1950).
154. Keder, W. E., J. C. Sheppard, and H. S. Wilson, *J. Inorg. Nucl. Chem.*, **12**, 327 (1960).
155. Kimura, K., *Bunseki Kagaku*, **6**, 637 (1957); translation available as *Comm. Energie At. (France) Rapport*, **CEA-Tr-X-154** (1959).
156. Kimura, K., *Bull. Chem. Soc. Japan*, **28**, 535 (1955).
157. Kimura, K., *J. At. Energy Soc. Japan*, **2**, 585 (1960).
158. Kimura, K., and T. Ishimori, "Some Studies on the Tracer Chemistry of Polonium," in *Proc. 2nd Intern. Conf. Peaceful Uses At. Energy, Geneva, 1958*, **28**, 151 (1958).

159. King, P., L. B. Lockhart, Jr., R. A. Baus, R. L. Patterson, Jr., H. Friedman, and I. H. Blifford, Jr., *Nucleonics*, **14**, No. 6, 78 (1956).
160. Kinsman, S., Ed., *Radiological Health Handbook*, U. S. Dept. of Health, Education, and Welfare, Washington, D. C., 1960.
161. Kirby, H. W., *Anal. Chem.*, **26**, 1063 (1954).
162. Kirby, H. W., *The Radiochemistry of Protactinium*, U. S. At. Energy Comm., **NAS-NS-3016** (1959).
163. Kirby, H. W., and R. M. Brodbeck, *U. S. At. Energy Comm.*, **MLM-1003** (1954).
164. Kirby, H. W., and P. E. Figgins, in H. W. Kirby, *The Radiochemistry of Protactinium*, U. S. At. Energy Comm., **NAS-NS-3016** (1959).
164a. Kourim, V., A. K. Lavrukhina, and S. S. Rodin, *J. Inorg. Nucl. Chem.*, **21**, 375 (1961).
165. Krasnov, K. S., and G. A. Krestov, *Radiokhimiya*, **2**, 671 (1960).
166. Kraus, K. A., and A. Garen, *U. S. At. Energy Comm.*, **ORNL-300** (1949).
167. Kraus, K. A., and A. Garen, in Hyde, E. K., "Radiochemical Separation Methods for the Actinide Elements," in *Proc. Intern. Conf. Peaceful Uses At. Energy, Geneva, 1955*, **7**, 281 (1956).
168. Kraus, K. A., and G. E. Moore, *J. Am. Chem. Soc.*, **73**, 2900 (1951).
169. Kraus, K. A., and G. E. Moore, *J. Am. Chem. Soc.*, **72**, 4293 (1950).
170. Kraus, K. A., and G. E. Moore, *J. Am. Chem. Soc.*, **77**, 1383 (1955).
172. Kraus, K. A., and Q. Van Winkle, "Extraction of Protactinium from Carbonate-Insoluble Residues of Uranium Ores," in L. I. Katzin, Ed., *Production and Separation of U^{233}*, U. S. At. Energy Comm., **TID-5223** (Pt. 1), Paper 6.2 (1952).
173. Krause, D. P., *U. S. At. Energy Comm.*, **ANL-5967** 126 (1959).
174. Kyi, R. T., *U. S. At. Energy Comm.*, **UCRL-8867**, 61 (1959).
175. Lange, R., *U. S. At. Energy Comm.*, **NYO-2301**, 44 (1961).
176. Larson, R. G., L. I. Katzin, and E. Hausman, "Extraction of Protactinium from Carbonate-Insoluble Residues of Uranium Ores," in L. I. Katzin, Ed., *Production and Separation of U^{233}*, U. S. At. Energy Comm., **TID-5223** (Pt. 1), Paper 6.13 (1952).
177. Lavrukhina, A. K., *Radiokhimiya*, **1**, 204 (1959).
178. Lavrukhina, A. K., A. A. Posdnyakov, and S. S. Rodin, *Intern. J. Appl. Radiation Isotopes*, **9**, 34 (1960).
179. Lavrukhina, A. K., A. A. Posdnyakov, and S. S. Rodin, *Doklady Akad. Nauk SSSR*, **130**, 88 (1960); through *English Translation of the Proceedings of the Academy of Sciences of the U.S.S.R.*, **130**, No. 1–6, Consultants Bureau, Inc., New York, Jan.–Feb., 1960.
180. Lecoin, M., M. Perey, and A. Pompei, *J. Chim. Phys.*, **46**, 158 (1949).
181. Lederer, M., *Compt. Rend.*, **236**, 200, 1557 (1953); *Anal. Chim. Acta*, **11**, 145 (1954).
182. Lederer, M., *J. Chromatog.*, **1**, 172 (1958).
183. Lederer, M., and J. Vernois, *Compt. Rend.*, **244**, 2388 (1957).
184. Lefort, M., G. Simonoff, and X. Tarrago, *Bull. Soc. Chim. France*, **1960**, 1726.
185. Levi, M. C., and J. Danon, *J. Chromatog.*, **3**, 584 (1960).
186. Lima, F. W., *J. Chem. Ed.*, **31**, 153 (1954).
187. Lindner, M., in P. E. Figgins, *The Radiochemistry of Polonium*, U. S. At. Energy Comm., **NAS-NS-3037**, 53–62 (1961).
188. Lindner, M., Ed., *U. S. At. Energy Comm.*, **UCRL-4377** (1954).
189. Mabuchi, H., *Bull. Chem. Soc. Japan*, **31**, 245 (1958).

190. Maddock, A. G., and G. L. Miles, *J. Chem. Soc.*, **1949**, S253.
191. Maddock, A. G., and W. Pugh, *J. Inorg. Nucl. Chem.*, **2**, 114 (1956).
192. Marckwald, W., *Chem. Ber.*, **35**, 4239 (1902).
193. Marckwald, W., *Chem. Ber.*, **36**, 2662 (1903).
194. Marechal-Cornil, J., and E. Picciotto, *Bull. Soc. Chim. Belges.*, **62**, 372 (1953).
195. Matsuura, W., and M. Haïssinsky, *J. Chim. Phys.*, **55**, 475 (1958).
196. Maxwell, C. R., *J. Chem. Phys.*, **17**, 1288 (1949).
197. Maxwell, C. R., and Beamer, W. H., *U. S. At. Energy Comm.*, **MDDC-721** (**LA-604**) (1946).
198. McCormack, J. J., F. H. Cripps, and W. H. Wiblin, *Anal. Chim. Acta*, **22**, 408 (1960).
199. McIssac, L. D., and E. C. Freiling, *Nucleonics*, **14**, No. 10, 65 (1956).
200. McLane, C. K., and S. Peterson, in G. T. Seaborg, J. J. Katz, and W. M. Manning, Eds., *The Transuranium Elements*, Part II, McGraw-Hill, New York, 1949, Paper 19.3.
201. McLane, C. K., and S. Peterson, *ibid.*, Paper 19.6.
202. McLane, C. K., and S. Peterson, *ibid.*, Paper 19.7.
203. Meggers, W. F., M. Fred, and F. S. Tomkins, *J. Opt. Soc. Am.*, **41**, 867 (1951).
204. Meinke, W. W., Ed., *U. S. At. Energy Comm.*, **AECD-2738** (1949).
205. Meinke, W. W., Ed., *U. S. At. Energy Comm.*, **AECD-2750** (1949).
206. Meitner, L., *Physik. Z.*, **12**, 1097 (1911).
207. Meyer, H. E., *U. S. At. Energy Comm.*, **WASH-736**, 7 (1957).
208. Miles, G. L., *Rev. Pure Appl. Chem.*, **2**, 163 (1952).
209. Minto, W. L., "Digestion of Tissues and Excreta," in R. M. Fink, Ed., *Biological Studies with Polonium, Radium, and Plutonium*, McGraw-Hill, New York, 1950, pp. 15–18.
210. Mitchell, R. F., *Anal. Chem.*, **32**, 326 (1960).
210a. Mjassoedou, M., and R. Muxart, *Bull. Soc. Chim. France*, **1962**, 237.
211. Moore, F. L., *Anal. Chem.*, **27**, 70 (1955).
212. Moore, F. L., *Anal. Chem.*, **29**, 1660 (1957).
213. Moore, F. L., *Metal Analysis with Thenoyltrifluoroacetone*, Am. Soc. Testing Mater., Spec. Tech. Publ. ,238 (1958).
214. Moore, F. L., *Anal. Chem.*, **32**, 1048 (1960).
215. Moore, F. L., *Liquid-Liquid Extraction with High-molecular-weight Amines*, U. S. At. Energy Comm., **NAS-NS-3101** (1960).
216. Moore, F. L., and S. A. Reynolds, *Anal. Chem.*, **29**, 1596 (1957).
217. Morrow, P. E., R. J. Della Rosa, L. J. Casarett, and G. J. Miller, *U. S. At. Energy Comm.*, **UR-363** (1954).
218. Moyer, H. V., Ed., *Polonium*, U. S. At. Energy Comm., **TID-5221** (1956).
219. Mulhaney, T. J., W. P. Norris, and W. T. Kisieleski, *U. S. At. Energy Comm.*, **ANL-4333**, 103 (1949).
220. Murbach, E. W., *U. S. At. Energy Comm.*, **NAA-SR-1988** (1957).
221. Muxart, R., M. Levi, and G. Bouissières, *Compt. Rend.*, **249**, 1000 (1959).
222. Nairn, J. S., D. A. Collins, H. A. C. McKay, and A. G. Maddock, "The Extraction of Actinide Elements from Wastes," in *Proc. 2nd Intern. Conf. Peaceful Uses At. Energy*, Geneva, 1958, 17, 216 (1958).
223. National Bureau of Standards *Handbook No. 59*, Government Printing Office, Washington, D. C., 1954.
224. Neumann, H. M., *J. Inorg. Nucl. Chem.*, **4**, 349 (1957).

225. Oelsner, G., and W. Forsling, *Arkiv. for Kemi*, **11**, 349 (1957).
226. Oliver, J. R., J. R. Meriwether, and R. H. Rainey, *U. S. At. Energy Comm.*, **ORNL-2668** (1959).
227. Orban, E., *U. S. At. Energy Comm.*, **MLM-973** (1954).
228. Parrott, M., W. M. Garrison, P. W. Durbin, M. Johnston, H. S. Powell, and J. G. Hamilton, *U. S. At. Energy Comm.*, **URCL-3065** (1955).
229. Peppard, D. F., G. W. Mason, P. R. Gray, and J. F. Mech, *J. Am. Chem. Soc.*, **74**, 6081 (1952).
230. Peppard, D. F., G. W. Mason, W. J. Driscoll, and R. J. Sironen, *J. Inorg. Nucl. Chem.*, **7**, 276 (1958).
231. Peppard, D. F., G. W. Mason, and M. V. Gergel, *J. Inorg. Nucl. Chem.*, **3**, 370 (1956).
232. Perey, M., *Compt. Rend.*, **208**, 97 (1939).
233. Perey, M., *J. Phys. Radium*, **10**, 435 (1939).
234. Perey, M., *Compt. Rend.*, **214**, 797 (1942).
235. Perey, M., *J. Chim. Phys.*, **43**, 155 (1946).
236. Perey, M., *J. Chim. Phys.*, **43**, 262 (1946).
237. Perey, M., *J. Chim. Phys.*, **43**, 269 (1946).
238. Perey, M., *Compt. Rend.*, **241**, 953 (1955).
239. Perey, M., *Compt. Rend.*, **242**, 2552 (1956).
240. Perey, M., *Compt. Rend.*, **243**, 1411 (1956).
241. Perey, M., *Compt. Rend.*, **243**, 1520 (1956).
242. Perey, M., "Francium," in P. Pascal, Ed., *Nouveau Traite de Chemie Minerale*, Masson et Cie, Paris, 1957, Vol. III, pp. 131–141.
243. Perey, M., and J. P. Adloff, *Compt. Rend.*, **236**, 1163 (1953).
244. Perey, M., and J. P. Adloff, *Compt. Rend.*, **239**, 1389 (1954).
245. Perey, M., and J. P. Adloff, in *Proceedings of the XVI Congrés International de Chemie Pure et Appliquée, Section de Chemie Minérale*, Paris, 1957, pp. 333–336.
246. Perey, M., and A. Chevallier, *Compt. Rend. Soc. Biol.*, **145**, 1208 (1951).
247. Peterson, M. D., and M. T. Kelley, *U. S. At. Energy Comm.*, **MonT-253** (1947).
248. Peterson, S., in G. T. Seaborg, J. J. Katz, and W. M. Manning, Eds., *The Transuranium Elements*, Part II, McGraw-Hill, New York, 1949, Paper 19.8.
249. Petrow, H. G., *U. S. At. Energy Comm.*, **TID-11282** (1961).
250. Petrow, H. G., R. J. Allen, R. Lindstrom, and B. Sohn, *U. S. At. Energy Comm.*, **TID-6870** (Part III), (1961).
251. Pluchet, E., and R. Muxart, *Bull. Soc. Chim. France*, **1961**, 372.
252. Poskanzer, A. M., and B. M. Foreman, Jr., *J. Inorg. Nucl. Chem.*, **16**, 323 (1961).
253. Potratz, H. A., and N. A. Bonner, in J. Kleinberg, Ed., *U. S. At. Energy Comm.*, **LA-1721** (2nd ed.), Pa-1 (1958).
254. Power, W. H., *U. S. At. Energy Comm.*, **MLM-909** (1953).
255. Radhakrishna, P., *J. Chim. Phys.*, **51**, 354 (1954).
256. Reynolds, S. A., in D. E. Ferguson and E. L. Nicholson, Eds., *U. S. At. Energy Comm.*, **ORNL-715** (1950).
257. Ripan, R., R. Paladi, and H. Hulubei, "A Natural Isotope of Element 84 with a Very Long Half Life," in *Proc. Intern. Conf. Peaceful Uses At. Energy, Geneva, 1955*, **7**, 392 (1956).
258. Rogers, N. E., and R. M. Watrous, *Anal. Chem.*, **27**, 2009 (1955).
259. Rosholt, J. N., *Anal. Chem.*, **26**, 1307 (1954).
260. Rosholt, J. N., *Anal. Chem.*, **29**, 1398 (1957).

261. Rosenblum, C., and E. W. Kaiser, *J. Phys. Chem.*, **39,** 797 (1935).
262. Rundo, J., *U. K. At. Energy Authority Rept.*, **AERE HP/A 627** (1950).
263. Sabol, W. W., and B. F. Rider, *U. S. At. Energy Comm.*, **KAPL-1477** (1956).
264. Sackett, W. M., *Science*, **132,** 1761 (1960).
265. Salutsky, M. L., K. Shaver, A. Elmlinger, and M. L. Curtis, *J. Inorg. Nucl. Chem.*, **3,** 289 (1956).
266. Salutsky, M. L., M. L. Curtis, K. Shaver, A. Elmlinger, and R. A. Miller, *Anal. Chem.*, **29,** 373 (1957).
267. Salutsky, M. L., and H. W. Kirby, *Anal. Chem.*, **28,** 1780 (1956).
268. Sanielevici, A. S., *J. Chim. Phys.*, **33,** 759 (1936).
269. Sakasi, Y., *Bull. Chem. Soc. Japan*, **28,** 89 (1955).
270. Schuler, F. W., F. L. Steahly, and R. W. Stoughton, "Dissolution of Thorium Metal and Thorium Dioxide," in L. I. Katzin, Ed., *Production and Separation of U^{233}, U. S. At. Energy Comm.*, **TID-5223,** Part 1, Paper 7.1 (1951).
271. Scott, R. G., and R. G. Stannard, *U. S. At. Energy Comm.*, **UR-235** (1953).
272. Sellers, P. A., S. Fried, R. E. Elson, and W. H. Zachariasen, *J. Am. Chem. Soc.*, **76,** 5935 (1954).
273. Shevchenko, V. B., S. I. Zolotu Kha, N. F. Kascheyev, S. A. Tsaryov, V. A. Mikhailov, and G. A. Toropchenova, "Complex Utilization of Uranium Ores," in *Proc. 2nd Intern. Conf. Peaceful Uses At. Energy, Geneva, 1958*, **4,** 40 (1958).
273a. Shevchenko, V. B., V. A. Mikhailov, and Y. P. Zavalskii, *J. Inorg. Chem. (U. S., S.R.)*, **8,** 1955, 1959 (1958).
274. Sill, C. W., *Anal. Chem.*, **33,** 1684 (1961).
275. Smales, A. A., L. Airey, J. Woodward, and D. Mopper, *U. K. At. Energy Authority Rept.*, **AERE C/R 2223** (1957).
276. Smith, F. A., R. J. Della Rosa, and L. J. Casarett, *U. S. At. Energy Comm.*, **UR-305** (1955).
277. Starik, I. E., A. P. Ratner, M. S. Pusvik, and L. D. Shleidina, in *Soviet Research on the Lanthanide and Actinide Elements, 1949–57*, Consultants Bureau, Ind., New York, Part II, 1959, p. 230.
278. Startizky, E., *U. S. At. Energy Comm.*, **LA-1286** (1951).
279. Stehney, A. F., and D. P. Krause, *U. S. At. Energy Comm.*, **ANL-5919,** 54 (1958).
280. Stevenson, P. C., and W. E. Nervik, *The Radiochemistry of the Rare Earths, Scandium, Yttrium, and Actinium, U. S. At. Energy Comm.*, **NAS-NS-3020** (1961).
281. Stites, J. G., M. L. Salutsky, and B. D. Stone, *J. Am. Chem. Soc.*, **77,** 237 (1955).
282. Strom, E. T., *U. S. At. Energy Comm.*, **UCRL-9372** (1960).
283. Strominger, D., J. M. Hollander, and G. T. Seaborg, *Rev. Mod. Phys.*, **30,** 585 (1958), Part II.
284. Sullivan, J. C., R. E. Elson, P. A. Sellers, E. R. John, E. C. Janda, and M. H. Studier, in *U. S. At. Energy Comm.*, **ANL-4545,** 11 (1950)
285. Thompson, R. C., *U. S. At. Energy Comm.*, **AECD-2488** (1946).
286. Thompson, R. C., "Protactinium Chemistry," in G. T. Seaborg, and L. I. Katzin, Eds., *Production and Separation of U^{233}, U. S. At. Energy Comm.*, **TID-5222** (1951), Chapter 6.
287. Thompson, R. C., Q. Van Winkle, and J. G. Malm, "Extraction of Protactinium from Carbonate-Insoluble Residues of Uranium Ores," in L. I. Katzin, Ed., *Production and Separation of U^{233}, U. S. At. Energy Comm.*, **TID-5223** (1952), Part 1, Paper 6.4.

288. Timma, D. L., D. J. Hunt, and G. Pish, *J. Opt. Soc. Am.*, **47**, 291 (1957).
289. Todt, F. Z., *Z. Physik. Chem.*, **113**, 329 (1924).
290. Tompkins, E. R., *U. S. At. Energy Comm.*, **UCRL-1294** (1951).
291. Treiman, L. H., and M. G. Bowman, *U. S. At. Energy Comm.*, **LA-1550** (1953).
292. Tsong, Y. J., *J. Chim. Phys.*, **47**, 805 (1950).
293. Tsong, Y. J., and M. Haïssinsky, *Bull. Soc. Chim. France*, **1949**, 546.
294. Van Winkle, Q., and R. John, in G. T. Seaborg and J. J. Katz, Eds., *The Actinide Elements*, McGraw-Hill, New York, 1954, p. 120.
295. Vernois, J., *J. Chromatog.*, **1**, 52 (1958).
296. Vernois, J., *J. Chromatog.*, **2**, 155 (1959).
297. Vittum, E. K., W. L. Minto, and R. M. Fink, "Plating Procedure," in R. M. Fink, Ed., *Biological Studies with Polonium, Radium, and Plutonium*, McGraw-Hill, New York, 1950, pp. 18–27.
298. Wahl, A. C., and N. A. Bonner, Eds., *Radioactivity Applied to Chemistry*, Wiley, New York, 1951.
299. Walter, G., and A. Coche, *J. Phys. Radium*, **21**, 477 (1960).
299a. Wang, Y., and V. Khalkin, *Radiokhimiya*, **3**, 662 (1961); through *Nucl. Sci. Abstr.*, **16**, 1180 (1962).
300. Warren, G. W., and R. W. Fink, in P. E. Figgins, *The Radiochemistry of Polonium*, U. S. At. Energy Comm., **NAS-NS-3037**, 46 (1961).
301. Weinstock, B., and C. L. Chernick, *J. Am. Chem. Soc.*, **82**, 4116 (1960).
302. Welch, G. A., quoted by A. T. Casey and A. G. Maddock, *J. Inorg. Nucl. Chem.*, **10**, 58 (1959).
303. Wyatt, E. J., H. P. Rosen, and W. S. Lyon, in *U. S. At. Energy Comm.*, **ORNL-1474**, 16 (1953).
304. Zachariasen, W. H., *Phys. Rev.*, 73, 1104 (1948).

GENERAL REFERENCES

Radioactivity Measurements

Cook, G. B., and J. F. Duncan, *Modern Radiochemical Practice*, Oxford Press, London, 1952.
Crouthamel, C., *Applied Gamma-Ray Spectrometry*, Pergamon Press, New York, 1960.
Friedlander, G., and J. W. Kennedy, *Nuclear and Radiochemistry*, Wiley, New York, 1955.
Jaffey, A. H., "Radiochemical Assay by Alpha and Fission Measurements," in G. T. Seaborg and J. J. Katz, Eds., *The Actinide Elements*, McGraw-Hill, New York, 1954, Chapter 16.
Kohman, T. P., *Anal. Chem.*, **21**, 352 (1949).
Price, W. J., *Nuclear Radiation Detection*, McGraw-Hill, New York, 1958.
Siegbahn, K., *Beta- and Gamma-Ray Spectroscopy*, Interscience, New York, 1955.
Steinberg, E. P., "Counting Methods for the Assay of Radioactive Samples," *U. S. At. Energy Comm.*, **ANL-6361** (1961).

Actinium

Bagnall, K. W., *Chemistry of the Rare Radioelements*, Academic Press, New York, 1957.
Bouissières, G., "Actinium," in P. Pascal, Ed., *Nouveau Traite de Chemie Minérale*, Masson et Cie, Paris, 1959, Tome VII, pp. 1413–1446.

Erbacher, O., "Actinium and Mesothor 2," in W. Fresenius and G. Jander, Eds., *Handbuch der Analytischen Chemie*, Springer Verlag, Berlin, 1956, Teil III, Band III aβ/IIIb, pp. 414–441.

Garner, C. S., "Radioactivity Applied to the Discovery and Investigation of The Newer Elements," in A. C. Wahl and N. A. Bonner, Eds., *Radioactivity Applied to Chemistry*, Wiley, New York, 1951, Chapter 7.

Hageman, F. T., "The Chemistry of Actinium," in G. T. Seaborg and J. J. Katz, Eds, *The Actinide Elements*, McGraw-Hill, New York, 1954, Chapter 2.

Herr, W., "Actinium and Isotope," in W. Fresenius and G. Jander, Eds., *Handbuch der Analytischen Chemie*, Springer Verlag, Berlin, 1956, Teil III, Band III aβ/IIIb, pp. 443–552.

Hyde, E. K., "Radiochemical Separation Methods for the Actinide Elements," in *Proc. Intern. Conf. Peaceful Uses At. Energy, Geneva, 1955*, **7**, 281 (1956).

Katz, J. J., and G. T. Seaborg, Eds., *The Chemistry of the Actinide Elements*, Methuen and Co., London, 1957.

Pietsch, E., Ed., "Actinium and Isotope," in *Gmelins Handbuch der Anorganischen Chemie*, Verlag Chemie, Berlin, 1942, System—Nummer 40.

Salutsky, M. L., "Actinium," in C. L. and D. W. Wilson, Eds., *Comprehensive Analytical Chemistry*, Elsevier Publ. Co., Amsterdam, 1962, Vol. Ic, pp. 492–496.

Seaborg, G. T., J. J. Katz, and W. M. Manning, Eds., *The Transuranium Elements*, McGraw-Hill, New York, 1949.

Stevenson, P. C., and W. E. Nervik, *The Radiochemistry of the Rare Earths, Scandium, Yttrium, and Actinium*, **NAS-NS-3020** (1961).

Astatine

Anders, E., "Technetium and Astatine Chemistry," in *Annual Reviews of Nuclear Science*, Annual Reviews, Inc., Palo Alto, Cal., 1959, Vol. 9, pp. 203–220.

Appelman, E. H., *The Radiochemistry of Astatine*, U. S. At. Energy Comm., **NAS-NS-3012** (1960).

Bagnall, K. W., *Chemistry of the Rare Radioelements*, Academic Press, New York, 1957.

Garner, C. S., "Radioactivity Applied to the Discovery and Investigation of the Newer Elements," in A. C. Wahl and N. A. Bonner, Eds., *Radioactivity Applied to Chemistry*, Wiley, New York, 1951, Chapter 7.

Haïssinsky, M., "Astate," in P. Pascal, Ed., *Nouveau Traite de Chemie Minérale*, Masson et Cie, Paris, 1960, Tome XVI, pp. 659–666.

Maddock, A. G., "Astatine," in *Mellor's Comprehensive Treatise on Inorganic and Theoretical Chemistry*, Supplement II, Part 1, Longmans, Green and Co., London, 1956, Chapter V, Section 44, pp. 1064–1079.

Francium

Bagnall, K. W., *Chemistry of the Rare Radioelements*, Academic Press, New York, 1957.

Garner, C. S., "Radioactivity Applied to the Discovery and Investigation of the Newer Elements," in A. C. Wahl and N. A. Bonner, Eds., *Radioactivity Applied to Chemistry*, Wiley, New York, 1951, Chapter 7.

Hyde, E. K., *The Radiochemistry of Francium*, U. S. At. Energy Comm., **NAS-NS-3003** (1960).

Perey, M., "Francium," in P. Pascal, Ed., *Nouveau Traite de Chemie Minérale*, Masson et Cie, Paris, 1957, Tome III, pp. 131–141.

Polonium

Bagnall, K. W., *Chemistry of the Rare Radioelements*, Academic Press, New York, 1957.

Figgins, P. E., *The Radiochemistry of Polonium*, U. S. At. Energy Comm., **NAS-NS-3037** (1961).

Garner, C. S., "Radioactivity Applied to the Discovery and Investigation of the Newer Elements," in A. C. Wahl and N. A. Bonner, Eds., *Radioactivity Applied to Chemistry*, Wiley, New York, 1951, Chapter 7.

Haïssinsky, M., "Polonium," in P. Pascal, Ed., *Nouveau Traite de Chemie Minérale*, Masson et Cie, Paris, 1960, Tome XIII, pp. 2041–2122.

Moyer, H. V., *Polonium*, U. S. At. Energy Comm., **TID-5221** (1956).

Pietsch, E., Ed., "Polonium and Isotope," in *Gmelins Handbuch der Anorganischen Chemie*, Verlag Chemie, Berlin, 1941, System—Nummer 12. Available in English translation as U. S. At. Energy Comm., **N-2167** (1945).

Protactinium

Elson, R. E., "The Chemistry of Protactinium," in G. T. Seaborg and J. J. Katz, Eds., *The Actinide Elements*, McGraw-Hill, New York, 1954, Chapter 5.

Garner, C. S., "Radioactivity Applied to the Discovery and Investigation of the Newer Elements," in A. C. Wahl and N. A. Bonner, Eds., *Radioactivity Applied to Chemistry*, Wiley, New York, 1951, Chapter 7.

Haïssinsky, M., and G. Bouissières, "Protactinium," in P. Pascal, Ed., *Nouveau Traite de Chimie Minérale*, Masson et Cie, Paris, 1958, Tome XII, pp. 617–680. Available in English translation by H. W. Kirby as U. S. At. Energy Comm., **MLM-1101-Tr**.

Hyde, E. K., "Radiochemical Separations Methods for the Actinide Elements," in *Proc. Intern. Conf. Peaceful Uses At. Energy, Geneva, 1955*, **7**, 281 (1956).

Jantsch, G., "Protactinium," in W. Fresenius and G. Jander, Eds., *Handbuch der Analytischen Chemie*, Springer-Verlag, 1956, Teil II, Band IVb, Va/b, pp. 746–756.

Katz, J. J., and G. T. Seaborg, Eds., *The Chemistry of the Actinide Elements*, Methuen and Co., London, 1957.

Katzin, L. I., Ed., "Production and Separation of U^{233}," U. S. At. Energy Comm., **TID 5223** (1952).

Kirby, H. W., *The Radiochemistry of Protactinium*, U. S. At. Energy Comm., **NAS-NS-3016** (1959).

Pietsch, E., Ed., "Protactinium and Isotope," in *Gmelins Handbuch der Anorganischen Chemie*, Verlag Chemie, Berlin, 1942, System—Number 51.

Salutsky, N. L., "Protactinium," in C. L. and D. W. Wilson, Eds., *Comprehensive Analytical Chemistry*, Elsevier Publ. Co., Amsterdam, 1962, Vol. Ic, pp. 570–580.

Seaborg, G. T., J. J. Katz, and W. M. Manning, Eds., *The Transuranium Elements*, McGraw-Hill, New York, 1949.

Subject Index

A

Actinium, absorption spectrum of, 453
 chemical properties of, 453–455
 chromatographic separation of, 465, 466
 coprecipitation behavior of, 454, 457, 458; table, 455
  with fluorides, 457, 458
  with hydroxides, 458, 459
  with oxalates, 458
  with sulfates, 459
 decay series of, 441; figure, 445
 detection of, 466–472
  by gamma-ray counting, 470–472; figures, 470, 471
  by growth of decay products, 467, 470; figures, 467, 469
  by use of Fr-233, 470
  of Pb-211, 469
  of Ra-223, 469
  of Th-227, 468, 469
 determination of, by active-deposit counting, 479
  by alpha-particle spectrometry, 474–476; figure, 476
  in bone, 482
  by calorimetric methods, 480, 481
  by coincidence counting of radon-219, 480, 481
  by counting, 474–477
  by decay-product separation, 477–480
  with di(2-ethylhexyl) phosphoric acid, 485, 486
  by emission spectroscopy, 480
  in ores, 481
  by TTA extraction, 481
  in uranium, irradiated, 482
  in urine, 482
  by use of Fr-223, 480, 481
 dissolution of samples of, 456
 electrochemical properties of, 451, 452
 emission spectrum of, 452, 453, 466; table, 453
 gamma-ray spectrometry of, 470–472, 545–547; figures, 470, 471, 545–547
 handling techniques for, 449, 450
 isolation of, 457–466, 472–474
 maximum exposure to, 448, 449; table, 449
 nuclear properties of, 451; table, 452
 occurrence of, 441–445; figures, 442–445
 optical properties of, 452, 453, 466; table, 453
 organic complex formation with, 454, 455
 physical properties of, 451
 radioactivity hazards of, 447, 448
 separation of, 457–466, 472–474
  from bismuth, 459
  from daughter elements, 459–461; table, 461
  by ion exchange, 463–465
   anion exchange, 465, 466
   cation exchange, 463–465
  from lanthanum, 461, 463, 464; table, 461
  from lead, 459
  by paper chromatography, 465, 466
  from polonium, 459
  from radium, 459–461; table, 461
  from thallium, 459
  from thorium, 459, 464, 465
  solvent extraction with alcohol, 462, 463
   with di(2-ethylhexyl) phosphoric acid, 462
   with TTA, 459–461; table, 461
   with tributyl phosphate, 461
 sources of, 445–447
 tracer chemistry of, 454, 455
 uses of, 447
Actinium compounds, crystal structures of, 451
 properties of, 451–454
Aquamarine, 4

611

Astatine, alpha counting of, 498–501
  chemical properties of, 489–492
  detection of, 497, 498
  determination of, 498–501
  discovery of, 487
  handling techniques for, 449, 450
  ionic radius of, 489
  ionization potential of, 489
  liquid–liquid extraction of, 492, 494, 496, 497
  maximum exposure to, 448, 449; table, 449
  nuclear properties of, 487, 488; table, 488
  occurrence of, 442
  oxidation-reduction potentials of, 491
  oxidation-reduction reactions of, table, 489
  oxidation states of, 489–491
  physical properties of, 487, 488
  radioactivity hazards of, 447, 448
  separation of, from biological materials, 497, 501
    from bismuth, 494
    by coprecipitation, 495, 496, 501
    by distillation, from bismuth, 492, 493, 496, 497
    by precipitation, 494, 495
    by spontaneous deposition, 494
  solvent extraction of, 492, 494, 496, 497
  sources of, 446, 447
  uses of, 447
  volatility properties of, 491; table, 491
  x-ray counting of, 499–501
Astatine isotopes, 487, 488
  energy of radiation of, table, 488
  half life of, table, 488
  sources of, 488
  type of decay, table, 488

B

n-Benzoylphenylhydroxylamine, 34
Bertrandite, 4
Beryl, 25, 26
  See also Beryllium aluminum silicate.
Beryllium, alpha-particle bombardment of, 25
  amphoterism of, 22
  complexation of, 23
  detection of, by colorimetry, 38
    by emission spectrography, 39, 60
    by precipitation, 39
    by radiochemistry, 39, 40, 50
  determination of, 40–50
    in air, 60
    in alloys, 58, 59
    by complexation, 23
    by gamma irradiation, 25
    by gravimetric methods, 40–42, 50, 51
    laboratory methods for, 50–60
    in minerals, 59, 60
    by nuclear activation, 50
    in ores, 59, 60
    oxygen in, 53–58; tables, 56, 57
    by photometric methods, 44–50
    by radiochemical methods, 50
    by spectrography, 48–50
    by spectrophotometry, 44–47
      with acetylacetone and EDTA, 44, 60
      with alkanan, 44
      with aluminon, 44, 45, 49
      with berillon, 45
      with Eriochrome Cyanine R, 45
      with morin, 60
      with Naphthachrome Azurine 2B, 45, 46
      with Naphthachrome Green, 45, 46
      with naphthazarin, 44
      with p-nitrophenylazoorcinol, 46, 57, 59, 60
      with other reagents, 47
      with 2-phenoxyquinizarin-3,4-sulfonic acid, 46
      with quinizarin-2-sulfonic acid, 46
      with solochrome cyanine, 45
      with Zenia, 46, 57, 59, 60
    by ultraviolet fluorescence, 47
    by volumetric methods, 42–50
      acid titration of hydroxide, 42, 50, 51
      complex fluoride titration, 42, 43
      volatilization with HCl, 57
  deuteron bombardment of, 11, 25
  equilibrium constants of, table, 23
  extraction as fluoride complex, 78
    as hydroxide, 5
    as sulfate, 5

SUBJECT INDEX

flake sampling of, 27
gamma-ray irradiation of, 24, 25
history of, 3
neutron liberation from, 24, 25
occurrence of, 3–5
optical properties of, 23, 24; table, 24
oxygen determination in, 53–58; tables, 56, 57
"pebbles," sampling of, 27
polarography of, 20, 21
production of by pyroreduction, 8, 9
proton bombardment of, 25
radiochemical-nuclear properties of, 24, 25; table, 25
reduction potential of, 23
separation of, 28–38
  by electrolysis, 37
  by ion exchange, 37, 38
  by precipitation, with ammonia, 33
    with n-benzoylphenylhydroxylamine, 34
    with cupferron, 34
    with 8-hydroxyquinoline, 33
    with nitrosophenylhydroxylamine, 34
    with other reagents, 34, 35
  by solvent extraction, with acetylacetone, 35, 36
    with butyrate, 36
    with chloroform, 33, 34
    with 8-hydroxyquinoline, 36
    with methyl isobutyl ketone, 36
    with tri-n-octyl phosphine oxide, 37
spectra of, table, 24
toxicity of, 10, 11
uses of, 11, 12
vacuum-cast and hot-pressed, 28
Beryllium alloys, 12, 13; table, 12
Beryllium aluminum silicate (beryl), 3–7, 25, 26
Beryllium chloride, electrolysis of, 4
Beryllium compounds, acetate, basic, 18
  ammonium beryllium fluoride, 14
  carbide, 18
  carbonate, basic, 18
  chloride, 16
  color of, 9, 13, 23
  fluoride, 16
  hydride, 19

hydroxide, 14, 15
hydrolysis of, 23
importance of, 13
intermetallics, 19, 23
iodide, 16
nitrate, 17
nitride, 17, 18
oxide, 16, 17
phosphate, 18, 19
polarograms of, 20, 21
reactions of, table, 23
sodium beryllium fluoride, 14
sulfate, 13, 14
thermodynamic constants of, 22
Beryllium fluoride, electrolysis of, 3, 4
  pyroreduction of, 4, 8
  reduction of by magnesium, 8, 9
Beryllium hydroxide, 5, 6, 14
  reaction of, with sodium hydroxide, 6
Beryllium ions, amphoterism of, 22
  complexation of, 23
  reduction potential of, 23
Beryllium isotopes, 24, 25; table, 25
Beryllium metal, 11, 12
  color of, 23
  determination of, by gravimetric methods, 50, 51
    by volumetric methods, 51–53, 57
  luster of, 21
  oxidation of, 21, 22
  production of, 8–10
  properties of, 19–24; tables, 20, 21, 24
  reactions of, 21–23; table, 23
  sampling of, 27, 28
Beryllium minerals and ores, determination of beryllium in, 59, 60
  opening of, 8
  sampling of, 25, 26
Beryllium oxide, reduction by carbon, 3
Beryllium salts, hydrolysis of, 23
Beryllium samples, dissolution of, 30, 31
Beryllium sulfate, 6, 13
Betafite, table, 186
Biplumbite ion, in determination of lead, 129
Bismutotantalite, table, 186

C

Columbite, 187; table, 186
Columbium. *See* Niobium.

SUBJECT INDEX

Copaux process, 7
Cupferron, 34

D

Diethyldithiocarbamate, 139, 140, 145; table, 101
Diphenylcarbazide, 120
Diphenylcarbazone, 120, 121
Dithizone, in lead analysis, 79, 96, 133–139, 144, 147, 148, 159–161; table, 101

E

EDTA, in beryllium analysis, 32, 36, 42, 47, 59
   in lead analysis, 85, 125–128, 132, 133, 144, 146, 149, 150, 154–157; table, 102
Electrolysis of lead, internal, 117
Emerald, 4
Eriochrome Black T, 126, 144
Eschynite, table, 186
Ethylenediaminetetraacetic acid. *See* EDTA.
Euxenite, 187; table, 186

F

Fabry-Perot interferometer, 94
Fergusonite, table, 186
Ferroniobium, 194
Ferroniobium–tantalum, preparation of, 194
Ferrotantalum alloy, decomposition of, 229
Francium, beta-ray counting of, 513
   coprecipitation reactions of, 504–506; table, 506
   chemical properties of, 503, 507
   decay curve of, 515, 516; figure, 515
   decay scheme of, 514
   detection of, 512, 513
   determination of, 513–517
     in actinium, 516, 517
     in thorium, 517
     in uranium, 517
   discovery of, 501, 503
   gamma-ray counting of, 513
   handling techniques for, 449, 450
   ion exchange behavior of, 510, 511
   maximum exposure to, 449
   nuclear properties of, 502, 503; table, 502
   occurrence of, 442, 502; figures, 443, 445
   preparation of, 502
   physical properties of, 503
   radioactivity hazards of, 447, 448
   separation of, from actinium, 511, 519, 520
     by chromatography, 509, 510
     by coprecipitation, 507–509
     by ion exchange, 510, 511
     by other methods, 511, 512
     from radium, 511, 512
     by solvent extraction, 511
     from thorium, 511, 512, 517–519
   sources of, 442, 446, 447; figures, 443, 445
   uses of, 447
   volatility of, 506, 507

G

Geological age determination, with lead isotopes, 94, 95

H

Halogenoniobates, 208–211
   boiling points of, table, 208
   bromides, 210
   chlorides, 208–210
   fluorides, 210, 211
   iodides, 210
   melting points of, table, 208
Halogenotantalates,
   boiling points of, table, 208
   borides, 213
   bromides, 210
   carbides, 213
   chemical properties of, 208
   chlorides, 208–210
   fluorides, 210, 211
   hydrides, 211, 212
   iodides, 210
   melting points of, table, 208
   nitrides, 212
   refractory binary compounds, properties of, 211
Hydrogen sulfide, action of, on lead salts, 104–106
8-Hydroxyquinoline, 33

J

Jones reductor, 80
  in niobium determination, 318

L

Lead, amperometric titration of, 132, 133, 163, 164
  chemical properties of, acetic acid effect on, 80
    corrosion of, 79
    hydrochloric acid effect on, 80
    hydrogen overpotential of, 79
    nitric acid effect on, 80
    reducing action of, 80
    sulfuric acid effect on, 80
  chromatographic separation of, tables, 102, 103
  consumption of, table, 73
  controlled-potential electrolysis of, 146
  coprecipitation of, table, 102
  detection of, in bismuth compounds, 108
    in gasoline, 108
    in water, 108
  determination of, by activation analysis, 143; table, 147
    in air, 147, 148
    in aluminum, 146, 147
    by amperometry, 132, 133
    in beryllium, 146
    in biological materials, 147
    in bismuth, 111, 131, 146
    in concentrates, 143, 144
    in copper, 140, 141, 146, 161–163
    by coulometry, 128
    by electrolysis, 114–117, 145, 157–159
    in food, 148
    in gasoline, 141, 142, 149, 150
    in glass, 150
    in magnesium, 146
    in meteorites, 143; table, 143
    in oil, 149
    in ores, 143, 144
    in organic materials, 97–99
    in paints, 150
    in rocks, 143, 144
    in rubber, 150
    in uranium, 131
    in urine, 147
    in water, 148, 149
    in zinc, 146, 147
  electrochemical properties of, reduction potentials, 78; table, 78
    storage battery, half-reactions of, 79
  electrodeposition of, 114–117; tables, 100, 115
  electrolytic determination of, 117, 157–159
  extraction of, with diethylammonium diethyldithiocarbamate, table, 101
    with diethyldithiocarbamate sodium, 139, 140, 145; table, 101
    with dithizone, table, 100
    by electrolytic refining, 74
    by flotation of lead ores, 73
    by Parke's refining process, 73, 74
    pure lead, requirements for, table, 74
    by roasting of lead ores, 73
    with sodium rhodizonate, table, 101
    by zone refining, 74
  flame photometric determination of, 140, 141
  fumes of, 148
  gravimetric determination of, as chloride, 118, 119
    as chromate, 112, 113, 153, 154
    by electrodeposition, 114–117
    by electrolysis, as PbO_2, 117, 157–159
    by internal electrolysis, 117
    as iodate, 118
    with mercaptobenzothiazole, 117
    as molybdate, 113, 114
    as nitrate, 118
    by precipitation from homogeneous solution, 112, 153, 154
    with salicylaldoxime, 117
    in steel, 117
    as sulfate, 110–112, 152, 153
    as sulfide, 118
    with thionalide, 117
    as triplumbic paraperiodate, 118
  ion-exchange separation of, table, 102
  mass spectrographic determination of, 92, 93
  metallic properties of, 79–81
  microdetection of, by crystal reactions, 106, 107

by potassium iodide, 106, 107
by "triple nitrate," 106
occurrence of, 72
optical determination of, 93, 94
optical properties of, table, 78
physical properties of, 77; table, 75
polarographic characteristics of, table, 129
polarographic determination of, by dropping mercury electrode, 130, 161–163
   by platinum electrode, 130
   in very low concentrations, 130–133, 143
precipitation of, from homogeneous solution, 112, 113
production of, 72
products of, industrial, 72, 73
pure, chemical requirements for, table, 74
qualitative analysis of, classical scheme, 104, 105
reactions of, 105, 106
   with sulfuric acid, 106
refining of, 73, 74
separation of, as chromate, table, 99
   as dioxide, table, 100
   as iodate, table, 100
   with mercaptobenzothiazole, table, 100
   as metal, table, 100
   as molybdate, table, 99
   from other elements, tables, 99–103
   as periodate, table, 99
   with salicylaldoxime, table, 100
   as sulfate, table, 99
   as sulfide, table, 99
   with thionalid, 146; table, 100
   with thiourea, 146; table, 100
spectrographic detection of, 103
spectrographic determination of, 141, 142, 145
spectrographic lines of, table, 104
spectrophotometric determination of, with diethylthiocarbamate, 139, 140, 145
   with dithizone, 133–139, 144, 147, 159–161
   with lead chloride, 140

with PAR-4, 139
with 4-(2-pyridylazo) resorcinol, 139
with tetramethyldiaminophenyl-methane, 139
spot tests for, with benzidine, 107
   with sodium rhodizonate, 107
   with p-tetramethyldiaminodiphenyl-methane, 107
thermal stability of weighing forms of, table, 109
titrimetric determination of, with alkali hydroxides, 121
   by alkalimetry, 127, 128
   with chromate ion, 120
   by coulometry, 128
   by EDTA complex formation, 125–128, 143, 144, 149, 150, 154, 155
   with ferrocyanide ion, 121
   with fluoride ion, 121
   with molybdate ion, 119, 120
   by oxidation-reduction, 122–125
      Bunsen method, 122
      chromate method, 124
      Diehl-Topf method, 123
      iodate method, 124
      Lux method, 123
      periodate method, 124
      Rupp-Siebler method, 123
   with phosphoric acid, 120
toxicity of, 74–77
volatilization of, table, 100
weighing forms of, table, 109
Lead alloys, 96, 97
   dissolution of, table, 97
Lead bromide, basic, determination of lead with, 121
Lead chloride, determination of lead with, 118, 119
Lead chlorofluoride, determination of lead with, 121
Lead chromate, determination of lead with, 120, 124, 163, 164
Lead compounds, chemical properties of, basic lead(II) salts, 85
   biplumbite ion, 82
   chlorocomplexes, 81
   halide complex ions, 83
   hydrolysis, 82, 83

instability products, table, 87–89
lead(II) acetate, 84
lead(II) cyanide, 85
lead(II) EDTA plumbate, 85
lead(II) halides, 83
lead hydride, 90
lead(II) hydroxide, 82
lead(II) nitrate, 83, 84
lead orthoplumbate, 91
lead(II) oxide, 81, 82
lead(IV) oxide, 81, 90
lead sesquioxide, 91
lead suboxide, 91
lead(II) sulfate, 84
lead(II) sulfide, 84
lead tetrachloride, 90
lead tetraethyl, 90
lead tetrafluoride, 90
litharge, 81, 82
other lead(II) salts, 87
oxidation states, 81
plumbite ion, 82
red lead, 91
solubility products, table, 86
detection of, with benzidine, 107, 152
as chloride, 150, 151
with sodium rhodizonate, 107
as sulfide, 151
with p-tetramethyldiaminodiphenyl methane, 107
by "triple nitrite" crystal reaction, 151, 152
determination of lead in, by alkalimetric methods, 85
by gravimetric methods, 85
by internal electrolysis, 117
by polarography, 128, 145, 146, 161–163
by titrimetric methods, 119–124
hydrolysis of lead(II) compounds, 82, 83
instability products of, table, 87–89
microdetection of, 83, 106, 107
microscopic detection of, 106, 107
physical properties of, lead(II) hydroxide, solubility of, 82
lead(II) oxide, 81
lead(IV) oxide, crystal structure of, 90
solubility products, table, 86

spectrographic detection of, 103
thermal stability of weighing forms of, table, 109
volatility of, 98
Lead dioxide, determination of lead with, 114–116
Lead ferrocyanide, determination of lead with, 121
Lead fumes, 148
Lead(II) hydroxide, amphoterism of, 82
thermogravimetry of, 82
Lead iodate, determination of lead with, 118
Lead isotopes, relative abundance, table, 76
decay type, table, 76
distribution of, 92
emission spectra of, 93
geological age determination by, 94, 95
half life, table, 76
Lead metal, chemical requirements for purity of, table, 74
Lead molybdate, determination of lead with, 119, 120, 145
Lead nitrate, determination of lead with, 119, 120, 145
Lead ores, dissolution of, 96
silicate rock, 96
sulfide minerals, 96
dithizone extraction of, 96
flotation of, 73
Lead oxides, determination of lead with, 122–125
Lead(IV) oxide, oxidizing properties of, 90
Lead phosphate, determination of lead with, 120, 121
Lead spectroscopy, by atomic absorption, 140, 141
Lead sulfide, determination of lead with, 118
Lead tetraethyl, detection of, 108
determination of, 149, 150
Lead tetraphenyl, detection of, 108
Lebeau, 3, 10

M

Mercaptobenzothiazole, in precipitation of lead, 117
Microlite, table, 186
Morin, in beryllium analysis, 31, 32, 35, 39

N

Neptunium, decay series of, 441; figure, 443
  discovery of, 441
  occurrence in uranium, 442
  production of, 442
Niobium, relative abundance of, table, 185
  chemical properties of, 203–213
  chemistry, comparative, of elements of fifth subgroup, 203, 204
  chromatographic separation of, 258–263; figures, 256, 257
    principles of, 258, 259
    procedures, 259–261
    sample preparation for, 259
  columbium, relation to, 183, 184
  corrosion resistance of, 201, 202
  detection of, by chemical procedures, 280
    by methylthymol blue method, 282, 283
    by paper chromatography, 282
    by reduction by zinc, 283
    by spectrography, 277, 278, 280; table, 279
      interferences, 278
      limits of, 280
    by tannin method, 280, 281
    by tartaric hydrolysis, 281
    by thiocyanate method, 283
  determination of, in concentrates, 376–380; figure, 377
    as impurity in tantalum, 368
    by neutron activation, 336–338; figure, 337
    in ores, 376–380; figure, 377
    by radioactive isotope dilution method, 336
    by radioactive tracer method, 336
  determination of metallic impurities in, 357–376
    of antimony, 374
    of arsenic, 375, 376
    of boron, 373, 374
    of cadmium, 374, 375
    of cobalt, 373
    of copper, 373, 374
    of iron, 368, 369
    of lead, 374, 375
    of molybdenum, 369–371
    of nickel, 373
    of phosphorus, 371
    of potassium, 375
    by radioactivation analysis, 362–366
      differential count method, 365
      nondestructive technique, 363; table, 364
      radiochemical separation technique, 365; table, 366
    of silicon, 371
    of sodium, 375
    by spectrography, with four-group method, 359, 360
      with germanium oxide, 359
      with lanthanum carbonate, 359
      with lithium carbonate-lanthanum, 358, 359
      with other procedures, 360, 361, 386–389; tables, 386, 388
      with silver chloride-barium fluoride, 359
      with strontium chloride-graphite, 359
    of tin, 374
    of titanium, 372, 373
    of tungsten, 369–371
    of zinc, 374
    of zirconium, 372
  earth acid of, 205, 206
  electrochemical properties of, 201, 202
  electronic configuration of, table, 203
  electrophoretic separation of, 263
  emission spectrum of, table, 202
  fission products of uranium-235 in, table, 201
  gravimetric determination of, by N-benzoyl-N-phenylhydroxylamine, 285
    by cupferron, 284
    by hydrolytic precipitation, 285
    by 8-hydroxyquinoline, 285, 286
    by 3,3,4,7-pentahydroxyflavone, 286
    by phenylarsonic acid, 285
    by tannic acid, 284
  history of, 183, 184
  industrial hygiene of, 194, 195

intermediate metallurgic products, decomposition of, 225
ion exchange separations of, in hydrochloric acid medium, 242–244
  in hydrochloric–oxalic acid medium, 255–258
  in hydrofluoric acid medium, 244, 245; figure, 245
  in mixed acid medium, 245–255
    applications of, 252–255
    chemistry of, 246–250; figures, 247–250
    separation principles, 250–252; figures, 251, 252
isotopic dilution, determination of, 336
mechanical properties of, 197; table, 195, 196
metallurgy of, extractive, 191–193
minerals, low-grade, decomposition of, 224, 225
neutron activation, determination of, 336–338; figure, 337
neutron cross section of, 188; tables, 196, 198, 199
occurrence, in columbite, 185, 187, 216, 217; table, 186
  in euxenite, 187, 216; table, 186
  fission products of uranium-235; table, 202
  in igneous rocks, 184; table, 184
  in minerals, 185, 187; table, 186
  in pyrochlore, 187; table, 186
  in tin mining by-products, 187, 188
optical properties of, 202; table, 202
oxidation of, 205
paper chromatographic separation of, 261–263
peroxy salts of, 207, 208
photometric determination of, by arsenazo method, 310
  by ascorbic acid, 310
  by disodium-1,2-dihydroxybenzene-3,5-disulfonate, 309
  by hydrochloric acid, 309
  by hydrogen peroxide, 297–301
    absorbance curve, figure, 297
    acid concentration, 298, 299
    application, 300, 301
    interfering ions, 299, 300; tables, 300
    procedure, 298
  by hydroquinone, 301–305
    absorbance curve, figure, 302
    application, 304, 305
    interfering ions, 303, 304
    procedure, 301, 302
    reagent preparation, 302, 303
    stability of colored complex, 303
  by molybdenum blue, 306, 307
    interfering ions, 307
  by niobium(III) method, 310
  by potassium carbonate, 310
  by pyrocatechol, 310
  by pyrogallol, 307–309
    interfering ions, 309
  by sulfosalicylic acid, 310
physical properties of, 195–197; table, 195, 196
polarographic determination of, in citrate system, 322, 323
  in EDTA system, 323, 324
  in hydrochloric acid system, 321
  in nitrate system, 320
  in other systems, 324
  in sulfuric acid system, 320, 321
precipitation procedures for, with cupferron, 235, 236
  by hydrolysis, 234, 235
  by miscellaneous methods, 239–241
  with phenylarsonic acid, 239
  with tannin, 236–239
preparation of, 193, 194, 202. *See also* separation of.
radioactive tracer determination of, 336
refractory binary compounds of, 211–213
sampling of, in concentrates, 213, 214
  in intermediate products, 214
  in metal, consolidated, 215
  in metal, powdered, 215
  in ores, 213, 214
separation of. *See also* chromatographic separation, ion-exchange separation, precipitation procedures, solvent extraction.
  with hydrogen sulfide precipitation of contaminants, 216
  "individuality," loss of, 215, 216

by solvent extraction, chemistry of, 263, 264
  with cyclohexanone, 268
  with diisobutyl carbinol, 268
  in fluoride medium, 265–269; figures, 266
  in hydrochloric acid system, 264, 265
  with miscellaneous solvents, 269
  with tributyl phosphate, 268
spectrographic determination of, for alloys, 329, 330
  for ores, high-grade, 324, 327; tables, 326, 327
  for ores, low-grade, 327–329
titrimetric determination of, by reduction in hydrofluoric acid, 317, 318
  by reduction in sulfuric acid, 318–320
toxicology of, 194, 195
uranium-235 fission products in, table, 201
uses of, in alloys, high-temperature, 189
  miscellaneous, 189
  in nuclear industry, 188
  in steel, 188, 189
x-ray fluorescence determination of, correction factors for, 333; table, 333
  fusion of sample, 333, 335
  instrumentation for, 331
  interferences with, 331, 332
  miscellaneous applications, 334, 335
  sample characteristics, 332
Niobium alloys, determination of niobium in, 380, 381
  preparation, of ferroniobium, 194
  of ferrotantalum–niobium, 194
  of niobium–chromium, 194
  of niobium–nickel, 194
Niobium alloys, high-temperature, decomposition of, 209–232
  miscellaneous, treatment of, 233, 234
Niobium borides, chemical properties of, 213
Niobium bromides, chemical properties of, 210
Niobium carbides, chemical properties of, 213
  preparation of, 194

Niobium chlorides, chemical properties of, 208–210
Niobium complexes, 246
Niobium compounds, electron probe measurement of, 335
  refractory binary, 211–213
Niobium concentrates, ion-exchange determination of niobium in, 376, 380; figure, 377
Niobium earth acids, 205, 206
Niobium fluorides, chemical properties of, 210, 211
Niobium hydrate, 205, 206
Niobium hydrides, chemical properties of, 211, 212
Niobium iodides, chemical properties of, 210
Niobium isotopes, activation product, table, 198, 199
  decay type, table, 198, 199
  half life of, 200, 201; table, 198, 199
  neutron cross section of, table, 198, 199
  niobium-93, 200, 201
  niobium-94, 200, 201; table, 201
  niobium-95, 200, 201; table, 201
  scattering cross-section of, table, 198, 199
Niobium metal, determination of tantalum in, 367, 368
  dissolution of, 228, 229
  electron probe measurement of, 335
  interstitial elements in, 349–357
    carbon determination, 357
    hydrogen determination, 353–355
      diffusion, 354
      spectrometry, 354, 355
      vacuum fusion, 354
    nitrogen determination, alkali fusion, 356, 357
      inert gas fusion, 355, 356
      Kjeldahl method, 356
      vacuum fusion, 355
  specifications for, table, 348
Niobium nitrides, chemical properties of, 212
Niobium ores and concentrates, decomposition of, by alkaline fusion, 219–221
  by borax, 221, 222

by chlorination, 223, 224
by hydrofluoric acid, 222, 223
by pyrosulfate, 217–219
Niobium ores, ion exchange determination of, niobium in, 376–380; figure, 377
   apparatus, 376, 377; figure, 377
   procedure, 377–380
   reagents, 376, 377
Niobium oxide, decomposition of, 226–228
   by alkaline attack, 227, 228
   by chlorination, 228
   by hydrofluoric acid, 228
   by pyrosulfate, 226, 227
   isolation of, metallurgic, 229–231
   treatment of, metallurgic, 231, 232
Niobium oxides, chemical properties of, 204, 205
Niobium silicides, chemical properties of, 213
Niobium slags, decomposition of, 225
Niobium steels, decomposition of, 209–232
   determination of niobium in, 380, 381
   determination of tantalum in, 380, 381
Niobium-95, 200, 201, 343–348; table, 201
   determination of, in aqueous solutions, 343, 344
      after cupferron extraction, 346
      with complement nuclide subtraction, 347, 348
      in fission products of plutonium fuels, 344, 345
      after ion-exchange separation, 347
      in uranium target, 346
      in uranyl nitrate from irradiated uranium, 345
   uses of, in determination of niobium, 336
Niobium–ferro alloys, decomposition of, 229
Niobium–plutonium alloys, treatment of, 232
Niobium–tantalum separation, 269–277
   by benzohydroxamic acid, 276
   by N-benzoyl-N-phenylhydroxylamine, 272–274
   by cellulose chromatography, 276, 277
   by cupferron, 275
   by diethyldithiocarbamate, 276
   by dihydroquercitin, 276

by EDTA, 275
by 8-hydroxyquinoline, 274, 275
by ion-exchange methods, 276
by Marignac method, 275, 276
by morin, 276
by peroxy method, 275
by phenylacetohydroxamic acid, 276
by phenylarsonic acid, 274
by Schoeller method, 270–272
by selenium oxychloride, 276
by selenous acid, 274
by solvent extraction, 277
by tannin precipitation, 270–272
by tetramethylene dithiocarbonate, 276
Niobium–titanium alloys, dissolution of, 232
Niobium–uranium alloys, treatment of, 232
p-Nitrobenzeneazoorcinol, 38
Nitron pertechnate, 426, 431
p-Nitrophenylazoorcinol, 38

O

Oxygen determination in beryllium, 53–58; tables, 56, 57

P

PAN in lead determination, 127, 156–158
Parke's process for lead extraction, 73, 74
Pegmatite dikes, 4
Pertechnetates, 415–417
   ammonium, 411, 427; table, 410
   conductance, of, 417
   coulometric reduction of, 416
   hydrogen sulfide action on, 415
   nitron, 426, 431
   potassium, 411; table, 410
   reduction of, 415
   tetraphenyl arsonium, 419, 426, 431
Pertechnetic acid, 411; table, 410
   conductance of, 417
   pK of, 414
   vapor pressure of, table, 414
Phenakite, 4
Polonium, absorption spectrum of, 527
   alpha counting of, 544, 545, 548, 549
      liquid scintillation, 548, 549
   chemical properties of, 525, 527–531
   chromatographic separations of, 542, 543

compounds of, 527–530
coprecipitation of, with hydroxides, 533
　with lead tellurate, 533
　with manganese dioxide, 533
　with other carriers, 533, 534, 544
　with selenium, 533
　with tellurium, 532, 533, 544
decay series of, figures, 442–445
deposition of, electrolytic, 536, 544
　spontaneous, 534–536, 547, 548
detection of, 543–547
determination of, 547–555
　in air, 552
　in biological materials, 549, 550, 555
　in bismuth, 550, 554, 555
　by deposition on silver foil, 553
　in ores, 551, 552
　in tungsten, 551
　in water, 552
discovery of, 520
dissolution of samples of, 531, 532
electrochemical properties of, 525, 526
electrolytic deposition of, 526
emission spectrum of, 526
gamma-ray counting of, 545–547; figures, 545, 547
handling techniques for, 449, 450
ion-exchange behavior, 539–541
isotopes of, table, 522, 523
　decay type, table, 522, 523
　energy of radiations, table, 522, 523
　half life of, table, 522, 523
　source of, table, 522, 523
liquid scintillation counting of, 548, 549
maximum exposure to, 448, 449; table, 449
nuclear properties of, 522, 524; table, 522, 523
occurrence of, 442, 443
oxidation states, properties of, 527–530
physical properties of, 521–525; table, 524
preparation of, 525
production of, 520
radioactivity hazards of, 447, 448
separation of, from bismuth, 554, 555
　by chromatography, 542, 543
　by coprecipitation, 532, 544
　by deposition, 534–536, 544, 547, 548

　by ion exchange, 539–541
　from radium-DEF, 554
　from selenium, 554, 555
　by solvent extraction, 536–539
　from tellurium, 554, 555
　by volatility methods, 541, 542
solvent extraction, by alcohols, 536
　by cupferron, 537
　by dithizone, 536, 537
　by ethers, 538
　by ketones, 538
　by sodium diethyldithiocarbamate, 537
　by thenoyltrifluoroacetone, 537
　by tri-n-benzylamine, 539
　by tributyl phosphate, 538, 539
sources of, 446, 447; table, 522, 523
spontaneous deposition of, 526
trace concentrations, chemical properties of, 530, 531
uses of, 447
volatility separations of, 541, 542
x-ray counting of, 547
Polycrase, table, 186
Protoactinium, absorption maxima of solutions of, table, 560
absorption spectrum of, 560, 561; table, 560
alpha counting of, 581, 585, 586; figures, 583, 584
alpha spectrometry of, 581, 585; figures, 583, 584
beta counting of, 582, 587
chemical properties of, 561–564
　protoactinium(IV), 563, 564
　protoactinium(V), 561–563
chromatographic separations, 580, 581; table, 581
coprecipitation of, with barium fluoride, 570
　with barium fluozirconate, 570
　with lanthanum fluoride, 570
　with lead dioxide, 567
　with manganese dioxide, 567
　with stannic oxide, 570
　with zirconium iodate, 570
　with zirconium mandelate, 570
　with zirconium phosphate, 570
coprecipitation reactions of, 566–570; tables, 566, 568, 569

decay series of, 441; tables, 443–445
deposition of, electrolytic, 559
　spontaneous, 557, 559
detection of, by alpha spectrometry, 581, 585; figures, 583, 584
　by beta counting, 582
　by color reactions, 583, 584
　by gamma-ray counting, 581; figure, 582
determination of, by alpha counting, 581, 585
　by alpha spectrometry, 585, 586
　by beta counting, 587
　by gamma-ray counting, 586, 587
　in geological samples, 593, 594
　by gravimetric methods, 585
　in ores, 587–591, 596
　by spectrophotometric methods, 560, 561; table, 560
　in thorium, 591–593
discovery of, 555, 556
dissolution of samples of, 564–566
electrochemical properties of, 557, 559
emission spectrum of, 559
gamma-ray counting of, 581, 586, 587; figure, 582
gamma-ray spectrometry of, 596, 597
handling techniques for, 449, 450
history of, 555, 556
ion-exchange behavior of, 578–580
maximum exposure to, 448, 449; table, 449
occurrence of, 441–445
optical properties of, 559–561
physical properties of, 557
polarography of, 559
radioactivity hazards of, 447, 448
reduction potential of, 559
separation of, 564–581; tables, 566, 568, 569, 572, 573, 581
solvent extraction of, 571–578; table, 572, 573
　fluoride effect on, 571
sources of, 445–447
uses of, 447
Protoactinium-223, determination of, 587
　preparation of, 598
Protoactinium isotopes, nuclear properties of, 557; table, 558

1-(2-Pyridylazo)-2-naphthol, 127, 156–158
Pyrocatechol violet in titrimetric determination of lead, 126
Pyrochlore, 187; table, 186
Pyroreduction method for beryllium production, 8, 9

Q

Quinalizarin, 38
Quinizarin, 39

R

Radiation exposure, maximum permissible, 448, 449
Radium, decay series of, 441; figures, 444, 445
Rhenium dioxide electrolytic preparation of, 415
Rhenium heptoxide, vapor pressure of, 414, 415
Rhenium metal, electrodeposition of, 416
Rhenium trioxide, 416

S

Salicylaldoxime, 117
Samarskite, table, 186
Simpsonite, table, 186
Sodium beryllate, reaction of with water, 6
Sodium beryllium fluoride, reaction of with sodium hydroxide, 7
Stibiocolumbite, table, 186
Stibotantalite, table, 186

T

Tantalates, chemical properties of, 206, 207
Tantalite, 185, 186; table, 186
Tantalum, abundance of, relative, 185; table, 185
　chemical properties of, 203–213
　chemistry, comparative, of members of fifth subgroup, 203, 204
　chromatographic separation of, principles of, 258, 259
　　procedures, 259–261
　　sample preparation, 259
　detection of, by chemical procedures, 280–283

by paper chromatography, 282
by reduction by tin, 283
by reduction by zinc, 283
by spectrography, 277–280; table, 279
  interferences, 279, 280
  limits of, 280
by tannin method, 281
by tartaric hydrolysis, 281
by thiocyanate method, 283
by x-ray emission, 280
determination of, by arsenazo method, 316
  by N-benzoyl-N-phenylhydroxylamine, 285
  by catechol, 316
  by cupferron, 284
  by gallic acid, 315
  by gravimetry, 284–286
  by hydrolytic precipitation, 285
  by 8-hydroxyquinoline, 285, 286
  by hydroquinone, 315
  by malachite green, 316
  by methyl violet, 316
  in niobium, 367, 368
  by neutron activation, 336–343
    decay scheme, figure, 339
    shelf-shielding effects, 341–343
    by 3,3,4,5,7-pentahydroxyflavone, 286
  by phenylarsonic acid, 285
  by phenylfluorone, 315, 316
  by photometry, 286–311
  by pyrogallol, 310–315; figure, 312
    acid medium effects, 311, 312
    interferences, 313–315; table, 314
    procedure, 311
    stability of colored complex, 313
  by radioactive isotope dilution, 336
  by radioactive tracers, 336
  by spectrophotometry, 286–311
  by tannic acid, 284
  by trihydroxy-6-fluorone, 315, 316
  by x-ray fluorescence, 331–335
    correction factors, 333; table, 333
    fusion of sample, 333, 334
    instrumentation, 331
    interferences, 331, 332
    applications, 334, 335

    sample characteristics, 332
determination of metallic impurities in, by spectrography, with four-group method, 359, 360
  with germanium oxide, 359
  with lanthanum carbonate, 359
  with lithium carbonate-lanthanum, 358
  with other reagents, 386–389; tables, 360, 361, 386, 388
  with silver chloride-barium fluoride, 359
  by x-ray fluorescence, 361, 362
electrochemical properties of, 200, 201
electronic configuration of, table, 203
electrophoretic separation of, 263
emission spectra of, table, 202
industrial hygiene of, 195
intermediate products of, decomposition of, 225
ion-exchange separation of, applications of, 252–255
  chemistry of, 246–250; figures, 247–250
  with hydrochloric acid system, 242–244; figure, 243
  with hydrochloric-oxalic acid system, 255–258; figures, 256, 257
  with hydrofluoric acid medium, 244, 245; figure, 245
  in mixed acid medium, 245, 246
    chemistry of, 246–250; figures, 247–250
    practical applications of, 252–255
    separation principles of, 250–252; figures, 251, 252
    tantalum complexes of, 246
  separation principles of, 250–252; figures, 251, 252
mechanical properties of, 195–197, 200; table, 195, 196
metallurgy of, extractive, 191–193
neutron cross-section of, table, 196
occurrence of, in euxenite, 187; table, 186
  in igneous rocks, 184, 185; table, 184
  in microlite, 186
  in minerals, 185–187; table, 185
  in pyrochlore, 187; table, 186

in tantalite, 185, 186; table, 186
in tin mining by-products, 187, 188
optical properties of, 202; table, 202
oxidation of, 205
paper chromatographic separation of, 261
peroxy salts of, chemical properties of, 207, 208
physical properties of, 195–197, 200; table, 195, 196
precipitation procedures for, with cupferron, 235, 236
　with hydrogen sulfide, 216
　by hydrolysis, 234, 235
　with miscellaneous reagents, 239–241
　with phenylarsonic acid, 239
　with tannin, 235, 236
preparation of, 193, 194
sampling of, concentrates, 213, 214
　intermediate products, 214
　metal, consolidated, 215
　metal, powdered, 215
　ore, 213, 214
separation of, 215–269
　by alkali fusion, 219–221
　from alloys of miscellaneous metals, 233
　　from plutonium, 232
　　from uranium, 232
　from borax, 221, 222
　chemistry of, 263
　by chlorination, 223, 224
　in fluoride medium, 265–268
　from high-grade ores and concentrates, 216–224
　in hydrochloric acid system, 264, 265
　by hydrofluoric acid, 222, 223
　"loss of individuality" in, 215, 216
　from niobium, by benzohydroxamic acid, 276
　　by N-benzoyl-N-phenylhydroxylamine, 272–274
　　by cellulose chromatography, 276, 277
　　by cupferron, 275
　　by diethyldithiocarbamate, 276
　　by dihydroquercitrin, 276
　　by 8-hydroxyquinoline, 274, 275
　　by hypophosphite, 276
　　by ion exchange, 276
　　by morin, 276
　　by peroxy method, 275
　　by phenylhydroxamic acid, 276
　　by phenylarsonic acid, 274
　　by Schoeller method, 270–272
　　by selenium, 276
　　by selenous acid, 274
　　by solvent extraction, 277
　　by tannin precipitation, 270
　　by tetramethylenedithiocarbonate, 276
　by pyrosulfate, 217–219
　by solvent extraction, with cyclohexanone, 268
　　with diisobutyl carbinol, 268
　　with methyl ethyl ketone, 268
　　with miscellaneous solvents, 269
　　with tributyl phosphate, 268
toxicology of, 195
uses of, in alloys, high-temperature, 190, 191
　in miscellaneous applications, 190, 191
　as pure metal, 190
　in steels, 188–191
x-ray spectra of, 202, 203; table, 202
Tantalum-182, 201
Tantalum alloys, determination of tantalum in, 329, 330, 380, 381
　dissolution of, 232
　preparation of, 194
Tantalum alloys, high-temperature, decomposition of, 229–232
　isolation of tantalum oxide, 229–231
　treatment of oxide, 231, 232
Tantalum carbide, preparation of, 194
Tantalum concentrates, determination of tantalum in, by ion exchange, 376–380
　apparatus, 376, 377; figure, 377
　procedure, 377–380
　reagents, 376, 377
Tantalum concentrates, high-grade, decomposition of, 216–224
Tantalum earth acids, chemical properties of, 205, 206
Tantalum isotopes, activation product, table, 199
　decay type, table, 199

determination of, 347, 348
half life, 199
neutron cross-section, table, 199
scattering cross-section, table, 199
tantalum-182, 201
Tantalum metal, carbon in, determination of, 357
  dissolution of, 228, 229
  hydrogen in, determination of, 353, 354
  interstitial elements in, determination of, carbon, 357
    nitrogen, 355–357
    oxygen, 349–351
  nonmetallic impurities in, determination of, 371
  specifications for, table, 349
Tantalum minerals, low-grade, decomposition of, 224, 225
Tantalum ores, determination of tantalum in, by ion exchange, 376, 380
  apparatus, 376, 377; figure, 377
  procedure, 377–380
  reagents, 376, 377
  high-grade, decomposition of, 216–224
    determination of tantalum in, by spectrophotometry, 324–327; tables, 326, 327
  low-grade, determination of tantalum in, by spectrophotometry, 327–329
Tantalum oxides, chemical properties of, 204, 205
  decomposition of, 226–228
    by alkaline attack, 227, 228
    by chlorination attack, 228
    by hydrofluoric acid attack, 228
    by pyrosulfate attack, 226, 227
Tantalum steels, decomposition of, 229–232
  isolation of tantalum oxide, 231, 232
  treatment of tantalum oxide, 231, 232
  determination of niobium in, 380, 381
  determination of tantalum in, 380, 381
Tapiolite, table, 186
Technetium, absorption spectra of, 412, 413; table, 413; figure, 413
  molar extinction coefficient, table, 413
  solvent, table, 413
  wavelength, table, 413
chemical properties of, 413–417

halogenation, 416
oxidation, 414
color of, 410
corrosion inhibition by, 409
detection of, 421–423; table, 422
  by emission spectroscopy, 423
  by radioactivity methods, 423
  by spot test, table, 423
determination of, as ammonium pertechnetate, 427
  analytical procedures for, 431, 432; tables, 425, 431
  by gravimetric methods, 424–427
  as metallic technetium, 424, 426
  by neutralization, 427
  as nitron pertechnetate, 426, 431
  by polarography, 428; table, 429
  by redox reactions, 427, 428
  by spectrophotometry, 428–430
  in technetium compounds, 424–431
  as technetium dioxide, 426
  as technetium sulfides, 427
  as technetium(V) thiocyanate complex, 431, 432
  as tetraphenyl arsonium pertechnetate, 426, 431
  as thioglycolic acid-Tc(VII) complex, 432
  by titrimetric methods, 427, 428
  in trace quantities, 429, 430
discovery of, 408
electrochemical properties of, 417, 418
electronic structure of, 410
magnetic susceptibility of, 410
natural occurrence of, 409
nuclear fission and, 408
nuclear properties of, 411, 412; table, 411
optical properties of, 412, 413; tables, 412, 413
oxidation potential of, 417; figure, 417
physical properties of, 409–411
polarography of, 417, 418
radioactivity of isotopes of, 423
sale of, 408
separation of, 418–421
  by anion-exchange procedure, 421
  by electroplating, 419
  from fission products, 418

by hydrogen peroxide oxidation, 418
by hydrogen sulfide precipitation, 418, 419
from molybdenum, 419
from rhenium, 419
from ruthenium, 419
by solvent extraction, with chloroform, 420, 421
  with hexone, 420
  with thiocyanate complex, 420
  with tributyl phosphate, 420
by tetraphenyl arsonium chloride, 419
by vaporization methods, 421
of technetium heptoxide, 418
spectral lines of, table, 412
in stars, 408, 409
toxicology of, 423, 424
  maximum permissible concentration, 424
  radioactivity of isotopes, 423
x-ray emission spectrum of, 412
Technetium compounds, 411; table, 410
Technetium dioxide, 415, 416
Technetium halides, 416
Technetium heptoxide, vapor pressure of, 414, 415; table, 414

Technetium isotopes, 411, 412
  radioactivity of, 423
Technetium sulfides, 415, 418, 419
Technetium trioxide, 416, 417
1,2,5,8-Tetrahydroxyanthraquinone, 38
Tetrahydroxyflavinol, 39
Tetraphenyl arsonium pertechnate, 419, 426, 431
Thionalide, 117
Thorium, decay series of, 441; figure, 442
Tracer solutions, 440
"Triple nitrite" crystal reaction for lead, 151
Triplumbic paraperiodate, 118

U V W X Y Z

Uranium, decay series of, 441–445; figures, 444, 445
Variamine blue in determination of lead, 121
Versene, 35
Wolframite, 4
Xylenol orange, 126, 155, 156
Yttrotantalite, table, 186
Zenia, in beryllium analysis, 38